A Visual Approach

Solve: $5 + \sqrt{x + 7} = x$.

Algebraic Solution

We first isolate the radical and then use the principle of powers.

$5 + \sqrt{x + 7} = x$

$\quad\quad \sqrt{x + 7} = x - 5$ Subtracting 5 on both sides

$\quad (\sqrt{x + 7})^2 = (x - 5)^2$ Using the principle of powers; squaring both sides

$\quad\quad\quad x + 7 = x^2 - 10x + 25$

$\quad\quad\quad\quad\quad 0 = x^2 - 11x + 18$ Subtracting x and 7

$\quad\quad\quad\quad\quad 0 = (x - 9)(x - 2)$ Factoring

$x - 9 = 0 \quad or \quad x - 2 = 0$

$\quad x = 9 \quad or \quad\quad\quad x = 2$

The possible solutions are 9 and 2.

CHECK: For 9:

$$\frac{5 + \sqrt{x + 7} = x}{}$$

$5 + \sqrt{9 + 7} \ ? \ 9$

$\quad 5 + \sqrt{16} \ \bigg|$

$\quad\quad 5 + 4 \ \bigg|$

$\quad\quad\quad\quad 9 \ \bigg| \ 9$ TRUE

For 2:

$$\frac{5 + \sqrt{x + 7} = x}{}$$

$5 + \sqrt{2 + 7} \ ? \ 2$

$\quad 5 + \sqrt{9} \ \bigg|$

$\quad\quad 5 + 3 \ \bigg|$

$\quad\quad\quad\quad 8 \ \bigg| \ 2$ FALSE

Since 9 checks but 2 does not, the only solution is 9.

Visualizing the Solution

When we graph $y = 5 + \sqrt{x + 7}$ and $y = x$, we find that the first coordinate of the point of intersection of the graphs is 9. Thus the solution of $5 + \sqrt{x + 7} = x$ is 9.

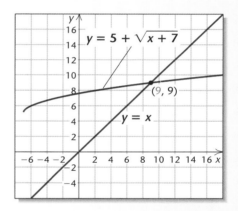

Note that the graphs show that the equation has only one solution.

Algebra and Trigonometry

Judith A. Beecher

Indiana University Purdue University Indianapolis

Judith A. Penna

Indiana University Purdue University Indianapolis

Marvin L. Bittinger

Indiana University Purdue University Indianapolis

Addison
Wesley

Boston   San Francisco   New York
London   Toronto   Sydney   Tokyo   Singapore   Madrid
Mexico City   Munich   Paris   Cape Town   Hong Kong   Montreal

Publisher	Greg Tobin
Acquisitions Editor	Jennifer Crum
Project Manager	Kari Heen
Associate Editor	Suzanne Alley
Assistant Editor	Greg Erb
Managing Editor	Ron Hampton
Production Supervisor	Kathleen A. Manley
Editorial and Production Services	Martha K. Morong/Quadrata, Inc. and Kathy Diamond
Art Editor	Geri Davis/The Davis Group, Inc.
Marketing Manager	Becky Anderson
Illustrator	Network Graphics
Prepress Supervisor	Caroline Fell
Compositor	The Beacon Group
Cover Designer	Dennis Schaefer
Interior Design	Geri Davis and Dennis Schaefer
Print Buyer	Evelyn Beaton
Media Producers	Ruth Berry and Rebecca Martin
Software Development	Malcolm Litowitz and Marty Wright

Library of Congress Cataloging-in-Publication Data

Beecher, Judith A.
    Algebra and trigonometry/Judith A. Beecher, Judith A. Penna, Marvin L. Bittinger.
        p. cm.
    Includes index.
    ISBN 0-201-74141-5
    1. Algebra. 2. Trigonometry. I. Penna, Judith A. II. Bittinger, Marvin L. III. Title.
QA152.3.B43 2002
512.13—dc21
                                                     2001045875

1 2 3 4 5 6 7 8 9 10—QWT—04030201

For our fathers

Floyd W. Swank

J. Keith Melvin

William Sterl Bittinger

Contents

5

The Trigonometric Functions

6

Trigonometric Identities, Inverse Functions, and Equations

7

Applications of Trigonometry

Preface

Note to Students

Our challenge, and our goal, when writing this textbook was to do everything possible to help you learn the concepts and skills contained between its covers. Every feature we have included was put here with this in mind. We realize that your time is both valuable and limited, so we communicate in a highly visual way that allows you to focus easily and learn quickly and efficiently. Take advantage of the side-by-side algebraic solutions and their visualizations, the Connecting the Concepts features, and the Study Tips. We included them to enable you to make the most of your study time and to be successful in this course.

Best wishes for a positive learning experience,

Judy Beecher
Judy Penna
Marv Bittinger

Content Features

Algebra and Trigonometry covers college-level algebra and trigonometry and is appropriate for a one- or two-term course in precalculus mathematics. We introduce functions earlier than most precalculus texts, and our approach is more visual as well. Although a course in intermediate algebra is a prerequisite for using this text, Chapter R, Basic Concepts of Algebra, provides sufficient review to unify the diverse mathematical backgrounds of most students.

• **Function Emphasis**   Functions are the core of this course and should be presented as a thread that runs throughout the course rather than as an isolated topic. We introduce functions in Chapter 1, whereas most traditional college algebra textbooks cover equation-solving in Chapter 1 (see pp. 56, 64–66, 137, and 153). Our approach introduces students to a relatively new concept at the beginning of the course rather than requiring them to begin with a review of what was previously covered in intermediate algebra. The concept of a function can be challenging for students. By repeatedly exposing them to the language, notation, and use of functions, demonstrating visually how functions relate to equations and graphs, and also showing how functions can be used to model real data, we hope that students will not only become comfortable with functions but also come to understand and appreciate them.

- **Zeros, Solutions, and x-Intercepts Theme** We find that when students understand the connections among the real zeros of a function, the solutions of its associated equation, and the first coordinates of the x-intercepts of its graph, a door opens to a new level of mathematical comprehension that increases the probability of success in this course. We emphasize zeros, solutions, and x-intercepts throughout the text by using consistent, precise terminology and including exceptional graphics. (See pp. 155, 165, and 468.)

- **Visual Emphasis** Our early introduction of functions allows graphs to be used to provide a visual aspect to solving equations and inequalities. For example, we are able to show the students both algebraically and visually that the solutions of a quadratic equation $ax^2 + bx + c = 0$ are the zeros of the quadratic function $f(x) = ax^2 + bx + c$ as well as the x-intercepts of the graph of that function. This makes it possible for students, particularly visual learners, to gain a quick understanding of these concepts. (See pp. 162 and 240.)

- **Side-by-Side Features** Many examples are presented in a side-by-side, two-column format in which the algebraic solution of an equation appears in the left column and a graphical interpretation of the solution appears in the right column (see pp. 155, 314, and 465). This enables students to visualize and comprehend the connections among the solutions of an equation, the zeros of a function, and the x-intercepts of the graph of a function.

- **Optional Review Chapter** Chapter R, Basic Concepts of Algebra, provides an optional review of intermediate algebra. Some or all of the topics in this chapter can be taught at the beginning of the course, or it can be used as a convenient source of information throughout the term for students who need a quick review of particular topics.

- **Connecting the Concepts** This feature highlights the importance of connecting concepts. When students are presented with concepts in a visual form—using graphs, an outline, or a chart—rather than merely in paragraphs of text, comprehension is streamlined and retention is maximized. The visual aspect of this feature invites students to stop and check their understanding of how concepts work together in one section or in several sections. This concept check in turn enhances student performance on homework assignments and exams. (See pp. 139, 280, and 458.)

- **Real-Data Applications** We encourage students to see and interpret the mathematics that appears every day in the world around them. Throughout the writing process, we conducted an energetic search for real-data applications, and the result is a variety of examples and exercises that connect the mathematical content with everyday life. Most of these applications feature source lines and frequently include charts and graphs. Many are drawn from the fields of health, business and economics, life and physical sciences, social science, and areas of general interest such as sports and travel. (See pp. 76, 277, 329, and 352.)

• **Study Tips** Appearing in the text margin, each Study Tip provides helpful study hints throughout the text and also briefly reminds students to use the electronic and print supplements that accompany the text. (See pp. 94 and 210.)

• **Review Icons** Placed next to the concept that a student is currently studying, a review icon references one or more sections of the text in which the student can find and review the topics on which the current concept is built. (See pp. 153 and 226.)

• **Technology Connections** This feature appears throughout the text to demonstrate how a graphing calculator can be used to solve problems. The technology is set apart from the traditional exposition, so that it does not intrude if no technology is desired. Although students might not be using graphing calculators, the graphing calculator windows that appear in the Technology Connection features enhance the visual element of the text, providing graphical interpretations of solutions of equations, zeros of functions, and x-intercepts of graphs of functions (see pp. 136, 162, 311, and 469). Exercises that are designed to be worked using a graphing calculator are grouped together in the exercise sets under the heading Technology Connection so they can be easily identified. A graphing calculator manual, providing keystroke-level instruction for six models of graphing calculators, is also available. (See Supplements for the Student.)

Pedagogical Features

• **Chapter Openers** Each chapter opens with an application relevant to the content of the chapter. Also included is a table of contents for the chapter, listing section titles. (See pp. 43, 193, and 335.)

• **Section Objectives** Content objectives are listed at the beginning of each section. Together with subheadings throughout the section, these objectives provide a useful outline of the section for both instructors and students. (See pp. 56, 153, and 357.)

• **Annotated Examples** Realizing that students quickly become discouraged and frustrated when a textbook does not prepare them adequately for the exercises in the exercise sets, we have included over 730 examples designed to prepare the student fully to do the exercises. Learning is carefully guided with the use of numerous color-coded art pieces and step-by-step annotations. Substitutions and annotations are highlighted in red for emphasis. (See pp. 157, 305, and 472.)

- **Use of Color** The text uses full color in an extremely functional way, as seen in the design elements and numerous pieces of art. The use of color has been carefully thought out so that it carries a consistent meaning that enhances students' ability to read and comprehend the exposition. (See pp. 120, 266, and 406.)

- **Art Package** The text contains over 1200 art pieces including photographs, situational art, and statistical graphs that not only highlight the abundance of real-world applications but also help students visualize the mathematics being discussed (see pp. 44, 235, 326, and 354). In particular, photorealistic art, in which mathematics is superimposed on a photograph, encourages students to see mathematics in familiar settings (see pp. 102 and 173).

- **Five-Step Problem-Solving Process** The basis for problem solving is a distinctive five-step process established early in the text (Section 2.1) to help students learn strategic ways to approach and solve applied problems. This process is then used consistently throughout the text to give students a consistent framework for problem solving. (See pp. 139 and 163.)

- **Variety of Exercises** There are over 5500 exercises in this text. The exercise sets are enhanced with real-data applications and source lines, detailed art pieces, tables, and graphs. In addition to the exercises that provide students practice with the concepts presented in the section, the exercise sets feature the following elements.

  Collaborative Discussion and Writing Exercises can be used in small groups or by the class as a whole to encourage students to talk and write about the key mathematical concepts in each section. (See pp. 167 and 212.)

  Skill Maintenance Exercises provide an ongoing review of concepts previously presented in the course, enhancing students' retention of these concepts. Answers to *all* Skill Maintenance exercises appear in the answer section at the back of the book along with a section reference that directs students quickly and efficiently to the appropriate section of the text if they need help with an exercise. (See pp. 124 and 224.)

  Synthesis Exercises appear at the end of each exercise set and encourage critical thinking by requiring students to synthesize concepts from several sections or to take a concept a step further than in the general exercises. (See pp. 213 and 276.)

  Technology Exercises are to be done using a graphing calculator. They are set apart from the non-calculator exercises, making it convenient for the instructor to avoid assigning them if graphing calculators are not being used in the course. (See pp. 69 and 316.)

- **Highlighted Information** Important definitions, properties, and rules are displayed in screened boxes, and summaries and procedures are listed in boxes outlined in color. This organization and presentation provides for efficient learning and review. (See pp. 218 and 233.)

- **Summary and Review** The Summary and Review at the end of each chapter contains an extensive set of review exercises along with a list of important properties and formulas covered in that chapter. This feature provides excellent preparation for chapter tests and the final examination. Answers to all review exercises appear in the answer section at the back of the book, along with corresponding section references. (See pp. 255 and 331.)

- **Chapter Tests** The test at the end of each chapter allows students to test themselves and target areas that need further study before taking the in-class test. Answers to all Chapter Test questions appear in the answer section at the back of the book, along with corresponding section references. (See pp. 130 and 334.)

Supplements for the Instructor

For more information on these and other helpful instructor supplements, please contact your Addison-Wesley sales representative.

Instructor's Edition *(ISBN 0-201-75162-3)*
This specially bound version of the student edition contains answers to both even- and odd-numbered exercises at the back of the text.

Instructor's Solutions Manual *(ISBN 0-201-75030-9)*
The *Instructor's Solutions Manual* by Judith A. Penna contains worked-out solutions to all exercises in the exercise sets, including the Collaborative Discussion and Writing exercises.

Printed Test Bank/Instructor's Resource Guide *(ISBN 0-201-75028-7)*
The *Printed Test Bank/Instructor's Resource Guide*, prepared by Laurie Hurley, contains the following:

- 4 free-response test forms for each chapter
- 2 multiple-choice test forms for each chapter
- 6 forms of a final examination, 4 with free-response questions and 2 with multiple-choice questions
- Index to the videotapes that accompany the text

TestGen-EQ with QuizMaster-EQ *(ISBN 0-201-75033-3)*
Available on a dual-platform Windows/Macintosh CD-ROM, this fully networkable software enables instructors to build, edit, print, and administer tests using a computerized test bank of questions organized according to the chapter content of the text. Tests can be printed or saved for on-line testing via a network or the Web, and the software can generate a variety of grading reports for tests and quizzes.

Supplements for the Student

For more information on these and other helpful student supplements, please contact your bookstore.

Student's Solutions Manual *(ISBN 0-201-75029-5)*

The *Student's Solutions Manual* by Judith A. Penna contains completely worked-out solutions with step-by-step annotations for all the odd-numbered exercises in the text, with the exception of the Collaborative Discussion and Writing exercises.

InterAct Math Tutorial CD-ROM *(ISBN 0-201-75032-5, stand-alone)*

This interactive tutorial software provides algorithmically generated practice exercises that correlate at the objective level to the odd-numbered exercises in the text. Each practice exercise is accompanied by an example and guided solution designed to involve students in the solution process. The software recognizes common student errors and provides appropriate feedback.

MyMathLab.com   www.mymathlab.com

A complete on-line course that integrates interactive multimedia instruction correlated to the textbook content, MyMathLab is easily customized to suit the needs of students and instructors and provides a comprehensive and efficient on-line course-management system that allows for diagnosis, assessment, and tracking of students' progress. Free when bundled with a new Addison-Wesley text.

AW Math Tutor Center *(ISBN 0-201-72170-8, stand-alone)*
www.aw.com/tutorcenter

The AW Math Tutor Center is staffed by qualified mathematics instructors who provide students with tutoring on examples and exercises from the textbook. Tutoring is available via toll-free telephone, fax, or e-mail five days a week, seven hours a day.

Videotapes *(ISBN 0-201-75034-1)*

Developed and produced especially for this text, these videotapes feature an engaging team of instructors presenting material and concepts in a format that stresses student interaction. There is a videotaped lesson for each section of the text. The lecturers' presentations include examples and exercises from the text and support an approach that emphasizes visualization and problem solving.

Digital Video Tutor *(ISBN 0-201-74997-1, stand-alone)*

The videos for this text are available on CD-ROM, making it easy, convenient, and affordable for students to watch video segments from a computer at home or on campus. The portability of this complete digitized video set makes it ideal for supplemental instruction or distance learning.

InterAct MathXL *(12-month registration ISBN 0-201-71630-5, stand-alone)*
www.mathxl.com

This Web-based diagnostic testing and tutorial system allows students to take practice tests correlated to the textbook and receive customized study plans based on their results. Each time a student takes a practice test, the resulting study plan identifies areas for improvement and links to appropriate practice exercises and tutorials generated by InterAct Math. A course-management feature allows instructors to view students' test results, study plans, and practice work.

Graphing Calculator Manual *(ISBN 0-201-75031-7)*

The *Graphing Calculator Manual* by Judith A. Penna, with the assistance of Daphne A. Bell, contains keystroke-level instruction for the Texas Instruments TI-82®, TI-83®, TI-83 Plus®, TI-85®, TI-86®, and TI-89® graphing calculators. This manual uses examples and exercises from the text to teach students to use the graphing calculator. The order of topics mirrors that of the text, providing a just-in-time mode of instruction.

Acknowledgments

We wish to express our heartfelt appreciation to a number of people who have contributed in special ways to the development of this textbook. Our editor, Jennifer Crum, encouraged our vision and spent many hours discussing it with us. Kari Heen, our executive project manager, deserves special recognition for overseeing every phase of the project and keeping it moving. We are very appreciative of the support that has been given to us by the entire Addison-Wesley Higher Education Group. We also thank Laurie Hurley, Barbara Johnson, and Patty Slipher for accuracy checking this textbook.

The following reviewers made invaluable contributions to the development of this text and we thank them for that:

Gerald Allen, *Howard College—San Angelo*

Donna Bailey, *Truman State University*

Randy Christ, *Creighton University*

Karin Deck, *University of North Carolina—Wilmington*

Tim Doyle, *Northern Illinois University*

Rebecca Easley, *Rose State College*

Jim Fischer, *Oregon Institute of Technology*

Marilyn Hasty, *Southern Illinois University—Edwardsville*

Maryann Justinger, *SUNY Erie Community College—South Campus*

Cathy Lovingier, *Linn Benton Community College*

Richard Nadel, *Florida International University*

Dana Nimic, *Southeast Community College*

Jane Pinnow, *University of Wisconsin—Parkside*

Joni Queen, *Pennsylvania State University*

Pascal Roubides, *Middle Georgia College*

Scott Satake, *North Idaho College*

Donald Solomon, *University of Wisconsin—Milwaukee*

Denise Widup, *University of Wisconsin—Parkside*

Kenneth Word, *Central Texas College—Killeen*

Henry Wyzinski, *Indiana University—Northwest*

J.A.B.
J.A.P.
M.L.B.

Feature Walkthrough

Functions, Equations, and Inequalities 2

*I*n this chapter, we examine functions and equations, placing particular emphasis on visualizing the connections among the zeros of a function, the *x*-intercepts of the graph of that function, and the solutions of the corresponding equation. We study linear, quadratic, rational, and radical equations as well as equations that are reducible to quadratic and equations with absolute value. We also study linear and quadratic functions and models. We end our work in this chapter by solving linear inequalities.

APPLICATION

In a recent study of 68 metropolitan areas, it was determined that in 1999 drivers in Indianapolis were delayed in traffic an average of 37 hr annually. This was 7 hr less than the annual delay in the mid-1990s. (*Source*: Texas Transportation Institute Urban Mobility Study) What was the annual delay in the mid-1990s?

This problem appears as Exercise 50 in Section 2.1.

Applied Chapter Openers

Each chapter begins with an application showing how chapter concepts are used in the real world. The chapter openers are not only motivating to students but also help them develop their critical-thinking skills.

ZEROS, SOLUTIONS, AND INTERCEPTS

FUNCTION	REAL ZEROS OF THE FUNCTION; SOLUTIONS OF THE EQUATION	x-INTERCEPTS OF THE GRAPH

Quadratic Polynomial
$g(x) = x^2 - 2x - 8$
$\quad = (x + 2)(x - 4),$
or
$\quad y = (x + 2)(x - 4)$

To find the **zeros** of $g(x)$, we solve $g(x) = 0$:

$$x^2 - 2x - 8 = 0$$
$$(x + 2)(x - 4) = 0$$
$$x + 2 = 0 \quad or \quad x - 4 = 0$$
$$x = -2 \quad or \quad x = 4.$$

The **solutions** of $x^2 - 2x - 8 = 0$ are -2 and 4. They are the zeros of the function $g(x)$. That is,

$$g(-2) = 0 \quad and \quad g(4) = 0.$$

The zeros of $g(x)$ are the x-coordinates of the **x-intercepts** of the graph of $y = g(x)$.

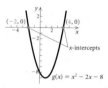

Cubic Polynomial
$h(x)$
$\quad = x^3 + 2x^2 - 5x - 6$
$\quad = (x + 3)(x + 1)(x - 2),$
or
$\quad y = (x + 3)(x + 1)(x - 2)$

To find the **zeros** of $h(x)$, we solve $h(x) = 0$:

$$x^3 + 2x^2 - 5x - 6 = 0$$
$$(x + 3)(x + 1)(x - 2) = 0$$
$$x + 3 = 0 \quad or \quad x + 1 = 0 \quad or \quad x - 2 = 0$$
$$x = -3 \quad or \quad x = -1 \quad or \quad x = 2.$$

The **solutions** of $x^3 + 2x^2 - 5x - 6 = 0$ are -3, -1, and 2. They are the zeros of the function $h(x)$. That is,

$$h(-3) = 0,$$
$$h(-1) = 0, \quad and$$
$$h(2) = 0.$$

The zeros of $h(x)$ are the x-coordinates of the **x-intercepts** of the graph of $y = h(x)$.

The connection between the zeros of a function and the x-intercepts of the graph of the function is easily seen in the examples above. If c is a real zero of a function (that is, $f(c) = 0$), then $(c, 0)$ is an x-intercept of the graph of the function.

Zeros, Solutions, and x-Intercepts Theme
There is a strong emphasis on visualizing and connecting the following concepts:

- the real zeros of a function
- the solutions of the associated equation
- the x-coordinates of the x-intercepts of the graph of the function.

Seeing the connections among these concepts increases student understanding. This theme is reinforced visually throughout the text.

Study Tips
Appearing in the text margins, each Study Tip provides helpful study hints throughout the text and briefly reminds students to use the electronic and print supplements that accompany the text.

Let $M = \log_a x$. Then $a^M = x$. Substituting $\log_a x$ for M, we obtain $a^{\log_a x} = x$. This also follows from the definition of a logarithm: $\log_a x$ is the power to which a is raised in order to get x.

STUDY TIP

Immediately after each quiz or chapter test, write out a step-by-step solution to the questions you missed. Visit your professor during office hours for help with problems that are still giving you trouble. When the week of the final examination arrives, you will be glad to have the excellent study guide these corrected tests provide.

A Base to a Logarithmic Power
For any base a and any positive real number x,

$$a^{\log_a x} = x.$$

(The number a raised to the power $\log_a x$ is x.)

EXAMPLE 10 Simplify each of the following.

a) $4^{\log_4 k}$ **b)** $e^{\ln 5}$ **c)** $10^{\log 7t}$

Solution
a) $4^{\log_4 k} = k$
b) $e^{\ln 5} = e^{\log_e 5} = 5$
c) $10^{\log 7t} = 10^{\log_{10} 7t} = 7t$

Side-by-Side Features

Many examples are presented in a side-by-side, two-column format in which the algebraic solution of an equation appears in the left column and a graphical interpretation of the solution appears in the right column. This enables students to visualize and comprehend the connections among the solutions of an equation, the zeros of a function, and the x-intercepts of the graph of the function.

EXAMPLE 6 Solve: $\log x + \log (x + 3) = 1$.

Algebraic Solution

In this case, we have common logarithms. Writing the base of 10 will help us understand the problem:

$$\log_{10} x + \log_{10} (x + 3) = 1$$
$$\log_{10} [x(x + 3)] = 1 \qquad \text{Using the product rule to obtain a single logarithm}$$
$$x(x + 3) = 10^1 \qquad \text{Writing an equivalent exponential equation}$$
$$x^2 + 3x = 10$$
$$x^2 + 3x - 10 = 0$$
$$(x - 2)(x + 5) = 0 \qquad \text{Factoring}$$
$$x - 2 = 0 \quad or \quad x + 5 = 0$$
$$x = 2 \quad or \qquad x = -5.$$

CHECK: For 2:
$$\log x + \log (x + 3) = 1$$

$$\log 2 + \log (2 + 3) \ ? \ 1$$
$$\log 2 + \log 5 \ |$$
$$\log (2 \cdot 5) \ |$$
$$\log 10 \ |$$
$$1 \ | \ 1 \quad \text{TRUE}$$

For -5:
$$\log x + \log (x + 3) = 1$$

$$\log (-5) + \log (-5 + 3) \ ? \ 1 \quad \text{FALSE}$$

The number -5 is not a solution because negative numbers do not have real-number logarithms. The solution is 2.

Visualizing the Solution

The solution of the equation
$$\log x + \log (x + 3) = 1$$
is the zero of the function
$$f(x) = \log x + \log (x + 3) - 1.$$
The solution is also the first coordinate of the x-intercept of the graph of the function.

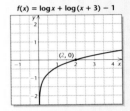

$$f(x) = \log x + \log (x + 3) - 1$$

The solution of the equation is 2. From the graph, we can easily see that there is only one solution.

Real-Data Applications

We encourage students to see and interpret the mathematics that appears every day in the world around them. Throughout the writing process, we conducted an energetic search for real-data applications, and the result is a variety of examples and exercises that connect the mathematical content with the real world. Most of these applications feature source lines and frequently include charts and graphs.

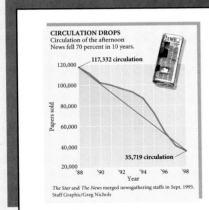

CIRCULATION DROPS
Circulation of the afternoon News fell 70 percent in 10 years.

117,332 circulation

35,719 circulation

Papers sold / Year

The Star and *The News* merged newsgathering staffs in Sept. 1995.
Staff Graphic/Greg Nichols

EXAMPLE 4 *Newspaper Circulation.* On September 30, 1999, the last edition of *The Indianapolis News*, an afternoon daily paper, rolled off the press. Circulation of the paper had dropped approximately 70% in the 10 preceding years, as shown in the graph at left. Find the average rate of change in the circulation from 1988 to 1998.

Solution We determine the coordinates of two points on the graph. In this case, we will use (1988, 117,332) and (1998, 35,719). Then we compute the slope, or average rate of change, as follows:

$$\text{Slope} = \text{Average rate of change} = \frac{\text{Change in } y}{\text{Change in } x}$$
$$= \frac{35,719 - 117,332}{1998 - 1988} = \frac{-81,613}{10} = -8161.3.$$

This result tells us that, on average, each year the circulation of the paper decreased by approximately 8161 copies. The *average rate of change* over the 10-year period was -8161 copies per year.

Connecting the Concepts

Comprehension is streamlined and retention is maximized when concepts are presented in a visual form. Combining design (usually in outline or chart form) and art, this feature emphasizes the importance of connecting concepts. Its visual aspect invites students to stop and check their understanding of how concepts work together within one or throughout several sections. The increased understanding gained from this feature enhances student performance on homework assignments.

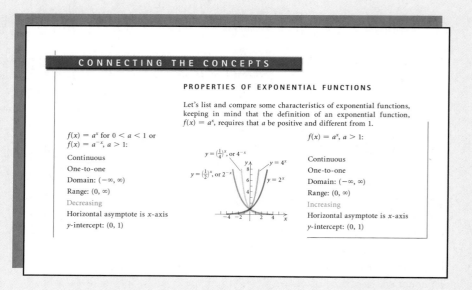

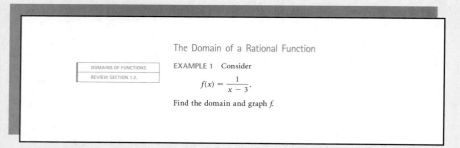

Review Icons

These icons prompt students to review topics necessary to understand a concept at hand, and the resulting review enhances students' comprehension. For easy navigation, the section number of the relevant review topic is given in the icon.

Technology Connections

This optional feature appears throughout the text to demonstrate how a graphing calculator can be used to solve problems. The technology is set apart from the traditional exposition, so that it can be easily omitted, if desired. Although students may not be using graphing calculators, the graphing calculator windows that appear in the Technology Connection features enhance the visual aspect of the text.

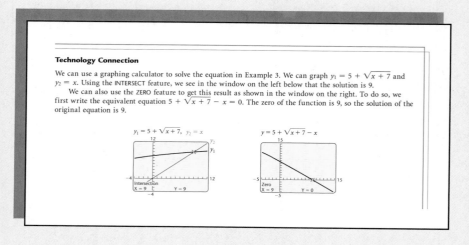

Exercise Sets

The exercise sets are enhanced with real-data applications and source lines, detailed art pieces, tables, and graphs. A variety of exercises appears in each exercise set.

Collaborative Discussion and Writing exercises encourage students both to consider and to write about key concepts in each section. These open-ended exercises are particularly suitable for use in class discussions or as collaborative activities.

Exercise Set **4.6**

1. *World Population Growth.* In 1999, the world population was 6.0 billion. The exponential growth rate was 1.3% per year.

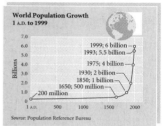

World Population Growth
1 A.D. to 1999

1999; 6 billion
1993; 5.5 billion
1975; 4 billion
1930; 2 billion
1850; 1 billion
1650; 500 million
200 million

Source: Population Reference Bureau

a) Find the exponential growth function.
b) Predict the population of the world in 2005 and 2010.
c) When will the world population be 8 billion?
d) Find the doubling time.

2. *Population Growth of Rabbits.* Under ideal conditions, a population of rabbits has an

4. *Female Olympic Athletes.* In 1985, the number of female athletes participating in Summer

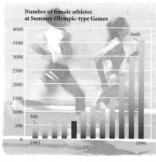

Number of female athletes at Summer Olympic-type Games

Olympic-type Games was 500. In 1996, about 3600 participated in the Summer Olympics in Atlanta. Assuming the exponential model applies:

Skill Maintenance *exercises review and reinforce skills previously presented in the text. These exercises provide excellent review for a final examination.*

Synthesis *exercises help build critical-thinking skills by requiring students to synthesize or combine learning objectives from several sections or to take a concept a step further than in the general exercises.*

Technology Connection *exercises are set apart from non-calculator exercises, making it convenient for the instructor to avoid assigning them if graphing calculators are not being used in the course.*

---

62. One leg of a right triangle is 7 cm less than the length of the other leg. The length of the hypotenuse is 13 cm. Find the lengths of the legs.

63. One number is 5 greater than another. The product of the numbers is 36. Find the numbers.

64. One number is 6 less than another. The product of the numbers is 72. Find the numbers.

65. *Box Construction.* An open box is made from a 10-cm by 20-cm piece of tin by cutting a square from each corner and folding up the edges. The area of the resulting base is 96 cm². What is the length of the sides of the squares?

10 cm

20 cm

66. *Picture Frame Dimensions.* The frame of a picture is 28 cm by 32 cm outside and is of uniform width. What is the width of the frame if 192 cm² of the picture shows?

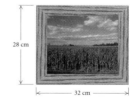

28 cm

32 cm

State whether the function is linear or quadratic.

67. $f(x) = 4 - 5x$ **68.** $f(x) = 4 - 5x^2$

69. $f(x) = 7x^2$ **70.** $f(x) = 23x + 6$

71. $f(x) = 1.2x - (3.6)^2$ **72.** $f(x) = 2 - x - x^2$

Technology Connection

Solve graphically. Round solutions to three decimal places, where appropriate.

73. $x^2 - 8x + 12 = 0$ **74.** $5x^2 + 42x + 16 = 0$

75. $7x^2 - 43x + 6 = 0$ **76.** $10x^2 - 23x + 12 = 0$

77. $6x + 1 = 4x^2$ **78.** $3x^2 + 5x = 3$

Use a graphing calculator to find the zeros of the function. Round to three decimal places.

79. $f(x) = 2x^2 - 5x - 4$ **80.** $f(x) = 4x^2 - 3x - 2$

81. $f(x) = 3x^2 + 2x - 4$ **82.** $f(x) = 9x^2 - 8x - 7$

83. $f(x) = 5.02x^2 - 4.19x - 2.057$

84. $f(x) = 1.21x^2 - 2.34x - 5.63$

Collaborative Discussion and Writing

85. Is it possible for a quadratic function to have one real zero and one imaginary zero? Why or why not?

86. The graph of a quadratic function can have 0, 1, or 2 x-intercepts. How can you predict the number of x-intercepts without drawing the graph or (completely) solving an equation?

Skill Maintenance

Determine whether the graph is symmetric with respect to the x-axis, the y-axis, and the origin.

87. $3x^2 + 4y^2 = 5$ **88.** $y^3 = 6x^2$

Determine whether the function is even, odd, or neither even nor odd.

89. $f(x) = 2x^3 - x$ **90.** $f(x) = 4x^2 + 2x - 3$

Synthesis

Solve.

91. $(x - 2)^3 = x^3 - 2$

92. $(x + 1)^3 = (x - 1)^3 + 26$

93. $(6x^3 + 7x^2 - 3x)(x^2 - 7) = 0$

94. $\left(x - \frac{1}{5}\right)\left(x^2 - \frac{1}{4}\right) + \left(x - \frac{1}{5}\right)\left(x^2 + \frac{1}{8}\right) = 0$

95. $x^2 + x - \sqrt{2} = 0$

96. $x^2 + \sqrt{5}x - \sqrt{3} = 0$

97. $2t^2 + (t - 4)^2 = 5t(t - 4) + 24$

98. $9t(t + 2) - 3t(t - 2) = 2(t + 4)(t + 6)$

99. $\sqrt{x - 3} - \sqrt[4]{x - 3} = 2$

100. $x^6 - 28x^3 + 27 = 0$

101. $\left(y + \frac{2}{y}\right)^2 + 3y + \frac{6}{y} = 4$

102. $x^2 + 3x + 1 - \sqrt{x^2 + 3x + 1} = 8$

103. Solve $\frac{1}{2}at^2 + v_0t + x_0 = 0$ for t.

4 Chapter Summary and Review

Important Properties and Formulas

The Composition of Two Functions:	$(f \circ g)(x) = f(g(x))$
One-to-One Function:	$f(a) = f(b) \rightarrow a = b$
Exponential Function:	$f(x) = a^x$
The Number e:	$e = 2.7182818284\ldots$
Logarithmic Function:	$f(x) = \log_a x$
A Logarithm Is an Exponent:	$\log_a x = y \longleftrightarrow x = a^y$
The Change-of-Base Formula:	$\log_b M = \dfrac{\log_a M}{\log_a b}$
The Product Rule:	$\log_a MN = \log_a M + \log_a N$
The Power Rule:	$\log_a M^p = p \log_a M$
The Quotient Rule:	$\log_a \dfrac{M}{N} = \log_a M - \log_a N$
Other Properties:	$\log_a a = 1, \qquad \log_a 1 = 0,$
	$\log_a a^x = x, \qquad a^{\log_a x} = x$
Base–Exponent Property:	$a^x = a^y \longleftrightarrow x = y,$ for $a > 0$, $a \neq 1$
Exponential Growth Model:	$P(t) = P_0 e^{kt}, \ k > 0$
Exponential Decay Model:	$P(t) = P_0 e^{-kt}, \ k > 0$
Interest Compounded Continuously:	$P(t) = P_0 e^{kt}, \ k > 0$
Limited Growth:	$P(t) = \dfrac{a}{1 + be^{-kt}}, \ k > 0$

Summary and Reviews
The Summary and Review at the end of each chapter contains an extensive set of review exercises along with a list of important properties and formulas covered in that chapter. This feature provides excellent preparation for chapter tests and the final examination.

REVIEW EXERCISES

In Exercises 1 and 2, for the pair of functions:
a) *Find the domain of $f \circ g$ and $g \circ f$.*
b) *Find $(f \circ g)(x)$ and $(g \circ f)(x)$.*

1. $f(x) = \dfrac{4}{x^2};\ g(x) = 3 - 2x$

2. $f(x) = 3x^2 + 4x;\ g(x) = 2x - 1$

Find $f(x)$ and $g(x)$ such that $h(x) = (f \circ g)x.$
3. $h(x) = \sqrt{5x + 2}$

4. $h(x) = 4(5x - 1)^2 + 9$

5. Find the inverse of the relation
$\{(1.3, -2.7),\ (8, -3),\ (-5, 3),\ (6, -3),\ (7, -5)\}.$

6. Find an equation of the inverse relation.

4 Chapter Test

1. For $f(x) = x - 5$ and $g(x) = x^2 + 1$, find $(f \circ g)(x)$ and $(g \circ f)(x)$.

2. Find the inverse of the relation
$\{(-2, 5),\ (4, 3),\ (0, -1),\ (-6, -3)\}.$

3. Determine whether the function shown below is one-to-one. Answer yes or no.

4. Find a formula for the inverse of the function $f(x) = x^3 + 1.$

Graph each of the following functions.
5. $f(x) = e^x - 3$ **6.** $f(x) = \ln(x + 2)$

7. Convert to an exponential equation: $\ln x = 4.$
8. Convert to a logarithmic equation: $3^x = 5.4.$

Solve.
9. $\log_{25} 5 = x$
10. $\log_3 x + \log_3 (x + 8) = 2$
11. $3^{4-x} = 27^x$
12. $e^x = 65$
13. Express $\ln \sqrt[5]{x^2 y}$ in terms of sums and differences of logarithms.
14. Given that $\log_a 2 = 0.328$ and $\log_a 8 = 0.984$, find $\log_a 4.$
15. Simplify: $\ln e^{-4t}.$
16. *Growth Rate.* A country's population doubled in 45 yr. What was the exponential growth rate?
17. *Compound Interest.* Suppose \$1000 is invested at interest rate k, compounded continuously, and grows to \$1144.54 in 3 yr.
  a) Find the interest rate.
  b) Find the exponential growth function.
  c) Find the balance after 8 yr.
  d) Find the doubling time.

Synthesis

18. Solve: $4^{\sqrt[3]{x}} = 8.$

Chapter Tests
The test at the end of each chapter allows students to test themselves and target areas that need further study before taking an in-class test.

Basic Concepts of Algebra R

*T*he algebraic concepts reviewed in this chapter provide the foundation for your further study in this text. They include the properties of the real-number system, the properties of exponents, and manipulations of algebraic expressions. You can continue to refer to this chapter to review these topics as needed as you proceed through the text.

APPLICATION

The formula

$$M = P\left[\frac{\frac{i}{12}\left(1 + \frac{i}{12}\right)^n}{\left(1 + \frac{i}{12}\right)^n - 1}\right]$$

gives the monthly mortgage payment M on a home loan of P dollars at interest rate i, where n is the total number of payments (12 times the number of years).

Suppose that a house costs $124,000, the down payment is $20,000, the interest rate is $7\frac{3}{4}\%$, and the loan period is 30 yr. What is the monthly mortgage payment?

This problem appears as Exercise 68 in Section R.2.

- *Identify various kinds of real numbers.*
- *Use interval notation to write a set of numbers.*
- *Identify the properties of real numbers.*
- *Find the absolute value of a real number.*

The Real-Number System

Real Numbers

In applications of algebraic concepts, we use real numbers to represent quantities such as distance, time, speed, area, profit, loss, and temperature. Some frequently used sets of real numbers and the relationships among them are shown below.

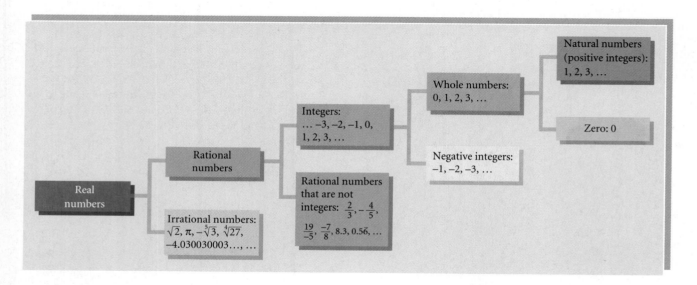

Numbers that can be expressed in the form p/q, where p and q are integers and $q \neq 0$, are **rational numbers.** Decimal notation for rational numbers either *terminates* (ends) or *repeats*. Each of the following is a rational number.

a) 0 $0 = \dfrac{0}{a}$ for any nonzero integer a

b) -7 $-7 = \dfrac{-7}{1}$, or $\dfrac{7}{-1}$, or $-\dfrac{7}{1}$

c) $\dfrac{1}{4} = 0.25$ Terminating decimal

d) $-\dfrac{5}{11} = -0.\overline{45}$ Repeating decimal

The real numbers that are not rational are **irrational numbers.** Decimal notation for irrational numbers neither terminates nor repeats.

Each of the following is an irrational number.

a) $\pi = 3.1415926535\ldots$ There is no repeating block of digits.

$\left(\frac{22}{7} \text{ and } 3.14 \text{ are rational } \textit{approximations} \text{ of the irrational number } \pi.\right)$

b) $\sqrt{2} = 1.414213562\ldots$ There is no repeating block of digits.

c) $-6.12122122212222\ldots$ Although there is a pattern, there is no repeating block of digits.

The set of all rational numbers combined with the set of all irrational numbers gives us the set of **real numbers.** The real numbers are *modeled* using a **number line,** as shown below.

Each point on the line represents a real number.

The order of the real numbers can be determined from the number line. If a number a is to the left of a number b, then a **is less than** b ($a < b$). Similarly, a **is greater than** b ($a > b$), if a is to the right of b on the number line. For example, we see from the number line above that $-2.9 < -\frac{3}{5}$, because -2.9 is to the left of $-\frac{3}{5}$. Also, $\frac{17}{4} > \sqrt{3}$, because $\frac{17}{4}$ is to the right of $\sqrt{3}$.

The statement $a \le b$, read "a is less than or equal to b," is true if either $a < b$ is true or $a = b$ is true.

The symbol $\in$ is used to indicate that a member, or **element**, belongs to a set. Thus if we let $\mathbb{Q}$ represent the set of rational numbers, we can see from the diagram on p. 2 that $0.\overline{56} \in \mathbb{Q}$. We can also write $\sqrt{2} \notin \mathbb{Q}$ to indicate that $\sqrt{2}$ is *not* an element of the set of rational numbers.

When *all* the elements of one set are elements of a second set, we say that the first set is a **subset** of the second set. The symbol $\subseteq$ is used to denote this. For instance, if we let $\mathbb{R}$ represent the set of real numbers, we can see from the diagram that $\mathbb{Q} \subseteq \mathbb{R}$ (read "$\mathbb{Q}$ is a subset of $\mathbb{R}$").

Interval Notation

Sets of real numbers can be expressed using **interval notation.** For example, for real numbers a and b such that $a < b$, the **open interval** (a, b) is the set of real numbers between, but not including, a and b. That is,

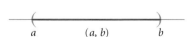

$$(a, b) = \{x \,|\, a < x < b\}.$$

The points a and b are the **endpoints** of the interval. The parentheses indicate that the endpoints are not included in the interval.

Some intervals extend without bound in one or both directions. The interval $[a, \infty)$, for example, begins at a and extends to the right without bound. That is,

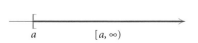

$$[a, \infty) = \{x \,|\, x \ge a\}.$$

The bracket indicates that a is included in the interval.

The various types of intervals are listed below.

Intervals: Types, Notation, and Graphs

TYPE	INTERVAL NOTATION	SET NOTATION	GRAPH
Open	(a, b)	$\{x \mid a < x < b\}$	
Closed	$[a, b]$	$\{x \mid a \leq x \leq b\}$	
Half-open	$[a, b)$	$\{x \mid a \leq x < b\}$	
Half-open	$(a, b]$	$\{x \mid a < x \leq b\}$	
Open	(a, ∞)	$\{x \mid x > a\}$	
Half-open	$[a, \infty)$	$\{x \mid x \geq a\}$	
Open	$(-\infty, b)$	$\{x \mid x < b\}$	
Half-open	$(-\infty, b]$	$\{x \mid x \leq b\}$	

The interval $(-\infty, \infty)$ names the set of all real numbers, $\mathbb{R}$.

EXAMPLE 1 Write interval notation for each set.

a) $\{x \mid -4 < x < 5\}$
b) $\{x \mid x \geq 1.7\}$
c) $\{x \mid -10 < x \leq -5\}$
d) $\{x \mid x < \sqrt{5}\}$

Solution

a) $\{x \mid -4 < x < 5\} = (-4, 5)$
b) $\{x \mid x \geq 1.7\} = [1.7, \infty)$
c) $\{x \mid -10 < x \leq -5\} = (-10, -5]$
d) $\{x \mid x < \sqrt{5}\} = (-\infty, \sqrt{5})$

Properties of the Real Numbers

The following properties can be used to manipulate algebraic expressions as well as real numbers.

Properties of the Real Numbers

For any real numbers a, b, and c:

$a + b = b + a$ and $ab = ba$	Commutative properties of addition and multiplication
$a + (b + c) = (a + b) + c$ and $a(bc) = (ab)c$	Associative properties of addition and multiplication
$a + 0 = 0 + a = a$	Additive identity property
$-a + a = a + (-a) = 0$	Additive inverse property
$a \cdot 1 = 1 \cdot a = a$	Multiplicative identity property
$a \cdot \dfrac{1}{a} = 1 \ (a \neq 0)$	Multiplicative inverse property
$a(b + c) = ab + ac$	Distributive property

Note that the distributive property is also true for subtraction since $a(b - c) = a[b + (-c)] = ab + a(-c) = ab - ac$.

EXAMPLE 2   State the property being illustrated in each sentence.

a) $8 \cdot 5 = 5 \cdot 8$ **b)** $5 + (m + n) = (5 + m) + n$

c) $14 + (-14) = 0$ **d)** $6 \cdot 1 = 1 \cdot 6 = 6$

e) $2(a - b) = 2a - 2b$

Solution

SENTENCE	PROPERTY
a) $8 \cdot 5 = 5 \cdot 8$	Commutative property of multiplication: $ab = ba$
b) $5 + (m + n) = (5 + m) + n$	Associative property of addition: $a + (b + c) = (a + b) + c$
c) $14 + (-14) = 0$	Additive inverse property: $a + (-a) = 0$
d) $6 \cdot 1 = 1 \cdot 6 = 6$	Multiplicative identity property: $a \cdot 1 = 1 \cdot a = a$
e) $2(a - b) = 2a - 2b$	Distributive property: $a(b + c) = ab + ac$

Absolute Value

The number line can be used to provide a geometric interpretation of *absolute value*. The **absolute value** of a number a, denoted $|a|$, is its distance from 0 on the number line. For example, $|-5| = 5$, because the distance of -5 from 0 is 5. Similarly, $\left|\frac{3}{4}\right| = \frac{3}{4}$, because the distance of $\frac{3}{4}$ from 0 is $\frac{3}{4}$.

Absolute Value

For any real number a,

$$|a| = \begin{cases} a, & \text{if } a \geq 0, \\ -a, & \text{if } a < 0. \end{cases}$$

It follows from the definition of absolute value that, for any real number a, $|a| \geq 0$.

Absolute value can be used to find the distance between two points on the number line.

$|a - b| = |b - a|$

Distance Between Two Points on the Number Line

For any real numbers a and b, the **distance between a and b** is $|a - b|$, or, equivalently, $|b - a|$.

EXAMPLE 3   Find the distance between -2 and 3.

Solution

$$|-2 - 3| = |-5| = 5, \quad \text{or, equivalently,}$$
$$|3 - (-2)| = |3 + 2| = |5| = 5.$$

Technology Connection

We can use the absolute-value operation on a graphing calculator to find the distance between two points. On many graphing calculators, absolute value is denoted "abs" and is found in the MATH NUM menu and also in the CATALOG.

```
abs (−2−3)
                        5
abs (3−(−2))
                        5
```

Exercise Set R.1

In Exercises 1–6, consider the numbers $-12, \sqrt{7}, 5.\overline{3},$ $-\frac{7}{3}, \sqrt[3]{8}, 0, 5.242242224\ldots, -\sqrt{14}, \sqrt[5]{5}, -1.96, 9, 4\frac{2}{3},$ $\sqrt{25}, \sqrt[4]{4}, \frac{5}{7}.$

1. Which are whole numbers?

2. Which are integers?

3. Which are irrational numbers?

4. Which are natural numbers?

5. Which are rational numbers?

6. Which are real numbers?

Find an example of each of the following. Answers may vary.

7. A rational number that is not an integer

8. A real number that is not a rational number

9. An integer that is not a whole number

10. A real number that is not an irrational number

Write interval notation.

11. $\{x \mid -3 \le x \le 3\}$ **12.** $\{x \mid -4 < x < 4\}$

13. $\{x \mid -14 \le x < -11\}$ **14.** $\{x \mid 6 < x \le 20\}$

15. $\{x \mid x \le -4\}$ **16.** $\{x \mid x > -5\}$

17. $\{x \mid x < 3.8\}$ **18.** $\{x \mid x \ge \sqrt{3}\}$

19. $\{x \mid 7 < x\}$ **20.** $\{x \mid -3 > x\}$

Write interval notation for the graph.

21.

22.

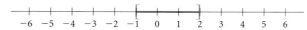

23.

24.

25.

26.

27.

28.

In Exercises 29–46, the following notation is used: $\mathbb{N}$ = the set of natural numbers, $\mathbb{W}$ = the set of whole numbers, $\mathbb{Z}$ = the set of integers, $\mathbb{Q}$ = the set of rational numbers, $\mathbb{I}$ = the set of irrational numbers, and $\mathbb{R}$ = the set of real numbers. Classify the statement as true or false.

29. $6 \in \mathbb{N}$ **30.** $0 \notin \mathbb{N}$

31. $3.2 \in \mathbb{Z}$ **32.** $-10.\overline{1} \in \mathbb{R}$

33. $-\dfrac{11}{5} \in \mathbb{Q}$ **34.** $-\sqrt{6} \in \mathbb{Q}$

35. $\sqrt{11} \notin \mathbb{R}$ **36.** $-1 \in \mathbb{W}$

37. $24 \notin \mathbb{W}$ **38.** $1 \in \mathbb{Z}$

39. $1.089 \notin \mathbb{I}$ **40.** $\mathbb{N} \subseteq \mathbb{W}$

41. $\mathbb{W} \subseteq \mathbb{Z}$ **42.** $\mathbb{Z} \subseteq \mathbb{N}$

43. $\mathbb{Q} \subseteq \mathbb{R}$ **44.** $\mathbb{Z} \subseteq \mathbb{Q}$

45. $\mathbb{R} \subseteq \mathbb{Z}$ **46.** $\mathbb{Q} \subseteq \mathbb{I}$

Name the property illustrated by the sentence.

47. $6x = x6$

48. $3 + (x + y) = (3 + x) + y$

49. $-3 \cdot 1 = -3$ **50.** $x + 4 = 4 + x$

51. $5(ab) = (5a)b$ **52.** $4(y - z) = 4y - 4z$

53. $2(a + b) = (a + b)2$ **54.** $-7 + 7 = 0$

55. $-6(m + n) = -6(n + m)$

56. $t + 0 = t$

57. $8 \cdot \dfrac{1}{8} = 1$

58. $9x + 9y = 9(x + y)$

Simplify.

59. $|-7.1|$ **60.** $|0|$

61. $\left| \dfrac{5}{4} \right|$

62. $\left| -\sqrt{3} \right|$

Find the distance between the given pair of points on the number line.

63. $-5, 6$

64. $0, -2.5$

65. $-2, -8$

66. $\dfrac{15}{8}, \dfrac{23}{12}$

67. $12.1, 6.7$

68. $-3, -14$

Collaborative Discussion and Writing

To the student and the instructor: The Collaborative Discussion and Writing exercises are meant to be answered with one or more sentences. These exercises can also be discussed and answered collaboratively by the entire class or by small groups. Because of their open-ended nature, the answers to these exercises do not appear at the back of the book. They are denoted by the words "Discussion and Writing."

69. How would you convince a classmate that division is not associative?

70. Under what circumstances is $\sqrt{a}$ a rational number?

Synthesis

To the student and the instructor: The Synthesis exercises found at the end of every exercise set challenge

students to combine concepts or skills studied in that section or in preceding parts of the text.

Between any two (different) real numbers there are many other real numbers. Find each of the following. Answers may vary.

71. An irrational number between 0.124 and 0.125

72. A rational number between $-\sqrt{2.01}$ and $-\sqrt{2}$

73. A rational number between $-\dfrac{1}{101}$ and $-\dfrac{1}{100}$

74. An irrational number between $\sqrt{5.99}$ and $\sqrt{6}$

75. The hypotenuse of an isosceles right triangle with legs of length 1 unit can be used to "measure" a value for $\sqrt{2}$ by using the Pythagorean theorem, as shown.

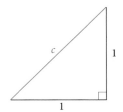

$$c^2 = 1^2 + 1^2$$
$$c^2 = 2$$
$$c = \sqrt{2}$$

Draw a right triangle that could be used to "measure" $\sqrt{10}$ units.

R.2

- Simplify expressions with integer exponents.
- Solve problems using scientific notation.
- Use the rules for order of operations.

Integer Exponents, Scientific Notation, and Order of Operations

Integers as Exponents

When a positive integer is used as an *exponent*, it indicates the number of times a factor appears in a product. For example, 7^3 means $7 \cdot 7 \cdot 7$ and 5^1 means 5.

For any positive integer n,

$$a^n = \underbrace{a \cdot a \cdot a \cdots a},$$
$$n \text{ factors}$$

where a is the **base** and n is the **exponent**.

Zero and negative-integer exponents are defined as follows.

For any nonzero real number a and any integer n,

$$a^0 = 1 \quad \text{and} \quad a^{-n} = \frac{1}{a^n}.$$

EXAMPLE 1   Simplify each of the following.

a) 6^0

b) $(-3.4)^0$

Solution

a) $6^0 = 1$

b) $(-3.4)^0 = 1$

EXAMPLE 2   Write each of the following with positive exponents.

a) 4^{-5}

b) $\dfrac{1}{(0.82)^{-7}}$

Solution

a) $4^{-5} = \dfrac{1}{4^5}$

b) $\dfrac{1}{(0.82)^{-7}} = (0.82)^{-(-7)} = (0.82)^7$

The following properties of exponents can be used to simplify expressions.

Properties of Exponents

For any real numbers a and b and any integers m and n, assuming 0 is not raised to a nonpositive power:

$a^m \cdot a^n = a^{m+n}$ Product rule

$\dfrac{a^m}{a^n} = a^{m-n} \ (a \neq 0)$ Quotient rule

$(a^m)^n = a^{mn}$ Power rule

$(ab)^m = a^m b^m$ Raising a product to a power

$\left(\dfrac{a}{b}\right)^m = \dfrac{a^m}{b^m} \ (b \neq 0)$ Raising a quotient to a power

EXAMPLE 3   Simplify each of the following.

a) $y^{-5} \cdot y^3$ 　　　　　　　　　　　**b)** $\dfrac{48x^{12}}{16x^4}$

c) $(t^{-3})^5$ 　　　　　　　　　　　**d)** $(2s^{-2})^5$

e) $\left(\dfrac{45x^{-4}y^2}{9z^{-8}}\right)^{-3}$

Solution

a) $y^{-5} \cdot y^3 = y^{-5+3} = y^{-2}$, or $\dfrac{1}{y^2}$

b) $\dfrac{48x^{12}}{16x^4} = \dfrac{48}{16}x^{12-4} = 3x^8$

c) $(t^{-3})^5 = t^{-3 \cdot 5} = t^{-15}$, or $\dfrac{1}{t^{15}}$

d) $(2s^{-2})^5 = 2^5(s^{-2})^5 = 32s^{-10}$, or $\dfrac{32}{s^{10}}$

e) $\left(\dfrac{45x^{-4}y^2}{9z^{-8}}\right)^{-3} = \left(\dfrac{5x^{-4}y^2}{z^{-8}}\right)^{-3}$

$\qquad = \dfrac{5^{-3}x^{12}y^{-6}}{z^{24}} = \dfrac{x^{12}y^{-6}}{125z^{24}}$, or $\dfrac{x^{12}}{125y^6z^{24}}$

Scientific Notation

We can use scientific notation to name very large and very small positive numbers and to perform computations.

Scientific Notation

Scientific notation for a number is an expression of the type

$$N \times 10^n,$$

where $1 \le N < 10$, N is in decimal notation, and n is an integer.

Keep in mind that in scientific notation, positive exponents are used for numbers greater than or equal to 10 and negative exponents for numbers between 0 and 1.

EXAMPLE 4   *Advertising.* Proctor and Gamble spent $1,703,053,300 on advertising in a recent year   (*Source*: Competitive Media Reporting and Publishers Information Bureau). Convert this number to scientific notation.

Solution We want the decimal point to be positioned between the 1 and the 7, so we move it 9 places to the left. Since the number to be converted is greater than 10, the exponent must be positive.

$$1{,}703{,}053{,}300 = 1.7030533 \times 10^9$$

EXAMPLE 5 *Mass of a Neutron.* The mass of a neutron is about 0.00000000000000000000000000167 kg. Convert this number to scientific notation.

Solution We want the decimal point to be positioned between the 1 and the 6, so we move it 27 places to the right. Since the number to be converted is between 0 and 1, the exponent must be negative.

$$0.00000000000000000000000000167 = 1.67 \times 10^{-27}$$

EXAMPLE 6 Convert each of the following to decimal notation.

a) 7.632×10^{-4}

b) 9.4×10^5

Solution

a) The exponent is negative, so the number is between 0 and 1. We move the decimal point 4 places to the left.

$$7.632 \times 10^{-4} = 0.0007632$$

b) The exponent is positive, so the number is greater than 10. We move the decimal point 5 places to the right.

$$9.4 \times 10^5 = 940{,}000$$

EXAMPLE 7 *Distance to a Star.* Alpha Centauri is about 4.3 light-years from Earth. One **light-year** is the distance that light travels in one year and is about 5.88×10^{12} miles. How many miles is it from Earth to Alpha Centauri? Express your answer in scientific notation.

Solution

$$\begin{aligned}
4.3 \times (5.88 \times 10^{12}) &= (4.3 \times 5.88) \times 10^{12} \\
&= 25.284 \times 10^{12} \qquad \text{This is not scientific notation.} \\
&= (2.5284 \times 10^1) \times 10^{12} \\
&= 2.5284 \times (10^1 \times 10^{12}) \\
&= 2.5284 \times 10^{13} \text{ miles} \qquad \text{Writing scientific notation}
\end{aligned}$$

Technology Connection

Most calculators make use of scientific notation. For example, the number 48,000,000,000,000 might be expressed in one of the ways shown below.

4.8E13

4.8 13

The computation in Example 7 can be performed on a calculator using scientific notation.

4.3∗5.88E12
2.5284E13

Order of Operations

Recall that to simplify the expression $3 + 4 \cdot 5$, we multiply 4 and 5 first to get 20 and then add 3 to get 23. Mathematicians have agreed on the following procedure, or rules for order of operations.

Rules for Order of Operations

1. Do all calculations within grouping symbols before operations outside. When nested grouping symbols are present, work from the inside out.

2. Evaluate all exponential expressions.

3. Do all multiplications and divisions in order from left to right.

4. Do all additions and subtractions in order from left to right.

EXAMPLE 8 Calculate each of the following.

a) $8(5 - 3)^3 - 20$

b) $\dfrac{10 \div (8 - 6) + 9 \cdot 4}{2^5 + 3^2}$

Solution

a)
$$8(5 - 3)^3 - 20 = 8 \cdot 2^3 - 20 \qquad \text{Doing the calculation within parentheses}$$
$$= 8 \cdot 8 - 20 \qquad \text{Evaluating the exponential expression}$$
$$= 64 - 20 \qquad \text{Multiplying}$$
$$= 44 \qquad \text{Subtracting}$$

b)
$$\frac{10 \div (8 - 6) + 9 \cdot 4}{2^5 + 3^2} = \frac{10 \div 2 + 9 \cdot 4}{32 + 9}$$
$$= \frac{5 + 36}{41} = \frac{41}{41} = 1$$

Note that fraction bars act as grouping symbols. That is, the given expression is equivalent to $[10 \div (8 - 6) + 9 \cdot 4] \div (2^5 + 3^2)$. ▬

EXAMPLE 9 *Compound Interest.* If a principal P is invested at an interest rate i, compounded n times per year, in t years it will grow to an amount A given by

$$A = P\left(1 + \frac{i}{n}\right)^{nt}.$$

Suppose that $1250 is invested at 4.6% interest, compounded quarterly. How much is in the account at the end of 8 years?

Solution We have $P = 1250$, $i = 4.6\%$, or 0.046, $n = 4$, and $t = 8$. Substituting, we find that the amount in the account at the end of 8 years is given by

$$A = 1250\left(1 + \frac{0.046}{4}\right)^{4 \cdot 8}.$$

Next, we evaluate this expression:

$$
\begin{aligned}
A &= 1250(1 + 0.0115)^{4 \cdot 8} && \text{Dividing} \\
&= 1250(1.0115)^{4 \cdot 8} && \text{Adding} \\
&= 1250(1.0115)^{32} && \text{Multiplying in the exponent} \\
&\approx 1250(1.441811175) && \text{Evaluating the exponential expression} \\
&\approx 1802.263969 && \text{Multiplying} \\
&\approx 1802.26. && \text{Rounding to the nearest cent}
\end{aligned}
$$

The amount in the account at the end of 8 years is $1802.26.

Exercise Set   R.2

Simplify.

1. 18^0

2. $\left(-\frac{4}{3}\right)^0$

3. $5^8 \cdot 5^{-6}$

4. $6^2 \cdot 6^{-7}$

5. $m^{-5} \cdot m^5$

6. $n^9 \cdot n^{-9}$

7. $7^3 \cdot 7^{-5} \cdot 7$

8. $3^6 \cdot 3^{-5} \cdot 3^4$

9. $2x^3 \cdot 3x^2$

10. $3y^4 \cdot 4y^3$

11. $(5a^2b)(3a^{-3}b^4)$

12. $(4xy^2)(3x^{-4}y^5)$

13. $(2x)^3(3x)^2$

14. $(4y)^2(3y)^3$

15. $\dfrac{b^{40}}{b^{37}}$

16. $\dfrac{a^{39}}{a^{32}}$

17. $\dfrac{x^2y^{-2}}{x^{-1}y}$

18. $\dfrac{x^3y^{-3}}{x^{-1}y^2}$

19. $\dfrac{32x^{-4}y^3}{4x^{-5}y^8}$

20. $\dfrac{20a^5b^{-2}}{5a^7b^{-3}}$

21. $(2ab^2)^3$

22. $(4xy^3)^2$

23. $(-2x^3)^4$

24. $(-3x^2)^4$

25. $(-5c^{-1}d^{-2})^{-2}$

26. $(-4x^{-5}z^{-2})^{-3}$

27. $\left(\dfrac{24a^{10}b^{-8}c^7}{3a^6b^{-3}c^5}\right)^5$

28. $\left(\dfrac{125p^{12}q^{-14}r^{22}}{25p^8q^6r^{-15}}\right)^{-4}$

Convert to scientific notation.

29. 405,000

30. 1,670,000

31. 0.00000039

32. 0.00092

33. One cubic inch is approximately equal to 0.000016 m^3.

34. The United States government received $828,597,000,000 in individual income taxes in a recent year (*Source*: U.S. Department of the Treasury).

Convert to decimal notation.

35. 8.3×10^{-5}

36. 4.1×10^6

37. 2.07×10^7

38. 3.15×10^{-6}

39. In 1997, the revenue of Microsoft Corp. was 1.1358×10^{10} (*Source: Fortune Magazine*).

40. The mass of a proton is about 1.67×10^{-24} g.

Compute. Write the answer using scientific notation.

41. $(3.1 \times 10^5)(4.5 \times 10^{-3})$

42. $(9.1 \times 10^{-17})(8.2 \times 10^3)$

43. $\dfrac{6.4 \times 10^{-7}}{8.0 \times 10^6}$

44. $\dfrac{1.1 \times 10^{-40}}{2.0 \times 10^{-71}}$

Solve. Write the answer using scientific notation.

45. *Chesapeake Bay Bridge-Tunnel.* The 17.6-mile-long Chesapeake Bay Bridge-Tunnel was completed in 1964. Construction costs were $210 million. Find the average cost per mile.

46. *Gambling Revenue.* Casino gambling revenue in Las Vegas and the surrounding county was $6,152,415,000 in 1997. That year 30,500,000 people visited the area. (*Source*: Las Vegas Convention and Visitors Authority) Find the average amount each visitor spent gambling.

47. *Nuclear Disintegration.* One gram of radium produces 37 billion disintegrations per second. How many disintegrations are produced in 1 hr?

48. *Length of Earth's Orbit.* The average distance from the earth to the sun is 93 million mi. About how far does the earth travel in a yearly orbit? (Assume a circular orbit.)

Calculate.

49. $3 \cdot 2 - 4 \cdot 2^2 + 6(3 - 1)$

50. $3[(2 + 4 \cdot 2^2) - 6(3 - 1)]$

51. $16 \div 4 \cdot 4 \div 2 \cdot 256$

52. $2^6 \cdot 2^{-3} \div 2^{10} \div 2^{-8}$

53. $\dfrac{4(8 - 6)^2 - 4 \cdot 3 + 2 \cdot 8}{3^1 + 19^0}$

54. $\dfrac{[4(8 - 6)^2 + 4](3 - 2 \cdot 8)}{2^2(2^3 + 5)}$

Compound Interest. Use the compound interest formula from Example 9 in Exercises 55–58. Round to the nearest cent.

55. Suppose that $2125 is invested at 6.2%, compounded semiannually. How much is in the account at the end of 5 yr?

56. Suppose that $9550 is invested at 5.4%, compounded semiannually. How much is in the account at the end of 7 yr?

57. Suppose that $6700 is invested at 4.5%, compounded quarterly. How much is in the account at the end of 6 yr?

58. Suppose that $4875 is invested at 5.8%, compounded quarterly. How much is in the account at the end of 9 yr?

Collaborative Discussion and Writing

59. Are parentheses necessary in the expression $4 \cdot 25 \div (10 - 5)$? Why or why not?

60. Is $x^{-2} < x^{-1}$ for any negative value(s) of x? Why or why not?

Synthesis

Simplify. Assume that all exponents are integers, all denominators are nonzero, and zero is not raised to a nonpositive power.

61. $(x^t \cdot x^{3t})^2$

62. $(x^y \cdot x^{-y})^3$

63. $(t^{a+x} \cdot t^{x-a})^4$

64. $(m^{x-b} \cdot n^{x+b})^x (m^b n^{-b})^x$

65. $\left[\dfrac{(3x^a y^b)^3}{(-3x^a y^b)^2} \right]^2$

66. $\left[\left(\dfrac{x^r}{y^t} \right) \left(\dfrac{x^{2r}}{y^{4t}} \right)^{-2} \right]^{-3}$

Mortgage Payments. The formula

$$M = P \left[\dfrac{\dfrac{i}{12}\left(1 + \dfrac{i}{12}\right)^n}{\left(1 + \dfrac{i}{12}\right)^n - 1} \right]$$

gives the monthly mortgage payment M on a home loan of P dollars at interest rate i, where n is the total number of payments (12 times the number of years). Use this formula in Exercises 67–70.

67. The cost of a house is $98,000. The down payment is $16,000, the interest rate is $8\frac{1}{2}\%$, and the loan period is 25 yr. What is the monthly mortgage payment?

68. The cost of a house is $124,000. The down payment is $20,000, the interest rate is $7\frac{3}{4}\%$, and the loan period is 30 yr. What is the monthly mortgage payment?

69. The cost of a house is $135,000. The down payment is $18,000, the interest rate is $7\frac{1}{2}\%$, and the loan period is 20 yr. What is the monthly mortgage payment?

70. The cost of a house is $151,000. The down payment is $21,000, the interest rate is $8\frac{1}{4}\%$, and the loan period is 25 yr. What is the monthly mortgage payment?

R.3

Addition, Subtraction, and Multiplication of Polynomials

- *Identify the terms, coefficients, and degree of a polynomial.*
- *Add, subtract, and multiply polynomials.*

Polynomials

Polynomials are a type of algebraic expression that you will often encounter in your study of algebra. Some examples of polynomials are

$$3x - 4y, \quad 5y^3 - \tfrac{7}{3}y^2 + 3y - 2, \quad 0, \quad -2.3a^4, \quad \text{and} \quad z^6 - \sqrt{5}.$$

All but the first are polynomials in one variable.

Polynomials in One Variable

A **polynomial in one variable** is any expression of the type

$$a_n x^n + a_{n-1} x^{n-1} + \cdots + a_2 x^2 + a_1 x + a_0,$$

where n is a nonnegative integer, $a_n, \ldots, a_0$ are real numbers, called **coefficients**, and $a_n \neq 0$. The parts of a polynomial separated by plus signs are called **terms**. The **degree** of the polynomial is n, the **leading coefficient** is a_n, and the **constant term** is a_0. The polynomial is said to be written in **descending order,** because the exponents decrease from left to right.

EXAMPLE 1 Identify the terms of the polynomial

$$2x^4 - 7.5x^3 + x - 12.$$

Solution We have

$$2x^4 - 7.5x^3 + x - 12 = 2x^4 + (-7.5x^3) + x + (-12),$$

so the terms are

$$2x^4, \quad -7.5x^3, \quad x, \quad \text{and} \quad -12.$$

A polynomial, like 23, consisting of only a nonzero constant term has degree 0. It is agreed that the polynomial consisting only of 0 has *no* degree.

EXAMPLE 2 Find the degree of each polynomial.

a) $2x^3 - 9$

b) $y^2 - \tfrac{3}{2} + 5y^4$

c) 7

Solution

POLYNOMIAL	DEGREE
a) $2x^3 - 9$	3
b) $y^2 - \frac{3}{2} + 5y^4 = 5y^4 + y^2 - \frac{3}{2}$	4
c) $7 = 7x^0$	0

Algebraic expressions like $3ab^3 - 8$ and $5x^4y^2 - 3x^3y^8 + 7xy^2 + 6$ are **polynomials in several variables.** The **degree of a term** is the sum of the exponents of the variables in that term. The **degree of a polynomial** is the degree of the term of highest degree.

EXAMPLE 3   Find the degree of the polynomial

$$7ab^3 - 11a^2b^4 + 8.$$

Solution   The degrees of the terms of $7ab^3 - 11a^2b^4 + 8$ are 4, 6, and 0, respectively, so the degree of the polynomial is 6.

A polynomial with just one term, like $-9y^6$, is a **monomial**. If a polynomial has two terms, like $x^2 + 4$, it is a **binomial**. A polynomial with three terms, like $4x^2 - 4xy + 1$, is a **trinomial**.

Expressions like

$$2x^2 - 5x + \frac{3}{x}, \qquad 9 - \sqrt{x}, \quad \text{and} \quad \frac{x+1}{x^4+5}$$

are not polynomials, because they cannot be written in the form $a_nx^n + a_{n-1}x^{n-1} + \cdots + a_1x + a_0$, where the exponents are all nonnegative integers and the coefficients are all real numbers.

Addition and Subtraction

If two terms of an expression have the same variables raised to the same powers, they are called **like terms,** or **similar terms.** We can **combine,** or **collect, like terms** using the distributive property. For example, $3y^2$ and $5y^2$ are like terms and

$$3y^2 + 5y^2 = (3 + 5)y^2$$
$$= 8y^2.$$

We add or subtract polynomials by combining like terms.

EXAMPLE 4   Add or subtract each of the following.

a) $(-5x^3 + 3x^2 - x) + (12x^3 - 7x^2 + 3)$

b) $(6x^2y^3 - 9xy) - (5x^2y^3 - 4xy)$

Solution

a) $(-5x^3 + 3x^2 - x) + (12x^3 - 7x^2 + 3)$

$\qquad = (-5x^3 + 12x^3) + (3x^2 - 7x^2) - x + 3$ Rearranging using the commutative and associative properties

$\qquad = (-5 + 12)x^3 + (3 - 7)x^2 - x + 3$ Using the distributive property

$\qquad = 7x^3 - 4x^2 - x + 3$

b) We can subtract by adding an opposite:

$(6x^2y^3 - 9xy) - (5x^2y^3 - 4xy)$

$\qquad = (6x^2y^3 - 9xy) + (-5x^2y^3 + 4xy)$ Adding the opposite of $5x^2y^3 - 4xy$

$\qquad = 6x^2y^3 - 9xy - 5x^2y^3 + 4xy$

$\qquad = x^2y^3 - 5xy.$ Combining like terms

Multiplication

Multiplication of polynomials is based on the distributive property. For example,

$(x + 4)(x + 3) = (x + 4)x + (x + 4)3$ Using the distributive property

$\qquad = x^2 + 4x + 3x + 12$ Using the distributive property two more times

$\qquad = x^2 + 7x + 12.$ Combining like terms

We can find the product of two binomials by multiplying the **F**irst terms, then the **O**uter terms, then the **I**nner terms, then the **L**ast terms. Then we collect like terms, if possible. This procedure is sometimes called **FOIL**.

EXAMPLE 5   Multiply: $(2x - 7)(3x + 4)$.

Solution   We have

$$(2x - 7)(3x + 4) = 6x^2 + 8x - 21x - 28$$
$$= 6x^2 - 13x - 28.$$

In general, to multiply two polynomials, we multiply each term of one by each term of the other and add the products. One way to do this is to use columns.

EXAMPLE 6   Multiply: $(4x^4y - 7x^2y + 3y)(2y - 3x^2y)$.

Solution   We can use columns to organize our work.

$$4x^4y - 7x^2y + 3y$$
$$2y - 3x^2y$$

$$\overline{}$$

$$-12x^6y^2 + 21x^4y^2 - 9x^2y^2 \qquad \text{Multiplying by } -3x^2y$$
$$8x^4y^2 - 14x^2y^2 + 6y^2 \qquad \text{Multiplying by } 2y$$

$$\overline{}$$

$$-12x^6y^2 + 29x^4y^2 - 23x^2y^2 + 6y^2 \qquad \text{Adding}$$

We can use FOIL to find some special products.

Special Products of Binomials

$(A + B)^2 = A^2 + 2AB + B^2$ Square of a sum

$(A - B)^2 = A^2 - 2AB + B^2$ Square of a difference

$(A + B)(A - B) = A^2 - B^2$ Product of a sum and a difference

---

EXAMPLE 7   Multiply each of the following.

a) $(4x + 1)^2$ **b)** $(3y^2 - 2)^2$ **c)** $(x^2 + 3y)(x^2 - 3y)$

Solution

a) $(4x + 1)^2 = (4x)^2 + 2 \cdot 4x \cdot 1 + 1^2 = 16x^2 + 8x + 1$

b) $(3y^2 - 2)^2 = (3y^2)^2 - 2 \cdot 3y^2 \cdot 2 + 2^2 = 9y^4 - 12y^2 + 4$

c) $(x^2 + 3y)(x^2 - 3y) = (x^2)^2 - (3y)^2 = x^4 - 9y^2$

Exercise Set

Determine the terms and the degree of the polynomial.

1. $-5y^4 + 3y^3 + 7y^2 - y - 4$

2. $2m^3 - m^2 - 4m + 11$

3. $3a^4b - 7a^3b^3 + 5ab - 2$

4. $6p^3q^2 - p^2q^4 - 3pq^2 + 5$

Perform the operations indicated.

5. $(5x^2y - 2xy^2 + 3xy - 5) +$
$(-2x^2y - 3xy^2 + 4xy + 7)$

6. $(6x^2y - 3xy^2 + 5xy - 3) +$
$(-4x^2y - 4xy^2 + 3xy + 8)$

7. $(2x + 3y + z - 7) + (4x - 2y - z + 8) +$
$(-3x + y - 2z - 4)$

8. $(2x^2 + 12xy - 11) + (6x^2 - 2x + 4) +$
$(-x^2 - y - 2)$

9. $(3x^2 - 2x - x^3 + 2) - (5x^2 - 8x - x^3 + 4)$

10. $(5x^2 + 4xy - 3y^2 + 2) - (9x^2 - 4xy + 2y^2 - 1)$

11. $(x^4 - 3x^2 + 4x) - (3x^3 + x^2 - 5x + 3)$

12. $(2x^4 - 3x^2 + 7x) - (5x^3 + 2x^2 - 3x + 5)$

13. $(a - b)(2a^3 - ab + 3b^2)$

14. $(n + 1)(n^2 - 6n - 4)$

15. $(x + 5)(x - 3)$

16. $(y - 4)(y + 1)$

17. $(x + 6)(x + 4)$

18. $(n - 5)(n - 8)$

19. $(2a + 3)(a + 5)$

20. $(3b + 1)(b - 2)$

21. $(2x + 3y)(2x + y)$ **22.** $(2a - 3b)(2a - b)$

23. $(y + 5)^2$ **24.** $(y + 7)^2$

25. $(x - 4)^2$ **26.** $(a - 6)^2$

27. $(5x - 3)^2$ **28.** $(3x - 2)^2$

29. $(2x + 3y)^2$ **30.** $(5x + 2y)^2$

31. $(2x^2 - 3y)^2$ **32.** $(4x^2 - 5y)^2$

33. $(a + 3)(a - 3)$ **34.** $(b + 4)(b - 4)$

35. $(2x - 5)(2x + 5)$

36. $(4y - 1)(4y + 1)$

37. $(3x - 2y)(3x + 2y)$

38. $(3x + 5y)(3x - 5y)$

39. $(2x + 3y + 4)(2x + 3y - 4)$

40. $(5x + 2y + 3)(5x + 2y - 3)$

41. $(x + 1)(x - 1)(x^2 + 1)$

42. $(y - 2)(y + 2)(y^2 + 4)$

Collaborative Discussion and Writing

43. Is the sum of two polynomials of degree n always a polynomial of degree n? Why or why not?

44. Explain how you would convince a classmate that $(A + B)^2 \neq A^2 + B^2$.

Synthesis

Multiply. Assume that all exponents are natural numbers.

45. $(a^n + b^n)(a^n - b^n)$

46. $(t^a + 4)(t^a - 7)$

47. $(a^n + b^n)^2$

48. $(x^{3m} - t^{5n})^2$

49. $(x - 1)(x^2 + x + 1)(x^3 + 1)$

50. $[(2x - 1)^2 - 1]^2$

51. $(x^{a-b})^{a+b}$

52. $(t^{m+n})^{m+n} \cdot (t^{m-n})^{m-n}$

53. $(a + b + c)^2$

R.4

Factoring

- *Factor polynomials by removing a common factor.*
- *Factor special products of polynomials.*

To factor a polynomial, we do the reverse of multiplying; that is, we find an equivalent expression that is written as a product.

Terms with Common Factors

When a polynomial is to be factored, we should always look first to factor out a factor that is common to all the terms using the distributive property. We usually look for the constant common factor with the largest absolute value and for variables with the largest exponent common to all the terms. In this sense, we factor out the "largest" common factor.

EXAMPLE 1   Factor each of the following.

a) $15 + 10x - 5x^2$

b) $12x^2y^2 - 20x^3y$

Solution

a) $15 + 10x - 5x^2 = 5 \cdot 3 + 5 \cdot 2x - 5 \cdot x^2 = 5(3 + 2x - x^2)$

We can check by multiplying: $5(3 + 2x - x^2) = 15 + 10x - 5x^2$.

b) $12x^2y^2 - 20x^3y = 4x^2y(3y - 5x)$

Note that there are several factors common to the terms of $12x^2y^2 - 20x^3y$, but $4x^2y$ is the "largest" of these.   ▬

In some polynomials, pairs of terms have a common factor that can be removed in a process called **factoring by grouping.**

EXAMPLE 2   Factor: $x^3 + 3x^2 - 5x - 15$.

Solution   We have

$$x^3 + 3x^2 - 5x - 15 = x^2(x + 3) - 5(x + 3)$$
$$= (x + 3)(x^2 - 5).$$ ▬

Trinomials

Some trinomials can be factored into the product of two binomials. To factor a trinomial of the form $x^2 + bx + c$, we look for two numbers with a product of c and a sum of b.

EXAMPLE 3   Factor each of the following.

a) $x^2 + 5x + 6$ **b)** $y^2 - 2y - 24$

Solution

a) We look for two numbers with a product of 6 and a sum of 5. They are 2 and 3. Then

$$x^2 + 5x + 6 = (x + 2)(x + 3).$$ To check, multiply.

b) We look for two numbers with a product of -24 and a sum of -2:

$$y^2 - 2y - 24 = (y - 6)(y + 4).$$ Note that $(-6)(4) = -24$ and $-6 + 4 = -2$.   ▬

The next example illustrates a method for factoring a trinomial of the form $ax^2 + bx + c$, $a \neq 1$.

EXAMPLE 4   Factor: $3x^2 - 10x - 8$.

Solution   We use FOIL in reverse, looking for two binomial factors of the form ($\blacksquare x + \quad$)($\blacksquare x + \quad$). The product of the First terms is $3x^2$, the product of the Last terms is -8, and the sum of the Inner and Outer products is $-10x$. By trial and error, we determine the factorization:

$$3x^2 - 10x - 8 = (3x + 2)(x - 4).$$

Special Factorizations

We reverse the equation $(A + B)(A - B) = A^2 - B^2$ to factor a **difference of squares.**

$$A^2 - B^2 = (A + B)(A - B)$$

EXAMPLE 5   Factor each of the following.

a) $x^2 - 16$ **b)** $9a^2 - 25$ **c)** $6x^4 - 6y^4$

Solution

a) $x^2 - 16 = (x + 4)(x - 4)$ To check, multiply.

b) $9a^2 - 25 = (3a + 5)(3a - 5)$

c) $6x^4 - 6y^4 = 6(x^4 - y^4)$

$\qquad\qquad\quad = 6(x^2 + y^2)(x^2 - y^2)$

$\qquad\qquad\quad = 6(x^2 + y^2)(x + y)(x - y)$ Because none of these factors can be factored further, we have *factored completely.*

The rules for squaring binomials can be reversed to factor trinomial squares.

$$A^2 + 2AB + B^2 = (A + B)^2$$
$$A^2 - 2AB + B^2 = (A - B)^2$$

EXAMPLE 6   Factor each of the following.

a) $x^2 + 8x + 16$ **b)** $25y^2 - 30y + 9$

Solution

a) $x^2 + 8x + 16 = (x + 4)^2$ To check, multiply.

b) $25y^2 - 30y + 9 = (5y - 3)^2$

We can use the following rules to factor a **sum** or a **difference of cubes:**

$$A^3 + B^3 = (A + B)(A^2 - AB + B^2);$$
$$A^3 - B^3 = (A - B)(A^2 + AB + B^2).$$

These rules can be verified by multiplying.

EXAMPLE 7   Factor each of the following.

a) $x^3 + 27$

b) $16y^3 - 250$

Solution

a) $x^3 + 27 = x^3 + 3^3$
$$= (x + 3)(x^2 - 3x + 9)$$

b) $16y^3 - 250 = 2(8y^3 - 125)$
$$= 2[(2y)^3 - 5^3]$$
$$= 2(2y - 5)(4y^2 + 10y + 25)$$

Not all polynomials can be factored into polynomials with integer coefficients. An example is $x^2 - x + 7$. There are no real factors of 7 whose sum is -1. In such a case we say that the polynomial is "not factorable," or **prime**.

CONNECTING THE CONCEPTS

A STRATEGY FOR FACTORING

A. Always factor out the largest common factor.

B. Look at the number of terms.

Two terms: Try factoring as a difference of squares first. Next, try factoring as a sum or a difference of cubes. Do *not* try to factor a *sum* of squares.

Three terms: Try factoring as a trinomial square. Next, try trial and error, using FOIL in reverse.

Four or more terms: Try factoring by grouping and factoring out a common binomial factor.

C. Always *factor completely*. If a factor with more than one term can itself be factored further, do so.

Exercise Set R.4

Factor out a common factor.

1. $2x - 10$

2. $7y + 42$

3. $3x^4 - 9x^2$

4. $20y^2 - 5y^5$

5. $4a^2 - 12a + 16$

6. $6n^2 + 24n - 18$

7. $a(b - 2) + c(b - 2)$

8. $a(x^2 - 3) - 2(x^2 - 3)$

Factor by grouping.

9. $x^3 + 3x^2 + 6x + 18$

10. $3x^3 + x^2 - 18x - 6$

11. $y^3 - 3y^2 - 4y + 12$

12. $p^3 - 2p^2 - 9p + 18$

13. $x^3 - x^2 + x - 1$

14. $x^3 - x^2 - x + 1$

Factor the trinomial.

15. $p^2 + 6p + 8$

16. $w^2 - 7w + 10$

17. $2n^2 + 9n - 56$

18. $3y^2 + 7y - 20$

19. $12x^2 + 11x + 2$

20. $6x^2 - 7x - 20$

21. $y^4 - 4y^2 - 21$

22. $m^4 - m^2 - 90$

Factor the difference of squares.

23. $m^2 - 4$

24. $z^2 - 81$

25. $9x^2 - 25$

26. $16x^2 - 9$

27. $4xy^4 - 4xz^2$

28. $5x^4y - 5yz^4$

Factor the trinomial square.

29. $y^2 - 6y + 9$

30. $x^2 + 8x + 16$

31. $4z^2 + 12z + 9$

32. $9z^2 - 12z + 4$

33. $1 - 8x + 16x^2$

34. $1 + 10x + 25x^2$

Factor the sum or difference of cubes.

35. $x^3 + 8$

36. $y^3 - 64$

37. $m^3 - 1$

38. $n^3 + 216$

Factor completely.

39. $18a^2b - 15ab^2$

40. $4x^2y + 12xy^2$

41. $x^3 - 4x^2 + 5x - 20$

42. $z^3 + 3z^2 - 3z - 9$

43. $8x^2 - 32$

44. $6y^2 - 6$

45. $4y^2 - 5$

46. $16x^2 - 7$

47. $m^2 - 9n^2$

48. $25t^2 - 16$

49. $x^2 + 9x + 20$

50. $y^2 + y - 6$

51. $y^2 - 6y + 5$

52. $x^2 - 4x - 21$

53. $2a^2 + 9a + 4$

54. $3b^2 - b - 2$

55. $6x^2 + 7x - 3$

56. $8x^2 + 2x - 15$

57. $y^2 - 18y + 81$

58. $n^2 + 2n + 1$

59. $9z^2 - 24z + 16$

60. $4z^2 + 20z + 25$

61. $x^2y^2 - 14xy + 49$

62. $x^2y^2 - 16xy + 64$

63. $4ax^2 + 20ax - 56a$

64. $21x^2y + 2xy - 8y$

65. $3z^3 - 24$

66. $4t^3 + 108$

67. $16a^7b + 54ab^7$

68. $24a^2x^4 - 375a^8x$

Collaborative Discussion and Writing

69. Under what circumstances can $A^2 + B^2$ be factored?

70. Explain how the rule for factoring a sum of cubes can be used to factor a difference of cubes.

Synthesis

Factor.

71. $y^4 - 84 + 5y^2$

72. $11x^2 + x^4 - 80$

73. $y^2 - \frac{8}{49} + \frac{2}{7}y$

74. $t^2 - 0.27 + 0.6t$

75. $(x + h)^3 - x^3$

76. $(x + 0.01)^2 - x^2$

77. $(y - 4)^2 + 5(y - 4) - 24$

78. $6(2p + q)^2 - 5(2p + q) - 25$

Factor. Assume that variables in exponents represent natural numbers.

79. $x^{2n} + 5x^n - 24$

80. $4x^{2n} - 4x^n - 3$

81. $x^2 + ax + bx + ab$

82. $bdy^2 + ady + bcy + ac$

83. $25y^{2m} - (x^{2n} - 2x^n + 1)$

84. $x^{6a} - t^{3b}$

85. $(y - 1)^4 - (y - 1)^2$

86. $x^6 - 2x^5 + x^4 - x^2 + 2x - 1$

R.5

Rational Expressions

- *Determine the domain of a rational expression.*
- *Simplify rational expressions.*
- *Multiply, divide, add, and subtract rational expressions.*
- *Simplify complex rational expressions.*

A **rational expression** is the quotient of two polynomials. For example,

$$\frac{3}{5}, \quad \frac{2}{x-3}, \quad \text{and} \quad \frac{x^2-4}{x^2-4x-5}$$

are rational expressions.

The Domain of a Rational Expression

The **domain** of an algebraic expression is the set of all real numbers for which the expression is defined. Since division by zero is not defined, any number that makes the denominator zero is not in the domain of a rational expression.

EXAMPLE 1   Find the domain of each of the following.

a) $\dfrac{2}{x-3}$ **b)** $\dfrac{x^2-4}{x^2-4x-5}$

Solution

a) Since $x-3$ is 0 when $x=3$, the domain of $2/(x-3)$ is the set of all real numbers except 3.

b) To determine the domain of $(x^2-4)/(x^2-4x-5)$, we first factor the denominator:

$$\frac{x^2-4}{x^2-4x-5} = \frac{x^2-4}{(x+1)(x-5)}.$$

The factor $x+1$ is 0 when $x=-1$, and the factor $x-5$ is 0 when $x=5$. Since $(x+1)(x-5)=0$ when $x=-1$ or $x=5$, the domain is the set of all real numbers except -1 and 5.   ▬

We can describe the domains found in Example 1 using *set-builder notation*. For example, we write "The set of all real numbers x such that x is not equal to 3" as

$$\{x \,|\, x \text{ is a real number } and \ x \neq 3\}.$$

Similarly, we write "The set of all real numbers x such that x is not equal to -1 and x is not equal to 5" as

$$\{x \,|\, x \text{ is a real number } and \ x \neq -1 \ and \ x \neq 5\}.$$

Simplifying, Multiplying, and Dividing Rational Expressions

To simplify rational expressions, we use the fact that

$$\frac{a \cdot c}{b \cdot c} = \frac{a}{b} \cdot \frac{c}{c} = \frac{a}{b} \cdot 1 = \frac{a}{b}.$$

EXAMPLE 2   Simplify: $\dfrac{9x^2 + 6x - 3}{12x^2 - 12}$.

Solution

$$\frac{9x^2 + 6x - 3}{12x^2 - 12} = \frac{3(x + 1)(3x - 1)}{3 \cdot 4(x + 1)(x - 1)} \quad \begin{array}{l}\text{Factoring the numerator}\\ \text{and the denominator}\end{array}$$

$$= \frac{3(x + 1)}{3(x + 1)} \cdot \frac{3x - 1}{4(x - 1)} \quad \begin{array}{l}\text{Factoring the rational}\\ \text{expression}\end{array}$$

$$= 1 \cdot \frac{3x - 1}{4(x - 1)} \qquad \frac{3(x + 1)}{3(x + 1)} = 1$$

$$= \frac{3x - 1}{4(x - 1)} \qquad \text{Removing a factor of 1}$$

Canceling is a shortcut that is often used to remove a factor of 1.

EXAMPLE 3   Simplify each of the following.

a) $\dfrac{4x^3 + 16x^2}{2x^3 + 6x^2 - 8x}$

b) $\dfrac{2 - x}{x^2 + x - 6}$

Solution

a) $\dfrac{4x^3 + 16x^2}{2x^3 + 6x^2 - 8x} = \dfrac{2 \cdot 2 \cdot x \cdot x(x + 4)}{2 \cdot x(x + 4)(x - 1)} \quad \begin{array}{l}\text{Factoring the numerator}\\ \text{and the denominator}\end{array}$

$$= \frac{\cancel{2} \cdot 2 \cdot \cancel{x} \cdot x(x\!\!\!\diagup\!\!+4)}{\cancel{2} \cdot \cancel{x}(x\!\!\!\diagup\!\!+4)(x - 1)} \quad \begin{array}{l}\text{Removing a factor of 1:}\\ \dfrac{2x(x + 4)}{2x(x + 4)} = 1\end{array}$$

$$= \frac{2x}{x - 1}$$

b) $\dfrac{2 - x}{x^2 + x - 6} = \dfrac{2 - x}{(x + 3)(x - 2)} \quad \text{Factoring the denominator}$

$$= \frac{-1(x - 2)}{(x + 3)(x - 2)} \qquad 2 - x = -1(x - 2)$$

$$= \frac{-1(x\!\!\!\diagup\!\!-2)}{(x + 3)(x\!\!\!\diagup\!\!-2)} \qquad \text{Removing a factor of 1: } \dfrac{x - 2}{x - 2} = 1$$

$$= \frac{-1}{x + 3}, \text{ or } -\frac{1}{x + 3}$$

In Example 3(b), we saw that

$$\frac{2 - x}{x^2 + x - 6} \quad \text{and} \quad -\frac{1}{x + 3}$$

are **equivalent expressions.** This means that they have the same value for all numbers that are in *both* domains. Note that -3 is not in the domain of *either* expression, whereas 2 is in the domain of $-1/(x + 3)$ but not in the domain of $(2 - x)/(x^2 + x - 6)$ and thus is not in the domain of *both* expressions.

To multiply rational expressions, we multiply numerators and multiply denominators and, if possible, simplify the result. To divide rational expressions, we multiply the dividend by the reciprocal of the divisor and, if possible, simplify the result—that is,

$$\frac{a}{b} \cdot \frac{c}{d} = \frac{ac}{bd} \quad \text{and} \quad \frac{a}{b} \div \frac{c}{d} = \frac{a}{b} \cdot \frac{d}{c} = \frac{ad}{bc}.$$

EXAMPLE 4 Multiply or divide and simplify each of the following.

a) $\dfrac{x + 4}{x - 3} \cdot \dfrac{x^2 - 9}{x^2 - x - 2}$

b) $\dfrac{y^3 - 1}{y^2 - 1} \div \dfrac{y^2 + y + 1}{y^2 + 2y + 1}$

Solution

a) $\dfrac{x + 4}{x - 3} \cdot \dfrac{x^2 - 9}{x^2 - x - 2} = \dfrac{(x + 4)(x^2 - 9)}{(x - 3)(x^2 - x - 2)}$

　　　　　　　　　　　　Multiplying the numerators and the denominators

$$= \frac{(x + 4)(x + 3)(x - 3)}{(x - 3)(x - 2)(x + 1)}$$

Factoring and removing a factor of 1: $\dfrac{x - 3}{x - 3} = 1$

$$= \frac{(x + 4)(x + 3)}{(x - 2)(x + 1)}$$

b) $\dfrac{y^3 - 1}{y^2 - 1} \div \dfrac{y^2 + y + 1}{y^2 + 2y + 1} = \dfrac{y^3 - 1}{y^2 - 1} \cdot \dfrac{y^2 + 2y + 1}{y^2 + y + 1}$

Multiplying by the reciprocal of the divisor

$$= \frac{(y^3 - 1)(y^2 + 2y + 1)}{(y^2 - 1)(y^2 + y + 1)}$$

$$= \frac{(y - 1)(y^2 + y + 1)(y + 1)(y + 1)}{(y + 1)(y - 1)(y^2 + y + 1)}$$

Factoring and removing a factor of 1

$$= y + 1$$

Adding and Subtracting Rational Expressions

When rational expressions have the same denominator, we can add or subtract by adding or subtracting the numerators and retaining the common denominator. If the denominators differ, we must find equivalent rational expressions that have a common denominator.

In general, it is most efficient to find the **least common denominator (LCD)** of the expressions. To do so, we factor each denominator and form the product that uses each factor the greatest number of times it occurs in any factorization.

EXAMPLE 5 Add or subtract and simplify each of the following.

a) $\dfrac{x^2 - 4x + 4}{2x^2 - 3x + 1} + \dfrac{x + 4}{2x - 2}$

b) $\dfrac{x}{x^2 + 11x + 30} - \dfrac{5}{x^2 + 9x + 20}$

Solution

a) $\dfrac{x^2 - 4x + 4}{2x^2 - 3x + 1} + \dfrac{x + 4}{2x - 2}$

$= \dfrac{x^2 - 4x + 4}{(2x - 1)(x - 1)} + \dfrac{x + 4}{2(x - 1)}$ Factoring the denominators

$\qquad$ The LCD is $(2x - 1)(x - 1)(2)$, or $2(2x - 1)(x - 1)$.

$= \dfrac{x^2 - 4x + 4}{(2x - 1)(x - 1)} \cdot \dfrac{2}{2} + \dfrac{x + 4}{2(x - 1)} \cdot \dfrac{2x - 1}{2x - 1}$ Multiplying each term by 1 to get the LCD

$= \dfrac{2x^2 - 8x + 8}{(2x - 1)(x - 1)(2)} + \dfrac{2x^2 + 7x - 4}{2(x - 1)(2x - 1)}$

$= \dfrac{4x^2 - x + 4}{2(2x - 1)(x - 1)}$ Adding the numerators

b) $\dfrac{x}{x^2 + 11x + 30} - \dfrac{5}{x^2 + 9x + 20}$

$= \dfrac{x}{(x + 5)(x + 6)} - \dfrac{5}{(x + 5)(x + 4)}$ Factoring the denominators

$\qquad$ The LCD is $(x + 5)(x + 6)(x + 4)$.

$= \dfrac{x}{(x + 5)(x + 6)} \cdot \dfrac{x + 4}{x + 4} - \dfrac{5}{(x + 5)(x + 4)} \cdot \dfrac{x + 6}{x + 6}$

$\qquad\qquad\qquad\qquad$ Multiplying each term by 1 to get the LCD

$= \dfrac{x^2 + 4x}{(x + 5)(x + 6)(x + 4)} - \dfrac{5x + 30}{(x + 5)(x + 4)(x + 6)}$

$= \dfrac{x^2 + 4x - 5x - 30}{(x + 5)(x + 6)(x + 4)}$ Be sure to change the sign of *every* term in the numerator of the expression being subtracted: $-(5x + 30) = -5x - 30$

$= \dfrac{x^2 - x - 30}{(x + 5)(x + 6)(x + 4)}$

$= \dfrac{(x + 5)(x - 6)}{(x + 5)(x + 6)(x + 4)}$ Factoring and removing a factor of 1: $\dfrac{x + 5}{x + 5} = 1$

$= \dfrac{x - 6}{(x + 6)(x + 4)}$

Complex Rational Expressions

A **complex rational expression** has rational expressions in its numerator or its denominator or both.

To simplify a complex rational expression:

Method 1. Find the LCD of all the denominators within the complex rational expression. Then multiply by 1 using the LCD as the numerator and the denominator of the expression for 1.

Method 2. First add or subtract, if necessary, to get a single rational expression in the numerator and in the denominator. Then divide by multiplying by the reciprocal of the denominator.

EXAMPLE 6 Simplify: $\dfrac{\dfrac{1}{a} + \dfrac{1}{b}}{\dfrac{1}{a^3} + \dfrac{1}{b^3}}$.

Solution

Method 1. The LCD of the four rational expressions in the numerator and the denominator is $a^3 b^3$.

$$\frac{\dfrac{1}{a} + \dfrac{1}{b}}{\dfrac{1}{a^3} + \dfrac{1}{b^3}} = \frac{\dfrac{1}{a} + \dfrac{1}{b}}{\dfrac{1}{a^3} + \dfrac{1}{b^3}} \cdot \frac{a^3 b^3}{a^3 b^3} \qquad \text{Multiplying by 1}$$
$$\text{using } \frac{a^3 b^3}{a^3 b^3}$$

$$= \frac{\left(\dfrac{1}{a} + \dfrac{1}{b}\right)(a^3 b^3)}{\left(\dfrac{1}{a^3} + \dfrac{1}{b^3}\right)(a^3 b^3)}$$

$$= \frac{a^2 b^3 + a^3 b^2}{b^3 + a^3}$$

$$= \frac{a^2 b^2 (b + a)}{(b + a)(b^2 - ba + a^2)} \qquad \text{Factoring and removing a}$$
$$\text{factor of 1: } \frac{b + a}{b + a} = 1$$

$$= \frac{a^2 b^2}{b^2 - ba + a^2}$$

Method 2. We add in the numerator and in the denominator.

$$\frac{\dfrac{1}{a} + \dfrac{1}{b}}{\dfrac{1}{a^3} + \dfrac{1}{b^3}} = \frac{\dfrac{1}{a} \cdot \dfrac{b}{b} + \dfrac{1}{b} \cdot \dfrac{a}{a}}{\dfrac{1}{a^3} \cdot \dfrac{b^3}{b^3} + \dfrac{1}{b^3} \cdot \dfrac{a^3}{a^3}} \quad \longleftarrow \text{ The LCD is } ab.$$
$$\longleftarrow \text{ The LCD is } a^3 b^3.$$

Then

$$= \frac{\dfrac{b}{ab} + \dfrac{a}{ab}}{\dfrac{b^3}{a^3b^3} + \dfrac{a^3}{a^3b^3}}$$

$$= \frac{\dfrac{b+a}{ab}}{\dfrac{b^3+a^3}{a^3b^3}} \qquad \text{We have a single rational expression in both the numerator and the denominator.}$$

$$= \frac{b+a}{ab} \cdot \frac{a^3b^3}{b^3+a^3} \qquad \text{Multiplying by the reciprocal of the denominator}$$

$$= \frac{(b+a)(a)(b)(a^2b^2)}{(a)(b)(b+a)(b^2 - ba + a^2)}$$

$$= \frac{a^2b^2}{b^2 - ba + a^2}.$$

Exercise Set

Find the domain of the rational expression.

1. $-\dfrac{3}{4}$

2. $\dfrac{5}{8-x}$

3. $\dfrac{3x-3}{x(x-1)}$

4. $\dfrac{15x-10}{2x(3x-2)}$

5. $\dfrac{x+5}{x^2+4x-5}$

6. $\dfrac{(x^2-4)(x+1)}{(x+2)(x^2-1)}$

7. $\dfrac{7x^2 - 28x + 28}{(x^2-4)(x^2+3x-10)}$

8. $\dfrac{7x^2 + 11x - 6}{x(x^2-x-6)}$

Multiply or divide and, if possible, simplify.

9. $\dfrac{x^2-y^2}{(x-y)^2} \cdot \dfrac{1}{x+y}$

10. $\dfrac{r-s}{r+s} \cdot \dfrac{r^2-s^2}{(r-s)^2}$

11. $\dfrac{x^2-2x-35}{2x^3-3x^2} \cdot \dfrac{4x^3-9x}{7x-49}$

12. $\dfrac{x^2+2x-35}{3x^3-2x^2} \cdot \dfrac{9x^3-4x}{7x+49}$

13. $\dfrac{a^2-a-6}{a^2-7a+12} \cdot \dfrac{a^2-2a-8}{a^2-3a-10}$

14. $\dfrac{a^2-a-12}{a^2-6a+8} \cdot \dfrac{a^2+a-6}{a^2-2a-24}$

15. $\dfrac{m^2-n^2}{r+s} \div \dfrac{m-n}{r+s}$

16. $\dfrac{a^2-b^2}{x-y} \div \dfrac{a+b}{x-y}$

17. $\dfrac{3x+12}{2x-8} \div \dfrac{(x+4)^2}{(x-4)^2}$

18. $\dfrac{a^2-a-2}{a^2-a-6} \div \dfrac{a^2-2a}{2a+a^2}$

19. $\dfrac{x^2-y^2}{x^3-y^3} \cdot \dfrac{x^2+xy+y^2}{x^2+2xy+y^2}$

20. $\dfrac{c^3+8}{c^2-4} \div \dfrac{c^2-2c+4}{c^2-4c+4}$

21. $\dfrac{(x-y)^2-z^2}{(x+y)^2-z^2} \div \dfrac{x-y+z}{x+y-z}$

22. $\dfrac{(a+b)^2-9}{(a-b)^2-9} \cdot \dfrac{a-b-3}{a+b+3}$

Add or subtract and, if possible, simplify.

23. $\dfrac{3}{2a + 3} + \dfrac{2a}{2a + 3}$

24. $\dfrac{a - 3b}{a + b} + \dfrac{a + 5b}{a + b}$

25. $\dfrac{y}{y - 1} + \dfrac{2}{1 - y}$

26. $\dfrac{a}{a - b} + \dfrac{b}{b - a}$

27. $\dfrac{x}{2x - 3y} - \dfrac{y}{3y - 2x}$

28. $\dfrac{3a}{3a - 2b} - \dfrac{2a}{2b - 3a}$

29. $\dfrac{3}{x + 2} + \dfrac{2}{x^2 - 4}$

30. $\dfrac{5}{a - 3} - \dfrac{2}{a^2 - 9}$

31. $\dfrac{y}{y^2 - y - 20} - \dfrac{2}{y + 4}$

32. $\dfrac{6}{y^2 + 6y + 9} - \dfrac{5}{y + 3}$

33. $\dfrac{3}{x + y} + \dfrac{x - 5y}{x^2 - y^2}$

34. $\dfrac{a^2 + 1}{a^2 - 1} - \dfrac{a - 1}{a + 1}$

35. $\dfrac{9x + 2}{3x^2 - 2x - 8} + \dfrac{7}{3x^2 + x - 4}$

36. $\dfrac{3y}{y^2 - 7y + 10} - \dfrac{2y}{y^2 - 8y + 15}$

37. $\dfrac{5a}{a - b} + \dfrac{ab}{a^2 - b^2} + \dfrac{4b}{a + b}$

38. $\dfrac{6a}{a - b} - \dfrac{3b}{b - a} + \dfrac{5}{a^2 - b^2}$

39. $\dfrac{7}{x + 2} - \dfrac{x + 8}{4 - x^2} + \dfrac{3x - 2}{4 - 4x + x^2}$

40. $\dfrac{6}{x + 3} - \dfrac{x + 4}{9 - x^2} + \dfrac{2x - 3}{9 - 6x + x^2}$

41. $\dfrac{1}{x + 1} + \dfrac{x}{2 - x} + \dfrac{x^2 + 2}{x^2 - x - 2}$

42. $\dfrac{x - 1}{x - 2} - \dfrac{x + 1}{x + 2} - \dfrac{x - 6}{4 - x^2}$

Simplify.

43. $\dfrac{\dfrac{x^2 - y^2}{xy}}{\dfrac{x - y}{y}}$

44. $\dfrac{\dfrac{a - b}{b}}{\dfrac{a^2 - b^2}{ab}}$

45. $\dfrac{\dfrac{x}{y} - \dfrac{y}{x}}{\dfrac{1}{y} + \dfrac{1}{x}}$

46. $\dfrac{\dfrac{a}{b} - \dfrac{b}{a}}{\dfrac{1}{a} - \dfrac{1}{b}}$

47. $\dfrac{c + \dfrac{8}{c^2}}{1 + \dfrac{2}{c}}$

48. $\dfrac{a - \dfrac{a}{b}}{b - \dfrac{b}{a}}$

49. $\dfrac{x^2 + xy + y^2}{\dfrac{x^2}{y} - \dfrac{y^2}{x}}$

50. $\dfrac{\dfrac{a^2}{b} + \dfrac{b^2}{a}}{a^2 - ab + b^2}$

51. $\dfrac{a - a^{-1}}{a + a^{-1}}$

52. $\dfrac{x^{-1} + y^{-1}}{x^{-3} + y^{-3}}$

53. $\dfrac{\dfrac{1}{x - 3} + \dfrac{2}{x + 3}}{\dfrac{3}{x - 1} - \dfrac{4}{x + 2}}$

54. $\dfrac{\dfrac{5}{x + 1} - \dfrac{3}{x - 2}}{\dfrac{1}{x - 5} + \dfrac{2}{x + 2}}$

55. $\dfrac{\dfrac{a}{1 - a} + \dfrac{1 + a}{a}}{\dfrac{1 - a}{a} + \dfrac{a}{1 + a}}$

56. $\dfrac{\dfrac{1 - x}{x} + \dfrac{x}{1 + x}}{\dfrac{1 + x}{x} + \dfrac{x}{1 - x}}$

57. $\dfrac{\dfrac{1}{a^2} + \dfrac{2}{ab} + \dfrac{1}{b^2}}{\dfrac{1}{a^2} - \dfrac{1}{b^2}}$

58. $\dfrac{\dfrac{1}{x^2} - \dfrac{1}{y^2}}{\dfrac{1}{x^2} - \dfrac{2}{xy} + \dfrac{1}{y^2}}$

Collaborative Discussion and Writing

59. When adding or subtracting rational expressions, we can always find a common denominator by forming the product of all the denominators. Explain why it is usually preferable to find the least common denominator.

60. How would you determine which method to use for simplifying a particular complex rational expression?

Synthesis

Simplify.

61. $\dfrac{(x + h)^2 - x^2}{h}$

62. $\dfrac{\dfrac{1}{x + h} - \dfrac{1}{x}}{h}$

63. $\dfrac{(x + h)^3 - x^3}{h}$

64. $\dfrac{\dfrac{1}{(x + h)^2} - \dfrac{1}{x^2}}{h}$

65. $\left[\dfrac{\dfrac{x + 1}{x - 1} + 1}{\dfrac{x + 1}{x - 1} - 1} \right]^5$

66. $1 + \dfrac{1}{1 + \dfrac{1}{1 + \dfrac{1}{1 + \dfrac{1}{x}}}}$

Perform the indicated operations and, if possible, simplify.

67. $\dfrac{n(n + 1)(n + 2)}{2 \cdot 3} + \dfrac{(n + 1)(n + 2)}{2}$

68. $\dfrac{n(n + 1)(n + 2)(n + 3)}{2 \cdot 3 \cdot 4} + \dfrac{(n + 1)(n + 2)(n + 3)}{2 \cdot 3}$

69. $\dfrac{x^2 - 9}{x^3 + 27} \cdot \dfrac{5x^2 - 15x + 45}{x^2 - 2x - 3} + \dfrac{x^2 + x}{4 + 2x}$

70. $\dfrac{x^2 + 2x - 3}{x^2 - x - 12} \div \dfrac{x^2 - 1}{x^2 - 16} - \dfrac{2x + 1}{x^2 + 2x + 1}$

R.6

Radical Notation and Rational Exponents

- *Simplify radical expressions.*
- *Rationalize denominators or numerators in rational expressions.*
- *Convert between exponential and radical notation.*
- *Simplify expressions with rational exponents.*

A number c is said to be a **square root** of a if $c^2 = a$. Thus, 3 is a square root of 9, because $3^2 = 9$, and -3 is also a square root of 9, because $(-3)^2 = 9$. Similarly, 5 is a third root (called a **cube root**) of 125, because $5^3 = 125$. The number 125 has no other real-number cube root.

nth Root

A number c is said to be an ***nth* root** of a if $c^n = a$.

The symbol $\sqrt{a}$ denotes the nonnegative square root of a, and the symbol $\sqrt[3]{a}$ denotes the real-number cube root of a. The symbol $\sqrt[n]{a}$ denotes the nth root of a, that is, a number whose nth power is a. The symbol $\sqrt[n]{}$ is called a **radical**, and the expression under the radical is called the **radicand**. The number n (which is omitted when it is 2) is called the **index**. Examples of roots for $n = 3$, 4, and 2, respectively, are

$$\sqrt[3]{125}, \qquad \sqrt[4]{16}, \quad \text{and} \quad \sqrt{3600}.$$

Any real number has only one real-number odd root. Any positive number has two square roots, one positive and one negative. The same is true for fourth roots or roots with any even index. The positive root is called the **principal root**. When an expression such as $\sqrt{4}$ or $\sqrt[6]{23}$ is used, it is understood to represent the principal (nonnegative) root. To denote a negative root, we use $-\sqrt{4}$, $-\sqrt[6]{23}$, and so on.

EXAMPLE 1   Simplify each of the following.

a) $\sqrt{36}$ b) $-\sqrt{36}$

c) $\sqrt[5]{\dfrac{32}{243}}$ d) $\sqrt[3]{-8}$

e) $\sqrt[4]{-16}$

Solution

a) $\sqrt{36} = 6$, because $6^2 = 36$.

b) $-\sqrt{36} = -6$, because $6^2 = 36$ and $-(\sqrt{36}) = -(6) = -6$.

c) $\sqrt[5]{\dfrac{32}{243}} = \dfrac{2}{3}$, because $\left(\dfrac{2}{3}\right)^5 = \dfrac{2^5}{3^5} = \dfrac{32}{243}$.

d) $\sqrt[3]{-8} = -2$, because $(-2)^3 = -8$.

e) $\sqrt[4]{-16}$ is not a real number, because we cannot find a real number that can be raised to the fourth power to get -16.   ▬

We can generalize Example 1(e) and say that when a is negative and n is even, $\sqrt[n]{a}$ is not a real number. For example, $\sqrt{-4}$ and $\sqrt[4]{-81}$ are not real numbers.

Technology Connection

We can find $\sqrt{36}$ and $-\sqrt{36}$ in Example 1 using the square-root feature on the keypad of a graphing calculator, and we can use the cube-root feature to find $\sqrt[3]{-8}$. We can use the xth-root feature to find higher roots.

When we try to find $\sqrt[4]{-16}$ on a graphing calculator set in REAL mode, we get an error message indicating that the answer is nonreal.

Simplifying Radical Expressions

Consider the expression $\sqrt{(-3)^2}$. This is equivalent to $\sqrt{9}$, or 3. Similarly, $\sqrt{3^2} = \sqrt{9} = 3$. This illustrates the first of several properties of radicals, listed below.

Properties of Radicals

Let a and b be any real numbers or expressions for which the given roots exist. For any natural numbers m and n ($n \neq 1$):

1. If n is even, $\sqrt[n]{a^n} = |a|$.

2. If n is odd, $\sqrt[n]{a^n} = a$.

3. $\sqrt[n]{a} \cdot \sqrt[n]{b} = \sqrt[n]{ab}$.

4. $\sqrt[n]{\dfrac{a}{b}} = \dfrac{\sqrt[n]{a}}{\sqrt[n]{b}}$ ($b \neq 0$).

5. $\sqrt[n]{a^m} = (\sqrt[n]{a})^m$.

EXAMPLE 2   Simplify each of the following.

a) $\sqrt{(-5)^2}$ **b)** $\sqrt[3]{(-5)^3}$ **c)** $\sqrt[4]{4} \cdot \sqrt[4]{5}$ **d)** $\sqrt{50}$

e) $\dfrac{\sqrt{72}}{\sqrt{6}}$ **f)** $\sqrt[3]{8^5}$ **g)** $\sqrt{216x^5y^3}$ **h)** $\sqrt{\dfrac{x^2}{16}}$

Solution

a) $\sqrt{(-5)^2} = |-5| = 5$ Using Property 1

b) $\sqrt[3]{(-5)^3} = -5$ Using Property 2

c) $\sqrt[4]{4} \cdot \sqrt[4]{5} = \sqrt[4]{4 \cdot 5} = \sqrt[4]{20}$ Using Property 3

d) $\sqrt{50} = \sqrt{25 \cdot 2} = \sqrt{25} \cdot \sqrt{2} = 5\sqrt{2}$ Using Property 3

e) $\dfrac{\sqrt{72}}{\sqrt{6}} = \sqrt{\dfrac{72}{6}}$ Using Property 4

$\quad\quad = \sqrt{12} = \sqrt{4 \cdot 3} = \sqrt{4}\,\sqrt{3}$ Using Property 3

$\quad\quad = 2\sqrt{3}$

f) $\sqrt[3]{8^5} = (\sqrt[3]{8})^5$ Using Property 5

$\quad\quad = 2^5 = 32$

g) $\sqrt{216x^5y^3} = \sqrt{36 \cdot 6 \cdot x^4 \cdot x \cdot y^2 \cdot y}$

$\quad\quad = \sqrt{36x^4y^2}\,\sqrt{6xy}$ Using Property 3

$\quad\quad = |6x^2y|\,\sqrt{6xy}$ Using Property 1

$\quad\quad = 6x^2|y|\,\sqrt{6xy}$ $6x^2$ cannot be negative, so absolute-value signs are not needed for it.

h) $\sqrt{\dfrac{x^2}{16}} = \dfrac{\sqrt{x^2}}{\sqrt{16}}$ Using Property 4

       $= \dfrac{|x|}{4}$ Using Property 1

In many situations, radicands are never formed by raising negative quantities to even powers. In such cases, absolute-value notation is not required. For this reason, **we will henceforth assume that no radicands are formed by raising negative quantities to even powers.** For example, we will write $\sqrt{x^2} = x$ and $\sqrt[4]{a^5b} = a\sqrt[4]{ab}$.

Radical expressions with the same index and the same radicand can be added or subtracted.

EXAMPLE 3   Perform the operations indicated.

a) $3\sqrt{8x^2} - 5\sqrt{2x^2}$

b) $(4\sqrt{3} + \sqrt{2})(\sqrt{3} - 5\sqrt{2})$

Solution

a) $3\sqrt{8x^2} - 5\sqrt{2x^2} = 3\sqrt{4x^2 \cdot 2} - 5\sqrt{x^2 \cdot 2}$

                 $= 3 \cdot 2x\sqrt{2} - 5x\sqrt{2}$

                 $= 6x\sqrt{2} - 5x\sqrt{2}$

                 $= (6x - 5x)\sqrt{2}$ Using the distributive property

                 $= x\sqrt{2}$

b) $(4\sqrt{3} + \sqrt{2})(\sqrt{3} - 5\sqrt{2}) = 4(\sqrt{3})^2 - 20\sqrt{6} + \sqrt{6} - 5(\sqrt{2})^2$

                                                         Multiplying

                                 $= 4 \cdot 3 + (-20 + 1)\sqrt{6} - 5 \cdot 2$

                                 $= 12 - 19\sqrt{6} - 10$

                                 $= 2 - 19\sqrt{6}$

An Application

The Pythagorean theorem relates the lengths of the sides of a right triangle. The side opposite the triangle's right angle is called the **hypotenuse**. The other sides are the **legs**.

The Pythagorean Theorem

The sum of the squares of the lengths of the legs of a right triangle is equal to the square of the length of the hypotenuse:

$$a^2 + b^2 = c^2.$$

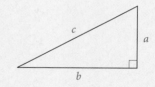

EXAMPLE 4 A surveyor places poles at points A, B, and C in order to measure the distance across a pond. The distances AC and BC are measured as shown. Find the distance AB across the pond.

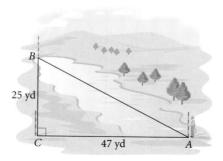

Solution We see that the lengths of the legs of a right triangle are given. Thus we use the Pythagorean theorem to find the length of the hypotenuse:

$$c^2 = a^2 + b^2$$
$$c = \sqrt{a^2 + b^2} \qquad \text{Solving for } c$$
$$= \sqrt{25^2 + 47^2}$$
$$= \sqrt{625 + 2209}$$
$$= \sqrt{2834}$$
$$\approx 53.2.$$

The distance across the pond is about 53.2 yd.

Rationalizing Denominators or Numerators

There are times when we need to remove the radicals in a denominator or a numerator. This is called **rationalizing the denominator** or **rationalizing the numerator.** It is done by multiplying by 1 in such a way as to obtain a perfect nth power.

EXAMPLE 5 Rationalize the denominator of each of the following.

a) $\sqrt{\dfrac{3}{2}}$

b) $\dfrac{\sqrt[3]{7}}{\sqrt[3]{9}}$

Solution

a) $\sqrt{\dfrac{3}{2}} = \sqrt{\dfrac{3}{2} \cdot \dfrac{2}{2}} = \sqrt{\dfrac{6}{4}} = \dfrac{\sqrt{6}}{\sqrt{4}} = \dfrac{\sqrt{6}}{2}$

b) $\dfrac{\sqrt[3]{7}}{\sqrt[3]{9}} = \dfrac{\sqrt[3]{7}}{\sqrt[3]{9}} \cdot \dfrac{\sqrt[3]{3}}{\sqrt[3]{3}} = \dfrac{\sqrt[3]{21}}{\sqrt[3]{27}} = \dfrac{\sqrt[3]{21}}{3}$

Pairs of expressions of the form $a\sqrt{b} + c\sqrt{d}$ and $a\sqrt{b} - c\sqrt{d}$ are called **conjugates**. The product of such a pair contains no radicals and can be used to rationalize a denominator or a numerator.

Technology Connection

We can perform the computation in Example 4 on a graphing calculator.

$$\sqrt{(25^2 + 47^2)}$$
$$53.23532662$$

EXAMPLE 6 Rationalize the numerator: $\dfrac{\sqrt{x} - \sqrt{y}}{5}$.

Solution

$$\frac{\sqrt{x} - \sqrt{y}}{5} = \frac{\sqrt{x} - \sqrt{y}}{5} \cdot \frac{\sqrt{x} + \sqrt{y}}{\sqrt{x} + \sqrt{y}} \qquad \text{The conjugate of } \sqrt{x} - \sqrt{y} \text{ is } \sqrt{x} + \sqrt{y}.$$

$$= \frac{(\sqrt{x})^2 - (\sqrt{y})^2}{5\sqrt{x} + 5\sqrt{y}}$$

$$= \frac{x - y}{5\sqrt{x} + 5\sqrt{y}}$$

Rational Exponents

We are motivated to define *rational exponents* so that the properties for integer exponents hold for them as well. For example, we must have

$$a^{1/2} \cdot a^{1/2} = a^{1/2 + 1/2} = a^1 = a.$$

Thus we are led to define $a^{1/2}$ to mean $\sqrt{a}$. Similarly, $a^{1/n}$ would mean $\sqrt[n]{a}$. Again, if the laws of exponents are to hold, we must have

$$(a^{1/n})^m = (a^m)^{1/n} = a^{m/n}.$$

Thus we are led to define $a^{m/n}$ to mean $(\sqrt[n]{a})^m$, or, equivalently, $\sqrt[n]{a^m}$.

Rational Exponents

For any real number a and any natural numbers m and n for which $\sqrt[n]{a}$ exists,

$$a^{1/n} = \sqrt[n]{a},$$

$$a^{m/n} = \sqrt[n]{a^m} = (\sqrt[n]{a})^m, \quad \text{and}$$

$$a^{-m/n} = \frac{1}{a^{m/n}}.$$

We can use the definition of rational exponents to convert between radical and exponential notation.

EXAMPLE 7 Convert to radical notation and, if possible, simplify each of the following.

a) $7^{3/4}$ **b)** $8^{-5/3}$

c) $m^{1/6}$ **d)** $(-32)^{2/5}$

Solution

a) $7^{3/4} = \sqrt[4]{7^3}$, or $(\sqrt[4]{7})^3$

b) $8^{-5/3} = \dfrac{1}{8^{5/3}} = \dfrac{1}{(\sqrt[3]{8})^5} = \dfrac{1}{2^5} = \dfrac{1}{32}$

c) $m^{1/6} = \sqrt[6]{m}$

d) $(-32)^{2/5} = \sqrt[5]{(-32)^2} = \sqrt[5]{1024} = 4,$ or
$\quad (-32)^{2/5} = (\sqrt[5]{-32})^2 = (-2)^2 = 4$

Technology Connection

Some graphing calculators will return an error message when an expression of the form $a^{m/n}$, with $a < 0$ and $m > 1$, such as $(-32)^{2/5}$, is entered. In this case, we enter the expression as $((a)^m)^{1/n}$, or $((a)^{1/n})^m$. For $(-32)^{2/5}$, for example, we enter $((-32)^2)^{1/5}$, or $((-32)^{1/5})^2$.

EXAMPLE 8 Convert to exponential notation and, if possible, simplify each of the following.

a) $(\sqrt[4]{7xy})^5$ **b)** $\sqrt[6]{x^3}$ **c)** $\sqrt[3]{\sqrt{7}}$

Solution

a) $(\sqrt[4]{7xy})^5 = (7xy)^{5/4}$
b) $\sqrt[6]{x^3} = x^{3/6} = x^{1/2}$
c) $\sqrt[3]{\sqrt{7}} = \sqrt[3]{7^{1/2}} = (7^{1/2})^{1/3} = 7^{1/6}$

We can use the laws of exponents to simplify exponential and radical expressions.

EXAMPLE 9 Simplify and then, if appropriate, write radical notation for each of the following.

a) $x^{5/6} \cdot x^{2/3}$ **b)** $(x + 3)^{5/2}(x + 3)^{-1/2}$

Solution

a) $x^{5/6} \cdot x^{2/3} = x^{5/6+2/3} = x^{9/6} = x^{3/2} = \sqrt{x^3} = \sqrt{x^2}\sqrt{x} = x\sqrt{x}$
b) $(x + 3)^{5/2}(x + 3)^{-1/2} = (x + 3)^{5/2-1/2} = (x + 3)^2$

Technology Connection

We can add and subtract rational exponents on a graphing calculator. The FRAC feature from the MATH menu allows us to express the result as a fraction. The addition of the exponents in Example 9(a) is shown here.

```
5/6+2/3▶Frac
                3/2
```

EXAMPLE 10 Write an expression containing a single radical: $a^{1/2}b^{5/6}$.

Solution

$$a^{1/2}b^{5/6} = a^{3/6}b^{5/6} = (a^3b^5)^{1/6} = \sqrt[6]{a^3b^5}$$

Exercise Set R.6

Simplify. Assume that variables can represent any real number.

1. $\sqrt{(-11)^2}$
2. $\sqrt{(-1)^2}$
3. $\sqrt{16y^2}$
4. $\sqrt{36t^2}$
5. $\sqrt{(b+1)^2}$
6. $\sqrt{(2c-3)^2}$
7. $\sqrt[3]{-27x^3}$
8. $\sqrt[3]{-8y^3}$
9. $\sqrt{x^2-4x+4}$
10. $\sqrt{x^2+16x+64}$
11. $\sqrt[5]{32}$
12. $\sqrt[5]{-32}$
13. $\sqrt{180}$
14. $\sqrt{48}$
15. $\sqrt[3]{54}$
16. $\sqrt[3]{135}$
17. $\sqrt{128c^2d^4}$
18. $\sqrt{162c^4d^6}$
19. $\sqrt[4]{48x^6y^4}$
20. $\sqrt[4]{243m^5n^{10}}$

Simplify. Assume that no radicands were formed by raising negative quantities to even powers.

21. $\sqrt{2x^3y}\,\sqrt{12xy}$
22. $\sqrt{3y^4z}\,\sqrt{20z}$
23. $\sqrt[3]{3x^2y}\,\sqrt[3]{36x}$
24. $\sqrt[5]{8x^3y^4}\,\sqrt[5]{4x^4y}$
25. $\sqrt[3]{2(x+4)}\,\sqrt[3]{4(x+4)^4}$
26. $\sqrt[3]{4(x+1)^2}\,\sqrt[3]{18(x+1)^2}$
27. $\sqrt[6]{\dfrac{m^{12}n^{24}}{64}}$
28. $\sqrt[8]{\dfrac{m^{16}n^{24}}{2^8}}$
29. $\dfrac{\sqrt[3]{40m}}{\sqrt[3]{5m}}$
30. $\dfrac{\sqrt{40xy}}{\sqrt{8x}}$
31. $\dfrac{\sqrt[3]{3x^2}}{\sqrt[3]{24x^5}}$
32. $\dfrac{\sqrt{128a^2b^4}}{\sqrt{16ab}}$
33. $\sqrt[3]{\dfrac{64a^4}{27b^3}}$
34. $\sqrt{\dfrac{9x^7}{16y^8}}$
35. $\sqrt{\dfrac{7x^3}{36y^6}}$
36. $\sqrt[3]{\dfrac{2yz}{250z^4}}$
37. $9\sqrt{50}+6\sqrt{2}$
38. $11\sqrt{27}-4\sqrt{3}$
39. $8\sqrt{2x^2}-6\sqrt{20x}-5\sqrt{8x^2}$
40. $2\sqrt[3]{8x^2}+5\sqrt[3]{27x^2}-3\sqrt{x^3}$
41. $(\sqrt{3}-\sqrt{2})(\sqrt{3}+\sqrt{2})$
42. $(\sqrt{8}+2\sqrt{5})(\sqrt{8}-2\sqrt{5})$
43. $(1+\sqrt{3})^2$
44. $(\sqrt{2}-5)^2$
45. $(2\sqrt{3}+\sqrt{5})(\sqrt{3}-3\sqrt{5})$
46. $(\sqrt{6}-4\sqrt{7})(3\sqrt{6}+2\sqrt{7})$

47. An airplane is flying at an altitude of 3700 ft. The slanted distance directly to the airport is 14,200 ft. How far horizontally is the airplane from the airport?

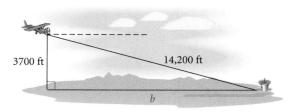

3700 ft   14,200 ft

b

48. During a summer heat wave, a 2-mi bridge expands 2 ft in length. Assuming that the bulge occurs straight up the middle, estimate the height of the bulge. (In reality, bridges are built with expansion joints to control such buckling.)

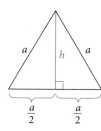

49. An *equilateral triangle* is shown below.

a h a

$\dfrac{a}{2}$ $\dfrac{a}{2}$

a) Find an expression for its height h in terms of a.
b) Find an expression for its area A in terms of a.

50. An isosceles right triangle has legs of length s. Find an expression for the length of the hypotenuse in terms of s.

51. The diagonal of a square has length $8\sqrt{2}$. Find the length of a side of the square.

52. The area of square $PQRS$ is 100 ft^2, and A, B, C, and D are the midpoints of the sides. Find the area of square $ABCD$.

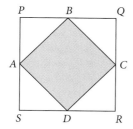

Rationalize the denominator.

53. $\sqrt{\dfrac{2}{3}}$

54. $\sqrt{\dfrac{3}{7}}$

55. $\dfrac{\sqrt[3]{5}}{\sqrt[3]{4}}$

56. $\dfrac{\sqrt[3]{7}}{\sqrt[3]{25}}$

57. $\sqrt[3]{\dfrac{16}{9}}$

58. $\sqrt[3]{\dfrac{3}{5}}$

59. $\dfrac{6}{3+\sqrt{5}}$

60. $\dfrac{2}{\sqrt{3}-1}$

61. $\dfrac{6}{\sqrt{m}-\sqrt{n}}$

62. $\dfrac{3}{\sqrt{v}+\sqrt{w}}$

Rationalize the numerator.

63. $\dfrac{\sqrt{12}}{5}$

64. $\dfrac{\sqrt{50}}{3}$

65. $\sqrt[3]{\dfrac{7}{2}}$

66. $\sqrt[3]{\dfrac{2}{5}}$

67. $\dfrac{\sqrt{11}}{\sqrt{3}}$

68. $\dfrac{\sqrt{5}}{\sqrt{2}}$

69. $\dfrac{9-\sqrt{5}}{3-\sqrt{3}}$

70. $\dfrac{8-\sqrt{6}}{5-\sqrt{2}}$

71. $\dfrac{\sqrt{a}+\sqrt{b}}{3a}$

72. $\dfrac{\sqrt{p}-\sqrt{q}}{1+\sqrt{q}}$

Convert to radical notation and simplify.

73. $x^{3/4}$

74. $y^{2/5}$

75. $16^{3/4}$

76. $4^{7/2}$

77. $125^{-1/3}$

78. $32^{-4/5}$

79. $a^{5/4}b^{-3/4}$

80. $x^{2/5}y^{-1/5}$

Convert to exponential notation and simplify.

81. $(\sqrt[4]{13})^5$

82. $\sqrt[5]{17^3}$

83. $\sqrt[3]{20^2}$

84. $(\sqrt[5]{12})^4$

85. $\sqrt[3]{\sqrt{11}}$

86. $\sqrt[5]{\sqrt[4]{7}}$

87. $\sqrt{5}\,\sqrt[3]{5}$

88. $\sqrt[3]{2}\,\sqrt{2}$

89. $\sqrt[5]{32^2}$

90. $\sqrt[3]{64^2}$

Simplify and then, if appropriate, write radical notation.

91. $(2a^{3/2})(4a^{1/2})$

92. $(3a^{5/6})(8a^{2/3})$

93. $\left(\dfrac{x^6}{9b^{-4}}\right)^{1/2}$

94. $\left(\dfrac{x^{2/3}}{4y^{-2}}\right)^{1/2}$

95. $\dfrac{x^{2/3}y^{5/6}}{x^{-1/3}y^{1/2}}$

96. $\dfrac{a^{1/2}b^{5/8}}{a^{1/4}b^{3/8}}$

Write an expression containing a single radical and simplify.

97. $\sqrt[3]{6}\,\sqrt{2}$

98. $\sqrt{2}\,\sqrt[4]{8}$

99. $\sqrt[4]{xy}\,\sqrt[3]{x^2y}$

100. $\sqrt[3]{ab^2}\,\sqrt{ab}$

101. $\sqrt[3]{a^4\sqrt{a^3}}$

102. $\sqrt{a^3\sqrt[3]{a^2}}$

103. $\dfrac{\sqrt{(a+x)^3}\,\sqrt[3]{(a+x)^2}}{\sqrt[4]{a+x}}$

104. $\dfrac{\sqrt[4]{(x+y)^2}\,\sqrt[3]{x+y}}{\sqrt{(x+y)^3}}$

Collaborative Discussion and Writing

105. Explain how you would convince a classmate that $\sqrt{a+b}$ is not equivalent to $\sqrt{a}+\sqrt{b}$, for positive real numbers a and b.

106. Explain how you would determine whether $10\sqrt{26}-50$ is positive or negative without carrying out the actual computation.

Synthesis

Simplify.

107. $\sqrt{1+x^2}+\dfrac{1}{\sqrt{1+x^2}}$

108. $\sqrt{1-x^2}-\dfrac{x^2}{2\sqrt{1-x^2}}$

109. $(\sqrt{a^{\sqrt{a}}})^{\sqrt{a}}$

110. $(2a^3b^{5/4}c^{1/7})^4 \div (54a^{-2}b^{2/3}c^{6/5})^{-1/3}$

Chapter Summary and Review

Important Properties and Formulas

Properties of the Real Numbers

Commutative: $a + b = b + a;$
     $ab = ba$

Associative: $a + (b + c) =$
     $(a + b) + c;$
     $a(bc) = (ab)c$

Additive Identity: $a + 0 = 0 + a = a$

Additive Inverse: $-a + a =$
     $a + (-a) = 0$

Multiplicative Identity: $a \cdot 1 = 1 \cdot a = a$

Multiplicative Inverse: $a \cdot \dfrac{1}{a} = 1 \quad (a \neq 0)$

Distributive: $a(b + c) = ab + ac$

Absolute Value

For any real number a,

$$|a| = \begin{cases} a, & \text{if } a \geq 0, \\ -a, & \text{if } a < 0. \end{cases}$$

Properties of Exponents

For any real numbers a and b and any integers m and n, assuming 0 is not raised to a nonpositive power:

The Product Rule: $a^m \cdot a^n = a^{m+n}$

The Quotient Rule: $\dfrac{a^m}{a^n} = a^{m-n} \quad (a \neq 0)$

The Power Rule: $(a^m)^n = a^{mn}$

Raising a Product to a Power:
   $(ab)^m = a^m b^m$

Raising a Quotient to a Power:
$$\left(\frac{a}{b}\right)^m = \frac{a^m}{b^m} \quad (b \neq 0)$$

Compound Interest Formula

$$A = P\left(1 + \frac{i}{n}\right)^{nt}$$

Special Products of Binomials

$$(A + B)^2 = A^2 + 2AB + B^2$$
$$(A - B)^2 = A^2 - 2AB + B^2$$
$$(A + B)(A - B) = A^2 - B^2$$

Sum or Difference of Cubes

$$A^3 + B^3 = (A + B)(A^2 - AB + B^2)$$
$$A^3 - B^3 = (A - B)(A^2 + AB + B^2)$$

Properties of Radicals

Let a and b be any real numbers or expressions for which the given roots exist. For any natural numbers m and n ($n \neq 1$):

If n is even, $\sqrt[n]{a^n} = |a|$.

If n is odd, $\sqrt[n]{a^n} = a$.

$\sqrt[n]{a} \cdot \sqrt[n]{b} = \sqrt[n]{ab}$.

$$\sqrt[n]{\frac{a}{b}} = \frac{\sqrt[n]{a}}{\sqrt[n]{b}} \quad (b \neq 0).$$

$\sqrt[n]{a^m} = (\sqrt[n]{a})^m$.

Rational Exponents

For any real number a and any natural numbers m and n for which $\sqrt[n]{a}$ exists,

$$a^{1/n} = \sqrt[n]{a},$$
$$a^{m/n} = \sqrt[n]{a^m} = (\sqrt[n]{a})^m, \quad \text{and}$$
$$a^{-m/n} = \frac{1}{a^{m/n}}.$$

Pythagorean Theorem

$$a^2 + b^2 = c^2$$

REVIEW EXERCISES

In Exercises 1–6, consider the numbers -43.89, 12, -3, $-\frac{1}{5}$, $\sqrt{7}$, $\sqrt[3]{10}$, -1, $-\frac{4}{3}$, $7\frac{2}{3}$, -19, 31, 0.

1. Which are integers?

2. Which are natural numbers?

3. Which are rational numbers?

4. Which are real numbers?

5. Which are irrational numbers?

6. Which are whole numbers?

7. Write interval notation for $\{x \mid -3 \le x < 5\}$.

Simplify.

8. $|-3.5|$ 9. $|16|$

10. Find the distance between -7 and 3 on the number line.

Calculate.

11. $5^3 - [2(4^2 - 3^2 - 6)]^3$ 12. $\dfrac{3^4 - (6 - 7)^4}{2^3 - 2^4}$

Convert to decimal notation.

13. 3.261×10^6 14. 4.1×10^{-4}

Convert to scientific notation.

15. 0.01432 16. $43{,}210$

Calculate. Write the answer using scientific notation.

17. $\dfrac{2.5 \times 10^{-8}}{3.2 \times 10^{13}}$

18. $(8.4 \times 10^{-17})(6.5 \times 10^{-16})$

Simplify.

19. $(7a^2 b^4)(-2a^{-4} b^3)$ 20. $\dfrac{54 x^6 y^{-4} z^2}{9 x^{-3} y^2 z^{-4}}$

21. $\sqrt[4]{81}$ 22. $\sqrt[5]{-32}$

23. $\dfrac{b - a^{-1}}{a - b^{-1}}$ 24. $\dfrac{\dfrac{x^2}{y} + \dfrac{y^2}{x}}{y^2 - xy + x^2}$

25. $(\sqrt{3} - \sqrt{7})(\sqrt{3} + \sqrt{7})$

26. $(5x^2 - \sqrt{2})^2$ 27. $8\sqrt{5} + \dfrac{25}{\sqrt{5}}$

28. $(x + t)(x^2 - xt + t^2)$ 29. $(5a + 4b)(2a - 3b)$

30. $(5xy^4 - 7xy^2 + 4x^2 - 3) -$
$(-3xy^4 + 2xy^2 - 2y + 4)$

Factor.

31. $x^3 + 2x^2 - 3x - 6$ 32. $12a^3 - 27ab^4$

33. $24x + 144 + x^2$ 34. $9x^3 + 35x^2 - 4x$

35. $8x^3 - 1$ 36. $27x^6 + 125y^6$

37. $6x^3 + 48$ 38. $4x^3 - 4x^2 - 9x + 9$

39. $9x^2 - 30x + 25$ 40. $18x^2 - 3x + 6$

41. $a^2 b^2 - ab - 6$

42. Divide and simplify:

$$\frac{3x^2 - 12}{x^2 + 4x + 4} \div \frac{x - 2}{x + 2}.$$

43. Subtract and simplify:

$$\frac{x}{x^2 + 9x + 20} - \frac{4}{x^2 + 7x + 12}.$$

Write an expression containing a single radical.

44. $\sqrt{y^5}\ \sqrt[3]{y^2}$ 45. $\dfrac{\sqrt{(a + b)^3}\ \sqrt[3]{a + b}}{\sqrt[6]{(a + b)^7}}$

46. Convert to radical notation: $b^{7/5}$.

47. Convert to exponential notation and simplify:

$$\sqrt[8]{\frac{m^{32} n^{16}}{3^8}}.$$

48. Rationalize the denominator:

$$\frac{\sqrt{x} - \sqrt{y}}{\sqrt{x} + \sqrt{y}}.$$

49. How long is a guy wire that reaches from the top of a 17-ft pole to a point on the ground 8 ft from the bottom of the pole?

Collaborative Discussion and Writing

50. Anya says that $15 - 6 \div 3 \cdot 4$ is 12. What mistake is she making?

51. A calculator indicates that $4^{21} = 4.398046511 \times 10^{12}$. How can you tell that this is an approximation?

Synthesis

Multiply. Assume that all exponents are integers.

52. $(x^n + 10)(x^n - 4)$

53. $(t^a + t^{-a})^2$

54. $(y^b - z^c)(y^b + z^c)$

55. $(a^n - b^n)^3$

Factor.

56. $y^{2n} + 16y^n + 64$

57. $x^{2t} - 3x^t - 28$

58. $m^{6n} - m^{3n}$

R Chapter Test

1. Consider the numbers

$$-8, \tfrac{11}{3}, \sqrt{15}, 0, -5.49, 36, \sqrt[3]{7}, 10\tfrac{1}{6}.$$

 a) Which are integers?
 b) Which are rational numbers?

2. Simplify: $|-1.2xy|$.

3. Write interval notation for $\{x \mid -3 < x \le 6\}$.

4. Compute: $32 \div 2^3 - 12 \div 4 \cdot 3$.

5. Compute and write scientific notation for the answer:

$$\frac{2.7 \times 10^4}{3.6 \times 10^{-3}}.$$

Simplify.

6. $(-3a^5b^{-4})(5a^{-1}b^3)$

7. $(3x^4 - 2x^2 + 6x) - (5x^3 - 3x^2 + x)$

8. $(x + 3)(2x - 5)$ **9.** $(2y - 1)^2$

10. $3\sqrt{75} + 2\sqrt{27}$ **11.** $\dfrac{\dfrac{x}{y} - \dfrac{y}{x}}{x + y}$

Factor.

12. $2n^2 + 5n - 12$

13. $8x^2 - 18$

14. $m^3 - 8$

15. Multiply and simplify:

$$\frac{x^2 + x - 6}{x^2 + 8x + 15} \cdot \frac{x^2 - 25}{x^2 - 4x + 4}.$$

16. Subtract and simplify: $\dfrac{x}{x^2 - 1} - \dfrac{3}{x^2 + 4x - 5}$.

17. Convert to radical notation: $t^{5/7}$.

18. Rationalize the denominator: $\dfrac{5}{7 - \sqrt{3}}$.

19. How long is a guy wire that reaches from the top of a 12-ft pole to a point on the ground 5 ft from the bottom of the pole?

Synthesis

20. Multiply: $(x - y - 1)^2$.

Graphs, Functions, and Models

1

*G*raphs help us *see* relationships among quantities. In this chapter, we graph many kinds of equations, emphasizing those that are straight lines. We also introduce the concept of a function and present models of data that enhance the relevance of functions in the real world. For linear equations and functions, we also develop the concepts of slope and intercepts.

APPLICATION

People in the United States create more solid waste than people in any other nation (*Source: Parade Magazine,* June 13, 1999).

YEAR	NUMBER OF TONS (IN MILLIONS)
1960	88
1970	121
1980	152
1990	205
1997	217

These data in the table are modeled with the linear function

$$f(x) = 3.65x - 7069.37,$$

where x is the year and y is the amount of solid waste, in millions of tons. We can use this function to predict amounts of solid waste in future years.

This problem appears as Exercise 37 in Section 1.4.

1.1

Introduction to Graphing

- *Plot points.*
- *Determine whether an ordered pair is a solution of an equation.*
- *Graph equations.*
- *Find the distance between two points in the plane and find the midpoint of a segment.*
- *Find an equation of a circle with a given center and radius, and given an equation of a circle, find the center and the radius.*
- *Graph equations of circles.*

Graphs

Graphs provide a means of displaying, interpreting, and analyzing data in a visual format. It is not uncommon to open a newspaper or magazine and encounter graphs. Shown below are examples of bar, circle, and line graphs.

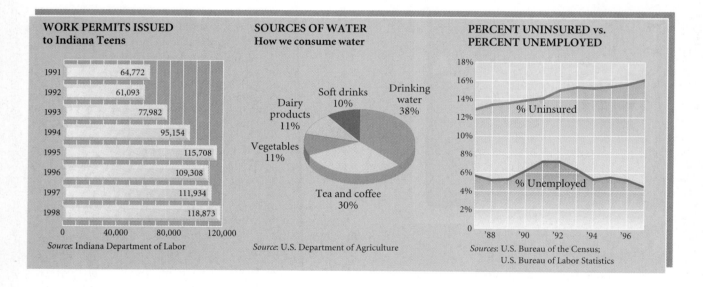

WORK PERMITS ISSUED to Indiana Teens

1991	64,772
1992	61,093
1993	77,982
1994	95,154
1995	115,708
1996	109,308
1997	111,934
1998	118,873

Source: Indiana Department of Labor

SOURCES OF WATER How we consume water

Drinking water 38%
Soft drinks 10%
Dairy products 11%
Vegetables 11%
Tea and coffee 30%

Source: U.S. Department of Agriculture

PERCENT UNINSURED vs. PERCENT UNEMPLOYED

% Uninsured
% Unemployed

Sources: U.S. Bureau of the Census; U.S. Bureau of Labor Statistics

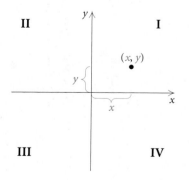

Many real-world situations can be modeled, or described mathematically, using equations in which two variables appear. We use a plane to graph a pair of numbers. To locate points on a plane, we use two perpendicular number lines, called **axes**, which intersect at (0, 0). We call this point the **origin**. The horizontal axis is called the **x-axis**, and the vertical axis is called the **y-axis**. (Other variables, such as *a* and *b*, can also be used.) The axes divide the plane into four regions, called **quadrants**, denoted by Roman numerals and numbered counterclockwise from the upper right. Arrows show the positive direction of each axis.

Each point (x, y) in the plane is called an **ordered pair.** The first number, *x*, indicates the point's horizontal location with respect to the *y*-axis, and the second number, y, indicates the point's vertical location with respect to the x-axis. We call x the **first coordinate, x-coordinate,** or **abscissa.** We call y the **second coordinate, y-coordinate,** or **ordinate.** Such a representation is called the **Cartesian coordinate system** in honor of the French mathematician and philosopher René Descartes (1596–1650).

EXAMPLE 1   Graph and label the points $(-3, 5)$, $(4, 3)$, $(3, 4)$, $(-4, -2)$, $(3, -4)$, $(0, 4)$, $(-3, 0)$, and $(0, 0)$.

Solution   To graph or **plot** $(-3, 5)$, we note that the x-coordinate tells us to move from the origin 3 units to the left of the y-axis. Then we move 5 units up from the x-axis. To graph the other points, we proceed in a similar manner. (See the graph at left.) Note that the origin, $(0, 0)$, is located at the intersection of the axes and that the point $(4, 3)$ is different from the point $(3, 4)$.

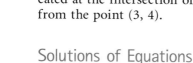

Solutions of Equations

Equations in two variables, like $2x + 3y = 18$, have solutions (x, y) that are ordered pairs such that when the first coordinate is substituted for x and the second coordinate is substituted for y, the result is a true equation.

EXAMPLE 2   Determine whether each ordered pair is a solution of $2x + 3y = 18$.

a) $(-5, 7)$ **b)** $(3, 4)$

Solution   We substitute the ordered pair into the equation and determine whether the resulting equation is true.

a) $2x + 3y = 18$

$2(-5) + 3(7) \ ? \ 18$ We substitute -5 for x and 7 for y (alphabetical order).

$-10 + 21$

$11 \ | \ 18$ FALSE

The equation $11 = 18$ is false, so $(-5, 7)$ is not a solution.

b) $2x + 3y = 18$

$2(3) + 3(4) \ ? \ 18$ We substitute 3 for x and 4 for y.

$6 + 12$

$18 \ | \ 18$ TRUE

The equation $18 = 18$ is true, so $(3, 4)$ is a solution.

Suggestions for Drawing Graphs

1. Calculate solutions and list the ordered pairs in a table.
2. Use graph paper.
3. Draw axes and label them with the variables.
4. Use arrows on the axes to indicate positive directions.
5. Scale the axes; that is, mark numbers on the axes. Consider the ordered pairs found in part (1) above.
6. Plot the ordered pairs, look for patterns, and complete the graph. Label the graph with the equation being graphed.

Graphs of Equations

The equation considered in Example 2 actually has an infinite number of solutions. Since we cannot list all the solutions, we will make a drawing, called a **graph**, that represents them. Shown at left are some suggestions for drawing graphs.

To Graph an Equation

To **graph an equation** is to make a drawing that represents the solutions of that equation.

Many equations of the type $Ax + By = C$ can be graphed conveniently using intercepts. The **x-intercept** of the graph of a linear equation or function is the point at which the graph crosses the x-axis. The **y-intercept** is the point at which the graph crosses the y-axis. We know from geometry that only one line can be drawn through two given points. Thus, if we know the intercepts, we can graph the line. To ensure that a computation error has not been made, it is a good idea to calculate a third point as a check.

---

x and y-Intercepts

An **x-intercept** is a point $(a, 0)$. To find a, let $y = 0$ and solve for x.

A **y-intercept** is a point $(0, b)$. To find b, let $x = 0$ and solve for y.

EXAMPLE 3   Graph: $2x + 3y = 18$.

Solution   To find ordered pairs that are solutions of this equation, we can replace either x or y with any number and then solve for the other variable. In this case, it is convenient to find the intercepts of the graph. For instance, if x is replaced with 0, then

$$2 \cdot 0 + 3y = 18$$
$$3y = 18$$
$$y = 6. \qquad \text{Dividing by 3}$$

Thus, $(0, 6)$ is a solution. It is the y-intercept of the graph. If y is replaced with 0, then

$$2x + 3 \cdot 0 = 18$$
$$2x = 18$$
$$x = 9. \qquad \text{Dividing by 2}$$

Thus, (9, 0) is a solution. It is the x-intercept of the graph. We find a third solution as a check. If x is replaced with 5, then

$$2 \cdot 5 + 3y = 18$$
$$10 + 3y = 18$$
$$3y = 8 \qquad \text{Subtracting 10}$$
$$y = \tfrac{8}{3}. \qquad \text{Dividing by 3}$$

Thus, $\left(5, \tfrac{8}{3}\right)$ is a solution.

We list the solutions in a table and then plot the points. Note that the points appear to lie on a straight line.

x	y	(x, y)
0	6	$(0, 6)$
9	0	$(9, 0)$
5	$\tfrac{8}{3}$	$\left(5, \tfrac{8}{3}\right)$

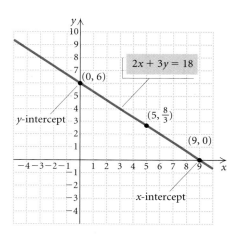

Were we to graph additional solutions of $2x + 3y = 18$, they would be on the same straight line. Thus, to complete the graph, we use a straightedge to draw a line as shown in the figure. This line represents all solutions of the equation.

EXAMPLE 4   Graph: $y = \tfrac{1}{2}x + 1$.

Solution   By choosing even integers for x, we can avoid fractional values when calculating y. For example, if we choose -6 for x, we get

$$y = \tfrac{1}{2}x + 1 = \tfrac{1}{2}(-6) + 1 = -3 + 1 = -2.$$

The table below lists a few more points.

x	y	(x, y)
-6	-2	$(-6, -2)$
-2	0	$(-2, 0)$
0	1	$(0, 1)$
4	3	$(4, 3)$

We plot the points and draw the graph.

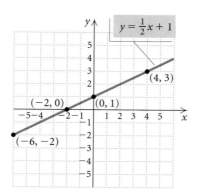

In the equation $y = \frac{1}{2}x + 1$, the value of y *depends* on the value chosen for x, so x is said to be the **independent variable** and y the **dependent variable.**

Technology Connection

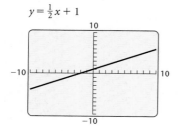

We can graph an equation on a graphing calculator. For the equation in Example 4, we enter $y = \frac{1}{2}x + 1$ on the equation-editor, or "$y =$", screen in the form $y = (1/2)x + 1$. Many calculators require an equation to be entered in the form "$y =$" as this one is written. If an equation is not initially given in this form, it must be solved for y before it is entered in the calculator.

Next, we determine the portion of the xy-plane that will appear on the calculator's screen. That portion of the plane is called the **viewing window.**

The notation used in this text to denote a window setting consists of four numbers [L, R, B, T], which represent the **L**eft and **R**ight endpoints of the x-axis and the **B**ottom and **T**op endpoints of the y-axis, respectively. The window with the settings $[-10, 10, -10, 10]$ is the **standard viewing window,** shown in the third window at left. On some graphing calculators, the standard window can be selected quickly using the ZSTANDARD feature from the ZOOM menu.

Xmin and Xmax are used to set the left and right endpoints of the x-axis, respectively; Ymin and Ymax are used to set the bottom and top endpoints of the y-axis. The settings Xscl and Yscl give the scales for the axes. For example, Xscl = 1 and Yscl = 1 means that there is 1 unit between tick marks on each of the axes. Generally in this text, Xscl and Yscl are both assumed to be 1 unless different values are given. Some exceptions will occur when the scaling factor is different from 1 but readily apparent.

After entering the equation and choosing a viewing window, we can then draw the graph.

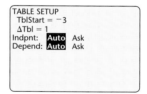
EXAMPLE 5   Graph: $y = x^2 - 9x - 12$.

Solution   We make a table of values, plot enough points to obtain an idea of the shape of the curve, and connect them with a smooth curve.

x	y	(x, y)
-3	24	$(-3, 24)$
-1	-2	$(-1, -2)$
0	-12	$(0, -12)$
2	-26	$(2, -26)$
4	-32	$(4, -32)$
5	-32	$(5, -32)$
10	-2	$(10, -2)$
12	24	$(12, 24)$

① Select values for x.

② Compute values for y.

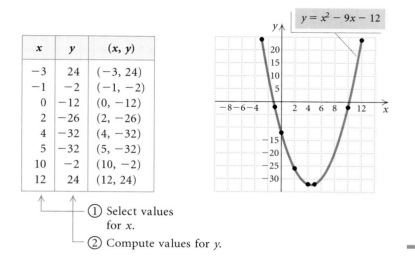

The Distance Formula

Suppose that a conservationist needs to determine the distance across an irregularly shaped pond. One way in which the conservationist might proceed is to measure two legs of a right triangle that is situated as shown below. The Pythagorean theorem, $a^2 + b^2 = c^2$, can then be used to find the length of the hypotenuse, which is the distance across the pond.

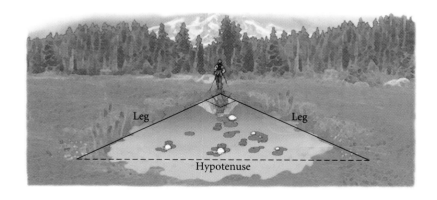

A similar strategy is used to find the distance between two points in a plane. For two points (x_1, y_1) and (x_2, y_2), we can draw a right triangle in which the legs have lengths $|x_2 - x_1|$ and $|y_2 - y_1|$.

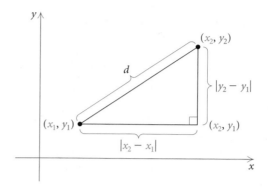

Using the Pythagorean theorem, we have

$$d^2 = |x_2 - x_1|^2 + |y_2 - y_1|^2.$$

Because we are squaring, parentheses can replace the absolute-value symbols:

$$d^2 = (x_2 - x_1)^2 + (y_2 - y_1)^2.$$

Taking the principal square root, we obtain the distance formula.

The Distance Formula

The **distance d** between any two points (x_1, y_1) and (x_2, y_2) is given by

$$d = \sqrt{(x_2 - x_1)^2 + (y_2 - y_1)^2}.$$

The subtraction of the x-coordinates can be done in any order, as can the subtraction of the y-coordinates. Although we derived the distance formula by considering two points not on a horizontal or a vertical line, the distance formula holds for *any* two points.

EXAMPLE 6 The point $(-2, 5)$ is on a circle that has $(3, -1)$ as its center. Find the length of the radius of the circle.

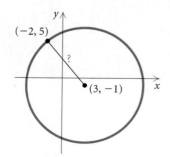

Solution Since the length of the radius is the distance from the center to a point on the circle, we substitute into the distance formula:

$$r = \sqrt{[3 - (-2)]^2 + [-1 - 5]^2} \qquad \text{Either point can serve as } (x_1, y_1).$$
$$= \sqrt{5^2 + (-6)^2} = \sqrt{61} \approx 7.8. \qquad \text{Rounded to the nearest tenth}$$

The radius of the circle is approximately 7.8.

Midpoints of Segments

The distance formula can be used to develop a way of determining the *midpoint* of a segment when the endpoints are known. We state the formula and leave its proof to the exercises.

The Midpoint Formula

If the endpoints of a segment are (x_1, y_1) and (x_2, y_2), then the coordinates of the **midpoint** are

$$\left(\frac{x_1 + x_2}{2}, \frac{y_1 + y_2}{2} \right).$$

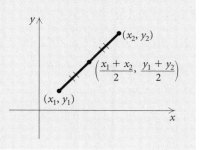

Note that we obtain the coordinates of the midpoint by averaging the coordinates of the endpoints. This is an easy way to remember the midpoint formula.

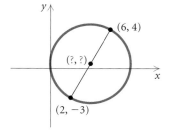

EXAMPLE 7 The diameter of a circle connects two points $(2, -3)$ and $(6, 4)$ on the circle. Find the coordinates of the center of the circle.

Solution Using the midpoint formula, we obtain

$$\left(\frac{2 + 6}{2}, \frac{-3 + 4}{2} \right), \quad \text{or} \quad \left(\frac{8}{2}, \frac{1}{2} \right), \quad \text{or} \quad \left(4, \frac{1}{2} \right).$$

The coordinates of the center are $\left(4, \frac{1}{2} \right)$. ▬

Circles

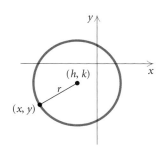

A **circle** is the set of all points in a plane that are a fixed distance r from a *center* (h, k). Thus if a point (x, y) is to be r units from the center, we must have

$$r = \sqrt{(x - h)^2 + (y - k)^2}. \qquad \textbf{Using the distance formula}$$

Squaring both sides gives an equation of a circle. The distance r is the length of a *radius* of the circle.

Technology Connection

Circles can be graphed using a graphing calculator. We show one method here. Another method will be discussed in Section 6.2.

We select the CIRCLE feature from the DRAW menu and enter the coordinates of the center and the length of the radius to draw the graph of the circle. The graph of the circle

$$(x - 2)^2 + (y + 1)^2 = 16$$

is shown here.

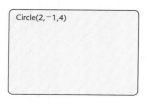

$(x - 2)^2 + (y + 1)^2 = 16$

The Equation of a Circle

The equation of a circle with center (h, k) and radius r, in standard form, is

$$(x - h)^2 + (y - k)^2 = r^2.$$

EXAMPLE 8   Find an equation of the circle having radius 5 and center $(3, -7)$.

Solution   Using the standard form, we have

$$[x - 3]^2 + [y - (-7)]^2 = 5^2 \qquad \text{Substituting}$$
$$(x - 3)^2 + (y + 7)^2 = 25.$$

EXAMPLE 9   Graph the circle $(x + 5)^2 + (y - 2)^2 = 16$.

Solution   We write the equation in standard form to determine the center and the radius:

$$[x - (-5)]^2 + [y - 2]^2 = 4^2.$$

The center is $(-5, 2)$ and the radius is 4. We locate the center and draw the circle using a compass.

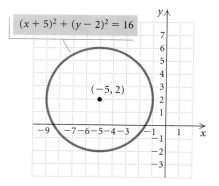

Technology Connection

When we graph a circle, we select a viewing window in which the distance between units is visually the same on both axes. This procedure is called **squaring the viewing window.** We do this so that the graph will not be distorted. A graph of the circle $x^2 + y^2 = 36$ in a nonsquared window is shown in Fig. 1.

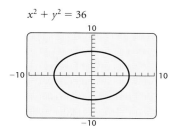

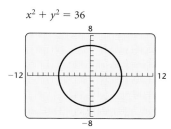

FIGURE 1   FIGURE 2

On many graphing calculators, the distance between units on the y-axis is about $\frac{2}{3}$ the distance between units on the x-axis. When we choose a window in which the length of the y-axis is $\frac{2}{3}$ the length of the x-axis, the window will be squared. The windows with dimensions $[-6, 6, -4, 4]$, $[-9, 9, -6, 6]$, and $[-12, 12, -8, 8]$ are examples of squared windows. A graph of the circle $x^2 + y^2 = 36$ in a squared window is shown in Fig. 2.

Many graphing calculators have an option on the ZOOM menu that squares the window automatically.

Exercise Set

Graph and label the given points.

1. $(4, 0)$, $(-3, -5)$, $(-1, 4)$, $(0, 2)$, $(2, -2)$

2. $(1, 4)$, $(-4, -2)$, $(-5, 0)$, $(2, -4)$, $(4, 0)$

3. $(-5, 1)$, $(5, 1)$, $(2, 3)$, $(2, -1)$, $(0, 1)$

4. $(4, 0)$, $(4, -3)$, $(-5, 2)$, $(-5, 0)$, $(-1, -5)$

Use substitution to determine whether the given ordered pairs are solutions of the given equation.

5. $(1, -1)$, $(0, 3)$; $y = 2x - 3$

6. $(2, 5)$, $(-2, -5)$; $y = 3x - 1$

7. $\left(\frac{2}{3}, \frac{3}{4}\right)$, $\left(1, \frac{3}{2}\right)$; $6x - 4y = 1$

8. $(1.5, 2.6)$, $(-3, 0)$; $x^2 + y^2 = 9$

9. $\left(-\frac{1}{2}, -\frac{4}{5}\right)$, $\left(0, \frac{3}{5}\right)$; $2a + 5b = 3$

10. $\left(0, \frac{3}{2}\right)$, $\left(\frac{2}{3}, 1\right)$; $3m + 4n = 6$

11. $(-0.75, 2.75)$, $(2, -1)$; $x^2 - y^2 = 3$

12. $(2, -4)$, $(4, -5)$; $5x + 2y^2 = 70$

Graph the equation.

13. $y = 3x + 5$ **14.** $y = -2x - 1$

15. $x - y = 3$

16. $x + y = 4$

17. $2x + y = 4$

18. $3x - y = 6$

19. $x - 4y = 5$

20. $x + 3y = 6$

21. $3x - 4y = 12$

22. $2x + 3y = -6$

23. $2x + 5y = -10$

24. $4x - 3y = 12$

25. $y = -x^2$

26. $y = x^2$

27. $y = x^2 - 3$

28. $y = 4 - x^2$

29. $y = x^2 - 2x - 3$

30. $y = x^2 + 2x - 1$

Find the distance between the pair of points. Give an exact answer and, where appropriate, an approximation to three decimal places.

31. $(4, 6)$ and $(5, 9)$

32. $(-3, 7)$ and $(2, 11)$

33. $(6, -1)$ and $(9, 5)$

34. $(-4, -7)$ and $(-1, 3)$

35. $(\sqrt{3}, -\sqrt{5})$ and $(-\sqrt{6}, 0)$

36. $(-\sqrt{2}, 1)$ and $(0, \sqrt{7})$

37. The points $(-3, -1)$ and $(9, 4)$ are the endpoints of the diameter of a circle. Find the length of the radius of the circle.

38. The point $(0, 1)$ is on a circle that has center $(-3, 5)$. Find the length of the diameter of the circle.

Use the distance formula and the Pythagorean theorem to determine whether the set of points could be vertices of a right triangle.

39. $(-4, 5)$, $(6, 1)$, and $(-8, -5)$

40. $(-3, 1)$, $(2, -1)$, and $(6, 9)$

41. The points $(-3, 4)$, $(0, 5)$, and $(3, -4)$ are all on the circle $x^2 + y^2 = 25$. Show that these three points are vertices of a right triangle.

42. The points $(-3, 4)$, $(2, -1)$, $(5, 2)$, and $(0, 7)$ are vertices of a quadrilateral. Show that the quadrilateral is a rectangle. (*Hint*: Show that the quadrilateral's opposite sides are the same length and that the two diagonals are the same length.)

43.–48. Find the midpoint of each segment having the endpoints given in Exercises 31–36, respectively.

49. Graph the rectangle described in Exercise 42. Then determine the coordinates of the midpoint for each of the four sides. Are the midpoints vertices of a rectangle?

50. Graph the square with vertices $(-5, -1)$, $(7, -6)$, $(12, 6)$, and $(0, 11)$. Then determine the midpoint for each of the four sides. Are the midpoints vertices of a square?

51. The points $(\sqrt{7}, -4)$ and $(\sqrt{2}, 3)$ are endpoints of the diameter of a circle. Determine the center of the circle.

52. The points $(-3, \sqrt{5})$ and $(1, \sqrt{2})$ are endpoints of the diagonal of a square. Determine the center of the square.

Find an equation for a circle satisfying the given conditions.

53. Center $(2, 3)$, radius of length $\frac{5}{3}$

54. Center $(4, 5)$, diameter of length 8.2

55. Center $(-1, 4)$, passes through $(3, 7)$

56. Center $(6, -5)$, passes through $(1, 7)$

57. The points $(7, 13)$ and $(-3, -11)$ are at either end of a diameter.

58. The points $(-9, 4)$, $(-2, 5)$, $(-8, -3)$, and $(-1, -2)$ are vertices of an inscribed square.

59. Center $(-2, 3)$, tangent (touching at one point) to the y-axis

60. Center $(4, -5)$, tangent to the x-axis

Find the center and the radius of the circle. Then graph the circle.

61. $x^2 + y^2 = 4$

62. $x^2 + y^2 = 81$

63. $x^2 + (y - 3)^2 = 16$

64. $(x + 2)^2 + y^2 = 100$

65. $(x - 1)^2 + (y - 5)^2 = 36$

66. $(x - 7)^2 + (y + 2)^2 = 25$

67. $(x + 4)^2 + (y + 5)^2 = 9$

68. $(x + 1)^2 + (y - 2)^2 = 64$

Technology Connection

In Exercises 69–72, use a graphing calculator to match the equation with one of the graphs (a)–(d), which follow.

a)

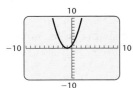

b)

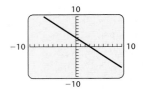

c)

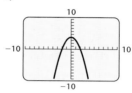

d)

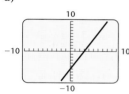

69. $y = 3 - x$

70. $y = 2x - 6$

71. $y = x^2 + 2x + 1$

72. $y = 4 - x^2$

Use a graphing calculator to graph the equation in the standard window.

73. $y = 2x + 1$

74. $y = 3x - 4$

75. $4x + y = 7$

76. $5x + y = -8$

77. $y = \frac{1}{3}x + 2$

78. $y = \frac{3}{2}x - 4$

79. $2x + 3y = -5$

80. $3x + 4y = 1$

81. $y = x^2 + 6$

82. $y = x^2 - 8$

83. $y = 2 - x^2$

84. $y = 5 - x^2$

85. $y = x^2 + 4x - 2$

86. $y = x^2 - 5x + 3$

Graph the equation in the standard window and in the given window. Determine which window better shows the shape of the graph and the x- and y-intercepts.

87. $y = x - 20$
$[-25, 25, -25, 25]$, with Xscl = 5 and Yscl = 5

88. $y = -2x + 24$
$[-15, 15, -10, 30]$, with Xscl = 3 and Yscl = 5

89. $y = 3x^2 - 6$
$[-4, 4, -4, 4]$

90. $y = 8 - x^2$
$[-3, 3, -3, 3]$

91. $y = -\frac{1}{6}x^2 + \frac{1}{12}$
$[-1, 1, -0.3, 0.3]$, with Xscl = 0.1 and Yscl = 0.1

92. $y = \frac{1}{2}x^2 - \frac{1}{4}x - \frac{1}{8}$
$[-0.8, 1.2, -0.5, 0.5]$, with Xscl = 0.1 and Yscl = 0.1

In Exercises 93 and 94, how would you change the window so that the circle is not distorted? Answers may vary.

93.

$(x + 3)^2 + (y - 2)^2 = 36$

94.

$(x - 4)^2 + (y + 5)^2 = 49$

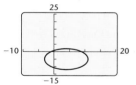

95.–102. Using a graphing calculator, graph each of the circles given in Exercises 61–68, respectively.

Collaborative Discussion and Writing

To the student and the instructor: The Collaborative Discussion and Writing exercises are meant to be answered with one or more sentences. They can be discussed and answered collaboratively by the entire class or by small groups. Because of their open-ended nature, the answers to these exercises do not appear at the back of the book. They are denoted by the words "Discussion and Writing."

103. Explain how the Pythagorean theorem is used to develop the equation of a circle in standard form.

104. Explain how you could find the coordinates of a point $\frac{7}{8}$ of the way from point A to point B.

Synthesis

To the student and the instructor: The Synthesis exercises found at the end of every exercise set challenge students to combine concepts or skills studied in that section or in preceding parts of the text.

Find the distance between the pair of points and find the midpoint of the segment having the given points as endpoints.

105. $(a, \sqrt{a})$ and $(a + h, \sqrt{a + h})$

106. $\left(a, \dfrac{1}{a}\right)$ and $\left(a + h, \dfrac{1}{a + h}\right)$

Find an equation of a circle satisfying the given conditions.

107. Center $(2, -7)$ with an area of 36π square units

108. Center $(-5, 8)$ with a circumference of 10π units

109. *Swimming Pool.* A swimming pool is being constructed in the corner of a yard, as shown. Before installation, the contractor needs to know measurements a_1 and a_2. Find them.

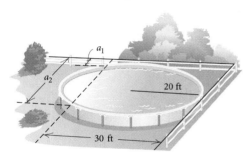

110. *An Arch of a Circle in Carpentry.*   Ace Carpentry needs to cut an arch for the top of an entranceway. The arch needs to be 8 ft wide and 2 ft high. To draw the arch, the carpenters will use a stretched string with chalk attached at an end as a compass.

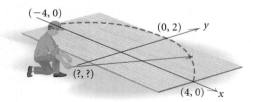

a) Using a coordinate system, locate the center of the circle.
b) What radius should the carpenters use to draw the arch?

*Determine whether each of the following points lies on the **unit circle**,* $x^2 + y^2 = 1$.

111. $\left(\dfrac{\sqrt{3}}{2}, -\dfrac{1}{2} \right)$

112. $(0, -1)$

113. $\left(-\dfrac{\sqrt{2}}{2}, \dfrac{\sqrt{2}}{2} \right)$

114. $\left(\dfrac{1}{2}, -\dfrac{\sqrt{3}}{2} \right)$

115. Find the point on the y-axis that is equidistant from the points $(-2, 0)$ and $(4, 6)$.

116. Consider any right triangle with base b and height h, situated as shown. Show that the midpoint of the hypotenuse P is equidistant from the three vertices of the triangle.

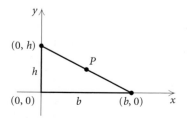

117. Prove the midpoint formula by showing that:
a) $\left(\dfrac{x_1 + x_2}{2}, \dfrac{y_1 + y_2}{2} \right)$ is equidistant from the points (x_1, y_1) and (x_2, y_2); and
b) the distance from (x_1, y_1) to the midpoint plus the distance from (x_2, y_2) to the midpoint equals the distance from (x_1, y_1) to (x_2, y_2).

1.2

Functions and Graphs

- *Determine whether a correspondence or a relation is a function.*
- *Find function values, or outputs, using a formula.*
- *Find the domain and the range of a function.*
- *Determine whether a graph is that of a function.*
- *Solve applied problems using functions.*

We now focus our attention on a concept that is fundamental to many areas of mathematics—the idea of a *function.*

Functions

We first consider an application.

Thunder Time Related to Lightning Distance. During a thunderstorm, it is possible to calculate how far away, y (in miles), lightning is when the sound of thunder arrives x seconds after the lightning has been sighted. It is known that the distance, in miles, is $\frac{1}{5}$ of the time, in seconds. If we hear the sound of thunder 15 seconds after we've seen the lightning, we know that the lightning is $\frac{1}{5} \cdot 15$, or 3 miles away. Similarly, 5 sec corresponds to 1 mi, 2 sec to $\frac{2}{5}$ mi, and so on. We can express this relationship with a set of ordered pairs, a graph, and an equation.

x	y	ORDERED PAIRS: (x, y)	CORRESPONDENCE
0	0	$(0, 0)$	$0 \longrightarrow 0$
1	$\frac{1}{5}$	$\left(1, \frac{1}{5}\right)$	$1 \longrightarrow \frac{1}{5}$
2	$\frac{2}{5}$	$\left(2, \frac{2}{5}\right)$	$2 \longrightarrow \frac{2}{5}$
5	1	$(5, 1)$	$5 \longrightarrow 1$
10	2	$(10, 2)$	$10 \longrightarrow 2$
15	3	$(15, 3)$	$15 \longrightarrow 3$

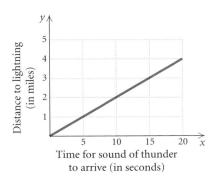

Time for sound of thunder
to arrive (in seconds)

The ordered pairs express a relationship, or correspondence, between the first and second coordinates. We can see this relationship in the graph as well. The equation that describes the correspondence is

$$y = \frac{1}{5}x.$$

This is an example of a *function*. In this case, distance is a function of time; that is, y is a function of x, where x is the independent variable and y is the dependent variable.

Let's consider some other correspondences before giving the definition of a function.

DOMAIN		RANGE
To each registered student	there corresponds	an I. D. number.
To each mountain bike sold	there corresponds	its price.
To each number between -3 and 3	there corresponds	the square of that number.

In each correspondence, the first set is called the **domain** and the second set is called the **range**. For each member, or **element**, in the domain, there is *exactly one* member in the range to which it corresponds. Thus each registered student has exactly *one* I. D. number, each mountain bike has exactly *one* price, and each number between -3 and 3 has exactly *one* square. Each correspondence is a *function*.

Function

A **function** is a correspondence between a first set, called the **domain**, and a second set, called the **range**, such that each member of the domain corresponds to *exactly one* member of the range.

It is important to note that not every correspondence between two sets is a function.

EXAMPLE 1   Determine whether each of the following correspondences is a function.

a) $-3 \longrightarrow 9$
    3
    $2 \longrightarrow 4$

b) Juan $\longrightarrow$ Casandra
    Boris $\longrightarrow$ Rebecca
    Nelson $\longrightarrow$ Helga
    Bernie $\longrightarrow$ Natasha

Solution

a) This correspondence *is* a function because each member of the domain corresponds to exactly one member of the range. Note that the definition of a function does allow more than one member of the domain to correspond to a member of the range.

b) This correspondence *is not* a function because there is a member of the domain (Boris) that is paired with two different members of the range (Rebecca and Helga).

EXAMPLE 2   Determine whether each of the following correspondences is a function.

DOMAIN	CORRESPONDENCE	RANGE
a) Years in which a presidential election occurs	The person elected	A set of presidents
b) The integers	Each integer's cube root	A subset of the real numbers
c) All states in the United States	A senator from that state	The set of all U.S. senators

Solution

a) This correspondence *is* a function, because in each presidential election *exactly one* president is elected.

b) This correspondence *is* a function, because each integer has *exactly one* cube root.

c) This correspondence *is not* a function, because each state can be paired with *two* different senators.

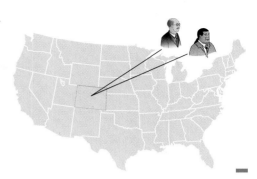

When a correspondence between two sets is not a function, it is still an example of a **relation**.

Relation

A **relation** is a correspondence between a first set, called the **domain**, and a second set, called the **range**, such that each member of the domain corresponds to *at least one* member of the range.

All the correspondences in Examples 1 and 2 are relations, but, as we have seen, not all are functions. Relations are sometimes written as sets of ordered pairs (as we saw earlier in the example on lightning) in which elements of the domain are the first coordinates of the ordered pairs and elements of the range are the second coordinates. For example, instead of writing $-3 \longrightarrow 9$, as we did in Example 1(a), we could write the ordered pair $(-3, 9)$.

EXAMPLE 3   Determine whether each of the following relations is a function. Identify the domain and the range.

a) $\{(-2, 5), (5, 7), (0, 1), (4, -2)\}$

b) $\{(9, -5), (9, 5), (2, 4)\}$

c) $\{(-5, 3), (0, 3), (6, 3)\}$

Solution

a) The relation *is* a function because *no* two ordered pairs have the same first coordinate and different second coordinates (see Fig. 1).

The domain is the set of all first coordinates: $\{-2, 5, 0, 4\}$.

The range is the set of all second coordinates: $\{5, 7, 1, -2\}$.

b) The relation *is not* a function because the ordered pairs $(9, -5)$ and $(9, 5)$ have the same first coordinate and different second coordinates (see Fig. 2).

The domain is the set of all first coordinates: $\{9, 2\}$.

The range is the set of all second coordinates: $\{-5, 5, 4\}$.

FIGURE 1

FIGURE 2

FIGURE 3

c) The relation *is* a function because *no* two ordered pairs have the same first coordinate and different second coordinates (see Fig. 3).

The domain is $\{-5, 0, 6\}$.

The range is $\{3\}$.

Notation for Functions

Functions used in mathematics are often given by equations. They generally require that certain calculations be performed in order to determine which member of the range is paired with each member of the domain. For example, in Section 1.1 we graphed the function $y = x^2 - 9x - 12$ by doing calculations like the following:

for $x = -2$, $y = (-2)^2 - 9(-2) - 12 = 10$,

for $x = 0$, $y = 0^2 - 9 \cdot 0 - 12 = -12$, and

for $x = 1$, $y = 1^2 - 9 \cdot 1 - 12 = -20$.

A more concise notation is often used. For $y = x^2 - 9x - 12$, the **inputs** (members of the domain) are values of x substituted into the equation. The **outputs** (members of the range) are the resulting values of y. If we call the function f, we can use x to represent an arbitrary *input* and $f(x)$—read "f of x," or "f at x," or "the value of f at x"—to represent the corresponding *output*. In this notation, the function given by $y = x^2 - 9x - 12$ is written as $f(x) = x^2 - 9x - 12$ and the above calculations would be

$$f(-2) = -2^2 - 9(-2) - 12 = 10,$$

$$f(0) = 0^2 - 9 \cdot 0 - 12 = -12,$$

$$f(1) = 1^2 - 9 \cdot 1 - 12 = -20.$$

Keep in mind that $f(x)$ *does not* mean $f \cdot x$.

Thus, instead of writing "when $x = -2$, the value of y is 10," we can simply write "$f(-2) = 10$," which can also be read as "f of -2 is 10" or "for the input -2, the output of f is 10." The letters g and h are also often used to name functions.

EXAMPLE 4 A function f is given by $f(x) = 2x^2 - x + 3$. Find each of the following.

a) $f(0)$ **b)** $f(-7)$

c) $f(5a)$ **d)** $f(a - 4)$

Solution We can think of this formula as follows:

$$f(\blacksquare) = 2(\blacksquare)^2 - (\blacksquare) + 3.$$

Then to find an output for a given input we think: "Whatever goes in the blank on the left goes in the blank(s) on the right." This gives us a "recipe" for finding outputs.

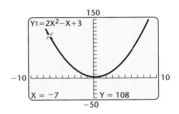
a) $f(0) = 2(0)^2 - 0 + 3 = 0 - 0 + 3 = 3$

b) $f(-7) = 2(-7)^2 - (-7) + 3 = 2 \cdot 49 + 7 + 3 = 108$

c) $f(5a) = 2(5a)^2 - 5a + 3 = 2 \cdot 25a^2 - 5a + 3 = 50a^2 - 5a + 3$

d) $f(a - 4) = 2(a - 4)^2 - (a - 4) + 3 = 2(a^2 - 8a + 16) - a + 4 + 3$
$$= 2a^2 - 16a + 32 - a + 4 + 3$$
$$= 2a^2 - 17a + 39$$

Graphs of Functions

We graph functions the same way we graph equations. We find ordered pairs (x, y) or $(x, f(x))$, plot points, and complete the graph.

EXAMPLE 5 Graph each of the following functions.

a) $f(x) = x^2 - 5$ b) $f(x) = x^3 - x$ c) $f(x) = \sqrt{x + 4}$

Solution We select values for x and find the corresponding values of $f(x)$. Then we plot the points and connect them with a smooth curve.

a) $f(x) = x^2 - 5$

x	$f(x)$	$(x, f(x))$
-3	4	$(-3, 4)$
-2	-1	$(-2, -1)$
-1	-4	$(-1, -4)$
0	-5	$(0, -5)$
1	-4	$(1, -4)$
2	-1	$(2, -1)$
3	4	$(3, 4)$

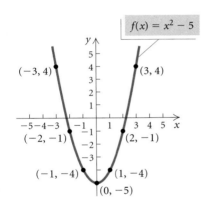

b) $f(x) = x^3 - x$

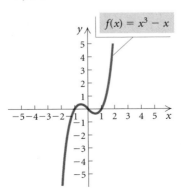

c) $f(x) = \sqrt{x + 4}$

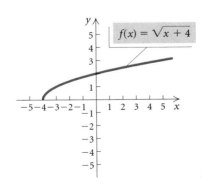

To find a function value, like $f(3)$, from a graph, we locate the input 3 on the horizontal axis, move vertically to the graph of the function, and then move horizontally to find the output on the vertical axis. For the function $f(x) = x^2 - 5$, we see that $f(3) = 4$.

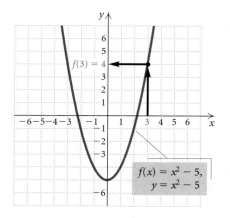

We know that when one member of the domain is paired with two or more different members of the range, the correspondence *is not* a function. Thus, when a graph contains two or more different points with the same first coordinate, the graph cannot represent a function (see the graph at left). Points sharing a common first coordinate are vertically above or below each other. This leads us to the *vertical-line test.*

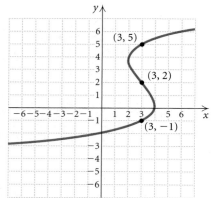

Since 3 is paired with more than one member of the range, the graph does not represent a function.

The Vertical-Line Test

If it is possible for a vertical line to cross a graph more than once, then the graph is *not* the graph of a function.

To apply the vertical-line test, we try to find a vertical line that crosses the graph more than once. If we succeed, then the graph is not that of a function. If we do not, then the graph is that of a function.

EXAMPLE 6 Which of graphs (a) through (f) (in red) are graphs of functions?

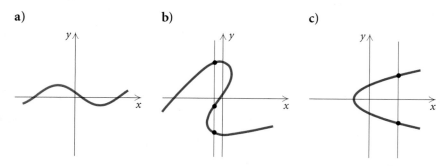

a)   b)   c)

d)

e)

f)

In graph (f), the solid dot shows that $(-1, 1)$ belongs to the graph. The open circle shows that $(-1, -2)$ does *not* belong to the graph.

Solution Graphs (a), (e), and (f) are graphs of functions because we cannot find a vertical line that crosses any of them more than once. In (b), the vertical line crosses the graph in three points and so it is not that of a function. Also, in (c) and (d), we can find a vertical line that crosses the graph more than once.

Finding Domains of Functions

When a function f, whose inputs and outputs are real numbers, is given by a formula, the *domain* is understood to be the set of all inputs for which the expression is defined as a real number. When a substitution results in an expression that is not defined as a real number, we say that the function value *does not exist* and that the number being substituted *is not* in the domain of the function.

EXAMPLE 7 Find the indicated function values.

a) $f(1)$ and $f(3)$, for $f(x) = \dfrac{1}{x - 3}$

b) $g(16)$ and $g(-7)$, for $g(x) = \sqrt{x} + 5$

Solution

a) $f(1) = \dfrac{1}{1 - 3} = \dfrac{1}{-2} = -\dfrac{1}{2}$;

Since $f(1)$ is defined, 1 is in the domain of f.

$f(3) = \dfrac{1}{3 - 3} = \dfrac{1}{0}$

Since division by 0 is not defined, the number 3 is not in the domain of f.

b) $g(16) = \sqrt{16} + 5 = 4 + 5 = 9$;

Since $g(16)$ is defined, 16 is in the domain of g.

$g(-7) = \sqrt{-7} + 5$

Since $\sqrt{-7}$ is not defined as a real number, the number -7 is not in the domain of g.

Technology Connection

When we use a graphing calculator to find function values and a function value does not exist, the calculator indicates this with an ERROR message.

In the tables below, we see in Example 7 that $f(3)$ for $f(x) = 1/(x - 3)$ and $g(-7)$ for $g(x) = \sqrt{x} + 5$ do not exist. Thus, 3 and -7 are *not* in the domains of the corresponding functions.

$y = 1/(x - 3)$

X	Y₁	
1	−.5	
3	ERROR	

X =

$y = \sqrt{x} + 5$

X	Y₁	
16	9	
−7	ERROR	

X =

Inputs that make a denominator 0 or the radicand in an even root negative are not in the domain of a function.

EXAMPLE 8   Find the domain of each of the following functions.

a) $f(x) = \dfrac{1}{x - 3}$ **b)** $g(x) = \sqrt{x} + 5$

c) $h(x) = \dfrac{3x^2 - x + 7}{x^2 + 2x - 3}$ **d)** $f(x) = x^3 + |x|$

Solution

INTERVAL NOTATION

REVIEW SECTION R.1.

a) The input 3 results in a denominator of 0. The domain is $\{x \mid x \neq 3\}$. We can also write the solution using interval notation and the symbol $\cup$ for the **union** or inclusion of both sets: $(-\infty, 3) \cup (3, \infty)$.

b) We can substitute any number for which the radicand is nonnegative, that is, for which $x \geq 0$. Thus the domain is $\{x \mid x \geq 0\}$, or $[0, \infty)$.

c) Although we can substitute any real number in the numerator, we must avoid inputs that make the denominator 0. To find those inputs, we solve $x^2 + 2x - 3 = 0$, or $(x + 3)(x - 1) = 0$. Thus the domain consists of the set of all real numbers except -3 and 1, or $\{x \mid x \neq -3$ *and* $x \neq 1\}$, or $(-\infty, -3) \cup (-3, 1) \cup (1, \infty)$.

d) All substitutions are suitable. The domain is the set of all real numbers, $\mathbb{R}$, or $(-\infty, \infty)$.

Visualizing Domain and Range

Keep the following in mind regarding the *graph* of a function:

Domain = the set of a function's inputs, found on the horizontal axis;

Range = the set of a function's outputs, found on the vertical axis.

By carefully examining the graph of a function, we may be able to determine the function's domain as well as its range. Consider the graph of $f(x) = \sqrt{4 - (x - 1)^2}$, shown at left. We look for the inputs on the x-axis that correspond to a point on the graph. We see that they extend from -1 to 3, inclusive. Thus the domain is $\{x \mid -1 \leq x \leq 3\}$, or $[-1, 3]$.

To find the range, we look for the outputs on the y-axis. We see that they extend from 0 to 2, inclusive. Thus the range of this function is $\{y \mid 0 \leq y \leq 2\}$, or $[0, 2]$.

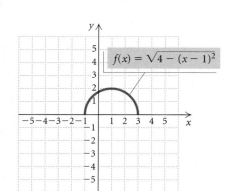

$f(x) = \sqrt{4 - (x - 1)^2}$

EXAMPLE 9   Graph each of the following functions. Then estimate the domain and the range of each.

a) $f(x) = \sqrt{x + 4}$ **b)** $f(x) = x^3 - x$

c) $f(x) = \dfrac{6}{x}$ **d)** $f(x) = x^4 - 2x^2 - 3$

Solution

a)

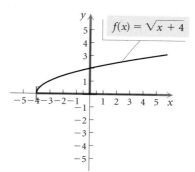

Domain = $[-4, \infty)$;
range = $[0, \infty)$

b)

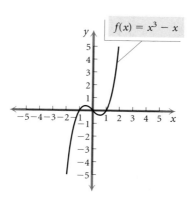

Domain = all real numbers,
$(-\infty, \infty)$; range = all real
numbers, $(-\infty, \infty)$

c)

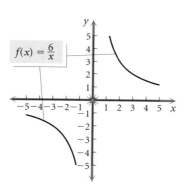

Since the graph does not touch or
cross either axis, 0 is excluded
both as an input and as an output.
Domain = $(-\infty, 0) \cup (0, \infty)$;
range = $(-\infty, 0) \cup (0, \infty)$

d)

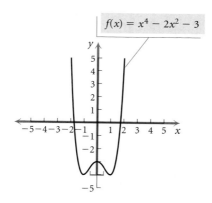

Domain = all real numbers,
$(-\infty, \infty)$; range = $[-4, \infty)$

Always consider adding the reasoning of Example 8 to a graphical
analysis. Think, "What can I substitute?" to find the domain. Think,
"What do I get out?" to find the range. Thus, in Example 9(d), it might
not look like the domain is all real numbers because the graph rises
steeply, but by examining the equation we see that we can indeed substitute any real number for x.

Applications of Functions

EXAMPLE 10   *Speed of Sound in Air.* The speed S of sound in air is a function of the temperature t, in degrees Fahrenheit, and is given by

$$S(t) = 1087.7\sqrt{\frac{5t + 2457}{2457}},$$

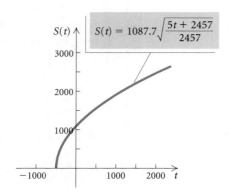

where S is in feet per second. Find the speed of sound in air when the temperature is $0°$, $32°$, $70°$, and $-10°$ Fahrenheit.

Solution   Using a calculator, we compute the function values. We find that

$$S(0) = 1087.7 \text{ ft/sec,}$$
$$S(32) \approx 1122.6 \text{ ft/sec,}$$
$$S(70) \approx 1162.6 \text{ ft/sec,} \quad \text{and}$$
$$S(-10) \approx 1076.6 \text{ ft/sec.}$$

We can visualize these values with the graph at left.   ▬

CONNECTING THE CONCEPTS

FUNCTION CONCEPTS

Formula for f: $f(x) = 5 + 2x^2 - x^4$.

For every input, there is exactly one output.

$(1, 6)$ is on the graph.

For the input 1, the output is 6.

$f(1) = 6$

Domain: set of all inputs $= (-\infty, \infty)$

Range: set of all outputs $= (-\infty, 6]$

GRAPH

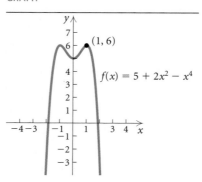

Exercise Set

In Exercises 1–14, determine whether the correspondence is a function.

1. $a \longrightarrow w$
$b \longrightarrow y$
$c \longrightarrow z$

2. $m \longrightarrow q$
$n r$
$o s$

3.
$-6 \longrightarrow 36$
$-2 4$
2

4.
$-3 \longrightarrow 2$
$1 4$
$5 \longrightarrow 6$
$9 \longrightarrow 8$

5. $m \longrightarrow A$
$n B$
$r C$
$s D$

6. $a \longrightarrow r$
$b s$
$c t$
d

7. WORLD'S TEN LARGEST EARTHQUAKES (1900–1999)

LOCATION AND DATE	MAGNITUDE
Chile (May 22, 1960)	9.5
Alaska (March 28, 1964)	9.2
Russia (Nov. 4, 1952)	9.0
Ecuador (Jan. 31, 1906)	8.8
Alaska (March 9, 1957)	8.8
Kuril Islands (Nov. 6, 1958)	8.7
Alaska (Feb. 4, 1965)	8.7
India (Aug. 15, 1950)	8.6
Argentina (Nov. 11, 1922)	8.5
Indonesia (Feb. 1, 1938)	8.5

(*Source*: National Earthquake Information Center, U.S. Geological Survey)

8.

ANIMAL	SPEED ON THE GROUND (IN MILES PER HOUR)
Cheetah	70
Lion	50
Reindeer	32
Giraffe	
Elephant	25
Giant tortoise	0.17

(*Source*: *Time Almanac*, 1999, p. 581)

	DOMAIN	CORRESPONDENCE	RANGE
9.	A set of cars in a parking lot	Each car's license number	A set of numbers
10.	A set of people in a town	A doctor a person uses	A set of doctors
11.	A set of members of a family	Each person's eye color	A set of colors
12.	A set of members of a rock band	An instrument each person plays	A set of instruments
13.	A set of students in a class	A student sitting in a neighboring seat	A set of students
14.	A set of bags of chips on a shelf	Each bag's weight	A set of weights

Determine whether the relation is a function. Identify the domain and the range.

15. $\{(2, 10), (3, 15), (4, 20)\}$

16. $\{(3, 1), (5, 1), (7, 1)\}$

17. $\{(-7, 3), (-2, 1), (-2, 4), (0, 7)\}$

18. $\{(1, 3), (1, 5), (1, 7), (1, 9)\}$

19. $\{(-2, 1), (0, 1), (2, 1), (4, 1), (-3, 1)\}$

20. $\{(5, 0), (3, -1), (0, 0), (5, -1), (3, -2)\}$

21. A graph of a function f is shown here. Find $f(-1)$, $f(0)$, and $f(1)$.

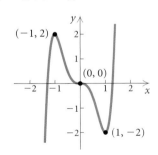

22. A graph of a function g is shown here. Find $g(-2)$, $g(0)$, and $g(2.4)$.

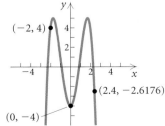

23. Given that $g(x) = 3x^2 - 2x + 1$, find each of the following.

    a) $g(0)$ **b)** $g(-1)$
    c) $g(3)$ **d)** $g(-x)$
    e) $g(1 - t)$

24. Given that $f(x) = 5x^2 + 4x$, find each of the following.

    a) $f(0)$ **b)** $f(-1)$
    c) $f(3)$ **d)** $f(t)$
    e) $f(t - 1)$

25. Given that $g(x) = x^3$, find each of the following.

    a) $g(2)$ **b)** $g(-2)$
    c) $g(-x)$ **d)** $g(3y)$
    e) $g(2 + h)$

26. Given that $f(x) = 2|x| + 3x$, find each of the following.

    a) $f(1)$ **b)** $f(-2)$
    c) $f(-x)$ **d)** $f(2y)$
    e) $f(2 - h)$

27. Given that $g(x) = \dfrac{x - 4}{x + 3}$, find each of the following.

    a) $g(5)$ **b)** $g(4)$
    c) $g(-3)$ **d)** $g(-16.25)$
    e) $g(x + h)$

28. Given that $f(x) = \dfrac{x}{2 - x}$, find each of the following.

    a) $f(2)$ **b)** $f(1)$
    c) $f(-16)$ **d)** $f(-x)$
    e) $f\left(-\dfrac{2}{3}\right)$

29. Find $g(0)$, $g(-1)$, $g(5)$, and $g\left(\frac{1}{2}\right)$ for

$$g(x) = \frac{x}{\sqrt{1 - x^2}}.$$

30. Find $h(0)$, $h(2)$, and $h(-x)$ for

$$h(x) = x + \sqrt{x^2 - 1}.$$

Find the domain of the function.

31. $f(x) = 7x + 4$

32. $f(x) = |3x - 2|$

33. $f(x) = 4 - \dfrac{2}{x}$

34. $f(x) = \dfrac{1}{x^4}$

35. $f(x) = \dfrac{x + 5}{2 - x}$

36. $f(x) = \dfrac{8}{x + 4}$

37. $f(x) = \dfrac{1}{x^2 - 4x - 5}$

38. $f(x) = \dfrac{x^4 - 2x^3 + 7}{3x^2 - 10x - 8}$

Graph the function. Then visually estimate the domain and the range.

39. $f(x) = |x|$ **40.** $f(x) = |x| - 2$
41. $f(x) = \sqrt{9 - x^2}$ **42.** $f(x) = -\sqrt{25 - x^2}$
43. $f(x) = (x - 1)^3 + 2$ **44.** $f(x) = (x - 2)^4 + 1$
45. $f(x) = \sqrt{7 - x}$ **46.** $f(x) = \sqrt{x + 8}$
47. $f(x) = -x^2 + 4x - 1$ **48.** $f(x) = 2x^2 - x^4 + 5$

In Exercises 49–56, determine whether the graph is that of a function. An open dot indicates that the point does not belong to the graph.

49.   **50.**

51. 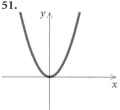   **52.**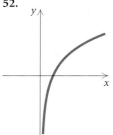

53. 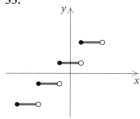   **54.**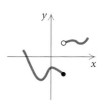

55. **56.**

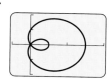 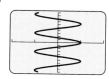

In Exercises 57–60, determine whether the graph is that of a function. Find the domain and the range.

57. **58.**

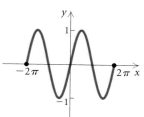 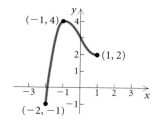

59. **60.**

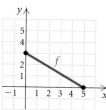 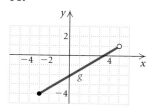

61. *Boiling Point and Elevation.* The elevation E, in meters, above sea level at which the boiling point of water is t degrees Celsius is given by the function

$$E(t) = 1000(100 - t) + 580(100 - t)^2.$$

At what elevation is the boiling point 99.5°? 100°?

62. *Average Price of a Movie Ticket.* The average price of a movie ticket, in dollars, can be estimated by the function P given by

$$P(x) = 0.1522x - 298.592$$

where x is the year. Thus, $P(2000)$ is the average price of a movie ticket in 2000. The price is lower than what might be expected due to lower prices for matinees, senior citizens' discounts, and so on.

a) Use the function to predict the average price in 2003 and 2010.

b) When will the average price be $8.00?

63. *Territorial Area of an Animal.* The territorial area of an animal is defined to be its defended, or exclusive region. For example, a lion has a certain region over which it is considered ruler. It has been shown that the territorial area T, in acres, of predatory animals is a function of body weight w, in pounds, and is given by the function

$$T(w) = w^{1.31}.$$

Find the territorial area of animals whose body weights are 0.5 lb, 10 lb, 20 lb, 100 lb, and 200 lb.

Technology Connection

In Exercises 64–66, use a graphing calculator and the TABLE *feature set in* ASK *mode.*

64. Given that

$$h(x) = 3x^4 - 10x^3 + 5x^2 - x + 6,$$

find $h(-11)$, $h(7)$, and $h(15)$.

65. Given that

$$g(x) = 0.06x^3 - 5.2x^2 - 0.8x,$$

find $g(-2.1)$, $g(5.08)$, and $g(10.003)$. Round answers to the nearest tenth.

66. Find the indicated function values if they exist.

a) $f(-4)$ and $f(-6)$, for $f(x) = \dfrac{4}{(x-4)(x+6)}$

b) $g(-5)$ and $g(1)$, for $g(x) = \sqrt{x-1} + 3$

Collaborative Discussion and Writing

67. Explain in your own words what a function is.

68. Explain in your own words the difference between the domain of a function and the range of a function.

Skill Maintenance

To the student and the instructor: The Skill Maintenance exercises review skills covered previously in the text. You can expect such exercises in every exercise set. They provide excellent review for a final examination. Answers to all skill maintenance exercises appear, along with section references, in the answer section at the back of the book.

Use substitution to determine whether the given ordered pairs are solutions of the given equation.

69. $(0, -7)$, $(8, 11)$; $y = 0.5x + 7$

70. $\left(\frac{4}{5}, -2\right)$, $\left(\frac{11}{5}, \frac{1}{10}\right)$; $15x - 10y = 32$

Graph the equation.

71. $y = (x - 1)^2$ **72.** $y = \frac{1}{3}x - 6$

73. $-2x - 5y = 10$

Synthesis

74. Draw a graph of a function for which the domain is $[-4, 4]$ and the range is $[1, 2] \cup [3, 5]$. Answers may vary.

75. Give an example of two different functions that have the same domain and the same range, but have no pairs in common. Answers may vary.

76. Draw a graph of a function for which the domain is $[-3, -1] \cup [1, 5]$ and the range is $\{1, 2, 3, 4\}$. Answers may vary.

77. Suppose that for some function f, $f(x - 1) = 5x$. Find $f(6)$.

1.3

Linear Functions, Slope, and Applications

• *Determine the slope of a line given two points on the line.*
• *Solve applied problems involving linear functions.*

In real-life situations, we often need to make decisions on the basis of limited information. When the given information is used to formulate an equation or inequality that at least approximates the situation mathematically, we have created a **model**. One of the most frequently used mathematical models is *linear*—the graph of a linear model is a straight line.

Linear Functions

Let's begin to examine the connections among equations, functions, and graphs that are straight lines. Compare the graphs of linear and non-linear functions shown here and on the next page.

LINEAR FUNCTIONS

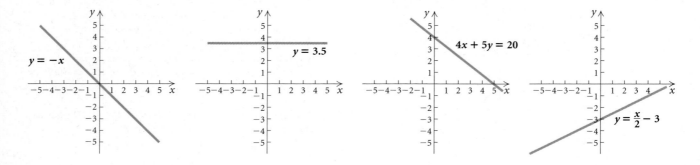

NONLINEAR FUNCTIONS

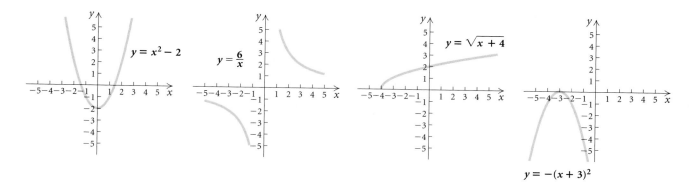

We have the following results and related terminology.

Linear Functions

A function f is a **linear function** if it can be written as

$$f(x) = mx + b,$$

where m and b are constants.

(If $m = 0$, the function is the **constant function** $f(x) = b$. If $m = 1$ and $b = 0$, the function is the **identity function** $f(x) = x$.)

Horizontal and Vertical Lines

Horizontal lines are given by equations of the type $y = b$ or $f(x) = b$. (They are functions.)

Vertical lines are given by equations of the type $x = a$. (They are *not* functions.)

Linear function: Identity function: Constant function, horizontal line: Not a function, vertical line:
$y = mx + b$ $y = x$ $y = b$ $x = a$

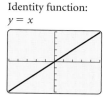
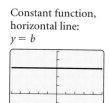

The Linear Function $f(x) = mx + b$ and Slope

To attach meaning to the constant m in the equation $f(x) = mx + b$, we first consider an application. FaxMax is an office machine business that currently has two stores in locations A and B in the same city. Their total costs for the same time period are given by two functions shown in the tables and graphs that follow. The variable x represents time, in months. The variable y represents total costs, in thousands of dollars, over that amount of time. Look for a pattern.

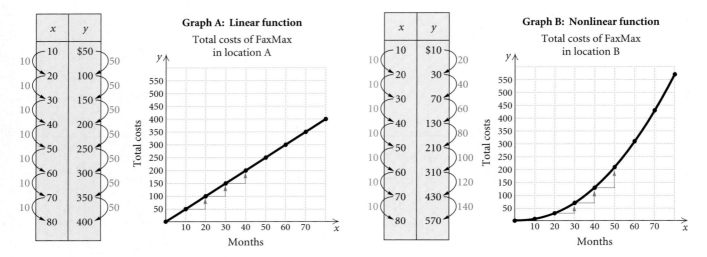

Graph A: Linear function

Total costs of FaxMax in location A

x	y
10	$50
20	100
30	150
40	200
50	250
60	300
70	350
80	400

Graph B: Nonlinear function

Total costs of FaxMax in location B

x	y
10	$10
20	30
30	70
40	130
50	210
60	310
70	430
80	570

We see in graph A that *every* change of 10 months results in a $50 thousand change in total costs. But in graph B, changes of 10 months do *not* result in constant changes in total costs. This is a way to distinguish linear from nonlinear functions. The rate at which a linear function changes, or the steepness of its graph, is constant.

Mathematically, we define a line's steepness, or **slope**, as the ratio of its vertical change (rise) to the corresponding horizontal change (run).

Slope

The **slope m** of a line containing points (x_1, y_1) and (x_2, y_2) is given by

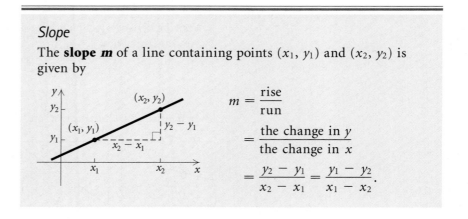

$$m = \frac{\text{rise}}{\text{run}}$$

$$= \frac{\text{the change in } y}{\text{the change in } x}$$

$$= \frac{y_2 - y_1}{x_2 - x_1} = \frac{y_1 - y_2}{x_1 - x_2}.$$

Technology Connection

Exploring the effect of the slope m in linear functions of the type $f(x) = mx$ becomes animated with the graphing calculator set in SEQUENTIAL mode. Graph the following equations:

$$y_1 = x, \qquad y_2 = 2x,$$
$$y_3 = 5x, \quad \text{and} \quad y_4 = 10x.$$

Try entering these equations as $y_1 = \{1, 2, 5, 10\}x$. What do you think the graph of $y = 128x$ will look like?

Clear the screen and graph the following equations:

$$y_1 = x, \qquad y_2 = 0.75x,$$
$$y_3 = 0.48x, \quad \text{and} \quad y_4 = 0.12x.$$

What do you think the graph of $y = 0.000029x$ will look like?

Again clear the screen and graph each set of equations:

$$y_1 = -x, \qquad y_2 = -2x,$$
$$y_3 = -4x, \quad \text{and} \quad y_4 = -10x$$

and

$$y_1 = -x, \qquad y_2 = -\tfrac{2}{3}x,$$
$$y_3 = -\tfrac{7}{20}x, \quad \text{and} \quad y_4 = -\tfrac{1}{10}x.$$

From your observations, what do you think the graphs of $y = -200x$ and $y = -\frac{17}{100,000}x$ will look like?

EXAMPLE 1 Graph the function $f(x) = -\frac{2}{3}x + 1$ and determine its slope.

Solution Since the equation for f is in the form $f(x) = mx + b$, we know it is a linear function. We can graph it by connecting two points on the graph with a straight line. We calculate two ordered pairs, plot the points, graph the function, and determine the slope:

$$f(3) = -\frac{2}{3} \cdot 3 + 1 = -2 + 1 = -1;$$

$$f(9) = -\frac{2}{3} \cdot 9 + 1 = -6 + 1 = -5;$$

Pairs: $(3, -1), (9, -5)$;

$$\text{Slope} = m = \frac{y_2 - y_1}{x_2 - x_1}$$

$$= \frac{-5 - (-1)}{9 - 3} = \frac{-4}{6} = -\frac{2}{3}.$$

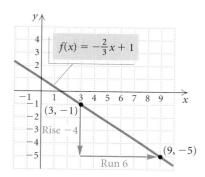

The slope is the same for any two points on a line. Thus, to check our work, note that $f(6) = -\frac{2}{3} \cdot 6 + 1 = -4 + 1 = -3$. Using the points $(6, -3)$ and $(3, -1)$, we have

$$m = \frac{-1 - (-3)}{3 - 6} = \frac{2}{-3} = -\frac{2}{3}.$$

We can also use the points in the opposite order when computing slope, so long as we are consistent:

$$m = \frac{-3 - (-1)}{6 - 3} = \frac{-2}{3} = -\frac{2}{3}.$$

Note also that the slope of the line is the number m in the equation for the function $f(x) = -\frac{2}{3}x + 1$. ▬

The slope of the line given by $y = mx + b$ is m.

If a line slants up from left to right, the change in x and the change in y have the same sign, so the line has a positive slope. The larger the slope is, the steeper the line. If a line slants down from left to right, the change in x and the change in y are of opposite signs, so the line has a negative slope. The larger the absolute value of the slope, the steeper the line. When $m = 0$, $y = 0x$, or $y = 0$. Note that this horizontal line is the x-axis.

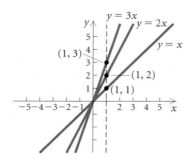

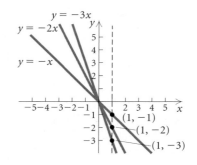

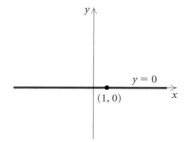

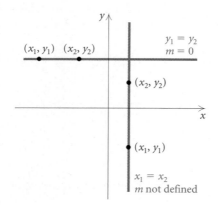

Horizontal and Vertical Lines

If a line is horizontal, the change in y for any two points is 0 and the change in x is nonzero. Thus a horizontal line has slope 0.

If a line is vertical, the change in y for any two points is nonzero and the change in x is 0. Thus the slope is *not defined* because we cannot divide by 0.

Note that zero slope and an undefined slope are two very different concepts.

EXAMPLE 2 Graph each linear equation and determine its slope.

a) $x = -2$ **b)** $y = \dfrac{5}{2}$

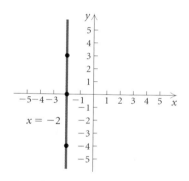

FIGURE 1

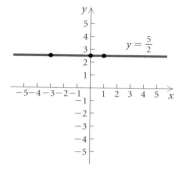

FIGURE 2

Solution

a) Since *y* is missing in $x = -2$, any value for *y* will do.

x	y
−2	0
−2	3
−2	−4

Choose any number for *y*; *x* must be −2.

The graph (see Fig. 1) is a *vertical line* 2 units to the left of the *y*-axis. The slope is not defined. The graph is not the graph of a function.

b) Since *x* is missing in $y = \frac{5}{2}$, any value for *x* will do.

x	y
0	$\frac{5}{2}$
−3	$\frac{5}{2}$
1	$\frac{5}{2}$

Choose any number for *x*; *y* must be $\frac{5}{2}$.

The graph (see Fig. 2) is a *horizontal line* $\frac{5}{2}$, or $2\frac{1}{2}$, units above the *x*-axis. The slope is 0. The graph is the graph of a function. ▬

Applications of Slope

Slope has many real-world applications. Numbers like 2%, 4%, and 7% are often used to represent the **grade** of a road. Such a number is meant to tell how steep a road is on a hill or mountain. For example, a 4% grade means that the road rises 4 ft for every horizontal distance of 100 ft.

The concept of grade is also used with a treadmill. During a treadmill test, a cardiologist might change the slope, or grade, of the treadmill to measure its effect on heart rate. Another example occurs in hydrology.

The strength or force of a river depends on how far the river falls vertically compared to how far it flows horizontally.

EXAMPLE 3 *Ramp for the Handicapped.* Construction laws regarding access ramps for the handicapped state that every vertical rise of 1 ft requires a horizontal run of 12 ft. What is the grade, or slope, of such a ramp?

Solution The grade, or slope, is given by $m = \frac{1}{12} \approx 0.083 \approx 8.3\%$.

Slope can also be considered as an **average rate of change.** To find the average rate of change between any two data points on a graph, we determine the slope of the line that passes through the two points.

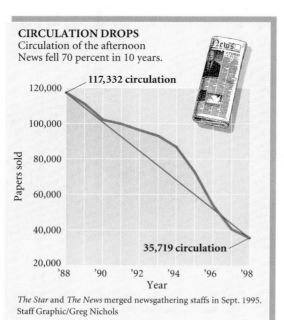

CIRCULATION DROPS
Circulation of the afternoon
News fell 70 percent in 10 years.

117,332 circulation

35,719 circulation

Papers sold

Year
'88   '90   '92   '94   '96   '98

The Star and *The News* merged newsgathering staffs in Sept. 1995.
Staff Graphic/Greg Nichols

EXAMPLE 4 *Newspaper Circulation.* On September 30, 1999, the last edition of *The Indianapolis News*, an afternoon daily paper, rolled off the press. Circulation of the paper had dropped approximately 70% in the 10 preceding years, as shown in the graph at left. Find the average rate of change in the circulation from 1988 to 1998.

Solution We determine the coordinates of two points on the graph. In this case, we will use (1988, 117,332) and (1998, 35,719). Then we compute the slope, or average rate of change, as follows:

$$\text{Slope} = \text{Average rate of change} = \frac{\text{Change in } y}{\text{Change in } x}$$

$$= \frac{35,719 - 117,332}{1998 - 1988} = \frac{-81,613}{10} = -8161.3.$$

This result tells us that, on average, each year the circulation of the paper decreased by approximately 8161 copies. The *average rate of change* over the 10-year period was −8161 copies per year.

Applications of Linear Functions

We now consider an application of linear functions.

Humerus

EXAMPLE 5 *Height Estimates.* An anthropologist can use linear functions to estimate the height of a male or a female, given the length of the humerus, the bone from the elbow to the shoulder. The height, in centimeters, of an adult male with a humerus of length x, in centimeters, is given by the function

$$M(x) = 2.89x + 70.64.$$

The height, in centimeters, of an adult female with a humerus of length x is given by the function

$$F(x) = 2.75x + 71.48.$$

A 26-cm humerus was uncovered in a ruins. Assuming it was from a female, how tall was she? What is the domain of this function?

Solution

We substitute into the function:

$$F(26) = 2.75(26) + 71.48 = 142.98.$$

Thus the female was 142.98 cm tall.

Theoretically, the domain of the function is the set of all real numbers. However, the context of the problem dictates a different domain. One could not find a bone with a length of 0 or less. Thus the domain consists of positive real numbers, that is, the interval $(0, \infty)$. (A more realistic domain might be 20 cm to 60 cm, the interval $[20, 60]$.)

Technology Connection

In Example 5, we can graph $F(x) = 2.75x + 71.48$ and use the VALUE feature to find $F(26)$.

$y = 2.75x + 71.48$

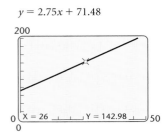

The female was 142.98 cm tall.

Exercise Set 1.3

In Exercises 1–4, the table of data contains input–output values for a function. Answer the following questions for each table.

a) Is the change in the inputs x the same?
b) Is the change in the outputs y the same?
c) Is the function linear?

1.

x	y
−3	7
−2	10
−1	13
0	16
1	19
2	22
3	25

2.

x	y
20	12.4
30	24.8
40	49.6
50	99.2
60	198.4
70	396.8
80	793.6

3.

x	y
11	3.2
26	5.7
41	8.2
56	9.3
71	11.3
86	13.7
101	19.1

4.

x	y
2	−8
4	−12
6	−16
8	−20
10	−24
12	−28
14	−36

Find the slope of the line containing the given points.

5.

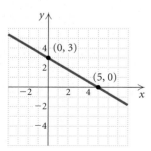

6.

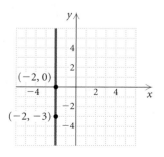

7.

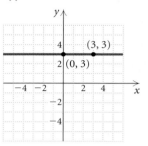

8.

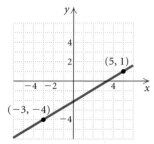

9. (9, 4) and (−1, 2)

10. (−3, 7) and (5, −1)

11. (4, −9) and (−5, 6)

12. (−6, −1) and (2, −13)

13. (π, −3) and (π, 2)

14. ($\sqrt{2}$, −4) and (0.56, −4)

15. (a, a^2) and ($a + h$, $(a + h)^2$)

16. (a, $3a + 1$) and ($a + h$, $3(a + h) + 1$)

17. *Road Grade.*   Using the following figure, find the road grade and an equation giving the height y as a function of the horizontal distance x.

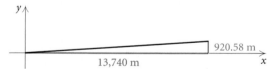

18. *Grade of Treadmills.*   A treadmill is 5 ft long and is set at an 8% grade. How high is the end of the treadmill?

19. Find the rate of change of the tuition and fees at public two-year colleges.

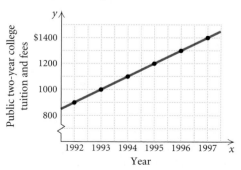

Source: Statistical Abstract of the United States

20. Find the rate of change of the cost of a formal wedding.

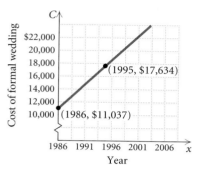

Source: *Modern Bride Magazine*

21. *Running Rate.* An Olympic marathoner passes the 10-km point of a race after 50 min and arrives at the 25-km point $1\frac{1}{2}$ hr later. Find the speed (average rate of change) of the runner.

22. *Work Rate.* As a typist resumes work on a project, $\frac{1}{6}$ of the paper has already been typed. Six hours later, the paper is $\frac{3}{4}$ done. Calculate the worker's typing rate.

23. *Ideal Weight.* One way to estimate the ideal weight of a woman in pounds is to multiply her height in inches by 3.5 and subtract 110. Let W = the ideal weight and h = height.

a) Express W as a linear function of h.
b) Find the ideal weight of a woman whose height is 62 in.
c) Find the domain of the function.

24. *Pressure at Sea Depth.* The function P, given by
$$P(d) = \frac{1}{33}d + 1,$$
gives the pressure, in atmospheres (atm), at a depth d, in feet, under the sea.

a) Find $P(0)$, $P(5)$, $P(10)$, $P(33)$, and $P(200)$.
b) Find the domain of the function.

25. *Stopping Distance on Glare Ice.* The stopping distance (at some fixed speed) of regular tires on glare ice is a function of the air temperature F, in degrees Fahrenheit. This function is estimated by
$$D(F) = 2F + 115,$$
where $D(F)$ is the stopping distance, in feet, when the air temperature is F, in degrees Fahrenheit.

a) Find $D(0°)$, $D(-20°)$, $D(10°)$, and $D(32°)$.
b) Explain why the domain should be restricted to $[-57.5°, 32°]$.

26. *Anthropology Estimates.* Consider Example 5 and the function
$$M(x) = 2.89x + 70.64$$
for estimating the height of a male.

a) If a 26-cm humerus from a male is found in an archeological dig, find the height of the male.
b) What is the domain of M?

27. *Reaction Time.* Suppose that while driving a car, you suddenly see a school crossing guard standing in the road. Your brain registers the information and sends a signal to your foot to hit the brake. The car travels a distance D, in feet, during this time, where D is a function of the speed r, in miles per hour, of the car when you see the crossing guard. That reaction distance is a linear function given by
$$D(r) = \frac{11r + 5}{10}.$$

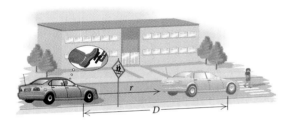

a) Find the slope of this line and interpret its meaning in this application.
b) Find $D(5)$, $D(10)$, $D(20)$, $D(50)$, and $D(65)$.
c) What is the domain of this function? Explain.

28. *Straight-line Depreciation.* A company buys a new color printer for $5200 to print banners for a sales campaign. The printer is purchased on January 1 and is expected to last 8 yr, at the end of which time its *trade-in*, or *salvage value*, will be $1100. If the company figures the decline or depreciation in value to be the same each year, then the salvage value V, after t years, is given by the linear function
$$V(t) = \$5200 - \$512.50t, \quad \text{for } 0 \le t \le 8.$$

a) Find $V(0)$, $V(1)$, $V(2)$, $V(3)$, and $V(8)$.
b) Find the domain and the range of this function.

In Exercises 29 and 30, express the slope as a ratio of two quantities.

29.

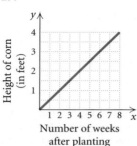

Number of weeks after planting

30.

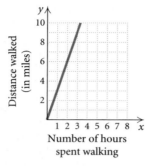

Number of hours spent walking

31. *Total Cost.* The Cellular Connection charges $60 for a phone and $40 per month under its economy plan. Write an equation that can be used to determine the total cost, $C(t)$, of operating a Cellular Connection phone for t months. Then find the total cost for 6 months.

32. *Total Cost.* Twin Cities Cable Television charges a $35 installation fee and $30 per month for "deluxe" service. Write an equation that can be used to determine the total cost, $C(t)$, for t months of deluxe cable television service. Then find the total cost for 8 months of service.

*In Exercises 33 and 34, the term **fixed costs** refers to the start-up costs of operating a business. This includes machinery and building costs. The term **variable costs** refers to what it costs a business to produce or service one item.*

33. Kara's Custom Tees experienced fixed costs of $800 and variable costs of $3 per shirt. Write an equation that can be used to determine the total costs encountered by Kara's Custom Tees. Then determine the total cost of producing 75 shirts.

34. It's My Racquet experienced fixed costs of $500 and variable costs of $2 for each tennis racquet that is restrung. Write an equation that can be used to determine the total costs encountered by It's My Racquet. Then determine the total cost of restringing 150 tennis racquets.

Technology Connection

35. Use the VALUE feature in the CALC menu to find the function values in Exercises 24(a), 25(a), 27(b), and 28(a).

Collaborative Discussion and Writing

36. Explain as you would to a fellow student how the numerical value of slope can be used to describe the slant and the steepness of a line.

37. Discuss why the graph of a vertical line $x = a$ cannot represent a function.

Skill Maintenance

If $f(x) = x^2 - 3x$, find each of the following.

38. $f(5)$ **39.** $f(-5)$

40. $f(-a)$ **41.** $f(a + h)$

Synthesis

42. *Fahrenheit and Celsius Temperatures.* Fahrenheit temperature F is a linear function of Celsius temperature C. When C is 0, F is 32; and when C is 100, F is 212. Use these data to express F as a function of C and to express C as a function of F.

Suppose that f is a linear function. Then $f(x) = mx + b$. Determine whether each of the following is true or false.

43. $f(cd) = f(c)f(d)$

44. $f(c + d) = f(c) + f(d)$

45. $f(c - d) = f(c) - f(d)$

46. $f(kx) = kf(x)$

Let $f(x) = mx + b$. Find a formula for $f(x)$ given each of the following.

47. $f(x + 2) = f(x) + 2$

48. $f(3x) = 3f(x)$

Equations of Lines and Modeling

- Determine equations of lines.
- Given the equations of two lines, determine whether their graphs are parallel or whether they are perpendicular.
- Model a set of data with a linear function.

Slope–Intercept Equations of Lines

Let's explore the effect of the constant b in linear equations of the type $f(x) = mx + b$. Compare the graphs of the equations

$$y = 3x$$

and

$$y = 3x - 2.$$

Note that the graph of $y = 3x - 2$ is a shift down of the graph of $y = 3x$, and that $y = 3x - 2$ has y-intercept $(0, -2)$. That is, the graph is parallel to $y = 3x$ and it crosses the y-axis at $(0, -2)$. The point $(0, -2)$ is the **y-intercept** of the graph.

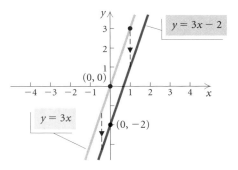

Technology Connection

We can explore the effect of the constant b in linear equations of the type $f(x) = mx + b$ with a graphing calculator. Begin with the graph of $y = x$. Now graph the lines $y = x + 3$ and $y = x - 4$ in the same viewing window. Try entering these equations as $y = x + \{0, 3, -4\}$ and compare the graphs. How do the last two lines differ from $y = x$? What do you think the line $y = x - 6$ will look like?

Try graphing $y = -0.5x$, $y = -0.5x - 4$, and $y = -0.5x + 3$ in the same viewing window. Describe what happens to the graph of $y = -0.5x$ when a number b is added.

The Slope–Intercept Equation

The linear function f given by

$$f(x) = mx + b$$

has a graph that is a straight line parallel to $y = mx$. The constant m is called the slope, and the y-intercept is $(0, b)$.

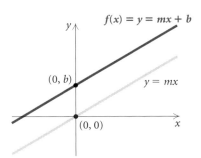

We know that a nonvertical line has slope m and y-intercept $(0, b)$ and that its equation can be written as $y = mx + b$. The advantage of writing the equation in the form $y = mx + b$ is that we can read the slope m and the y-intercept $(0, b)$ directly from the equation.

EXAMPLE 1 Find the slope and the y-intercept of the line with equation $y = -0.25x - 3.8$.

Solution

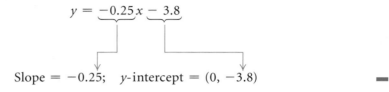

$$y = \underbrace{-0.25}x \underbrace{- 3.8}$$

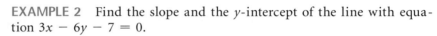

Slope $= -0.25$; y-intercept $= (0, -3.8)$ ▬

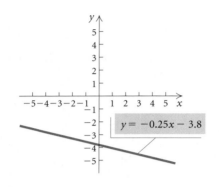

$y = -0.25x - 3.8$

Any equation whose graph is a straight line is a **linear equation.** To find the slope and the y-intercept of the graph of a linear equation, we can solve for y, and then read the information from the equation.

EXAMPLE 2 Find the slope and the y-intercept of the line with equation $3x - 6y - 7 = 0$.

Solution   We solve for y:

$$3x - 6y - 7 = 0$$
$$-6y = -3x + 7 \qquad \text{Adding } -3x \text{ and 7 on both sides}$$
$$-\tfrac{1}{6}(-6y) = -\tfrac{1}{6}(-3x + 7) \quad \text{Multiplying by } -\tfrac{1}{6}$$
$$y = \tfrac{1}{2}x - \tfrac{7}{6}.$$

Thus the slope is $\tfrac{1}{2}$, and the y-intercept is $\left(0, -\tfrac{7}{6}\right)$.   ▬

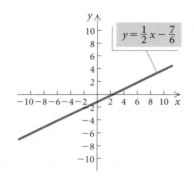

$y = \tfrac{1}{2}x - \tfrac{7}{6}$

EXAMPLE 3 A line has slope $-\tfrac{7}{9}$ and y-intercept $(0, 16)$. Find an equation of the line.

Solution   We use the slope–intercept equation and substitute $-\tfrac{7}{9}$ for m and 16 for b:

$$y = mx + b$$
$$y = -\tfrac{7}{9}x + 16.$$ ▬

Point–Slope Equations of Lines

Suppose that we have a nonvertical line and that the coordinates of point P_1 are (x_1, y_1). We can think of P_1 as fixed and imagine a movable point P on the line with coordinates (x, y). Thus the slope is given by

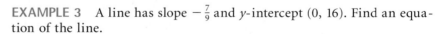

$$\frac{y - y_1}{x - x_1} = m.$$

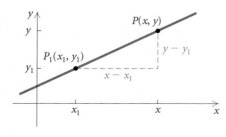

Multiplying on both sides by $x - x_1$, we get the *point–slope equation* of the line:

$$(x - x_1) \cdot \frac{y - y_1}{x - x_1} = m \cdot (x - x_1)$$

$$y - y_1 = m(x - x_1).$$

Point–Slope Equation

The **point–slope equation** of the line with slope m passing through (x_1, y_1) is

$$y - y_1 = m(x - x_1).$$

Thus if we know the slope of a line and the coordinates of one point on the line, we can find an equation of the line.

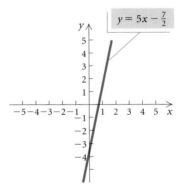

EXAMPLE 4   Find an equation of the line containing the point $\left(\frac{1}{2}, -1\right)$ and with slope 5.

Solution   If we substitute in $y - y_1 = m(x - x_1)$, we get

$$y - (-1) = 5\left(x - \tfrac{1}{2}\right),$$

which simplifies as follows:

$$y + 1 = 5x - \tfrac{5}{2}$$
$$y = 5x - \tfrac{5}{2} - 1$$
$$y = 5x - \tfrac{7}{2}. \qquad \text{Slope–intercept equation}$$

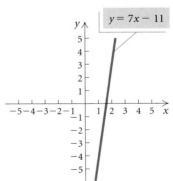

EXAMPLE 5   Find an equation of the line containing the points $(2, 3)$ and $(1, -4)$.

Solution   We first determine the slope:

$$m = \frac{-4 - 3}{1 - 2}$$
$$= \frac{-7}{-1}$$
$$= 7.$$

Then we substitute 7 for m and either of the points $(2, 3)$ or $(1, -4)$ for (x_1, y_1) in the point–slope equation. In this case, we use $(2, 3)$. We get

$$y - 3 = 7(x - 2),$$

which simplifies to

$$y - 3 = 7x - 14$$
$$y = 7x - 11.$$

Parallel Lines

Can we determine whether the graphs of two linear equations are parallel without graphing them? Let's observe three pairs of equations and their graphs.

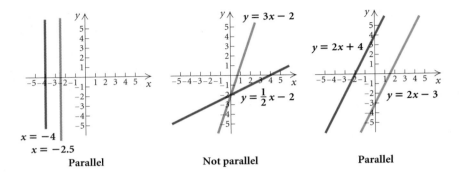

Parallel   Not parallel   Parallel

If two different lines are vertical, then they are parallel. Thus two equations such as $x = a_1$, $x = a_2$, where $a_1 \neq a_2$, have graphs that are *parallel lines*. Two nonvertical lines such as $y = mx + b_1$, $y = mx + b_2$, where the slopes are the same and $b_1 \neq b_2$, also have graphs that are *parallel lines*.

Parallel Lines

Vertical lines are **parallel**. Nonvertical lines are **parallel** if and only if they have the same slope and different y-intercepts.

Perpendicular Lines

Can we examine a pair of equations to determine whether their graphs are perpendicular without graphing the equations? Let's observe the following pairs of equations and their graphs.

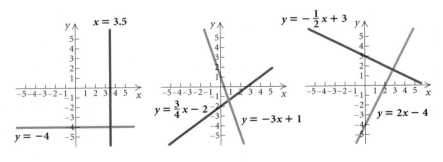

Perpendicular   Not perpendicular   Perpendicular

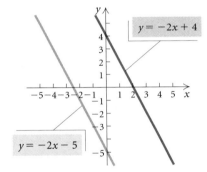

FIGURE 1

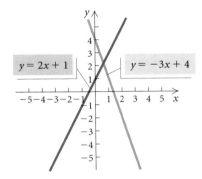

FIGURE 2

Perpendicular Lines

Two lines with slopes m_1 and m_2 are **perpendicular** if and only if the product of their slopes is -1:

$$m_1 m_2 = -1.$$

Lines are also **perpendicular** if one is vertical ($x = a$) and the other is horizontal ($y = b$).

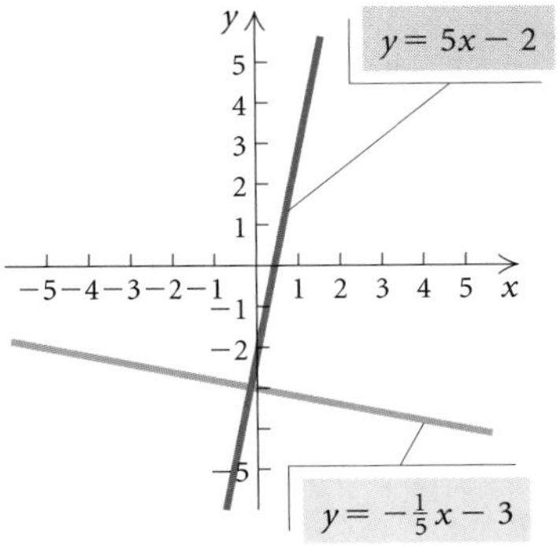

If a line has slope m_1, the slope m_2 of a line perpendicular to it is $-1/m_1$ (the slope of one line is the opposite of the reciprocal of the other).

EXAMPLE 6 Determine whether each of the following pairs of lines is parallel, perpendicular, or neither.

a) $y + 2 = 5x$, $\quad 5y + x = -15$
b) $2y + 4x = 8$, $\quad 5 + 2x = -y$
c) $2x + 1 = y$, $\quad y + 3x = 4$

Solution We use the slopes of the lines to determine whether the lines are parallel or perpendicular.

a) We solve each equation for y:

$$y = 5x - 2, \qquad y = -\tfrac{1}{5}x - 3.$$

The slopes are 5 and $-\tfrac{1}{5}$. Their product is -1, so the lines are perpendicular (see Fig. 1).

b) Solving each equation for y, we get

$$y = -2x + 4, \qquad y = -2x - 5.$$

We see that $m_1 = -2$ and $m_2 = -2$. Since the slopes are the same and the y-intercepts, 4 and -5, are different, the lines are parallel (see Fig. 2).

c) Solving each equation for y, we get

$$y = 2x + 1, \qquad y = -3x + 4.$$

Since $m_1 = 2$ and $m_2 = -3$, we know that $m_1 \neq m_2$ and that $m_1 m_2 \neq -1$. It follows that the lines are neither parallel nor perpendicular (see Fig. 3).

FIGURE 3

EXAMPLE 7 Write equations of the lines (a) parallel and (b) perpendicular to the graph of the line $4y - x = 20$ and containing the point $(2, -3)$.

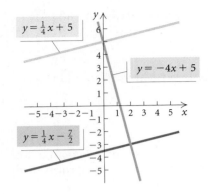

Solution We first solve $4y - x = 20$ for y to get $y = \frac{1}{4}x + 5$. Thus the slope of the given line is $\frac{1}{4}$.

a) The line parallel to the given line will have slope $\frac{1}{4}$. We use the point–slope equation for a line with slope $\frac{1}{4}$ and containing the point $(2, -3)$:

$$y - y_1 = m(x - x_1)$$
$$y - (-3) = \tfrac{1}{4}(x - 2)$$
$$y + 3 = \tfrac{1}{4}x - \tfrac{1}{2}$$
$$y = \tfrac{1}{4}x - \tfrac{7}{2}.$$

b) The slope of the perpendicular line is the opposite of the reciprocal of $\frac{1}{4}$, or -4. Now we use the point–slope equation to write an equation for a line with slope -4 and containing the point $(2, -3)$:

$$y - y_1 = m(x - x_1)$$
$$y - (-3) = -4(x - 2)$$
$$y + 3 = -4x + 8$$
$$y = -4x + 5.$$

Summary of Terminology About Lines

TERMINOLOGY	MATHEMATICAL INTERPRETATION
Slope	$m = \dfrac{y_2 - y_1}{x_2 - x_1}$, or $\dfrac{y_1 - y_2}{x_1 - x_2}$
Slope–intercept equation	$y = mx + b$
Point–slope equation	$y - y_1 = m(x - x_1)$
Horizontal lines	$y = b$
Vertical lines	$x = a$
Parallel lines	$m_1 = m_2,\ b_1 \neq b_2$
Perpendicular lines	$m_1 m_2 = -1$

Mathematical Models

When a real-world problem can be described in mathematical language, we have a **mathematical model.** For example, the natural numbers constitute a mathematical model for situations in which counting is essential. Situations in which algebra can be brought to bear often require the use of functions as models.

Mathematical models are abstracted from real-world situations. Procedures within the mathematical model give results that allow one to predict what will happen in that real-world situation. If the predictions are inaccurate or the results of experimentation do not conform to the model, the model needs to be changed or discarded.

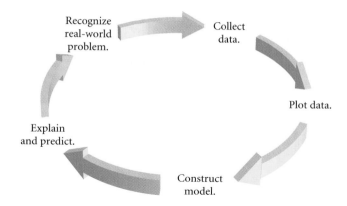

Mathematical modeling can be an ongoing process. For example, finding a mathematical model that will enable an accurate prediction of population growth is not a simple problem. Any population model that one might devise would need to be reshaped as further information is acquired.

Curve Fitting

We will develop and use many kinds of mathematical models in this text. In this chapter, we have used *linear* functions as models. Other types of functions, such as quadratic, cubic, or exponential functions, can also model data. These functions are *nonlinear.* Modeling with quadratic and cubic functions will be discussed in Chapter 3. Modeling with exponential functions will be discussed in Chapter 4.

Quadratic function:
$y = ax^2 + bx + c, a > 0$

Cubic function:
$y = ax^3 + bx^2 + cx + d, a > 0$

Exponential function:
$y = ab^x, a, b > 0$

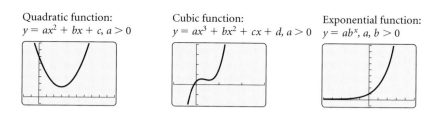

In general, we try to find a function that fits, as well as possible, observations (data), theoretical reasoning, and common sense. We call this **curve fitting;** it is one aspect of mathematical modeling.

Let's look at some data and related graphs or **scatterplots** and determine whether any of the above functions, linear or nonlinear, seem to fit either set of data.

Number of Apartment Households in the United States

YEAR, x	NUMBER OF APARTMENT HOUSEHOLDS, y (IN MILLIONS)	SCATTERPLOT	
1970, 0 1975, 5 1980, 10 1985, 15 1990, 20 1997, 27	8.5 9.9 10.8 12.9 14.2 14.5		It appears that the data points can be represented or modeled by a linear function. The graph is **linear**.

Sources: U.S. Bureau of the Census; National Multi-Housing Council

Number of Cellular Phone Subscribers

YEAR, x	NUMBER OF CELLULAR PHONE SUBSCRIBERS, y (IN MILLIONS)	SCATTERPLOT	
1985, 0 1986, 1 1987, 2 1988, 3 1989, 4 1990, 5 1991, 6 1992, 7 1993, 8 1994, 9 1995, 10 1996, 11 1997, 12	0.3 0.7 1.2 2.1 3.5 5.3 7.6 11.0 16.0 24.1 33.8 44.0 55.3		It appears that the data points cannot be modeled by a linear function. The graph is **nonlinear**.

Source: Cellular Telecommunications Industry Association

Looking at the scatterplots, we see that the apartment households data seem to be rising in a manner to suggest that a *linear function* might fit, although a "perfect" straight line cannot be drawn through the data points. A linear function does not seem to fit the cellular phone data.

EXAMPLE 8 *Apartment Households.* Model the data in the apartment household table on the preceding page with two different linear functions. Then with each function, predict the number of apartment households in 2003. Of the two models, which appears to be the better fit?

Solution We can choose any two of the data points to determine an equation. Let's use (10, 10.8) and (20, 14.2) for our first model.

We first determine the slope of the line:

$$m = \frac{14.2 - 10.8}{20 - 10} = \frac{3.4}{10} = 0.34.$$

Then we substitute 0.34 for m and either of the points (10, 10.8) or (20, 14.2) for (x_1, y_1) in the point–slope equation. In this case, we use (10, 10.8). We get

$$y - 10.8 = 0.34(x - 10),$$

which simplifies to

$$y = 0.34x + 7.4. \qquad \text{Model I}$$

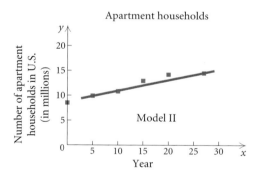

For the second model, let's use (5, 9.9) and (27, 14.5). Again we first determine the slope:

$$m = \frac{14.5 - 9.9}{27 - 5} = \frac{4.6}{22} \approx 0.21.$$

Substuting 0.21 for m and (5, 9.9) for (x_1, y_1), we get

$$y - 9.9 = 0.21(x - 5),$$

which simplifies to

$$y = 0.21x + 8.85. \qquad \text{Model II}$$

Now we can predict the number of apartment households in 2003 by substituting 33 for x (2003 − 1970 = 33) in each model:

$$\text{Model I:} \quad y = 0.34x + 7.4$$
$$y = 0.34(33) + 7.4$$
$$\approx 18.6;$$

$$\text{Model II:} \quad y = 0.21x + 8.85$$
$$y = 0.21(33) + 8.85$$
$$\approx 15.8.$$

Thus, using model I, we predict that there will be 18.6 million apartment households in 2003 and using model II, we predict 15.8 million.

Since it appears from the graphs above that model II fits the data more closely, we would choose model II over model I.

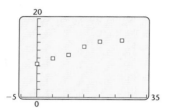

FIGURE 1

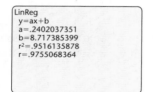

FIGURE 2

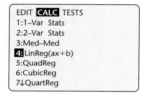

FIGURE 3

FIGURE 4

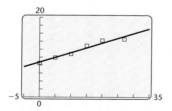

FIGURE 5

Technology Connection

The Regression Line

We now consider **linear regression,** a procedure that can be used to model a set of data using a linear function. Although discussion leading to a complete understanding of this method belongs in a statistics course, we present the procedure here because we can carry it out easily using technology. The graphing calculator gives us the powerful capability to find linear models and to make predictions using them.

Consider the data presented before Example 8 on the number of apartment households in the United States. We can fit a regression line of the form $y = mx + b$ to the data using the LINEAR REGRESSION feature on a graphing calculator.

First, we enter the data in lists on the calculator. We enter the values of the independent variable x in list L1 and the corresponding values of the dependent variable y in L2 (see Fig. 1). The graphing calculator can then create a scatterplot of the data, as shown in Fig. 2.

When we select the LINEAR REGRESSION feature from the STAT CALC menu, we find the linear equation that best models the data. It is

$$y = 0.2402037351x + 8.717385399 \qquad \text{Regression line}$$

(see Figs. 3 and 4). We can then graph the regression line on the same graph as the scatterplot shown in Fig. 5.

To predict the number of apartment households in 2003, we evaluate the function found in part (a) for $x = 33$.

Using this model, we see that the number of apartment households in 2003 is predicted to be about 16.6 million.

Y1(33)	
	16.64410866

The Correlation Coefficient

On some graphing calculators with DIAGNOSTIC turned on, a constant r between -1 and 1, called the **coefficient of linear correlation,** appears with the equation of the regression line. Though we cannot develop a formula for calculating r in this text, keep in mind that it is used to describe the strength of the linear relationship between x and y. The closer $|r|$ is to 1, the better the correlation. A positive value of r also indicates that the regression line has a positive slope, and a negative value of r indicates that the regression line has a negative slope. For the apartment household data just discussed,

$$r = 0.9755068364,$$

which indicates a very good linear correlation.

Exercise Set 1.4

Find the slope and the y-intercept of the equation.

1. $y = \frac{3}{5}x - 7$ **2.** $f(x) = -2x + 3$

3. $f(x) = 5 - \frac{1}{2}x$ **4.** $y = 2 + \frac{3}{7}x$

5. $3x + 2y = 10$ **6.** $2x - 3y = 12$

7. $4x - 3f(x) - 15 = 6$ **8.** $9 = 3 + 5x - 2f(x)$

Write a slope–intercept equation for a line with the given characteristics.

9. $m = \frac{2}{9}$, *y*-intercept $(0, 4)$

10. $m = -\frac{3}{8}$, *y*-intercept $(0, 5)$

11. $m = -4$, *y*-intercept $(0, -7)$

12. $m = \frac{2}{7}$, *y*-intercept $(0, -6)$

13. $m = -4.2$, *y*-intercept $\left(0, \frac{3}{4}\right)$

14. $m = -4$, *y*-intercept $\left(0, -\frac{3}{2}\right)$

15. $m = \frac{2}{9}$, passes through $(3, 7)$

16. $m = -\frac{3}{8}$, passes through $(5, 6)$

17. $m = 3$, passes through $(1, -2)$

18. $m = -2$, passes through $(-5, 1)$

19. $m = -\frac{3}{5}$, passes through $(-4, -1)$

20. $m = \frac{2}{3}$, passes through $(-4, -5)$

21. Passes through $(-1, 5)$ and $(2, -4)$

22. Passes through $(2, -1)$ and $(7, -11)$

23. Passes through $(7, 0)$ and $(-1, 4)$

24. Passes through $(-3, 7)$ and $(-1, -5)$

Determine whether the pair of lines is parallel, perpendicular, or neither.

25. $x + 2y = 5$,
$2x + 4y = 8$

26. $2x - 5y = -3$,
$2x + 5y = 4$

27. $y = 4x - 5$,
$4y = 8 - x$

28. $y = 7 - x$,
$y = x + 3$

Write a slope–intercept equation for a line passing through the given point that is parallel to the given line. Then write a second equation for a line passing through the given point that is perpendicular to the given line.

29. $(3, 5)$, $y = \frac{2}{7}x + 1$

30. $(-1, 6)$, $f(x) = 2x + 9$

31. $(-7, 0)$, $y = -0.3x + 4.3$

32. $(-4, -5)$, $2x + y = -4$

33. $(3, -2)$, $3x + 4y = 5$

34. $(8, -2)$, $y = 4.2(x - 3) + 1$

35. $(3, -3)$, $x = -1$ **36.** $(4, -5)$, $y = -1$

37. *Garbage Production.* America creates more garbage than any other nation. According to Denis Hayes, president of Seattle's nonprofit Bullitt Foundation and one of the founders of Earth Day in 1970, "We need to be an Heirloom Society instead of a Throw-Away Society." The Environmental Protection Agency estimates that, on average, we each produce 4.4 lb of garbage a day (*Source*: "Take Out the Trash, and Put It—Where?" by Bernard Bavzer, *Parade Magazine*, June 13, 1999). The following graph illustrates the annual production of municipal solid waste in the United States for certain years.

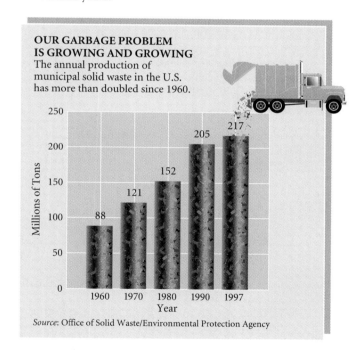

OUR GARBAGE PROBLEM IS GROWING AND GROWING
The annual production of municipal solid waste in the U.S. has more than doubled since 1960.

Source: Office of Solid Waste/Environmental Protection Agency

a) Model the data with two linear functions. Let the independent variable represent the number of years after 1960. That is, the data points are $(0, 88)$, $(10, 121)$, and so on. Answers will vary depending on the data points used.

b) With each function found in part (a), predict the amount of municipal solid waste in 2005.

c) Which of the two models appears to fit the data more closely (that is, provides the more realistic prediction)?

38. *The Cost of Tuition at State Universities.* Model the data in the chart below with a linear function and predict the costs of college tuition in 2004–2005, 2006–2007, and 2010–2011. Answers will vary depending on the data points used.

COLLEGE YEAR, *x*	ESTIMATED TUITION, *y*
1997–1998, 0	$ 9,838
1998–1999, 1	10,424
1999–2000, 2	11,054
2000–2001, 3	11,717
2001–2002, 4	12,420

Source: College Board, Senate Labor Committee

Technology Connection

39. *Maximum Heart Rate.* A person who is exercising should not exceed his or her maximum heart rate, which is determined on the basis of that person's sex, age, and resting heart rate. The following table relates resting heart rate and maximum heart rate for a 20-yr-old man.

RESTING HEART RATE, *H* (IN BEATS PER MINUTE)	MAXIMUM HEART RATE, *M* (IN BEATS PER MINUTE)
50	166
60	168
70	170
80	172

Source: American Heart Association

a) Use a graphing calculator to model the data with a linear function.
b) Estimate the maximum heart rate if the resting heart rate is 40, 65, 76, and 84.
c) What is the correlation coefficient? How confident are you about using the regression line as an estimate?

40. *Study Time versus Grades.* A math instructor asked her students to keep track of how much time each spent studying a chapter on functions in her algebra–trigonometry course. She collected the information together with test scores from that chapter's test. The data are listed in the table at the top of the next column.

STUDY TIME, *x* (IN HOURS)	TEST GRADE, *y* (IN PERCENT)
23	81
15	85
17	80
9	75
21	86
13	80
16	85
11	93

a) Use a graphing calculator to model the data with a linear function.
b) Predict a student's score if he or she studies 24 hr, 6 hr, and 18 hr.
c) What is the correlation coefficient? How confident are you about using the regression line as a predictor?

41. a) Use a graphing calculator to fit a regression line to the data in Exercise 37.
b) Predict the amount of municipal solid waste in 2005 and compare the result with the result found with the model in Exercise 37.
c) Find the correlation coefficient for the regression line and determine whether the line fits the data closely.

42. a) Use a graphing calculator to fit a regression line to the data in Exercise 38.
b) Predict the tuition costs in 2004–2005, 2006–2007, and 2010–2011, and compare the values with the results found in Exercise 38.
c) Find the correlation coefficient for the regression line and determine whether the line fits the data closely.

Collaborative Discussion and Writing

43. If one line has a slope of $-\frac{3}{4}$ and another has a slope of $\frac{2}{5}$, which line is steeper? Why?

44. A company offers its new employees a starting salary with a guaranteed 5% increase each year. Can a linear function be used to express the yearly salary as a function of the number of years an employee has worked? Why or why not?

Skill Maintenance

Find the slope of the line containing the given points.

45. $(2, -8)$ and $(-5, -1)$ **46.** $(5, 7)$ and $(5, -7)$

Find an equation for a circle satisfying the given conditions.

47. Center $(-7, -1)$, radius of length $\frac{9}{5}$

48. Center $(0, 3)$, diameter of length 5

Synthesis

49. Find k so that the line containing the points $(-3, k)$ and $(4, 8)$ is parallel to the line containing the points $(5, 3)$ and $(1, -6)$.

50. Find an equation of the line passing through the point $(4, 5)$ perpendicular to the line passing through the points $(-1, 3)$ and $(2, 9)$.

More on Functions

- *Graph functions, looking for intervals on which the function is increasing, decreasing, or constant, and estimate relative maxima and minima.*
- *Given an application, find a function that models the application; find the domain of the function and function values, and then graph the function.*
- *Graph functions defined piecewise.*
- *Find the sum, the difference, the product, and the quotient of two functions, and determine their domains.*

Because functions occur in so many real-world situations, it is important to be able to analyze them carefully.

Increasing, Decreasing, and Constant Functions

On a given interval, if the graph of a function rises from left to right, it is said to be **increasing** on that interval. If the graph drops from left to right, it is said to be **decreasing**. If the function stays the same from left to right, it is said to be **constant**.

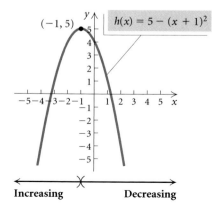

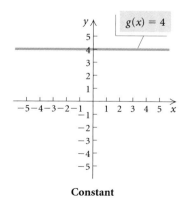

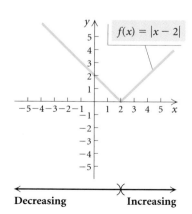

We are led to the following definitions.

Increasing, Decreasing, and Constant Functions

A function f is said to be **increasing** on an open interval I, if for all a and b in that interval, $a < b$ implies $f(a) < f(b)$.

A function f is said to be **decreasing** on an open interval I, if for all a and b in that interval, $a < b$ implies $f(a) > f(b)$.

A function f is said to be **constant** on an open interval I, if for all a and b in that interval, $f(a) = f(b)$.

EXAMPLE 1   Determine the intervals on which the function in the figure at left is (**a**) increasing; (**b**) decreasing; (**c**) constant.

Solution

a) The function is increasing on the interval $(3, 5)$.

b) The function is decreasing on the intervals $(-\infty, -1)$ and $(5, \infty)$.

c) The function is constant on the interval $(-1, 3)$.

Relative Maximum and Minimum Values

Consider the graph shown below. Note the "peaks" and "valleys" at the points c_1, c_2, and c_3. The function value $f(c_2)$ is called a **relative maximum** (plural, **maxima**). Each of the function values $f(c_1)$ and $f(c_3)$ is called a **relative minimum** (plural, **minima**).

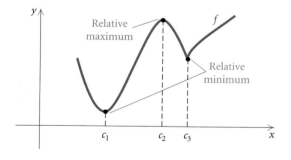

Relative Maxima and Minima

Suppose that f is a function for which $f(c)$ exists for some c in the domain of f. Then:

  $f(c)$ is a **relative maximum** if there exists an open interval I containing c such that $f(c) > f(x)$, for all x in I where $x \neq c$; and

  $f(c)$ is a **relative minimum** if there exists an open interval I containing c such that $f(c) < f(x)$, for all x in I where $x \neq c$.

Technology Connection

We can approximate relative maximum and minimum values with MAXIMUM and MINIMUM features on the CALC menu on a graphing calculator.

$y = 0.1x^3 - 0.6x^2 - 0.1x + 2$

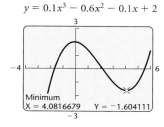

$y = 0.1x^3 - 0.6x^2 - 0.1x + 2$

Simply stated, $f(c)$ is a relative maximum if $f(c)$ is the highest point in some open interval, and $f(c)$ is a relative minimum if $f(c)$ is the lowest point in some open interval.

If you take a calculus course, you will learn a method for determining exact values of relative maxima and minima. In Section 2.4, we will find exact maximum and minimum values of quadratic functions algebraically. MAXIMUM or MINIMUM features on a graphing calculator can be used to approximate relative maxima or minima. In this section, we will read the values shown on a graph.

EXAMPLE 2   Using the graph shown below, determine any relative maxima or minima of the function $f(x) = 0.1x^3 - 0.6x^2 - 0.1x + 2$ and intervals on which the function is increasing or decreasing.

Solution   We see that the relative maximum value of the function is 2.004. It occurs when $x = -0.082$. We also see the relative minimum: -1.604 at $x = 4.082$.

We note that the graph starts rising, or increasing, from the left and stops increasing at the relative maximum. From this point, the graph decreases to the relative minimum and then begins to rise again. The function is *increasing* on the intervals

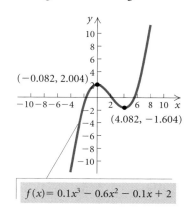

$f(x) = 0.1x^3 - 0.6x^2 - 0.1x + 2$

$$(-\infty, -0.082) \quad \text{and} \quad (4.082, \infty)$$

and *decreasing* on the interval

$$(-0.082, 4.082).$$

Applications of Functions

Many real-world situations can be modeled by functions.

EXAMPLE 3   *Car Distance.* Jamie and Ruben drive away from a restaurant at right angles to each other. Jamie's speed is 65 mph and Ruben's is 55 mph.

a) Express the distance between the cars as a function of time.

b) Find the domain of the function.

Solution

a) Suppose 1 hr goes by. At that time, Jamie has traveled 65 mi and Ruben has traveled 55 mi. We can use the Pythagorean theorem then to find the distance between them. This distance would be the length of the hypotenuse of a triangle with legs measuring 65 mi and 55 mi. After 2 hr, the triangle's legs would measure 130 mi and 110 mi. Observing that the distances will always be changing, we make a drawing and let t = the time, in hours, that Jamie and Ruben have been driving since leaving the restaurant.

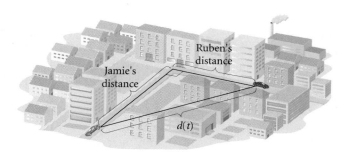

After t hours, Jamie has traveled $65t$ miles and Ruben $55t$ miles. We can use the Pythagorean theorem:

$$[d(t)]^2 = (65t)^2 + (55t)^2.$$

Because distance must be nonnegative, we need only consider the positive square root when solving for $d(t)$:

$$
\begin{aligned}
d(t) &= \sqrt{(65t)^2 + (55t)^2} \\
&= \sqrt{4225t^2 + 3025t^2} \\
&= \sqrt{7250t^2} \\
&= \sqrt{7250}\,\sqrt{t^2} \\
&\approx 85.15|t| \qquad \text{Approximating the root to two decimal places} \\
&\approx 85.15t. \qquad \text{Since } t \geq 0,\ |t| = t.
\end{aligned}
$$

Thus, $d(t) = 85.15t$, $t \geq 0$.

b) Since the time traveled must be nonnegative, the domain is the set of nonnegative real numbers $[0, \infty)$. ▬

EXAMPLE 4   *Storage Area.* The Sound Shop has 20 ft of dividers with which to set off a rectangular area for the storage of overstock. If a corner of the store is used, the partition need only form two sides of a rectangle.

a) Express the floor area of the storage space as a function of the length of the partition.

b) Find the domain of the function.

c) With the graph shown on the following page, determine the dimensions that maximize the area of the floor.

Solution

a) Note that the dividers will form two sides of a rectangle. If, for example, 14 ft of dividers are used for the length of the rectangle, that would leave $20 - 14$, or 6 ft of dividers for the width. Thus if x = the length, in feet, of the rectangle, then $20 - x$ = the width. We represent this information in a sketch, as shown at left.

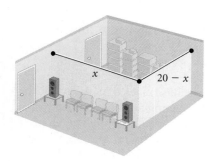

The area, $A(x)$, is given by

$$A(x) = x(20 - x) \qquad \text{Area} = \text{length} \cdot \text{width}$$
$$= 20x - x^2.$$

The function $A(x) = 20x - x^2$ can be used to express the rectangle's area as a function of the length.

b) Because the rectangle's length must be positive and only 20 ft of dividers is available, we restrict the domain of A to $\{x \mid 0 < x < 20\}$, that is, the interval $(0, 20)$.

c) On the graph of the function, the maximum value of the area function on the interval $(0, 20)$ appears to be 100 when $x = 10$. Thus the dimensions that maximize the area are

$$\text{Length} = x = 10 \text{ ft} \quad \text{and}$$
$$\text{Width} = 20 - x = 20 - 10 = 10 \text{ ft.}$$

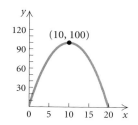

Functions Defined Piecewise

Sometimes functions are defined **piecewise** using different output formulas for different parts of the domain.

EXAMPLE 5 Graph the function defined as

$$f(x) = \begin{cases} 4, & \text{for } x \le 0, \\ 4 - x^2, & \text{for } 0 < x \le 2, \\ 2x - 6, & \text{for } x > 2. \end{cases}$$

Solution We create the graph in three parts, as shown and described below.

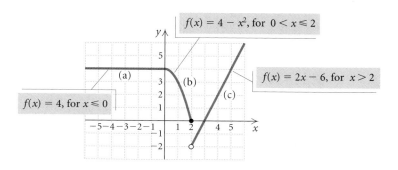

a) We graph $f(x) = 4$ *only* for inputs x less than or equal to 0 (that is, on the interval $(-\infty, 0]$).

b) We graph $f(x) = 4 - x^2$ *only* for inputs x greater than 0 and less than or equal to 2 (that is, on the interval $(0, 2]$).

c) We graph $f(x) = 2x - 6$ *only* for inputs x greater than 2 (that is, on the interval $(2, \infty)$).

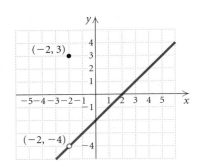

EXAMPLE 6 Graph the function defined as

$$f(x) = \begin{cases} \dfrac{x^2 - 4}{x + 2}, & \text{for } x \neq -2, \\ 3, & \text{for } x = -2. \end{cases}$$

Solution When $x \neq -2$, the denominator of $(x^2 - 4)/(x + 2)$ is nonzero, so we can simplify:

$$\frac{x^2 - 4}{x + 2} = \frac{(x + 2)(x - 2)}{x + 2} = x - 2.$$

Thus,

$$f(x) = x - 2, \quad \text{for } x \neq -2.$$

The graph of this part of the function consists of a line with a "hole" at the point $(-2, -4)$, indicated by the open dot. At $x = -2$, we have $f(-2) = 3$, so the point $(-2, 3)$ is plotted above the open dot. ▬

A piecewise function with importance in calculus and computer programming is the *greatest integer function, f,* denoted as $f(x) = [\![x]\!]$, or int(x).

Greatest Integer Function
$f(x) = [\![x]\!] =$ the greatest integer less than or equal to x.

The greatest integer function pairs each input with the greatest integer less than or equal to that input. Thus, 1, 1.2, and 1.9 are all paired with the y-value 1. Similarly, 2, 2.3, and 2.8 are all paired with 2, and -3.4, -3.06, and -3.99027 are all paired with -4, and so on.

EXAMPLE 7 Graph $f(x) = [\![x]\!]$ and determine its domain and range.

Solution The greatest integer function can also be defined as a piecewise function with an infinite number of statements.

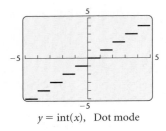
$$f(x) = [\![x]\!] = \begin{cases} \vdots \\ -3 & \text{for } -3 \leq x < -2 \\ -2 & \text{for } -2 \leq x < -1 \\ -1 & \text{for } -1 \leq x < 0 \\ 0 & \text{for } 0 \leq x < 1 \\ 1 & \text{for } 1 \leq x < 2 \\ 2 & \text{for } 2 \leq x < 3 \\ 3 & \text{for } 3 \leq x < 4 \\ \vdots \end{cases}$$

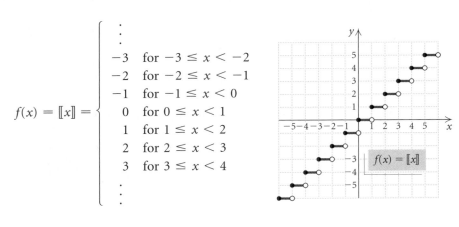

We see that the domain of $f(x)$ is the set of all real numbers, $(-\infty, \infty)$, and the range is the set of all integers, $\{\ldots, -3, -2, -1, 0, 1, 2, 3, \ldots\}$. ▬

The Algebra of Functions: Sums, Differences, Products, and Quotients

We now use addition, subtraction, multiplication, and division to combine functions and obtain new functions.

Consider the following two functions f and g:

$$f(x) = x + 2 \quad \text{and} \quad g(x) = x^2 + 1.$$

Since $f(3) = 3 + 2 = 5$ and $g(3) = 3^2 + 1 = 10$, we have

$$f(3) + g(3) = 5 + 10 = 15,$$
$$f(3) - g(3) = 5 - 10 = -5,$$
$$f(3) \cdot g(3) = 5 \cdot 10 = 50, \quad \text{and} \quad \frac{f(3)}{g(3)} = \frac{5}{10} = \frac{1}{2}.$$

In fact, so long as x is in the domain of both f and g, we can easily compute $f(x) + g(x)$, $f(x) - g(x)$, $f(x) \cdot g(x)$, and, assuming $g(x) \neq 0$, $f(x)/g(x)$. Notation has been developed to facilitate this work.

Sums, Differences, Products, and Quotients of Functions

If f and g are functions and x is in the domain of each function, then

$(f + g)(x) = f(x) + g(x)$, $\quad (f - g)(x) = f(x) - g(x)$,

$(fg)(x) = f(x) \cdot g(x)$, $\quad (f/g)(x) = f(x)/g(x)$,

provided $g(x) \neq 0$.

EXAMPLE 8 Given that $f(x) = x + 1$ and $g(x) = \sqrt{x + 3}$, find each of the following.

a) $(f + g)(x)$ **b)** $(f + g)(6)$ **c)** $(f + g)(-4)$

Solution

a) $(f + g)(x) = f(x) + g(x)$
$$= x + 1 + \sqrt{x + 3} \qquad \text{This cannot be simplified.}$$

b) We can find $(f + g)(6)$ provided 6 is in the domain of each function. The domain of f is all real numbers. The domain of g is all real numbers x for which $x + 3 \geq 0$, or $x \geq -3$. This is the interval $[-3, \infty)$. Thus, 6 is in both domains, so we have

$$f(6) = 6 + 1 = 7, \qquad g(6) = \sqrt{6 + 3} = \sqrt{9} = 3,$$
$$(f + g)(6) = f(6) + g(6) = 7 + 3 = 10.$$

Another method is to use the formula found in part (a):

$$(f + g)(6) = 6 + 1 + \sqrt{6 + 3} = 7 + \sqrt{9} = 7 + 3 = 10.$$

c) To find $(f + g)(-4)$, we must first determine whether -4 is in the domain of each function. We note that -4 is not in the domain of g, $[-3, \infty)$. That is, $\sqrt{-4 + 3}$ is not a real number. Thus, $(f + g)(-4)$ does not exist.   ▬

It is useful to view the concept of the sum of two functions graphically. In the graph below, we see the graphs of two functions f and g and their sum, $f + g$. Consider finding $(f + g)(4) = f(4) + g(4)$. We can locate $g(4)$ on the graph of g and use a compass to measure it. Then we add that length on top of $f(4)$ on the graph of f. The sum gives us $(f + g)(4)$.

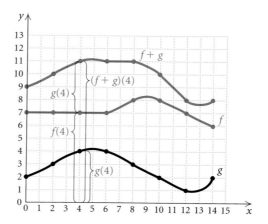

With this in mind, let's view Example 8 from a graphical perspective. Let's look at the graphs of

$$f(x) = x + 1, \qquad g(x) = \sqrt{x + 3}, \quad \text{and}$$
$$(f + g)(x) = x + 1 + \sqrt{x + 3}.$$

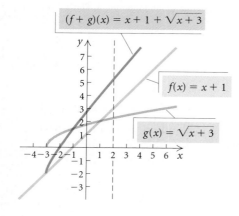

See the graph at left. Note that because the domain of $g(x)$ is $[-3, \infty)$, the graph of the sum exists only on $[-3, \infty)$. We can confirm that the y-coordinates of the graph of $(f + g)(x)$ are the sums of the corresponding y-coordinates of the graphs of $f(x)$ and $g(x)$. Here we confirm it for $x = 2$.

$$f(x) = x + 1 \qquad\qquad g(x) = \sqrt{x + 3}$$
$$f(2) = 2 + 1 = 3; \qquad\qquad g(2) = \sqrt{2 + 3} = \sqrt{5};$$
$$(f + g)(x) = x + 1 + \sqrt{x + 3}$$
$$(f + g)(2) = 2 + 1 + \sqrt{2 + 3}$$
$$\qquad = 3 + \sqrt{5}.$$

EXAMPLE 9   Given that $f(x) = x^2 - 4$ and $g(x) = x + 2$, find each of the following.

a) $(f + g)(x)$ **b)** $(f - g)(x)$

c) $(fg)(x)$ 　　　　　　　　　　　　**d)** $(f/g)(x)$

e) $(gg)(x)$

f) The domain of $f + g$, $f - g$, fg, and f/g

Solution

a) $(f + g)(x) = f(x) + g(x) = (x^2 - 4) + (x + 2) = x^2 + x - 2$

b) $(f - g)(x) = f(x) - g(x) = (x^2 - 4) - (x + 2) = x^2 - x - 6$

c) $(fg)(x) = f(x) \cdot g(x) = (x^2 - 4)(x + 2) = x^3 + 2x^2 - 4x - 8$

d) $(f/g)(x) = \dfrac{f(x)}{g(x)}$

$$= \frac{x^2 - 4}{x + 2} \qquad \text{Note that } (f/g)(x) \text{ is}$$
$$\text{undefined when } x = -2.$$

$$= \frac{(x + 2)(x - 2)}{x + 2} \qquad \text{Factoring}$$

$$= x - 2 \qquad \text{Removing a factor of 1: } \frac{x + 2}{x + 2} = 1$$

Thus, $(f/g)(x) = x - 2$ with the added stipulation that $x \neq -2$ since -2 is not in the domain of $(f/g)(x)$.

e) $(gg)(x) = g(x) \cdot g(x) = [g(x)]^2 = (x + 2)^2 = x^2 + 4x + 4$

f) The domain of f is the set of all real numbers. The domain of g is also the set of all real numbers. The domain of $f + g$, $f - g$, and fg is the set of numbers in the intersection of the domains—that is, the set of numbers in both domains, which is again the set of real numbers. For f/g, we must exclude -2, since $g(-2) = 0$. Thus the domain of f/g is the set of real numbers excluding -2, or $(-\infty, -2) \cup (-2, \infty)$. ▬

Difference Quotients

It is important for students preparing for calculus to be able to simplify rational expressions like

$$\frac{f(x + h) - f(x)}{h}.$$

In calculus, this process is referred to as finding the **difference quotient.**

EXAMPLE 10 For the function f given by $f(x) = 2x^2 - x - 3$, find the difference quotient

$$\frac{f(x + h) - f(x)}{h}.$$

SOLUTION We first find $f(x + h)$:

$$f(x + h) = 2(x + h)^2 - (x + h) - 3$$
$$= 2[x^2 + 2xh + h^2] - x - h - 3$$
$$= 2x^2 + 4xh + 2h^2 - x - h - 3.$$

Then

$$\frac{f(x + h) - f(x)}{h}$$

$$= \frac{[2x^2 + 4xh + 2h^2 - x - h - 3] - [2x^2 - x - 3]}{h}$$

$$= \frac{2x^2 + 4xh + 2h^2 - x - h - 3 - 2x^2 + x + 3}{h}$$

$$= \frac{4xh + 2h^2 - h}{h}$$

$$= \frac{h(4x + 2h - 1)}{h \cdot 1} = \frac{h}{h} \cdot \frac{4x + 2h - 1}{1} = 4x + 2h - 1.$$

Exercise Set 1.5

Determine intervals on which the function is (a) increasing, (b) decreasing, and (c) constant.

1.

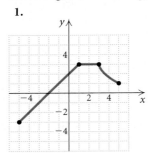

2.

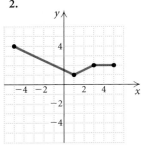

3.

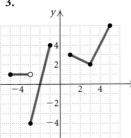

4.

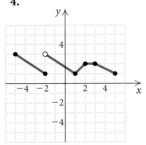

Graph the function. Estimate the intervals on which the function is increasing or decreasing and any relative maxima or minima.

5. $f(x) = x^2$

6. $f(x) = 4 - x^2$

7. $f(x) = 5 - |x|$

8. $f(x) = |x + 3| - 5$

9. $f(x) = x^2 - 6x + 10$

10. $f(x) = -x^2 - 8x - 9$

11. *Rising Balloon.* A hot-air balloon rises straight up from the ground at a rate of 120 ft/min. The balloon is tracked from a rangefinder on the ground at point P, which is 400 ft from the release point Q of the balloon. Let d = the distance from the balloon to the rangefinder at point P and t = the time, in minutes, since the balloon was released. Express d as a function of t.

12. *Airplane Distance.* An airplane is flying at an altitude of 3700 ft. The slanted distance directly to the airport is *d* feet. Express the horizontal distance *h* as a function of *d*.

3700 ft

13. *Garden Area.* Yardbird Landscaping has 48 m of fencing with which to enclose a rectangular garden. If the garden is *x* meters long, express the garden's area as a function of the length.

14. *Triangular Flag.* A scout troop is designing a triangular flag so that the length of its base, in inches, is 7 less than twice the height, *h*. Express the area of the flag as a function of the height.

15. *Inscribed Rhombus.* A rhombus is inscribed in a rectangle that is *w* meters wide with a perimeter of 40 m. Each vertex of the rhombus is a midpoint of a side of the rectangle. Express the area of the rhombus as a function of the rectangle's width. (*Hint*: Consider the area of the rectangle.)

w

16. *Tablecloth Area.* A tailor uses 16 ft of lace to trim the edges of a rectangular tablecloth. If the tablecloth is *w* feet wide, express its area as a function of the width.

17. *Golf Distance Finder.* A device used in golf to estimate the distance *d*, in yards, to a hole measures the size *s*, in inches, that the 7-ft pin appears to be in a viewfinder. Express the distance *d* as a function of *s*.

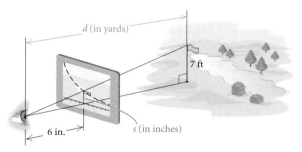

d (in yards)

7 ft

6 in.   *s* (in inches)

18. *Gas Tank Volume.* A gas tank has ends that are hemispheres of radius *r* feet. The cylindrical midsection is 6 ft long. Express the volume of the tank as a function of *r*.

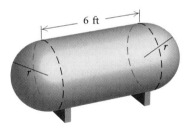

6 ft

r

19. *Play Space.* A daycare center has 30 ft of dividers with which to enclose a rectangular play space in a corner of a large room. The sides against the wall require no partition. Suppose the play space is *x* feet long.

a) Express the area of the play space as a function of *x*.
b) Find the domain of the function.
c) Using the graph shown below, determine the dimensions that yield the maximum area.

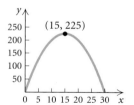

y

(15, 225)

250
200
150
100
50

0   5 10 15 20 25 30   *x*

20. *Corral Design.* A rancher has 360 yd of fencing with which to enclose two adjacent rectangular corrals, one for sheep and one for cattle. A river forms one side of the corrals. Suppose the width of each corral is *x* yards.

a) Express the total area of the two corrals as a function of *x*.
b) Find the domain of the function.

c) Using the graph shown below, determine the dimensions that yield the maximum area.

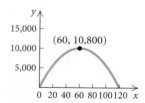

21. *Volume of a Box.*   From a 12-cm by 12-cm piece of cardboard, square corners are cut out so that the sides can be folded up to make a box.

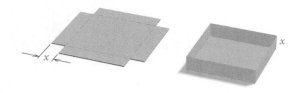

a) Express the volume of the box as a function of the length x, in centimeters, of a cut-out square.
b) Find the domain of the function.
c) Using the graph shown below, determine the dimensions that yield the maximum volume.

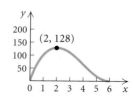

22. *Molding Plastics.*   Perfect Plastics plans to produce a one-component vertical file by bending the long side of an 8-in. by 14-in. sheet of plastic along two lines to form a U shape.

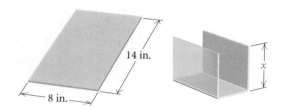

a) Express the volume of the file as a function of the height x, in inches, of the file.
b) Find the domain of the function.

c) Using the graph shown below, determine how tall the file should be in order to maximize the volume the file can hold.

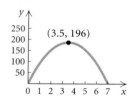

Graph each of the following.

23. $f(x) = \begin{cases} \frac{1}{2}x, & \text{for } x < 0, \\ x + 3, & \text{for } x \geq 0 \end{cases}$

24. $f(x) = \begin{cases} -\frac{1}{3}x + 2, & \text{for } x \leq 0, \\ x - 5, & \text{for } x > 0 \end{cases}$

25. $f(x) = \begin{cases} -\frac{3}{4}x + 2, & \text{for } x < 4, \\ -1, & \text{for } x \geq 4 \end{cases}$

26. $f(x) = \begin{cases} 4, & \text{for } x \leq -2, \\ x + 1, & \text{for } -2 < x < 3, \\ -x, & \text{for } x \geq 3 \end{cases}$

27. $f(x) = \begin{cases} x + 1, & \text{for } x \leq -3, \\ -1, & \text{for } -3 < x < 4, \\ \frac{1}{2}x, & \text{for } x \geq 4 \end{cases}$

28. $f(x) = \begin{cases} \dfrac{x^2 - 9}{x + 3}, & \text{for } x \neq -3, \\ 5, & \text{for } x = -3 \end{cases}$

29. $f(x) = \begin{cases} 2, & \text{for } x = 5, \\ \dfrac{x^2 - 25}{x - 5}, & \text{for } x \neq 5 \end{cases}$

30. $f(x) = \begin{cases} \dfrac{x^2 + 3x + 2}{x + 1}, & \text{for } x \neq -1, \\ 7, & \text{for } x = -1 \end{cases}$

31. $f(x) = [\![x]\!]$

32. $f(x) = 2[\![x]\!]$

33. $g(x) = 1 + [\![x]\!]$

34. $h(x) = \frac{1}{2}[\![x]\!] - 2$

Given that $f(x) = x^2 - 3$ and $g(x) = 2x + 1$, find each of the following, if it exists.

35. $(f + g)(5)$

36. $(f - g)(3)$

37. $(f - g)(-1)$

38. $(fg)(0)$

39. $(fg)(2)$

40. $(fg)(-2)$

41. $(f/g)(-1)$

42. $(f/g)\left(-\frac{1}{2}\right)$

43. $(fg)\left(-\frac{1}{2}\right)$

44. $(f/g)(-\sqrt{3})$

45. $(f/g)(\sqrt{3})$

46. $(f - g)(0)$

For each pair of functions in Exercises 47–50:

a) Find the domain of f, g, $f + g$, $f - g$, fg, ff, f/g, and g/f.

b) Find $(f + g)(x)$, $(f - g)(x)$, $(fg)(x)$, $(ff)(x)$, $(f/g)(x)$, and $(g/f)(x)$.

47. $f(x) = x - 3$, $g(x) = \sqrt{x + 4}$

48. $f(x) = x^2 - 1$, $g(x) = 2x + 5$

49. $f(x) = x^3$, $g(x) = 2x^2 + 5x - 3$

50. $f(x) = x^2$, $g(x) = \sqrt{x}$

For each pair of functions in Exercises 51–54, find $(f + g)(x)$, $(f - g)(x)$, $(fg)(x)$, and $(f/g)(x)$.

51. $f(x) = x^2 - 4$, $g(x) = x^2 + 2$

52. $f(x) = x^2 + 2$, $g(x) = 4x - 7$

53. $f(x) = \sqrt{x - 7}$, $g(x) = x^2 - 25$

54. $f(x) = \sqrt{x - 1}$, $g(x) = \sqrt{3x - 8}$

In Exercises 55–60, consider the functions F and G as shown in the following graph.

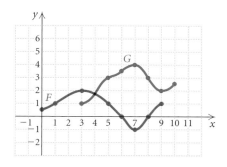

55. Find the domain of F, the domain of G, and the domain of $F + G$.

56. Find the domain of $F - G$, FG, and F/G.

57. Find the domain of G/F.

58. Graph $F + G$.

59. Graph $G - F$.

60. Graph $F - G$.

61. *Total Cost, Revenue, and Profit.* In economics, functions that involve revenue, cost, and profit are used. For example, suppose that $R(x)$ and $C(x)$ denote the total revenue and the total cost, respectively, of producing a new kind of tool for King Hardware Stores. Then the difference

$$P(x) = R(x) - C(x)$$

represents the total profit for producing x tools. Given

$$R(x) = 60x - 0.4x^2 \quad \text{and} \quad C(x) = 3x + 13,$$

find each of the following.

a) $P(x)$

b) $R(100)$, $C(100)$, and $P(100)$

62. *Total Cost, Revenue, and Profit.* Given that

$$R(x) = 200x - x^2 \quad \text{and} \quad C(x) = 5000 + 8x,$$

for a new radio produced by Clear Communication, find each of the following. (See Exercise 61.)

a) $P(x)$

b) $R(175)$, $C(175)$, and $P(175)$

For each function f, construct and simplify the difference quotient

$$\frac{f(x + h) - f(x)}{h}.$$

63. $f(x) = 3x^2 - 2x + 1$

64. $f(x) = 5x^2 + 4x$

65. $f(x) = x^3$

66. $f(x) = 2|x| + 3x$

67. $f(x) = \dfrac{x - 4}{x + 3}$

68. $f(x) = \dfrac{x}{2 - x}$

Technology Connection

69. *Temperature During an Illness.* The temperature of a patient during an illness is given by the function

$$T(t) = -0.1t^2 + 1.2t + 98.6, \quad 0 \le t \le 12,$$

where T is the temperature, in degrees Fahrenheit, at time t, in days, after the onset of the illness.

a) Graph the function using a graphing calculator.

b) Use the MAXIMUM feature to determine at what time the patient's temperature was the highest. What was the highest temperature?

70. *Advertising Effect.* A software firm estimates that it will sell N units of a new CD-ROM video game after spending a dollars on advertising, where

$$N(a) = -a^2 + 300a + 6, \quad 0 \le a \le 300,$$

and a is measured in thousands of dollars.

a) Graph the function using a graphing calculator.
b) Use the MAXIMUM feature to find the relative maximum.
c) For what advertising expenditure will the greatest number of games be sold? How many games will be sold for that amount?

Use a graphing calculator to find the intervals on which each function is increasing or decreasing. Consider the entire set of real numbers if no domain is given.

71. $f(x) = \dfrac{8x}{x^2 + 1}$

72. $f(x) = \dfrac{-4}{x^2 + 1}$

73. $f(x) = x\sqrt{4 - x^2}, \quad$ for $-2 \le x \le 2$

74. $f(x) = -0.8x\sqrt{9 - x^2}, \quad$ for $-3 \le x \le 3$

75. *Area of an Inscribed Rectangle.* A rectangle that is x feet wide is inscribed in a circle of radius 8 ft.

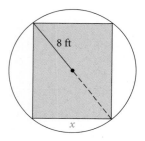

a) Express the area of the rectangle as a function of x.
b) Find the domain of the function.
c) Graph the function with a graphing calculator.
d) What dimensions maximize the area of the rectangle?

76. *Cost of Material.* A rectangular box with volume 320 ft³ is built with a square base and top. The cost is \$1.50/ft² for the bottom, \$2.50/ft² for the sides, and \$1/ft² for the top. Let x = the length of the base, in feet.

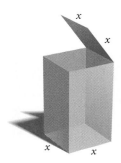

a) Express the cost of the box as a function of x.
b) Find the domain of the function.
c) Graph the function with a graphing calculator.
d) What dimensions minimize the cost of the box?

77. Using a graphing calculator, graph the three functions in Exercise 61 in the viewing window [0, 160, 0, 3000].

78. Using a graphing calculator, graph the three functions in Exercise 62 in the viewing window [0, 200, 0, 10,000].

Collaborative Discussion and Writing

79. Describe a real-world situation that could be modeled by a function that is, in turn, increasing, then constant, and finally decreasing.

80. Simply stated, a *continuous function* is a function whose graph can be drawn without lifting the pencil from the paper. Examine several functions in this exercise set to see if they are continuous. Then explore the continuous functions to estimate the relative maxima and minima. For continuous functions, how can you connect the ideas of increasing and decreasing on an interval to relative maxima and minima?

Skill Maintenance

Find the domain of the function.

81. $f(x) = -2x - 3$

82. $g(x) = \dfrac{1}{(x - 3)^2}$

83. $h(x) = \dfrac{x}{|x + 5|}$

84. $f(x) = -5$

Synthesis

85. *Parking Costs.* A parking garage charges $2 for up to (but not including) 1 hr of parking, $4 for up to 2 hr of parking, $6 for up to 3 hr of parking, and so on. Let $C(t) =$ the cost of parking for t hours.

a) Graph the function.

b) Write an equation for $C(t)$ using the greatest integer notation $[\![t]\!]$.

86. If $[\![x + 2]\!] = -3$, what are the possible inputs for x?

87. If $([\![x]\!])^2 = 25$, what are the possible inputs for x?

88. *Minimizing Power Line Costs.* A power line is constructed from a power station at point A to an island at point C, which is 1 mi directly out in the water from a point B on the shore. Point B is 4 mi downshore from the power station at A. It costs $5000 per mile to lay the power line under water and $3000 per mile to lay the power line under ground. The line comes to the shore at point S downshore from A. Let $x =$ the distance from B to S.

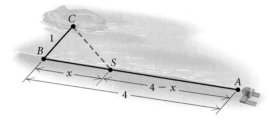

a) Express the cost C of laying the line as a function of x.

b) At what distance x should the line come to shore in order to minimize cost?

89. *Volume of an Inscribed Cylinder.* A right circular cylinder of height h and radius r is inscribed in a right circular cone with a height of 10 ft and a base with radius 6 ft.

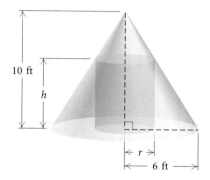

a) Express the height h of the cylinder as a function of r.

b) Express the volume V of the cylinder as a function of r.

c) Express the volume V of the cylinder as a function of h.

1.6

Symmetry and Transformations

- *Determine whether a graph is symmetric with respect to the x-axis, the y-axis, and the origin.*
- *Determine whether a function is even, odd, or neither even nor odd.*
- *Given the graph of a function, graph its transformation under translations, reflections, stretchings, and shrinkings.*

Symmetry

Symmetry occurs often in nature and in art. For example, when viewed from the front, the bodies of most animals are at least approximately symmetric. This means that each eye is the same distance from the center of the bridge of the nose, each shoulder is the same distance from the center of the chest, and so on. Architects have used symmetry for thousands of years to enhance the beauty of buildings.

   A knowledge of symmetry in mathematics helps us graph and analyze equations and functions.

   Consider the points (4, 2) and (4, −2) that appear in the graph of $x = y^2$ (see Fig. 1 at the top of the following page). Points like these have the same x-value but opposite y-values and are **reflections** of each other across the x-axis. If for any point (x, y) on a graph the point (x, −y) is also on the graph, then the graph is said to be **symmetric with respect to the x-axis.** If we fold the graph on the x-axis, the parts above and below the x-axis will coincide.

   Consider the points (3, 4) and (−3, 4) that appear in the graph of $y = x^2 − 5$ (see Fig. 2 at the top of the following page). Points like these have the same y-value but opposite x-values and are **reflections** of each other across the y-axis. If for any point (x, y) on a graph the point

$(-x, y)$ is also on the graph, then the graph is said to be **symmetric with respect to the y-axis.** If we fold the graph on the y-axis, the parts to the left and right of the y-axis will coincide.

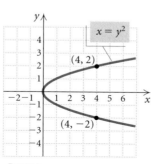

FIGURE 1

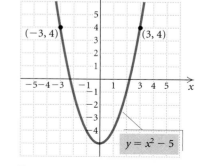

FIGURE 2

Consider the points $(3, \sqrt{7})$ and $(-3, -\sqrt{7})$ that appear in the graph of $x^2 = y^2 + 2$ (see Fig. 3). Note that if we take the opposites of the coordinates of one pair, we get the other pair. If for any point (x, y) on a graph the point $(-x, -y)$ is also on the graph, then the graph is said to be **symmetric with respect to the origin.** Visually, if we rotate the graph $180°$ about the origin, the resulting figure coincides with the original.

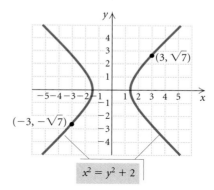

FIGURE 3

Algebraic Tests of Symmetry

x-axis: If replacing y with $-y$ produces an equivalent equation, then the graph is *symmetric with respect to the x-axis.*

y-axis: If replacing x with $-x$ produces an equivalent equation, then the graph is *symmetric with respect to the y-axis.*

Origin: If replacing x with $-x$ and y with $-y$ produces an equivalent equation, then the graph is *symmetric with respect to the origin.*

EXAMPLE 1   Test $y = x^2 + 2$ for symmetry with respect to the x-axis, the y-axis, and the origin.

Algebraic Solution

x-AXIS

We replace y with $-y$:

$$y = x^2 + 2$$

$$\downarrow$$

$$-y = x^2 + 2$$

$$y = -x^2 - 2. \qquad \text{Multiplying by } -1 \\ \text{on both sides}$$

The resulting equation is not equivalent to the original equation, so the graph *is not* symmetric with respect to the x-axis.

y-AXIS

We replace x with $-x$:

$$y = x^2 + 2$$

$$\searrow$$

$$y = (-x)^2 + 2$$

$$y = x^2 + 2. \qquad \text{Simplifying}$$

The resulting equation is equivalent to the original equation, so the graph *is* symmetric with respect to the y-axis.

ORIGIN

We replace x with $-x$ and y with $-y$:

$$y = x^2 + 2$$

$$\downarrow \qquad \searrow$$

$$-y = (-x)^2 + 2$$

$$-y = x^2 + 2 \qquad \text{Simplifying}$$

$$y = -x^2 - 2.$$

The resulting equation is not equivalent to the original equation, so the graph *is not* symmetric with respect to the origin.

Visualizing the Solution

Let's look at the graph of $y = x^2 + 2$.

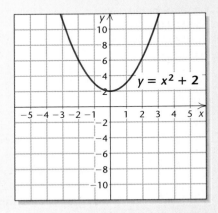

$y = x^2 + 2$

Note that if the graph were folded on the x-axis, the parts above and below the x-axis would not coincide. If it were folded on the y-axis, the parts to the left and right of the y-axis would coincide. If we rotated it 180°, the resulting graph would not coincide with the original graph.

Thus we see that the graph *is not* symmetric with respect to the x-axis or the origin. The graph *is* symmetric with respect to the y-axis.

EXAMPLE 2 Test $x^2 + y^4 = 5$ for symmetry with respect to the x-axis, the y-axis, and the origin.

Algebraic Solution

x-AXIS

We replace y with $-y$:

$$x^2 + y^4 = 5$$
$$x^2 + (-y)^4 = 5$$
$$x^2 + y^4 = 5.$$

The resulting equation is equivalent to the original equation. Thus the graph is symmetric with respect to the x-axis.

We leave it to the student to verify algebraically that the graph is also symmetric with respect to the y-axis and the origin.

Visualizing the Solution

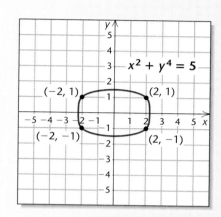

From the graph of the equation, we see symmetry with respect to both axes and with respect to the origin.

Even and Odd Functions

Now we relate symmetry to graphs of functions.

Even and Odd Functions

If the graph of a function f is symmetric with respect to the y-axis, we say that it is an **even function**. That is, for each x in the domain of f, $f(x) = f(-x)$.

If the graph of a function f is symmetric with respect to the origin, we say that it is an **odd function**. That is, for each x in the domain of f, $f(-x) = -f(x)$.

Algebraic Procedure for Determining Even and Odd Functions

Given the function $f(x)$:

1. Find $f(-x)$ and simplify. If $f(x) = f(-x)$, then f is even.
2. Find $-f(x)$, simplify, and compare with $f(-x)$ from step (1). If $f(-x) = -f(x)$, then f is odd.

Except for the function $f(x) = 0$, a function cannot be *both* even and odd. Thus if we see in step (1) that $f(x) = f(-x)$ (that is, f is even), we need not continue.

An algebraic procedure for determining even and odd functions is shown at left.

EXAMPLE 3 Determine whether each of the following functions is even, odd, or neither.

a) $f(x) = 5x^7 - 6x^3 - 2x$ **b)** $h(x) = 5x^6 - 3x^2 - 7$

a) Algebraic Solution

$f(x) = 5x^7 - 6x^3 - 2x$

1. $f(-x) = 5(-x)^7 - 6(-x)^3 - 2(-x)$

$\qquad = 5(-x^7) - 6(-x^3) + 2x$

$\qquad\qquad (-x)^7 = (-1 \cdot x)^7 = (-1)^7 x^7 = -x^7$

$\qquad = -5x^7 + 6x^3 + 2x$

We see that $f(x) \neq f(-x)$. Thus, f is *not* even.

2. $-f(x) = -(5x^7 - 6x^3 - 2x)$

$\qquad\quad = -5x^7 + 6x^3 + 2x$

We see that $f(-x) = -f(x)$. Thus, f is odd.

Visualizing the Solution

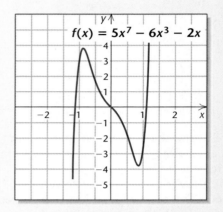

We see that the graph appears to be symmetric with respect to the origin. The function is odd.

b) Algebraic Solution

$h(x) = 5x^6 - 3x^2 - 7$

1. $h(-x) = 5(-x)^6 - 3(-x)^2 - 7$

$\qquad\quad = 5x^6 - 3x^2 - 7$

We see that $h(x) = h(-x)$. Thus the function is even.

Visualizing the Solution

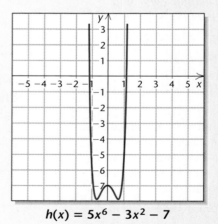

$h(x) = 5x^6 - 3x^2 - 7$

We see that the graph appears to be symmetric with respect to the y-axis. The function is even.

Transformations of Functions

The graphs of some basic functions are shown below. Others can be seen on the inside back cover.

Squaring function:
$y = x^2$

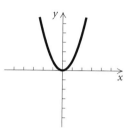

Square root function:
$y = \sqrt{x}$

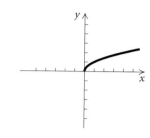

Cubing function:
$y = x^3$

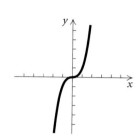

Cube root function:
$y = \sqrt[3]{x}$

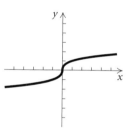

Reciprocal function:
$y = \dfrac{1}{x}$

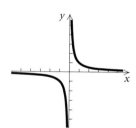

Absolute-value function:
$y = |x|$

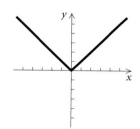

$y = \frac{1}{5}x^4 + 5$

$y = \frac{1}{5}x^4$

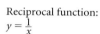

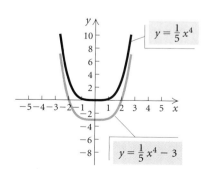

These functions can be considered building blocks for many other functions. We can create graphs of new functions by shifting them horizontally or vertically, stretching or shrinking them, and reflecting them across an axis. We now consider these *transformations*.

Vertical and Horizontal Translations

Suppose that we have a function given by $y = f(x)$. Let's explore the graphs of the new functions $y = f(x) + b$ and $y = f(x) - b$, for $b > 0$.

Consider the functions $y = \frac{1}{5}x^4$, $y = \frac{1}{5}x^4 + 5$, and $y = \frac{1}{5}x^4 - 3$ and compare their graphs shown at left. What pattern do you see? Test it with some other graphs.

The effect of adding a constant to or subtracting a constant from $f(x)$ in $y = f(x)$ is a shift of $f(x)$ up or down. Such a shift is called a **vertical translation.**

Vertical Translation

For $b > 0$,

the graph of $y = f(x) + b$ is the graph of $y = f(x)$ shifted *up* b units;

the graph of $y = f(x) - b$ is the graph of $y = f(x)$ shifted *down* b units.

Suppose that we have a function given by $y = f(x)$. Let's explore the graphs of the new functions $y = f(x - d)$ and $y = f(x + d)$, for $d > 0$.

Consider the functions $y = \frac{1}{5}x^4$, $y = \frac{1}{5}(x - 3)^4$, and $y = \frac{1}{5}(x + 7)^4$ and compare their graphs shown below. What pattern do you observe? Test it with some other graphs.

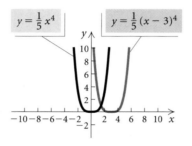
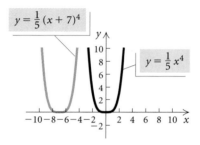

The effect of subtracting a constant from the x-value or adding a constant to the x-value in $y = f(x)$ is a shift of $f(x)$ to the right or left. Such a shift is called a **horizontal translation.**

Horizontal Translation

For $d > 0$:

the graph of $y = f(x - d)$ is the graph of $y = f(x)$ shifted *right d* units;

the graph of $y = f(x + d)$ is the graph of $y = f(x)$ shifted *left d* units.

EXAMPLE 4   Graph each of the following. Before doing so, describe how each graph can be obtained from one of the basic graphs shown on the preceding page.

a) $g(x) = x^2 - 6$

b) $g(x) = \sqrt{x + 2}$

c) $g(x) = |x - 4|$

d) $h(x) = \sqrt{x + 2} - 3$

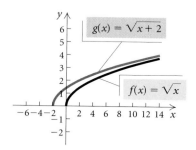

FIGURE 1

Solution

a) To graph $g(x) = x^2 - 6$, think of the graph of $f(x) = x^2$. Since $g(x) = f(x) - 6$, the graph of $g(x) = x^2 - 6$ is the graph of $f(x) = x^2$, shifted, or translated, *down* 6 units (see Fig. 1).

b) To graph $g(x) = \sqrt{x + 2}$, think of the graph of $f(x) = \sqrt{x}$. Since $g(x) = f(x + 2)$, the graph of $g(x) = \sqrt{x + 2}$ is the graph of $f(x) = \sqrt{x}$, shifted *left* 2 units (see Fig. 2).

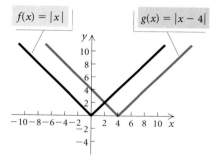

FIGURE 3

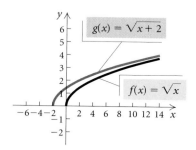

FIGURE 2

c) To graph $g(x) = |x - 4|$, think of the graph of $f(x) = |x|$. Since $g(x) = f(x - 4)$, the graph of $g(x) = |x - 4|$ is the graph of $f(x) = |x|$ shifted *right* 4 units (see Fig. 3).

d) To graph $h(x) = \sqrt{x + 2} - 3$, think of the graph of $f(x) = \sqrt{x}$. In part (b), we found that the graph of $g(x) = \sqrt{x + 2}$ is the graph of $f(x) = \sqrt{x}$ shifted *left* 2 units. Since $h(x) = g(x) - 3$, we shift the graph of $g(x) = \sqrt{x + 2}$ *down* 3 units. Together, the graph of $f(x) = \sqrt{x}$ is shifted *left* 2 units and *down* 3 units (see Fig. 4).

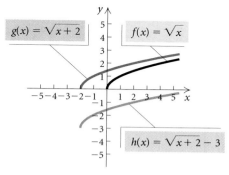

FIGURE 4

Reflections

Suppose that we have a function given by $y = f(x)$. Let's explore the graphs of the new functions $y = -f(x)$ and $y = f(-x)$.

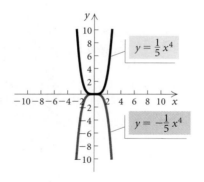

Compare the functions $y = f(x)$ and $y = -f(x)$ by observing the graphs of $y = \frac{1}{5}x^4$ and $y = -\frac{1}{5}x^4$ shown at left. What do you see? Test your observation with some other functions y_1 and y_2 where $y_2 = -y_1$.

Compare the functions $y = f(x)$ and $y = f(-x)$ by observing the graphs of $y = 2x^3 - x^4 + 5$ and $y = 2(-x)^3 - (-x)^4 + 5$ shown below. What do you see? Test your observation with some other functions in which x is replaced with $-x$.

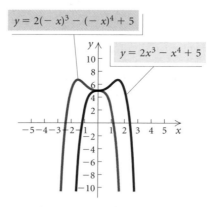

Given the graph of $y = f(x)$, we can reflect each point across the x-axis to obtain the graph of $y = -f(x)$. We can reflect each point of y across the y-axis to obtain the graph of $y = f(-x)$. The new graphs are called **reflections** of $f(x)$.

Reflections

The graph of $y = -f(x)$ is the **reflection** of the graph of $y = f(x)$ across the x-axis.

The graph of $y = f(-x)$ is the **reflection** of the graph of $y = f(x)$ across the y-axis.

EXAMPLE 5   Graph each of the following. Before doing so, describe how each graph can be obtained from the graph of $f(x) = x^3 - 4x^2$.

a) $g(x) = (-x)^3 - 4(-x)^2$ **b)** $h(x) = 4x^2 - x^3$

Solution

a) We first note that

$$f(-x) = (-x)^3 - 4(-x)^2 = g(x).$$

Thus the graph of g is a reflection of the graph of f across the y-axis (see Fig. 1).

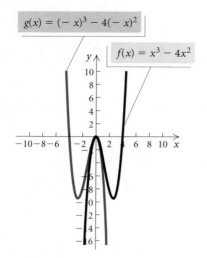

FIGURE 1

b) We first note that

$$-f(x) = -(x^3 - 4x^2)$$
$$= -x^3 + 4x^2$$
$$= 4x^2 - x^3$$
$$= h(x).$$

Thus the graph of h is a reflection of the graph of f across the x-axis (see Fig. 2).

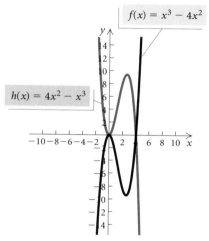

$f(x) = x^3 - 4x^2$

$h(x) = 4x^2 - x^3$

FIGURE 2

Vertical and Horizontal Stretchings and Shrinkings

Suppose that we have a function given by $y = f(x)$. Let's explore the graphs of the new functions $y = af(x)$ and $y = f(cx)$.

Consider the functions $y = x^3 - x$, $y = \frac{1}{10}(x^3 - x)$, $y = 2(x^3 - x)$, and $y = -2(x^3 - x)$ and compare their graphs. What pattern do you observe? Test it with some other graphs.

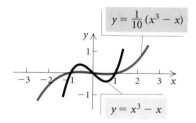

$y = \frac{1}{10}(x^3 - x)$

$y = x^3 - x$

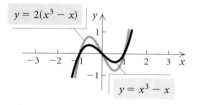

$y = 2(x^3 - x)$

$y = x^3 - x$

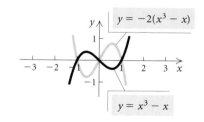

$y = -2(x^3 - x)$

$y = x^3 - x$

Consider any function f given by $y = f(x)$. Multiplying on the right by any constant a, where $|a| > 1$, to obtain $g(x) = af(x)$ will *stretch* the graph vertically away from the x-axis. If $0 < |a| < 1$, then the graph will be flattened or *shrunk* vertically toward the x-axis. If $a < 0$, the graph is also reflected across the x-axis.

Vertical Stretching and Shrinking

The graph of $y = af(x)$ can be obtained from the graph of $y = f(x)$ by

> stretching vertically for $|a| > 1$, or
>
> shrinking vertically for $0 < |a| < 1$.

For $a < 0$, the graph is also reflected across the x-axis.

Consider the functions $y = x^3 - x$, $y = (2x)^3 - (2x)$, $y = \left(\frac{1}{2}x\right)^3 - \left(\frac{1}{2}x\right)$, and $y = \left(-\frac{1}{2}x\right)^3 - \left(-\frac{1}{2}x\right)$ and compare their graphs. What pattern do you observe? Test it with some other graphs.

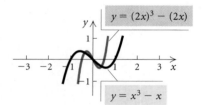

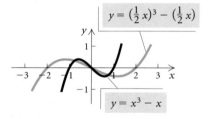

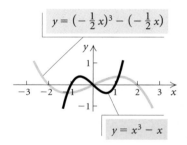

The constant c in the equation $g(x) = f(cx)$ will *stretch* the graph of $y = f(x)$ horizontally away from the y-axis if $0 < |c| < 1$. If $|c| > 1$, the graph will be *shrunk* horizontally toward the y-axis. If $c < 0$, the graph is also reflected across the y-axis.

Horizontal Stretching and Shrinking

The graph of $y = f(cx)$ can be obtained from the graph of $y = f(x)$ by

> shrinking horizontally for $|c| > 1$, or
>
> stretching horizontally for $0 < |c| < 1$.

For $c < 0$, the graph is also reflected across the y-axis.

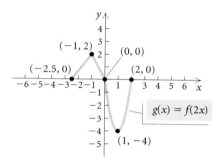

It is instructive to use these concepts to create transformations of a given graph.

EXAMPLE 6   Shown at left is a graph of $y = f(x)$ for some function f. No formula for f is given. Graph each of the following.

a) $g(x) = f(2x)$ **b)** $h(x) = f\left(\frac{1}{2}x\right)$ **c)** $t(x) = f\left(-\frac{1}{2}x\right)$

Solution

a) Since $|2| > 1$, the graph of $g(x) = f(2x)$ is a horizontal shrinking of the graph of $y = f(x)$. We can consider the key points $(-5, 0)$, $(-2, 2)$, $(0, 0)$, $(2, -4)$, and $(4, 0)$ on the graph of $y = f(x)$. The transformation divides each x-coordinate by 2 to obtain the key points $(-2.5, 0)$, $(-1, 2)$, $(0, 0)$, $(1, -4)$, and $(2, 0)$ of the graph of $g(x) = f(2x)$. The graph is shown below.

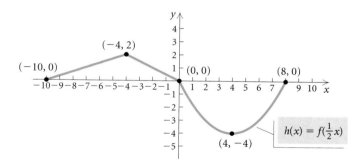

b) Since $\left|\frac{1}{2}\right| < 1$, the graph of $h(x) = f\left(\frac{1}{2}x\right)$ is a horizontal stretching of the graph of $y = f(x)$. We again consider the key points $(-5, 0)$, $(-2, 2)$, $(0, 0)$, $(2, -4)$, and $(4, 0)$ on the graph of $y = f(x)$. The transformation divides each x-coordinate by $\frac{1}{2}$ (which is the same as multiplying by 2) to obtain the key points $(-10, 0)$, $(-4, 2)$, $(0, 0)$, $(4, -4)$, and $(8, 0)$ of the graph of $h(x) = f\left(\frac{1}{2}x\right)$. The graph is shown below.

c) The graph of $t(x) = f\left(-\frac{1}{2}x\right)$ can be obtained by reflecting the graph in part (b) across the y-axis. It is shown at the top of the following page.

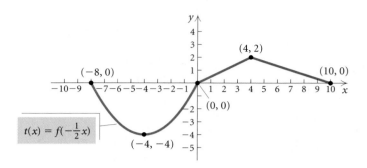

EXAMPLE 7 Use the graph of $y = f(x)$ given in Example 6 to graph $y = -2f(x - 3) + 1$.

Solution

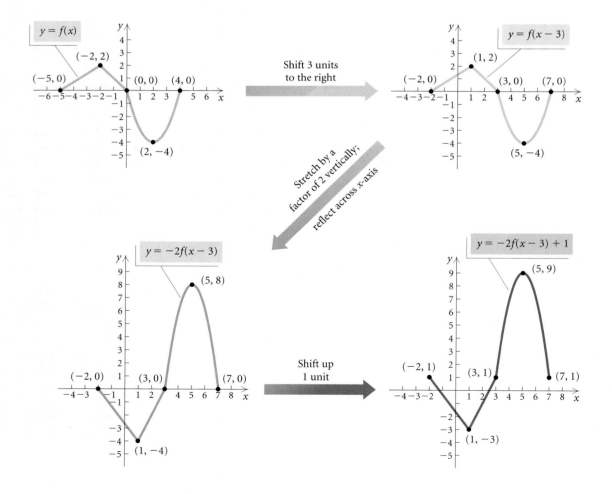

Summary of Transformations of $y = f(x)$

Vertical Translation: $y = f(x) \pm b$

For $b > 0$,

> the graph of $y = f(x) + b$ is the graph of $y = f(x)$ shifted *up* b units;
>
> the graph of $y = f(x) - b$ is the graph of $y = f(x)$ shifted *down* b units.

Horizontal Translation: $y = f(x \mp d)$

For $d > 0$,

> the graph of $y = f(x - d)$ is the graph of $y = f(x)$ shifted *right* d units;
>
> the graph of $y = f(x + d)$ is the graph of $y = f(x)$ shifted *left* d units.

Reflections

Across the x-axis: The graph of $y = -f(x)$ is the reflection of the graph of $y = f(x)$ across the x-axis.

Across the y-axis: The graph of $y = f(-x)$ is the reflection of the graph of $y = f(x)$ across the y-axis.

Vertical Stretching or Shrinking: $y = af(x)$

The graph of $y = af(x)$ can be obtained from the graph of $y = f(x)$ by

> stretching vertically for $|a| > 1$, or
>
> shrinking vertically for $0 < |a| < 1$.

For $a < 0$, the graph is also reflected across the x-axis.

Horizontal Stretching or Shrinking: $y = f(cx)$

The graph of $y = f(cx)$ can be obtained from the graph of $y = f(x)$ by

> shrinking horizontally for $|c| > 1$, or
>
> stretching horizontally for $0 < |c| < 1$.

For $c < 0$, the graph is also reflected across the y-axis.

Exercise Set 1.6

Determine visually whether the graph is symmetric with respect to the x-axis, the y-axis, and/or the origin.

1.

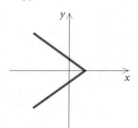

2.

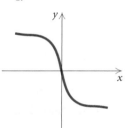

3.

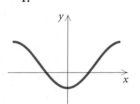

4.

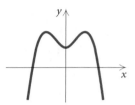

5.

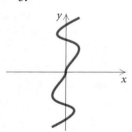

6.

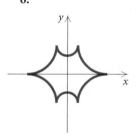

Determine whether the graph is symmetric with respect to the x-axis, the y-axis, and/or the origin.

7. $5x - 5y = 0$

8. $6x + 7y = 0$

9. $3x^2 - 2y^2 = 3$

10. $5y = 7x^2 - 2x$

11. $y = |2x|$

12. $y^3 = 2x^2$

13. $2x^4 + 3 = y^2$

14. $2y^2 = 5x^2 + 12$

15. $3y^3 = 4x^3 + 2$

16. $3x = |y|$

17. $xy = 12$

18. $xy - x^2 = 3$

Determine visually whether the function is even, odd, or neither even nor odd.

19.

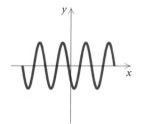

20.

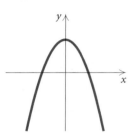

21.

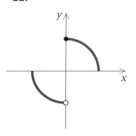

22.

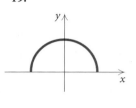

23.

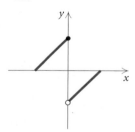

24.

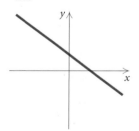

Determine whether the function is even, odd, or neither even nor odd.

25. $f(x) = -3x^3 + 2x$

26. $f(x) = 7x^3 + 4x - 2$

27. $f(x) = 5x^2 + 2x^4 - 1$

28. $f(x) = x + \dfrac{1}{x}$

29. $f(x) = x^{17}$

30. $f(x) = \sqrt[3]{x}$

31. $f(x) = \dfrac{1}{x^2}$

32. $f(x) = x - |x|$

33. $f(x) = 8$

34. $f(x) = \sqrt{x^2 + 1}$

Describe how the graph of the function can be obtained from one of the basic graphs on page 113.

35. $f(x) = x^2 + 1$

36. $g(x) = |3x|$

37. $g(x) = (x + 5)^3$

38. $f(x) = \frac{1}{2}\sqrt[3]{x}$

39. $f(x) = -x^2$

40. $f(x) = |x - 3| - 4$

41. $f(x) = 3\sqrt{x} - 5$

42. $f(x) = 5 - \dfrac{1}{x}$

43. $g(x) = \left|\frac{1}{3}x\right| - 4$

44. $f(x) = \frac{2}{3}x^3 - 4$

45. $f(x) = (x + 5)^2 - 4$

46. $f(x) = (-x)^3 - 5$

47. $f(x) = -\frac{1}{4}(x - 5)^2$

48. $g(x) = \sqrt{-x} + 5$

49. $f(x) = \dfrac{1}{x + 3} + 2$

50. $f(x) = 3(x + 4)^2 - 3$

Write an equation for a function that has a graph with the given characteristics.

51. The shape of $y = x^2$, but upside-down and shifted right 8 units

52. The shape of $y = \sqrt{x}$, but shifted left 6 units and down 5 units

53. The shape of $y = |x|$, but shifted left 7 units and up 2 units

54. The shape of $y = x^3$, but upside-down and shifted right 5 units

55. The shape of $y = 1/x$, but shrunk vertically by a factor of $\frac{1}{2}$ and shifted down 3 units

56. The shape of $y = x^2$, but shifted right 6 units and up 2 units

57. The shape of $y = x^2$, but upside-down and shifted right 3 units and up 4 units

58. The shape of $y = |x|$, but stretched horizontally by a factor of 2 and shifted down 5 units

59. The shape of $y = \sqrt{x}$, but reflected across the y-axis and shifted left 2 units and down 1 unit

60. The shape of $y = 1/x$, but reflected across the x-axis and shifted up 1 unit

61. The shape of $y = x^3$, but shifted left 4 units and shrunk vertically by a factor of 0.83

A graph of $y = f(x)$ follows. No formula for f is given. In Exercises 62–67, graph the given equation.

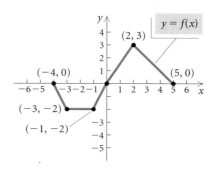

62. $g(x) = \frac{1}{2}f(x)$

63. $g(x) = -2f(x)$

64. $g(x) = f(2x)$

65. $g(x) = f\left(-\frac{1}{2}x\right)$

66. $g(x) = -3f(x + 1) - 4$

67. $g(x) = -\frac{1}{2}f(x - 1) + 3$

The graph of the function f is shown in figure (a). In Exercises 68–75, match the function g with one of the graphs (a)–(h), which follow. Some graphs may be used more than once.

a)

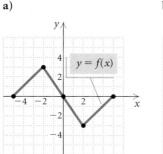

b)

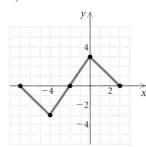

c)

d)

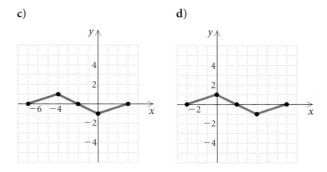

e)

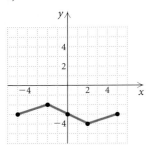

f)

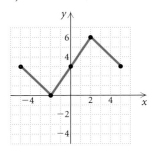

g)

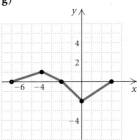

h)

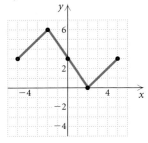

68. $g(x) = f(x) + 3$

69. $g(x) = f(-x) + 3$

70. $g(x) = -f(-x)$

71. $g(x) = -f(x) + 3$

72. $g(x) = \frac{1}{3}f(x) - 3$

73. $g(x) = \frac{1}{3}f(x - 2)$

74. $g(x) = -f(x + 2)$

75. $g(x) = \frac{1}{3}f(x + 2)$

For each pair of functions, determine if $g(x) = f(-x)$.

76. $f(x) = \frac{1}{4}x^4 + \frac{1}{5}x^3 - 81x^2 - 17$,
$\quad g(x) = \frac{1}{4}x^4 + \frac{1}{5}x^3 + 81x^2 - 17$

77. $f(x) = 2x^4 - 35x^3 + 3x - 5$,
$\quad g(x) = 2x^4 + 35x^3 - 3x - 5$

A graph of the function $f(x) = x^3 - 3x^2$ *is shown below. Exercises 78–81 show graphs of functions transformed from this one. Find a formula for each function.*

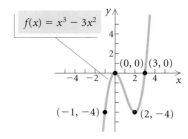

78.

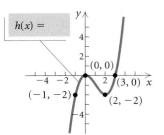

79.

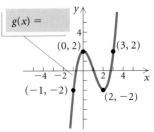

80.

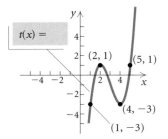

81.

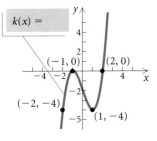

Collaborative Discussion and Writing

82. Describe conditions under which you would know whether a polynomial function $f(x) = a_n x^n + a_{n-1}x^{n-1} + \cdots + a_2 x^2 + a_1 x + a_0$ is even or odd without using an algebraic procedure. Explain.

83. Consider the constant function $f(x) = 0$. Determine whether the graph of this function is symmetric with respect to the x-axis, the y-axis, and/or the origin. Determine whether this function is even or odd.

84. Without drawing the graph, describe what the graph of $f(x) = |x^2 - 9|$ looks like.

85. Explain in your own words why the graph of $y = f(-x)$ is a reflection of the graph of $y = f(x)$ across the y-axis.

Skill Maintenance

86. Given $f(x) = 4x^3 - 5x$, find each of the following.
  a) $f(2)$ **b)** $f(-2)$ **c)** $f(a)$ **d)** $f(-a)$

87. Given $f(x) = 5x^2 - 7$, find each of the following.
  a) $f(-3)$ **b)** $f(3)$ **c)** $f(a)$ **d)** $f(-a)$

88. Find the slope and the y-intercept of the line with equation $2x - 9y + 1 = 0$.

89. Write an equation of the line perpendicular to the graph of the line $8x - y = 10$ and containing the point $(-1, 1)$.

Synthesis

Determine whether the function is even, odd, or neither even nor odd.

90. $f(x) = \dfrac{x^2 + 1}{x^3 - 1}$

91. $f(x) = x\sqrt{10 - x^2}$

Determine whether the graph is symmetric with respect to the x-axis, the y-axis, and/or the origin.

92.

$x^3 = y^2(2 - x)$

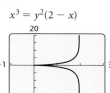

93.

$x^4 = y^2 + x^6$

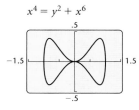

94.

$(x^2 + y^2)^2 = 2xy$

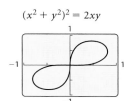

95.

$y^2 + 4xy^2 - y^4 =$
$x^4 - 4x^3 + 3x^2 + 2x^2y^2$

96. The graph of $f(x) = |x|$ passes through the points $(-3, 3)$, $(0, 0)$, and $(3, 3)$. Transform this function to one whose graph passes through the points $(5, 1)$, $(8, 4)$, and $(11, 1)$.

97. If $(-1, 5)$ is a point on the graph of $y = f(x)$, find b such that $(2, b)$ is on the graph of $y = f(x - 3)$.

State whether each of the following is true or false.

98. The product of two odd functions is odd.

99. The sum of two even functions is even.

100. The product of an even function and an odd function is odd.

101. Show that if f is *any* function, then the function E defined by

$$E(x) = \frac{f(x) + f(-x)}{2}$$

is even.

102. Show that if f is *any* function, then the function O defined by

$$O(x) = \frac{f(x) - f(-x)}{2}$$

is odd.

103. Consider the functions E and O of Exercises 101 and 102.

a) Show that $f(x) = E(x) + O(x)$. This means that every function can be expressed as the sum of an even and an odd function.

b) Let $f(x) = 4x^3 - 11x^2 + \sqrt{x} - 10$. Express f as a sum of an even function and an odd function.

Chapter Summary and Review

Important Properties and Formulas

The Distance Formula

$$d = \sqrt{(x_2 - x_1)^2 + (y_2 - y_1)^2}$$

The Midpoint Formula

$$\left(\frac{x_1 + x_2}{2}, \frac{y_1 + y_2}{2} \right)$$

Equation of a Circle

$$(x - h)^2 + (y - k)^2 = r^2$$

Terminology about Lines

Slope: $m = \dfrac{y_2 - y_1}{x_2 - x_1}$

The Slope–Intercept Equation:
$$y = mx + b$$

The Point–Slope Equation:
$$y - y_1 = m(x - x_1)$$

Horizontal Lines: $y = b$

Vertical Lines: $x = a$

Parallel Lines: $m_1 = m_2, \; b_1 \neq b_2$

Perpendicular Lines:
$$m_1 m_2 = -1, \text{ or}$$
$$x = a, \; y = b$$

The Algebra of Functions

The Sum of Two Functions:
$$(f + g)(x) = f(x) + g(x)$$

The Difference of Two Functions:
$$(f - g)(x) = f(x) - g(x)$$

The Product of Two Functions:
$$(fg)(x) = f(x) \cdot g(x)$$

The Quotient of Two Functions:
$$(f/g)(x) = f(x)/g(x), \; g(x) \neq 0$$

Tests for Symmetry

x-axis: If replacing y with $-y$ produces an equivalent equation, then the graph is symmetric with respect to the x-axis.

y-axis: If replacing x with $-x$ produces an equivalent equation, then the graph is symmetric with respect to the y-axis.

Origin: If replacing x with $-x$ and y with $-y$ produces an equivalent equation, then the graph is symmetric with respect to the origin.

Even Function: $f(-x) = f(x)$

Odd Function: $f(-x) = -f(x)$

Transformations

Vertical Translation: $y = f(x) \pm b$

Horizontal Translation: $y = f(x \mp d)$

Reflection across the x-axis: $y = -f(x)$

Reflection across the y-axis: $y = f(-x)$

Vertical Stretching or Shrinking:
$$y = af(x)$$

Horizontal Stretching or Shrinking:
$$y = f(cx)$$

REVIEW EXERCISES

Use substitution to determine whether the given ordered pairs are solutions of the given equation.

1. $\left(3, \frac{24}{9}\right)$, $(0, -9)$; $2x - 9y = -18$

2. $(0, 7)$, $(7, 1)$; $y = 7$

Graph the equation.

3. $y = -\frac{2}{3}x + 1$

4. $2x - 4y = 8$

5. Find the distance between $(3, 7)$ and $(-2, 4)$.

6. Find the midpoint of the segment with endpoints $(3, 7)$ and $(-2, 4)$.

7. Find an equation of the circle with center $(-2, 6)$ and radius $\sqrt{13}$.

8. Find the center and the radius of the circle
$$(x + 1)^2 + (y - 3)^2 = 16.$$

9. Find an equation of the circle having a diameter with endpoints $(-3, 5)$ and $(7, 3)$.

Determine whether the relation is a function. Identify the domain and the range.

10. $\{(3, 1), (5, 3), (7, 7), (3, 5)\}$

11. $\{(2, 7), (-2, -7), (7, -2), (0, 2), (1, -4)\}$

Determine whether the graph is that of a function.

12.

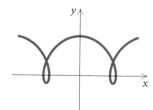

13.

14.

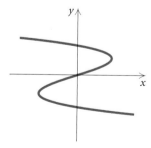

15.

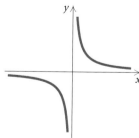

Find the domain of the function.

16. $f(x) = 4 - 5x + x^2$

17. $f(x) = \dfrac{3}{x} + 2$

18. $f(x) = \dfrac{1}{x^2 - 6x + 5}$

19. $f(x) = \dfrac{-5x}{|16 - x^2|}$

Graph the function. Then visually estimate the domain and the range.

20. $f(x) = \sqrt{16 - x^2}$

21. $g(x) = |x - 5|$

22. $f(x) = x^3 - 7$

23. $h(x) = x^4 + x^2$

24. Given that $f(x) = x^2 - x - 3$, find each of the following.

  a) $f(0)$

  b) $f(-3)$

  c) $f(a - 1)$

  d) $\dfrac{f(x + h) - f(x)}{h}$

In Exercises 25 and 26, the table of data contains input–output values for a function. Answer the following questions.

a) *Is the change in the inputs, x, the same?*

b) *Is the change in the outputs, y, the same?*

c) *Is the function linear?*

25.

x	y
-3	8
-2	11
-1	14
0	17
1	20
2	22
3	26

26.

x	y
20	11.8
30	24.2
40	36.6
50	49.0
60	61.4
70	73.8
80	86.2

Find the slope of the line containing the given points.

27. $(2, -11)$, $(5, -6)$

28. $(5, 4)$, $(-3, 4)$

29. $\left(\frac{1}{2}, 3\right)$, $\left(\frac{1}{2}, 0\right)$

30. Find the slope and the y-intercept of the graph of $-2x - y = 7$.

Write a point–slope equation for a line with the following characteristics.

31. $m = -\frac{2}{3}$, y-intercept $(0, -4)$

32. $m = 3$, passes through $(-2, -1)$

33. Passes through $(4, 1)$ and $(-2, -1)$

Determine whether the lines are parallel, perpendicular, or neither.

34. $3x - 2y = 8$,
  $6x - 4y = 2$

35. $y - 2x = 4$,
  $2y - 3x = -7$

36. $y = \frac{3}{2}x + 7$,
  $y = -\frac{2}{3}x - 4$

Given the point $(1, -1)$ *and the line* $2x + 3y = 4$:

37. Find an equation of the line containing the given point and parallel to the given line.

38. Find an equation of the line containing the given point and perpendicular to the given line.

39. *Total Cost.* Clear County Cable Television charges a $25 installation fee and $20 per month for "basic" service. Write an equation that can be used to determine the total cost, $C(t)$, of t months of basic cable television service. Find the total cost of 6 months of service.

40. *Temperature and Depth of the Earth.* The function T given by $T(d) = 10d + 20$ can be used to determine the temperature T, in degrees Celsius, at a depth d, in kilometers, inside the earth.

a) Find $T(5)$, $T(20)$, and $T(1000)$.
b) The radius of the earth is about 5600 km. Use this fact to determine the domain of the function.

41. *Motor Vehicle Production.* The data in the following table shows the increase in world motor vehicle production since 1950.

YEAR, x	NUMBER OF VEHICLES PRODUCED IN THE WORLD, y (IN MILLIONS)
1950, 0	10.6
1960, 10	16.5
1970, 20	29.4
1980, 30	38.6
1990, 40	48.6
1995, 45	50.0

Source: The Wall Street Journal Almanac, 1999

Model the data with a linear function and, using this function, estimate the number of vehicles that will be produced in 2010. Answers will vary depending on the data points used.

Graph of each of the following.

42. $f(x) = \begin{cases} x^3, & \text{for } x < -2, \\ |x|, & \text{for } -2 \le x \le 2, \\ \sqrt{x - 1}, & \text{for } x > 2 \end{cases}$

43. $f(x) = \begin{cases} \dfrac{x^2 - 1}{x + 1}, & \text{for } x \ne -1, \\ 3, & \text{for } x = -1 \end{cases}$

44. $f(x) = [\![x]\!]$

45. $f(x) = [\![x - 3]\!]$

46. *Inscribed Rectangle.* A rectangle is inscribed in a semicircle of radius 2, as shown. The variable $x =$ half the length of the rectangle. Express the area of the rectangle as a function of x.

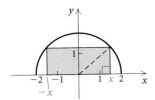

47. *Minimizing Surface Area.* A container firm is designing an open-top rectangular box, with a square base, that will hold 108 in³. Let $x =$ the length of a side of the base.

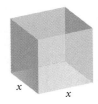

a) Express the surface area as a function of x.
b) Find the domain of the function.
c) Using the graph shown below, determine the dimensions that will minimize the surface area of the box.

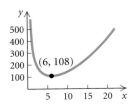

First, graph the equation and determine visually whether it is symmetric with respect to the x-axis, the y-axis, and/or the origin. Then verify your assertion algebraically.

48. $x^2 + y^2 = 4$

49. $y^2 = x^2 + 3$

50. $x + y = 3$

51. $y = x^2$

52. $y = x^3$

53. $y = x^4 - x^2$

Determine visually whether the function is even, odd, or neither even nor odd.

54.

55.

56.

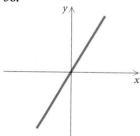

57.

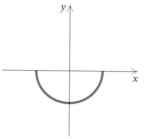

Test whether the function is even, odd, or neither even nor odd.

58. $f(x) = 9 - x^2$

59. $f(x) = x^3 - 2x + 4$

60. $f(x) = x^7 - x^5$

61. $f(x) = |x|$

62. $f(x) = \sqrt{16 - x^2}$

63. $f(x) = \dfrac{10x}{x^2 + 1}$

Write an equation for a function that has a graph with the given characteristics.

64. The shape of $y = x^2$, but shifted left 3 units

65. The shape of $y = \sqrt{x}$, but upside-down and shifted right 3 units and up 4 units

66. The shape of $y = |x|$, but stretched vertically by a factor of 2 and shifted right 3 units

A graph of $y = f(x)$ is shown below. No formula for f is given. Graph each of the following.

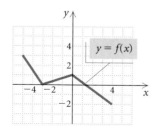

67. $y = f(x - 1)$

68. $y = f(2x)$

69. $y = -2f(x)$

70. $y = 3 + f(x)$

For each pair of functions in Exercises 71 and 72:

a) *Find the domain of f, g, $f + g$, $f - g$, fg, and f/g.*

b) *Find $(f + g)(x)$, $(f - g)(x)$, $fg(x)$, and $(f/g)(x)$.*

71. $f(x) = \dfrac{4}{x^2}$; $g(x) = 3 - 2x$

72. $f(x) = 3x^2 + 4x$; $g(x) = 2x - 1$

73. Given the total-revenue and total-cost functions
$$R(x) = 120x - 0.5x^2 \quad \text{and} \quad C(x) = 15x + 6,$$
find the total profit function $P(x)$.

Technology Connection

74. a) Fit a regression line to the data in Exercise 41 and use it to estimate the number of vehicles produced in 2010.

b) What is the correlation coefficient for the regression line? How close a fit is the regression line?

Collaborative Discussion and Writing

75. Given that $f(x) = 4x^3 - 2x + 7$, find each of the following. Then discuss how each expression differs from the other.

a) $f(x) + 2$ **b)** $f(x + 2)$ **c)** $f(x) + f(2)$

76. Given the graph of $y = f(x)$, explain and contrast the effect of the constant c on the graphs of $y = f(cx)$ and $y = cf(x)$.

77. a) Graph several functions of the type $y_1 = f(x)$ and $y_2 = |f(x)|$. Describe a procedure, involving transformations, for creating a graph of y_2 from y_1.

b) Describe a procedure, involving transformations, for creating a graph of $y_2 = f(|x|)$ from $y_1 = f(x)$.

Synthesis

Find the domain.

78. $f(x) = \dfrac{\sqrt{1 - x}}{x - |x|}$

79. $f(x) = (x - 9x^{-1})^{-1}$

80. Prove that the sum of two odd functions is odd.

81. Describe how the graph of $y = -f(-x)$ is obtained from the graph of $y = f(x)$.

Chapter Test

1. Graph: $5x - 2y = -10$.

2. Find the distance between $(5, 8)$ and $(-1, 5)$.

3. Find the midpoint of the segment with endpoints $(-2, 6)$ and $(-4, 3)$.

4. Find an equation of the circle with center $(-1, 2)$ and radius $\sqrt{5}$.

5. a) Determine whether the relation

$$\{(-4, 7), (3, 0), (1, 5), (0, 7)\}$$

is a function. Answer yes or no.
b) Find the domain of the relation.
c) Find the range of the relation.

6. Given that $f(x) = 2x^2 - x + 5$, find each of the following.

a) $f(-1)$
b) $f(a + 2)$

7. a) Graph: $f(x) = |x - 2| + 3$.
b) Visually estimate the domain of $f(x)$.
c) Visually estimate the range of $f(x)$.

8. Determine whether each graph is that of a function. Answer yes or no.

a)

b)

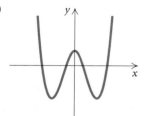

9. Graph $f(x)$

$$f(x) = \begin{cases} x^2, & \text{for } x < -1, \\ |x|, & \text{for } -1 \le x \le 1, \\ \sqrt{x - 1}, & \text{for } x > 1 \end{cases}$$

10. Find the slope and the y-intercept of the graph of $-3x + 2y = 5$.

11. Write an equation for the line that passes through $(-5, 4)$ and $(3, -2)$.

12. Find an equation of the line containing the point $(-1, 3)$ and parallel to the line $x + 2y = -6$.

13. The table below shows the per capita consumption of tea in the United States in several years.

YEAR, x	PER CAPITA CONSUMPTION OF TEA (IN GALLONS)
1993, 0	8.4
1994, 1	8.2
1995, 2	8.0
1996, 3	8.0

Model the data listed in the table with a linear function and use that function to predict the per capita tea consumption in 2005. Answers will vary depending on the data points used.

14. Determine whether the graph of $f(x) = x^4 - 2x^2$ is symmetric with respect to the x-axis, the y-axis, and/or the origin.

15. Test whether the function

$$f(x) = \frac{2x}{x^2 + 1}$$

is even, odd, or neither even nor odd. Show your work.

16. Write an equation for a function that has the shape of $y = x^2$, but shifted right 2 units and down 1 unit.

17. Write an equation for a function that has the shape of $y = x^2$ but shifted left 2 units and down 3 units.

18. The graph of a function $y = f(x)$ is shown below. No formula for f is given. Graph $y = -\frac{1}{2} f(x)$.

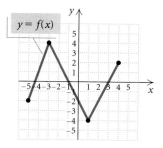

19. For $f(x) = x^2$ and $g(x) = \sqrt{x - 3}$, find each of the following.
   a) The domain of f
   b) The domain of g
   c) $(f - g)(x)$
   d) $(fg)(x)$
   e) The domain of $(f/g)(x)$

Synthesis

20. If $(-3, 1)$ is a point on the graph of $y = f(x)$, what point do you know is on the graph of $y = f(3x)$?

Functions, Equations, and Inequalities 2

In this chapter, we examine functions and equations, placing particular emphasis on visualizing the connections among the zeros of a function, the x-intercepts of the graph of that function, and the solutions of the corresponding equation. We study linear, quadratic, rational, and radical equations as well as equations that are reducible to quadratic and equations with absolute value. We also study linear and quadratic functions and models. We end our work in this chapter by solving linear inequalities.

APPLICATION

In a recent study of 68 metropolitan areas, it was determined that in 1999 drivers in Indianapolis were delayed in traffic an average of 37 hr annually. This was 7 hr less than the annual delay in the mid-1990s. (*Source*: Texas Transportation Institute Urban Mobility Study) What was the annual delay in the mid-1990s?

This problem appears as Exercise 50 in Section 2.1.

2.1

Linear Equations, Functions, and Models

- *Solve linear equations.*
- *Solve a formula for a given variable.*
- *Find zeros of linear functions.*
- *Solve applied problems using linear models.*

An **equation** is a statement that two expressions are equal. To **solve** an equation in one variable is to find all the values of the variable that make the equation true. Each of these numbers is a **solution** of the equation. The set of all solutions of an equation is its **solution set.** Equations that have the same solution set are called **equivalent equations.**

Linear Equations

A **linear equation in one variable** is an equation that is equivalent to one of the form $ax + b = 0$, where a and b are real numbers and $a \neq 0$.

The following principles allow us to solve linear equations.

Equation-Solving Principles

For any real numbers a, b, and c,

The Addition Principle: If $a = b$ is true, then $a + c = b + c$ is true.

The Multiplication Principle: If $a = b$ is true, then $ac = bc$ is true.

EXAMPLE 1   Solve: $6x - 4 = 1$.

Algebraic Solution

We have

$6x - 4 = 1$

$\quad\quad 6x = 5$ Using the addition principle to
                        add 4 on both sides

$\quad\quad\quad x = \frac{5}{6}.$ Using the multiplication principle
                        to multiply by $\frac{1}{6}$, or divide by 6,
                        on both sides

CHECK: $\dfrac{6x - 4 = 1}{}$

$\quad\quad 6 \cdot \frac{5}{6} - 4 \; ? \; 1$ Substituting $\frac{5}{6}$ for x

$\quad\quad\quad\quad 5 - 4 \; \Big|$

$\quad\quad\quad\quad\quad\quad 1 \; \Big| \; 1$ TRUE

The solution is $\frac{5}{6}$.

Visualizing the Solution

We graph $y = 6x - 4$ and $y = 1$. The first coordinate of the point of intersection of the graphs is the value of x for which $6x - 4 = 1$ and is thus the solution of the equation.

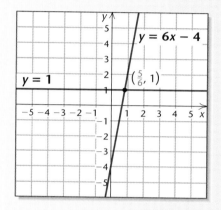

The solution is $\frac{5}{6}$.

Technology Connection

We can use the INTERSECT feature on a graphing calculator to solve equations. We call this the **Intersect method.** To use the Intersect method to solve the equation in Example 1, for instance, we graph $y_1 = 6x - 4$ and $y_2 = 1$. The value of x for which $y_1 = y_2$ is the solution of the equation. This value of x is the first coordinate of the point of intersection of the graphs of y_1 and y_2. Using the INTERSECT feature, we find that the first coordinate of this point is approximately 0.83333333.

If the solution is a rational number, we can find fractional notation for the exact solution by using the ▶FRAC feature. The solution is $\frac{5}{6}$.

$y_1 = 6x - 4, \;\; y_2 = 1$

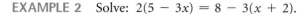

Technology Connection

We can use the TABLE feature on a graphing calculator, set in ASK mode, to check the solutions of equations. In Example 2, for instance, let $y_1 = 2(5 - 3x)$ and $y_2 = 8 - 3(x + 2)$. When $\frac{8}{3}$ is entered for x, we see that y_1 and y_2 are both -6. (Note that the calculator converts $\frac{8}{3}$ to decimal notation.)

X	Y1	Y2
2.6667	−6	−6

X =

EXAMPLE 2   Solve: $2(5 - 3x) = 8 - 3(x + 2)$.

Solution

We have

$$2(5 - 3x) = 8 - 3(x + 2)$$
$$10 - 6x = 8 - 3x - 6 \qquad \text{Using the distributive property}$$
$$10 - 6x = 2 - 3x \qquad \text{Combining like terms}$$
$$10 = 2 + 3x \qquad \text{Using the addition principle to add } 6x \text{ on both sides}$$
$$8 = 3x \qquad \text{Using the addition principle to add } -2, \text{ or subtract 2, on both sides}$$
$$\frac{8}{3} = x. \qquad \text{Using the multiplication principle to multiply by } \frac{1}{3}, \text{ or divide by 3, on both sides}$$

CHECK:
$$\frac{2(5 - 3x) = 8 - 3(x + 2)}{}$$

$$2\left(5 - 3 \cdot \tfrac{8}{3}\right) \; ? \; 8 - 3\left(\tfrac{8}{3} + 2\right) \qquad \text{Substituting } \tfrac{8}{3} \text{ for } x$$
$$2(5 - 8) \;\Big|\; 8 - 3\left(\tfrac{14}{3}\right)$$
$$2(-3) \;\Big|\; 8 - 14$$
$$-6 \;\Big|\; -6 \qquad\qquad \text{TRUE}$$

The solution is $\frac{8}{3}$.

Formulas

A **formula** is an equation that can be used to *model* a situation. For example, the formula $P = 2l + 2w$ gives the perimeter of a rectangle with length l and width w. The equation-solving principles presented earlier can be used to solve a formula for a given variable.

EXAMPLE 3   Solve $P = 2l + 2w$ for l.

Solution

$$P = 2l + 2w$$
$$P - 2w = 2l \qquad \text{Subtracting } 2w \text{ on both sides}$$
$$\frac{P - 2w}{2} = l \qquad \text{Dividing by 2 on both sides}$$

The formula $l = \dfrac{P - 2w}{2}$ can be used to determine a rectangle's length if we are given its perimeter and its width.

On many graphing calculators, an equation must be solved for y before it can be entered. Solving for y is the same as solving a formula for a given variable.

EXAMPLE 4   Solve $3x + 4y = 12$ for y.

Solution   We have

$$3x + 4y = 12$$
$$4y = -3x + 12 \qquad \text{Subtracting } 3x \text{ on both sides}$$
$$y = \frac{-3x + 12}{4}. \qquad \text{Dividing by 4 on both sides}$$

If we were to graph the equation $3x + 4y = 12$ on a graphing calculator, we would enter it as $y = (-3x + 12)/4$.

Zeros of Linear Functions

An input for which a function's output is 0 is called a **zero** of the function.

Zeros of Functions

An input c of a function f is called a **zero** of the function, if the output for c is 0. That is, $f(c) = 0$.

LINEAR FUNCTIONS

REVIEW SECTION 1.3.

We will restrict our attention to zeros of linear functions in this section. Recall that a linear function is given by $f(x) = mx + b$, where m and b are constants.

For the linear function $f(x) = 2x - 4$, we have $f(2) = 2 \cdot 2 - 4 = 0$, so 2 is a **zero** of the function. In fact, 2 is the *only* zero of this function. In general, a **linear function $f(x) = mx + b$, with $m \neq 0$, has exactly one zero.**

Consider the graph of $f(x) = 2x - 4$, shown here.

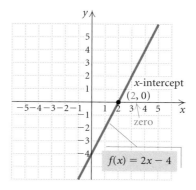

We see from the graph that the zero, 2, is the first coordinate of the point at which the graph crosses the x-axis. This point, $(2, 0)$, is the **x-intercept** of the graph. Thus when we find the zero of a linear function, we are also finding the first coordinate of the x-intercept of the graph of the function.

For every linear function $f(x) = mx + b$, there is an associated linear equation $mx + b = 0$. When we find the zero of a function $f(x) = mx + b$, we are also finding the solution of the equation $mx + b = 0$.

EXAMPLE 5 Find the zero of $f(x) = 5x - 9$.

Algebraic Solution

We find the value of x for which $f(x) = 0$:

$5x - 9 = 0$	Setting $f(x) = 0$
$5x - 9 + 9 = 0 + 9$	Adding 9 on both sides
$5x = 9$	
$\dfrac{5x}{5} = \dfrac{9}{5}$	Dividing by 5 on both sides
$x = \dfrac{9}{5}$, or 1.8.	

The zero is 9/5, or 1.8. Note that the *zero* of the function $f(x) = 5x - 9$ is the *solution* of the equation $5x - 9 = 0$.

Visualizing the Solution

We graph $f(x) = 5x - 9$.

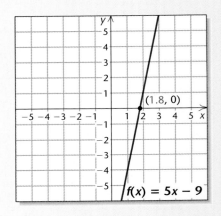

The x-intercept of the graph is $\left(\frac{9}{5}, 0\right)$, or $(1.8, 0)$. Thus, $\frac{9}{5}$, or 1.8, is the zero of the function.

Technology Connection

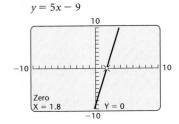

We can use the ZERO feature on a graphing calculator to find the zeros of a function $f(x)$ and to solve the corresponding equation $f(x) = 0$. We call this the **Zero method.** To use the Zero method in Example 5, for instance, we graph $y = 5x - 9$ and use the ZERO feature to find the coordinates of the x-intercept of the graph. Note that the x-intercept must appear in the window when the ZERO feature is used. We see that the zero of the function is 1.8.

CONNECTING THE CONCEPTS

ZEROS, SOLUTIONS, AND INTERCEPTS

The zero of a linear function $f(x) = mx + b$, with $m \neq 0$, is the solution of the linear equation $mx + b = 0$ and is the first coordinate of the x-intercept of the graph of $f(x) = mx + b$. To find the zero of $f(x) = mx + b$, we solve $f(x) = 0$, or $mx + b = 0$.

FUNCTION	ZERO OF THE FUNCTION; SOLUTION OF THE EQUATION	X-INTERCEPT OF THE GRAPH

Linear Function

$f(x) = 2x - 4$, or

$\quad y = 2x - 4$

To find the **zero** of $f(x)$, we solve $f(x) = 0$:

$$2x - 4 = 0$$
$$2x = 4$$
$$x = 2.$$

The **solution** of $2x - 4 = 0$ is 2. This is the zero of the function $f(x) = 2x - 4$. That is, $f(2) = 0$.

The zero of $f(x)$ is the first coordinate of the **x-intercept** of the graph of $y = f(x)$.

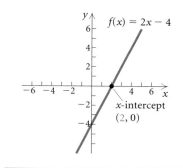

Applications Using Linear Models

Mathematical techniques are used to answer questions arising from real-world situations. Functions and equations of the form $f(x) = mx + b$ and $g(x) = h(x)$, where g and h are linear functions, *model* many of these situations.

The following strategy is of great assistance in problem solving.

Five Steps for Problem Solving

1. **Familiarize** yourself with the problem situation. If the problem is presented in words, then, of course, this means to read carefully. Some or all of the following can also be helpful.

   a) Make a drawing, if it makes sense to do so.

   (continued)

b) Make a written list of the known facts and a list of what you wish to find out.

c) Assign variables to represent unknown quantities.

d) Organize the information in a chart or a table.

e) Find further information. Look up a formula, consult a reference book or an expert in the field, or do research on the Internet.

f) Guess or estimate the answer and check your guess or estimate.

2. **Translate** the problem situation to mathematical language or symbolism. For most of the problems you will encounter in algebra, this means to write one or more equations, but sometimes an inequality or some other mathematical symbolism may be appropriate.

3. **Carry out** some type of mathematical manipulation. Use your mathematical skills to find a possible solution. In algebra, this usually means to solve an equation, an inequality, or a system of equations.

4. **Check** to see whether your possible solution actually fits the problem situation and is thus really a solution of the problem. You might be able to solve an equation, but the solution(s) of the equation might or might not be solution(s) of the original problem.

5. **State** the answer clearly using a complete sentence.

EXAMPLE 6 *Library Acquisitions.* In 1998, college and university libraries in the United States spent $1.9 billion on books, periodicals, and other materials for their collections. This amount is 5.6% more than the amount spent in 1997. (*Source*: *Book Industry Trends 1998*, Book Industry Study Group, Inc.) How much was spent in 1997?

Solution

1. **Familiarize.** Let's estimate that $1.5 billion was spent in 1997. Then the amount spent in 1998, in billions of dollars, would have been

$$1.5 + 5.6\% \text{ of } 1.5 = 1(1.5) + 0.056(1.5)$$
$$= 1.056(1.5) = \$1.584 \text{ billion.}$$

Since $1.9 billion was the actual amount spent in 1998, the estimate is too low. Nevertheless, the calculations performed in making the estimate indicate how we can translate this problem to an equation. We let $a =$ the amount that libraries spent, in billions of dollars, adding to their collections in 1997. Then $a + 5.6\%a$, or $1.056a$, is the amount spent in 1998.

2. **Translate.** We translate to an equation.

$$\underbrace{\text{The amount spent in 1998}}_{1.056a} \quad \underset{=}{\text{is}} \quad \underset{1.9}{\text{\$1.9 billion.}}$$

3. **Carry out.** We solve the equation:

$$1.056a = 1.9$$

$$\frac{1.056a}{1.056} = \frac{1.9}{1.056} \qquad \text{Dividing by 1.056 on both sides}$$

$$a \approx 1.8.$$

4. **Check.** We see that 5.6% of 1.8 is 0.1008 and $1.8 + 0.1008 = 1.9008 \approx 1.9$, so the answer checks. (Remember that we rounded the value of a.)

5. **State.** Colleges and universities in the United States spent about $1.8 billion adding books, periodicals, and other materials to their collections in 1997.

In some applications, we need to use a formula that describes the relationships among variables. When a situation involves distance, speed, and time, for example, we need to recall the **motion formula:**

$d = rt$, where $d =$ distance, $r =$ rate (or speed), and $t =$ time.

EXAMPLE 7 *Airplane Speed.* Continental Airlines' fleet includes 727s with a cruising speed of 545 mph and B1900Ds with a cruising speed of 285 mph (*Source*: Continental Airlines). Suppose that a B1900D takes off and travels at its cruising speed. One hour later a 727 takes off and follows the same route, traveling at its cruising speed. How long will it take the 727 to overtake the B1900D?

Solution

1. **Familiarize.** We make a drawing showing both the known and the unknown information. We let $t =$ the time, in hours, that the 727 travels before it overtakes the B1900D. Since the B1900D takes off 1 hr before the 727, it will travel for $t + 1$ hr before being overtaken. The planes will have traveled the same distance, d, when one overtakes the other.

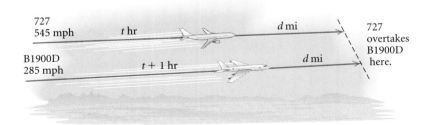

We can also organize the information in a table, as follows.

$$d = r \cdot t$$

	DISTANCE	RATE	TIME
727	d	545	t	$\longrightarrow d = 545t$
B1900D	d	285	$t + 1$	$\longrightarrow d = 285(t + 1)$

2. **Translate.**   Using the formula $d = rt$ in each row of the table, we get
   two expressions for d:

   $$d = 545t \quad \text{and} \quad d = 285(t + 1).$$

   Substituting $545t$ for d in the second equation, we have

   $$545t = 285(t + 1).$$

3. **Carry out.**   We solve the equation:

$545t = 285(t + 1)$	
$545t = 285t + 285$	Using the distributive property
$260t = 285$	Subtracting $285t$ on both sides
$t \approx 1.1.$	Dividing by 260 on both sides

4. **Check.**   If the 727 travels for about 1.1 hr, then the B1900D travels
   for about $1.1 + 1$, or 2.1 hr. In 2.1 hr, the B1900D travels $285(2.1)$,
   or 598.5 mi; and in 1.1 hr, the 727 travels $545(1.1)$, or 599.5 mi.
   Since 598.5 mi $\approx$ 599.5 mi, the answer checks. (Remember that we
   rounded the value of t.)

5. **State.**   About 1.1 hr after the 727 has taken off, it will overtake the
   B1900D.

Technology Connection

CONNECTING THE CONCEPTS

THE INTERSECT AND ZERO METHODS

An equation $f(x) = g(x)$ can be solved using the Intersect method by graphing $y_1 = f(x)$ and $y_2 = g(x)$ and using the INTERSECT feature to find the first coordinate of the point of intersection of the graphs. The equation can also be solved using the Zero method by rewriting it with 0 on one side of the equals sign, if necessary, and then using the ZERO feature.

Solve: $x - 1 = 2x - 6$.

The Intersect Method

Graph

$f(x) = y_1 = x - 1$

and

$g(x) = y_2 = 2x - 6$.

The Zero Method

First add $-2x$ and 6 on both sides of the equation to get 0 on one side.

$$x - 1 = 2x - 6$$
$$x - 1 - 2x + 6 = 0$$

Graph

$$y = x - 1 - 2x + 6.$$

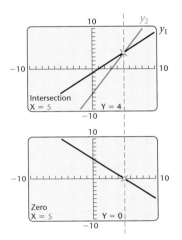

Exercise Set

Solve.

1. $4x + 5 = 21$

2. $2y - 1 = 3$

3. $4x + 3 = 0$

4. $3x - 16 = 0$

5. $y + 1 = 2y - 7$

6. $5 - 4x = x - 13$

7. $2x + 7 = x + 3$

8. $5x - 4 = 2x + 5$

9. $3x - 5 = 2x + 1$

10. $4x + 3 = 2x - 7$

11. $5x - 2 + 3x = 2x + 6 - 4x$

12. $5x - 17 - 2x = 6x - 1 - x$

13. $7(3x + 6) = 11 - (x + 2)$

14. $4(5y + 3) = 3(2y - 5)$

15. $3(x + 1) = 5 - 2(3x + 4)$

16. $4(3x + 2) - 7 = 3(x - 2)$

17. $2(x - 4) = 3 - 5(2x + 1)$

18. $3(2x - 5) + 4 = 2(4x + 3)$

Solve.

19. $A = \frac{1}{2}bh$, for b
   (Area of a triangle)

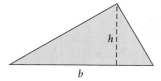

20. $A = \pi r^2$, for π
   (Area of a circle)

21. $P = 2l + 2w$, for w
   (Perimeter of a rectangle)

22. $A = P + Prt$, for P
   (Simple interest)

23. $A = \frac{1}{2}h(b_1 + b_2)$, for h
   (Area of a trapezoid)

24. $A = \frac{1}{2}h(b_1 + b_2)$, for b_2

25. $V = \frac{4}{3}\pi r^3$, for π
   (Volume of a sphere)

26. $V = \frac{4}{3}\pi r^3$, for r^3

27. $F = \frac{9}{5}C + 32$, for C
   (Temperature conversion)

28. $Ax + By = C$, for y
   (Standard linear equation)

Find the zero of the linear function.

29. $f(x) = x + 5$ | **30.** $f(x) = 5x + 20$

31. $f(x) = -x + 18$ | **32.** $f(x) = 8 + x$

33. $f(x) = 16 - x$ | **34.** $f(x) = -2x + 7$

35. $f(x) = x + 12$ | **36.** $f(x) = 8x + 2$

37. $f(x) = -x + 6$ | **38.** $f(x) = 4 + x$

39. $f(x) = 20 - x$ | **40.** $f(x) = -3x + 13$

41. $f(x) = x - 6$ | **42.** $f(x) = 3x - 9$

43. $f(x) = -x + 15$ | **44.** $f(x) = 4 - x$

In Exercises 45 and 46, use the given graph to find each of the following:

a) *the x-intercept;*
b) *the zero of the function.*

45.

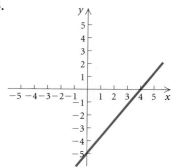

46.

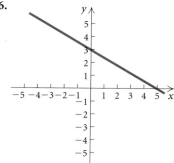

47. *Tourism.*   France was the leading destination of foreign tourists in 1998, with 70 million arrivals. This exceeded the number of foreign-tourist arrivals in the United States by 22.9 million. (*Source*: World Tourism Organization) How many foreign tourists visited the United States in 1998?

48. *Express Delivery.*   DHL Worldwide Express has determined that its couriers in Cairo walk an average of 11 mi on a single route when delivering packages. This is about six times the average distance walked by its couriers in Amsterdam. (*Source*: DHL Worldwide Express) How far, on average, does a courier in Amsterdam walk on a route?

49. *Direct Marketing.* In a recent year, 8000 catalog companies mailed 14 billion catalogs to the 112 million households in the United States (*Source*: W. A. Dean and Associates). On average, how many catalogs were received in each household?

50. *Traffic Delays.* In a recent study of 68 metropolitan areas, it was determined that in 1999 drivers in Indianapolis were delayed in traffic an average of 37 hr annually. This was 7 hr less than the annual delay in the mid-1990s. (*Source*: Texas Transportation Institute Urban Mobility Study) What was the annual delay in the mid-1990s?

51. *Braille Reading Material.* With an increasing reliance on tape recorders, letter magnifiers, and computer voice translators, the percentage of legally blind students learning Braille, a system that translates language into a code of raised dots, has dropped from 53% in 1963 to 10% in 1997. Nevertheless, an increasing number of Braille materials are being published, because those who do read Braille continue to seek reading material. In 1997, the Braille Press published 5.8 million pages. This was about three times the number of pages published in 1977. (*Source*: The American Foundation for the Blind) How many Braille pages were published by the Braille Press in 1977?

52. *Corporate Fax Expense.* The average *Fortune 500* company spends $13 million annually for local and domestic long-distance fax service. The amount spent for local faxes is about one third the amount spent for domestic long-distance faxes. (*Source*: Gallup survey of *Fortune 500* companies for Pitney Bowes) Find the amount spent for each type of fax service.

53. *Test Plot Dimensions.* A flower seed company has a rectangular test plot with a perimeter of 322 m.

The length is 25 m more than the width. Find the dimensions of the plot.

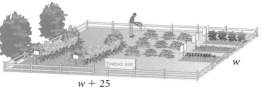

$w + 25$

w

54. *Garden Dimensions.* The children at Tiny Tots Day Care plant a rectangular vegetable garden with a perimeter of 39 m. The length is twice the width. Find the dimensions of the garden.

55. *Angle Measure.* In triangle *ABC*, angle *B* is five times as large as angle *A*. The measure of angle *C* is 2° less than that of angle *A*. Find the measures of the angles. (*Hint*: The sum of the angle measures is 180°.)

56. *Angle Measure.* In triangle *ABC*, angle *B* is twice as large as angle *A*. Angle *C* measures 20° more than angle *A*. Find the measures of the angles.

57. *HMO Enrollment.* The number of enrollees in health maintenance organizations (HMOs) in Kentucky in a recent year was 416,300. This was a 48.4% increase over the previous year. (*Source*: Marion Merrell Dow Managed Care Department) What was the former number of enrollees?

58. *Moving Costs.* The average cost of moving a distance of 1255 mi is $1035 if you handle the move yourself. This is about 52% of the cost of hiring a professional moving firm. (*Source*: American Movers Conference) What is the cost of the professionally handled move?

59. *Train Speeds.* The speed of an Amtrak passenger train is 14 mph faster than the speed of a Central Railway freight train. The passenger train travels 400 mi in the same time it takes the freight train to travel 330 mi. Find the speed of each train.

60. *Distance Traveled.* A private airplane leaves Midway airport and flies due east at a speed of 180 km/h. Two hours later, a jet leaves Midway and flies due east at a speed of 900 km/h. How far from the airport will the jet overtake the private plane?

61. *Water Weight.* Water accounts for 50% of a woman's weight (*Source*: National Institute for Fitness and Sport). Jocelyn weighs 135 lb. How much of her body weight is water?

62. *Water Weight.* Water accounts for 60% of a man's weight (*Source:* National Institute for Fitness and Sport). Reggie weighs 186 lb. How much of his body weight is water?

63. *Cardiovascular Procedures.* About 6 million Americans had cardiovascular procedures performed in 1999. This was about 9% more than the number who had such procedures performed in 1996. (*Source: Parade Magazine,* July 4, 1999) How many Americans had cardiovascular procedures performed in 1996?

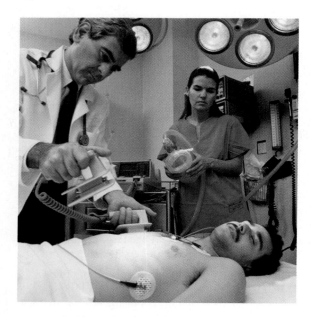

64. *Cost of Commercials.* In the fall of 1998, the average cost of a 30-sec commercial on NBC television was $168,000. This was a 10.2% decrease from the cost a year earlier. (*Source: Advertising Age*) What was the cost of a 30-sec commercial on NBC in the fall of 1997?

65. *Programming Cost.* Rights to the medical drama *ER* cost NBC television $13 million per episode. NBC charges advertisers $565,000 per 30-sec commercial on *ER*. (*Source: Indianapolis Star,* October 1, 1998) How many 30-sec commercials must the network sell in order to break even on the show?

66. *Book Buyers.* 15% of book buyers are 40–44 years old. This is two and one-half times the number who are 18–24 years old. (*Source:* Consumer Research Study on Book Purchasing) What percent of book buyers are 18–24 years old?

67. *Erosion.* Because of erosion, Horseshoe Falls, one of the two falls that make up Niagara Falls, is migrating upstream at a rate of 2 ft per year (*Source: Indianapolis Star,* February 14, 1999). At this rate, how long will it take the falls to move one-fourth mile?

68. *Volcanic Activity.* A volcano that is currently about one-half mile below the surface of the Pacific Ocean near the Big Island of Hawaii will eventually become a new Hawaiian island, Loihi. The volcano will break the surface of the ocean in about 50,000 yr. (*Source:* U.S. Geological Survey) On average, how many inches does the volcano rise in a year?

Technology Connection

69. Use a graphing calculator to solve the equations in Exercises 1–18.

70. Use a graphing calculator to find the zeros of the functions in Exercises 29–44.

Collaborative Discussion and Writing

71. The formula $A = P + Prt$ gives the amount A in an account when a principal P is invested at interest rate r for t years. Under what circumstances would it be helpful to solve this formula for P?

72. Explain in your own words why a linear function $f(x) = mx + b$, with $m \neq 0$, has exactly one zero.

Skill Maintenance

73. Write a slope–intercept equation for the line containing the point $(-1, 4)$ and parallel to the line $3x + 4y = 7$.

74. Write an equation of the line containing the points $(-5, 4)$ and $(3, -2)$.

Given that $f(x) = 2x - 1$ and $g(x) = x^2 - 5$, find each of the following.

75. $(f + g)(x)$ **76.** $(fg)(-1)$

Synthesis

Solve.

77. $2x - \{x - [3x - (6x + 5)]\} = 4x - 1$

78. $14 - 2[3 + 5(x - 1)] = 3\{x - 4[1 + 6(2 - x)]\}$

State whether each of the following is a linear function.

79. $f(x) = 7 - \dfrac{3}{2}x$

80. $f(x) = \dfrac{3}{2x} + 5$

81. $f(x) = x^2 + 1$

82. $f(x) = \dfrac{3}{4}x - (2.4)^2$

2.2

The Complex Numbers

• *Perform computations involving complex numbers.*

Some functions have zeros that are not real numbers. In order to find the zeros of such functions, we must consider the **complex-number system.**

The Complex-Number System

We know that the square root of a negative number is not a real number. For example, $\sqrt{-1}$ is not a real number because there is no real number x such that $x^2 = -1$. This means that certain equations, like $x^2 = -1$, or $x^2 + 1 = 0$, do not have real-number solutions and certain functions, like $f(x) = x^2 + 1$, do not have real-number zeros. Consider the graph of $f(x) = x^2 + 1$.

RADICAL EXPRESSIONS

REVIEW SECTION R.6.

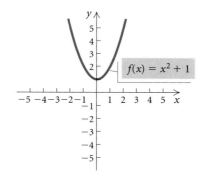

We see that the graph does not cross the x-axis and thus has no x-intercepts. This illustrates that the function $f(x) = x^2 + 1$ has no real-number zeros. Thus there are no real-number solutions of the corresponding equation $x^2 + 1 = 0$.

We can define a number that is a solution of the equation $x^2 + 1 = 0$.

The Number i

The number i is defined such that

$$i = \sqrt{-1} \quad \text{and} \quad i^2 = -1.$$

To express roots of negative numbers in terms of i, we can use the fact that

$$\sqrt{-p} = \sqrt{-1 \cdot p} = \sqrt{-1}\,\sqrt{p} = i\sqrt{p}$$

when p is a positive real number.

STUDY TIP

Don't be hesitant to ask questions in class at appropriate times. Most instructors welcome questions and encourage students to ask them. Other students in your class probably have the same questions you do.

EXAMPLE 1   Express each number in terms of i.

a) $\sqrt{-7}$ **b)** $\sqrt{-16}$
c) $-\sqrt{-13}$ **d)** $-\sqrt{-64}$
e) $\sqrt{-48}$

Solution

a) $\sqrt{-7} = \sqrt{-1 \cdot 7} = \sqrt{-1} \cdot \sqrt{7}$
$= i\sqrt{7}$, or $\sqrt{7}i$ ← i is *not* under the radical.

b) $\sqrt{-16} = \sqrt{-1 \cdot 16} = \sqrt{-1} \cdot \sqrt{16}$
$= i \cdot 4 = 4i$

c) $-\sqrt{-13} = -\sqrt{-1 \cdot 13} = -\sqrt{-1} \cdot \sqrt{13}$
$= -i\sqrt{13}$, or $-\sqrt{13}i$

d) $-\sqrt{-64} = -\sqrt{-1 \cdot 64} = -\sqrt{-1} \cdot \sqrt{64}$
$= -i \cdot 8 = -8i$

e) $\sqrt{-48} = \sqrt{-1 \cdot 48} = \sqrt{-1} \cdot \sqrt{48}$
$= i\sqrt{16 \cdot 3} = i \cdot 4\sqrt{3}$
$= 4i\sqrt{3}$, or $4\sqrt{3}i$

The complex numbers are formed by adding real numbers and multiples of i.

Complex Numbers

A **complex number** is a number of the form $a + bi$, where a and b are real numbers. The number a is said to be the **real part** of $a + bi$ and the number b is said to be the **imaginary part** of $a + bi$.*

Note that either a or b or both can be 0. When $b = 0$, $a + bi = a + 0i = a$, so every real number is a complex number. A complex number like $3 + 4i$ or $17i$, for which $b \neq 0$, is called an **imaginary number.** A complex number like $17i$ or $-4i$, in which $a = 0$ and $b \neq 0$, is some-

*Sometimes bi is considered to be the imaginary part.

times called a **pure imaginary number.** The relationships among various types of complex numbers are shown in the figure below.

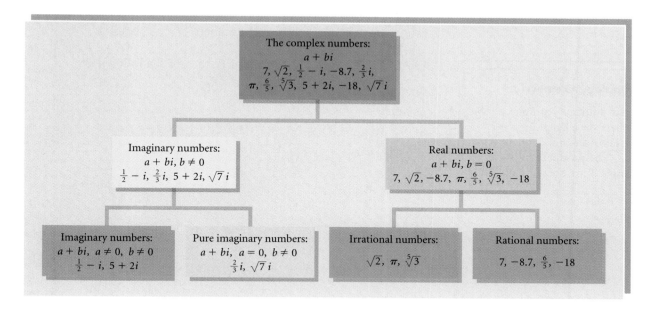

Addition and Subtraction

The complex numbers obey the commutative, associative, and distributive laws. Thus we can add and subtract them as we do binomials.

EXAMPLE 2 Add or subtract and simplify each of the following.

a) $(8 + 6i) + (3 + 2i)$ **b)** $(4 + 5i) - (6 - 3i)$

Solution

a) $(8 + 6i) + (3 + 2i) = (8 + 3) + (6i + 2i)$

Collecting the real parts and the imaginary parts

$$= 11 + (6 + 2)i = 11 + 8i$$

b) $(4 + 5i) - (6 - 3i) = (4 - 6) + [5i - (-3i)]$

Note that 6 and $-3i$ are both being subtracted.

$$= -2 + 8i$$

Multiplication

When $\sqrt{a}$ and $\sqrt{b}$ are real numbers, $\sqrt{a} \cdot \sqrt{b} = \sqrt{ab}$, but this is not true when $\sqrt{a}$ and $\sqrt{b}$ are not real numbers. Thus,

$$\sqrt{-2} \cdot \sqrt{-5} = \sqrt{-1} \cdot \sqrt{2} \cdot \sqrt{-1} \cdot \sqrt{5}$$
$$= i\sqrt{2} \cdot i\sqrt{5}$$
$$= i^2\sqrt{10} = -1\sqrt{10} = -\sqrt{10} \quad \text{is correct!}$$

Technology Connection

When set in $a + bi$ mode, most graphing calculators perform operations on complex numbers. The operations in Example 2 are shown in the window below. Some calculators will express a complex number in the form (a, b) rather than $a + bi$.

```
(8+6i)+(3+2i)
                    11+8i
(4+5i)−(6−3i)
                    −2+8i
```

But

$$\sqrt{-2} \cdot \sqrt{-5} = \sqrt{(-2)(-5)} = \sqrt{10} \quad \text{is wrong!}$$

Keeping this and the fact that $i^2 = -1$ in mind, we multiply in much the same way that we do with real numbers.

EXAMPLE 3   Multiply and simplify each of the following.

a) $\sqrt{-16} \cdot \sqrt{-25}$ b) $(1 + 2i)(1 + 3i)$ c) $(3 - 7i)^2$

Solution

a) $\sqrt{-16} \cdot \sqrt{-25} = \sqrt{-1} \cdot \sqrt{16} \cdot \sqrt{-1} \cdot \sqrt{25}$

$$= i \cdot 4 \cdot i \cdot 5$$

$$= i^2 \cdot 20$$

$$= -1 \cdot 20 \qquad i^2 = -1$$

$$= -20$$

b) $(1 + 2i)(1 + 3i) = 1 + 3i + 2i + 6i^2$ Multiplying each term of one number by every term of the other (FOIL)

$$= 1 + 3i + 2i - 6 \qquad i^2 = -1$$

$$= -5 + 5i \qquad \text{Collecting like terms}$$

c) $(3 - 7i)^2 = 3^2 - 2 \cdot 3 \cdot 7i + (7i)^2$ Recall that $(A - B)^2 = A^2 - 2AB + B^2$

$$= 9 - 42i + 49i^2$$

$$= 9 - 42i - 49 \qquad i^2 = -1$$

$$= -40 - 42i$$

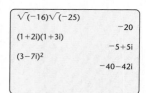
Recall that -1 raised to an *even* power is 1, and -1 raised to an *odd* power is -1. Simplifying powers of i can then be done by using the fact that $i^2 = -1$ and expressing the given power of i in terms of i^2. Consider the following:

$i = \sqrt{-1}$, $i^5 = i^4 \cdot i = (i^2)^2 \cdot i = (-1)^2 \cdot i = i$,

$i^2 = -1$, $i^6 = (i^2)^3 = (-1)^3 = -1$,

$i^3 = i^2 \cdot i = (-1)i = -i$, $i^7 = (i^2)^3 \cdot i = (-1)^3 \cdot i = -i$,

$i^4 = (i^2)^2 = (-1)^2 = 1$, $i^8 = (i^2)^4 = (-1)^4 = 1$.

Note that the powers of i cycle through the values i, -1, $-i$, and 1.

EXAMPLE 4   Simplify each of the following.

a) i^{37} b) i^{58} c) i^{75} d) i^{80}

Solution

a) $i^{37} = i^{36} \cdot i = (i^2)^{18} \cdot i = (-1)^{18} \cdot i = 1 \cdot i = i$

b) $i^{58} = (i^2)^{29} = (-1)^{29} = -1$

c) $i^{75} = i^{74} \cdot i = (i^2)^{37} \cdot i = (-1)^{37} \cdot i = -1 \cdot i = -i$

d) $i^{80} = (i^2)^{40} = (-1)^{40} = 1$

Conjugates and Divison

Conjugates of complex numbers are defined as follows.

Conjugate of a Complex Number

The **conjugate** of a complex number $a + bi$ is $a - bi$. The numbers $a + bi$ and $a - bi$ are **complex conjugates.**

Each of the following pairs of numbers are complex conjugates: $-3 + 7i$ and $-3 - 7i$; $14 - 5i$ and $14 + 5i$; and $8i$ and $-8i$.

The product of a complex number and its conjugate is a real number.

EXAMPLE 5 Multiply each of the following.

a) $(5 + 7i)(5 - 7i)$ **b)** $(8i)(-8i)$

Solution

a)
$$
\begin{aligned}
(5 + 7i)(5 - 7i) &= 5^2 - (7i)^2 \quad \text{Using } (A + B)(A - B) = A^2 - B^2\\
&= 25 - 49i^2\\
&= 25 - 49(-1)\\
&= 74
\end{aligned}
$$

b)
$$
\begin{aligned}
(8i)(-8i) &= -64i^2\\
&= -64(-1)\\
&= 64
\end{aligned}
$$

—

When we are dividing complex numbers, conjugates are used.

EXAMPLE 6 Divide $2 - 5i$ by $1 - 6i$.

Solution We write fractional notation and then multiply by 1, using the conjugate of the denominator to form the symbol for 1.

$$
\begin{aligned}
\frac{2 - 5i}{1 - 6i} &= \frac{2 - 5i}{1 - 6i} \cdot \frac{1 + 6i}{1 + 6i} \quad \begin{array}{l}\text{Note that } 1 + 6i \text{ is the conjugate}\\ \text{of the divisor.}\end{array}\\[2mm]
&= \frac{(2 - 5i)(1 + 6i)}{(1 - 6i)(1 + 6i)}\\[2mm]
&= \frac{2 + 7i - 30i^2}{1 - 36i^2}\\[2mm]
&= \frac{2 + 7i + 30}{1 + 36} \quad i^2 = -1\\[2mm]
&= \frac{32 + 7i}{37}\\[2mm]
&= \frac{32}{37} + \frac{7}{37}i.
\end{aligned}
$$

—

Technology Connection

With a graphing calculator set in $a + bi$ mode, we can divide complex numbers and express the real and imaginary parts in fractional form, just as we did in Example 6.

```
(2−5i)/(1−6i) ▶Frac
                32/37+7/37i
```

Exercise Set 2.2

Simplify. Write answers in the form a + bi, where a and b are real numbers.

1. $(-5 + 3i) + (7 + 8i)$

2. $(-6 - 5i) + (9 + 2i)$

3. $(4 - 9i) + (1 - 3i)$

4. $(7 - 2i) + (4 - 5i)$

5. $(3 + \sqrt{-16}) + (2 + \sqrt{-25})$

6. $(7 - \sqrt{-36}) + (2 + \sqrt{-9})$

7. $(10 + 7i) - (5 + 3i)$

8. $(-3 - 4i) - (8 - i)$

9. $(13 + 9i) - (8 + 2i)$

10. $(-7 + 12i) - (3 - 6i)$

11. $(1 + 3i)(1 - 4i)$

12. $(1 - 2i)(1 + 3i)$

13. $(2 + 3i)(2 + 5i)$

14. $(3 - 5i)(8 - 2i)$

15. $7i(2 - 5i)$

16. $3i(6 + 4i)$

17. $(3 + \sqrt{-16})(2 + \sqrt{-25})$

18. $(7 - \sqrt{-16})(2 + \sqrt{-9})$

19. $(5 - 4i)(5 + 4i)$

20. $(5 + 9i)(5 - 9i)$

21. $(4 + 2i)^2$

22. $(5 - 4i)^2$

23. $(-2 + 7i)^2$

24. $(-3 + 2i)^2$

25. $\dfrac{2 + \sqrt{3}i}{5 - 4i}$

26. $\dfrac{\sqrt{5} + 3i}{1 - i}$

27. $\dfrac{4 + i}{-3 - 2i}$

28. $\dfrac{5 - i}{-7 + 2i}$

29. $\dfrac{3}{5 - 11i}$

30. $\dfrac{i}{2 + i}$

31. $\dfrac{1 + i}{(1 - i)^2}$

32. $\dfrac{1 - i}{(1 + i)^2}$

33. $\dfrac{4 - 2i}{1 + i} + \dfrac{2 - 5i}{1 + i}$

34. $\dfrac{3 + 2i}{1 - i} + \dfrac{6 + 2i}{1 - i}$

Simplify.

35. i^{11}

36. i^7

37. i^{35}

38. i^{24}

39. i^{64}

40. i^{42}

41. $(-i)^{71}$

42. $(-i)^6$

43. $(5i)^4$

44. $(2i)^5$

Technology Connection

45. Use a graphing calculator to perform the operations in Exercises 1, 5, 7, 9, 11, 15, 21, 25, and 27.

46. Use a graphing calculator to perform the operations in Exercises 2, 6, 8, 10, 12, 16, 22, 26, and 28.

Collaborative Discussion and Writing

47. Is the sum of two imaginary numbers always an imaginary number? Explain your answer.

48. Is the product of two imaginary numbers always an imaginary number? Explain your answer.

Skill Maintenance

49. Write a slope–intercept equation for the line containing the point $(3, -5)$ and perpendicular to the line $3x - 6y = 7$.

Given that $f(x) = x^2 + 4$ and $g(x) = 3x + 5$, find each of the following.

50. $(f - g)(x)$

51. $(f/g)(2)$

52. For the function $f(x) = x^2 - 3x + 4$, construct and simplify the difference quotient

$$\frac{f(x + h) - f(x)}{h}.$$

Synthesis

Determine whether each of the following is true or false.

53. The sum of two numbers that are conjugates of each other is always a real number.

54. The conjugate of a sum is the sum of the conjugates of the individual complex numbers.

55. The conjugate of a product is the product of the conjugates of the individual complex numbers.

Let $z = a + bi$ and $\bar{z} = a - bi$.

56. Find a general expression for $1/z$.

57. Find a general expression for $z\bar{z}$.

58. Solve $z + 6\bar{z} = 7$ for z.

2.3

Quadratic Equations, Functions, and Models

- *Find zeros of quadratic functions and solve quadratic equations by using the principle of zero products, by using the principle of square roots, by completing the square, and by using the quadratic formula.*
- *Solve equations that are reducible to quadratic.*
- *Solve applied problems.*

Quadratic Equations and Quadratic Functions

In this section, we will explore the relationship between the solutions of quadratic equations and the zeros of quadratic functions. We define quadratic equations and functions as follows.

Quadratic Equations

A **quadratic equation** is an equation equivalent to

$$ax^2 + bx + c = 0, \quad a \neq 0,$$

where a, b, and c are real numbers.

Quadratic Functions

A **quadratic function** f is a second-degree polynomial function

$$f(x) = ax^2 + bx + c, \quad a \neq 0,$$

where a, b, and c are real numbers.

ZEROS OF A FUNCTION

REVIEW SECTION 2.1.

The zeros of a quadratic function $f(x) = ax^2 + bx + c$ are the solutions of the associated quadratic equation $ax^2 + bx + c = 0$. (These solutions are sometimes called *roots* of the equation.) Quadratic functions can have real-number or imaginary-number zeros and quadratic equations can have real-number or imaginary-number solutions. If the zeros or solutions are real numbers, they are also the first coordinates of the x-intercepts of the graph of the quadratic function or equation.

The following principles allow us to solve many quadratic equations.

Equation-Solving Principles

The Principle of Zero Products: If $ab = 0$ is true, then $a = 0$ or $b = 0$, and if $a = 0$ or $b = 0$, then $ab = 0$.

The Principle of Square Roots: If $x^2 = k$, then $x = \sqrt{k}$ or $x = -\sqrt{k}$.

EXAMPLE 1   Solve: $2x^2 - x = 3$.

Algebraic Solution

We have

$$2x^2 - x = 3$$
$$2x^2 - x - 3 = 0 \qquad \text{Subtracting 3 on both sides}$$
$$(x + 1)(2x - 3) = 0 \qquad \text{Factoring}$$
$$x + 1 = 0 \quad or \quad 2x - 3 = 0 \qquad \text{Using the principle of zero products}$$
$$x = -1 \quad or \qquad 2x = 3$$
$$x = -1 \quad or \qquad x = \tfrac{3}{2}.$$

CHECK: For $x = -1$:

$$\frac{2x^2 - x = 3}{2(-1)^2 - (-1) \;\overset{?}{}\; 3}$$
$$2 \cdot 1 + 1 \;\Big|$$
$$3 \;\Big|\; 3 \qquad \text{TRUE}$$

For $x = \tfrac{3}{2}$:

$$\frac{2x^2 - x = 3}{2\left(\tfrac{3}{2}\right)^2 - \tfrac{3}{2} \;\overset{?}{}\; 3}$$
$$2 \cdot \tfrac{9}{4} - \tfrac{3}{2} \;\Big|$$
$$\tfrac{9}{2} - \tfrac{3}{2} \;\Big|$$
$$3 \;\Big|\; 3 \qquad \text{TRUE}$$

The solutions are -1 and $\tfrac{3}{2}$.

Visualizing the Solution

The solutions of the equation $2x^2 - x = 3$ are the first coordinates of the points of intersection of the graphs of $y = 2x^2 - x$ and $y = 3$.

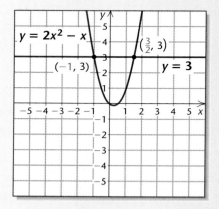

The solutions are -1 and $\tfrac{3}{2}$.

EXAMPLE 2 Solve: $2x^2 - 10 = 0$.

Algebraic Solution

We have

$2x^2 - 10 = 0$

$2x^2 = 10$ Adding 10 on both sides

$x^2 = 5$ Dividing by 2 on both sides

$x = \sqrt{5}$ *or* $x = -\sqrt{5}$.   Using the principle of square roots

CHECK: $2x^2 - 10 = 0$

     $2(\pm\sqrt{5})^2 - 10 \; ? \; 0$ We can check both solutions at once.

        $2 \cdot 5 - 10$

         $10 - 10$

           $0 \; | \; 0$ TRUE

The solutions are $\sqrt{5}$ and $-\sqrt{5}$, or $\pm\sqrt{5}$.

Visualizing the Solution

The solutions of the equation $2x^2 - 10 = 0$ are the zeros of the function $f(x) = 2x^2 - 10$. Note that they are also the first coordinates of the x-intercepts of the graph of $f(x) = 2x^2 - 10$.

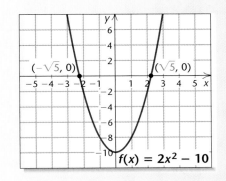

The solutions are $-\sqrt{5}$ and $\sqrt{5}$.

Completing the Square

Quadratic equations like $x^2 - 3x - 4 = 0$ and $x^2 - 6x + 9 = 0$ can be solved by factoring and using the principle of zero products:

$x^2 - 3x - 4 = 0$

$(x + 1)(x - 4) = 0$ Factoring

$x + 1 = 0$ *or* $x - 4 = 0$ Using the principle of zero products

    $x = -1$ *or* $x = 4$.

The equation $x^2 - 3x - 4 = 0$ has *two real-number* solutions, -1 and 4. These are the zeros of the associated quadratic function $f(x) = x^2 - 3x - 4$ and the first coordinates of the x-intercepts of the graph of this function. (See Fig. 1.)

Next, we have

$x^2 - 6x + 9 = 0$

$(x - 3)(x - 3) = 0$ Factoring

$x - 3 = 0$ *or* $x - 3 = 0$ Using the principle of zero products

    $x = 3$ *or* $x = 3$.

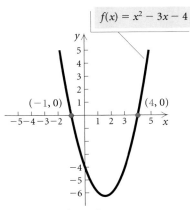

Two real-number zeros
Two x-intercepts

FIGURE 1

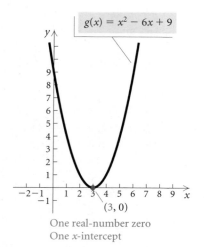

$g(x) = x^2 - 6x + 9$

$(3, 0)$

One real-number zero
One x-intercept

FIGURE 2

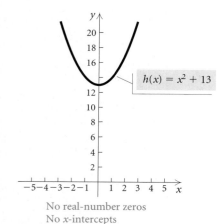

$h(x) = x^2 + 13$

No real-number zeros
No x-intercepts

FIGURE 3

The equation $x^2 - 6x + 9 = 0$ has *one real-number* solution, 3. It is the zero of the quadratic function $g(x) = x^2 - 6x + 9$ and the first coordinate of the x-intercept of the graph of this function. (See Fig. 2.)

The principle of square roots can be used to solve quadratic equations like $x^2 + 13 = 0$:

$$x^2 + 13 = 0$$
$$x^2 = -13$$
$$x = \pm\sqrt{-13} \qquad \text{Using the principle of square roots}$$
$$x = \pm\sqrt{13}i.$$

The equation has *two imaginary-number* solutions, $-\sqrt{13}i$ and $\sqrt{13}i$. These are the zeros of the associated quadratic function $h(x) = x^2 + 13$. Since the zeros are not real numbers, the graph of the function has no x-intercepts. (See Fig. 3.)

Neither the principle of zero products nor the principle of square roots would yield the *exact* zeros of a function like $f(x) = x^2 - 6x - 10$ or the *exact* solutions of the associated equation $x^2 - 6x - 10 = 0$. If we wish to do that, we can use a procedure called **completing the square** and then use the principle of square roots.

EXAMPLE 3   Find the zeros of $f(x) = x^2 - 6x - 10$ by completing the square.

Solution   We find the values of x for which $f(x) = 0$. That is, we solve the associated equation $x^2 - 6x - 10 = 0$. Our goal is to find an equivalent equation of the form $x^2 + bx + c = d$ in which $x^2 + bx + c$ is a perfect square. Since

$$x^2 + bx + \left(\frac{b}{2}\right)^2 = \left(x + \frac{b}{2}\right)^2,$$

the number c is found by taking half the coefficient of the x-term and squaring it. Then for the equation $x^2 - 6x - 10 = 0$, we have

$$x^2 - 6x - 10 = 0$$
$$x^2 - 6x \qquad = 10 \qquad \text{Adding 10}$$
$$x^2 - 6x + 9 = 10 + 9 \qquad \text{Adding 9 to complete the square:}$$
$$\left(\frac{b}{2}\right)^2 = \left(\frac{-6}{2}\right)^2 = 9$$
$$x^2 - 6x + 9 = 19.$$

Because $x^2 - 6x + 9$ is a perfect square, we are able to write it as $(x - 3)^2$, the square of a binomial. We can then use the principle of square roots to finish the solution:

$$(x - 3)^2 = 19 \qquad \text{Factoring}$$
$$x - 3 = \pm\sqrt{19} \qquad \text{Using the principle of square roots}$$
$$x = 3 \pm \sqrt{19}. \qquad \text{Adding 3}$$

Technology Connection

Approximations for the zeros of the quadratic function $f(x) = x^2 - 6x - 10$ in Example 3 can be found using the Zero method.

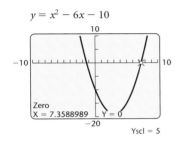

$y = x^2 - 6x - 10$

Zero
X = 7.3588989   Y = 0

Yscl = 5

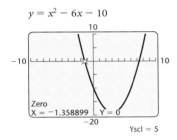

$y = x^2 - 6x - 10$

Zero
X = -1.358899   Y = 0

Yscl = 5

STUDY TIP

The examples in the text are carefully chosen to prepare you for success with the exercise sets. Study the step-by-step solutions of the examples, noting that substitutions and explanations appear in red. The time you spend studying the examples will save you valuable time when you do your homework.

Thus the solutions of the equation are $3 + \sqrt{19}$ and $3 - \sqrt{19}$, or simply $3 \pm \sqrt{19}$. The zeros of $f(x) = x^2 - 6x - 10$ are also $3 + \sqrt{19}$ and $3 - \sqrt{19}$, or $3 \pm \sqrt{19}$.

Decimal approximations for $3 \pm \sqrt{19}$ can be found using a calculator:

$$3 + \sqrt{19} \approx 7.359 \quad \text{and} \quad 3 - \sqrt{19} \approx -1.359.$$

The zeros are approximately 7.359 and -1.359. ▬

Before we can complete the square, the coefficient of the x^2-term must be 1. When it is not, we divide by the x^2-coefficient on both sides of the equation.

EXAMPLE 4 Solve: $2x^2 - 1 = 3x$.

Solution We have

$$2x^2 - 1 = 3x$$

$$2x^2 - 3x - 1 = 0 \qquad \text{Subtracting } 3x. \text{ We are unable to factor the result.}$$

$$2x^2 - 3x \quad\;\; = 1 \qquad \text{Adding 1}$$

$$x^2 - \frac{3}{2}x \quad\;\; = \frac{1}{2} \qquad \text{Dividing by 2 to make the } x^2\text{-coefficient 1}$$

$$x^2 - \frac{3}{2}x + \frac{9}{16} = \frac{1}{2} + \frac{9}{16} \qquad \text{Completing the square: } \frac{1}{2}\left(-\frac{3}{2}\right) = -\frac{3}{4} \text{ and } \left(-\frac{3}{4}\right)^2 = \frac{9}{16}; \text{ adding } \frac{9}{16}$$

$$\left(x - \frac{3}{4}\right)^2 = \frac{17}{16} \qquad \text{Factoring and simplifying}$$

$$x - \frac{3}{4} = \pm\frac{\sqrt{17}}{4} \qquad \text{Using the principle of square roots and the quotient rule for radicals}$$

$$x = \frac{3 \pm \sqrt{17}}{4}. \qquad \text{Adding } \frac{3}{4}$$

The solutions are

$$\frac{3 + \sqrt{17}}{4} \quad \text{and} \quad \frac{3 - \sqrt{17}}{4}, \quad \text{or} \quad \frac{3 \pm \sqrt{17}}{4}. \qquad \text{▬}$$

To solve a quadratic equation by completing the square:

1. Isolate the terms with variables on one side of the equation and arrange them in descending order.
2. Divide by the coefficient of the squared term if that coefficient is not 1.
3. Complete the square by taking half the coefficient of the first-degree term and adding its square on both sides of the equation.
4. Express one side of the equation as the square of a binomial.
5. Use the principle of square roots.
6. Solve for the variable.

Using the Quadratic Formula

Because completing the square works for *any* quadratic equation, it can be used to solve the general quadratic equation $ax^2 + bx + c = 0$ for x. The result will be a formula that can be used to solve any quadratic equation quickly.

Consider any quadratic equation in standard form:

$$ax^2 + bx + c = 0, \quad a \neq 0.$$

For now, we assume that $a > 0$ and solve by completing the square. As the steps are carried out, compare them with those of Example 4.

$$ax^2 + bx + c = 0 \qquad \text{Standard form}$$

$$ax^2 + bx = -c \qquad \text{Adding } -c$$

$$x^2 + \frac{b}{a}x = -\frac{c}{a} \qquad \text{Dividing by } a$$

Half of $\dfrac{b}{a}$ is $\dfrac{b}{2a}$ and $\left(\dfrac{b}{2a}\right)^2 = \dfrac{b^2}{4a^2}$. Thus we add $\dfrac{b^2}{4a^2}$:

$$x^2 + \frac{b}{a}x + \frac{b^2}{4a^2} = -\frac{c}{a} + \frac{b^2}{4a^2} \qquad \text{Adding } \frac{b^2}{4a^2} \text{ to complete the square}$$

$$\left(x + \frac{b}{2a}\right)^2 = -\frac{4ac}{4a^2} + \frac{b^2}{4a^2} \qquad \begin{array}{l}\text{Factoring and finding a common} \\ \text{denominator:} \\ -\dfrac{c}{a} = -\dfrac{4a}{4a} \cdot \dfrac{c}{a} = -\dfrac{4ac}{4a^2}\end{array}$$

$$\left(x + \frac{b}{2a}\right)^2 = \frac{b^2 - 4ac}{4a^2}$$

$$x + \frac{b}{2a} = \pm\frac{\sqrt{b^2 - 4ac}}{2a} \qquad \begin{array}{l}\text{Using the principle of square roots} \\ \text{and the quotient rule for radicals.} \\ \text{Since } a > 0, \sqrt{4a^2} = 2a.\end{array}$$

$$x = \frac{-b \pm \sqrt{b^2 - 4ac}}{2a}. \qquad \text{Adding } -\frac{b}{2a}$$

It can also be shown that this result holds if $a < 0$.

The Quadratic Formula

The solutions of $ax^2 + bx + c = 0$, $a \neq 0$, are given by

$$x = \frac{-b \pm \sqrt{b^2 - 4ac}}{2a}.$$

EXAMPLE 5 Solve $3x^2 + 2x = 7$. Find exact solutions and approximate solutions rounded to the nearest thousandth.

Solution

After finding standard form, we are unable to factor, so we identify a, b, and c in order to use the quadratic formula:

$$3x^2 + 2x - 7 = 0;$$
$$a = 3, \quad b = 2, \quad c = -7.$$

We then use the quadratic formula:

$$x = \frac{-b \pm \sqrt{b^2 - 4ac}}{2a}$$

$$= \frac{-2 \pm \sqrt{2^2 - 4(3)(-7)}}{2(3)} \qquad \text{Substituting}$$

$$= \frac{-2 \pm \sqrt{4 + 84}}{6}$$

$$= \frac{-2 \pm \sqrt{88}}{6}$$

$$= \frac{-2 \pm \sqrt{4 \cdot 22}}{6}$$

$$= \frac{-2 \pm 2\sqrt{22}}{6}$$

$$= \frac{2}{2} \cdot \frac{-1 \pm \sqrt{22}}{3}$$

$$= \frac{-1 \pm \sqrt{22}}{3}.$$

The exact solutions are

$$\frac{-1 - \sqrt{22}}{3} \quad \text{and} \quad \frac{-1 + \sqrt{22}}{3}.$$

Using a calculator, we approximate the solutions to be -1.897 and 1.230.

Technology Connection

We can solve the equation $3x^2 + 2x = 7$ in Example 5 using the Intersect method. We graph $y_1 = 3x^2 + 2x$ and $y_2 = 7$ and use the INTERSECT feature to find the coordinates of the points of intersection. The first coordinates of these points are the solutions of the equation $y_1 = y_2$, or $3x^2 + 2x = 7$.

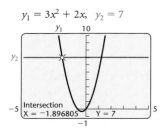

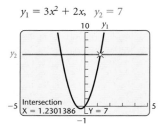

The solutions are approximately -1.897 and 1.230. We could also write the equation in standard form, $3x^2 + 2x - 7 = 0$, and use the Zero method.

EXAMPLE 6   Solve: $x^2 + 5x + 8 = 0$.

Algebraic Solution

To find the solutions, we use the quadratic formula. Here

$$a = 1, \quad b = 5, \quad c = 8;$$

$$x = \frac{-b \pm \sqrt{b^2 - 4ac}}{2a}$$

$$= \frac{-5 \pm \sqrt{5^2 - 4(1)(8)}}{2 \cdot 1} \qquad \text{Substituting}$$

$$= \frac{-5 \pm \sqrt{-7}}{2} \qquad \text{Simplifying}$$

$$= \frac{-5 \pm \sqrt{7}i}{2}.$$

The solutions are $-\dfrac{5}{2} - \dfrac{\sqrt{7}}{2}i$ and $-\dfrac{5}{2} + \dfrac{\sqrt{7}}{2}i$.

Visualizing the Solution

The graph of the function $f(x) = x^2 + 5x + 8$ has no x-intercepts.

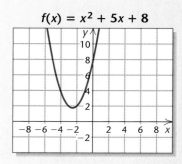

Thus the function has no real-number zeros and there are no real-number solutions of the associated equation $x^2 + 5x + 8 = 0$.

The Discriminant

From the quadratic formula, we know that the solutions x_1 and x_2 of a quadratic equation are given by

$$x_1 = \frac{-b + \sqrt{b^2 - 4ac}}{2a} \quad \text{and} \quad x_2 = \frac{-b - \sqrt{b^2 - 4ac}}{2a}.$$

The expression $b^2 - 4ac$ shows the nature of the solutions. This expression is called the **discriminant**. If it is 0, then it makes no difference whether we choose the plus or the minus sign in the formula. There is just one solution. In this case, we sometimes say that there is one repeated real solution. If the discriminant is positive, there will be two real solutions. If it is negative, we will be taking the square root of a negative number; hence there will be two imaginary-number solutions, and they will be complex conjugates.

Discriminant

For $ax^2 + bx + c = 0$:

$b^2 - 4ac = 0 \longrightarrow$ One real-number solution;

$b^2 - 4ac > 0 \longrightarrow$ Two different real-number solutions;

$b^2 - 4ac < 0 \longrightarrow$ Two different imaginary-number solutions, complex conjugates.

In Example 5, the discriminant, 88, is positive, indicating that there are two different real-number solutions. If the discriminant is negative, as it is in Example 6, we know that there are two different imaginary-number solutions.

Equations Reducible to Quadratic

Some equations can be treated as quadratic, provided that we make a suitable substitution. For example, consider the following:

$$x^4 - 5x^2 + 4 = 0$$
$$(x^2)^2 - 5x^2 + 4 = 0 \qquad x^4 = (x^2)^2$$
$$u^2 - 5u + 4 = 0. \qquad \text{Substituting } u \text{ for } x^2$$

The equation $u^2 - 5u + 4 = 0$ can be solved for u by factoring or using the quadratic formula. Then we can reverse the substitution, replacing u with x^2, and solve for x. Equations like the one above are said to be **reducible to quadratic,** or **quadratic in form.**

EXAMPLE 7   Solve: $x^4 - 5x^2 + 4 = 0$.

Algebraic Solution

We let $u = x^2$ and substitute:

$u^2 - 5u + 4 = 0$ Substituting u for x^2

$(u - 1)(u - 4) = 0$ Factoring

$u - 1 = 0$ or $u - 4 = 0$ Using the principle of zero products

      $u = 1$ or $u = 4$.

Don't stop here! We must solve for the original variable. We substitute x^2 for u and solve for x:

$x^2 = 1$ or $x^2 = 4$

$x = \pm 1$ or $x = \pm 2$.   Using the principle of square roots

The solutions are -1, 1, -2, and 2.

Visualizing the Solution

The solutions of the given equation are the zeros of $y = x^4 - 5x^2 + 4$. Observe that the zeros occur at the x-values -2, -1, 1, and 2.

$y = x^4 - 5x^2 + 4$

Technology Connection

We can use the Zero method to solve the equation in Example 7, $x^4 - 5x^2 + 4 = 0$. We graph the function $y = x^4 - 5x^2 + 4$ and use the ZERO feature to find the zeros.

$y = x^4 - 5x^2 + 4$

Zero
X = -2   Y = 0

The leftmost zero is -2. Using the ZERO feature three more times, we find that the other zeros are -1, 1, and 2.

Applications

Some applied problems can be translated to quadratic equations.

EXAMPLE 8 *Time of a Free Fall.* The Petronas Towers in Kuala Lumpur, Malaysia are 1482 ft tall. How long would it take an object dropped from the top to reach the ground?

Solution

1. **Familiarize.** The formula $s = 16t^2$ is used to approximate the distance s, in feet, that an object falls freely from rest in t seconds. In this case, the distance is 1482 ft.

2. **Translate.** We substitute 1482 for s in the formula:

   $$1482 = 16t^2.$$

3. **Carry out.** We use the principle of square roots:

   $$1482 = 16t^2$$
   $$\frac{1482}{16} = t^2 \qquad \text{Dividing by 16}$$
   $$\sqrt{\frac{1482}{16}} = t \qquad \text{Taking the positive square root. Time cannot be negative in this application.}$$
   $$9.624 \approx t.$$

4. **Check.** In 9.624 sec, a dropped object would travel a distance of $16(9.624)^2$, or about 1482 ft. The answer checks.

5. **State.** It would take about 9.624 sec for an object dropped from the top of the Petronas Towers to reach the ground. ▬

EXAMPLE 9 *Bicycling Speed.* Castulo and Linette leave a campsite, Castulo biking due north and Linette biking due east. Castulo bikes 7 km/h slower than Linette. After 4 hr, they are 68 km apart. Find the speed of each bicyclist.

Solution

1. **Familiarize.** We let r = Linette's speed, in kilometers per hour. Then $r - 7$ = Castulo's speed, in kilometers per hour. We will use the motion formula $d = rt$, where d is the distance, r is the rate (or

speed), and t is the time. Then, after 4 hr, Linette has traveled $4r$ km and Castulo has traveled $4(r - 7)$ km. We add these distances to the drawing, as shown below.

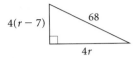

2. **Translate.** We use the Pythagorean theorem, $a^2 + b^2 = c^2$, where a and b are the lengths of the legs of a right triangle and c is the length of the hypotenuse:

$$(4r)^2 + [4(r - 7)]^2 = 68^2.$$

3. **Carry out.** We solve the equation:

$$(4r)^2 + [4(r - 7)]^2 = 68^2$$
$$16r^2 + 16(r^2 - 14r + 49) = 4624$$
$$16r^2 + 16r^2 - 224r + 784 = 4624$$
$$32r^2 - 224r - 3840 = 0 \qquad \text{Subtracting 4624}$$
$$r^2 - 7r - 120 = 0 \qquad \text{Dividing by 32}$$
$$(r + 8)(r - 15) = 0 \qquad \text{Factoring}$$
$$r + 8 = 0 \quad \text{or} \quad r - 15 = 0 \qquad \text{Principle of zero products}$$
$$r = -8 \quad \text{or} \qquad r = 15.$$

4. **Check.** Since speed cannot be negative, we need to check only 15. If Linette's speed is 15 km/h, then Castulo's speed is $15 - 7$, or 8 km/h. In 4 hr, Linette travels $4 \cdot 15$, or 60 km, and Castulo travels $4 \cdot 8$, or 32 km. Then they are $60^2 + 32^2$, or 4624 km apart. Since $4624 = 68^2$, the answer checks.

5. **State.** Linette's speed is 15 km/h and Castulo's speed is 8 km/h.

CONNECTING THE CONCEPTS

ZEROS, SOLUTIONS, AND INTERCEPTS

The zeros of a function $y = f(x)$ are also the solutions of the equation $f(x) = 0$ and the first coordinates of the x-intercepts of the graph of the function.

FUNCTION	ZEROS OF THE FUNCTION; SOLUTIONS OF THE EQUATION	X-INTERCEPTS OF THE GRAPH
Linear Function $f(x) = 2x - 4$, or $y = 2x - 4$	To find the **zero** of $f(x)$, we solve $f(x) = 0$: $$2x - 4 = 0$$ $$2x = 4$$ $$x = 2.$$ The **solution** of $2x - 4 = 0$ is 2. This is the zero of the function $f(x) = 2x - 4$. That is, $f(2) = 0$.	The zero of $f(x)$ is the first coordinate of the **x-intercept** of the graph of $y = f(x)$. 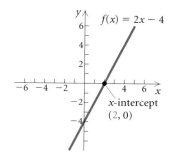
Quadratic Function $g(x) = x^2 - 3x - 4$, or $y = x^2 - 3x - 4$	To find the **zeros** of $g(x)$, we solve $g(x) = 0$: $$x^2 - 3x - 4 = 0$$ $$(x + 1)(x - 4) = 0$$ $$x + 1 = 0 \quad or \quad x - 4 = 0$$ $$x = -1 \quad or \quad x = 4.$$ The **solutions** of $x^2 - 3x - 4 = 0$ are -1 and 4. They are the zeros of the function $g(x)$. That is, $g(-1) = 0$ and $g(4) = 0$.	The zeros of $g(x)$ are the first coordinates of the **x-intercepts** of the graph of $y = g(x)$. 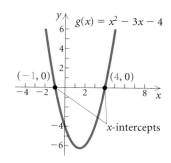

Exercise Set 2.3

Solve.

1. $(2x - 3)(3x - 2) = 0$
2. $(5x - 2)(2x + 3) = 0$
3. $3x^2 + x - 2 = 0$
4. $10x^2 - 16x + 6 = 0$
5. $4x^2 - 12 = 0$
6. $6x^2 = 36$
7. $2x^2 = 6x$
8. $18x + 9x^2 = 0$
9. $3y^3 - 5y^2 - 2y = 0$
10. $3t^3 + 2t = 5t^2$
11. $7x^3 + x^2 - 7x - 1 = 0$
    (*Hint*: Factor by grouping.)
12. $3x^3 + x^2 - 12x - 4 = 0$
    (*Hint*: Factor by grouping.)

In Exercises 13 and 14, use the given graph to find each of the following:

a) *the x-intercepts and*
b) *the zeros of the function.*

13.

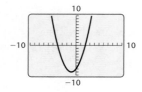

14.

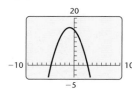

Solve by completing the square to obtain exact solutions.

15. $x^2 + 6x = 7$
16. $x^2 + 8x = -15$
17. $x^2 = 8x - 9$
18. $x^2 = 22 + 10x$
19. $x^2 + 8x + 25 = 0$
20. $x^2 + 6x + 13 = 0$
21. $3x^2 + 5x - 2 = 0$
22. $2x^2 - 5x - 3 = 0$

Use the quadratic formula to find exact solutions.

23. $x^2 - 2x = 15$
24. $x^2 + 4x = 5$
25. $5m^2 + 3m = 2$
26. $2y^2 - 3y - 2 = 0$
27. $3x^2 + 6 = 10x$
28. $3t^2 + 8t + 3 = 0$
29. $x^2 + x + 2 = 0$
30. $x^2 + 1 = x$
31. $5t^2 - 8t = 3$
32. $5x^2 + 2 = x$
33. $3x^2 + 4 = 5x$
34. $2t^2 - 5t = 1$

For each of the following, consider only $b^2 - 4ac$, the discriminant of the quadratic formula, to determine whether imaginary solutions exist.

35. $4x^2 = 8x + 5$
36. $4x^2 - 12x + 9 = 0$
37. $x^2 + 3x + 4 = 0$
38. $x^2 - 2x + 4 = 0$
39. $5t^2 - 7t = 0$
40. $5t^2 - 4t = 11$

Find the zeros of the function.

41. $f(x) = x^2 + 6x + 5$
42. $f(x) = x^2 - x - 2$
43. $f(x) = x^2 - 3x - 3$
44. $f(x) = 3x^2 + 8x + 2$
45. $f(x) = x^2 - 5x + 1$
46. $f(x) = x^2 - 3x - 7$
47. $f(x) = x^2 + 2x - 5$
48. $f(x) = x^2 - x - 4$

Solve.

49. $x^4 - 3x^2 + 2 = 0$
50. $x^4 + 3 = 4x^2$
51. $x - 3\sqrt{x} - 4 = 0$
    (*Hint*: Let $u = \sqrt{x}$.)
52. $2x - 9\sqrt{x} + 4 = 0$
53. $m^{2/3} - 2m^{1/3} - 8 = 0$
    (*Hint*: Let $u = m^{1/3}$.)
54. $t^{2/3} + t^{1/3} - 6 = 0$
55. $(2x - 3)^2 - 5(2x - 3) + 6 = 0$
    (*Hint*: Let $u = 2x - 3$.)
56. $(3x + 2)^2 + 7(3x + 2) - 8 = 0$
57. $(2t^2 + t)^2 - 4(2t^2 + t) + 3 = 0$
58. $12 = (m^2 - 5m)^2 + (m^2 - 5m)$

Time of a Free Fall.   The formula $s = 16t^2$ is used to approximate the distance s, in feet, that an object falls freely from rest in t seconds.

59. The Warszawa Radio Mast in Poland, at 2120 ft, is the world's tallest structure (*Source: The Cambridge Fact Finder*). How long would it take an object falling freely from the top to reach the ground?

60. The tallest structure in the United States, at 2063 ft, is the KTHI-TV tower in North Dakota (*Source: The Cambridge Fact Finder*). How long would it take an object falling freely from the top to reach the ground?

61. The length of a rectangular rug is 1 ft more than the width and a diagonal of the rug is 5 ft. Find the length and the width.

62. One leg of a right triangle is 7 cm less than the length of the other leg. The length of the hypotenuse is 13 cm. Find the lengths of the legs.

63. One number is 5 greater than another. The product of the numbers is 36. Find the numbers.

64. One number is 6 less than another. The product of the numbers is 72. Find the numbers.

65. *Box Construction.* An open box is made from a 10-cm by 20-cm piece of tin by cutting a square from each corner and folding up the edges. The area of the resulting base is 96 cm². What is the length of the sides of the squares?

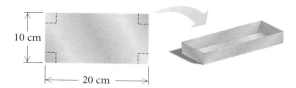

66. *Picture Frame Dimensions.* The frame of a picture is 28 cm by 32 cm outside and is of uniform width. What is the width of the frame if 192 cm² of the picture shows?

State whether the function is linear or quadratic.

67. $f(x) = 4 - 5x$

68. $f(x) = 4 - 5x^2$

69. $f(x) = 7x^2$

70. $f(x) = 23x + 6$

71. $f(x) = 1.2x - (3.6)^2$

72. $f(x) = 2 - x - x^2$

Technology Connection

Solve graphically. Round solutions to three decimal places, where appropriate.

73. $x^2 - 8x + 12 = 0$

74. $5x^2 + 42x + 16 = 0$

75. $7x^2 - 43x + 6 = 0$

76. $10x^2 - 23x + 12 = 0$

77. $6x + 1 = 4x^2$

78. $3x^2 + 5x = 3$

Use a graphing calculator to find the zeros of the function. Round to three decimal places.

79. $f(x) = 2x^2 - 5x - 4$

80. $f(x) = 4x^2 - 3x - 2$

81. $f(x) = 3x^2 + 2x - 4$

82. $f(x) = 9x^2 - 8x - 7$

83. $f(x) = 5.02x^2 - 4.19x - 2.057$

84. $f(x) = 1.21x^2 - 2.34x - 5.63$

Collaborative Discussion and Writing

85. Is it possible for a quadratic function to have one real zero and one imaginary zero? Why or why not?

86. The graph of a quadratic function can have 0, 1, or 2 x-intercepts. How can you predict the number of x-intercepts without drawing the graph or (completely) solving an equation?

Skill Maintenance

Determine whether the graph is symmetric with respect to the x-axis, the y-axis, and the origin.

87. $3x^2 + 4y^2 = 5$

88. $y^3 = 6x^2$

Determine whether the function is even, odd, or neither even nor odd.

89. $f(x) = 2x^3 - x$

90. $f(x) = 4x^2 + 2x - 3$

Synthesis

Solve.

91. $(x - 2)^3 = x^3 - 2$

92. $(x + 1)^3 = (x - 1)^3 + 26$

93. $(6x^3 + 7x^2 - 3x)(x^2 - 7) = 0$

94. $\left(x - \frac{1}{5}\right)\left(x^2 - \frac{1}{4}\right) + \left(x - \frac{1}{5}\right)\left(x^2 + \frac{1}{8}\right) = 0$

95. $x^2 + x - \sqrt{2} = 0$

96. $x^2 + \sqrt{5}x - \sqrt{3} = 0$

97. $2t^2 + (t - 4)^2 = 5t(t - 4) + 24$

98. $9t(t + 2) - 3t(t - 2) = 2(t + 4)(t + 6)$

99. $\sqrt{x - 3} - \sqrt[4]{x - 3} = 2$

100. $x^6 - 28x^3 + 27 = 0$

101. $\left(y + \frac{2}{y}\right)^2 + 3y + \frac{6}{y} = 4$

102. $x^2 + 3x + 1 - \sqrt{x^2 + 3x + 1} = 8$

103. Solve $\frac{1}{2}at^2 + v_0 t + x_0 = 0$ for t.

2.4

Analyzing Graphs of Quadratic Functions

- *Find the vertex, the line of symmetry, and the maximum or minimum value of a quadratic function using the method of completing the square.*
- *Graph quadratic functions.*
- *Solve applied problems involving maximum and minimum function values.*

Graphing Quadratic Functions of the Type $f(x) = a(x - h)^2 + k$

The graph of a quadratic function is called a **parabola**. The graph of every parabola evolves from the graph of the squaring function $f(x) = x^2$ using transformations.

We get the graph of $f(x) = a(x - h)^2 + k$ from the graph of $f(x) = x^2$ as follows:

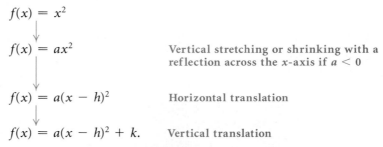

$$f(x) = x^2$$

$$f(x) = ax^2 \qquad \text{Vertical stretching or shrinking with a reflection across the } x\text{-axis if } a < 0$$

$$f(x) = a(x - h)^2 \qquad \text{Horizontal translation}$$

$$f(x) = a(x - h)^2 + k. \qquad \text{Vertical translation}$$

Consider the following graphs of the form $f(x) = a(x - h)^2 + k$. The point (h, k) at which the graph turns is called the **vertex**. The maximum or minimum value of $f(x)$ occurs at the vertex. Each graph has a line $x = h$ that is called the **line of symmetry.**

TRANSFORMATIONS
REVIEW SECTION 1.6.

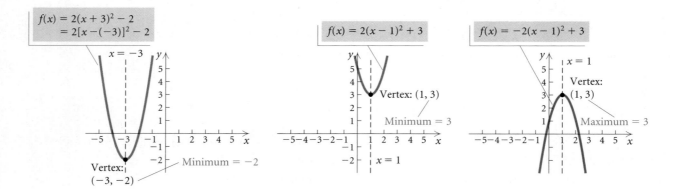

CONNECTING THE CONCEPTS

GRAPHING QUADRATIC FUNCTIONS

The graph of $f(x) = a(x - h)^2 + k$ is a parabola that:

opens up if $a > 0$ and down if $a < 0$;

has (h, k) as the vertex;

has $x = h$ as the line of symmetry;

has k as a minimum value (output) if $a > 0$ and has k as a maximum value if $a < 0$.

If the equation is in the form $f(x) = a(x - h)^2 + k$, we can learn a great deal about the graph without graphing.

FUNCTION	$f(x) = 3\left(x - \frac{1}{4}\right)^2 - 2$ $= 3\left(x - \frac{1}{4}\right)^2 + (-2)$	$g(x) = -3(x + 5)^2 + 7$ $= -3[x - (-5)]^2 + 7$
VERTEX	$\left(\frac{1}{4}, -2\right)$	$(-5, 7)$
LINE OF SYMMETRY	$x = \frac{1}{4}$	$x = -5$
MAXIMUM	No: $3 > 0$, graph opens up.	Yes, 7 $(-3 < 0,$ graph opens down.)
MINIMUM	Yes, -2 $(3 > 0,$ graph opens up.)	No: $-3 < 0$, graph opens down.

Note that the vertex (h, k) is used to find the maximum or the minimum value of the function. The maximum or minimum value is the number k, *not* the ordered pair (h, k).

Graphing Quadratic Functions of the Type $f(x) = ax^2 + bx + c, a \neq 0$

We now use a modification of the method of completing the square as an aid in graphing and analyzing quadratic functions of the form $f(x) = ax^2 + bx + c, a \neq 0$.

EXAMPLE 1 Complete the square to find the vertex, the line of symmetry, and the maximum or minimum value of $f(x) = x^2 + 10x + 23$. Then graph the function.

Solution To express $f(x) = x^2 + 10x + 23$ in the form $f(x) = a(x - h)^2 + k$, we construct a trinomial square. To do so, we take half the coefficient of x and square it, obtaining $(10/2)^2$, or 25. We now add and subtract that number on the right-hand side. We can think of this as adding $25 - 25$, which is 0.

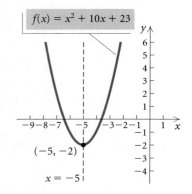

$$f(x) = x^2 + 10x + 23$$

Note that 25 completes the square for $x^2 + 10x$.

$$= x^2 + 10x + 25 - 25 + 23$$

Adding $25 - 25$, or 0, to the right side

$$= (x^2 + 10x + 25) - 25 + 23$$

$$= (x + 5)^2 - 2$$

Factoring and simplifying

$$= [x - (-5)]^2 + (-2)$$

Writing in the form $f(x) = a(x - h)^2 + k$

From this form of the function, we know the following:

Vertex: $(-5, -2)$;

Line of symmetry: $x = -5$;

Minimum value of the function: -2.

The graph of f, shown at left, is a shift of the graph of $y = x^2$ left 5 units and down 2 units. ▬

Keep in mind that the line of symmetry is not part of the graph; it is a characteristic of the graph. If you fold the graph on its line of symmetry, the two halves of the graph will coincide.

EXAMPLE 2   Complete the square to find the vertex, the line of symmetry, and the maximum or minimum value of $g(x) = x^2/2 - 4x + 8$. Then graph the function.

Solution   To complete the square, we factor $\frac{1}{2}$ out of the first two terms. This makes the coefficient of x^2 within the parentheses 1:

$$g(x) = \frac{x^2}{2} - 4x + 8$$

$$= \frac{1}{2}(x^2 - 8x) + 8.$$

Factoring $\frac{1}{2}$ out of the first two terms

Now we complete the square inside the parentheses: Half of -8 is -4, and $(-4)^2 = 16$. We add and subtract 16 inside the parentheses:

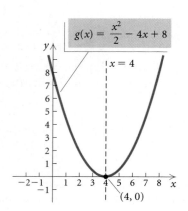

$$g(x) = \tfrac{1}{2}(x^2 - 8x + 16 - 16) + 8$$

$$= \tfrac{1}{2}(x^2 - 8x + 16) - \tfrac{1}{2} \cdot 16 + 8$$

Using the distributive law to remove -16 from within the parentheses

$$= \tfrac{1}{2}(x - 4)^2 + 0, \text{ or } \tfrac{1}{2}(x - 4)^2.$$

Factoring and simplifying

We know the following:

Vertex: $(4, 0)$;

Line of symmetry: $x = 4$;

Minimum value of the function: 0.

The graph of g is a vertical shrinking of the graph of $y = x^2$ along with a shift 4 units to the right.   ▬

EXAMPLE 3   Complete the square to find the vertex, the line of symmetry, and the maximum or minimum value of $f(x) = -2x^2 + 10x - \frac{23}{2}$. Then graph the function.

Solution   We have

$$f(x) = -2x^2 + 10x - \frac{23}{2}$$

$$= -2(x^2 - 5x) - \frac{23}{2} \qquad \text{Factoring } -2 \text{ out of the first two terms}$$

$$= -2\left(x^2 - 5x + \frac{25}{4} - \frac{25}{4}\right) - \frac{23}{2} \qquad \text{Completing the square inside the parentheses}$$

$$= -2\left(x^2 - 5x + \frac{25}{4}\right) - 2\left(-\frac{25}{4}\right) - \frac{23}{2} \qquad \text{Removing } -\frac{25}{4} \text{ from within the parentheses}$$

$$= -2\left(x - \frac{5}{2}\right)^2 + \frac{25}{2} - \frac{23}{2}$$

$$= -2\left(x - \frac{5}{2}\right)^2 + 1.$$

This form of the function yields the following:

Vertex: $\left(\frac{5}{2}, 1\right)$;

Line of symmetry: $x = \frac{5}{2}$;

Maximum value of the function: 1.

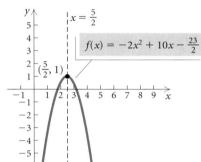

The graph is found by shifting the graph of $f(x) = x^2$ to the right $\frac{5}{2}$ units, reflecting it across the x-axis, stretching it vertically, and shifting it up 1 unit.   ▬

In many situations, we want to find the coordinates of the vertex directly from the equation $f(x) = ax^2 + bx + c$ using a formula. One way to develop such a formula is to observe that the x-coordinate of the vertex is centered between the x-intercepts, or zeros, of the function. By averaging the two solutions of $ax^2 + bx + c = 0$, we find a formula for the x-coordinate of the vertex:

$$\begin{aligned}
\frac{x\text{-coordinate}}{\text{of vertex}} &= \frac{\dfrac{-b - \sqrt{b^2 - 4ac}}{2a} + \dfrac{-b + \sqrt{b^2 - 4ac}}{2a}}{2} \\[2mm]
&= \frac{\dfrac{-2b}{2a}}{2} = \frac{-\dfrac{b}{a}}{2} \\[2mm]
&= -\frac{b}{a} \cdot \frac{1}{2} = -\frac{b}{2a}.
\end{aligned}$$

We use this value of x to find the y-coordinate of the vertex,

$$f\left(-\frac{b}{2a}\right).$$

The Vertex of a Parabola

The **vertex** of the graph of $f(x) = ax^2 + bx + c$ is

$$\left(-\frac{b}{2a}, f\left(-\frac{b}{2a}\right)\right).$$

We calculate the   We substitute to
x-coordinate.   find the y-coordinate.

EXAMPLE 4   For the function $f(x) = -x^2 + 14x - 47$:

a) Find the vertex.

b) Determine whether there is a maximum or minimum value and find that value.

c) Find the range.

d) On what intervals is the function increasing? decreasing?

Solution   There is no need to graph the function.

a) The x-coordinate of the vertex is

$$-\frac{b}{2a} = -\frac{14}{2(-1)}, \text{ or } 7.$$

Since

$$f(7) = -7^2 + 14 \cdot 7 - 47 = 2,$$

the vertex is $(7, 2)$.

b) Since a is negative $(a = -1)$, the graph opens down so the second co-ordinate of the vertex, 2, is the maximum value of the function.

c) The range is $(-\infty, 2]$.

d) Since the graph opens down, function values increase to the left of the vertex and decrease to the right of the vertex. Thus the function is increasing on the interval $(-\infty, 7)$ and decreasing on $(7, \infty)$.   ▬

Applications

Many real-world situations involve finding the maximum or minimum value of a quadratic function.

EXAMPLE 5   *Maximizing Area.* A stone mason has enough stones to enclose a rectangular patio with 60 ft of stone wall. If the house forms one side of the rectangle, what is the maximum area that the mason can enclose? What should the dimensions of the patio be in order to yield this area?

Technology Connection

We can use a graphing calculator to do Example 4. Once we have graphed $y = -x^2 + 14x - 47$, we see that the graph opens down and thus has a maximum value. We can use the MAXIMUM feature to find the coordinates of the vertex. Using these coordinates, we can then find the maximum value and the range of the function along with the intervals on which the function is increasing or decreasing.

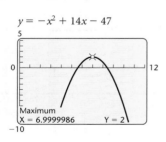

$y = -x^2 + 14x - 47$

Maximum
X = 6.9999986   Y = 2

PROBLEM-SOLVING STRATEGY

REVIEW SECTION 2.1.

Technology Connection

As a more complete check in Example 5, assuming that the function $A(w)$ is correct, we could examine a table of values for

$$A(w) = (60 - 2w)w$$

and/or examine its graph.

X	Y₁
14.7	449.82
14.8	449.92
14.9	449.98
15	**450**
15.1	449.98
15.2	449.92
15.3	449.82

X = 15

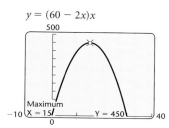

Solution   We will use the five-step problem-solving strategy.

1. **Familiarize.** We make a drawing of the situation, using w to represent the width of the patio, in feet. Then $(60 - 2w)$ feet of stone is available for the length. Suppose the patio were 10 ft wide. It would then be $60 - 2 \cdot 10 = 40$ ft long. The area would be $(10 \text{ ft})(40 \text{ ft}) = 400 \text{ ft}^2$. If the patio were 12 ft wide, it would be $60 - 2 \cdot 12 = 36$ ft long. The area would be $(12 \text{ ft})(36 \text{ ft}) = 432 \text{ ft}^2$. If it were 16 ft wide, it would be $60 - 2 \cdot 16 = 28$ ft long and the area would be $(16 \text{ ft})(28 \text{ ft}) = 448 \text{ ft}^2$. There are more combinations of length and width than we could possibly try. Instead we will find a function that represents the area and then determine the maximum value of the function.

2. **Translate.** Since the area of a rectangle is given by length times width, we have

$$A(w) = (60 - 2w)w \qquad A = lw;\ l = 60 - 2w$$
$$= -2w^2 + 60w,$$

   where $A(w)$ is the area of the patio, in square feet, as a function of the width, w.

3. **Carry out.** To solve this problem, we need to determine the maximum value of $A(w)$ and find the dimensions for which that maximum occurs. Since A is a quadratic function and w^2 has a negative coefficient, we know that the function has a maximum value that occurs at the vertex of the graph of the function. The first coordinate of the vertex, $(w, A(w))$, is

$$w = -\frac{b}{2a} = -\frac{60}{2(-2)} = 15 \text{ ft.}$$

   Thus, if $w = 15$ ft, then the length $l = 60 - 2 \cdot 15 = 30$ ft; and the area is $15 \cdot 30$, or 450 ft^2.

4. **Check.** As a partial check, we note that $450 \text{ ft}^2 > 448 \text{ ft}^2$, which is the largest area we found in a guess in the *Familiarize* step.

5. **State.** The maximum possible area is 450 ft^2 when the patio is 15 ft wide and 30 ft long.

EXAMPLE 6   *Finding the Depth of a Well.* Two seconds after a chlorine tablet has been dropped into a well, a splash is heard. The speed of sound is 1100 ft/sec. How far is the top of the well from the water?

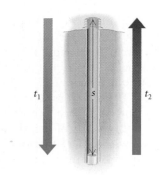

Solution

1. **Familiarize.** We first make a drawing and label it with known and unknown information. We let s = the depth of the well, in feet, t_1 = the time, in seconds, that it takes for the tablet to hit the water, and t_2 = the time, in seconds, that it takes for the sound to reach the top of the well. This gives us the equation

$$t_1 + t_2 = 2. \tag{1}$$

2. **Translate.** Can we find any relationship between the two times and the distance s? Often in problem solving you may need to look up related formulas in a physics book, another mathematics book, or on the Internet. We find that the formula

$$s = 16t^2$$

gives the distance, in feet, that a dropped object falls in t seconds. The time t_1 that it takes the tablet to hit the water can be found as follows:

$$s = 16t_1^2, \quad \text{or} \quad t_1 = \frac{\sqrt{s}}{4}. \tag{2}$$

To find an expression for t_2, the time it takes the sound to travel to the top of the well, recall that *Distance = Rate · Time*. Thus,

$$s = 1100t_2, \quad \text{or} \quad t_2 = \frac{s}{1100}. \tag{3}$$

We now have expressions for t_1 and t_2, both in terms of s. Substituting into equation (1), we obtain

$$t_1 + t_2 = 2, \quad \text{or} \quad \frac{\sqrt{s}}{4} + \frac{s}{1100} = 2. \tag{4}$$

3. **Carry out.** We solve equation (4) for s. Multiplying by 1100, we get

$$275\sqrt{s} + s = 2200, \quad \text{or} \quad s + 275\sqrt{s} - 2200 = 0.$$

This equation is reducible to quadratic with $u = \sqrt{s}$. Substituting, we get

$$u^2 + 275u - 2200 = 0.$$

Using the quadratic formula, we can solve for u:

$$u = \frac{-b \pm \sqrt{b^2 - 4ac}}{2a}$$

$$= \frac{-275 + \sqrt{275^2 - 4 \cdot 1 \cdot (-2200)}}{2 \cdot 1} \qquad \text{We want only the positive solution.}$$

$$= \frac{-275 + \sqrt{84{,}425}}{2}$$

$$\approx 7.78.$$

Since $u \approx 7.78$, we have

$$\sqrt{s} = 7.78$$

$$s \approx 60.5. \qquad \text{Squaring both sides}$$

4. Check. To check, we can substitute 60.5 for s in equation (4) and see that $t_1 + t_2 \approx 2$. We leave the mathematics for the student.

5. State. The top of the well is about 60.5 ft above the water.

Exercise Set 2.4

In Exercises 1 and 2, use the given graph to find each of the following:

a) the vertex;
b) the line of symmetry; and
c) the maximum or minimum value of the function.

1.

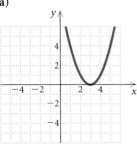

2.

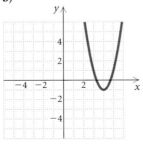

In Exercises 3–10, complete the square to:

a) find the vertex;
b) find the line of symmetry; and
c) determine whether there is a maximum or minimum value and find that value.

3. $f(x) = x^2 - 8x + 12$ **4.** $g(x) = x^2 + 7x - 8$

5. $f(x) = x^2 - 7x + 12$ **6.** $g(x) = x^2 - 5x + 6$

7. $g(x) = 2x^2 + 6x + 8$ **8.** $f(x) = 2x^2 - 10x + 14$

9. $g(x) = -2x^2 + 2x + 1$ **10.** $f(x) = -3x^2 - 3x + 1$

In Exercises 11–18, match the equation with one of the figures (a)–(h), which follow.

a)

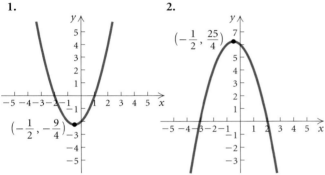

b)

c)

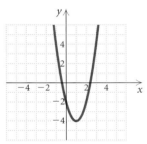

d)

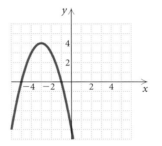

e)

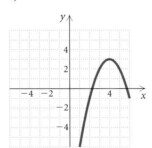

f)

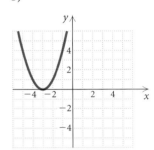

g)

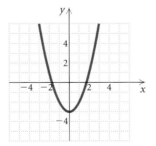

h)

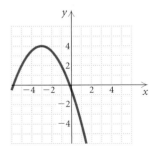

11. $y = (x + 3)^2$ **12.** $y = -(x - 4)^2 + 3$

13. $y = 2(x - 4)^2 - 1$ **14.** $y = x^2 - 3$

15. $y = -\frac{1}{2}(x + 3)^2 + 4$ **16.** $y = (x - 3)^2$

17. $y = -(x + 3)^2 + 4$ **18.** $y = 2(x - 1)^2 - 4$

In Exercises 19–26:

a) *Find the vertex.*
b) *Determine whether there is a maximum or minimum value and find that value.*
c) *Find the range.*
d) *Find the intervals on which the function is increasing and the intervals on which the function is decreasing.*

19. $f(x) = x^2 - 6x + 5$ **20.** $f(x) = x^2 + 4x - 5$

21. $f(x) = 2x^2 + 4x - 16$ **22.** $f(x) = \frac{1}{2}x^2 - 3x + \frac{5}{2}$

23. $f(x) = -\frac{1}{2}x^2 + 5x - 8$

24. $f(x) = -2x^2 - 24x - 64$

25. $f(x) = 3x^2 + 6x + 5$

26. $f(x) = -3x^2 + 24x - 49$

27. *Maximizing Volume.* Plasco Manufacturing plans to produce a one-compartment vertical file by bending the long side of an 10-in. by 18-in. sheet of plastic along two lines to form a U-shape. How tall should the file be in order to maximize the volume that the file can hold?

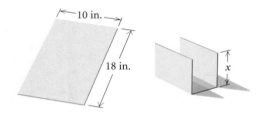

28. *Maximizing Area.* A fourth-grade class decides to enclose a rectangular garden, using the side of the school as one side of the rectangle. What is the maximum area that the class can enclose with 32 ft of fence? What should the dimensions of the garden be in order to yield this area?

29. *Maximizing Area.* The sum of the base and the height of a triangle is 20 cm. Find the dimensions for which the area is a maximum.

30. *Maximizing Area.* The sum of the base and the height of a parallelogram is 69 cm. Find the dimensions for which the area is a maximum.

31. *Finding the Height of an Elevator Shaft.* Jenelle dropped a screwdriver from the top of an elevator shaft. Exactly 5 sec later, she hears the sound of the screwdriver hitting the bottom of the shaft. How tall is the elevator shaft? (*Hint*: See Example 6.)

32. *Finding the Height of a Cliff.* A water balloon is dropped from a cliff. Exactly 3 sec later, the sound of the balloon hitting the ground reaches the top of the cliff. How high is the cliff? (*Hint*: See Example 6.)

33. *Minimizing Cost.* Aki's Bicycle Designs has determined that when x hundred bicycles are built, the average cost per bicycle is given by

$$C(x) = 0.1x^2 - 0.7x + 2.425,$$

where $C(x)$ is in hundreds of dollars. How many bicycles should be built in order to minimize the average cost per bicycle?

Maximizing Profit. *In business, profit is the difference between revenue and cost, that is,*

$$Total\ profit = Total\ revenue - Total\ cost,$$
$$P(x) = R(x) - C(x),$$

where x is the number of units sold. Find the maximum profit and the number of units that must be sold in order to yield the maximum profit for each of the following.

34. $R(x) = 5x,\ C(x) = 0.001x^2 + 1.2x + 60$

35. $R(x) = 50x - 0.5x^2,\ C(x) = 10x + 3$

36. $R(x) = 20x - 0.1x^2,\ C(x) = 4x + 2$

37. *Maximizing Area.* A rancher needs to enclose two adjacent rectangular corrals, one for sheep and one for cattle. If a river forms one side of the corrals and 240 yd of fencing is available, what is the largest total area that can be enclosed?

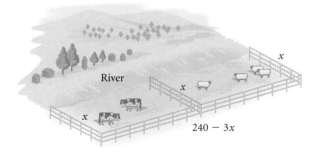

38. *Norman Window.*   A Norman window is a rectangle with a semicircle on top. Sky Blue Windows is designing a Norman window that will require 24 ft of trim. What dimensions will allow the maximum amount of light to enter a house?

A Norman window

Technology Connection

Use a graphing calculator to check the answers for each of the following.

39. Exercises 3 and 9

40. Exercises 4 and 10

41. Exercises 19 and 23

42. Exercises 20 and 24

Collaborative Discussion and Writing

43. Write a problem for a classmate to solve. Design it so that it is a maximum or minimum problem using a quadratic function.

44. Discuss two ways in which we used completing the square in this chapter.

45. Suppose that the graph of $f(x) = ax^2 + bx + c$ has x-intercepts $(x_1, 0)$ and $(x_2, 0)$. What are the x-intercepts of $g(x) = -ax^2 - bx - c$? Explain.

Skill Maintenance

For each function f, construct and simplify the difference quotient
$$\frac{f(x + h) - f(x)}{h}.$$

46. $f(x) = 3x - 7$

47. $f(x) = 2x^2 - x + 4$

A graph of $y = f(x)$ follows. No formula is given for f. Make a hand-drawn graph of each of the following.

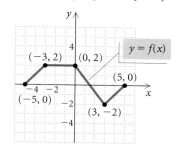

48. $g(x) = f(2x)$

49. $g(x) = -2f(x)$

Synthesis

For each equation in Exercises 50–53, under the given condition:

a) *Find k.*

b) *Find a second solution.*

50. $kx^2 - 2x + k = 0$; one solution is -3

51. $kx^2 - 17x + 33 = 0$; one solution is 3

52. $x^2 - (6 + 3i)x + k = 0$; one solution is 3

53. $x^2 - kx + 2 = 0$; one solution is $1 + i$

54. Find b such that
$$f(x) = -4x^2 + bx + 3$$
has a maximum value of 50.

55. Find c such that
$$f(x) = -0.2x^2 - 3x + c$$
has a maximum value of -225.

56. Find a quadratic function with vertex $(4, -5)$ and containing the point $(-3, 1)$.

57. Graph: $f(x) = (|x| - 5)^2 - 3$.

58. *Minimizing Area.*   A 24-in. piece of string is cut into two pieces. One piece is used to form a circle while the other is used to form a square. How should the string be cut so that the sum of the areas is a minimum?

2.5

More Equation Solving

• *Solve rational and radical equations and equations with absolute value.*

Rational Equations

Equations containing rational expressions are called **rational equations.** Solving such equations involves multiplying on both sides by the least common denominator (LCD).

EXAMPLE 1   Solve: $\dfrac{x-8}{3} + \dfrac{x-3}{2} = 0.$

Algebraic Solution

We have

$$\dfrac{x-8}{3} + \dfrac{x-3}{2} = 0 \qquad \text{The LCD is } 3 \cdot 2, \text{ or 6.}$$

$$6\left(\dfrac{x-8}{3} + \dfrac{x-3}{2}\right) = 6 \cdot 0 \qquad \begin{array}{l}\text{Multiplying by}\\ \text{the LCD on}\\ \text{both sides}\end{array}$$

$$6 \cdot \dfrac{x-8}{3} + 6 \cdot \dfrac{x-3}{2} = 0$$

$$2(x-8) + 3(x-3) = 0$$

$$2x - 16 + 3x - 9 = 0$$

$$5x - 25 = 0$$

$$5x = 25$$

$$x = 5.$$

The possible solution is 5.

CHECK:
$$\dfrac{x-8}{3} + \dfrac{x-3}{2} = 0$$

$$\overline{\dfrac{5-8}{3} + \dfrac{5-3}{2}} \ ? \ 0$$

$$\dfrac{-3}{3} + \dfrac{2}{2} \ \Big|$$

$$-1 + 1 \ \Big|$$

$$0 \ \Big| \ 0 \quad \text{TRUE}$$

The solution is 5.

Visualizing the Solution

The solution of the given equation is the zero of the function

$$f(x) = \dfrac{x-8}{3} + \dfrac{x-3}{2}.$$

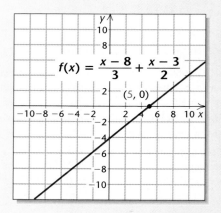

The zero of the function is 5. Thus the solution of the equation is 5.

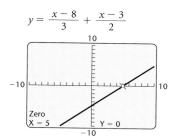

$$y = \frac{x-8}{3} + \frac{x-3}{2}$$

We can use the Zero method to solve the equation in Example 1. We find the zero of the function

$$f(x) = \frac{x-8}{3} + \frac{x-3}{2}.$$

The zero of the function is 5. Thus the solution of the equation is 5.

When we use the multiplication principle to multiply (or divide) both sides of an equation by an expression with a variable, we might not obtain an equivalent equation. We must check possible solutions obtained in this manner by substituting in the original equation. The next example illustrates this.

EXAMPLE 2 Solve: $\dfrac{x^2}{x-3} = \dfrac{9}{x-3}$.

Solution The LCD is $x - 3$.

$$(x-3) \cdot \frac{x^2}{x-3} = (x-3) \cdot \frac{9}{x-3}$$

$$x^2 = 9$$

$$x = -3 \quad or \quad x = 3 \qquad \text{Using the principle of square roots}$$

The possible solutions are -3 and 3. We check.

CHECK: For -3:

$$\frac{x^2}{x-3} = \frac{9}{x-3}$$

$$\frac{(-3)^2}{-3-3} \; ? \; \frac{9}{-3-3}$$

$$\frac{9}{-6} \; \bigg| \; \frac{9}{-6} \qquad \text{TRUE}$$

For 3:

$$\frac{x^2}{x-3} = \frac{9}{x-3}$$

$$\frac{3^2}{3-3} \; ? \; \frac{9}{3-3}$$

$$\frac{9}{0} \; \bigg| \; \frac{9}{0} \qquad \text{UNDEFINED}$$

The number -3 checks, so it is a solution. Since division by 0 is undefined, 3 is not a solution. Note that 3 is not in the domain of either $x^2/(x-3)$ or $9/(x-3)$.

Radical Equations

A **radical equation** is an equation in which variables appear in one or more radicands. For example,

$$\sqrt{2x-5} - \sqrt{x-3} = 1$$

is a radical equation. The following principle is used to solve such equations.

The Principle of Powers

For any positive integer n:

If $a = b$ is true, then $a^n = b^n$ is true.

EXAMPLE 3 Solve: $5 + \sqrt{x + 7} = x$.

Algebraic Solution

We first isolate the radical and then use the principle of powers.

$$5 + \sqrt{x + 7} = x$$

$\sqrt{x + 7} = x - 5$ Subtracting 5 on both sides

$(\sqrt{x + 7})^2 = (x - 5)^2$ Using the principle of powers; squaring both sides

$x + 7 = x^2 - 10x + 25$

$0 = x^2 - 11x + 18$ Subtracting x and 7

$0 = (x - 9)(x - 2)$ Factoring

$x - 9 = 0$ *or* $x - 2 = 0$

$x = 9$ *or* $x = 2$

The possible solutions are 9 and 2.

CHECK: For 9:

$$5 + \sqrt{x + 7} = x$$

$$5 + \sqrt{9 + 7} \ ? \ 9$$
$$5 + \sqrt{16}$$
$$5 + 4$$
$$9 \ \bigg| \ 9 \quad \text{TRUE}$$

For 2:

$$5 + \sqrt{x + 7} = x$$

$$5 + \sqrt{2 + 7} \ ? \ 2$$
$$5 + \sqrt{9}$$
$$5 + 3$$
$$8 \ \bigg| \ 2 \quad \text{FALSE}$$

Since 9 checks but 2 does not, the only solution is 9.

Visualizing the Solution

When we graph $y = 5 + \sqrt{x + 7}$ and $y = x$, we find that the first coordinate of the point of intersection of the graphs is 9. Thus the solution of $5 + \sqrt{x + 7} = x$ is 9.

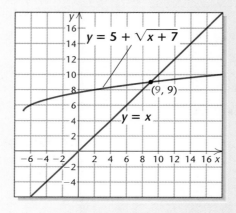

Note that the graphs show that the equation has only one solution.

Technology Connection

We can use a graphing calculator to solve the equation in Example 3. We can graph $y_1 = 5 + \sqrt{x + 7}$ and $y_2 = x$. Using the INTERSECT feature, we see in the window on the left below that the solution is 9.

We can also use the ZERO feature to get this result as shown in the window on the right. To do so, we first write the equivalent equation $5 + \sqrt{x + 7} - x = 0$. The zero of the function is 9, so the solution of the original equation is 9.

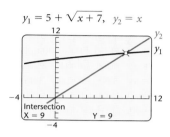

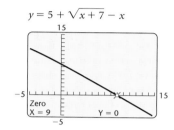

When we raise both sides of an equation to an even power, the resulting equation can have solutions that the original equation does not. This is because the converse of the principle of powers is not necessarily true. That is, if $a^n = b^n$ is true, we do not know that $a = b$ is true. For example, $(-2)^2 = 2^2$, but $-2 \neq 2$. Thus, as we see in Example 3, it is necessary to check the possible solutions in the original equation when the principle of powers is used to raise both sides of an equation to an even power.

When a radical equation has two radical terms on one side, we isolate one of them and then use the principle of powers. If, after doing so, a radical term remains, we repeat these steps.

EXAMPLE 4 Solve: $\sqrt{x - 3} + \sqrt{x + 5} = 4$.

Solution We have

$$\sqrt{x - 3} = 4 - \sqrt{x + 5} \qquad \text{Isolating one radical}$$

$$(\sqrt{x - 3})^2 = (4 - \sqrt{x + 5})^2 \qquad \text{Using the principle of powers; squaring both sides}$$

$$x - 3 = 16 - 8\sqrt{x + 5} + (x + 5)$$

$$x - 3 = 21 - 8\sqrt{x + 5} + x \qquad \text{Combining like terms}$$

$$-24 = -8\sqrt{x + 5} \qquad \text{Isolating the remaining radical; subtracting } x \text{ and 21 on both sides}$$

$$3 = \sqrt{x + 5} \qquad \text{Dividing by } -8 \text{ on both sides}$$

$$3^2 = (\sqrt{x + 5})^2 \qquad \text{Using the principle of powers; squaring both sides}$$

$$9 = x + 5$$

$$4 = x.$$

The number 4 checks and is the solution.

Equations with Absolute Value

Recall that the absolute value of a number is its distance from 0 on the number line. We use this concept to solve equations with absolute value.

For $a > 0$:

$$|X| = a \text{ is equivalent to } X = -a \text{ or } X = a.$$

EXAMPLE 5 Solve each of the following.

a) $|x| = 5$ **b)** $|x - 3| = 2$

a)

Algebraic Solution

We have

$$|x| = 5$$
$$x = -5 \quad or \quad x = 5. \qquad \text{Writing an equivalent statement}$$

The solutions are -5 and 5.

Visualizing the Solution

The first coordinates of the points of intersection of the graphs of $y = |x|$ and $y = 5$ are -5 and 5. These are the solutions of the equation $|x| = 5$.

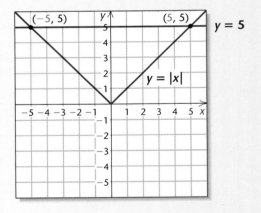

b) $\qquad |x - 3| = 2$
$$x - 3 = -2 \quad or \quad x - 3 = 2 \qquad \text{Writing an equivalent statement}$$
$$x = 1 \qquad or \qquad x = 5 \qquad \text{Adding 3}$$

The solutions are 1 and 5.

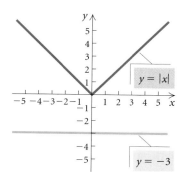

When $a = 0$, $|X| = a$ is equivalent to $x = 0$. Note that for $a < 0$, $|X| = a$ has no solution, because the absolute value of an expression is never negative. We can use a graph to illustrate the last statement for a specific value of a. For example, we let $a = -3$ and graph $y = |x|$ and $y = -3$. The graphs do not intersect, as shown at left. Thus the equation $|x| = -3$ has no solution. The solution set is the **empty set,** denoted $\varnothing$.

Exercise Set

Solve.

1. $\dfrac{1}{4} + \dfrac{1}{5} = \dfrac{1}{t}$

2. $\dfrac{1}{3} - \dfrac{5}{6} = \dfrac{1}{x}$

3. $\dfrac{x + 2}{4} - \dfrac{x - 1}{5} = 15$

4. $\dfrac{t + 1}{3} - \dfrac{t - 1}{2} = 1$

5. $\dfrac{1}{2} + \dfrac{2}{x} = \dfrac{1}{3} + \dfrac{3}{x}$

6. $\dfrac{1}{t} + \dfrac{1}{2t} + \dfrac{1}{3t} = 5$

7. $\dfrac{3x}{x + 2} + \dfrac{6}{x} = \dfrac{12}{x^2 + 2x}$

8. $\dfrac{5x}{x - 4} - \dfrac{20}{x} = \dfrac{80}{x^2 - 4x}$

9. $\dfrac{4}{x^2 - 1} - \dfrac{2}{x - 1} = \dfrac{3}{x + 1}$

10. $\dfrac{3y + 5}{y^2 + 5y} + \dfrac{y + 4}{y + 5} = \dfrac{y + 1}{y}$

11. $\dfrac{490}{x^2 - 49} = \dfrac{5x}{x - 7} - \dfrac{35}{x + 7}$

12. $\dfrac{3}{m + 2} + \dfrac{2}{m} = \dfrac{4m - 4}{m^2 - 4}$

13. $\dfrac{1}{x - 6} - \dfrac{1}{x} = \dfrac{6}{x^2 - 6x}$

14. $\dfrac{8}{x^2 - 4} = \dfrac{x}{x - 2} - \dfrac{2}{x + 2}$

15. $\dfrac{8}{x^2 - 2x + 4} = \dfrac{x}{x + 2} + \dfrac{24}{x^3 + 8}$

16. $\dfrac{18}{x^2 - 3x + 9} - \dfrac{x}{x + 3} = \dfrac{81}{x^3 + 27}$

17. $\sqrt{3x - 4} = 1$

18. $\sqrt[3]{2x + 1} = -5$

19. $\sqrt[4]{x^2 - 1} = 1$

20. $\sqrt{m + 1} - 5 = 8$

21. $\sqrt{y - 1} + 4 = 0$

22. $\sqrt[5]{3x + 4} = 2$

23. $\sqrt[3]{6x + 9} + 8 = 5$

24. $\sqrt{6x + 7} = x + 2$

25. $\sqrt{x - 3} + \sqrt{x + 2} = 5$

26. $\sqrt{x} - \sqrt{x - 5} = 1$

27. $\sqrt{3x - 5} + \sqrt{2x + 3} + 1 = 0$

28. $\sqrt{2m - 3} = \sqrt{m + 7} - 2$

29. $\sqrt{x} - \sqrt{3x - 3} = 1$

30. $\sqrt{2x + 1} - \sqrt{x} = 1$

31. $\sqrt{2y - 5} - \sqrt{y - 3} = 1$

32. $\sqrt{4p + 5} + \sqrt{p + 5} = 3$

33. $x^{1/3} = -2$

34. $t^{1/5} = 2$

35. $t^{1/4} = 3$

36. $m^{1/2} = -7$

Solve.

37. $|x| = 7$

38. $|x| = 4.5$

39. $|x| = -10.7$

40. $|x| = -\frac{3}{5}$

41. $|x - 1| = 4$

42. $|x - 7| = 5$

43. $|3x| = 1$

44. $|5x| = 4$

45. $|x| = 0$

46. $|6x| = 0$

47. $|3x + 2| = 1$

48. $|7x - 4| = 8$

49. $\left|\frac{1}{2}x - 5\right| = 17$

50. $\left|\frac{1}{3}x - 4\right| = 13$

51. $|x - 1| + 3 = 6$

52. $|x + 2| - 5 = 9$

Solve.

53. $\dfrac{P_1 V_1}{T_1} = \dfrac{P_2 V_2}{T_2}$, for T_1
(A chemistry formula for gases)

54. $\dfrac{1}{F} = \dfrac{1}{m} + \dfrac{1}{p}$, for F
(A formula from optics)

55. $\dfrac{1}{R} = \dfrac{1}{R_1} + \dfrac{1}{R_2}$, for R_2
(Resistance)

56. $A = P(1 + i)^2$, for i
(Compound interest)

57. $\dfrac{1}{F} = \dfrac{1}{m} + \dfrac{1}{p}$, for p
(A formula from optics)

Technology Connection

58. Use a graphing calculator to do Exercises 4, 16, 24, 32, and 52.

59. Use a graphing calculator to do Exercises 5, 15, 29, 31, and 51.

Collaborative Discussion and Writing

60. Explain why it is necessary to check the possible solutions of a rational equation.

61. Explain in your own words why it is necessary to check the possible solutions when the principle of powers is used to solve an equation.

Skill Maintenance

Find the zero of the function.

62. $f(x) = -3x + 9$

63. $f(x) = 15 - 2x$

64. In 1991, the U.S. per-capita consumption of bananas was 25.1 lb. In 1997, this number rose to 27.7 lb. (*Source*: U.S. Department of Agriculture) Find the percent of increase.

65. In 1998, 506,000 adults earned high school equivalency diplomas by passing the General Educational Development (GED) test. This was 25,000 more than the number who passed the test in 1997. (*Source: Indianapolis Star,* August 7, 1999) How many adults passed the GED in 1997?

Synthesis

Solve.

66. $\dfrac{x + 3}{x + 2} - \dfrac{x + 4}{x + 3} = \dfrac{x + 5}{x + 4} - \dfrac{x + 6}{x + 5}$

67. $(x - 3)^{2/3} = 2$

68. $\sqrt{15 + \sqrt{2x + 80}} = 5$

69. $\sqrt{x + 5} + 1 = \dfrac{6}{\sqrt{x + 5}}$

70. $x^{2/3} = x$

2.6

Solving Linear Inequalities

• *Solve linear inequalities, using interval notation to express solution sets.*
• *Solve compound inequalities.*
• *Solve inequalities with absolute value.*

An **inequality** is a sentence with $<$, $>$, $\leq$, or $\geq$ as its verb. An example is $3x - 5 < 6 - 2x$. To **solve** an inequality is to find all values of the variable that make the inequality true. Each of these numbers is a **solution** of the inequality, and the set of all such solutions is its **solution**

set. Inequalities that have the same solution set are called **equivalent inequalities.**

Linear Inequalities

The principles for solving inequalities are similar to those for solving equations.

Principles for Solving Inequalities

For any real numbers a, b, and c:

The Addition Principle for Inequalities: If $a < b$ is true, then $a + c < b + c$ is true.

The Multiplication Principle for Inequalities: If $a < b$ and $c > 0$ are true, then $ac < bc$ is true. If $a < b$ and $c < 0$ are true, then $ac > bc$ is true.

Similar statements hold for $a \leq b$.

> When both sides of an inequality are multiplied by a negative number, we must reverse the inequality sign.

First-degree inequalities with one variable, like those in Example 1 below, are **linear inequalities.**

EXAMPLE 1   Solve each of the following.

a) $3x - 5 < 6 - 2x$ 　　　　　　　　　**b)** $13 - 7x \geq 10x - 4$

Solution

a) $3x - 5 < 6 - 2x$

$\qquad 5x - 5 < 6$ 　　　Using the addition principle for inequalities; adding $2x$

$\qquad\qquad 5x < 11$ 　　　Using the addition principle for inequalities; adding 5

$\qquad\qquad\quad x < \tfrac{11}{5}$ 　　　Using the multiplication principle for inequalities; multiplying by $\tfrac{1}{5}$ or dividing by 5

Any number less than $\tfrac{11}{5}$ is a solution. The solution set is $\left\{x \mid x < \tfrac{11}{5}\right\}$, or $\left(-\infty, \tfrac{11}{5}\right)$.

b) 　$13 - 7x \geq 10x - 4$

$\qquad 13 - 17x \geq -4$ 　　　Subtracting $10x$

$\qquad\qquad -17x \geq -17$ 　　　Subtracting 13

$\qquad\qquad\quad x \leq 1$ 　　　Dividing by -17 and reversing the inequality sign

The solution set is $\{x \mid x \leq 1\}$, or $(-\infty, 1]$. 　　—

Technology Connection

To check Example 1(a) graphically, we graph $y_1 = 3x - 5$ and $y_2 = 6 - 2x$. The graph shows that for $x < 2.2$, or $x < \tfrac{11}{5}$, we have $y_1 < y_2$.

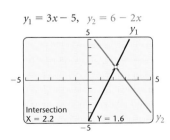

$y_1 = 3x - 5, \quad y_2 = 6 - 2x$

Compound Inequalities

When two inequalities are joined by the word *and* or the word *or*, a **compound inequality** is formed. A compound inequality like $-3 < 2x + 5$ *and* $2x + 5 \leq 7$ is called a **conjunction**, because it uses the word *and*. The sentence $-3 < 2x + 5 \leq 7$ is an abbreviation for the preceding conjunction.

Compound inequalities can be solved using the addition and multiplication principles for inequalities.

EXAMPLE 2   Solve $-3 < 2x + 5 \leq 7$. Then graph the solution set.

Solution   We have

$$-3 < 2x + 5 \leq 7$$
$$-8 < 2x \leq 2 \qquad \text{Subtracting 5}$$
$$-4 < x \leq 1. \qquad \text{Dividing by 2}$$

The solution set is $\{x \mid -4 < x \leq 1\}$, or $(-4, 1]$. The graph of the solution set is as shown below.

Technology Connection

We can perform a partial check of the solution of Example 2 graphically using operations from the TEST menu of a graphing calculator. We graph $y_1 = (-3 < 2x + 5)$ *and* $(2x + 5 \leq 7)$ in DOT mode. The calculator graphs a segment 1 unit above the x-axis for the values of x for which this expression for y is true. Here the number 1 corresponds to "true."

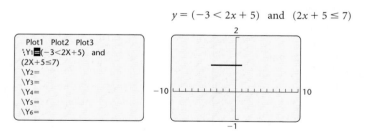

The segment extends from -4 to 1, confirming that all x-values from -4 to 1 are in the solution set. The algebraic solution indicates that the endpoint 1 is also in the solution set.

A compound inequality like $2x - 5 \leq -7$ *or* $2x - 5 > 1$ is called a **disjunction**, because it contains the word *or*. Unlike some conjunctions, it cannot be abbreviated; that is, it cannot be written without the word *or*.

Technology Connection

To check Example 3
graphically, we graph
$y_1 = 2x - 5$, $y_2 = -7$, and
$y_3 = 1$. Note that for
$\{x \mid x \leq -1 \text{ or } x > 3\}$,
$y_1 < y_2 \text{ or } y_1 > y_3$.

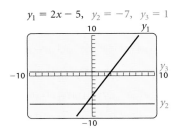

$y_1 = 2x - 5$, $y_2 = -7$, $y_3 = 1$

EXAMPLE 3   Solve $2x - 5 \leq -7$ or $2x - 5 > 1$. Then graph the solution set.

Solution   We have

$$2x - 5 \leq -7 \quad or \quad 2x - 5 > 1$$
$$2x \leq -2 \quad or \quad 2x > 6 \qquad \text{Adding 5}$$
$$x \leq -1 \quad or \quad x > 3. \qquad \text{Dividing by 2}$$

The solution set is $\{x \mid x \leq -1 \text{ or } x > 3\}$. We can also write the solution using interval notation and the symbol $\cup$ for the **union** or inclusion of both sets: $(-\infty, -1] \cup (3, \infty)$. The graph of the solution set is shown below.

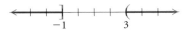

Inequalities with Absolute Value

Inequalities sometimes contain absolute-value notation. The following properties are used to solve them.

For $a > 0$:

$\quad |X| < a$ is equivalent to $-a < X < a$.
$\quad |X| > a$ is equivalent to $X < -a$ or $X > a$.

Similar statements hold for $|X| \leq a$ and $|X| \geq a$.

EXAMPLE 4   Solve each of the following.

a) $|3x + 2| < 5$
b) $|5 - 2x| \geq 1$

Solution

a) $|3x + 2| < 5$

$$-5 < 3x + 2 < 5 \qquad \text{Writing an equivalent inequality}$$
$$-7 < 3x < 3 \qquad \text{Subtracting 2}$$
$$-\tfrac{7}{3} < x < 1 \qquad \text{Dividing by 3}$$

The solution is $\left\{ x \mid -\tfrac{7}{3} < x < 1 \right\}$, or $\left(-\tfrac{7}{3}, 1 \right)$.

b) $|5 - 2x| \geq 1$

$$5 - 2x \leq -1 \quad or \quad 5 - 2x \geq 1 \qquad \text{Writing an equivalent inequality}$$
$$-2x \leq -6 \quad or \quad -2x \geq -4 \qquad \text{Subtracting 5}$$
$$x \geq 3 \quad or \quad x \leq 2 \qquad \text{Dividing by } -2 \text{ and reversing the inequality signs}$$

The solution set is $\{x \mid x \leq 2 \text{ or } x \geq 3\}$, or $(-\infty, 2] \cup [3, \infty)$.

Exercise Set 2.6

Solve.

1. $x + 6 < 5x - 6$
2. $3 - x < 4x + 7$
3. $3x - 3 + 2x \geq 1 - 7x - 9$
4. $5y - 5 + y \leq 2 - 6y - 8$
5. $14 - 5y \leq 8y - 8$
6. $8x - 7 < 6x + 3$
7. $-\frac{3}{4}x \geq -\frac{5}{8} + \frac{2}{3}x$
8. $-\frac{5}{6}x \leq \frac{3}{4} + \frac{8}{3}x$
9. $4x(x - 2) < 2(2x - 1)(x - 3)$
10. $(x + 1)(x + 2) > x(x + 1)$

Solve. Write interval notation.

11. $-2 \leq x + 1 < 4$
12. $-3 < x + 2 \leq 5$
13. $5 \leq x - 3 \leq 7$
14. $-1 < x - 4 < 7$
15. $-3 \leq x + 4 \leq 3$
16. $-5 < x + 2 < 15$
17. $-2 < 2x + 1 < 5$
18. $-3 \leq 5x + 1 \leq 3$
19. $-4 \leq 6 - 2x < 4$
20. $-3 < 1 - 2x \leq 3$
21. $-5 < \frac{1}{2}(3x + 1) \leq 7$
22. $\frac{2}{3} \leq -\frac{4}{5}(x - 3) < 1$
23. $3x \leq -6 \text{ or } x - 1 > 0$
24. $2x < 8 \text{ or } x + 3 \geq 10$
25. $2x + 3 \leq -4 \text{ or } 2x + 3 \geq 4$
26. $3x - 1 < -5 \text{ or } 3x - 1 > 5$
27. $2x - 20 < -0.8 \text{ or } 2x - 20 > 0.8$
28. $5x + 11 \leq -4 \text{ or } 5x + 11 \geq 4$
29. $x + 14 \leq -\frac{1}{4} \text{ or } x + 14 \geq \frac{1}{4}$
30. $x - 9 < -\frac{1}{2} \text{ or } x - 9 > \frac{1}{2}$

Solve and graph the solution set.

31. $|x| < 7$
32. $|x| \leq 4.5$
33. $|x| \geq 4.5$
34. $|x| > 7$

Solve.

35. $|x + 8| < 9$
36. $|x + 6| \leq 10$
37. $|x + 8| \geq 9$
38. $|x + 6| > 10$
39. $\left|x - \frac{1}{4}\right| < \frac{1}{2}$
40. $|x - 0.5| \leq 0.2$
41. $|3x| < 1$
42. $|5x| \leq 4$
43. $|2x + 3| \leq 9$
44. $|3x + 4| < 13$
45. $|x - 5| > 0.1$
46. $|x - 7| \geq 0.4$

47. $|6 - 4x| \leq 8$
48. $|5 - 2x| > 10$
49. $\left|x + \frac{2}{3}\right| \leq \frac{5}{3}$
50. $\left|x + \frac{3}{4}\right| < \frac{1}{4}$
51. $\left|\dfrac{2x + 1}{3}\right| > 5$
52. $\left|\dfrac{2x - 1}{3}\right| \geq \dfrac{5}{6}$
53. $|2x - 4| < -5$
54. $|3x + 5| < 0$

Technology Connection

55. Use a graphing calculator to check your answers to Exercises 9, 21, and 41.
56. Use a graphing calculator to check your answers to Exercises 10, 22, and 42.

Collaborative Discussion and Writing

57. Explain why $|x| < p$ has no solution for $p \leq 0$.
58. Explain why all real numbers are solutions of $|x| > p$, for $p < 0$.

Skill Maintenance

Simplify. Write the answer in the form $a + bi$, where a and b are real numbers.

59. $(6 - 4i) - (-4 + 3i)$
60. $(-1 + 2i) + (3 - i)$
61. $\dfrac{5 + 2i}{3 - 4i}$
62. $(2 - 3i)(5 + i)$

Synthesis

Solve.

63. $2x \leq 5 - 7x < 7 + x$
64. $x \leq 3x - 2 \leq 2 - x$
65. $|3x - 1| > 5x - 2$
66. $|x + 2| \leq |x - 5|$
67. $|p - 4| + |p + 4| < 8$
68. $|x| + |x + 1| < 10$
69. $|x - 3| + |2x + 5| > 6$

2 | Chapter Summary and Review

Important Properties and Formulas

Equation-Solving Principles

The Addition Principle:

If $a = b$ is true, then $a + c = b + c$ is true.

The Multiplication Principle:

If $a = b$ is true, then $ac = bc$ is true.

The Principle of Zero Products:

If $ab = 0$ is true, then $a = 0$ or $b = 0$, and
if $a = 0$ or $b = 0$, then $ab = 0$.

The Principle of Square Roots:

If $x^2 = k$, then $x = \sqrt{k}$ or $x = -\sqrt{k}$.

The Principle of Powers:

For any positive integer n, if $a = b$ is true, then $a^n = b^n$ is true.

Zero of a Function:

An input c of a function f is a zero of f if $f(c) = 0$.

Complex Number: $a + bi$, a, b real, $i^2 = -1$

Imaginary Number: $a + bi$, $b \neq 0$

Complex Conjugates: $a + bi$, $a - bi$

Quadratic Equation:

$ax^2 + bx + c = 0$, $a \neq 0$, a, b, c real

Quadratic Function:

$f(x) = ax^2 + bx + c$, $a \neq 0$, a, b, c real

Quadratic Formula:

For $ax^2 + bx + c = 0$, $a \neq 0$,
$$x = \frac{-b \pm \sqrt{b^2 - 4ac}}{2a}.$$

Five Steps for Problem Solving

1. Familiarize.
2. Translate.
3. Carry out.
4. Check.
5. State.

Principles for Solving Inequalities

The Addition Principle for Inequalities:

If $a < b$ is true, then $a + c < b + c$ is true.

The Multiplication Principle for Inequalities:

If $a < b$ and $c > 0$ are true, then $ac < bc$ is true.
If $a < b$ and $c < 0$ are true, then $ac > bc$ is true.

Similar statements hold for $\leq$.

Equations and Inequalities with Absolute Value

For $a > 0$,

$|X| = a$	$X = -a$ or $X = a$
$|X| < a$	$-a < X < a$
$|X| > a$	$X < -a$ or $X > a$

REVIEW EXERCISES

Solve.

1. $4y - 5 = 1$

2. $3x - 4 = 5x + 8$

3. $5(3x + 1) = 2(x - 4)$

4. $2(n - 3) = 3(n + 5)$

5. $(2y + 5)(3y - 1) = 0$

6. $x^2 + 4x - 5 = 0$

7. $3x^2 + 2x = 8$

8. $5x^2 = 15$

9. $x^2 - 10 = 0$

Find the zero(s) of the function.

10. $f(x) = 6x - 18$

11. $f(x) = x - 4$

12. $f(x) = 2 - 10x$

13. $f(x) = 8 - 2x$

14. $f(x) = x^2 - 2x + 1$

15. $f(x) = x^2 + 2x - 15$

16. $f(x) = 2x^2 - x - 5$

17. $f(x) = 3x^2 + 2x - 3$

Solve.

18. $\dfrac{5}{2x + 3} + \dfrac{1}{x - 6} = 0$

19. $\dfrac{3}{8x + 1} + \dfrac{8}{2x + 5} = 1$

20. $\sqrt{5x + 1} - 1 = \sqrt{3x}$

21. $\sqrt{x - 1} - \sqrt{x - 4} = 1$

22. $|x - 4| = 3$

23. $|2y + 7| = 9$

24. $-3 \le 3x + 1 < 5$

25. $-2 < 5x - 4 \le 6$

26. $2x < -1 \text{ or } x + 3 > 0$

27. $3x + 7 \le 2 \text{ or } 2x + 3 \ge 5$

28. $|6x - 1| < 5$

29. $|x + 4| \ge 2$

30. Solve $V = lwh$ for h.

31. Solve $M = n + 0.3s$ for s.

32. Solve $v = \sqrt{2gh}$ for h.

33. Solve $\dfrac{1}{a} + \dfrac{1}{b} = \dfrac{1}{t}$ for t.

Express in terms of i.

34. $-\sqrt{-40}$

35. $\sqrt{-12} \cdot \sqrt{-20}$

36. $\dfrac{\sqrt{-49}}{-\sqrt{-64}}$

Simplify each of the following. Write the answer in the form $a + bi$, where a and b are real numbers.

37. $(6 + 2i)(-4 - 3i)$

38. $\dfrac{2 - 3i}{1 - 3i}$

39. $(3 - 5i) - (2 - i)$

40. $(6 + 2i) + (-4 - 3i)$

41. i^{23}

42. $(-3i)^{28}$

Solve by completing the square to obtain exact solutions. Show your work.

43. $x^2 - 3x = 18$

44. $3x^2 - 12x - 6 = 0$

Solve. Give exact solutions.

45. $3x^2 + 10x = 8$

46. $r^2 - 2r + 10 = 0$

47. $x^2 = 18 + 3x$

48. $x = 2\sqrt{x} - 1$

49. $y^4 - 3y^2 + 1 = 0$

50. $(x^2 - 1)^2 - (x^2 - 1) - 2 = 0$

51. $(p - 3)(3p + 2)(p + 2) = 0$

52. $x^3 + 5x^2 - 4x - 20 = 0$

In Exercises 53 and 54, complete the square to:

a) *find the vertex;*

b) *find the line of symmetry; and*

c) *determine whether there is a maximum or minimum value and find that value.*

53. $f(x) = -4x^2 + 3x - 1$ 54. $f(x) = 5x^2 - 10x + 3$

In Exercises 55–58, match the equation with one of the figures (a)–(d), which follow.

a)

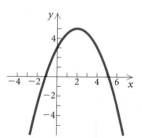

b)

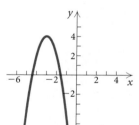

c)

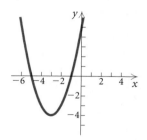

d)

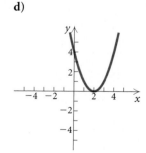

55. $y = (x - 2)^2$

56. $y = (x + 3)^2 - 4$

57. $y = -2(x + 3)^2 + 4$

58. $y = -\frac{1}{2}(x - 2)^2 + 5$

59. *Legs of a Right Triangle.* The hypotenuse of a right triangle is 50 ft. One leg is 10 ft longer than the other. What are the lengths of the legs?

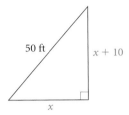

60. *Motion.* A Riverboat Cruise Line boat travels 8 mi upstream and 8 mi downstream. The total time for both parts of the trip is 3 hr. The speed of the stream is 2 mph. What is the speed of the boat in still water?

61. *Motion.* Two freight trains leave the same city at right angles. The first train travels at a speed of 60 km/h. In 1 hr, the trains are 100 km apart. How fast is the second train traveling?

62. *Sidewalk Width.* A 60-ft by 80-ft parking lot is torn up to install a sidewalk of uniform width around its perimeter. The new area of the parking lot is two thirds of the old area. How wide is the sidewalk?

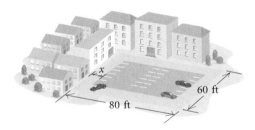

63. *Maximizing Volume.* The Berniers have 24 ft of flexible fencing with which to build a rectangular "toy corral." If the fencing is 2 ft high, what dimensions should the corral have in order to maximize its volume?

64. *Dimensions of a Box.* An open box is made from a 10-cm by 20-cm piece of aluminum by cutting a square from each corner and folding up the edges. The area of the resulting base is 90 cm². What is the length of the sides of the squares?

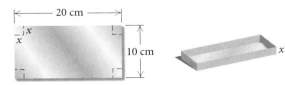

65. *Maximizing Box Dimensions.* An open box is to be made from a 10-cm by 20-cm piece of aluminum by cutting a square from each corner and folding up the edges. What should the length of the sides of the squares be in order to maximize the volume?

Technology Connection

Use a graphing calculator to do each of the following.

66. Exercises 2, 6, 18, 20, and 22

67. Exercises 1, 7, 19, 21, and 23

68. Exercises 10 and 14

69. Exercises 13 and 15

Collaborative Discussion and Writing

70. As the first step in solving

$$3x - 1 = 8,$$

Juliet multiplies by $\frac{1}{3}$ on both sides. What advice would you give her about the procedure for solving equations?

71. If the graphs of

$$f(x) = a_1(x - h_1)^2 + k_1$$

and

$$g(x) = a_2(x - h_2)^2 + k_2$$

have the same shape, what, if anything, can you conclude about the a's, the h's, and the k's? Explain your answer.

Synthesis

Solve.

72. Solve: $\sqrt{\sqrt{\sqrt{x}}} = 2$.

73. $(x - 1)^{2/3} = 4$

74. $(t - 4)^{4/5} = 3$

75. $\sqrt{x + 2} + \sqrt[4]{x + 2} - 2 = 0$

76. $(2y - 2)^2 + y - 1 = 5$

77. Find b such that $f(x) = -3x^2 + bx - 1$ has a maximum value of 2.

78. At the beginning of the year, $3500 was deposited in a savings account. One year later, $4000 was deposited in another account. The interest rate was the same for both accounts. At the end of the second year, there was a total of $8518.35 in the accounts. What was the annual interest rate?

2 Chapter Test

Solve. Find exact solutions.

1. $6x + 7 = 1$

2. $(2x - 1)(x + 5) = 0$

3. $x^2 - 2x - 3 = 0$

4. $6x^2 - 36 = 0$

5. $2t^2 - 3t + 4 = 0$

6. $x + 5\sqrt{x} - 36 = 0$

7. $\dfrac{3}{3x + 4} + \dfrac{2}{x - 1} = 2$

8. $\sqrt{x + 4} - 2 = 1$

9. $|4y - 3| = 5$

10. $-7 < 2x + 3 < 9$

11. $|x + 5| > 2$

12. Solve $V = \frac{2}{3}\pi r^2 h$ for h.

13. Solve $R = \sqrt{3np}$ for n.

Find the zero(s) of each function.

14. $f(x) = 3x + 9$

15. $f(x) = 4x^2 - 11x - 3$

16. $f(x) = 2x^2 - x - 7$

17. Solve $x^2 + 4x = 1$ by completing the square. Find the exact solutions. Show your work.

Express in terms of i.

18. $\sqrt{-43}$

19. $-\sqrt{-25}$

Simplify.

20. $(3 + 4i)(2 - i)$

21. i^{33}

22. For the graph of the function $f(x) = -x^2 + 2x + 8$:

  a) Find the vertex.

  b) Find the line of symmetry.

  c) State whether there is a maximum or minimum value and find that value.

  d) Find the range.

23. *Maximizing Area.* A homeowner wants to fence a rectangular play yard using 60 ft of fencing. The side of the house will be used as one side of the rectangle. Find the dimensions for which the area is a maximum.

Synthesis

24. Find a such that $f(x) = ax^2 - 4x + 3$ has a maximum value of 12.

Polynomial and Rational Functions 3

A *polynomial function* is a function that can be defined using a polynomial expression. A *rational function* is a function that can be defined using a quotient of two polynomials. We will study both kinds of functions and examine zeros of functions in greater depth. In addition, we will study the graphs of rational functions and use the graphs of both polynomial and rational functions to solve related inequalities.

APPLICATION

Under certain conditions, the power P, in watts per hour, generated by a windmill with winds blowing v miles per hour is given by the polynomial function

$$P(v) = 0.015v^3.$$

We can use this function to find the power generated by specific wind speeds.

This problem appears as Exercise 22 in Section 3.1.

3.1

Polynomial Functions and Models

- *Determine the behavior of the graph of a polynomial function using the leading-term test.*
- *Determine whether a function has a real zero between two given real numbers.*
- *Solve applied problems using polynomial models.*

There are many different kinds of functions. The constant, linear, and quadratic functions that we studied in Chapters 1 and 2 are part of a larger group of functions called *polynomial functions*.

Polynomial Function

A **polynomial function** P is given by

$$P(x) = a_n x^n + a_{n-1} x^{n-1} + a_{n-2} x^{n-2} + \cdots + a_1 x + a_0,$$

where the coefficients a_n, a_{n-1}, ..., a_1, a_0 are real numbers and the exponents are whole numbers.

The first nonzero coefficient, a_n, is called the **leading coefficient.** The term $a_n x^n$ is called the **leading term.** The **degree** of the polynomial function is n. Some examples of types of polynomial functions and their names are as follows.

POLYNOMIAL FUNCTION	DEGREE	EXAMPLE
Constant	0	$f(x) = 3$
Linear	1	$f(x) = \frac{2}{3}x + 1$
Quadratic	2	$f(x) = 2x^2 - x + 3$
Cubic	3	$f(x) = x^3 + 2x^2 + x - 5$
Quartic	4	$f(x) = -x^4 - 1.1x^3 + 0.3x^2 - 2.8x - 1.7$

The function $f(x) = 0$ can be described in many ways: $f(x) = 0 = 0x^2 = 0x^{15} = 0x^{48}$, and so on. For this reason, we say that the constant function $f(x) = 0$ has no degree.

From our study of functions in Chapters 1 and 2, we know how to find or at least estimate many characteristics of a polynomial function. Let's consider two examples for review.

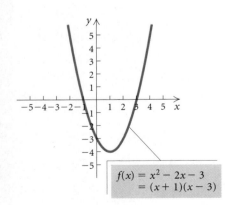

Function: $f(x) = x^2 - 2x - 3$
           $= (x + 1)(x - 3)$

Zeros: -1, 3

x-intercepts: $(-1, 0)$, $(3, 0)$

y-intercept: $(0, -3)$

Relative minimum: -4 at $x = 1$

Relative maximum: None

Domain: All real numbers, $(-\infty, \infty)$

Range: $[-4, \infty)$

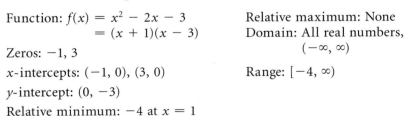
$f(x) = x^2 - 2x - 3$
$= (x + 1)(x - 3)$

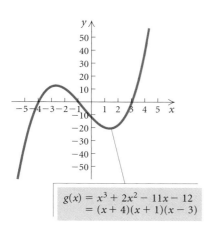

$g(x) = x^3 + 2x^2 - 11x - 12$
$= (x + 4)(x + 1)(x - 3)$

Function: $g(x) = x^3 + 2x^2 - 11x - 12$
$= (x + 4)(x + 1)(x - 3)$

Zeros: $-4, -1, 3$

x-intercepts: $(-4, 0)$, $(-1, 0)$, $(3, 0)$

y-intercept: $(0, -12)$

Relative minimum: -20.7 at $x = 1.4$

Relative maximum: 12.6 at $x = -2.7$

Domain: All real numbers, $(-\infty, \infty)$

Range: All real numbers, $(-\infty, \infty)$

All graphs of polynomial functions have some characteristics in common. Compare the following graphs. How do the graphs of polynomial functions differ from the graphs of nonpolynomial functions? Describe some characteristics of the graphs of polynomial functions that you observe.

POLYNOMIAL FUNCTIONS

$f(x) = x^2 + 3x + 1$

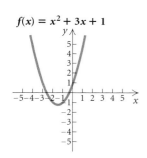

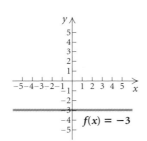

$f(x) = -3$

$f(x) = 2x^3 + x^2 + x - 1$

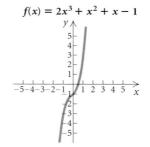

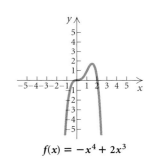

$f(x) = -x^4 + 2x^3$

NONPOLYNOMIAL FUNCTIONS

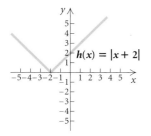

$h(x) = |x + 2|$

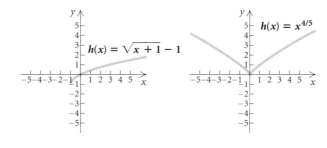

$h(x) = \sqrt{x + 1} - 1$

$h(x) = x^{4/5}$

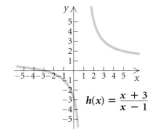

$h(x) = \dfrac{x + 3}{x - 1}$

You probably noted that the graph of a polynomial function is continuous, that is, it has no holes or breaks. It is also smooth; there are no sharp corners. Furthermore, the *domain* of a polynomial function is the set of all real numbers, $(-\infty, \infty)$.

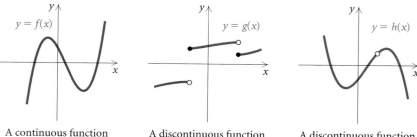

A continuous function   A discontinuous function   A discontinuous function

The Leading-Term Test

It would seem reasonable that as x becomes very large $(x \to \infty)$ or as x becomes very small $(x \to -\infty)$, the behavior of the graph of a polynomial function would be determined by the *leading term*. Let's see if we can discover some general patterns with the graphs shown below.

EVEN DEGREE

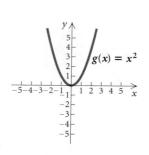

$g(x) = x^2$

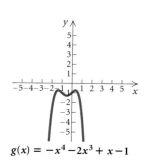

$g(x) = -x^4 - 2x^3 + x - 1$

$g(x) = \frac{1}{2}x^6 + 3$

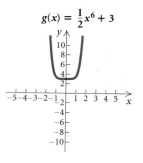

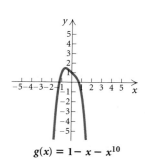

$g(x) = 1 - x - x^{10}$

ODD DEGREE

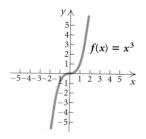

$f(x) = x^3$

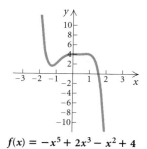

$f(x) = -x^5 + 2x^3 - x^2 + 4$

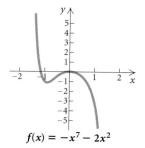

$f(x) = -x^7 - 2x^2$

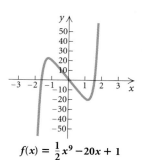

$f(x) = \frac{1}{2}x^9 - 20x + 1$

We can summarize our observations as follows.

The Leading-Term Test

If $a_n x^n$ is the leading term of a polynomial, then the behavior of the graph as $x \to \infty$ or as $x \to -\infty$ can be described in one of the four following ways.

If n is even, and $a_n > 0$:

If n is even, and $a_n < 0$:

If n is odd, and $a_n > 0$:

If n is odd, and $a_n < 0$:

The $\bf{www}$ portion of the graph is not determined by this test.

EXAMPLE 1 Using the leading-term test, match each of the following functions with one of the graphs A–D, which follow.

a) $f(x) = 3x^4 - 2x^3 + 3$

b) $f(x) = -5x^3 - x^2 + 4x + 2$

c) $f(x) = x^5 + \frac{1}{4}x + 1$

d) $f(x) = -x^6 + x^5 - 4x^3$

A.

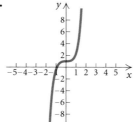

B.

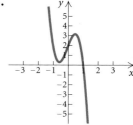

C.

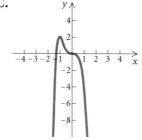

D.

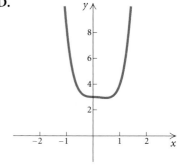

Solution

	LEADING TERM	DEGREE OF LEADING TERM	SIGN OF LEADING COEFFICIENT	GRAPH
a)	$3x^4$	Even	Positive	D
b)	$-5x^3$	Odd	Negative	B
c)	x^5	Odd	Positive	A
d)	$-x^6$	Even	Negative	C

In addition to the leading-term test, it is helpful to consider the following facts when graphing polynomial functions.

If $P(x)$ is a polynomial function of degree n, the graph of the function:

- has at most n x-intercepts, and thus at most n real zeros;
- has at most $n - 1$ turning points.

(Turning points on a graph occur when the function changes from decreasing to increasing or from increasing to decreasing.)

Polynomial functions P are continuous, hence their graphs are unbroken. All polynomial functions have $(-\infty, \infty)$ as the domain. Suppose two function values $P(a)$ and $P(b)$ have opposite signs. Since P is continuous, its graph must be a curve from $(a, P(a))$ to $(b, P(b))$ without a break. Then it follows that the curve must cross the x-axis at some point c between a and b—that is, the function has a zero at c between a and b.

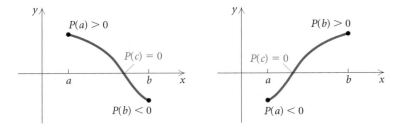

The Intermediate Value Theorem

For any polynomial function $P(x)$ with real coefficients, suppose that for $a \neq b$, $P(a)$ and $P(b)$ are of opposite signs. Then the function has a real zero between a and b.

EXAMPLE 2 Using the intermediate value theorem, determine, if possible, whether the function has a real zero between a and b.

a) $f(x) = x^3 + x^2 - 6x$; $a = -4$, $b = -2$

b) $f(x) = x^3 + x^2 - 6x$; $a = -1$, $b = 3$

c) $g(x) = \frac{1}{3}x^4 - x^3$; $a = -\frac{1}{2}$, $b = \frac{1}{2}$

d) $g(x) = \frac{1}{3}x^4 - x^3$; $a = 1$, $b = 2$

Solution We find $f(a)$ and $f(b)$ or $g(a)$ and $g(b)$ and determine whether they differ in sign. The graphs of $f(x)$ and $g(x)$ at left provide a visual check of the solutions.

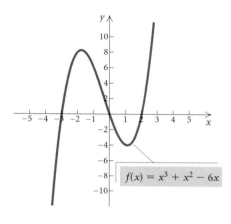

$f(x) = x^3 + x^2 - 6x$

a) $f(-4) = (-4)^3 + (-4)^2 - 6(-4) = -24$,
$f(-2) = (-2)^3 + (-2)^2 - 6(-2) = 8$

Note that $f(-4)$ is negative and $f(-2)$ is positive. By the intermediate value theorem, since $f(-4)$ and $f(-2)$ have opposite signs, then $f(x)$ has a zero between -4 and -2.

b) $f(-1) = (-1)^3 + (-1)^2 - 6(-1) = 6$,
$f(3) = 3^3 + 3^2 - 6(3) = 18$

Both $f(-1)$ and $f(3)$ are positive. This does not necessarily mean that there is not a zero between -1 and 3. In fact, the graph of $f(x)$ shows that there are two zeros between -1 and 3. However, in this case, the function values 6 and 18 do not allow us to determine this.

c) $g\left(-\frac{1}{2}\right) = \frac{1}{3}\left(-\frac{1}{2}\right)^4 - \left(-\frac{1}{2}\right)^3 = \frac{7}{48}$,
$g\left(\frac{1}{2}\right) = \frac{1}{3}\left(\frac{1}{2}\right)^4 - \left(\frac{1}{2}\right)^3 = -\frac{5}{48}$

Since $g\left(-\frac{1}{2}\right)$ and $g\left(\frac{1}{2}\right)$ have opposite signs, $g(x)$ has a zero between $-\frac{1}{2}$ and $\frac{1}{2}$.

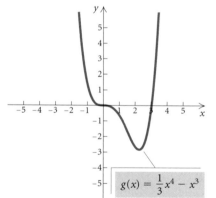

$g(x) = \frac{1}{3}x^4 - x^3$

d) $g(1) = \frac{1}{3}(1)^4 - 1^3 = -\frac{2}{3}$,
$g(2) = \frac{1}{3}(2)^4 - 2^3 = -\frac{8}{3}$

Both $g(1)$ and $g(2)$ are negative. This does not necessarily mean that there is not a zero between 1 and 2. The graph of $g(x)$ does show that there are no zeros between 1 and 2, but the function values $-\frac{2}{3}$ and $-\frac{8}{3}$ do not allow us to determine this. ▬

Polynomial Models

Polynomial functions have many uses as models in science, engineering, and business. The simplest use of polynomial functions in applied problems occurs when we merely evaluate a polynomial function. In such cases, a model has already been developed.

EXAMPLE 3 *Ibuprofen in the Bloodstream.* Ibuprofen is a pain relief medication. The polynomial function

$$M(t) = 0.5t^4 + 3.45t^3 - 96.65t^2 + 347.7t$$

can be used to estimate the number of milligrams of ibuprofen in the bloodstream t hours after 400 mg of the medication has been taken. Find the number of milligrams in the bloodstream at $t = 0$, 0.5, 1, 1.5, and so on, up to 6 hr. Round the function values to the nearest tenth.

Solution Using a calculator, we compute the function values.

$M(0) = 0,$ $M(3.5) = 255.9,$

$M(0.5) = 150.2,$ $M(4) = 193.2,$

$M(1) = 255,$ $M(4.5) = 126.9,$

$M(1.5) = 318.3,$ $M(5) = 66,$

$M(2) = 344.4,$ $M(5.5) = 20.2,$

$M(2.5) = 338.6,$ $M(6) = 0.$

$M(3) = 306.9,$

The domain of a polynomial function, unless restricted by a statement of the function, is $(-\infty, \infty)$. The implications of the application in Example 3 restrict the domain of the function. If we assume that a patient had not taken any of the medication before, it seems reasonable that $M(0) = 0$; that is, at time 0, there is 0 mg of the medication in the bloodstream. After the medication has been taken, $M(t)$ will be positive for a period of time and eventually decrease back to 0 when $t = 6$ and not increase again (unless another dose is taken). Thus the restricted domain is [0, 6].

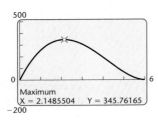

X	Y₁
0	**0**
.5	150.15
1	255
1.5	318.26
2	344.4
2.5	338.63
3	306.9

X = 0

500

0   6

Maximum
X = 2.1485504   Y = 345.76165

−200

Technology Connection

We can evaluate the function in Example 3 with the TABLE feature of a graphing calculator. Let's start at 0 and use a step-value of 0.5.

Using the restriction $0 \le t \le 6$, we find that the restricted domain of $M(t)$ is [0, 6]. To determine the range, we find the relative maximum value of the function using the MAXIMUM feature.

The maximum is about 345.76 mg. It occurs approximately 2.15 hr, or 2 hr 9 min, after the initial dose has been taken. The range is about [0, 345.76].

Technology Connection

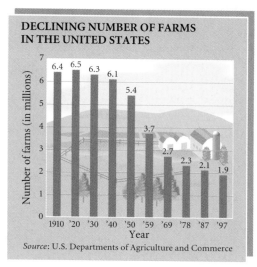

**DECLINING NUMBER OF FARMS
IN THE UNITED STATES**

Source: U.S. Departments of Agriculture and Commerce

Previously, we used regression to model data with linear functions. We now expand that procedure to include quadratic, cubic, and quartic models.

Declining Number of Farms in the United States. Today U.S. farm acreage is about the same as it was in the early part of the twentieth century, but the number of farms has shrunk.

Looking at the graph at left, we note that the data could be modeled with a cubic or a quartic function. Here we use the quartic model. Using the REGRESSION feature with DIAGNOSTIC turned on, we get the following.

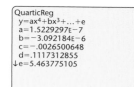
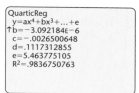

Note that *a* and *b* are given in scientific notation on the graphing calculator, but we convert to decimal notation when we write the function.

$$f(x) = 0.00000015229297x^4 - 0.000003092184x^3$$
$$- 0.0026500648x^2 + 0.1117312855x + 5.463775105$$

The graph is shown at left. With this function, we can estimate that there were about 5.5 million farms in 1945 and about 2.5 million in 1975. Looking at the bar graph shown above, we see that these estimates appear to be fairly accurate. But if we use the function to estimate the number of farms in 2003, we get an unrealistic number, about 2.6 million.

Since the number of farms will probably continue to decrease, the estimate of 2.6 million for 2003 is probably not a good prediction. The quartic model has a high value for R^2, approximately 0.984, over the domain of the data, but this number does not reflect the degree of accuracy for extended values. It is always important when using regression to evaluate predictions with common sense and knowledge of current trends.

Exercise Set 3.1

Classify the polynomial as constant, linear, quadratic, cubic, or quartic and determine the leading term, the leading coefficient, and the degree of the polynomial.

1. $g(x) = \frac{1}{2}x^3 - 10x + 8$

2. $f(x) = 15x^2 - 10 + 0.11x^4 - 7x^3$

3. $h(x) = 0.9x - 0.13$

4. $f(x) = -6$

5. $g(x) = 305x^4 + 4021$

6. $h(x) = 2.4x^3 + 5x^2 - x + \frac{7}{8}$

In Exercises 7–12, use the leading-term test and your knowledge of y-intercepts to match the function with one of graphs (a)–(f), which follow.

a)

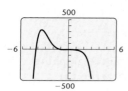

b)

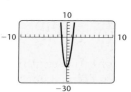

c)

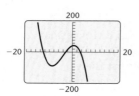

d)

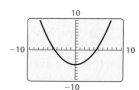

e)

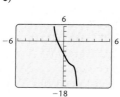

f)

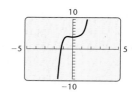

7. $f(x) = \frac{1}{4}x^2 - 5$

8. $f(x) = -0.5x^6 - x^5 + 4x^4 - 5x^3 - 7x^2 + x - 3$

9. $f(x) = x^5 - x^4 + x^2 + 4$

10. $f(x) = -\frac{1}{3}x^3 - 4x^2 + 6x + 42$

11. $f(x) = x^4 - 2x^3 + 12x^2 + x - 20$

12. $f(x) = -0.3x^7 + 0.11x^6 - 0.25x^5 + x^4 + x^3 - 6x - 5$

Using the intermediate value theorem, determine if possible whether the function f has a real zero between a and b.

13. $f(x) = x^3 + 3x^2 - 9x - 13;\ a = -5,\ b = -4$

14. $f(x) = x^3 + 3x^2 - 9x - 13;\ a = 1,\ b = 2$

15. $f(x) = 3x^2 - 2x - 11;\ a = -3,\ b = -2$

16. $f(x) = 3x^2 - 2x - 11;\ a = 2,\ b = 3$

17. $f(x) = x^4 - 2x^2 - 6;\ a = 2,\ b = 3$

18. $f(x) = 2x^5 - 7x + 1;\ a = 1,\ b = 2$

19. *Vertical Leap.* A formula relating an athlete's vertical leap V, in inches, to hang time T, in seconds, is

$$V = 48T^2.$$

Anfernee Hardaway of the Phoenix Suns has a vertical leap of 36 in. (*Source:* National Basketball Association). What is his hang time?

20. *Projectile Motion.* A stone thrown downward with an initial velocity of 34.3 m/sec will travel a distance of s meters, where

$$s(t) = 4.9t^2 + 34.3t$$

and t is in seconds. If a stone is thrown downward at 34.3 m/sec from a height of 294 m, how long will it take the stone to hit the ground?

21. *Games in a Sports League.* If there are x teams in a sports league and all the teams play each other twice, a total of $N(x)$ games are played, where

$$N(x) = x^2 - x.$$

A softball league has 9 teams, each of which plays the others twice. If the league pays $45 per game for the field and umpires, how much will it cost to play the entire schedule?

22. *Windmill Power.* Under certain conditions, the power P, in watts per hour, generated by a windmill with winds blowing v miles per hour is given by

$$P(v) = 0.015v^3.$$

a) Find the power generated by 15-mph winds.

b) How fast must the wind blow in order to generate 120 watts of power in 1 hr?

23. *Interest Compounded Annually.* When P dollars is invested at interest rate i, compounded annually, for t years, the investment grows to A dollars, where

$$A = P(1 + i)^t.$$

a) Find the interest rate i if \$2560 grows to \$3610 in 2 yr.

b) Find the interest rate i if \$10,000 grows to \$13,310 in 3 yr.

24. *Threshold Weight.* In a study performed by Alvin Shemesh, it was found that the **threshold weight W**, defined as the weight above which the risk of death rises dramatically, is given by

$$W(h) = \left(\frac{h}{12.3}\right)^3,$$

where W is in pounds and h is a person's height, in inches. Find the threshold weight of a person who is 5 ft, 7 in. tall.

Technology Connection

Using a graphing calculator, estimate the relative maxima and minima and the range of the polynomial function.

25 $g(x) = x^3 - 1.2x + 1$

26. $h(x) = -\frac{1}{2}x^4 + 3x^3 - 5x^2 + 3x + 6$

27. $f(x) = x^6 - 3.8$

28. $h(x) = 2x^3 - x^4 + 20$

29. $f(x) = x^2 + 10x - x^5$

30. $f(x) = 2x^4 - 5.6x^2 + 10$

31. *U.S. Farms.* As the number of farms has decreased in the United States, the average size of the remaining farms has grown larger, as shown below.

YEAR	AVERAGE ACREAGE PER FARM
1910	139
1920	149
1930	157
1940	175
1950	216
1959	303
1969	390
1978	449
1987	462
1997	487

Source: U.S. Departments of Agriculture and of Commerce

a) Use a graphing calculator to fit linear, quadratic, cubic, and power functions to the data. Let x represent the number of years since 1900.

b) With each function found in part (a), estimate the average acreage in 2000 and 2010 and determine which function gives the most realistic estimates.

32. *Funeral Costs.* Since the Federal Trade Commission began regulating funeral directors in 1984, the average cost for the funeral of an adult has greatly increased, as shown in the following table.

YEAR	AVERAGE COST OF FUNERAL
1980	\$1809
1985	2737
1991	3742
1995	4624
1996	4782
1998	5020

Source: National Funeral Directors Association

a) Use a graphing calculator to fit quadratic, cubic, and quartic functions to the data. Let x represent the number of years since 1900.

b) With each function found in part (a), estimate the cost of a funeral in 2000 and 2010 and determine which function gives the most realistic estimates.

33. *Consumer Debt.* Nonmortgage consumer debt is mounting in the United States, as shown in the table below.

YEAR	NONMORTGAGE DEBT (IN BILLIONS)
1989	\$ 762
1990	789
1991	783
1992	775
1993	804
1994	902
1995	1038
1996	1161
1997	1216
1998	1266

Source: Federal Reserve

a) Use a graphing calculator to fit linear, quadratic, cubic, and quartic functions to the data. Let x represent the number of years since 1989.

b) Use the functions found in part (a) to estimate the debt in 2003. Compare the estimates and determine which model gives the most realistic estimate.

34. Find a cubic function that fits the data shown in the graph, noting that $R^2 = 1$. Let x represent the number of years since 1979. Then, using this model, estimate the number of gunfire deaths in 2000 and discuss the limitations of the model.

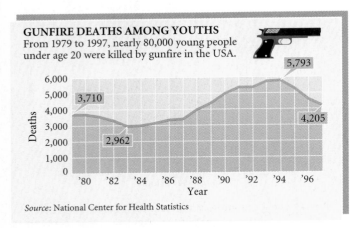

GUNFIRE DEATHS AMONG YOUTHS
From 1979 to 1997, nearly 80,000 young people under age 20 were killed by gunfire in the USA.

Source: National Center for Health Statistics

Collaborative Discussion and Writing

35. How is the range of a polynomial function related to the degree of the polynomial?

36. Polynomial functions are continuous. Discuss what "continuous" means in terms of the domains of the functions and the characteristics of their graphs.

Skill Maintenance

Find the distance between the pair of points.

37. $(3, -5)$ and $(0, -1)$

38. $(4, 2)$ and $(-2, -4)$

39. Find the center and the radius of the circle
$$(x - 3)^2 + (y + 5)^2 = 49.$$

40. The diameter of a circle connects the points $(-6, 5)$ and $(-2, 1)$ on the circle. Find the coordinates of the center of the circle and the length of the radius.

Synthesis

41. In early 1995, $2000 was deposited at a certain interest rate compounded annually. One year later, $1200 was deposited in another account at the same rate. At the end of that year, there was a total of $3573.80 in both accounts. What is the annual interest rate?

3.2

Polynomial Division; The Remainder and Factor Theorems

- *Perform long division with polynomials and determine whether one polynomial is a factor of another.*
- *Use synthetic division to divide a polynomial by $x - c$.*
- *Use the remainder theorem to find a function value $f(c)$.*
- *Use the factor theorem to determine whether $x - c$ is a factor of $f(x)$.*

In general, finding exact zeros of many polynomial functions is neither easy nor straightforward. In this section and the one that follows, we develop concepts that help us find exact zeros of certain polynomial functions with degree 3 or greater.

Let's review the meaning of the real zeros of a function and their connection to the x-intercepts of the function's graph.

CONNECTING THE CONCEPTS

ZEROS, SOLUTIONS, AND INTERCEPTS

FUNCTION	REAL ZEROS OF THE FUNCTION; SOLUTIONS OF THE EQUATION	x–INTERCEPTS OF THE GRAPH

Quadratic Polynomial

$$g(x) = x^2 - 2x - 8$$
$$= (x + 2)(x - 4),$$

or

$$y = (x + 2)(x - 4)$$

To find the **zeros** of $g(x)$, we solve $g(x) = 0$:

$$x^2 - 2x - 8 = 0$$
$$(x + 2)(x - 4) = 0$$
$$x + 2 = 0 \quad or \quad x - 4 = 0$$
$$x = -2 \quad or \quad x = 4.$$

The **solutions** of $x^2 - 2x - 8 = 0$ are -2 and 4. They are the zeros of the function $g(x)$. That is,

$$g(-2) = 0 \quad and \quad g(4) = 0.$$

The zeros of $g(x)$ are the x-coordinates of the **x-intercepts** of the graph of $y = g(x)$.

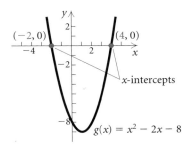

Cubic Polynomial

$$h(x)$$
$$= x^3 + 2x^2 - 5x - 6$$
$$= (x + 3)(x + 1)(x - 2),$$

or

$$y = (x + 3)(x + 1)(x - 2)$$

To find the **zeros** of $h(x)$, we solve $h(x) = 0$:

$$x^3 + 2x^2 - 5x - 6 = 0$$
$$(x + 3)(x + 1)(x - 2) = 0$$
$$x + 3 = 0 \quad or \quad x + 1 = 0 \quad or \quad x - 2 = 0$$
$$x = -3 \quad or \quad x = -1 \quad or \quad x = 2.$$

The **solutions** of $x^3 + 2x^2 - 5x - 6 = 0$ are -3, -1, and 2. They are the zeros of the function $h(x)$. That is,

$$h(-3) = 0,$$
$$h(-1) = 0, \quad and$$
$$h(2) = 0.$$

The zeros of $h(x)$ are the x-coordinates of the **x-intercepts** of the graph of $y = h(x)$.

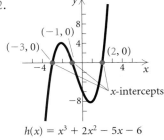

The connection between the zeros of a function and the x-intercepts of the graph of the function is easily seen in the examples above. If c is a real zero of a function (that is, $f(c) = 0$), then $(c, 0)$ is an x-intercept of the graph of the function.

EXAMPLE 1   Consider $P(x) = x^3 + x^2 - 17x + 15$. Determine whether each of the numbers 2 and -5 is a zero of $P(x)$.

Solution   We have

$$P(2) = (2)^3 + (2)^2 - 17(2) + 15 = -7.$$ Substituting 2 into the polynomial

Since $P(2) \neq 0$, we know that 2 is *not* a zero of the polynomial.
We also have

$$P(-5) = (-5)^3 + (-5)^2 - 17(-5) + 15 = 0.$$ Substituting -5 into the polynomial

Since $P(-5) = 0$, we know that -5 is a zero of $P(x)$.

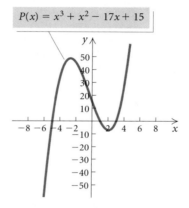

Let's take a closer look at the polynomial

$$h(x) = x^3 + 2x^2 - 5x - 6$$

(see Connecting the Concepts on p. 205). Is there a connection between the factors of the polynomial and the zeros of the function? For $h(x)$, the factors are

$$x + 3, \qquad x + 1, \quad \text{and} \quad x - 2,$$

and the zeros are

$$-3, \qquad -1, \quad \text{and} \quad 2.$$

We note that when the polynomial is expressed in factored form, each factor determines a zero of the function. Thus if we know the factors of a polynomial, we can easily find the zeros. We now show how this idea can be "reversed" so that if we know the zeros of a polynomial function, we can find the factors of the polynomial.

Division and Factors

When we divide one polynomial by another, we obtain a quotient and a remainder. If the remainder is 0, then the divisor is a factor of the dividend.

EXAMPLE 2   Divide to determine whether $x + 1$ and $x - 3$ are factors of

$$x^3 + 2x^2 - 5x - 6.$$

Solution   We have

$$
\begin{array}{r}
\overbrace{x^2 + x - 6}^{\text{Quotient}} \\
x + 1 \overline{)\, x^3 + 2x^2 - 5x - 6} \quad \leftarrow \text{ Dividend} \\
\underline{x^3 + x^2} \\
x^2 - 5x \\
\underline{x^2 + x} \\
-6x - 6 \\
\underline{-6x - 6} \\
0 \leftarrow \text{ Remainder}
\end{array}
$$

Since the remainder is 0, we know that $x + 1$ is a factor of $x^3 + 2x^2 - 5x - 6$. In fact, we know that $x^3 + 2x^2 - 5x - 6 = (x + 1)(x^2 + x - 6)$.

We also have

$$
\begin{array}{r}
x^2 + 5x + 10 \\
x - 3 \overline{)\, x^3 + 2x^2 - 5x - 6} \\
\underline{x^3 - 3x^2} \\
5x^2 - 5x \\
\underline{5x^2 - 15x} \\
10x - 6 \\
\underline{10x - 30} \\
24 \leftarrow \text{ Remainder}
\end{array}
$$

Since the remainder is not 0, we know that $x - 3$ is *not* a factor of $x^3 + 2x^2 - 5x - 6$.

EXAMPLE 3   Divide to determine whether $x^2 + 3x - 1$ is a factor of $x^4 - 81$.

Solution   We have

$$
\begin{array}{r}
x^2 - 3x + 10 \\
x^2 + 3x - 1 \overline{)\, x^4 + 0x^3 + 0x^2 + 0x - 81} \\
\underline{x^4 + 3x^3 - x^2} \\
-3x^3 + x^2 + 0x \\
\underline{-3x^3 - 9x^2 + 3x} \\
10x^2 - 3x - 81 \\
\underline{10x^2 + 30x - 10} \\
-33x - 71
\end{array}
$$

Note that terms in which the coefficient is 0 have been inserted for the missing terms. Spaces can also be used to indicate missing terms.

Since the remainder is not 0, we know that $x^2 + 3x - 1$ is *not* a factor of $x^4 - 81$.

When we divide a polynomial $P(x)$ by a divisor $d(x)$, a polynomial $Q(x)$ is the quotient and a polynomial $R(x)$ is the remainder. The remainder must either be 0 or have degree less than that of $d(x)$.

As in arithmetic, to check a division, we multiply the quotient by the divisor and add the remainder, to see if we get the dividend. Thus these polynomials are related as follows:

$$P(x) = d(x) \cdot Q(x) + R(x)$$

Dividend   Divisor   Quotient   Remainder

For example, if $P(x) = x^3 + 2x^2 - 5x - 6$ and $d(x) = x + 1$, as in Example 2, then $Q(x) = x^2 + x - 6$ and $R(x) = 0$, and

$$\underbrace{x^3 + 2x^2 - 5x - 6}_{P(x)} = \underbrace{(x + 1)}_{d(x)} \cdot \underbrace{(x^2 + x - 6)}_{Q(x)} + \underbrace{0}_{R(x).}$$

If $P(x) = x^4 - 81$ and $d(x) = x^2 + 3x - 1$, as in Example 3, then $Q(x) = x^2 - 3x + 10$ and $R(x) = -33x - 71$, and

$$\underbrace{x^4 - 81}_{P(x)} = \underbrace{(x^2 + 3x - 1)}_{d(x)} \cdot \underbrace{(x^2 - 3x + 10)}_{Q(x)} + \underbrace{(-33x - 71)}_{R(x).}$$

The Remainder Theorem and Synthetic Division

Consider the function

$$h(x) = x^3 + 2x^2 - 5x - 6.$$

When we divided $h(x)$ by $x + 1$ and $x - 3$ in Example 2, the remainders were 0 and 24, respectively. Let's now find the function values $h(-1)$ and $h(3)$:

$$h(-1) = (-1)^3 + 2(-1)^2 - 5(-1) - 6 = 0,$$
$$h(3) = (3)^3 + 2(3)^2 - 5(3) - 6 = 24.$$

Note that the function values are the same as the remainders. This suggests the following theorem.

The Remainder Theorem

If a number c is substituted for x in the polynomial $f(x)$, then the result $f(c)$ is the remainder that would be obtained by dividing $f(x)$ by $x - c$. That is, if $f(x) = (x - c) \cdot Q(x) + R$, then $f(c) = R$.

PROOF (OPTIONAL). The equation $f(x) = d(x) \cdot Q(x) + R(x)$, where $d(x) = x - c$, is the basis of this proof. If we divide $f(x)$ by $x - c$, we obtain a quotient $Q(x)$ and a remainder $R(x)$ related as follows:

$$f(x) = (x - c) \cdot Q(x) + R(x).$$

The remainder $R(x)$ must either be 0 or have degree less than $x - c$. Thus, $R(x)$ must be a constant. Let's call this constant R. The equation above is true for any replacement of x, so we replace x with c. We get

$$\begin{aligned} f(c) &= (c - c) \cdot Q(c) + R \\ &= 0 \cdot Q(c) + R \\ &= R. \end{aligned}$$

Thus the function value $f(c)$ is the remainder obtained when we divide $f(x)$ by $x - c$.

The remainder theorem motivates us to find a rapid way of dividing by $x - c$, in order to find function values. To streamline division, we can arrange the work so that duplicate and unnecessary writing is avoided. Consider the following.

A.
$$\begin{array}{r} 4x^2 + 5x + 11 \\ x - 2\overline{)4x^3 - 3x^2 + x + 7} \\ \underline{4x^3 - 8x^2} \\ 5x^2 + x \\ \underline{5x^2 - 10x} \\ 11x + 7 \\ \underline{11x - 22} \\ 29 \end{array}$$

B.
$$\begin{array}{r} 4 \quad 5 \quad 11 \\ 1 - 2\overline{)4 - 3 + 1 + 7} \\ \underline{4 - 8} \\ 5 + 1 \\ \underline{5 - 10} \\ 11 + 7 \\ \underline{11 - 22} \\ 29 \end{array}$$

The division in (B) is the same as that in (A), but we wrote only the coefficients. The color numerals are duplicated, so we look for an arrangement in which they are not duplicated. In place of the divisor in the form $x - c$, we can simply use c and then add rather than subtract. When the procedure is "collapsed," we have the algorithm known as **synthetic division.**

C. *Synthetic Division*

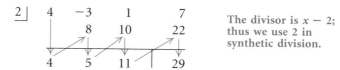

The divisor is $x - 2$; thus we use 2 in synthetic division.

We "bring down" the 4. Then we multiply it by the 2 to get 8 and add to get 5. We then multiply 5 by 2 to get 10, add, and so on. The last number, 29, is the remainder. The others, 4, 5, and 11, are the coefficients of the quotient.

When using synthetic division, we write a 0 for a missing term in the dividend.

EXAMPLE 4 Use synthetic division to find the quotient and the remainder:

$$(2x^3 + 7x^2 - 5) \div (x + 3).$$

Solution   First, we note that $x + 3 = x - (-3)$.

$$
\begin{array}{r|rrrr}
-3 & 2 & 7 & 0 & -5 \\
   &   & -6 & -3 & 9 \\
\hline
   & 2 & 1 & -3 & \;\;\Big|\;\; 4
\end{array}
$$

Note: We must write a 0 for the missing term.

The quotient is $2x^2 + x - 3$. The remainder is 4. (Note that the degree of the quotient is 1 less than the degree of the dividend when the degree of the divisor is 1.)

We can now use synthetic division to find polynomial function values.

EXAMPLE 5   Given that $f(x) = 2x^5 - 3x^4 + x^3 - 2x^2 + x - 8$, find $f(10)$.

Solution   By the remainder theorem, $f(10)$ is the remainder when $f(x)$ is divided by $x - 10$. We use synthetic division to find that remainder.

$$
\begin{array}{r|rrrrrr}
10 & 2 & -3 & 1 & -2 & 1 & -8 \\
   &   & 20 & 170 & 1710 & 17{,}080 & 170{,}810 \\
\hline
   & 2 & 17 & 171 & 1708 & 17{,}081 & \;\Big|\; 170{,}802
\end{array}
$$

Thus, $f(10) = 170{,}802$.

Compare the computations in Example 5 with those in a direct substitution:

$$
\begin{aligned}
f(10) &= 2(10)^5 - 3(10)^4 + (10)^3 - 2(10)^2 + 10 - 8 \\
&= 170{,}802.
\end{aligned}
$$

EXAMPLE 6   Determine whether -4 is a zero of $f(x)$, where

$$f(x) = x^3 + 8x^2 + 8x - 32.$$

Solution   We use synthetic division and the remainder theorem to find $f(-4)$.

$$
\begin{array}{r|rrrr}
-4 & 1 & 8 & 8 & -32 \\
   &   & -4 & -16 & 32 \\
\hline
   & 1 & 4 & -8 & \;\Big|\; 0
\end{array}
$$

Since $f(-4) = 0$, the number -4 is a zero of $f(x)$.

Finding Factors of Polynomials

We now consider a useful result that follows from the remainder theorem.

The Factor Theorem

For a polynomial $f(x)$, if $f(c) = 0$, then $x - c$ is a factor of $f(x)$.

PROOF (OPTIONAL). If we divide $f(x)$ by $x - c$, we obtain a quotient and a remainder, related as follows:

$$f(x) = (x - c) \cdot Q(x) + f(c).$$

Then if $f(c) = 0$, we have

$$f(x) = (x - c) \cdot Q(x),$$

so $x - c$ is a factor of $f(x)$.

The factor theorem is very useful in factoring polynomials, and hence in solving polynomial equations.

Technology Connection

In Example 7, if $x - c$ is a factor of

$$f(x) = x^3 + 2x^2 - 5x - 6,$$

then $f(c) = 0$. Evaluating $f(c)$ with a graphing calculator is more efficient than using synthetic division to find function values. Here, we check the solutions, -1, -3, and 2, of $f(x) = 0$ by evaluating $f(-1)$, $f(-3)$, and $f(2)$.

X	Y₁
-1	0
-3	0
2	0
X = 2	

EXAMPLE 7 Let $f(x) = x^3 + 2x^2 - 5x - 6$. Factor $f(x)$ and solve the equation $f(x) = 0$.

Solution We look for linear factors of the form $x - c$. Let's try $x - 1$. (In the next section, we will learn a method for choosing the numbers to try for c.) We use synthetic division to determine whether $f(1) = 0$.

$$
\begin{array}{r|rrrr}
1 & 1 & 2 & -5 & -6 \\
  &   & 1 & 3 & -2 \\
\hline
  & 1 & 3 & -2 & \,|\; -8
\end{array}
$$

Since $f(1) \neq 0$, we know that $x - 1$ is *not a factor* of $f(x)$. We try $x + 1$ or $x - (-1)$.

$$
\begin{array}{r|rrrr}
-1 & 1 & 2 & -5 & -6 \\
   &   & -1 & -1 & 6 \\
\hline
   & 1 & 1 & -6 & \,|\; 0
\end{array}
$$

Since $f(-1) = 0$, we know that $x + 1$ *is one factor* and the quotient, $x^2 + x - 6$, is another. Thus,

$$f(x) = (x + 1)(x^2 + x - 6).$$

The trinomial is easily factored in this case, so we have

$$f(x) = (x + 1)(x + 3)(x - 2).$$

Our goal is to solve the equation $f(x) = 0$. To do so, we use the principle of zero products.

$$x + 1 = 0 \quad \text{or} \quad x + 3 = 0 \quad \text{or} \quad x - 2 = 0$$
$$x = -1 \quad \text{or} \qquad x = -3 \quad \text{or} \qquad x = 2$$

The solutions are -1, -3, and 2.

Exercise Set 3.2

1. Use substitution to determine whether 4, 5, and -2 are zeros of
$$f(x) = x^3 - 9x^2 + 14x + 24.$$

2. Use substitution to determine whether 2, 3, and -1 are zeros of
$$f(x) = 2x^3 - 3x^2 + x - 1.$$

3. For $f(x)$ in Exercise 1, use long division to determine which of the following are factors of $f(x)$.

a) $x - 4$ **b)** $x - 5$ **c)** $x + 2$

4. For $f(x)$ in Exercise 2, use long division to determine which of the following are factors of $f(x)$.

a) $x - 2$ **b)** $x - 3$ **c)** $x + 1$

In each of the following, a polynomial $P(x)$ and a divisor $d(x)$ are given. Use long division to find the quotient $Q(x)$ and the remainder $R(x)$ when $P(x)$ is divided by $d(x)$, and express $P(x)$ in the form $d(x) \cdot Q(x) + R(x)$.

5. $P(x) = x^3 - 8$,
$d(x) = x + 2$

6. $P(x) = 2x^3 - 3x^2 + x - 1$,
$d(x) = x - 3$

7. $P(x) = x^4 + 9x^2 + 20$,
$d(x) = x^2 + 4$

8. $P(x) = x^4 + x^2 + 2$,
$d(x) = x^2 + x + 1$

Use synthetic division to find the quotient and the remainder.

9. $(2x^4 + 7x^3 + x - 12) \div (x + 3)$

10. $(x^3 - 7x^2 + 13x + 3) \div (x - 2)$

11. $(x^3 - 2x^2 - 8) \div (x + 2)$

12. $(x^3 - 3x + 10) \div (x - 2)$

13. $(x^4 - 1) \div (x - 1)$

14. $(x^5 + 32) \div (x + 2)$

15. $(2x^4 + 3x^2 - 1) \div \left(x - \frac{1}{2}\right)$

16. $(3x^4 - 2x^2 + 2) \div \left(x - \frac{1}{4}\right)$

17. $(x^4 - y^4) \div (x - y)$

18. $(x^3 + 3ix^2 - 4ix - 2) \div (x + i)$

Use synthetic division to find the function values.

19. $f(x) = x^3 - 6x^2 + 11x - 6$; find $f(1)$, $f(-2)$, and $f(3)$.

20. $f(x) = x^3 + 7x^2 - 12x - 3$; find $f(-3)$, $f(-2)$, and $f(1)$.

21. $f(x) = 2x^5 - 3x^4 + 2x^3 - x + 8$; find $f(20)$ and $f(-3)$.

22. $f(x) = x^5 - 10x^4 + 20x^3 - 5x - 100$; find $f(-10)$ and $f(5)$.

23. $f(x) = x^4 - 16$; find $f(2)$, $f(-2)$, $f(3)$, and $f(1 - \sqrt{2})$.

24. $f(x) = x^5 + 32$; find $f(2)$, $f(-2)$, $f(3)$, and $f(2 + 3i)$.

Using synthetic division, determine whether the numbers are zeros of the polynomials.

25. $-3, 2$; $f(x) = 3x^3 + 5x^2 - 6x + 18$

26. $-4, 2$; $f(x) = 3x^3 + 11x^2 - 2x + 8$

27. $-3, \frac{1}{2}$; $f(x) = x^3 - \frac{7}{2}x^2 + x - \frac{3}{2}$

28. $i, -i, -2$; $f(x) = x^3 + 2x^2 + x + 2$

Factor the polynomial $f(x)$. Then solve the equation $f(x) = 0$.

29. $f(x) = x^3 + 4x^2 + x - 6$

30. $f(x) = x^3 + 5x^2 - 2x - 24$

31. $f(x) = x^3 - 6x^2 + 3x + 10$

32. $f(x) = x^3 + 2x^2 - 13x + 10$

33. $f(x) = x^3 - x^2 - 14x + 24$

34. $f(x) = x^3 - 3x^2 - 10x + 24$

35. $f(x) = x^4 - x^3 - 19x^2 + 49x - 30$

36. $f(x) = x^4 + 11x^3 + 41x^2 + 61x + 30$

Technology Connection

37. Use a graphing calculator to find the function values in Exercises 19–24.

Collaborative Discussion and Writing

38. Can an nth-degree polynomial function have more than n zeros? Why or why not?

39. Let $P(x)$ be a polynomial function with $p(x)$ a factor of $P(x)$. If $p(2) = 0$, does it follow that $P(2) = 0$? Why or why not? If $P(2) = 0$, does it follow that $p(2) = 0$? Why or why not?

Skill Maintenance

Find exact solutions.

40. $2x - 7 = 5x + 8$

41. $2x^2 + 12 = 5x$

42. $7x^2 + 4x = 3$

43. The sum of the base and the height of a triangle is 30 in. Find the dimensions for which the area is a maximum.

Synthesis

In Exercises 44 and 45, a graph of a polynomial function is given. On the basis of the graph:

a) *Find as many factors of the polynomial as you can.*
b) *Construct a polynomial function with the zeros shown in the graph.*
c) *Can you find any other polynomial functions with the given zeros?*
d) *Can you find any other polynomial functions with the given zeros and the same graph?*

44.

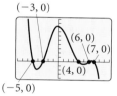

45.

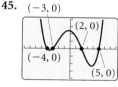

46. For what values of k will the remainder be the same when $x^2 + kx + 4$ is divided by $x - 1$ and $x + 1$?

47. Find k such that $x + 2$ is a factor of $x^3 - kx^2 + 3x + 7k$.

48. *Beam Deflection.* A beam rests at two points A and B and has a concentrated load applied to its center. Let $y = $ the deflection, in feet, of the beam at a distance of x feet from A. Under certain conditions, this deflection is given by

$$y = \frac{1}{13}x^3 - \frac{1}{14}x.$$

Find the zeros of the polynomial in the interval $[0, 2]$.

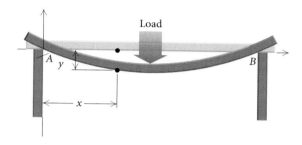

Solve.

49. $\dfrac{2x^2}{x^2 - 1} + \dfrac{4}{x + 3} = \dfrac{32}{x^3 + 3x^2 - x - 3}$

50. $\dfrac{6x^2}{x^2 + 11} + \dfrac{60}{x^3 - 7x^2 + 11x - 77} = \dfrac{1}{x - 7}$

51. Find a 15th-degree polynomial for which $x - 1$ is a factor. Answers may vary.

Use synthetic division to divide.

52. $(x^2 - 4x - 2) \div [x - (3 + 2i)]$

53. $(x^2 - 3x + 7) \div (x - i)$

3.3

Theorems about Zeros of Polynomial Functions

- *Factor polynomial functions and find the zeros and their multiplicities.*
- *Find a polynomial with specified zeros.*
- *For a polynomial function with integer coefficients, find the rational zeros and the other zeros, if possible.*
- *Use Descartes' rule of signs to find information about the number of real zeros of a polynomial function with real coefficients.*

We will now allow the coefficients of a polynomial to be complex numbers. In certain cases, we will restrict the coefficients to be real numbers, rational numbers, or integers, as shown in the following examples.

POLYNOMIAL	TYPE OF COEFFICIENT
$5x^3 - 3x^2 + (2 + 4i)x + i$	Complex
$5x^3 - 3x^2 + \sqrt{2}\,x - \pi$	Real
$5x^3 - 3x^2 + \frac{2}{3}x - \frac{7}{4}$	Rational
$5x^3 - 3x^2 + 8x - 11$	Integer

The Fundamental Theorem of Algebra

A linear, or first-degree, polynomial function $f(x) = mx + b$ (where $m \neq 0$) has just one zero, $-b/m$. It can be shown that any quadratic polynomial function with complex numbers for coefficients has at least one, and at most two, complex zeros. The following theorem is a generalization. No proof is given.

The Fundamental Theorem of Algebra

Every polynomial function of degree n, with $n \geq 1$, has at least one zero in the system of complex numbers.

Note that although the fundamental theorem of algebra guarantees that a zero exists, it does not tell how to find it. Recall that the zeros of a polynomial function $f(x)$ are the solutions of the polynomial equation $f(x) = 0$. We now develop some concepts that can help in finding zeros. First, we consider one of the results of the fundamental theorem of algebra.

Every polynomial function f of degree n, with $n \geq 1$, can be factored into n linear factors (not necessarily unique); that is,
$$f(x) = a_n(x - c_1)(x - c_2) \cdots (x - c_n).$$

Finding Zeros of Factored Polynomial Functions

When a polynomial function is factored into a product of linear factors, it is easy to find the zeros by solving the equation $f(x) = 0$ using the principle of zero products.

EXAMPLE 1   Find the zeros of

$$f(x) = 5(x - 2)(x - 2)(x - 2)(x + 1).$$

Solution   To solve the equation $f(x) = 0$, we use the principle of zero products. The zeros of $f(x)$ are 2 and -1.   ▬

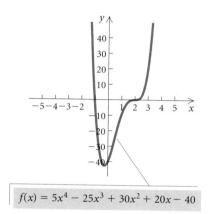

$f(x) = 5x^4 - 25x^3 + 30x^2 + 20x - 40$

In Example 1, the factor $x - 2$ occurs three times. In a case like this, we sometimes say that the zero we obtain from this factor, 2, has a **multiplicity** of 3. If we multiply out the expression for $f(x)$, we obtain

$$f(x) = 5x^4 - 25x^3 + 30x^2 + 20x - 40.$$

Had we started with this form of the function, we might have had trouble finding the zeros. Some polynomials, however, can be factored using techniques we already know, such as factoring by grouping.

EXAMPLE 2   Find the zeros of

$$f(x) = x^3 - 2x^2 - 9x + 18.$$

Solution   We factor by grouping, as follows:

$$\begin{aligned}
f(x) &= x^3 - 2x^2 - 9x + 18 \\
&= x^2(x - 2) - 9(x - 2) \\
&= (x - 2)(x^2 - 9) \\
&= (x - 2)(x + 3)(x - 3).
\end{aligned}$$

Then, by the principle of zero products, the solutions of the equation $f(x) = 0$ are 2, -3, and 3. These are the zeros of $f(x)$.   ▬

Other factoring techniques can also be used.

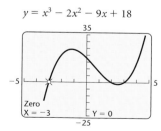

$y = x^3 - 2x^2 - 9x + 18$

Zero
X = -3   Y = 0

Technology Connection

Using the Zero method, we can determine the zeros of the function in Example 2.

The window at left shows the calculator display when we find the leftmost zero. The other zeros, 2 and 3, can be found in the same manner.

EXAMPLE 3   Find the zeros of

$$f(x) = x^4 + 4x^2 - 45.$$

Solution   We factor as follows:

$$f(x) = x^4 + 4x^2 - 45$$
$$= (x^2 - 5)(x^2 + 9).$$

We now solve the equation $f(x) = 0$ to determine the zeros. We use the principle of zero products:

$$(x^2 - 5)(x^2 + 9) = 0$$
$$x^2 - 5 = 0 \quad or \quad x^2 + 9 = 0$$
$$x^2 = 5 \quad or \quad x^2 = -9$$
$$x = \pm\sqrt{5} \quad or \quad x = \pm\sqrt{-9} = \pm 3i.$$

The solutions are $\pm\sqrt{5}$ and $\pm 3i$. These are the zeros of $f(x)$.   ▬

Only the real-number zeros of a function can be seen on its graph. The real-number zeros $\sqrt{5}$ and $-\sqrt{5}$ of the function in Example 3 can be seen on the graph below. The nonreal zeros, $3i$ and $-3i$, do not appear on the graph.

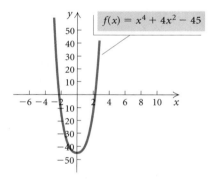

> Every polynomial function of degree n, with $n \geq 1$, has at least one zero and at most n zeros.

This is often stated as follows: "Every polynomial function of degree n, with $n \geq 1$, has *exactly* n zeros." This statement is compatible with the preceding statement, if one takes multiplicities into account.

Finding Polynomials with Given Zeros

Given several numbers, we can find a polynomial function with those numbers as its zeros.

EXAMPLE 4   Find a polynomial function of degree 3, having the zeros 1, $3i$, and $-3i$.

Solution   Such a polynomial has factors $x - 1$, $x - 3i$, and $x + 3i$, so we have

$$f(x) = a_n(x - 1)(x - 3i)(x + 3i).$$

The number a_n can be any nonzero number. The simplest polynomial will be obtained if we let it be 1. If we then multiply the factors, we obtain

$$f(x) = (x - 1)(x^2 + 9)$$
$$= x^3 - x^2 + 9x - 9.$$

EXAMPLE 5   Find a polynomial function of degree 5 with -1 as a zero of multiplicity 3, 4 as a zero of multiplicity 1, and 0 as a zero of multiplicity 1.

Solution   Proceeding as in Example 4, letting $a_n = 1$, we obtain

$$f(x) = (x + 1)^3(x - 4)(x - 0)$$
$$= x^5 - x^4 - 9x^3 - 11x^2 - 4x.$$

Zeros of Polynomial Functions with Real Coefficients

Consider the quadratic equation $x^2 - 2x + 2 = 0$, with real coefficients. Its solutions are $1 + i$ and $1 - i$. Note that they are complex conjugates. This generalizes to any polynomial with real coefficients.

> If a complex number $a + bi$, $b \neq 0$, is a zero of a polynomial function $f(x)$ with *real* coefficients, then its conjugate, $a - bi$, is also a zero. (Nonreal zeros occur in conjugate pairs.)

For the preceding to be true, it is essential that the coefficients be real numbers.

Rational Coefficients

When a polynomial has rational numbers for coefficients, certain irrational zeros also occur in pairs, as described in the following theorem.

> If $a + c\sqrt{b}$, a and c rational, b not a square, is a zero of a polynomial function $f(x)$ with *rational* coefficients, then $a - c\sqrt{b}$ is also a zero.

EXAMPLE 6   Suppose that a polynomial function of degree 6 with rational coefficients has $-2 + 5i$, $-2i$, and $1 - \sqrt{3}$ as three of its zeros. Find the other zeros.

Solution   The other zeros are $-2 - 5i$, $2i$, and $1 + \sqrt{3}$. There are no other zeros because a polynomial of degree 6 can have at most 6 zeros.

EXAMPLE 7   Find a polynomial function of lowest degree with rational coefficients that has $1 - \sqrt{2}$ and $1 + 2i$ as two of its zeros.

Solution   The function must also have the zeros $1 + \sqrt{2}$ and $1 - 2i$. Because we want to find the polynomial function of lowest degree with the given zeros, we will not include additional zeros. That is, we will write a polynomial function of degree 4. Thus the polynomial function is

$$f(x) = [x - (1 - \sqrt{2})][x - (1 + \sqrt{2})][x - (1 + 2i)][x - (1 - 2i)]$$
$$= (x^2 - 2x - 1)(x^2 - 2x + 5)$$
$$= x^4 - 4x^3 + 8x^2 - 8x - 5.$$

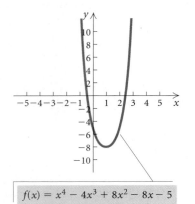

$f(x) = x^4 - 4x^3 + 8x^2 - 8x - 5$

Integer Coefficients and the Rational Zeros Theorem

It is not always easy to find the zeros of a polynomial function. However, if a polynomial function has integer coefficients, there is a procedure that will yield all the rational zeros.

The Rational Zeros Theorem

Let

$$P(x) = a_n x^n + a_{n-1} x^{n-1} + \cdots + a_1 x + a_0,$$

where all the coefficients are integers. Consider a rational number denoted by p/q, where p and q are relatively prime (having no common factor besides -1 and 1). If p/q is a zero of $P(x)$, then p is a factor of a_0 and q is a factor of a_n.

EXAMPLE 8   Given $f(x) = 3x^4 - 11x^3 + 10x - 4$:

a) Find the rational zeros, and then the other zeros; that is, solve $f(x) = 0$.

b) Factor $f(x)$ into linear factors.

Solution

a) Because the degree of $f(x)$ is 4, there are at most 4 distinct zeros. The rational zeros theorem says that if p/q is a zero of $f(x)$, then p must

We can narrow the list of possibilities for the rational zeros of a function by examining its graph.

$$y = 3x^4 - 11x^3 + 10x - 4$$

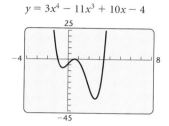

From the graph of the function in Example 8, we see that of the possibilities for rational zeros, only the numbers

$$-1, \ \tfrac{1}{3}, \ \tfrac{2}{3}$$

might actually be rational zeros. We can check these with the TABLE feature. Note that the graphing calculator converts $\tfrac{1}{3}$ and $\tfrac{2}{3}$ to decimal notation.

⋏	Y1	
−1	0	
.33333	−1.037	
.66667	0	
X =		

Since $f(-1) = 0$ and $f\left(\tfrac{2}{3}\right) = 0$, -1 and $\tfrac{2}{3}$ are zeros. Since $f\left(\tfrac{1}{3}\right) \neq 0$, $\tfrac{1}{3}$ is not a zero.

be a factor of -4 and q must be a factor of 3. Thus the possibilities for p/q are

$$\frac{\text{Possibilities for } p}{\text{Possibilities for } q}: \ \frac{\pm 1, \ \pm 2, \ \pm 4}{\pm 1, \ \pm 3}.$$

Possibilities for p/q: $\ 1, \ -1, \ 2, \ -2, \ 4, \ -4, \ \tfrac{1}{3}, \ -\tfrac{1}{3}, \ \tfrac{2}{3}, \ -\tfrac{2}{3}, \ \tfrac{4}{3}, \ -\tfrac{4}{3}.$

To find which are zeros, we could use substitution, but synthetic division is usually more efficient. It is easier to consider first the integers. Then we consider the fractions, if the integers do not produce all the zeros.

We try 1:

$$\begin{array}{r|rrrr} 1 & 3 & -11 & 0 & 10 & -4 \\ & & 3 & -8 & -8 & 2 \\ \hline & 3 & -8 & -8 & 2 & \big| \ -2. \end{array}$$

Since $f(1) = -2$, 1 is not a zero.

We try -1:

$$\begin{array}{r|rrrr} -1 & 3 & -11 & 0 & 10 & -4 \\ & & -3 & 14 & -14 & 4 \\ \hline & 3 & -14 & 14 & -4 & \big| \ 0. \end{array}$$

We have $f(-1) = 0$, so -1 is a zero. Thus, $x + 1$ is a factor. Using the results of the second division above, we can express $f(x)$ as

$$f(x) = (x + 1)(3x^3 - 14x^2 + 14x - 4).$$

We now consider the factor $3x^3 - 14x^2 + 14x - 4$ and check the other possible zeros. We use synthetic division again, to determine whether -1 is a zero of multiplicity 2 of $f(x)$:

$$\begin{array}{r|rrr} -1 & 3 & -14 & 14 & -4 \\ & & -3 & 17 & -31 \\ \hline & 3 & -17 & 31 & \big| \ -35. \end{array}$$

We see that -1 is not a double zero. We leave it to the student to verify that $2, -2, 4,$ and -4 are not zeros. There are no other zeros that are integers, so we start checking the fractions.

Let's now try $\tfrac{2}{3}$.

$$\begin{array}{r|rrr} 2/3 & 3 & -14 & 14 & -4 \\ & & 2 & -8 & 4 \\ \hline & 3 & -12 & 6 & \big| \ 0 \end{array}$$

Since the remainder is 0, we know that $x - \tfrac{2}{3}$ is a factor of $3x^3 - 14x^2 + 14x - 4$ and is also a factor of $f(x)$. Thus, $\tfrac{2}{3}$ is a zero of $f(x)$.

Using the results of the synthetic division, we can factor further:

$$f(x) = (x + 1)\left(x - \tfrac{2}{3}\right)(3x^2 - 12x + 6) \qquad \text{Using the results of the last synthetic division}$$

$$= (x + 1)\left(x - \tfrac{2}{3}\right) \cdot 3 \cdot (x^2 - 4x + 2). \qquad \text{Removing a factor of 3}$$

The quadratic formula can be used to find the values of x for which $x^2 - 4x + 2 = 0$. Those values are also zeros of $f(x)$:

$$x = \frac{-b \pm \sqrt{b^2 - 4ac}}{2a}$$

$$= \frac{-(-4) \pm \sqrt{(-4)^2 - 4 \cdot 1 \cdot 2}}{2 \cdot 1}$$

$$= \frac{4 \pm \sqrt{8}}{2} = \frac{4 \pm 2\sqrt{2}}{2} = \frac{2(2 \pm \sqrt{2})}{2}$$

$$= 2 \pm \sqrt{2}.$$

The rational zeros are -1 and $\frac{2}{3}$. The other zeros are $2 \pm \sqrt{2}$.

b) The complete factorization of $f(x)$ is

$$f(x) = (x + 1)\left(x - \tfrac{2}{3}\right) \cdot 3 \cdot [x - (2 - \sqrt{2})][x - (2 + \sqrt{2})].$$

EXAMPLE 9   Given $f(x) = 2x^5 - x^4 - 4x^3 + 2x^2 - 30x + 15$:

a) Find the rational zeros, and then the other zeros; that is, solve $f(x) = 0$.

b) Factor $f(x)$ into linear factors.

Solution

a) Because the degree of $f(x)$ is 5, there are at most 5 distinct zeros. According to the rational zeros theorem, any rational zero of f must be of the form p/q, where p is a factor of 15 and q is a factor of 2. The possibilities are

$$\frac{\textit{Possibilities for } p}{\textit{Possibilities for } q} : \quad \frac{\pm 1, \pm 3, \pm 5, \pm 15}{\pm 1, \pm 2};$$

Possibilities for p/q: $1, -1, 3, -3, 5, -5, 15, -15, \frac{1}{2}, -\frac{1}{2}, \frac{3}{2},$
$-\frac{3}{2}, \frac{5}{2}, -\frac{5}{2}, \frac{15}{2}, -\frac{15}{2}.$

We use synthetic division to check each of the possibilities. Let's try 1 and -1.

$$\begin{array}{r|rrrrrr}
1 & 2 & -1 & -4 & 2 & -30 & 15 \\
  &   & 2 & 1 & -3 & -1 & -31 \\
\hline
  & 2 & 1 & -3 & -1 & -31 & -16
\end{array}$$

$$\begin{array}{r|rrrrrr}
-1 & 2 & -1 & -4 & 2 & -30 & 15 \\
   &   & -2 & 3 & 1 & -3 & 33 \\
\hline
   & 2 & -3 & -1 & 3 & -33 & 48
\end{array}$$

Since $f(1) = -16$ and $f(-1) = 48$, neither 1 nor -1 is a zero of the function. We leave it to the student to verify that the other integer

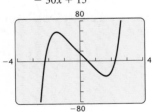

possibilities are not zeros. We now check $\frac{1}{2}$.

$$\frac{1}{2} \begin{array}{|rrrrrr} 2 & -1 & -4 & 2 & -30 & 15 \\ & 1 & 0 & -2 & 0 & -15 \\ \hline 2 & 0 & -4 & 0 & -30 & 0 \end{array}$$

This means that $x - \frac{1}{2}$ is a factor of $f(x)$. We write the factorization and try to factor further:

$$\begin{aligned} f(x) &= \left(x - \tfrac{1}{2}\right)(2x^4 - 4x^2 - 30) \\ &= \left(x - \tfrac{1}{2}\right) \cdot 2 \cdot (x^4 - 2x^2 - 15) \qquad \text{Factoring out the 2} \\ &= \left(x - \tfrac{1}{2}\right) \cdot 2 \cdot (x^2 - 5)(x^2 + 3). \qquad \text{Factoring the trinomial} \end{aligned}$$

We now solve the equation $f(x) = 0$ to determine the zeros. We use the principle of zero products:

$$\left(x - \tfrac{1}{2}\right) \cdot 2 \cdot (x^2 - 5)(x^2 + 3) = 0$$

$$\begin{array}{ccccc} x - \tfrac{1}{2} = 0 & or & x^2 - 5 = 0 & or & x^2 + 3 = 0 \\ x = \tfrac{1}{2} & or & x^2 = 5 & or & x^2 = -3 \\ x = \tfrac{1}{2} & or & x = \pm\sqrt{5} & or & x = \pm\sqrt{3}\,i. \end{array}$$

There is only one rational zero, $\frac{1}{2}$. The other zeros are $\pm\sqrt{5}$ and $\pm\sqrt{3}i$.

b) The factorization into linear factors is

$$f(x) = 2\left(x - \tfrac{1}{2}\right)(x + \sqrt{5})(x - \sqrt{5})(x + \sqrt{3}\,i)(x - \sqrt{3}\,i). \qquad \blacksquare$$

Descartes' Rule of Signs

The development of a rule that helps determine the number of positive real zeros and the number of negative real zeros of a polynomial function is credited to the French mathematician René Descartes. To use the rule, we must have the polynomial arranged in descending or ascending order, with no zero terms written in, the leading coefficient positive, and the *constant term* not 0. The **constant term** is the term that does not contain a variable. Then we determine the number of *variations of sign*, that is, the number of times, in reading through the polynomial, that successive coefficients are of different signs.

EXAMPLE 10 Determine the number of variations of sign in the polynomial function $P(x) = 2x^5 - 3x^2 + x + 4$.

Solution We have

$$P(x) = \underbrace{2x^5 - }\underbrace{3x^2 + }\underbrace{x + } 4.$$

From positive to negative; a variation

From negative to positive; a variation

Both positive; no variation

The number of variations of sign is 2. $\quad\blacksquare$

Note the following:

$$P(-x) = 2(-x)^5 - 3(-x)^2 + (-x) + 4$$
$$= -2x^5 - 3x^2 - x + 4.$$

We see that the number of variations of sign in $P(-x)$ is 1.

We now state Descartes' rule, without proof.

Descartes' Rule of Signs

Let $P(x)$ be a polynomial function with real coefficients and a nonzero constant term. The number of positive real zeros of $P(x)$ is either:

1. The same as the number of variations of sign in $P(x)$, or
2. Less than the number of variations of sign in $P(x)$ by a positive even integer.

The number of negative real zeros of $P(x)$ is either:

3. The same as the number of variations of sign in $P(-x)$, or
4. Less than the number of variations of sign in $P(-x)$ by a positive even integer.

A zero of multiplicity m must be counted m times.

In each of Examples 11–13, what does Descartes' rule of signs tell you about the number of positive real zeros and the number of negative real zeros?

EXAMPLE 11 $P(x) = 2x^5 - 5x^2 - 3x + 6$

Solution   The number of variations of sign in $P(x)$ is 2. Therefore, the number of positive real zeros is either 2 or less than 2 by 2, 4, 6, and so on. Thus the number of positive real zeros is either 2 or 0, since a negative number of zeros has no meaning.

$$P(-x) = -2x^5 - 5x^2 + 3x + 6$$

The number of variations of sign in $P(-x)$ is 1. Thus there is exactly 1 negative real zero. Since nonreal, complex conjugates occur in pairs, we also know the possible ways in which zeros might occur, as summarized in the table shown at left.

TOTAL NUMBER OF ZEROS	5	5
POSITIVE REAL	2	0
NEGATIVE REAL	1	1
NONREAL	2	4

EXAMPLE 12 $P(x) = 5x^4 - 3x^3 + 7x^2 - 12x + 4$

Solution   There are 4 variations of sign. Thus the number of positive real zeros is either

$$4 \quad \text{or} \quad 4 - 2 \quad \text{or} \quad 4 - 4.$$

That is, the number of positive real zeros is 4, 2, or 0.

$$P(-x) = 5x^4 + 3x^3 + 7x^2 + 12x + 4$$

There are 0 changes in sign, so there are no negative real zeros.

EXAMPLE 13 $P(x) = 6x^6 - 2x^2 - 5x$

Solution   As stated, the polynomial does not satisfy the conditions of Descartes' rule of signs because the constant term is 0. But because x is a factor of every term, we know that the polynomial has 0 as a zero. We can then factor as follows:

$$P(x) = x(6x^5 - 2x - 5).$$

Now we analyze $Q(x) = 6x^5 - 2x - 5$ and $Q(-x) = -6x^5 + 2x - 5$. The number of variations of sign in $Q(x)$ is 1. Therefore, there is exactly 1 positive real zero. The number of variations of sign in $Q(-x)$ is 2. Thus the number of negative real zeros is 2 or 0. The same results apply to $P(x)$.

Exercise Set

Find the zeros of the polynomial function and state the multiplicity of each.

1. $f(x) = (x + 3)^2(x - 1)$

2. $f(x) = -8(x - 3)^2(x + 4)^3 x^4$

3. $f(x) = x^3(x - 1)^2(x + 4)$

4. $f(x) = (x^2 - 5x + 6)^2$

5. $f(x) = x^4 - 4x^2 + 3$

6. $f(x) = x^4 - 10x^2 + 9$

7. $f(x) = x^3 + 3x^2 - x - 3$

8. $f(x) = x^3 - x^2 - 2x + 2$

Find a polynomial function of degree 3 with the given numbers as zeros.

9. $-2, 3, 5$

10. $2, i, -i$

11. $-3, 2i, -2i$

12. $1 + 4i, 1 - 4i, -1$

13. $\sqrt{2}, -\sqrt{2}, \sqrt{3}$

14. $-3, 0, \frac{1}{2}$

15. Find a polynomial function of degree 5 with -1 as a zero of multiplicity 3, 0 as a zero of multiplicity 1, and 1 as a zero of multiplicity 1.

16. Find a polynomial function of degree 4 with -2 as a zero of multiplicity 1, 3 as a zero of multiplicity 2, and -1 as a zero of multiplicity 1.

Suppose that a polynomial function of degree 5 with rational coefficients has the given numbers as zeros. Find the other zeros.

17. $6, -3 + 4i, 4 - \sqrt{5}$ **18.** $-2, 3, 4, 1 - i$

Find a polynomial function of lowest degree with rational coefficients that has the given numbers as some of its zeros.

19. $1 + i, 2$

20. $2 - i, -1$

21. $-4i, 5$

22. $2 - \sqrt{3}, 1 + i$

23. $\sqrt{5}, -3i$

24. $-\sqrt{2}, 4i$

Given that the polynomial function has the given zero, find the other zeros.

25. $f(x) = x^4 - 5x^3 + 7x^2 - 5x + 6;$ $-i$

26. $f(x) = x^4 - 16;$ $2i$

27. $f(x) = x^3 - 6x^2 + 13x - 20;$ 4

28. $f(x) = x^3 - 8;$ 2

List all possible rational zeros.

29. $f(x) = x^5 - 3x^2 + 1$

30. $f(x) = x^7 + 37x^5 - 6x^2 + 12$

31. $f(x) = 15x^6 + 47x^2 + 2$

32. $f(x) = 10x^{25} + 3x^{17} - 35x + 6$

For each polynomial function:

a) *Find the rational zeros, and then the other zeros; that is, solve* $f(x) = 0$.

b) *Factor* $f(x)$ *into linear factors.*

33. $f(x) = x^3 + 3x^2 - 2x - 6$

34. $f(x) = x^3 - x^2 - 3x + 3$

35. $f(x) = x^3 - 3x + 2$

36. $f(x) = x^3 - 2x + 4$

37. $f(x) = x^3 - 5x^2 + 11x + 17$

38. $f(x) = 2x^3 + 7x^2 + 2x - 8$

39. $f(x) = 5x^4 - 4x^3 + 19x^2 - 16x - 4$

40. $f(x) = 3x^4 - 4x^3 + x^2 + 6x - 2$

41. $f(x) = x^4 - 3x^3 - 20x^2 - 24x - 8$

42. $f(x) = x^4 + 5x^3 - 27x^2 + 31x - 10$

43. $f(x) = x^3 - 4x^2 + 2x + 4$

44. $f(x) = x^3 - 8x^2 + 17x - 4$

45. $f(x) = x^3 + 8$

46. $f(x) = x^3 - 8$

47. $f(x) = \frac{1}{3}x^3 - \frac{1}{2}x^2 - \frac{1}{6}x + \frac{1}{6}$

48. $f(x) = \frac{2}{3}x^3 - \frac{1}{2}x^2 + \frac{2}{3}x - \frac{1}{2}$

Find only the rational zeros.

49. $f(x) = x^4 + 32$

50. $f(x) = x^6 + 8$

51. $f(x) = x^3 - x^2 - 4x + 3$

52. $f(x) = 2x^3 + 3x^2 + 2x + 3$

53. $f(x) = x^4 + 2x^3 + 2x^2 - 4x - 8$

54. $f(x) = x^4 + 6x^3 + 17x^2 + 36x + 66$

55. $f(x) = x^5 - 5x^4 + 5x^3 + 15x^2 - 36x + 20$

56. $f(x) = x^5 - 3x^4 - 3x^3 + 9x^2 - 4x + 12$

What does Descartes' rule of signs tell you about the number of positive real zeros and the number of negative real zeros of the function?

57. $f(x) = 3x^5 - 2x^2 + x - 1$

58. $g(x) = 5x^6 - 3x^3 + x^2 - x$

59. $h(x) = 6x^7 + 2x^2 + 5x + 4$

60. $P(x) = -3x^5 - 7x^3 - 4x - 5$

61. $F(p) = 3p^{18} + 2p^4 - 5p^2 + p + 3$

62. $H(t) = 5t^{12} - 7t^4 + 3t^2 + t + 1$

63. $C(x) = 7x^6 + 3x^4 - x - 10$

64. $g(z) = -z^{10} + 8z^7 + z^3 + 6z - 1$

65. $h(t) = -4t^5 - t^3 + 2t^2 + 1$

66. $P(x) = x^6 + 2x^4 - 9x^3 - 4$

67. $f(y) = y^4 + 13y^3 - y + 5$

68. $Q(x) = x^4 - 2x^2 + 12x - 8$

69. $r(x) = x^4 - 6x^2 + 20x - 24$

70. $f(x) = x^5 - 2x^3 - 8x$

71. $R(x) = 3x^5 - 5x^3 - 4x$

72. $f(x) = x^4 - 9x^2 - 6x + 4$

Technology Connection

73. Use a graphing calculator to narrow the list of possible rational zeros in Exercises 38–41. Then determine the rational zeros by evaluating the function using the TABLE feature.

Collaborative Discussion and Writing

74. If $Q(x) = -P(x)$, do $P(x)$ and $Q(x)$ have the same zeros? Why or why not?

75. Is it possible for a third-degree polynomial with rational coefficients to have no real zeros? Why or why not?

Skill Maintenance

For Exercises 76 and 77, complete the square to:

a) *find the vertex;*

b) *find the line of symmetry; and*

c) *determine whether there is a maximum or minimum value and find that value.*

76. $f(x) = 3x^2 - 6x - 1$

77. $f(x) = x^2 - 8x + 10$

Find the zeros of the function.

78. $g(x) = x^2 - 8x - 33$

79. $f(x) = -\frac{4}{5}x + 8$

Synthesis

80. Use the rational zeros theorem and the equation $x^4 - 12 = 0$ to show that $\sqrt[4]{12}$ is irrational.

81. Consider $f(x) = 2x^3 - 5x^2 - 4x + 3$. Find the solutions of each equation.

a) $f(x) = 0$ b) $f(x - 1) = 0$

c) $f(x + 2) = 0$ d) $f(2x) = 0$

Find the rational zeros.

82. $P(x) = x^6 - 6x^5 - 72x^4 - 81x^2 + 486x + 5832$

83. $P(x) = 2x^5 - 33x^4 - 84x^3 + 2203x^2 - 3348x - 10,080$

3.4

Rational Functions

- *Graph a rational function, identifying all asymptotes.*
- *Solve applied problems involving rational functions.*

Now we turn our attention to functions that represent the quotient of two polynomials. Whereas, the sum, difference, or product of two polynomials is a polynomial, in general the quotient of two polynomials is *not* itself a polynomial.

A *rational number* can be expressed as the quotient of two integers, p/q, where $q \neq 0$. A *rational function* is formed by the quotient of two polynomials, $p(x)/q(x)$, where $q(x) \neq 0$. Here are some examples of rational functions and their graphs.

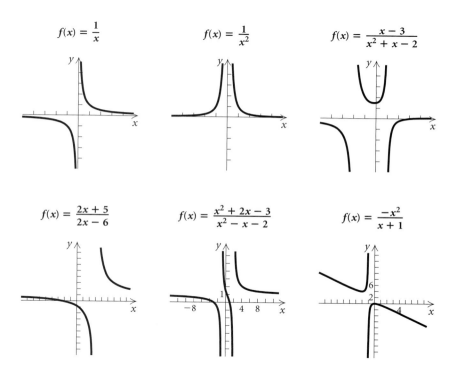

$$f(x) = \frac{1}{x} \qquad f(x) = \frac{1}{x^2} \qquad f(x) = \frac{x-3}{x^2+x-2}$$

$$f(x) = \frac{2x+5}{2x-6} \qquad f(x) = \frac{x^2+2x-3}{x^2-x-2} \qquad f(x) = \frac{-x^2}{x+1}$$

Rational Function

A **rational function** is a function f that is a quotient of two polynomials, that is,

$$f(x) = \frac{p(x)}{q(x)},$$

where $p(x)$ and $q(x)$ are polynomials and where $q(x)$ is not the zero polynomial. The domain of f consists of all inputs x for which $q(x) \neq 0$.

The Domain of a Rational Function

DOMAINS OF FUNCTIONS

REVIEW SECTION 1.2.

EXAMPLE 1   Consider

$$f(x) = \frac{1}{x - 3}.$$

Find the domain and graph f.

Solution   The only input that results in a denominator of 0 is 3. Thus the domain is

$$\{x \mid x \neq 3\}, \text{ or } (-\infty, 3) \cup (3, \infty).$$

The graph of this function is the graph of $y = 1/x$ translated right 3 units.

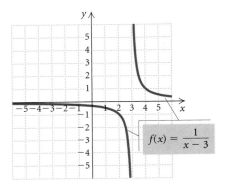

Technology Connection

The graphs at left show the function $f(x) = 1/(x - 3)$ graphed in CONNECTED mode and in DOT mode. Using CONNECTED mode can lead to an incorrect graph. In CONNECTED mode, a graphing calculator connects plotted points with line segments. In DOT mode, it simply plots unconnected points. In the first graph, the graphing calculator has connected the points plotted on either side of the x-value 3 with a line that appears to be the vertical line $x = 3$. (It is not actually vertical since it connects the last point to the left of $x = 3$ with the first point to the right of $x = 3$.) Since 3 is not in the domain of the function, the vertical line $x = 3$ cannot be part of the graph. We will see later in this section that vertical lines like $x = 3$, although not part of the graph, are important in the construction of graphs. If you have a choice when graphing rational functions, use DOT mode.

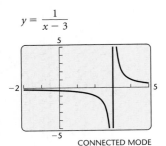

$$y = \frac{1}{x - 3}$$

CONNECTED MODE

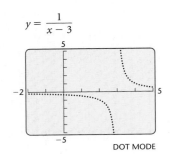

$$y = \frac{1}{x - 3}$$

DOT MODE

EXAMPLE 2 Determine the domain of each of the functions illustrated at the beginning of this section.

Solution The domain of each rational function will be the set of all real numbers except those values that make the denominator 0. To determine those exceptions, we set the denominator equal to 0 and solve for x.

FUNCTION	DOMAIN
$f(x) = \dfrac{1}{x}$	$\{x \mid x \neq 0\}$, or $(-\infty, 0) \cup (0, \infty)$
$f(x) = \dfrac{1}{x^2}$	$\{x \mid x \neq 0\}$, or $(-\infty, 0) \cup (0, \infty)$
$f(x) = \dfrac{x - 3}{x^2 + x - 2} = \dfrac{x - 3}{(x + 2)(x - 1)}$	$\{x \mid x \neq -2 \text{ and } x \neq 1\}$, or $(-\infty, -2) \cup (-2, 1) \cup (1, \infty)$
$f(x) = \dfrac{2x + 5}{2x - 6} = \dfrac{2x + 5}{2(x - 3)}$	$\{x \mid x \neq 3\}$, or $(-\infty, 3) \cup (3, \infty)$
$f(x) = \dfrac{x^2 + 2x - 3}{x^2 - x - 2} = \dfrac{x^2 + 2x - 3}{(x + 1)(x - 2)}$	$\{x \mid x \neq -1 \text{ and } x \neq 2\}$, or $(-\infty, -1) \cup (-1, 2) \cup (2, \infty)$
$f(x) = \dfrac{-x^2}{x + 1}$	$\{x \mid x \neq -1\}$, or $(-\infty, -1) \cup (-1, \infty)$

As a partial check of the domains, we can observe the discontinuities (breaks) in the graphs of these functions (see p. 225). ▬

Asymptotes

Look at the graph of $f(x) = 1/(x - 3)$ in Example 1. Let's explore what happens as x-values get closer and closer to 3 from the left. We then explore what happens as x-values get closer and closer to 3 from the right.

From left:

x	2	$2\frac{1}{2}$	$2\frac{99}{100}$	$2\frac{9999}{10,000}$	$2\frac{999,999}{1,000,000}$	$\longrightarrow 3$
$f(x)$	-1	-2	-100	$-10,000$	$-1,000,000$	$\longrightarrow -\infty$

From right:

x	4	$3\frac{1}{2}$	$3\frac{1}{100}$	$3\frac{1}{10,000}$	$3\frac{1}{1,000,000}$	$\longrightarrow 3$
$f(x)$	1	2	100	10,000	1,000,000	$\longrightarrow \infty$

We see that as x-values get closer and closer to 3 from the left, the function values (y-values) decrease without bound. Similarly, as the x-values approach 3 from the right, the function values increase without bound. We write this as

$$f(x) \to -\infty \text{ as } x \to 3^- \quad \text{and} \quad f(x) \to \infty \text{ as } x \to 3^+.$$

We read "$f(x) \to -\infty$ as $x \to 3^-$" as "$f(x)$ decreases without bound as x approaches 3 from the left." We read "$f(x) \to \infty$ as $x \to 3^+$" as "$f(x)$ increases without bound as x approaches 3 from the right." The vertical line $x = 3$ is said to be a **vertical asymptote** for this curve.

Vertical Asymptote

The line $x = a$ is a **vertical asymptote** for the graph of f if any of the following is true:

$$f(x) \to \infty \text{ as } x \to a^- \quad \text{or} \quad f(x) \to -\infty \text{ as } x \to a^-, \quad \text{or}$$
$$f(x) \to \infty \text{ as } x \to a^+ \quad \text{or} \quad f(x) \to -\infty \text{ as } x \to a^+.$$

The following figures show the four ways in which a vertical asymptote can occur.

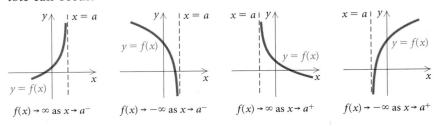

The vertical asymptotes of a rational function $f(x) = p(x)/q(x)$ are found by determining the zeros of $q(x)$ that are not also zeros of $p(x)$. If $p(x)$ and $q(x)$ are polynomials with no common factors other than constants, we need determine only the zeros of the denominator $q(x)$.

Determining Vertical Asymptotes

For a rational function $f(x) = p(x)/q(x)$, where $p(x)$ and $q(x)$ are polynomials with no common factors other than constants, if a is a zero of the denominator, then the line $x = a$ is a vertical asymptote for the graph of the function.

EXAMPLE 3   Determine the vertical asymptotes of each of the following functions.

a) $f(x) = \dfrac{2x - 11}{x^2 + 2x - 8}$

b) $g(x) = \dfrac{x - 2}{x^3 - 5x}$

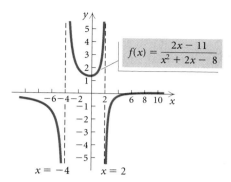

FIGURE 1

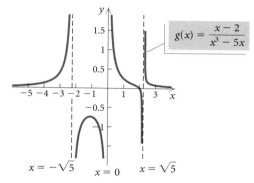

FIGURE 2

Solution

a) We factor to find the zeros of the denominator:

$$x^2 + 2x - 8 = (x + 4)(x - 2).$$

The zeros of the denominator are -4 and 2. Thus the vertical asymptotes are the lines $x = -4$ and $x = 2$ (see Fig. 1).

b) We factor to find the zeros of the denominator:

$$x^3 - 5x = x(x^2 - 5).$$

Setting $x(x^2 - 5) = 0$ and solving for x, we get

$$x = 0 \quad or \quad x^2 - 5 = 0$$
$$x = 0 \quad or \quad x^2 = 5$$
$$x = 0 \quad or \quad x = \pm\sqrt{5}.$$

The zeros of the denominator are 0, $\sqrt{5}$, and $-\sqrt{5}$. Thus the vertical asymptotes are the lines $x = 0$, $x = \sqrt{5}$, and $x = -\sqrt{5}$ (see Fig. 2).

Referring again to Example 1, let's explore what happens to $f(x) = 1/(x - 3)$ as x increases without bound and as x decreases without bound.

x increases without bound:

x	100	5000	1,000,000	$\longrightarrow \infty$
$f(x)$	≈ 0.0103	≈ 0.0002	≈ 0.000001	$\longrightarrow 0$

x decreases without bound:

x	-300	-8000	$-1,000,000$	$\longrightarrow -\infty$
$f(x)$	≈ -0.0033	≈ -0.0001	≈ -0.000001	$\longrightarrow 0$

We see that

$$\frac{1}{x - 3} \to 0 \text{ as } x \to \infty \quad \text{and} \quad \frac{1}{x - 3} \to 0 \text{ as } x \to -\infty.$$

Since $y = 0$ is the equation of the x-axis, we say that the curve approaches the x-axis asymptotically and that the x-axis is a horizontal asymptote for the curve.

Horizontal Asymptote

The line $y = b$ is a **horizontal asymptote** for the graph of f if either or both of the following are true:

$$f(x) \to b \text{ as } x \to \infty \quad \text{or} \quad f(x) \to b \text{ as } x \to -\infty.$$

The following figures illustrate four ways in which horizontal asymptotes can occur. In each case, the curve gets close to the line $y = b$ either as $x \to \infty$ or as $x \to -\infty$. Keep in mind that the symbols ∞ and $-\infty$ convey the idea of increasing without bound and decreasing without bound, respectively.

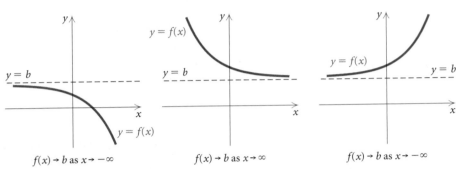

$f(x) \to b$ as $x \to \infty$ $f(x) \to b$ as $x \to -\infty$ $f(x) \to b$ as $x \to \infty$ $f(x) \to b$ as $x \to -\infty$

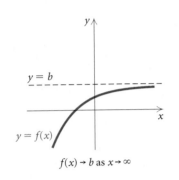

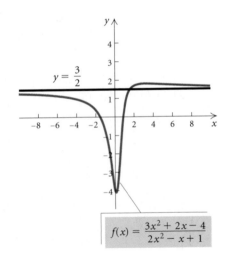

$$f(x) = \frac{3x^2 + 2x - 4}{2x^2 - x + 1}$$

How can we determine a horizontal asymptote? As x gets very large or very small, the value of the polynomial function $p(x)$ is dominated by the function's leading term. Because of this, if $p(x)$ and $q(x)$ have the *same* degree, the value of $p(x)/q(x)$ as $x \to \infty$ or as $x \to -\infty$ is dominated by the ratio of the numerator's leading coefficient to the denominator's leading coefficient.

For $f(x) = (3x^2 + 2x - 4)/(2x^2 - x + 1)$, we see that

$$\frac{3x^2 + 2x - 4}{2x^2 - x + 1} \to \frac{3}{2}, \text{ or } 1.5, \text{ as } x \to \infty, \quad \text{and}$$

$$\frac{3x^2 + 2x - 4}{2x^2 - x + 1} \to \frac{3}{2}, \text{ or } 1.5, \text{ as } x \to -\infty.$$

We say that the curve approaches the horizontal line $y = \frac{3}{2}$ asymptotically and that $y = \frac{3}{2}$ is a horizontal asymptote for the curve.

It follows that when the numerator and the denominator of a rational function have the same degree, the line $y = a/b$ is the horizontal asymptote, where a and b are the leading coefficients of the numerator and the denominator, respectively.

EXAMPLE 4   Find the horizontal asymptote: $f(x) = \dfrac{-7x^4 - 10x^2 + 1}{11x^4 + x - 2}$.

Solution   The numerator and the denominator have the same degree, so the line $y = -\frac{7}{11}$, or $-0.\overline{63}$, is a horizontal asymptote.   ▬

To check Example 4, we could evaluate the function for a very large and a very small value of x. Another check, one that is useful in calculus,

We can check Example 4 by using a graphing calculator to evaluate the function for a very large and a very small value of x.

X	Y₁
100000	−.6364
−80000	−.6364

X =

For each number, the function value is close to −0.63.

is to multiply by 1, using $(1/x^4)/(1/x^4)$:

$$f(x) = \frac{-7x^4 - 10x^2 + 1}{11x^4 + x - 2} \cdot \frac{\dfrac{1}{x^4}}{\dfrac{1}{x^4}}$$

$$= \frac{-7 - \dfrac{10}{x^2} + \dfrac{1}{x^4}}{11 + \dfrac{1}{x^3} - \dfrac{2}{x^4}}.$$

As $|x|$ becomes very large, each expression with a power of x in the denominator tends toward 0. Specifically, as $x \to \infty$ or as $x \to -\infty$, we have

$$f(x) \to \frac{-7 - 0 + 0}{11 + 0 - 0}, \quad \text{or} \quad f(x) \to -\frac{7}{11}.$$

The horizontal asymptote is $y = -\frac{7}{11}$.

We now investigate the occurrence of a horizontal asymptote when the degree of the numerator is less than the degree of the denominator. We saw in Example 1 that $y = 0$, the x-axis, is the horizontal asymptote of $f(x) = 1/(x - 3)$.

EXAMPLE 5 Find the horizontal asymptote: $f(x) = \dfrac{2x + 3}{x^3 - 2x^2 + 4}$.

Solution We let $p(x) = 2x + 3$, $q(x) = x^3 - 2x^2 + 4$, and $f(x) = p(x)/q(x)$. Note that as x grows large, the value of $q(x)$ grows much faster than the value of $p(x)$. Because of this, the ratio $p(x)/q(x)$ shrinks toward 0. As $x \to -\infty$, the ratio $p(x)/q(x)$ behaves in a similar manner. The horizontal asymptote is $y = 0$, the x-axis. ▬

The following statements describe the two ways in which a horizontal asymptote occurs.

Determining a Horizontal Asymptote
- When the numerator and the denominator of a rational function have the same degree, the line $y = a/b$ is the horizontal asymptote, where a and b are the leading coefficients of the numerator and the denominator, respectively.
- When the degree of the numerator of a rational function is less than the degree of the denominator, the x-axis, or $y = 0$, is the horizontal asymptote.
- When the degree of the numerator of a rational function is greater than the degree of the denominator, there is no horizontal asymptote.

The following statements are also true.

> The graph of a rational function never crosses a vertical asymptote.

> The graph of a rational function may or may not cross a horizontal asymptote.

EXAMPLE 6   Graph

$$g(x) = \frac{2x^2 + 1}{x^2}.$$

Include and label all asymptotes.

Solution   Since 0 is the zero of the denominator, the y-axis, $x = 0$, is the vertical asymptote. Note also that the degree of the numerator is the same as the degree of the denominator. Thus, $y = 2/1$, or 2, is the horizontal asymptote.

To draw the graph, we first draw the asymptotes with dashed lines. Then we compute and plot some ordered pairs and draw the two branches of the curve.

x	$g(x)$
-2	$2\frac{1}{4}$
$-1\frac{1}{2}$	$2\frac{4}{9}$
-1	3
$-\frac{1}{2}$	6
$\frac{1}{2}$	6
1	3
$1\frac{1}{2}$	$2\frac{4}{9}$
2	$2\frac{1}{4}$

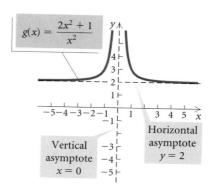

Sometimes a line that is neither horizontal nor vertical is an asymptote. Such a line is called an **oblique asymptote,** or a **slant asymptote.**

EXAMPLE 7   Find all the asymptotes of

$$f(x) = \frac{2x^2 - 3x - 1}{x - 2}.$$

Solution   The line $x = 2$ is the vertical asymptote because 2 is the zero of the denominator. There is no horizontal asymptote because the degree of the numerator is greater than the degree of the denominator. When the

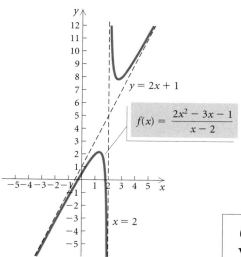

degree of the numerator is 1 greater than the degree of the denominator, we divide to find an equivalent expression:

$$\frac{2x^2 - 3x - 1}{x - 2} = (2x + 1) + \frac{1}{x - 2}.$$

$$\begin{array}{r} 2x + 1 \\ x - 2 \overline{\smash{\big)}\ 2x^2 - 3x - 1} \\ \underline{2x^2 - 4x} \\ x - 1 \\ \underline{x - 2} \\ 1 \end{array}$$

Now we see that when $x \to \infty$ or $x \to -\infty$, $1/(x - 2) \to 0$ and the value of $f(x) \to 2x + 1$. This means that as $|x|$ becomes very large, the graph of $f(x)$ gets very close to the graph of $y = 2x + 1$. Thus the line $y = 2x + 1$ is the oblique asymptote. ▬

Occurrence of Lines as Asymptotes of Rational Functions

Vertical asymptotes of a rational function $p(x)/q(x)$, where $p(x)$ and $q(x)$ have no common factors other than constants, occur at any x-values that make the denominator 0.

The x-axis is the horizontal asymptote when the degree of the numerator is less than the degree of the denominator.

A horizontal asymptote other than the x-axis occurs when the numerator and the denominator have the same degree.

An oblique asymptote occurs when the degree of the numerator is 1 greater than the degree of the denominator.

There can be only one horizontal asymptote or one oblique asymptote and never both.

An asymptote is *not* part of the graph of the function.

The following is an outline of a procedure that we can follow to create accurate graphs of rational functions.

To graph a rational function $f(x) = p(x)/q(x)$, where $p(x)$ and $q(x)$ have no common factor:

1. Find the real zeros of the denominator. Determine the domain of the function and sketch the vertical asymptotes.
2. Find the horizontal or the oblique asymptote, if there is one, and sketch it.
3. Find the zeros of the function. These are the first coordinates of the x-intercepts of the function. The zeros are found by determining the zeros of the numerator.
4. Find $f(0)$. This gives the y-intercept of the function.
5. Find other function values to determine the general shape. Then draw the graph.

EXAMPLE 8   Graph: $f(x) = \dfrac{2x + 3}{3x^2 + 7x - 6}$.

Solution

1. We find the zeros of the denominator by solving $3x^2 + 7x - 6 = 0$. Since

$$3x^2 + 7x - 6 = (3x - 2)(x + 3),$$

the zeros are $\frac{2}{3}$ and -3. Thus the domain excludes $\frac{2}{3}$ and -3 and is

$$(-\infty, -3) \cup \left(-3, \tfrac{2}{3}\right) \cup \left(\tfrac{2}{3}, \infty\right).$$

The graph has vertical asymptotes $x = -3$ and $x = \frac{2}{3}$. We sketch these as dashed lines.

2. Because the degree of the numerator is less than the degree of the denominator, the x-axis is the horizontal asymptote.

3. To find the zeros of the numerator, we solve $2x + 3 = 0$ and get $x = -\frac{3}{2}$. Thus, $-\frac{3}{2}$ is the zero of the function, and the pair $\left(-\frac{3}{2}, 0\right)$ is the x-intercept.

4. We find $f(0)$:

$$f(0) = \frac{2 \cdot 0 + 3}{3 \cdot 0^2 + 7 \cdot 0 - 6}$$

$$= \frac{3}{-6} = -\frac{1}{2}.$$

Thus, $\left(0, -\frac{1}{2}\right)$ is the y-intercept.

5. We find other function values to determine the general shape and then draw the graph. Note that the graph of this function crosses its horizontal asymptote at $x = -\frac{3}{2}$. ▬

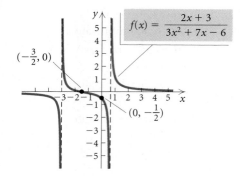

EXAMPLE 9   Graph: $g(x) = \dfrac{x^2 - 1}{x^2 + x - 6}$.

Solution

1. We find the zeros of the denominator by solving $x^2 + x - 6 = 0$. Since

$$x^2 + x - 6 = (x + 3)(x - 2),$$

the zeros are -3 and 2. Thus the domain excludes the x-values -3 and 2 and is $(-\infty, -3) \cup (-3, 2) \cup (2, \infty)$. The graph has vertical asymptotes $x = -3$ and $x = 2$. We sketch these as dashed lines.

2. The numerator and the denominator have the same degree, so the horizontal asymptote is determined by the ratio of the leading coefficients: $1/1$, or 1. Thus, $y = 1$ is the horizontal asymptote. We sketch it with a dashed line.

3. To find the zeros of the numerator, we solve $x^2 - 1 = 0$. The solutions are -1 and 1. Thus, -1 and 1 are the zeros of the function and the pairs $(-1, 0)$ and $(1, 0)$ are the x-intercepts.

4. We find $g(0)$:

$$g(0) = \frac{0^2 - 1}{0^2 + 0 - 6} = \frac{-1}{-6} = \frac{1}{6}.$$

Thus, $\left(0, \frac{1}{6}\right)$ is the y-intercept.

5. We find other function values to determine the general shape and then draw the graph.

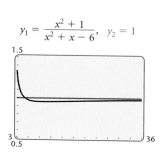

$y_1 = \dfrac{x^2 + 1}{x^2 + x - 6}, \quad y_2 = 1$

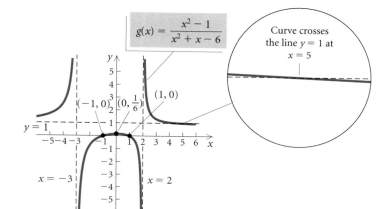

The magnified portion of the graph in Example 9 shows another situation in which a graph can cross its horizontal asymptote. Let's observe the behavior of the curve after it crosses the horizontal asymptote at $x = 5$. (See the graph at left.) It continues to decrease for a short interval and then begins to increase, getting closer and closer to $y = 1$ as $x \to \infty$.

Graphs of rational functions can also cross an oblique asymptote. The graph of

$$f(x) = \frac{2x^3}{x^2 + 1}$$

shown at left crosses its oblique asymptote $y = 2x$. Remember, graphs can cross horizontal or oblique asymptotes, but they cannot cross vertical asymptotes.

Let's now graph a rational function $p(x)/q(x)$, where $p(x)$ and $q(x)$ have a common factor.

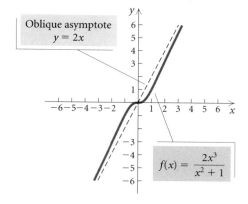

EXAMPLE 10 Graph: $g(x) = \dfrac{x - 2}{x^2 - x - 2}$.

Solution We first express the denominator in factored form:

$$g(x) = \frac{x - 2}{x^2 - x - 2} = \frac{x - 2}{(x + 1)(x - 2)}.$$

The domain of the function is $\{x \,|\, x \neq -1 \ and \ x \neq 2\}$, or $(-\infty, -1) \cup (-1, 2) \cup (2, \infty)$. The zeros of the denominator are

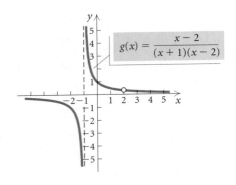

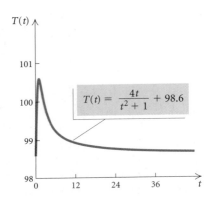

-1 and 2, and the zero of the numerator is 2. Thus the graph of the function has $x = -1$ as its only vertical asymptote, since -1 is the only zero of the denominator that is not a zero of the numerator. The degree of the numerator is less than the degree of the denominator, so $y = 0$ is the horizontal asymptote. There are no zeros of the function and thus no x-intercepts, because 2 is the only zero of the numerator and 2 is not in the domain of the function. Since $g(0) = 1$, $(0, 1)$ is the y-intercept. We draw the graph indicating the "hole" when $x = 2$ with an open circle.

The rational expression $(x - 2)/((x + 1)(x - 2))$ can be simplified. Thus,

$$g(x) = \frac{x - 2}{(x + 1)(x - 2)} = \frac{1}{x + 1}, \quad \text{where } x \neq -1 \text{ and } x \neq 2.$$

The graph of $g(x)$ is the graph of $y = 1/(x + 1)$ with the point $\left(2, \frac{1}{3}\right)$ missing.

Applications

EXAMPLE 11   *Temperature During an Illness.* The temperature T, in degrees Fahrenheit, of a person during an illness is given by the function

$$T(t) = \frac{4t}{t^2 + 1} + 98.6,$$

where time t is given in hours since the onset of the illness. The graph of this function is shown at left.

a) Find the temperature at $t = 0, 1, 2, 5, 12$, and 24.

b) Find the horizontal asymptote of the graph of $T(t)$. Complete:

$$T(t) \rightarrow \boxed{} \text{ as } t \rightarrow \infty.$$

c) Give the meaning of the answer to part (b) in terms of the application.

Solution

a) We have

$$T(0) = 98.6, \quad T(1) = 100.6, \quad T(2) = 100.2,$$
$$T(5) \approx 99.369, \quad T(12) \approx 98.931, \quad \text{and} \quad T(24) \approx 98.766.$$

b) Since

$$T(t) = \frac{4t}{t^2 + 1} + 98.6$$
$$= \frac{98.6t^2 + 4t + 98.6}{t^2 + 1},$$

the horizontal asymptote is $y = 98.6/1$, or 98.6. Then it follows that $T(t) \rightarrow 98.6$ as $t \rightarrow \infty$.

c) As time goes on, the temperature returns to "normal," which is 98.6°.

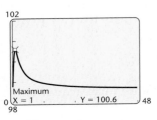

Exercise Set 3.4

In Exercises 1–6, use your knowledge of asymptotes and intercepts to match the equation with one of the graphs (a)–(f), which follow. List all asymptotes.

a)

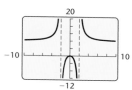

b)

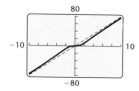

c)

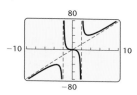

d)

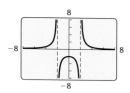

e)

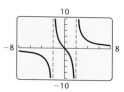

f)

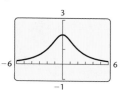

1. $f(x) = \dfrac{8}{x^2 - 4}$

2. $f(x) = \dfrac{8}{x^2 + 4}$

3. $f(x) = \dfrac{8x}{x^2 - 4}$

4. $f(x) = \dfrac{8x^2}{x^2 - 4}$

5. $f(x) = \dfrac{8x^3}{x^2 - 4}$

6. $f(x) = \dfrac{8x^3}{x^2 + 4}$

Graph each of the following. Be sure to label all the asymptotes. List the x- and y-intercepts.

7. $f(x) = \dfrac{1}{x + 3}$

8. $f(x) = \dfrac{1}{x - 5}$

9. $f(x) = \dfrac{-2}{x - 5}$

10. $f(x) = \dfrac{3}{3 - x}$

11. $f(x) = \dfrac{2x + 1}{x}$

12. $f(x) = \dfrac{3x - 1}{x}$

13. $f(x) = \dfrac{1}{(x - 2)^2}$

14. $f(x) = \dfrac{-2}{(x - 3)^2}$

15. $f(x) = \dfrac{1}{x^2}$

16. $f(x) = \dfrac{1}{3x^2}$

17. $f(x) = \dfrac{1}{x^2 + 3}$

18. $f(x) = \dfrac{-1}{x^2 + 2}$

19. $f(x) = \dfrac{x^2 - 4}{x - 2}$

20. $f(x) = \dfrac{x^2 - 9}{x + 3}$

21. $f(x) = \dfrac{x - 1}{x + 2}$

22. $f(x) = \dfrac{x - 2}{x + 1}$

23. $f(x) = \dfrac{x + 3}{2x^2 - 5x - 3}$

24. $f(x) = \dfrac{3x}{x^2 + 5x + 4}$

25. $f(x) = \dfrac{x^2 - 9}{x + 1}$

26. $f(x) = \dfrac{x^2 - 4}{x - 1}$

27. $f(x) = \dfrac{x^2 + x - 2}{2x^2 + 1}$

28. $f(x) = \dfrac{x^2 - 2x - 3}{3x^2 + 2}$

29. $g(x) = \dfrac{3x^2 - x - 2}{x - 1}$

30. $f(x) = \dfrac{2x + 1}{2x^2 - 5x - 3}$

31. $f(x) = \dfrac{x - 1}{x^2 - 2x - 3}$

32. $f(x) = \dfrac{x + 2}{x^2 + 2x - 15}$

33. $f(x) = \dfrac{x - 3}{(x + 1)^3}$

34. $f(x) = \dfrac{x + 2}{(x - 1)^3}$

35. $f(x) = \dfrac{x^3 + 1}{x}$

36. $f(x) = \dfrac{x^3 - 1}{x}$

37. $f(x) = \dfrac{x^3 + 2x^2 - 15x}{x^2 - 5x - 14}$

38. $f(x) = \dfrac{x^3 + 2x^2 - 3x}{x^2 - 25}$

39. $f(x) = \dfrac{5x^4}{x^4 + 1}$

40. $f(x) = \dfrac{x + 1}{x^2 + x - 6}$

41. $f(x) = \dfrac{x^2}{x^2 - x - 2}$

42. $f(x) = \dfrac{x^2 - x - 2}{x + 2}$

Find a rational function that satisfies the given conditions for each of the following. Answers may vary, but try to give the simplest answer possible.

43. Vertical asymptotes $x = -4$, $x = 5$

44. Vertical asymptotes $x = -4$, $x = 5$; x-intercept $(-2, 0)$

45. Vertical asymptotes $x = -4$, $x = 5$; horizontal asymptote $y = \frac{3}{2}$; x-intercept $(-2, 0)$

46. Oblique asymptote $y = x - 1$

47. *Medical Dosage.* The function

$$N(t) = \frac{0.8t + 1000}{5t + 4}, \quad t \geq 15$$

gives the body concentration $N(t)$, in parts per million, of a certain dosage of medication after time t, in hours.

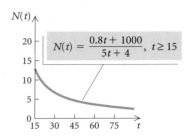

a) Find the horizontal asymptote of the graph and complete the following:

$$N(t) \to \boxed{} \text{ as } t \to \infty.$$

b) Explain the meaning of the answer to part (a) in terms of the application.

48. *Average Cost.* The average cost per tape, in dollars, for a company to produce x videotapes on exercising is given by the function

$$A(x) = \frac{13x + 100}{x}, \quad x > 0.$$

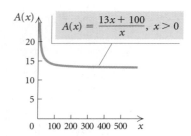

a) Find the horizontal asymptote of the graph and complete the following:

$$A(x) \to \boxed{} \text{ as } x \to \infty.$$

b) Explain the meaning of the answer to part (a) in terms of the application.

49. *Population Growth.* The population P, in thousands, of Lordsburg is given by

$$P(t) = \frac{500t}{2t^2 + 9},$$

where t is the time, in months.

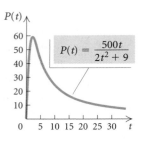

a) Find the population at $t = 0, 1, 3,$ and 8 months.

b) Find the horizontal asymptote of the graph and complete the following:

$$P(t) \to \boxed{} \text{ as } t \to \infty.$$

c) Explain the meaning of the answer to part (b) in terms of the application.

Technology Connection

50. *Minimizing Surface Area.* The Hold-It Container Co. is designing an open-top rectangular box, with a square base, that will hold 108 cubic centimeters.

a) Express the surface area S as a function of the length x of a side of the base.

b) Graph the function on the interval $(0, \infty)$.

c) Estimate the minimum surface area and the value of x that will yield it.

51. For Exercise 49, find the maximum population and the value of t that will yield it.

52. Graph

$$y_1 = \frac{x^3 + 4}{x} \quad \text{and} \quad y_2 = x^2$$

using the same viewing window. Explain how the parabola $y_2 = x^2$ can be thought of as a nonlinear asymptote for y_1.

Collaborative Discussion and Writing

53. Under what circumstances will a rational function have a domain consisting of all real numbers?

54. Explain why the graph of a rational function cannot have both a horizontal and an oblique asymptote.

Skill Maintenance

Solve.

55. $2y - 3 \geq 1 - y + 5$

56. $(x - 2)(x + 5) > x(x - 3)$

57. $|x + 6| \geq 7$ **58.** $\left| x + \frac{1}{4} \right| \leq \frac{2}{3}$

Synthesis

Find the nonlinear asymptotes of the function.

59. $f(x) = \dfrac{x^5 + 2x^3 + 4x^2}{x^2 + 2}$

60. $f(x) = \dfrac{x^4 + 3x^2}{x^2 + 1}$

Graph the function.

61. $f(x) = \dfrac{2x^3 + x^2 - 8x - 4}{x^3 + x^2 - 9x - 9}$

62. $f(x) = \dfrac{x^3 + 4x^2 + x - 6}{x^2 - x - 2}$

Find the domain.

63. $f(x) = \sqrt{\dfrac{72}{x^2 - 4x - 21}}$

64. $f(x) = \sqrt{x^2 - 4x - 21}$

3.5

Polynomial and Rational Inequalities

• *Solve polynomial and rational inequalities.*

We will use a combination of algebraic and graphical methods to solve polynomial and rational inequalities.

Polynomial Inequalities

Just as a quadratic equation can be written in the form $ax^2 + bx + c = 0$, a **quadratic inequality** can be written in the form $ax^2 + bx + c \ \square \ 0$, where $\square$ is $<$, $>$, $\leq$, or $\geq$. Here are some examples of quadratic inequalities:

$$3x^2 - 2x - 5 > 0, \qquad -\tfrac{1}{2}x^2 + 4x - 7 \leq 0.$$

Quadratic inequalities are one type of **polynomial inequality.** Other examples of polynomial inequalities are

$$-2x^4 + x^2 - 3 < 7, \qquad \tfrac{2}{3}x + 4 \geq 0, \quad \text{and} \quad 4x^3 - 2x^2 > 5x + 7.$$

When the inequality symbol in a polynomial inequality is replaced with an equals sign, a **related equation** is formed. Polynomial inequalities can be easily solved once the related equation has been solved.

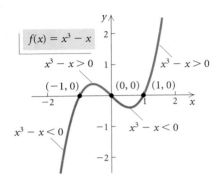

$f(x) = x^3 - x$

$x^3 - x > 0$

$(-1, 0)$ $(0, 0)$ $(1, 0)$

$x^3 - x > 0$

$x^3 - x < 0$ $x^3 - x < 0$

EXAMPLE 1   Solve: $x^3 - x > 0$.

Solution   We are asked to find all x-values for which $x^3 - x > 0$. To locate these values, we graph $f(x) = x^3 - x$. Then we note that whenever the function changes sign, its graph passes through an x-intercept. Thus to solve $x^3 - x > 0$, we first solve the related equation $x^3 - x = 0$ to find all zeros of the function:

$$x^3 - x = 0$$
$$x(x^2 - 1) = 0$$
$$x(x + 1)(x - 1) = 0.$$

The zeros are -1, 0, and 1. Thus the x-intercepts of the graph are $(-1, 0)$, $(0, 0)$, and $(1, 0)$, as shown in the figure at left. The zeros divide the x-axis into four intervals:

$$(-\infty, -1), \qquad (-1, 0), \qquad (0, 1), \quad \text{and} \quad (1, \infty).$$

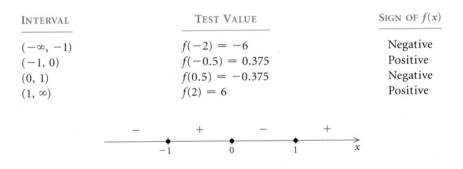

For all x-values within a given interval, the sign of $x^3 - x$ must be either *positive* or *negative*. To determine which, we choose a test value for x from each interval and find $f(x)$. We can also determine the sign of $f(x)$ in each interval by simply looking at the graph of the function.

INTERVAL	TEST VALUE	SIGN OF $f(x)$
$(-\infty, -1)$	$f(-2) = -6$	Negative
$(-1, 0)$	$f(-0.5) = 0.375$	Positive
$(0, 1)$	$f(0.5) = -0.375$	Negative
$(1, \infty)$	$f(2) = 6$	Positive

Since we are solving $x^3 - x > 0$, the solution set consists of only two of the four intervals, those in which the sign of $f(x)$ is *positive*. We see that the solution set is $(-1, 0) \cup (1, \infty)$, or $\{x \,|\, -1 < x < 0 \text{ or } x > 1\}$. ▬

Shown in the box at left is a method for solving polynomial inequalities.

EXAMPLE 2   Solve: $3x^4 + 10x \leq 11x^3 + 4$.

By subtracting $11x^3 + 4$, we form the equivalent inequality

$$3x^4 - 11x^3 + 10x - 4 \leq 0.$$

To solve a polynomial inequality:

1. Find an equivalent inequality with 0 on one side.

2. Solve the related polynomial equation.

3. Use the solutions to divide the x-axis into intervals. Then select a test value from each interval and determine the polynomial's sign on the interval.

4. Determine the intervals for which the inequality is satisfied and write interval notation or set-builder notation for the solution set. Include the endpoints of the intervals in the solution set if the inequality symbol is $\leq$ or $\geq$.

Algebraic Solution

To solve the related equation

$$3x^4 - 11x^3 + 10x - 4 = 0,$$

we need to use the theorems of Section 3.3 (see Example 8 in Section 3.3). The solutions are

$$-1, \quad 2 - \sqrt{2}, \quad \tfrac{2}{3}, \quad \text{and} \quad 2 + \sqrt{2},$$

or approximately

$$-1, \quad 0.586, \quad 0.667, \quad \text{and} \quad 3.414.$$

These numbers divide the x-axis into five intervals: $(-\infty, -1)$, $(-1, 2 - \sqrt{2})$, $\left(2 - \sqrt{2}, \tfrac{2}{3}\right)$, $\left(\tfrac{2}{3}, 2 + \sqrt{2}\right)$, and $(2 + \sqrt{2}, \infty)$.

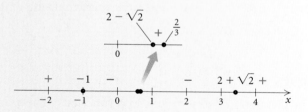

We then let $f(x) = 3x^4 - 11x^3 + 10x - 4$ and, using test values for $f(x)$, determine the sign of $f(x)$ in each interval:

INTERVAL	TEST VALUE	SIGN OF $f(x)$
$(-\infty, -1)$	$f(-2) = 112$	$+$
$(-1, 2 - \sqrt{2})$	$f(0) = -4$	$-$
$(2 - \sqrt{2}, 2/3)$	$f(0.6) = 0.128$	$+$
$(2/3, 2 + \sqrt{2})$	$f(1) = -2$	$-$
$(2 + \sqrt{2}, \infty)$	$f(4) = 100$	$+$

Function values are negative in the intervals $(-1, 2 - \sqrt{2})$ and $\left(\tfrac{2}{3}, 2 + \sqrt{2}\right)$. Since the inequality sign is $\leq$, we include the endpoints of the intervals in the solution set. The solution set is

$$\left[-1, 2 - \sqrt{2}\right] \cup \left[\tfrac{2}{3}, 2 + \sqrt{2}\right], \quad \text{or}$$
$$\left\{x \mid -1 \leq x \leq 2 - \sqrt{2} \; or \; \tfrac{2}{3} \leq x \leq 2 + \sqrt{2}\right\}.$$

Visualizing the Solution

Observing the graph of the function

$$f(x) = 3x^4 - 11x^3 + 10x - 4$$

and a closeup view of the graph on the interval $(0, 1)$, we see the intervals on which $f(x) \leq 0$. The values of $f(x)$ are less than or equal to 0 in two intervals.

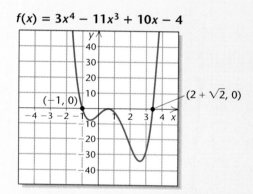

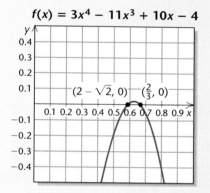

The solution set of the inequality

$$3x^4 - 11x^3 + 10x - 4 \leq 0$$

is

$$\left[-1, 2 - \sqrt{2}\right] \cup \left[\tfrac{2}{3}, 2 + \sqrt{2}\right].$$

Technology Connection

Polynomial inequalities can be solved quickly with a graphing calculator. Consider the inequality in Example 2:

$$3x^4 + 10x \le 11x^3 + 4, \quad \text{or}$$
$$3x^4 - 11x^3 + 10x - 4 \le 0.$$

We graph the related equation

$$y = 3x^4 - 11x^3 + 10x - 4$$

and use the ZERO feature. We see in the window on the left below that two of the zeros are -1 and approximately 3.414 ($2 + \sqrt{2} \approx 3.414$). However, this window leaves us uncertain about the number of zeros of the function in the interval $[0, 1]$.

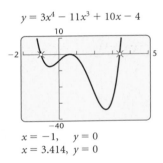

$y = 3x^4 - 11x^3 + 10x - 4$

$x = -1, \quad y = 0$
$x = 3.414, \quad y = 0$

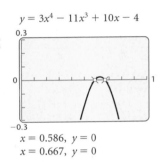

$y = 3x^4 - 11x^3 + 10x - 4$

$x = 0.586, \quad y = 0$
$x = 0.667, \quad y = 0$

The window on the right above shows another view of the zeros in the interval $[0, 1]$. Those zeros are about 0.586 and 0.667 $\left(2 - \sqrt{2} \approx 0.586; \frac{2}{3} \approx 0.667\right)$. The intervals to be considered are $(-\infty, -1), (-1, 0.586), (0.586, 0.667), (0.667, 3.414)$, and $(3.414, \infty)$. We note on the graph where the function is negative. Then, including appropriate endpoints, we find that the solution set is approximately

$$[-1, 0.586] \cup [0.667, 3.414].$$

Rational Inequalities

Some inequalities involve rational expressions and functions. These are called **rational inequalities.** To solve rational inequalities, we need to make some adjustments to the preceding method.

EXAMPLE 3   Solve: $\dfrac{x - 3}{x + 4} \ge \dfrac{x + 2}{x - 5}$.

Solution   We first subtract $(x + 2)/(x - 5)$ in order to find an equivalent inequality with 0 on one side:

$$\frac{x - 3}{x + 4} - \frac{x + 2}{x - 5} \ge 0.$$

Algebraic Solution

We look for all values of x for which the related function

$$f(x) = \frac{x-3}{x+4} - \frac{x+2}{x-5}$$

is not defined or is 0. These are called **critical values.**

A look at the denominators shows that $f(x)$ is not defined for $x = -4$ and $x = 5$. Next, we solve $f(x) = 0$:

$$\frac{x-3}{x+4} - \frac{x+2}{x-5} = 0$$

$$(x+4)(x-5)\left(\frac{x-3}{x+4} - \frac{x+2}{x-5}\right) = (x+4)(x-5) \cdot 0$$

$$(x-5)(x-3) - (x+4)(x+2) = 0$$

$$(x^2 - 8x + 15) - (x^2 + 6x + 8) = 0$$

$$-14x + 7 = 0$$

$$x = \tfrac{1}{2}.$$

The critical values are -4, $\tfrac{1}{2}$, and 5. These values divide the x-axis into four intervals:

$$(-\infty, -4), \quad \left(-4, \tfrac{1}{2}\right), \quad \left(\tfrac{1}{2}, 5\right), \quad \text{and} \quad (5, \infty).$$

We then use a test value to determine the sign of $f(x)$ in each interval.

INTERVAL	TEST VALUE	SIGN OF $f(x)$
$(-\infty, -4)$	$f(-5) = 7.7$	$+$
$\left(-4, \tfrac{1}{2}\right)$	$f(-2) = -2.5$	$-$
$\left(\tfrac{1}{2}, 5\right)$	$f(3) = 2.5$	$+$
$(5, \infty)$	$f(6) = -7.7$	$-$

Function values are positive in the intervals $(-\infty, -4)$ and $\left(\tfrac{1}{2}, 5\right)$. Since $f\left(\tfrac{1}{2}\right) = 0$ and the inequality symbol is $\geq$, we know that $\tfrac{1}{2}$ must be in the solution set. Note that since neither -4 nor 5 is in the domain of f, they cannot be part of the solution set.

The solution set is $(-\infty, -4) \cup \left[\tfrac{1}{2}, 5\right)$.

Visualizing the Solution

The graph of the related function

$$f(x) = \frac{x-3}{x+4} - \frac{x+2}{x-5}$$

confirms the three critical values found algebraically: -4 and 5 where $f(x)$ is not defined and $\tfrac{1}{2}$ where $f(x) = 0$.

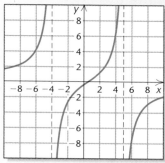

The graph shows where $f(x)$ is positive and where it is negative. Note that -4 and 5 cannot be in the solution set since y is not defined for these values. We do include $\tfrac{1}{2}$, however, since the inequality symbol is $\geq$ and $f\left(\tfrac{1}{2}\right) = 0$. The solution set is

$$(-\infty, -4) \cup \left[\tfrac{1}{2}, 5\right).$$

The following is a method for solving rational inequalities.

To solve a rational inequality:

1. Find an equivalent inequality with 0 on one side.

2. Change the inequality symbol to an equals sign and solve the related equation.

3. Find values of the variable for which the related rational function is not defined.

4. The numbers found in steps (2) and (3) are called critical values. Use the critical values to divide the x-axis into intervals. Then test an x-value from each interval to determine the function's sign in that interval.

5. Select the intervals for which the inequality is satisfied and write interval notation or set-builder notation for the solution set. If the inequality symbol is $\leq$ or $\geq$, then the solutions to step (2) should be included in the solution set. The x-values found in step (3) are never included in the solution set.

Exercise Set 3.5

In Exercises 1–4, a related function is graphed. Solve the given inequality.

1. $x^3 + 6x^2 < x + 30$

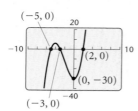

2. $x^4 - 27x^2 - 14x + 120 \geq 0$

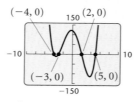

3. $\dfrac{8x}{x^2 - 4} \geq 0$

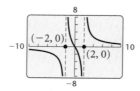

4. $\dfrac{8}{x^2 - 4} < 0$

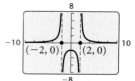

Solve.

5. $(x - 1)(x + 4) < 0$

6. $(x + 3)(x - 5) < 0$

7. $(x - 4)(x + 2) \geq 0$

8. $(x - 2)(x + 1) \geq 0$

9. $x^2 + x - 2 > 0$

10. $x^2 - x - 6 > 0$

11. $x^2 > 25$

12. $x^2 \leq 1$

13. $4 - x^2 \leq 0$

14. $11 - x^2 \geq 0$

15. $6x - 9 - x^2 < 0$

16. $x^2 + 2x + 1 \leq 0$

17. $x^2 + 12 < 4x$

18. $x^2 - 8 > 6x$

19. $4x^3 - 7x^2 \leq 15x$

20. $2x^3 - x^2 < 5x$

21. $x^3 + 3x^2 - x - 3 \geq 0$

22. $x^3 + x^2 - 4x - 4 \geq 0$

23. $x^3 - 2x^2 < 5x - 6$

24. $x^3 + x \leq 6 - 4x^2$

25. $x^5 + x^2 \geq 2x^3 + 2$

26. $x^5 + 24 > 3x^3 + 8x^2$

27. $2x^3 + 6 \leq 5x^2 + x$

28. $2x^3 + x^2 < 10 + 11x$

29. $x^3 + 5x^2 - 25x \leq 125$

30. $x^3 - 9x + 27 \geq 3x^2$

31. $\dfrac{1}{x + 4} > 0$

32. $\dfrac{1}{x - 3} \leq 0$

33. $\dfrac{-4}{2x + 5} < 0$

34. $\dfrac{-2}{5 - x} \geq 0$

35. $\dfrac{x - 4}{x + 3} - \dfrac{x + 2}{x - 1} \leq 0$

36. $\dfrac{x + 1}{x - 2} + \dfrac{x - 3}{x - 1} < 0$

37. $\dfrac{2x - 1}{x + 3} \geq \dfrac{x + 1}{3x + 1}$

38. $\dfrac{x + 5}{x - 4} > \dfrac{3x + 2}{2x + 1}$

39. $\dfrac{x + 1}{x - 2} \geq 3$

40. $\dfrac{x}{x - 5} < 2$

41. $x - 2 > \dfrac{1}{x}$

42. $4 \geq \dfrac{4}{x} + x$

43. $\dfrac{2}{x^2 - 4x + 3} \leq \dfrac{5}{x^2 - 9}$

44. $\dfrac{3}{x^2 - 4} \leq \dfrac{5}{x^2 + 7x + 10}$

45. $\dfrac{3}{x^2 + 1} \geq \dfrac{6}{5x^2 + 2}$

46. $\dfrac{4}{x^2 - 9} < \dfrac{3}{x^2 - 25}$

47. $\dfrac{5}{x^2 + 3x} < \dfrac{3}{2x + 1}$

48. $\dfrac{2}{x^2 + 3} > \dfrac{3}{5 + 4x^2}$

49. $\dfrac{5x}{7x - 2} > \dfrac{x}{x + 1}$

50. $\dfrac{x^2 - x - 2}{x^2 + 5x + 6} < 0$

51. $\dfrac{x}{x^2 + 4x - 5} + \dfrac{3}{x^2 - 25} \leq \dfrac{2x}{x^2 - 6x + 5}$

52. $\dfrac{2x}{x^2 - 9} + \dfrac{x}{x^2 + x - 12} \geq \dfrac{3x}{x^2 + 7x + 12}$

53. *Temperature During an Illness.* The temperature T, in degrees Fahrenheit, of a person during an illness is given by the function

$$T(t) = \frac{4t}{t^2 + 1} + 98.6,$$

where t is the time, in hours. Find the interval on which the temperature was over 100°. (See Example 11 in Section 3.4.)

54. *Population Growth.* The population P, in thousands, of Lordsburg is given by

$$P(t) = \frac{500t}{2t^2 + 9},$$

where t is the time, in months. Find the interval on which the population was 40 thousand or greater. (See Exercise 49 in Exercise Set 3.4.)

55. *Total Profit.* Flexl, Inc., determines that its total profit is given by the function

$$P(x) = -3x^2 + 630x - 6000.$$

a) Flexl makes a profit for those nonnegative values of x for which $P(x) > 0$. Find the values of x for which Flexl makes a profit.

b) Flexl loses money for those nonnegative values of x for which $P(x) < 0$. Find the values of x for which Flexl loses money.

56. *Height of a Thrown Object.*   The function

$$S(t) = -16t^2 + 32t + 1920$$

gives the height S, in feet, of an object thrown from a cliff that is 1920 ft high. Here t is the time, in seconds, that the object is in the air.

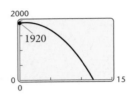

a) For what times is the height greater than 1920 ft?
b) For what times is the height less than 640 ft?

57. *Number of Diagonals.*   A polygon with n sides has D diagonals, where D is given by the function

$$D(n) = \frac{n(n-3)}{2}.$$

Find the number of sides n if

$$27 \le D \le 230.$$

58. *Number of Handshakes.*   There are n people in a room. The number N of possible handshakes by all the people in the room is given by the function

$$N(n) = \frac{n(n-1)}{2}.$$

For what number n of people is

$$66 \le N \le 300?$$

Technology Connection

59. Use a graphing calculator and the ZERO feature to check your answers to Exercises 5–52.

Collaborative Discussion and Writing

60. Why, when solving rational inequalities, do we need to find values for which the function is undefined as well as zeros of the function?

61. Under what circumstances would a quadratic inequality have a solution set that is a closed interval? Under what circumstances would a quadratic inequality have an empty solution set?

Skill Maintenance

Find an equation for a circle satisfying the given conditions.

62. Center: $(0, -3)$; diameter of length $\frac{7}{2}$

63. Center: $(-2, 4)$; radius of length 3

In Exercises 64 and 65:

a) *Find the vertex.*
b) *Determine whether there is a maximum or minimum value and find that value.*
c) *Find the range.*

64. $g(x) = x^2 - 10x + 2$

65. $h(x) = -2x^2 + 3x - 8$

Synthesis

Solve.

66. $x^4 - 6x^2 + 5 > 0$

67. $x^2 + 9 \le 6x$

68. $\left| \dfrac{x+3}{x-4} \right| < 2$

69. $x^4 + 3x^2 > 4x - 15$

70. $(7 - x)^{-2} < 0$

71. $|x^2 - 5| = 5 - x^2$

72. $|x|^2 - 4|x| + 4 \ge 9$

73. $2|x|^2 - |x| + 2 \le 5$

74. $\left| 2 - \dfrac{1}{x} \right| \le 2 + \left| \dfrac{1}{x} \right|$

75. $\left| 1 + \dfrac{1}{x} \right| < 3$

76. $|1 + 5x - x^2| \ge 5$

77. $|x^2 + 3x - 1| < 3$

78. Write a polynomial inequality for which the solution set is $[-4, 3] \cup [7, \infty)$.

79. Write a quadratic inequality for which the solution set is $(-4, 3)$.

3.6

Variation and Applications

- *Find equations of direct, inverse, and combined variation given values of the variables.*
- *Solve applied problems involving variation.*

We now extend our study of formulas and functions by considering applications involving variation.

Direct Variation

An executive chef earns $23 per hour. In 1 hr, $23 is earned; in 2 hr, $46 is earned; in 3 hr, $69 is earned; and so on. This gives rise to a set of ordered pairs of numbers:

$$(1, 23), \quad (2, 46), \quad (3, 69), \quad (4, 92),$$

and so on. Note that the ratio of the second coordinate to the first is the same number for each pair:

$$\frac{23}{1} = 23, \quad \frac{46}{2} = 23, \quad \frac{69}{3} = 23, \quad \frac{92}{4} = 23, \quad \text{and so on.}$$

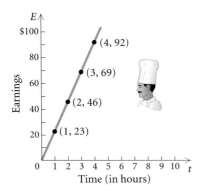

Whenever a situation produces pairs of numbers in which the *ratio is constant*, we say that there is **direct variation.** Here the amount earned E varies directly as the time worked t:

$$\frac{E}{t} = 23 \text{ (a constant)}, \quad \text{or} \quad E = 23t,$$

or, using function notation, $E(t) = 23t$. This equation is an equation of **direct variation.** The coefficient, 23 in the situation above, is called the **variation constant.** In this case, it is the rate of change of earnings with respect to time.

The graph of $y = kx$, $k > 0$, always goes through the origin and rises from left to right. Note that as x increases, y increases; that is, the function is increasing on the interval $(0, \infty)$. The constant k is also the slope of the line.

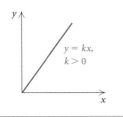

Direct Variation

If a situation gives rise to a linear function $f(x) = kx$, or $y = kx$, where k is a positive constant, we say that we have **direct variation,** or that **y varies directly as x,** or that **y is directly proportional to x.** The number k is called the **variation constant,** or **constant of proportionality.**

EXAMPLE 1 Find the variation constant and an equation of variation in which y varies directly as x, and $y = 32$ when $x = 2$.

Solution   We know that (2, 32) is a solution of $y = kx$. Thus,

$$y = kx$$
$$32 = k \cdot 2 \qquad \text{Substituting}$$
$$\frac{32}{2} = k, \text{ or } k = 16. \qquad \text{Solving for } k$$

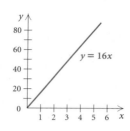

The variation constant, 16, is the rate of change of y with respect to x. The equation of variation is $y = 16x$.

EXAMPLE 2   *Water from Melting Snow.* The number of centimeters W of water produced from melting snow varies directly as S, the number of centimeters of snow. Meteorologists have found that 150 cm of snow will melt to 16.8 cm of water. To how many centimeters of water will 200 cm of snow melt?

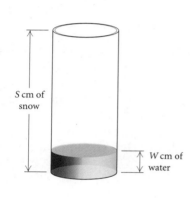

Solution   We can express the amount of water as a function of the amount of snow. Thus, $W(S) = kS$, where k is the variation constant. We first find k using the given data and then find an equation of variation:

$$W(S) = kS \qquad \text{W varies directly as S.}$$
$$W(150) = k \cdot 150 \qquad \text{Substituting 150 for S}$$
$$16.8 = k \cdot 150 \qquad \text{Replacing $W(150)$ with 16.8}$$
$$\frac{16.8}{150} = k \qquad \text{Solving for k}$$
$$0.112 = k. \qquad \text{This is the variation constant.}$$

The equation of variation is $W(S) = 0.112S$.

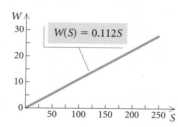

Next, we use the equation to find how many centimeters of water will result from melting 200 cm of snow:

$$W(S) = 0.112S$$
$$W(200) = 0.112(200) \qquad \text{Substituting}$$
$$= 22.4.$$

Thus, 200 cm of snow will melt to 22.4 cm of water.

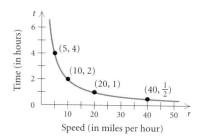

Speed (in miles per hour)

Inverse Variation

A bus is traveling a distance of 20 mi. At a speed of 5 mph, the trip will take 4 hr; at 10 mph, it will take 2 hr; at 20 mph, it will take 1 hr; at 40 mph, it will take $\frac{1}{2}$ hr; and so on. We plot this information on a graph, using speed as the first coordinate and time as the second coordinate to determine a set of ordered pairs:

$$(5, 4), \quad (10, 2), \quad (20, 1), \quad \left(40, \tfrac{1}{2}\right), \quad \text{and so on.}$$

Note that the products of the coordinates are all the same number:

$$5 \cdot 4 = 20, \quad 10 \cdot 2 = 20, \quad 20 \cdot 1 = 20, \quad 40 \cdot \tfrac{1}{2} = 20, \quad \text{and so on.}$$

Whenever a situation produces pairs of numbers in which the *product is constant*, we say that there is **inverse variation.** Here the time varies inversely as the speed:

$$rt = 20 \text{ (a constant)}, \quad \text{or} \quad t = \frac{20}{r},$$

or, using function notation, $t(r) = 20/r$. This equation is an equation of **inverse variation.** The coefficient, 20 in the situation above, is called the **variation constant.** Note that as the first number increases, the second number decreases.

It is helpful to look at the graph of $y = k/x$, $k > 0$. The graph is like the one shown below for positive values of x. Note that as x increases, y decreases; that is, the function is decreasing on the interval $(0, \infty)$.

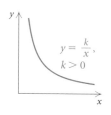

Inverse Variation

If a situation gives rise to a function $f(x) = k/x$, or $y = k/x$, where k is a positive constant, we say that we have **inverse variation,** or that **y varies inversely as x,** or that **y is inversely proportional to x.** The number k is called the **variation constant,** or **constant of proportionality.**

EXAMPLE 3   Find the variation constant and an equation of variation in which y varies inversely as x, and $y = 16$ when $x = 0.3$.

Solution   We know that $(0.3, 16)$ is a solution of $y = k/x$. We substitute:

$$y = \frac{k}{x}$$

$$16 = \frac{k}{0.3} \qquad \text{Substituting}$$

$$(0.3)16 = k \qquad \text{Solving for } k$$

$$4.8 = k.$$

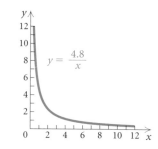

The variation constant is 4.8. The equation of variation is $y = 4.8/x$.

There are many problems that translate to an equation of inverse variation.

EXAMPLE 4 *Building a House.* The time t required to do a job varies inversely as the number of people P who work on the job (assuming that all work at the same rate). If it takes 32 hr for 30 people to build a house, how long will it take 20 people to complete the same job?

Solution We can express the amount of time required, in hours, as a function of the number of people working. Thus we have $t(P) = k/P$. We first find k using the given information and then find an equation of variation:

$$t(P) = \frac{k}{P} \qquad t \text{ varies inversely as } P.$$

$$t(30) = \frac{k}{30} \qquad \text{Substituting 30 for } P$$

$$32 = \frac{k}{30} \qquad \text{Replacing } t(30) \text{ with 32}$$

$$30 \cdot 32 = k \qquad \text{Solving for } k$$

$$960 = k. \qquad \text{This is the variation constant.}$$

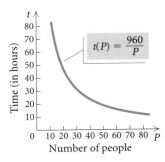

The equation of variation is $t(P) = 960/P$.

Next, we use the equation to find the time that it would take 20 people to do the job. We compute $t(20)$:

$$t(P) = \frac{960}{P} \qquad t \text{ varies inversely as } P.$$

$$t(20) = \frac{960}{20} \qquad \text{Substituting}$$

$$t = 48.$$

Thus it would take 48 hr for 20 people to complete the job. ▬

Combined Variation

We now look at other kinds of variation.

y varies **directly as the *n*th power of *x*** if there is some positive constant k such that

$$y = kx^n.$$

y varies **inversely as the *n*th power of *x*** if there is some positive constant k such that

$$y = \frac{k}{x^n}.$$

y varies **jointly as *x* and *z*** if there is some positive constant k such that

$$y = kxz.$$

There are other types of combined variation as well. Consider the formula $V = \pi r^2 h$, in which V, r, and h are variables and π is a constant. We say that V varies jointly as h and the square of r.

EXAMPLE 5 Find an equation of variation in which y varies directly as the square of x, and $y = 12$ when $x = 2$.

Solution We write an equation of variation and find k:

$$y = kx^2$$
$$12 = k \cdot 2^2 \quad \text{Substituting}$$
$$12 = k \cdot 4$$
$$3 = k.$$

Thus, $y = 3x^2$.

EXAMPLE 6 Find an equation of variation in which y varies jointly as x and z, and $y = 42$ when $x = 2$ and $z = 3$.

Solution We have

$$y = kxz$$
$$42 = k \cdot 2 \cdot 3 \quad \text{Substituting}$$
$$42 = k \cdot 6$$
$$7 = k.$$

Thus, $y = 7xz$.

EXAMPLE 7 Find an equation of variation in which y varies jointly as x and z and inversely as the square of w, and $y = 105$ when $x = 3$, $z = 20$, and $w = 2$.

Solution We have

$$y = k \cdot \frac{xz}{w^2}$$
$$105 = k \cdot \frac{3 \cdot 20}{2^2} \quad \text{Substituting}$$
$$105 = k \cdot 15$$
$$7 = k.$$

Thus, $y = 7\frac{xz}{w^2}$.

Many applied problems can be modeled using equations of combined variation.

EXAMPLE 8 *Volume of a Tree.* The volume of wood V in a tree varies jointly as the height h and the square of the girth g (girth is distance around). If the volume of a redwood tree is 216 m^3 when the height is 30 m and the girth is 1.5 m, what is the height of a tree whose volume is 960 m^3 and girth is 2 m?

Solution   We first find k using the first set of data. Then we solve for h using the second set of data.

$$V = khg^2$$
$$216 = k \cdot 30 \cdot 1.5^2$$
$$216 = k \cdot 30 \cdot 2.25$$
$$216 = k \cdot 67.5$$
$$3.2 = k$$

Then the equation of variation is $V = 3.2hg^2$. We substitute the second set of data into the equation:

$$960 = 3.2 \cdot h \cdot 2^2$$
$$960 = 12.8 \cdot h$$
$$75 = h.$$

Thus the height of the tree is 75 m.   ▬

Exercise Set   3.6

Find the variation constant and an equation of variation for the given situation.

1. y varies directly as x, and $y = 54$ when $x = 12$

2. y varies directly as x, and $y = 0.1$ when $x = 0.2$

3. y varies inversely as x, and $y = 3$ when $x = 12$

4. y varies inversely as x, and $y = 12$ when $x = 5$

5. y varies directly as x, and $y = 1$ when $x = \frac{1}{4}$

6. y varies inversely as x, and $y = 0.1$ when $x = 0.5$

7. y varies inversely as x, and $y = 32$ when $x = \frac{1}{8}$

8. y varies directly as x, and $y = 3$ when $x = 33$

9. y varies directly as x, and $y = \frac{3}{4}$ when $x = 2$

10. y varies inversely as x, and $y = \frac{1}{5}$ when $x = 35$

11. y varies inversely as x, and $y = 1.8$ when $x = 0.3$

12. y varies directly as x, and $y = 0.9$ when $x = 0.4$

13. *Work Rate.*   The time T required to do a job varies inversely as the number of people P working. It takes 5 hr for 7 bricklayers to build a park wall (see

the graph below). How long will it take 10 bricklayers to complete the job?

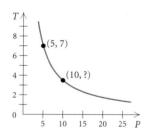

14. *Weekly Allowance.*   According to Fidelity Investments *Investment Vision Magazine*, the average weekly allowance A of children varies directly as their grade level G. It is known that the average allowance of a 9th-grade student is $9.66

per week. What then is the average allowance of a 4th-grade student?

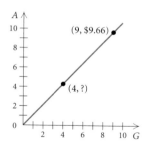

15. *Fat Intake.* The maximum number of grams of fat that should be in a diet varies directly as a person's weight. A person weighing 120 lb should have no more than 60 g of fat per day. What is the maximum daily fat intake for a person weighing 180 lb?

16. *Rate of Travel.* The time t required to drive a fixed distance varies inversely as the speed r. It takes 5 hr at a speed of 80 km/h to drive a fixed distance. How long will it take to drive the same distance at a speed of 70 km/h?

17. *Beam Weight.* The weight W that a horizontal beam can support varies inversely as the length L of the beam. Suppose an 8-m beam can support 1200 kg. How many kilograms can a 14-m beam support?

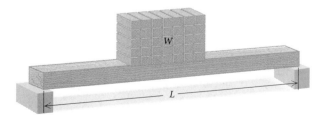

18. *House of Representatives.* The number of representatives N that each state has varies directly as the number of people P living in the state. If New York, with 18,137,226 residents, has 31 representatives, how many representatives does Colorado, with a population of 3,892,644, have?

19. *Weight on Mars.* The weight M of an object on Mars varies directly as its weight E on Earth. A person who weighs 95 lb on Earth weighs 38 lb on Mars. How much would a 100-lb person weigh on Mars?

20. *Pumping Rate.* The time t required to empty a tank varies inversely as the rate r of pumping. If a pump can empty a tank in 45 min at the rate of 600 kL/min, how long will it take the pump to empty the same tank at the rate of 1000 kL/min?

21. *Hooke's Law.* Hooke's law states that the distance d that a spring will stretch varies directly as the mass m of an object hanging from the spring. If a 3-kg mass stretches a spring 40 cm, how far will a 5-kg mass stretch the spring?

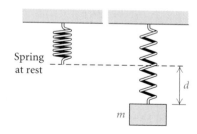

22. *Relative Aperture.* The relative aperture, or f-stop, of a 23.5-mm diameter lens is directly proportional to the focal length F of the lens. If a 150-mm focal length has an f-stop of 6.3, find the f-stop of a 23.5-mm diameter lens with a focal length of 80 mm.

23. *Musical Pitch.* The pitch P of a musical tone varies inversely as its wavelength W. One tone has a pitch of 330 vibrations per second and a wavelength of 3.2 ft. Find the wavelength of another tone that has a pitch of 550 vibrations per second.

24. *Lead Pollution.* The average U.S. community of population 12,500 released about 385 tons of lead into the environment in a recent year (*Source: Conservation Matters,* Autumn 1995 issue. Boston: Conservation Law Foundation, p. 30). How many tons were released nationally? Use 250,000,000 as the U.S. population.

Find an equation of variation for the given situation.

25. y varies inversely as the square of x, and $y = 0.15$ when $x = 0.1$

26. y varies inversely as the square of x, and $y = 6$ when $x = 3$

27. y varies directly as the square of x, and $y = 0.15$ when $x = 0.1$

28. y varies directly as the square of x, and $y = 6$ when $x = 3$

29. y varies jointly as x and z, and $y = 56$ when $x = 7$ and $z = 8$

30. y varies directly as x and inversely as z, and $y = 4$ when $x = 12$ and $z = 15$

31. y varies jointly as x and the square of z, and $y = 105$ when $x = 14$ and $z = 5$

32. y varies jointly as x and z and inversely as w, and $y = \frac{3}{2}$ when $x = 2$, $z = 3$, and $w = 4$

33. y varies jointly as x and z and inversely as the product of w and p, and $y = \frac{3}{28}$ when $x = 3$, $z = 10$, $w = 7$, and $p = 8$

34. y varies jointly as x and z and inversely as the square of w, and $y = \frac{12}{5}$ when $x = 16$, $z = 3$, and $w = 5$

35. *Intensity of Light.* The intensity I of light from a light bulb varies inversely as the square of the distance d from the bulb. Suppose that I is 90 W/m^2 (watts per square meter) when the distance is 5 m. How much *farther* would it be to a point where the intensity is 40 W/m^2?

36. *Atmospheric Drag.* Wind resistance, or atmospheric drag, tends to slow down moving objects. Atmospheric drag varies jointly as an object's surface area A and velocity v. If a car traveling at a speed of 40 mph with a surface area of 37.8 ft^2 experiences a drag of 222 N (Newtons), how fast must a car with 51 ft^2 of surface area travel in order to experience a drag force of 430 N?

37. *Stopping Distance of a Car.* The stopping distance d of a car after the brakes have been applied varies directly as the square of the speed r. If a car traveling 60 mph can stop in 200 ft, how fast can a car travel and still stop in 72 ft?

38. *Weight of an Astronaut.* The weight W of an object varies inversely as the square of the distance d from the center of the earth. At sea level (3978 mi from the center of the earth), an astronaut weighs 220 lb. Find his weight when he is 200 mi above the surface of the earth and the spacecraft is not in motion.

39. *Earned-Run Average.* A pitcher's earned-run average E varies directly as the number R of earned runs allowed and inversely as the number I of innings pitched. In a recent year, Shawn Estes of the San Francisco Giants had an earned-run average of 3.18. He gave up 71 earned runs in 201 innings. How many earned runs would he have given up had he pitched 300 innings with the same average? Round to the nearest whole number.

40. *Boyle's Law.* The volume V of a given mass of a gas varies directly as the temperature T and inversely as the pressure P. If $V = 231$ cm^3 when $T = 42°$ and $P = 20$ kg/cm^2, what is the volume when $T = 30°$ and $P = 15$ kg/cm^2?

Collaborative Discussion and Writing

41. If y varies directly as x^2, explain why doubling x would not cause y to be doubled as well.

42. If y varies directly as x and x varies inversely as z, how does y vary with regard to z? Why?

Skill Maintenance

43. Graph: $f(x) = \begin{cases} x - 2, & \text{for } x \le -1, \\ 3, & \text{for } -1 < x \le 2, \\ x, & \text{for } x > 2. \end{cases}$

Determine algebraically whether the graph is symmetric with respect to the x-axis, the y-axis, and the origin.

44. $y = 3x^4 - 3$

45. $y^2 = x$

46. $2x - 5y = 0$

Synthesis

47. *Volume and Cost.* A peanut butter jar in the shape of a right circular cylinder is 4 in. high and 3 in. in diameter and sells for $1.20. If we assume that cost is directly proportional to volume, how much should a jar 6 in. high and 6 in. in diameter cost?

48. In each of the following equations, state whether y varies directly as x, inversely as x, or neither directly nor inversely as x.

a) $7xy = 14$

b) $x - 2y = 12$

c) $-2x + 3y = 0$

d) $x = \frac{3}{4}y$

e) $\frac{x}{y} = 2$

49. *Area of a Circle.* The area of a circle varies directly as the square of the length of a diameter. What is the variation constant?

50. Describe in words the variation given by the equation

$$Q = \frac{kp^2}{q^3}.$$

3

Chapter Summary and Review

Important Properties and Formulas

Polynomial Function:

$$P(x) = a_n x^n + a_{n-1} x^{n-1} + a_{n-2} x^{n-2} + \cdots + a_1 x + a_0$$

The Leading-Term Test: If $a_n x^n$ is the leading term of a polynomial, then the behavior of the graph as $x \to \infty$ and as $x \to -\infty$ can be described in one of the four following ways.

If n is even, and $a_n > 0$:

If n is even, and $a_n < 0$:

If n is odd, and $a_n > 0$:

If n is odd, and $a_n < 0$:

The Intermediate Value Theorem: For any polynomial function $P(x)$ with real coefficients, suppose that $a \neq b$ and that $P(a)$ and $P(b)$ are of opposite signs. Then the polynomial has a real zero between a and b.

Polynomial Division:

$$P(x) = D(x) \cdot Q(x) + R(x)$$

Dividend   Divisor   Quotient   Remainder

The Remainder Theorem: The remainder found by dividing $P(x)$ by $x - c$ is $P(c)$.

The Factor Theorem: $P(c) = 0 \leftrightarrow x - c$ is a factor of $P(x)$.

The Fundamental Theorem of Algebra: Every polynomial of degree n, $n \geq 1$, with complex coefficients has at least one complex-number zero.

The Rational Zeros Theorem: Consider the polynomial equation

$$a_n x^n + a_{n-1} x^{n-1} + a_{n-2} x^{n-2} + \cdots + a_1 x + a_0 = 0,$$

where all the coefficients are integers and $n \geq 1$. Also, consider a rational number p/q, where p and q have no common factor other than -1 and 1. If p/q is a solution of the polynomial equation, then p is a factor of a_0 and q is a factor of a_n.

(continued)

Descartes' Rule of Signs

Let $P(x)$ be a polynomial function with real coefficients and a nonzero constant term. The number of positive real zeros of $P(x)$ is either:

1. The same as the number of variations of sign in $P(x)$, or
2. Less than the number of variations of sign in $P(x)$ by a positive even integer.

The number of negative real zeros of $P(x)$ is either:

3. The same as the number of variations of sign in $P(-x)$, or
4. Less than the number of variations of sign in $P(-x)$ by a positive even integer.

A zero of multiplicity m must be counted m times.

Rational Function:

$$f(x) = \frac{p(x)}{q(x)},$$

where $p(x)$ and $q(x)$ are polynomials and where $q(x)$ is not the zero polynomial, and the domain of $f(x)$ consists of all x for which $q(x) \neq 0$.

Occurrence of Lines as Asymptotes

For a rational function $p(x)/q(x)$, where $p(x)$ and $q(x)$ have no common factors other than constants:

Vertical asymptotes occur at any x-values that make the denominator 0.

The x-axis is the horizontal asymptote when the degree of the numerator is less than the degree of the denominator.

A horizontal asymptote other than the x-axis occurs when the numerator and the denominator have the same degree.

An oblique asymptote occurs when the degree of the numerator is 1 greater than the degree of the denominator.

Variation

Direct: $y = kx$

Inverse: $y = \dfrac{k}{x}$

Joint: $y = kxz$

REVIEW EXERCISES

Classify the polynomial as constant, linear, quadratic, cubic, or quartic and determine the leading term, the leading coefficient, and the degree of the polynomial.

1. $f(x) = 7x^2 - 5 + 0.45x^4 - 3x^3$
2. $h(x) = -25$
3. $g(x) = 6 - 0.5x$
4. $f(x) = \frac{1}{3}x^3 - 2x + 3$

Using the intermediate value theorem, determine whether the function f has a zero between a and b.

5. $f(x) = 4x^2 - 5x - 3; \ a = 1, b = 2$

6. $f(x) = x^3 - 4x^2 + \frac{1}{2}x + 2; \ a = 0, b = 1$

7. *Interest Compounded Annually.* When P dollars is invested at interest rate i, compounded annually, for t years, the investment grows to A dollars, where

$$A = P(1 + i)^t.$$

   a) Find the interest rate i if \$6250 grows to \$6760 in 2 yr.
   b) Find the interest rate i if \$1,000,000 grows to \$1,215,506.25 in 4 yr.

Do the long division. Check by multiplying.

8. $(20x^3 - 6x^2 - 9x + 10) \div (5x + 2)$
9. $(2x^5 - 3x^3 + x^2 + 4) \div (x^2 - 2)$

In Exercises 10 and 11, a polynomial equation is given along with information about factors. Use the information to solve the equation.

10. $x^3 + 2x^2 - 13x + 10 = 0$;
$x + 5$ is a factor of $x^3 + 2x^2 - 13x + 10$

11. $x^4 + 5x^3 - 23x^2 - 87x + 140 = 0$;
$x^2 + x - 20$ is a factor of
$x^4 + 5x^3 - 23x^2 - 87x + 140$

Use synthetic division to find the quotient and the remainder.

12. $(x^3 + 2x^2 - 13x + 10) \div (x - 5)$

13. $(x^4 + 3x^3 + 3x^2 + 3x + 2) \div (x + 2)$

Use synthetic division to find the indicated function value.

14. $P(x) = x^3 + 2x^2 - 13x + 10$; $P(5)$

15. $P(x) = x^4 - 16$; $P(-2)$

Determine whether the given values are zeros of $P(x)$. Then factor and find any other zeros that exist.

16. $P(x) = x^3 + 7x^2 + 7x - 15$; -1, 2, and -3

17. $P(x) = x^4 + 9x^3 + 19x^2 + 9x - 2$; 1, -4, and -2

Find a polynomial equation of lowest degree with integer coefficients and the following as some of its zeros.

18. 4, -1, -2

19. -4, $-\sqrt{3}$, $\sqrt{3}$, 1

20. 3, $2 - 3i$, $4 + \sqrt{5}$, -2

Use the rational zeros theorem and/or factoring to solve the equation.

21. $x^3 - 4x^2 + x + 6 = 0$

22. $x^3 + 7x^2 + 7x = 15$

23. $x^4 + 36x + 63 = 4x^3 + 16x^2$

In Exercises 24 and 25, for the polynomial function:
a) Solve $P(x) = 0$.
b) Express $P(x)$ as a product of linear factors.

24. $P(x) = x^3 + 8x^2 - 19x + 10$

25. $P(x) = x^6 + x^5 - 28x^4 - 16x^3 + 192x^2$

What does Descartes' rule of signs tell you about the number of positive real zeros and the number of negative real zeros of each of the following polynomial functions?

26. $f(x) = 2x^6 - 7x^3 + x^2 - x$

27. $h(x) = -x^8 + 6x^5 - x^3 + 2x - 2$

28. $g(x) = 5x^5 - 4x^2 + x - 1$

Graph each of the following. Be sure to label all the asymptotes.

29. $f(x) = \dfrac{x^2 - 5}{x + 2}$

30. $f(x) = \dfrac{5}{(x - 2)^2}$

31. $f(x) = \dfrac{x^2 + x - 6}{x^2 - x - 20}$

32. $f(x) = \dfrac{x - 2}{x^2 - 2x - 15}$

In Exercises 33 and 34, find a rational function that satisfies the given conditions. Answers may vary, but try to give the simplest answer possible.

33. Vertical asymptotes $x = -2$, $x = 3$

34. Vertical asymptotes $x = -2$, $x = 3$; horizontal asymptote $y = 4$; x-intercept $(-3, 0)$

35. *Medical Dosage.* The function

$$N(t) = \frac{0.7t + 2000}{8t + 9}, \quad t \geq 5$$

gives the body concentration $N(t)$, in parts per million, of a certain dosage of medication after time t, in hours.

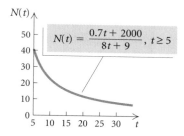

a) Find the horizontal asymptote of the graph and complete the following:

$$N(t) \rightarrow \boxed{} \text{ as } t \rightarrow \infty.$$

b) Explain the meaning of the answer to part (a) in terms of the application.

Solve.

36. $x^2 - 9 < 0$

37. $2x^2 > 3x + 2$

38. $(1 - x)(x + 4)(x - 2) \leq 0$

39. $\dfrac{x - 2}{x + 3} < 4$

40. $\dfrac{x - 3}{x^2 + x - 20} \geq \dfrac{4}{x^2 - 4}$

41. *Height of a Rocket.* The function
$$S(t) = -16t^2 + 80t + 224$$
gives the height S, in feet, of a model rocket launched with a velocity of 80 ft/sec from a hill that is 224 ft high, where t is the time, in seconds.

a) Determine when the rocket reaches the ground.
b) On what interval is the height greater than 320 ft?

42. *Population Growth.* The population P, in thousands, of Novi is given by
$$P(t) = \frac{8000t}{4t^2 + 10},$$
where t is the time, in months. Find the interval on which the population was 400,000 or greater.

43. Find an equation of variation in which y varies directly as x, and $y = 100$ when $x = 25$.

44. Find an equation of variation in which y varies inversely as x, and $y = 100$ when $x = 25$.

45. *Pumping Time.* The time t required to empty a tank varies inversely as the rate r of pumping. If a pump can empty a tank in 35 min at the rate of 800 kL per minute, how long will it take the pump to empty the same tank at the rate of 1400 kL per minute?

46. *Power of Electric Current.* The power P expended by heat in an electric circuit of fixed resistance varies directly as the square of the current C in the circuit. A circuit expends 180 watts when a current of 6 amperes is flowing. What is the amount of heat expended when the current is 10 amperes?

47. *Test Score.* The score N on a test varies directly as the number of correct responses a. Ellen answers 28 questions correctly and earns a score of 87. What would Ellen's score have been if she had answered 25 questions correctly?

Technology Connection

Use a graphing calculator to graph the polynomial function. Then estimate the function's (a) zeros; (b) relative maxima; (c) relative minima; (d) domain and range.

48. $f(x) = -2x^2 - 3x + 6$

49. $f(x) = x^3 + 3x^2 - 2x - 6$

50. $f(x) = x^4 - 3x^3 + 2x^2$

51. *Cholesterol Level and the Risk of Heart Attack.* The data in the following table show the relationship of cholesterol level in men to the risk of a heart attack.

CHOLESTEROL LEVEL	NUMBER OF MEN, PER 10,000, WHO SUFFER A HEART ATTACK
100	30
200	65
250	100
275	130

Source: Nutrition Action Healthletter

a) Use regression on a graphing calculator to fit linear, quadratic, and cubic functions to the data.
b) It is also known that 180 of 10,000 men have a heart attack with a cholesterol level of 300. Which function in part (a) would best make this prediction?
c) Use the answer to part (b) to predict the heart attack rate for men with cholesterol levels of 350 and of 400.

Collaborative Discussion and Writing

52. Explain the difference between a polynomial function and a rational function.

53. Explain and contrast the three types of asymptotes considered for rational functions.

Synthesis

54. *Interest Rate.* In early 2000, $3500 was deposited at a certain interest rate. One year later, $4000 was deposited in another account at the same rate. At the end of that year, there was a total of $8518.35 in both accounts. What is the annual interest rate?

Solve.

55. $x^2 \geq 5 - 2x$

56. $\left| 1 - \dfrac{1}{x^2} \right| < 3$

57. $x^4 - 2x^3 + 3x^2 - 2x + 2 = 0$

58. $(x - 2)^{-3} < 0$

59. Express $x^3 - 1$ as a product of linear factors.

60. Find k such that $x + 3$ is a factor of $x^3 + kx^2 + kx - 15$.

61. When $x^2 - 4x + 3k$ is divided by $x + 5$, the remainder is 33. Find the value of k.

Find the domain of the function.

62. $f(x) = \sqrt{x^2 + 3x - 10}$

63. $f(x) = \sqrt{x^2 - 3.1x + 2.2} + 1.75$

64. $f(x) = \dfrac{1}{\sqrt{5 - |7x + 2|}}$

3 Chapter Test

Classify the polynomial as constant, linear, quadratic, cubic, or quartic and determine the leading term, the leading coefficient, and the degree of the polynomial.

1. $f(x) = 2x^3 + 6x^2 - x^4 + 11$

2. $h(x) = -4.7x + 29$

Using the intermediate value theorem, determine whether the function has a zero between a and b.

3. $f(x) = -5x^2 + 3$; $a = 0$, $b = 2$

4. $g(x) = 2x^3 + 6x^2 - 3$; $a = -2$, $b = -1$

5. *Interest Compounded Annually.* When P dollars is invested at interest rate i, compounded annually, for t years, the investment grows to A dollars, where
$$A = P(1 + i)^t.$$
Find the interest rate i if \$1500 grows to \$1858.24 in 4 yr.

6. Use long division to find the quotient and remainder. Show your work.
$$(x^4 + 3x^3 + 2x - 5) \div (x^2 - 1)$$

7. Use synthetic division to find the quotient and remainder. Show your work.
$$(3x^3 - 12x + 7) \div (x - 5)$$

8. Use synthetic division to determine whether -2 is a zero of $f(x) = x^3 + 4x^2 + x - 6$. Answer yes or no. Show your work.

9. Use synthetic division to find $P(-3)$ for $P(x) = 2x^3 - 6x^2 + x - 4$. Show your work.

10. Suppose that a polynomial function of degree 5 with rational coefficients has 1, $\sqrt{3}$, and $2 - i$ as zeros. Find the other zeros.

11. For the polynomial function
$$P(x) = x^4 + 2x^3 - 4x^2 - 5x + 6:$$
a) Solve $P(x) = 0$.
b) Express $P(x)$ as a product of linear factors.

12. What does Descartes' rule of signs tell you about the number of positive real zeros and the number of negative real zeros of the following function?
$$g(x) = -x^8 + 2x^6 - 4x^3 - 1$$

13. Graph
$$f(x) = \frac{2}{(x - 3)^2}.$$
Label all the asymptotes.

14. Find a rational function that has vertical asymptotes $x = -1$ and $x = 2$ and x-intercept $(-4, 0)$.

Solve.

15. $2x^2 > 5x + 3$

16. $\dfrac{x + 1}{x - 4} \le 3$

17. The function $S(t) = -16t^2 + 64t + 192$ gives the height S, in feet, of a model rocket launched with a velocity of 64 ft/sec from a hill that is 192 ft high.
a) Determine how long it will take the rocket to reach the ground.
b) Find the interval on which the height of the rocket is greater than 240 ft.

18. Find an equation of variation in which y varies inversely as x, and $y = 5$ when $x = 6$.

19. The stopping distance d of a car after the brakes have been applied varies directly as the square of the speed r. If a car traveling 60 mph can stop in 200 ft, how long will it take a car traveling 30 mph to stop?

Synthesis

20. Find the domain of $f(x) = \sqrt{x^2 + x - 12}$.

Exponential and Logarithmic Functions 4

*I*n this chapter, we will consider two kinds of closely related functions. The first type, *exponential functions,* has a variable in the exponent. Such functions have many applications to the growth of populations, commodities, and investments.

Recall that a function takes an input to an output. Suppose that we can reverse the process and take the output back to an input. That process produces the *inverse* of the original function. The inverses of exponential functions, called *logarithmic functions,* or *logarithm functions,* are also important in many applications such as earthquake magnitude, sound level, and chemical pH.

APPLICATION

The magnitude R, measured on the Richter scale, of an earthquake of intensity I is defined as

$$R = \log \frac{I}{I_0},$$

where I_0 is a minimum intensity used for comparison. The earthquake in Ahmedabad, India on January 26, 2001, had an intensity of $10^{7.9} \cdot I_0$. What is its magnitude on the Richter scale?

This problem appears as Example 12 in Section 4.3.

4.1

Composite and Inverse Functions

- Find the composition of two functions and the domain of the composition; decompose a function as a composition of two functions.
- Determine whether a function is one-to-one, and if it is, find a formula for its inverse.
- Simplify expressions of the type $(f \circ f^{-1})(x)$ and $(f^{-1} \circ f)(x)$.

The Composition of Functions

In the real world, it is not uncommon for a function's output to depend on some input that is itself an output of another function. For instance, the amount a person pays as state income tax usually depends on the amount of adjusted gross income on the person's federal tax return, which, in turn, depends on his or her annual earnings. Such functions are called **composite functions.**

To illustrate how composite functions work, suppose a chemistry student needs a formula to convert Fahrenheit temperatures to Kelvin units. The formula

$$c(t) = \tfrac{5}{9}(t - 32)$$

gives the Celsius temperature $c(t)$ that corresponds to the Fahrenheit temperature t. The formula

$$k(c) = c + 273$$

gives the Kelvin temperature $k(c)$ that corresponds to the Celsius temperature c. Thus, 50° Fahrenheit corresponds to

$$c(50) = \tfrac{5}{9}(50 - 32)$$
$$= \tfrac{5}{9}(18) = 10° \text{ Celsius}$$

and 10° Celsius corresponds to

$$k(10) = 10 + 273$$
$$= 283° \text{ Kelvin.}$$

We see that 50° Fahrenheit is the same as 283° Kelvin. This two-step procedure can be used to convert any Fahrenheit temperature to Kelvin units.

Technology Connection

With the TABLE feature, we can convert Fahrenheit temperatures, x, to Celsius temperatures, y_1, using

$$y_1 = \frac{5}{9}(x - 32).$$

We can also convert Celsius temperatures to Kelvin units, y_2, using

$$y_2 = y_1 + 273.$$

$$y_1 = \frac{5}{9}(x - 32), \quad y_2 = y_1 + 273$$

X	Y₁	Y₂
50	10	283
59	15	288
68	20	293
77	25	298
86	30	303
95	35	308
104	40	313

X = 50

	F°	C°	K°
Boiling point of water	212°	100°	373°
	50° →	10° →	283°
Freezing point of water	32°	0°	273°
Absolute zero	−460°	−273°	0°

A student making numerous conversions might look for a formula that converts directly from Fahrenheit to Kelvin. Such a formula can be found by substitution:

$$k(c(t)) = c(t) + 273$$

$$= \frac{5}{9}(t - 32) + 273 \qquad \text{Substituting}$$

$$= \frac{5t + 2297}{9}. \qquad \text{Simplifying}$$

Since the last equation expresses the Kelvin temperature as a new function, K, of the Fahrenheit temperature, t, we can write

$$K(t) = \frac{5t + 2297}{9},$$

where $K(t)$ is the Kelvin temperature corresponding to the Fahrenheit temperature, t. Here we have $K(t) = k(c(t))$. The new function K is called the **composition** of k and c and can be denoted $k \circ c$ (read "k composed with c," "the composition of k and c," or "k circle c").

Composition of Functions

The **composite function** $f \circ g$, the **composition** of f and g, is defined as

$$(f \circ g)(x) = f(g(x)),$$

where x is in the domain of g and $g(x)$ is in the domain of f.

EXAMPLE 1 Given that $f(x) = 2x - 5$ and $g(x) = x^2 - 3x + 8$, find each of the following.

a) $(f \circ g)(x)$ and $(g \circ f)(x)$ **b)** $(f \circ g)(7)$ and $(g \circ f)(7)$

Solution Consider each function separately:

$$f(x) = 2x - 5 \qquad \text{This function multiplies each input by 2 and subtracts 5.}$$

and

$$g(x) = x^2 - 3x + 8. \qquad \text{This function squares an input, subtracts 3 times the input from the result, and then adds 8.}$$

a) To find $(f \circ g)(x)$, we substitute $g(x)$ for x in the equation for $f(x)$:

$$(f \circ g)(x) = f(g(x)) = f(x^2 - 3x + 8)$$

$$\qquad\qquad \text{Substituting } x^2 - 3x + 8 \text{ for } g(x)$$

$$= 2(x^2 - 3x + 8) - 5$$

$$= 2x^2 - 6x + 16 - 5$$

$$= 2x^2 - 6x + 11.$$

Technology Connection

We can check our work in Example 1(b) using a graphing calculator. We enter the following on the equation-editor screen:

$$y_1 = 2x - 5$$

and

$$y_2 = x^2 - 3x + 8.$$

Then, on the home screen, we find $(f \circ g)(7)$ and $(g \circ f)(7)$ using the function notations $Y_1(Y_2(7))$ and $Y_2(Y_1(7))$, respectively.

$$y_1 = 2x - 5, \quad y_2 = x^2 - 3x + 8$$

Y1(Y2(7))	
	67
Y2(Y1(7))	
	62

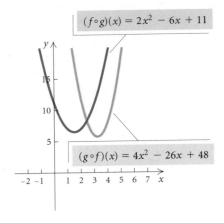

$(f \circ g)(x) = 2x^2 - 6x + 11$

$(g \circ f)(x) = 4x^2 - 26x + 48$

To find $(g \circ f)(x)$, we substitute $f(x)$ for x in the equation for $g(x)$:

$$(g \circ f)(x) = g(f(x)) = g(2x - 5) \qquad \text{Substituting } 2x - 5 \text{ for } f(x)$$
$$= (2x - 5)^2 - 3(2x - 5) + 8$$
$$= 4x^2 - 20x + 25 - 6x + 15 + 8$$
$$= 4x^2 - 26x + 48.$$

b) To find $(f \circ g)(7)$, we first find $g(7)$. Then we use $g(7)$ as an input for f:

$$(f \circ g)(7) = f(g(7)) = f(7^2 - 3 \cdot 7 + 8)$$
$$= f(36) = 2 \cdot 36 - 5$$
$$= 67.$$

To find $(g \circ f)(7)$, we first find $f(7)$. Then we use $f(7)$ as an input for g:

$$(g \circ f)(7) = g(f(7)) = g(2 \cdot 7 - 5)$$
$$= g(9) = 9^2 - 3 \cdot 9 + 8$$
$$= 62.$$

We could also find $(f \circ g)(7)$ and $(g \circ f)(7)$ by substituting 7 for x in the equations that we found in part (a):

$$(f \circ g)(x) = 2x^2 - 6x + 11$$
$$(f \circ g)(7) = 2 \cdot 7^2 - 6 \cdot 7 + 11 = 67;$$

$$(g \circ f)(x) = 4x^2 - 26x + 48$$
$$(g \circ f)(7) = 4 \cdot 7^2 - 26 \cdot 7 + 48 = 62.$$

Note in Example 1 that, as a rule, $(f \circ g)(x) \neq (g \circ f)(x)$. We can see this graphically, as shown in the graphs at left.

EXAMPLE 2 Given that $f(x) = \sqrt{x}$ and $g(x) = x - 3$:

a) Find $h(x)$ and $k(x)$ if $h = f \circ g$ and $k = g \circ f$.
b) Find the domains of h and k.

Solution

a) $h(x) = (f \circ g)(x) = f(g(x)) = f(x - 3) = \sqrt{x - 3}$

$k(x) = (g \circ f)(x) = g(f(x)) = g(\sqrt{x}) = \sqrt{x} - 3$

b) The domain of f is $\{x \mid x \geq 0\}$, or the interval $[0, \infty)$. The domain of g is $(-\infty, \infty)$.

To find the domain of $h = f \circ g$, we must consider that the outputs for g will serve as inputs for f. Since the inputs for f cannot be negative, we must have

$$g(x) = x - 3, \quad \text{where } x - 3 \geq 0, \text{ or } x \geq 3.$$

Thus the domain of $h = \{x \mid x \geq 3\}$, or the interval $[3, \infty)$, as the graph in Fig. 1 on the next page confirms.

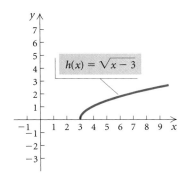

$h(x) = \sqrt{x} - 3$

FIGURE 1

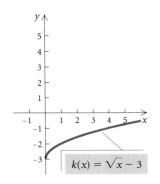

$k(x) = \sqrt{x - 3}$

FIGURE 2

To find the domain of $k = g \circ f$, we must consider that the outputs for f will serve as inputs for g. Since g can accept *any* real number as an input, any output from f is acceptable. Thus the domain of $k = \{x \mid x \geq 0\}$, as the graph in Fig. 2 confirms.

Decomposing a Function as a Composition

In calculus, one often needs to recognize how a function can be expressed as the composition of two functions. In this way, we are "decomposing" the function.

EXAMPLE 3   If $h(x) = (2x - 3)^5$, find $f(x)$ and $g(x)$ such that $h(x) = (f \circ g)(x)$.

Solution   The function $h(x)$ raises $(2x - 3)$ to the 5th power. Two functions that can be used for the composition are

$$f(x) = x^5 \quad \text{and} \quad g(x) = 2x - 3.$$

We can check by forming the composition:

$$h(x) = (f \circ g)(x) = f(g(x)) = f(2x - 3) = (2x - 3)^5.$$

This is the most "obvious" answer to the question. There can be other less obvious answers. For example, if

$$f(x) = (x + 7)^5 \quad \text{and} \quad g(x) = 2x - 10,$$

then

$$
\begin{aligned}
h(x) &= (f \circ g)(x) = f(g(x)) \\
&= f(2x - 10) \\
&= [(2x - 10) + 7]^5 = (2x - 3)^5.
\end{aligned}
$$

Inverses

When we go from an output of a function back to its input or inputs, we get an inverse relation. When that relation is a function, we have an inverse function.

Consider the relation h given as follows:

$$h = \{(-8, 5), (4, -2), (-7, 1), (3.8, 6.2)\}.$$

Suppose we *interchange* the first and second coordinates. The relation we obtain is called the **inverse** of the relation h and is given as follows:

$$\text{Inverse of } h = \{(5, -8), (-2, 4), (1, -7), (6.2, 3.8)\}.$$

Inverse Relation

Interchanging the first and second coordinates of each ordered pair in a relation produces the **inverse relation.**

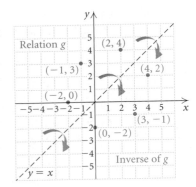

Relation g

(2, 4)

(−1, 3)

(4, 2)

(−2, 0)

(3, −1)

(0, −2)

Inverse of g

y = x

EXAMPLE 4   Consider the relation g given by

$$g = \{(2, 4), (-1, 3), (-2, 0)\}.$$

Graph the relation in blue. Find the inverse and graph it in red.

Solution   The relation g is shown in blue in the figure at left. The inverse of the relation is

$$\{(4, 2), (3, -1), (0, -2)\}$$

and is shown in red. The pairs in the inverse are reflections across the line $y = x$.   —

Inverse Relation

If a relation is defined by an equation, interchanging the variables produces an equation of the **inverse relation.**

EXAMPLE 5   Find an equation for the inverse of the relation:

$$y = x^2 - 5x.$$

Solution   We interchange x and y and obtain an equation of the inverse:

$$x = y^2 - 5y.$$ —

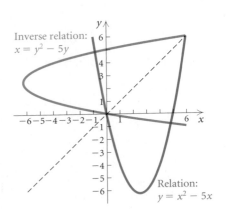

Inverse relation:
$x = y^2 - 5y$

Relation:
$y = x^2 - 5x$

If a relation is given by an equation, then the solutions of the inverse can be found from those of the original equation by interchanging the first and second coordinates of each ordered pair. Thus the graphs of a relation and its inverse are always reflections of each other across the line $y = x$. This is illustrated with the equations of Example 5 in the graph at left. We will explore inverses and their graphs later in this section.

Inverses and One-to-One Functions

Let's consider the following two functions.

YEAR (DOMAIN)	FIRST-CLASS POSTAGE COST, IN CENTS (RANGE)
1978	15
1983	20
1984	
1989	25
1991	29
1995	32
1999	33
2001	34

Source: U.S. Postal Service

NUMBER (DOMAIN)	CUBE (RANGE)
−3	−27
−2	−8
−1	−1
0	0
1	1
2	8
3	27

Suppose we reverse the arrows. Are these inverse relations functions?

Year (Range)	First-Class Postage Cost, in Cents (Domain)
1978 ←	15
1983 ←	20
1984 ←	
1989 ←	25
1991 ←	29
1995 ←	32
1999 ←	33
2001 ←	34

Number (Range)	Cube (Domain)
−3 ←	−27
−2 ←	−8
−1 ←	−1
0 ←	0
1 ←	1
2 ←	8
3 ←	27

We see that the inverse of the postage function is not a function. Like all functions, each input in the postage function has exactly one output. However, the outputs for both 1983 and 1984 are the same, 20. Thus in the inverse of the postage function, the input 20 has *two* outputs, 1983 and 1984. When the same output of a function comes from two or more different inputs, the inverse cannot be a function. In the cubing function, each output corresponds to exactly one input, so its inverse is also a function. The cubing function is an example of a **one-to-one function.** If the inverse of a function f is also a function, it is named f^{-1} (read "f-inverse").

The −1 in f^{-1} is *not* an exponent!

One-to-One Functions and Inverses

A function f is **one-to-one** if different inputs have different outputs—that is,

if $a \neq b$, then $f(a) \neq f(b)$. Or,

a function f is **one-to-one** if when the outputs are the same, the inputs are the same—that is,

if $f(a) = f(b)$, then $a = b$.

If a function is one-to-one, then its inverse is a function.

The domain of a one-to-one function f is the range of the inverse f^{-1}.

The range of a one-to-one function f is the domain of the inverse f^{-1}.

EXAMPLE 6 Given the function f described by $f(x) = 2x - 3$, prove that f is one-to-one (that is, it has an inverse that is a function).

Solution   To show that f is one-to-one, we show that if $f(a) = f(b)$, then $a = b$. Assume that $f(a) = f(b)$ for any numbers a and b in the domain of f. Since $f(a) = 2a - 3$ and $f(b) = 2b - 3$, we have

$$2a - 3 = 2b - 3$$
$$2a = 2b \qquad \text{Adding 3}$$
$$a = b. \qquad \text{Dividing by 2}$$

Thus, if $f(a) = f(b)$, then $a = b$. This shows that f is one-to-one.   ▄

EXAMPLE 7   Given the function g described by $g(x) = x^2$, prove that g is not one-to-one.

Solution   We can prove that g is not one-to-one by finding two numbers a and b for which $a \neq b$ and $g(a) = g(b)$. Two such numbers are -3 and 3, because $-3 \neq 3$ and $g(-3) = g(3) = 9$. Thus g is not one-to-one.   ▄

The following graph shows a function, in blue, and its inverse, in red. To determine whether the inverse is a function, we can apply the vertical-line test to its graph. By reflecting each such vertical line back across the line $y = x$, we obtain an equivalent **horizontal-line test** for the original function.

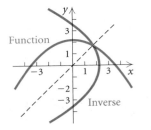

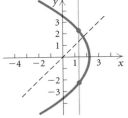

The vertical-line test shows that the inverse is not a function.

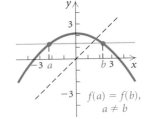

The horizontal-line test shows that the function is not one-to-one.

Horizontal-Line Test

If it is possible for a horizontal line to intersect the graph of a function more than once, then the function is *not* one-to-one and its inverse is *not* a function.

EXAMPLE 8   From the graph shown, determine whether each function is one-to-one and thus has an inverse that is a function.

a)

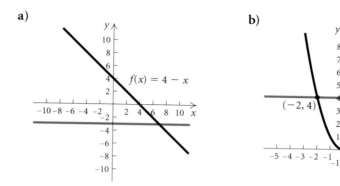

b)

c)

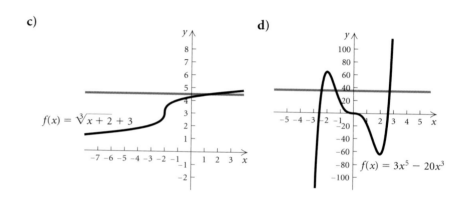

d)

Solution   In each, we apply the horizontal-line test.

a) No horizontal line intersects the graph more than once, so the function is one-to-one. It has an inverse that is a function.

b) There are many horizontal lines that intersect the graph more than once. Note that where the line $y = 4$ intersects the graph, the first coordinates are -2 and 2. Although these are different inputs, they have the same output, 4. Thus the function is not one-to-one. The inverse is not a function.

c) No horizontal line intersects the graph more than once, so the function is one-to-one. It has an inverse that is a function.

d) There are many horizontal lines that intersect the graph more than once. Thus the function is not one-to-one. The inverse is not a function.

Finding Formulas for Inverses

Suppose that a function is described by a formula. If it has an inverse that is a function, we proceed as follows to find a formula for f^{-1}.

Obtaining a Formula for an Inverse

If a function f is one-to-one, a formula for its inverse can generally be found as follows:

1. Replace $f(x)$ with y.
2. Interchange x and y.
3. Solve for y.
4. Replace y with $f^{-1}(x)$.

EXAMPLE 9 Determine whether the function $f(x) = 2x - 3$ is one-to-one, and if it is, find a formula for $f^{-1}(x)$.

Solution The graph of f is shown at left. It passes the horizontal-line test. Thus it is one-to-one and its inverse is a function. We also proved that f is one-to-one in Example 6.

1. Replace $f(x)$ with y: $\qquad y = 2x - 3$
2. Interchange x and y: $\qquad x = 2y - 3$
3. Solve for y: $\qquad\qquad x + 3 = 2y$

$$\frac{x + 3}{2} = y$$

4. Replace y with $f^{-1}(x)$: $\quad f^{-1}(x) = \dfrac{x + 3}{2}.$

Consider

$$f(x) = 2x - 3 \quad \text{and} \quad f^{-1}(x) = \frac{x + 3}{2}$$

from Example 9. For the input 5, we have

$$f(5) = 2 \cdot 5 - 3 = 10 - 3 = 7.$$

The output is 7. Now we use 7 for the input in the inverse:

$$f^{-1}(7) = \frac{7 + 3}{2}$$

$$= \frac{10}{2} = 5.$$

The function f takes the number 5 to 7. The inverse function f^{-1} takes the number 7 back to 5.

EXAMPLE 10 Graph

$$f(x) = 2x - 3 \quad \text{and} \quad f^{-1}(x) = \frac{x + 3}{2}$$

using the same set of axes. Then compare the two graphs.

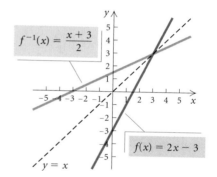

$f^{-1}(x) = \dfrac{x+3}{2}$

$f(x) = 2x - 3$

$y = x$

Solution The graphs of *f* and f^{-1} are shown at left. Note that the graph of f^{-1} can be drawn by reflecting the graph of *f* across the line $y = x$. That is, if we were to graph $f(x) = 2x - 3$ in wet ink and fold along the line $y = x$, the graph of $f^{-1}(x) = (x + 3)/2$ would be formed by the ink transferred from *f*.

When we interchange *x* and *y* in finding a formula for the inverse of $f(x) = 2x - 3$, we are in effect reflecting the graph of that function across the line $y = x$. For example, when the coordinates of the *y*-intercept, $(0, -3)$, of the graph of *f* are reversed, we get the *x*-intercept, $(-3, 0)$, of the graph of f^{-1}.

> The graph of f^{-1} is a reflection of the graph of *f* across the line $y = x$.

EXAMPLE 11 Consider $g(x) = x^3 + 2$.

a) Determine whether the function is one-to-one.

b) If it is one-to-one, find a formula for its inverse.

c) Graph the function and its inverse.

Solution

a) The graph of $g(x) = x^3 + 2$ is shown below. It passes the horizontal-line test and thus has an inverse that is a function.

Technology Connection

On some graphing calculators, we can graph the inverse of a function after graphing the function itself by accessing a drawing feature. Consult your user's manual or the *Graphing Calculator Manual* that accompanies this text for the procedure.

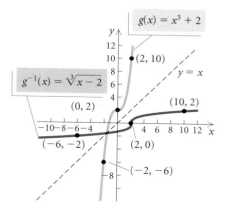

$g(x) = x^3 + 2$

$g^{-1}(x) = \sqrt[3]{x} - 2$

$y = x$

$(2, 10)$

$(0, 2)$

$(10, 2)$

$(-6, -2)$

$(2, 0)$

$(-2, -6)$

b) We follow the procedure for finding an inverse.

    1. Replace $g(x)$ with *y*: $y = x^3 + 2$

    2. Interchange *x* and *y*: $x = y^3 + 2$

    3. Solve for *y*: $x - 2 = y^3$

                             $\sqrt[3]{x - 2} = y$

    4. Replace *y* with $g^{-1}(x)$: $g^{-1}(x) = \sqrt[3]{x - 2}$.

c) To find the graph, we reflect the graph of $g(x) = x^3 + 2$ across the line $y = x$. This can be done by plotting points, as shown in part (a) above.   ▬

Inverse Functions and Composition

Suppose that we were to use some input a for a one-to-one function f and find its output, $f(a)$. The function f^{-1} would then take that output back to a. Similarly, if we began with an input b for the function f^{-1} and found its output, $f^{-1}(b)$, the original function f would then take that output back to b. This is summarized as follows.

If a function f is one-to-one, then f^{-1} is the unique function such that each of the following holds:

$$(f^{-1} \circ f)(x) = f^{-1}(f(x)) = x, \quad \text{for each } x \text{ in the domain of } f, \text{ and}$$

$$(f \circ f^{-1})(x) = f(f^{-1}(x)) = x, \quad \text{for each } x \text{ in the domain of } f^{-1}.$$

EXAMPLE 12   Given that $f(x) = 5x + 8$, use composition of functions to show that $f^{-1}(x) = (x - 8)/5$.

Solution   We find $(f^{-1} \circ f)(x)$ and $(f \circ f^{-1})(x)$ and check to see that each is x:

$$(f^{-1} \circ f)(x) = f^{-1}(f(x))$$

$$= f^{-1}(5x + 8) = \frac{(5x + 8) - 8}{5}$$

$$= \frac{5x}{5} = x;$$

$$(f \circ f^{-1})(x) = f(f^{-1}(x))$$

$$= f\left(\frac{x - 8}{5}\right) = 5\left(\frac{x - 8}{5}\right) + 8$$

$$= x - 8 + 8 = x.$$ ▬

Restricting a Domain

In the case in which the inverse of a function is not a function, the domain of the function can be restricted to allow the inverse to be a function. We saw in Examples 7 and 8(b) that $f(x) = x^2$ is not one-to-one. The graph is shown at left.

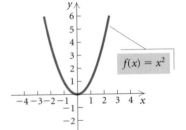

Suppose that we had tried to find a formula for the inverse as follows:

$$y = x^2 \qquad \text{Replacing } f(x) \text{ with } y$$
$$x = y^2 \qquad \text{Interchanging } x \text{ and } y$$
$$\pm\sqrt{x} = y. \qquad \text{Solving for } y$$

This is not the equation of a function. An input of, say, 4 would yield two outputs, -2 and 2. In such cases, it is convenient to consider "part" of the function by restricting the domain of $f(x)$. For example, if we restrict the domain of $f(x) = x^2$ to nonnegative numbers, then its inverse is a function as shown with the graphs of $f(x) = x^2$, $x \geq 0$, and $f^{-1}(x) = \sqrt{x}$, $x \geq 0$ below.

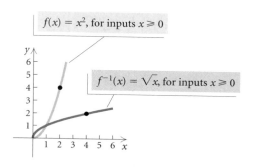

$f(x) = x^2$, for inputs $x \geqslant 0$

$f^{-1}(x) = \sqrt{x}$, for inputs $x \geqslant 0$

Exercise Set

Find $(f \circ g)(x)$ *and* $(g \circ f)(x)$.

1. $f(x) = x + 3$, $g(x) = x - 3$

2. $f(x) = \frac{4}{5}x$, $g(x) = \frac{5}{4}x$

3. $f(x) = 3x - 7$, $g(x) = \dfrac{x + 7}{3}$

4. $f(x) = \frac{2}{3}x - \frac{4}{5}$, $g(x) = 1.5x + 1.2$

5. $f(x) = 20$, $g(x) = 0.05$

6. $f(x) = x^4$, $g(x) = \sqrt[4]{x}$

7. $f(x) = \sqrt{x + 5}$, $g(x) = x^2 - 5$

8. $f(x) = x^5 - 2$, $g(x) = \sqrt[5]{x + 2}$

9. $f(x) = \dfrac{1 - x}{x}$, $g(x) = \dfrac{1}{1 + x}$

10. $f(x) = \dfrac{x^2 - 1}{x^2 + 1}$, $g(x) = \dfrac{3x - 4}{5x - 2}$

11. $f(x) = x^3 - 5x^2 + 3x + 7$, $g(x) = x + 1$

12. $f(x) = x - 1$, $g(x) = x^3 + 2x^2 - 3x - 9$

Find $f(x)$ *and* $g(x)$ *such that* $h(x) = (f \circ g)(x)$. *Answers may vary.*

13. $h(x) = (4 + 3x)^5$

14. $h(x) = \sqrt[3]{x^2 - 8}$

15. $h(x) = \dfrac{1}{(x - 2)^4}$

16. $h(x) = \dfrac{1}{\sqrt{3x + 7}}$

17. $h(x) = \dfrac{x^3 - 1}{x^3 + 1}$

18. $h(x) = |9x^2 - 4|$

19. $h(x) = \left(\dfrac{2 + x^3}{2 - x^3}\right)^6$

20. $h(x) = (\sqrt{x} - 3)^4$

21. $h(x) = \sqrt{\dfrac{x - 5}{x + 2}}$

22. $h(x) = \sqrt{1 + \sqrt{1 + x}}$

23. $h(x) = (x + 2)^3 - 5(x + 2)^2 + 3(x + 2) - 1$

24. $h(x) = 2(x - 1)^{5/3} + 5(x - 1)^{2/3}$

25. *Dress Sizes.* A dress that is size x in France is size $s(x)$ in the United States, where $s(x) = x - 32$. A dress that is size x in the United States is size $y(x)$ in Italy, where $y(x) = 2(x + 12)$. Find a function that will convert French dress sizes to Italian dress sizes.

26. *Ripple Spread.* A stone is thrown into a pond. A circular ripple is spreading over the pond in such a way that the radius is increasing at the rate of 3 ft/sec.

a) Find a function $r(t)$ for the radius in terms of t.
b) Find a function $A(r)$ for the area of the ripple in terms of the radius r.
c) Find $(A \circ r)(t)$. Explain the meaning of this function.

Find the inverse of the relation.

27. $\{(7, 8), (-2, 8), (3, -4), (8, -8)\}$

28. $\{(0, 1), (5, 6), (-2, -4)\}$

29. $\{(-1, -1), (-3, 4)\}$

30. $\{(-1, 3), (2, 5), (-3, 5), (2, 0)\}$

Find an equation of the inverse relation.

31. $y = 4x - 5$

32. $2x^2 + 5y^2 = 4$

33. $x^3y = -5$

34. $y = 3x^2 - 5x + 9$

Graph the equation by substituting and plotting points. Then reflect the graph across the line $y = x$ to obtain the graph of its inverse.

35. $x = y^2 - 3$

36. $y = x^2 + 1$

37. $y = |x|$

38. $x = |y|$

Using the horizontal-line test, determine whether the function is one-to-one.

39. $f(x) = 2.7^x$

40. $f(x) = 2^{-x}$

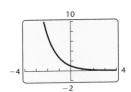

41. $f(x) = 4 - x^2$

42. $f(x) = x^3 - 3x + 1$

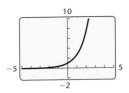

43. $f(x) = \dfrac{8}{x^2 - 4}$

44. $f(x) = \sqrt{\dfrac{10}{4 + x}}$

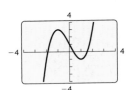

45. $f(x) = \sqrt[3]{x + 2} - 2$

46. $f(x) = \dfrac{8}{x}$

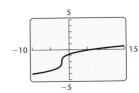
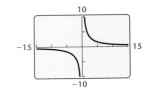
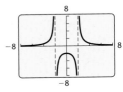
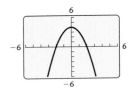
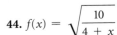
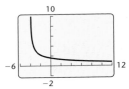
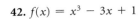

Determine whether the function is one-to-one using the tests on p. 269.

47. $f(x) = 5x - 8$

48. $f(x) = 3 + 4x$

49. $f(x) = 1 - x^2$

50. $f(x) = |x| - 2$

51. $f(x) = |x + 2|$

52. $f(x) = -0.8$

53. $f(x) = -\dfrac{4}{x}$

54. $f(x) = \dfrac{2}{x + 3}$

In Exercises 55–70, for each function:

a) *Sketch the graph and use the graph to determine whether it is one-to-one.*

b) *If it is one-to-one, find a formula for the inverse.*

55. $f(x) = x + 4$

56. $f(x) = 7 - x$

57. $f(x) = 2x - 1$

58. $f(x) = 5x + 8$

59. $f(x) = \dfrac{4}{x + 7}$

60. $f(x) = -\dfrac{3}{x}$

61. $f(x) = \dfrac{x + 4}{x - 3}$

62. $f(x) = \dfrac{5x - 3}{2x + 1}$

63. $f(x) = x^3 - 1$

64. $f(x) = (x + 5)^3$

65. $f(x) = x\sqrt{4 - x^2}$

66. $f(x) = 2x^2 - x - 1$

67. $f(x) = 5x^2 - 2, \ x \geq 0$

68. $f(x) = 4x^2 + 3, \ x \geq 0$

69. $f(x) = \sqrt{x + 1}$

70. $f(x) = \sqrt[3]{x - 8}$

Find the inverse by thinking about the operations of the function and then reversing, or undoing, them. Check your work algebraically.

FUNCTION	INVERSE
71. $f(x) = 3x$	$f^{-1}(x) = $
72. $f(x) = \frac{1}{4}x + 7$	$f^{-1}(x) = $
73. $f(x) = -x$	$f^{-1}(x) = $
74. $f(x) = \sqrt[3]{x} - 5$	$f^{-1}(x) = $
75. $f(x) = \sqrt[3]{x - 5}$	$f^{-1}(x) = $
76. $f(x) = x^{-1}$	$f^{-1}(x) = $

For the function f, use composition of functions to show that f^{-1} is as given.

77. $f(x) = \frac{7}{8}x, \ f^{-1}(x) = \frac{8}{7}x$

78. $f(x) = \dfrac{x + 5}{4}, \ f^{-1}(x) = 4x - 5$

79. $f(x) = \dfrac{1 - x}{x}, \ f^{-1}(x) = \dfrac{1}{x + 1}$

80. $f(x) = \sqrt[3]{x + 4}, \ f^{-1}(x) = x^3 - 4$

81. Find $f(f^{-1}(5))$ and $f^{-1}(f(a))$:
$$f(x) = x^3 - 4.$$

82. Find $f^{-1}(f(p))$ and $f(f^{-1}(1253))$:
$$f(x) = \sqrt[5]{\dfrac{2x - 7}{3x + 4}}.$$

83. *Dress Sizes in the United States and Italy.* A function that will convert dress sizes in the United States to those in Italy is
$$g(x) = 2(x + 12).$$

a) Find the dress sizes in Italy that correspond to sizes 6, 8, 10, 14, and 18 in the United States.

b) Find a formula for the inverse of the function.

c) Use the inverse function to find the dress sizes in the United States that correspond to 36, 40, 44, 52, and 60 in Italy.

84. *Bus Chartering.* An organization determines that the cost per person of chartering a bus is given by the formula
$$C(x) = \dfrac{100 + 5x}{x},$$

where x is the number of people in the group and $C(x)$ is in dollars. Determine $C^{-1}(x)$ and explain what it represents.

Technology Connection

Graph the function and its inverse using a graphing calculator. Use an inverse drawing feature, if available. Find the domain and the range of f. Find the domain and the range of the inverse f^{-1}.

85. $f(x) = 0.8x + 1.7$

86. $f(x) = 2.7 - 1.08x$

87. $f(x) = \frac{1}{2}x - 4$

88. $f(x) = x^3 - 1$

89. $f(x) = \sqrt{x - 3}$

90. $f(x) = -\dfrac{2}{x}$

91. $f(x) = x^2 - 4, \ x \geq 0$

92. $f(x) = 3 - x^2, \ x \geq 0$

93. $f(x) = (3x - 9)^3$

94. $f(x) = \sqrt[3]{\dfrac{x - 3.2}{1.4}}$

95. *Reaction Distance.* You are driving a car when a deer suddenly darts across the road in front of you. Your brain registers the emergency and sends a signal to your foot to hit the brake. The car travels a distance D, in feet, during this time, where D is a function of the speed r, in miles per hour, that the car is traveling when you see the deer. That reaction distance D is a linear function given by

$$D(r) = \frac{11r + 5}{10}.$$

a) Find $D(0)$, $D(10)$, $D(20)$, $D(50)$, and $D(65)$.
b) Find $D^{-1}(r)$ and explain what it represents.
c) Graph the function and its inverse.

96. *Bread Consumption.* The number N of 1-lb loaves of bread consumed per person per year t years after 1995 is given by the function

$$N(t) = 0.6514t + 53.1599$$

(*Source*: U.S. Department of Agriculture).

a) Find the consumption of bread per person in 1998 and 2000.
b) Graph the function and its inverse.
c) Explain what the inverse represents.

Collaborative Discussion and Writing

97. If you graph a function using a list of ordered pairs and see that it is one-to-one, how could you then use that list of ordered pairs to graph the inverse?

98. The following formulas for the conversion between Fahrenheit and Celsius temperatures have been considered several times in this text:

$$C = \tfrac{5}{9}(F - 32)$$

and

$$F = \tfrac{9}{5}C + 32.$$

Discuss these formulas from the standpoint of inverses.

Skill Maintenance

Graph.

99. $y = \dfrac{3}{2}x - 4$

100. $(x - 1)^2 + y^2 = 9$

101. $y = \dfrac{x + 3}{x^2 - x - 2}$

102. $y = -x^2 + x - 4$

Synthesis

103. The function $f(x) = x^2 - 3$ is not one-to-one. Restrict the domain of f so that its inverse is a function. Find the inverse and state the restriction on the domain of the inverse.

104. Consider the function f given by

$$f(x) = \begin{cases} x^3 + 2, & x \le -1, \\ x^2, & -1 < x < 1, \\ x + 1, & x \ge 1. \end{cases}$$

Does f have an inverse that is a function? Why or why not?

105. Find three examples of functions that are their own inverses, that is, $f = f^{-1}$.

4.2

Exponential Functions and Graphs

- *Graph exponential equations and functions.*
- *Solve applied problems involving exponential functions and their graphs.*

We now turn our attention to the study of a set of functions that are very rich in application. Consider the following graphs. Each one illustrates an *exponential function.* In this section, we consider such functions, their inverses, and some important applications of exponential functions.

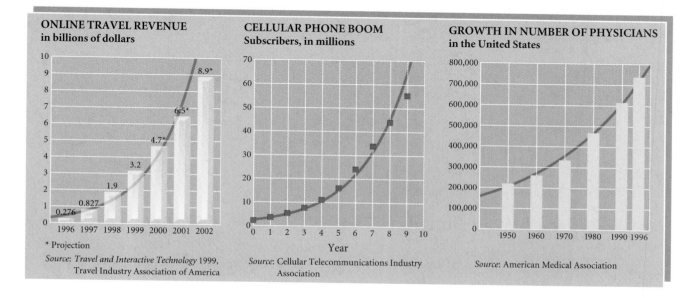

ONLINE TRAVEL REVENUE
in billions of dollars

* Projection
Source: Travel and Interactive Technology 1999, Travel Industry Association of America

CELLULAR PHONE BOOM
Subscribers, in millions

Year

Source: Cellular Telecommunications Industry Association

GROWTH IN NUMBER OF PHYSICIANS
in the United States

Source: American Medical Association

Graphing Exponential Functions

EXPONENTS

REVIEW SECTIONS R.2 AND R.6.

We now define exponential functions. We assume that a^x has meaning for any real number x and any positive real number a and that the laws of exponents still hold, though we will not prove them here.

Exponential Function

The function $f(x) = a^x$, where x is a real number, $a > 0$ and $a \neq 1$, is called the **exponential function,** base a.

We require the **base** to be positive in order to avoid the complex numbers that would occur by taking even roots of negative numbers—an example is $(-1)^{1/2}$, the square root of -1, which is not a real number. The restriction $a \neq 1$ is made to exclude the constant function $f(x) = 1^x = 1$, which does not have an inverse because it is not one-to-one.

The following are examples of exponential functions:

$$f(x) = 2^x, \qquad f(x) = \left(\frac{1}{2}\right)^x, \qquad f(x) = (3.57)^x.$$

Note that, in contrast to functions like $f(x) = x^5$ and $f(x) = x^{1/2}$ in which the variable is the base of an exponential expression, the variable in an exponential function is *in the exponent*.

Let's now consider graphs of exponential functions.

EXAMPLE 1   Graph the exponential function

$$y = f(x) = 2^x.$$

Solution   We compute some function values and list the results in a table (at left).

$$f(0) = 2^0 = 1;$$
$$f(1) = 2^1 = 2;$$
$$f(2) = 2^2 = 4;$$
$$f(3) = 2^3 = 8;$$
$$f(-1) = 2^{-1} = \frac{1}{2^1} = \frac{1}{2};$$
$$f(-2) = 2^{-2} = \frac{1}{2^2} = \frac{1}{4};$$
$$f(-3) = 2^{-3} = \frac{1}{2^3} = \frac{1}{8}.$$

x	$y = f(x) = 2^x$	(x, y)
0	1	$(0, 1)$
1	2	$(1, 2)$
2	4	$(2, 4)$
3	8	$(3, 8)$
-1	$\frac{1}{2}$	$\left(-1, \frac{1}{2}\right)$
-2	$\frac{1}{4}$	$\left(-2, \frac{1}{4}\right)$
-3	$\frac{1}{8}$	$\left(-3, \frac{1}{8}\right)$

Next, we plot these points and connect them with a smooth curve. Be sure to plot enough points to determine how steeply the curve rises.

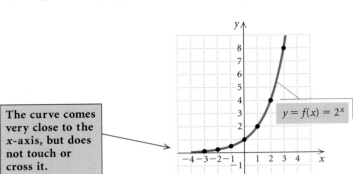

The curve comes very close to the *x*-axis, but does not touch or cross it.

Note that as x increases, the function values increase without bound. As x decreases, the function values decrease, getting close to 0. That is, as $x \to -\infty$, $y \to 0$. Thus the *x*-axis, or the line $y = 0$, is a horizontal asymptote. As the *x*-inputs decrease, the curve gets closer and closer to this line, but does not cross it.

HORIZONTAL ASYMPTOTES

REVIEW SECTION 3.4.

EXAMPLE 2   Graph the exponential function $y = f(x) = \left(\frac{1}{2}\right)^x$.

x	y $y = f(x) = 2^{-x}$	(x, y)
0	1	$(0, 1)$
1	$\frac{1}{2}$	$\left(1, \frac{1}{2}\right)$
2	$\frac{1}{4}$	$\left(2, \frac{1}{4}\right)$
3	$\frac{1}{8}$	$\left(3, \frac{1}{8}\right)$
−1	2	$(-1, 2)$
−2	4	$(-2, 4)$
−3	8	$(-3, 8)$

Solution   We compute some function values and list the results in a table (at left). Before we plot these points and draw the curve, note that

$$y = f(x) = \left(\frac{1}{2}\right)^x = (2^{-1})^x = 2^{-x}.$$

This tells us, before we begin graphing, that this graph is a reflection of the graph of $y = 2^x$ across the y-axis.

$$f(0) = 2^{-0} = 1; \qquad\qquad f(-1) = 2^{-(-1)} = 2^1 = 2;$$

$$f(1) = 2^{-1} = \frac{1}{2^1} = \frac{1}{2}; \qquad f(-2) = 2^{-(-2)} = 2^2 = 4;$$

$$f(2) = 2^{-2} = \frac{1}{2^2} = \frac{1}{4}; \qquad f(-3) = 2^{-(-3)} = 2^3 = 8.$$

$$f(3) = 2^{-3} = \frac{1}{2^3} = \frac{1}{8};$$

Next, we plot these points and connect them with a smooth curve.

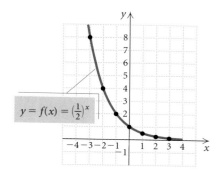

Note that as x increases, the function values decrease, getting close to 0. As x decreases, the function values increase without bound.   ▬

Observe the following graphs of exponential functions and look for patterns in the graphs.

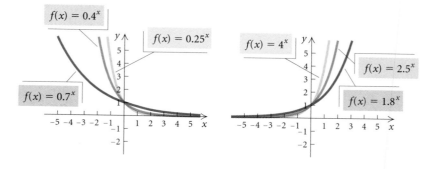

What relationship do you see between the base a and the shape of the resulting graph of $f(x) = a^x$? What do all the graphs have in common? How do they differ?

CONNECTING THE CONCEPTS

PROPERTIES OF EXPONENTIAL FUNCTIONS

Let's list and compare some characteristics of exponential functions, keeping in mind that the definition of an exponential function, $f(x) = a^x$, requires that a be positive and different from 1.

$f(x) = a^x$ for $0 < a < 1$ or
$f(x) = a^{-x}$, $a > 1$:

Continuous

One-to-one

Domain: $(-\infty, \infty)$

Range: $(0, \infty)$

Decreasing

Horizontal asymptote is x-axis

y-intercept: $(0, 1)$

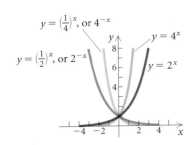

$f(x) = a^x$, $a > 1$:

Continuous

One-to-one

Domain: $(-\infty, \infty)$

Range: $(0, \infty)$

Increasing

Horizontal asymptote is x-axis

y-intercept: $(0, 1)$

TRANSFORMATIONS
OF FUNCTIONS

REVIEW SECTION 1.6.

To graph other types of exponential functions, keep in mind the ideas of translation, stretching, and reflection. All these concepts allow us to visualize the graph before drawing it.

EXAMPLE 3 Graph each of the following. Before doing so, describe how each graph can be obtained from the graph of $f(x) = 2^x$.

a) $f(x) = 2^{x-2}$

b) $f(x) = 2^x - 4$

c) $f(x) = 5 - 2^{-x}$

Solution

a) The graph of this function is the graph of $y = 2^x$ shifted *right* 2 units.

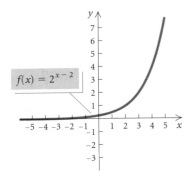

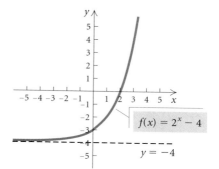

b) The graph is the graph of $y = 2^x$ shifted *down* 4 units (see the graph at left).

c) The graph is a reflection of the graph of $y = 2^x$ across the y-axis, followed by a reflection across the x-axis and then a shift *up* 5 units.

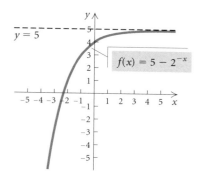

Graphs of Inverses of Exponential Functions

We have noted that every exponential function (with $a > 0$ and $a \neq 1$) is one-to-one. Thus such a function has an inverse that is a function. In the next section, we will name these inverse functions and use them in applications. For now, we draw their graphs by interchanging x and y.

EXAMPLE 4 Graph: $x = 2^y$.

Solution Note that x is alone on one side of the equation. We can find ordered pairs that are solutions by choosing values for y and then computing the corresponding x-values.

For $y = 0$, $x = 2^0 = 1$.

For $y = 1$, $x = 2^1 = 2$.

For $y = 2$, $x = 2^2 = 4$.

For $y = 3$, $x = 2^3 = 8$.

For $y = -1$, $x = 2^{-1} = \dfrac{1}{2^1} = \dfrac{1}{2}$.

For $y = -2$, $x = 2^{-2} = \dfrac{1}{2^2} = \dfrac{1}{4}$.

For $y = -3$, $x = 2^{-3} = \dfrac{1}{2^3} = \dfrac{1}{8}$.

x		
$x = 2^y$	y	(x, y)
1	0	$(1, 0)$
2	1	$(2, 1)$
4	2	$(4, 2)$
8	3	$(8, 3)$
$\dfrac{1}{2}$	-1	$\left(\dfrac{1}{2}, -1\right)$
$\dfrac{1}{4}$	-2	$\left(\dfrac{1}{4}, -2\right)$
$\dfrac{1}{8}$	-3	$\left(\dfrac{1}{8}, -3\right)$

(1) Choose values for y.
(2) Compute values for x.

We plot the points and connect them with a smooth curve. Note that the curve does not touch or cross the y-axis. The y-axis is a vertical asymptote.

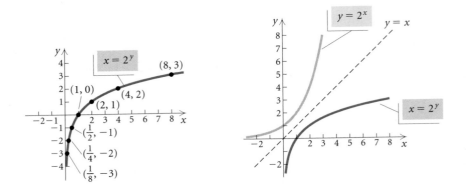

Note too that this curve looks just like the graph of $y = 2^x$, except that it is reflected across the line $y = x$, as we would expect for an inverse. The inverse of $y = 2^x$ is $x = 2^y$.

Applications

One of the most frequent applications of exponential functions occurs with compound interest.

EXAMPLE 5 *Compound Interest.* The amount of money A that a principal P will be worth after t years at interest rate i, compounded n times per year, is given by the formula

$$A = P\left(1 + \frac{i}{n}\right)^{nt}.$$

Suppose that $100,000 is invested at 8% interest, compounded semi-annually.

a) Find a function for the amount of money after t years.

b) Find the amount of money in the account at $t = 0$, 4, 8, and 10 yr.

c) Graph the function.

Solution

a) Since $P = \$100{,}000$, $i = 8\% = 0.08$, and $n = 2$, we can substitute these values and write the following function:

$$A(t) = 100{,}000\left(1 + \frac{0.08}{2}\right)^{2 \cdot t} = \$100{,}000(1.04)^{2t}.$$

Technology Connection

We can find the function values in Example 5(b) using the VALUE feature from the CALC menu.

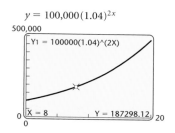

We could also use the TABLE feature.

b) We can compute function values with a calculator:

$$A(0) = 100,000(1.04)^{2 \cdot 0} = \$100,000;$$
$$A(4) = 100,000(1.04)^{2 \cdot 4} \approx \$136,856.91;$$
$$A(8) = 100,000(1.04)^{2 \cdot 8} \approx \$187,298.12;$$
$$A(10) = 100,000(1.04)^{2 \cdot 10} \approx \$219,112.31.$$

c) We use the function values computed in part (b) and others if we wish, and draw the graph as follows. Note that the axes are scaled differently because of the large numbers and that t is restricted to non-negative values, because negative time values have no meaning here.

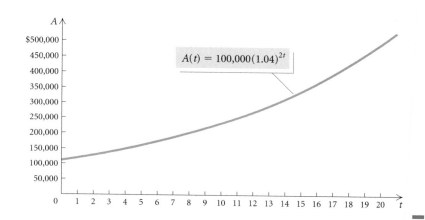

The Number e

We now consider a very special number in mathematics. In 1741, Leonhard Euler named this number e. Though you may not have encountered it before, you will see here and in future mathematics that it has many important applications. To explain this number, we use the compound interest formula $A = P(1 + i/n)^{nt}$ in Example 5. Suppose that \$1 is an initial investment at 100% interest for 1 yr. (No bank would pay this.) The formula above becomes a function A defined in terms of the number of compounding periods n. Since $P = 1$, $i = 1$, and $t = 1$,

$$A(n) = \left(1 + \frac{1}{n}\right)^n.$$

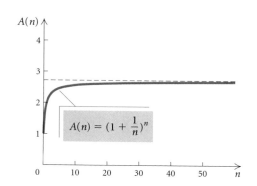

Let's visualize this function with its graph shown at left and explore the values of $A(n)$ as $n \to \infty$. Consider the graph for larger and larger values of n. Does this function have a horizontal asymptote?

Let's find some function values using a calculator.

n, NUMBER OF COMPOUNDING PERIODS	$A(n) = \left(1 + \dfrac{1}{n}\right)^n$
1 (compounded annually)	$2.00
2 (compounded semiannually)	$2.25
3	$2.3704
4 (compounded quarterly)	$2.4414
5	$2.4883
100	$2.7048
365 (compounded daily)	$2.7146
8760 (compounded hourly)	$2.7181

It appears from these values that the graph does have a horizontal asymptote, $y \approx 2.7$. As the values of n get larger and larger, the function values get closer and closer to the number Euler named e. Its decimal representation does not terminate or repeat; it is irrational.

$$e = 2.7182818284\ldots$$

EXAMPLE 6 Find each value of e^x, to four decimal places, using the $\boxed{e^x}$ key on a calculator.

a) e^3 **b)** $e^{-0.23}$

c) e^0 **d)** e^1

Solution

FUNCTION VALUE	READOUT	ROUNDED
a) e^3	e^(3) 20.08553692	20.0855
b) $e^{-0.23}$	e^(−.23) .7945336025	0.7945
c) e^0	e^(0) 1	1
d) e^1	e^(1) 2.718281828	2.7183

Graphs of Exponential Functions, Base e

We demonstrate ways in which to graph exponential functions.

EXAMPLE 7 Graph $f(x) = e^x$ and $g(x) = e^{-x}$.

Solution We can compute points for each equation using the $\boxed{e^x}$ key on a calculator. Then we plot these points and draw the graphs of the functions.

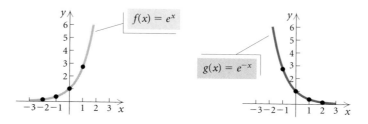

Note that the graph of g is a reflection of the graph of f across the y-axis.

EXAMPLE 8 Graph each of the following. Before doing so, describe how each graph can be obtained from the graph of $y = e^x$.

a) $f(x) = e^{-0.5x}$ **b)** $f(x) = 1 - e^{-2x}$ **c)** $f(x) = e^{x+3}$

Solution

a) We note that the graph of this function is a horizontal stretching of the graph of $y = e^x$ followed by a reflection across the y-axis. (See Fig. 1.)

b) The graph is a horizontal shrinking of the graph of $y = e^x$, followed by a reflection across the y-axis, then across the x-axis, followed by a translation up 1 unit. (See Fig. 2.)

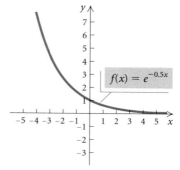

FIGURE 1

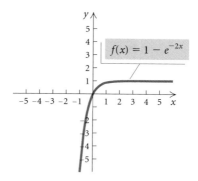

FIGURE 2

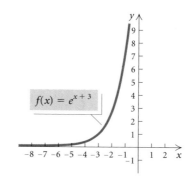

FIGURE 3

c) The graph is a translation of the graph of $y = e^x$ to the left 3 units. (See Fig. 3.)

Exercise Set 4.2

Find each of the following, to four decimal places, using a calculator.

1. e^4

2. e^{10}

3. $e^{-2.458}$

4. $\left(\dfrac{1}{e^3}\right)^2$

Graph the function.

5. $f(x) = 3^x$

6. $f(x) = 5^x$

7. $f(x) = 6^x$

8. $f(x) = 3^{-x}$

9. $f(x) = \left(\dfrac{1}{4}\right)^x$

10. $f(x) = \left(\dfrac{2}{3}\right)^x$

11. $x = 3^y$

12. $x = 4^y$

13. $x = \left(\dfrac{1}{2}\right)^y$

14. $x = \left(\dfrac{4}{3}\right)^y$

15. $y = \dfrac{1}{4}e^x$

16. $y = 2e^{-x}$

17. $f(x) = 1 - e^{-x}$

18. $f(x) = e^x - 2$

Sketch the graph of the function. Describe how each graph can be obtained from the graph of a basic exponential function.

19. $f(x) = 2^{x+1}$

20. $f(x) = 2^{x-1}$

21. $f(x) = 2^x - 3$

22. $f(x) = 2^x + 1$

23. $f(x) = 4 - 3^{-x}$

24. $f(x) = 2^{x-1} - 3$

25. $f(x) = \left(\dfrac{3}{2}\right)^{x-1}$

26. $f(x) = 3^{4-x}$

27. $f(x) = 2^{x+3} - 5$

28. $f(x) = -3^{x-2}$

29. $f(x) = e^{2x}$

30. $f(x) = e^{-0.2x}$

31. $y = e^{-x+1}$

32. $y = e^{2x} + 1$

33. $f(x) = 2(1 - e^{-x})$

34. $f(x) = 1 - e^{-0.01x}$

35. *Growth of AIDS.* The total number of AIDS cases reported in the United States is approximated by the exponential function

$$N(t) = 274,390(1.195)^t,$$

where $t = 0$ corresponds to 1992 (*Source*: U.S. Centers for Disease Control and Prevention, HIV/AID Surveillance Reports).

a) According to this function, how many AIDS cases had been reported by 1998?

b) Estimate the number of AIDS cases that will have been reported by 2001.

36. *Growth of Bacteria* Escherichia coli.   The bacteria *Escherichia coli* are commonly found in the human bladder. Suppose that 3000 of the bacteria are present at time $t = 0$. Then under certain conditions, t minutes later, the number of bacteria present is

$$N(t) = 3000(2)^{t/20}.$$

a) How many bacteria will be present after 10 min? 20 min? 30 min? 40 min? 60 min?

b) Graph the function.

37. *Recycling Aluminum Cans.*   It is estimated that two thirds of all aluminum cans distributed will be recycled each year (*Source*: Alcoa Corporation). A beverage company distributes 350,000 cans. The number still in use after time t, in years, is given by the exponential function

$$N(t) = 350,000\left(\dfrac{2}{3}\right)^t.$$

a) How many cans are still in use after 0 yr? 1 yr? 4 yr? 10 yr?

b) Graph the function.

38. *Interest in a College Trust Fund.*   Following the birth of a child, Juan deposits $10,000 in a college trust fund where interest is 6.4%, compounded semiannually.

a) Find a function for the amount in the account after t years.

b) Find the amount of money in the account at $t = 0$, 4, 8, 10, and 18 yr.

39. *Salvage Value.*   A top-quality fax–copying machine is purchased for $5800. Its value each year is about 80% of the value of the preceding year. After t years, its value, in dollars, is given by the exponential function

$$V(t) = 5800(0.8)^t.$$

Find the value of the machine after 0 yr, 1 yr, 2 yr, 5 yr, and 10 yr.

40. *Cellular Phone Boom.*   The number of people using cellular phones has grown exponentially in recent years. The total number of subscribers C, in millions, is given by

$$C(t) = 0.4703477987(1.523963441)^x,$$

where t is the number of years since 1985.

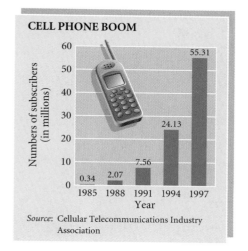

CELL PHONE BOOM

Source: Cellular Telecommunications Industry Association

Find the total number of subscribers in 1995, 1999, 2000, and 2004.

41. *Timber Demand.* World demand for timber is increasing exponentially. The demand N, in billions of cubic feet purchased, is given by

$$N(t) = 46.6(1.018)^t,$$

where t is the number of years since 1981 (*Sources*: U.N. Food and Agricultural Organization, American Forest and Paper Association). Find the demand for timber in 2005 and 2010.

42. *Typing Speed.* Sarah is taking keyboarding at a community college. After she practices for t hours, her speed, in words per minute, is given by the function

$$S(t) = 200[1 - (0.86)^t].$$

What is Sarah's speed after practicing for 10 hr? 20 hr? 40 hr? 100 hr?

43. *Advertising.* A company begins a radio advertising campaign in New York City to market a new CD-ROM video game. The percentage of the target market that buys a game is generally a function of the length of the advertising campaign. The estimated percentage is given by

$$f(t) = 100(1 - e^{-0.04t}),$$

where t is the number of days of the campaign. Find $f(25)$, the percentage of the target market that has bought the product after a 25-day advertising campaign.

44. *Growth of a Stock.* The value of a stock is given by the function

$$V(t) = 58(1 - e^{-1.1t}) + 20,$$

where V is the value of the stock after time t, in months. Find $V(1)$, $V(2)$, $V(4)$, $V(6)$, and $V(12)$.

Technology Connection

In Exercises 45–58, use a graphing calculator to match the equation with one of figures (a)–(n), which follow.

a)
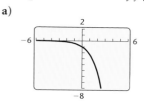

b)
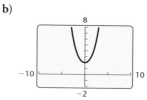

c)
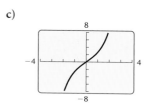

d)
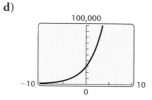

e)
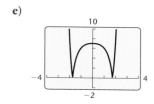

f)
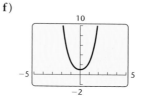

g)
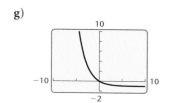

h)
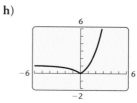

i)
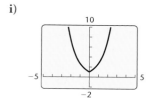

j)
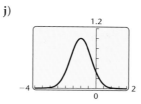

k)
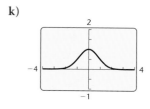

l)
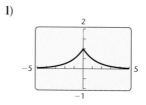

m)

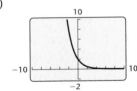

n)

45. $y = 3^x - 3^{-x}$

46. $y = 3^{-(x+1)^2}$

47. $f(x) = -2.3^x$

48. $f(x) = 30,000(1.4)^x$

49. $y = 2^{-|x|}$

50. $y = 2^{-(x-1)}$

51. $f(x) = (0.58)^x - 1$

52. $y = 2^x + 2^{-x}$

53. $g(x) = e^{|x|}$

54. $f(x) = |2^x - 1|$

55. $y = 2^{-x^2}$

56. $y = |2^{x^2} - 8|$

57. $g(x) = \dfrac{e^x - e^{-x}}{2}$

58. $f(x) = \dfrac{e^x + e^{-x}}{2}$

Collaborative Discussion and Writing

59. Describe the differences between the graphs of $f(x) = x^3$ and $g(x) = 3^x$.

60. Suppose that \$10,000 is invested for 8 yr at 6.4% interest, compounded annually. In what year will the most interest be earned? Why?

Graph the pair of equations using the same set of axes. Then compare the results.

61. $y = 3^x, \ x = 3^y$

62. $y = 1^x, \ x = 1^y$

Skill Maintenance

Find the zeros of the function.

63. $h(x) = x^4 - x^2$

64. $g(x) = x^3 + x^2 - 12x$

Solve.

65. $x^3 + 6x^2 - 16x = 0$

66. $3x^2 - 6 = 5x$

Synthesis

67. Which is larger, 7^π or π^7? 70^{80} or 80^{70}?

4.3

Logarithmic Functions and Graphs

- *Graph logarithmic functions.*
- *Convert between exponential and logarithmic equations.*
- *Find common and natural logarithms using a calculator.*

We now consider *logarithmic*, or *logarithm*, *functions.* These functions are inverses of exponential functions and have many applications.

Logarithmic Functions

Consider the function $f(x) = 2^x$ graphed at right. We see from the graph that this function passes the horizontal-line test and is therefore one-to-one. Thus, f has an inverse that is a function.

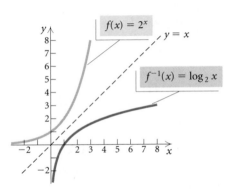

STUDY TIP

Try being a tutor for a fellow student. Understanding and retention of the concepts can be maximized when you explain the material to someone else.

To find a formula for f^{-1} when $f(x) = 2^x$, we try to use the method of Section 4.1:

1. Replace $f(x)$ with y:

$$f(x) = 2^x$$
$$y = 2^x$$

2. Interchange x and y: $\qquad x = 2^y$

3. Solve for y: $\qquad y =$ the power to which we raise 2 to get x

4. Replace y with $f^{-1}(x)$: $\quad f^{-1}(x) =$ the power to which we raise 2 to get x.

Mathematicians have defined a new symbol to replace the words "the power to which we raise 2 to get x." That symbol is "$\log_2 x$," read "the logarithm, base 2, of x."

Logarithmic Function, Base 2

"$\log_2 x$," read "the logarithm, base 2, of x," means "the power to which we raise 2 to get x."

Thus if $f(x) = 2^x$, then $f^{-1}(x) = \log_2 x$. For example,

$$f^{-1}(8) = \log_2 8 = 3,$$

because

3 is the power to which we raise 2 to get 8.

Similarly, $\log_2 13$ is the power to which we raise 2 to get 13. As yet, we have no simpler way to say this other than

"$\log_2 13$ is the power to which we raise 2 to get 13."

Later, however, we will learn how to approximate this expression using a calculator.

For any exponential function $f(x) = a^x$, its inverse is called a **logarithmic function, base *a*.** The graph of the inverse can be obtained by reflecting the graph of $y = a^x$ across the line $y = x$, to obtain $x = a^y$. Then $x = a^y$ is equivalent to $y = \log_a x$. We read $\log_a x$ as "the logarithm, base a, of x."

The inverse of $f(x) = a^x$ is given by $f^{-1}(x) = \log_a x.$

Logarithmic Function, Base a

We define $y = \log_a x$ as that number y such that $x = a^y$, where $x > 0$ and a is a positive constant other than 1.

Let's look at the graphs of $f(x) = a^x$ and $f^{-1}(x) = \log_a x$ for $a > 1$ and $0 < a < 1$.

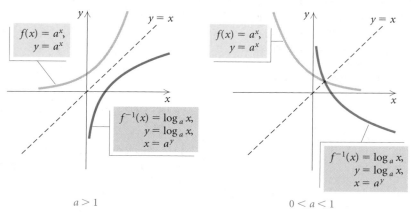

$a > 1$ $0 < a < 1$

Note that the graphs of $f(x)$ and $f^{-1}(x)$ are reflections of each other across the line $y = x$.

CONNECTING THE CONCEPTS

COMPARING EXPONENTIAL AND LOGARITHMIC FUNCTIONS

Usually, we use a number that is greater than 1 for the logarithmic base. In the following table, we compare exponential and logarithmic functions with bases a greater than 1. Similar statements would be made for a, where $0 < a < 1$. It is helpful to visualize the differences by carefully observing the graphs.

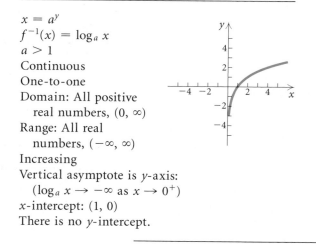

EXPONENTIAL FUNCTION

$y = a^x$
$f(x) = a^x$
$a > 1$
Continuous
One-to-one
Domain: All real
  numbers, $(-\infty, \infty)$
Range: All positive
  real numbers, $(0, \infty)$
Increasing
Horizontal asymptote is x-axis:
  $(a^x \to 0$ as $x \to -\infty)$
y-intercept: $(0, 1)$
There is no x-intercept.

LOGARITHMIC FUNCTION

$x = a^y$
$f^{-1}(x) = \log_a x$
$a > 1$
Continuous
One-to-one
Domain: All positive
  real numbers, $(0, \infty)$
Range: All real
  numbers, $(-\infty, \infty)$
Increasing
Vertical asymptote is y-axis:
  $(\log_a x \to -\infty$ as $x \to 0^+)$
x-intercept: $(1, 0)$
There is no y-intercept.

Converting Between Exponential and Logarithmic Equations

It is helpful in dealing with logarithmic functions to remember that a logarithm of a number is an *exponent*. It is the exponent y in $x = a^y$. You might think to yourself, "the logarithm, base a, of a number x is the power to which a must be raised to get x."

We are led to the following. (The symbol $\longleftrightarrow$ means that the two statements are equivalent; that is, when one is true, the other is true. The words "if and only if" can be used in place of $\longleftrightarrow$.)

$$\log_a x = y \longleftrightarrow x = a^y \qquad \text{A logarithm is an exponent!}$$

EXAMPLE 1 Convert each of the following to a logarithmic equation.

a) $16 = 2^x$ **b)** $10^{-3} = 0.001$ **c)** $e^t = 70$

Solution

                  The exponent is the logarithm.

a) $16 = 2^x \qquad \log_2 16 = x$

                  The base remains the same.

b) $10^{-3} = 0.001 \rightarrow \log_{10} 0.001 = -3$

c) $e^t = 70 \rightarrow \log_e 70 = t$

EXAMPLE 2 Convert each of the following to an exponential equation.

a) $\log_2 32 = 5$ **b)** $\log_a Q = 8$ **c)** $x = \log_t M$

Solution

                  The logarithm is the exponent.

a) $\log_2 32 = 5 \qquad 2^5 = 32$

                  The base remains the same.

b) $\log_a Q = 8 \rightarrow a^8 = Q$

c) $x = \log_t M \rightarrow t^x = M$

Finding Certain Logarithms

Let's use the definition of logarithms to find some logarithmic values.

EXAMPLE 3 Find each of the following logarithms.

a) $\log_{10} 10{,}000$ **b)** $\log_{10} 0.01$

c) $\log_2 8$ **d)** $\log_9 3$

e) $\log_6 1$ **f)** $\log_8 8$

Solution

a) The exponent to which we raise 10 to obtain 10,000 is 4, thus $\log_{10} 10{,}000 = 4$.

b) We have $0.01 = \dfrac{1}{100} = \dfrac{1}{10^2} = 10^{-2}$. The exponent to which we raise 10 to get 0.01 is -2, so $\log_{10} 0.01 = -2$.

c) $8 = 2^3$. The exponent to which we raise 2 to get 8 is 3, so $\log_2 8 = 3$.

d) $3 = \sqrt{9} = 9^{1/2}$. The exponent to which we raise 9 to get 3 is $\frac{1}{2}$, so $\log_9 3 = \frac{1}{2}$.

e) $1 = 6^0$. The exponent to which we raise 6 to get 1 is 0, so $\log_6 1 = 0$.

f) $8 = 8^1$. The exponent to which we raise 8 to get 8 is 1, so $\log_8 8 = \underline{1}$.

Examples 3(e) and 3(f) illustrate two important properties of logarithms. The property $\log_a 1 = 0$ follows from the fact that $a^0 = 1$. Thus, $\log_5 1 = 0$, $\log_{10} 1 = 0$, and so on. The property $\log_a a = 1$ follows from the fact that $a^1 = a$. Thus, $\log_5 5 = 1$, $\log_{10} 10 = 1$, and so on.

$$\log_a 1 = 0 \quad \text{and} \quad \log_a a = 1, \quad \text{for any logarithmic base } a.$$

Finding Logarithms on a Calculator

Before calculators became so widely available, base-10 logarithms, or **common logarithms,** were used extensively to simplify complicated calculations. In fact, that is why logarithms were invented. The abbreviation **log**, with no base written, is used to represent common logarithms, or base-10 logarithms. Thus,

$$\log 29 \quad \text{means} \quad \log_{10} 29.$$

Let's compare log 29 with log 10 and log 100:

$$\left.\begin{aligned} \log 10 &= \log_{10} 10 = 1 \\ \log 29 &= ? \\ \log 100 &= \log_{10} 100 = 2 \end{aligned}\right\}$$ Since 29 is between 10 and 100, it seems reasonable that log 29 is between 1 and 2.

On a calculator, the key for common logarithms is generally marked $\boxed{\text{LOG}}$. Using that key, we find that

$$\log 29 \approx 1.462397998 \approx 1.4624$$

rounded to four decimal places. Since $1 < 1.4624 < 2$, our answer seems reasonable. This also tells us that $10^{1.4624} \approx 29$.

EXAMPLE 4 Find each of the following common logarithms on a calculator. If you are using a graphing calculator, set the calculator in REAL mode. Round to four decimal places.

a) log 645,778   **b)** log 0.0000239   **c)** log (-3)

Solution

FUNCTION VALUE	READOUT	ROUNDED
a) log 645,778	log(645778) 5.810083246	5.8101
b) log 0.0000239	log(0.0000239) −4.621602099	−4.6216
c) log (-3)	ERR:NONREAL ANS	Does not exist

Since 5.810083246 is the power to which we raise 10 to get 645,778, we can check part (a) by finding $10^{5.810083246}$. We can check part (b) in a similar manner. In part (c), log (-3) does not exist as a real number because there is no real-number power to which we can raise 10 to get -3. The number 10 raised to any real-number power is positive. The common logarithm of a negative number does not exist as a real number. Recall that the domain of $f(x) = \log_a x$ is $(0, \infty)$.

Natural Logarithms

Logarithms, base e, are called **natural logarithms.** The abbreviation "ln" is generally used for natural logarithms. Thus,

> ln 53   means $\log_e 53$.

On a calculator, the key for natural logarithms is generally marked $\boxed{\text{LN}}$. Using that key, we find that

> ln 53 ≈ 3.970291914

> ≈ 3.9703

rounded to four decimal places. This also tells us that $e^{3.9703} \approx 53$.

EXAMPLE 5   Find each of the following natural logarithms on a calculator. If you are using a graphing calculator, set the calculator in REAL mode. Round to four decimal places.

a) ln 645,778   **b)** ln 0.0000239

c) ln (-5) **d)** ln e

e) ln 1

---

*If the graphing calculator is set in $a + bi$ mode, the readout is .4771212547 + 1.364376354i.

Solution

FUNCTION VALUE	READOUT	ROUNDED
a) ln 645,778	ln(645778) 13.37821107	13.3782
b) ln 0.0000239	ln(0.0000239) −10.6416321	−10.6416
c) ln (−5)	ERR:NONREAL ANS *	Does not exist
d) ln *e*	ln(e) 1	1
e) ln 1	ln(1) 0	0

Since 13.37821107 is the power to which we raise e to get 645,778, we can check part (a) by finding $e^{13.37821107}$. We can check parts (b), (d), and (e) in a similar manner. In parts (d) and (e), note that $\ln e = \log_e e = 1$ and $\ln 1 = \log_e 1 = 0$.

Changing Logarithmic Bases

Most calculators give the values of both common logarithms and natural logarithms. To find a logarithm with a base other than 10 or e, we can use the following conversion formula.

The Change-of-Base Formula

For any logarithmic bases a and b, and any positive number M,

$$\log_b M = \frac{\log_a M}{\log_a b}.$$

We will prove this result in the next section.

---

*If the graphing calculator is set in $a + bi$ mode, the readout is 1.609437912 + 3.141592654i.

EXAMPLE 6 Find $\log_5 8$ using common logarithms.

Solution First, we let $a = 10$, $b = 5$, and $M = 8$. Then we substitute into the change-of-base formula:

$$\log_5 8 = \frac{\log_{10} 8}{\log_{10} 5} \qquad \text{Substituting}$$

$$\approx 1.2920. \qquad \text{Using a calculator}$$

We can also use base e for a conversion.

EXAMPLE 7 Find $\log_5 8$ using natural logarithms.

Solution Substituting e for a, 5 for b, and 8 for M, we have

$$\log_5 8 = \frac{\log_e 8}{\log_e 5}$$

$$= \frac{\ln 8}{\ln 5} \approx 1.2920.$$

Since $\log_5 8$ is the power to which we raise 5 to get 8, we would expect this power to be greater than 1 ($5^1 = 5$) and less than 2 ($5^2 = 25$).

Graphs of Logarithmic Functions

Let's now consider graphs of logarithmic functions.

EXAMPLE 8 Graph: $y = f(x) = \log_5 x$.

Solution

The equation $y = \log_5 x$ is equivalent to $x = 5^y$. We can find ordered pairs that are solutions by choosing values for y and computing the x-values. We then plot points, remembering that x still is the first coordinate.

For $y = 0$, $x = 5^0 = 1$.
For $y = 1$, $x = 5^1 = 5$.
For $y = 2$, $x = 5^2 = 25$.
For $y = 3$, $x = 5^3 = 125$.
For $y = -1$, $x = 5^{-1} = \dfrac{1}{5}$.
For $y = -2$, $x = 5^{-2} = \dfrac{1}{25}$.

x, or 5^y	y
1	0
5	1
25	2
125	3
$\dfrac{1}{5}$	-1
$\dfrac{1}{25}$	-2

(1) Select y.
(2) Compute x.

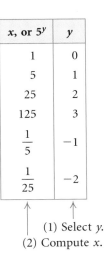

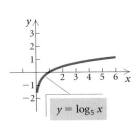

$y = \log_5 x$

To graph $y = \log_5 x$ in Example 8 with a graphing calculator, we must first change the base. Here we change from base 5 to base e:

$$y = \log_5 x = \frac{\ln x}{\ln 5}.$$

The graph is shown below.

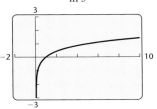

$$y = \log_5 x = \frac{\ln x}{\ln 5}$$

Some graphing calculators can graph inverses without the need to first find an equation of the inverse. If we begin with $y_1 = 5^x$, the graphs of both y_1 and its inverse, $y_2 = \log_5 x$, will be drawn as shown below.

$y_1 = 5^x$, $y_2 = \log_5 x$

EXAMPLE 9   Graph: $g(x) = \ln x$.

Solution   To graph $y = g(x) = \ln x$, we select values for x and use a calculator to find the corresponding values of $\ln x$. We then plot points and
draw the curve.

	$g(x)$
x	$g(x) = \ln x$
0.5	-0.7
1	0
2	0.7
3	1.1
4	1.4
5	1.6

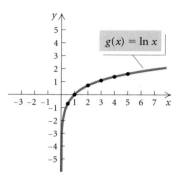

We could also write $g(x) = \ln x$, or $y = \ln x$, as $x = e^y$, select
values for y, and use a calculator to find the corresponding values of x.

Recall that the graph of $f(x) = \log_a x$, for any base a, has the
x-intercept $(1, 0)$. The domain is the set of positive real numbers, and the
range is the set of all real numbers. The y-axis is the vertical asymptote.

EXAMPLE 10   Graph each of the following. Describe how each graph
can be obtained from the graph of $y = \ln x$. Give the domain and the
vertical asymptote of each function.

a) $f(x) = \ln (x + 3)$
b) $f(x) = 3 - \frac{1}{2} \ln x$
c) $f(x) = |\ln (x - 1)|$

Solution

a) The graph is a shift of the graph of $y = \ln x$ left 3 units (see
Fig. 1). The domain is the set of all real numbers greater than -3,
$(-3, \infty)$. The line $x = -3$ is the vertical asymptote.

b) The graph is a vertical shrinking of the graph of $y = \ln x$, followed
by a reflection across the x-axis, and then a translation up 3 units (see
Fig. 2). The domain is the set of all positive real numbers, $(0, \infty)$. The
y-axis is the vertical asymptote.

c) The graph is a translation of the graph of $y = \ln x$, 1 unit to the right.
Then the absolute value has the effect of reflecting negative outputs
across the x-axis (see Fig. 3). The domain is the set of all real numbers
greater than 1, $(1, \infty)$. The line $x = 1$ is the vertical asymptote.

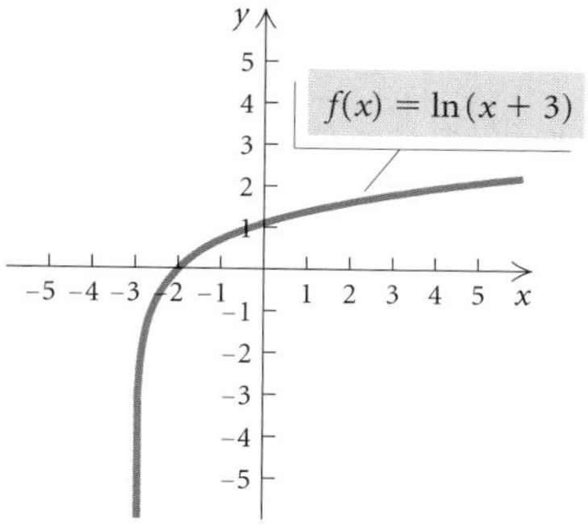

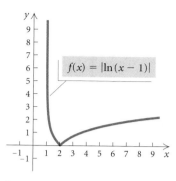

FIGURE 1

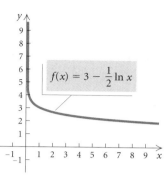

FIGURE 2

FIGURE 3

Applications

EXAMPLE 11 *Walking Speed.* In a study by psychologists Bornstein and Bornstein, it was found that the average walking speed w, in feet per second, of a person living in a city of population P, in thousands, is given by the function

$$w(P) = 0.37 \ln P + 0.05$$

(*Source*: *International Journal of Psychology*).

a) The population of San Diego, California, is 1,171,000. Find the average walking speed of people living in San Diego.

b) The population of Pueblo, Colorado, is 103,600. Find the average walking speed of people living in Pueblo.

Solution

a) We substitute 1171 for P, since P is in thousands:

$$w(1171) = 0.37 \ln 1171 + 0.05 \qquad \text{Substituting}$$
$$\approx 2.7 \text{ ft/sec.} \qquad \text{Finding the natural logarithm and simplifying}$$

The average walking speed of people living in San Diego is about 2.7 ft/sec.

b) We substitute 103.6 for P:

$$w(103.6) = 0.37 \ln 103.6 + 0.05 \qquad \text{Substituting}$$
$$\approx 1.8 \text{ ft/sec.}$$

The average walking speed of people living in Pueblo is about 1.8 ft/sec.

EXAMPLE 12 *Earthquake Magnitude.* The magnitude R, measured on the Richter scale, of an earthquake of intensity I is defined as

$$R = \log \frac{I}{I_0},$$

where I_0 is a minimum intensity used for comparison. We can think of I_0 as a threshold intensity that is the weakest earthquake that can be recorded on a seismograph. If one earthquake is 10 times as intense as another, its magnitude on the Richter scale is 1 greater than that of the other. If one earthquake is 100 times as intense as another, its magnitude on the Richter scale is 2 higher, and so on. Thus an earthquake whose magnitude is 7 on the Richter scale is 10 times as intense as an earthquake whose magnitude is 6. Earthquake intensities can be interpreted as multiples of the minimum intensity I_0.

The earthquake in Ahmedabad, India on January 26, 2001, had an intensity of $10^{7.9} \cdot I_0$. What was its magnitude on the Richter scale?

Solution We substitute into the formula:

$$R = \log \frac{I}{I_0} = \log \frac{10^{7.9} I_0}{I_0}$$
$$= \log 10^{7.9} = 7.9.$$

The magnitude of the earthquake was 7.9 on the Richter scale.

Exercise Set

Graph each of the following.

1. $y = \log_3 x$ **2.** $y = \log_4 x$

3. $f(x) = \log x$ **4.** $f(x) = \ln x$

Find each of the following. Do not use a calculator.

5. $\log_2 16$ **6.** $\log_3 9$

7. $\log_5 125$ **8.** $\log_2 64$

9. $\log 0.001$ **10.** $\log 100$

11. $\log_2 \frac{1}{4}$ **12.** $\log_8 2$

13. $\ln 1$ **14.** $\ln e$

15. $\log 10$ **16.** $\log 1$

Convert to a logarithmic equation.

17. $10^3 = 1000$ **18.** $5^{-3} = \frac{1}{125}$

19. $8^{1/3} = 2$ **20.** $10^{0.3010} = 2$

21. $e^3 = t$ **22.** $Q^t = x$

23. $e^2 = 7.3891$ **24.** $e^{-1} = 0.3679$

25. $p^k = 3$ **26.** $e^{-t} = 4000$

Convert to an exponential equation.

27. $\log_5 5 = 1$ **28.** $t = \log_4 7$

29. $\log 0.01 = -2$ **30.** $\log 7 = 0.845$

31. $\ln 30 = 3.4012$

32. $\ln 0.38 = -0.9676$

33. $\log_a M = -x$

34. $\log_t Q = k$

35. $\log_a T^3 = x$

36. $\ln W^5 = t$

Find each of the following using a calculator. Round to four decimal places.

37. $\log 3$ **38.** $\log 8$

39. $\log 532$ **40.** $\log 93{,}100$

41. $\log 0.57$ **42.** $\log 0.082$

43. $\log (-2)$ **44.** $\ln 50$

45. $\ln 2$ **46.** $\ln (-4)$

47. $\ln 809.3$ **48.** $\ln 0.00037$

49. $\ln (-1.32)$ **50.** $\ln 0$

Find the logarithm using the change-of-base formula.

51. $\log_4 100$ **52.** $\log_3 20$

53. $\log_{100} 0.3$ **54.** $\log_\pi 100$

55. $\log_{200} 50$ **56.** $\log_{5.3} 1700$

For each of the following functions, briefly describe how the graph can be obtained from a basic logarithmic function. Then graph the function. Give the domain and the vertical asymptote of each function.

57. $f(x) = \log_2 (x + 3)$ **58.** $f(x) = \log_3 (x - 2)$

59. $y = \log_3 x - 1$ **60.** $y = 3 + \log_2 x$

61. $f(x) = 4 \ln x$ **62.** $f(x) = \frac{1}{2} \ln x$

63. $y = 2 - \ln x$ **64.** $y = \ln (x + 1)$

Graph the function and its inverse using the same set of axes.

65. $f(x) = 3^x$, $f^{-1}(x) = \log_3 x$

66. $f(x) = \log_4 x$, $f^{-1}(x) = 4^x$

67. $f(x) = \log x$, $f^{-1}(x) = 10^x$

68. $f(x) = e^x$, $f^{-1}(x) = \ln x$

69. *Walking Speed.* Refer to Example 11. Various cities and their populations are given below. Find the average walking speed in each city.

    a) Albuquerque, New Mexico: 419,681
    b) Chicago, Illinois: 2,721,547
    c) Pittsburgh, Pennsylvania: 350,363
    d) Durham, North Carolina: 149,799
    e) Green Bay, Wisconsin: 94,466

70. *Earthquake Magnitude.* Refer to Example 12. Various locations of earthquakes and their intensities are given below. What was the magnitude on the Richter scale?

    a) Mexico City, 1978: $10^{7.85} \cdot I_0$
    b) San Francisco, 1906: $10^{8.25} \cdot I_0$
    c) Chile, 1960: $10^{9.6} \cdot I_0$
    d) Italy, 1980: $10^{7.85} \cdot I_0$
    e) San Francisco, 1989: $10^{6.9} \cdot I_0$

71. *Forgetting.* Students in an accounting class took a final exam. They took equivalent forms of the exam in monthly intervals thereafter. The average score $S(t)$, as a percent, after t months was found to be given by the function

$$S(t) = 78 - 15 \log (t + 1), \quad t \geq 0.$$

    a) What was the average score when they initially took the test, $t = 0$?
    b) What was the average score after 4 months? 24 months?

72. *pH of Substances in Chemistry.* In chemistry, the pH of a substance is defined as

$$pH = -\log [H^+],$$

where H^+ is the hydrogen ion concentration, in moles per liter. Find the pH of each substance.

Litmus paper is used to test pH.

SUBSTANCE	HYDROGEN ION CONCENTRATION
a) Pineapple juice	1.6×10^{-4}
b) Hair rinse	0.0013
c) Mouthwash	6.3×10^{-7}
d) Eggs	1.6×10^{-8}
e) Tomatoes	6.3×10^{-5}

73. Find the hydrogen ion concentration of each substance, given the pH (see Exercise 72). Express the answer in scientific notation.

SUBSTANCE	pH
a) Tap water	7
b) Rainwater	5.4
c) Orange juice	3.2
d) Wine	4.8

74. *Advertising.* A model for advertising response is given by the function

$$N(a) = 1000 + 200 \ln a, \quad a \geq 1,$$

where $N(a)$ is the number of units sold when a is the amount spent on advertising, in thousands of dollars.

    a) How many units were sold after spending $1000 ($a = 1$) on advertising?
    b) How many units were sold after spending $5000?

75. *Loudness of Sound.* The **loudness L**, in bels (after Alexander Graham Bell), of a sound of intensity I is defined to be

$$L = \log \frac{I}{I_0},$$

where I_0 is the minimum intensity detectable by the human ear (such as the tick of a watch at 20 ft under quiet conditions). If a sound is 10 times as intense as another, its loudness is 1 bel greater than that of the other. If a sound is 100 times as intense as another, its loudness is 2 bels greater, and so on. The bel is a large unit, so a subunit, the **decibel**, is generally used. For L, in decibels, the formula is

$$L = 10 \log \frac{I}{I_0}.$$

Find the loudness, in decibels, of each sound with the given intensity.

SOUND	INTENSITY
a) Library	$2510 \cdot I_0$
b) Dishwasher	$2{,}500{,}000 \cdot I_0$
c) Conversational speech	$10^6 \cdot I_0$
d) Heavy truck	$10^9 \cdot I_0$

Technology Connection

76. Graph the functions in each of Exercises 65–68 along with their inverses. Use a graphing calculator that can graph inverses without the need to first find an equation of the inverse.

Collaborative Discussion and Writing

77. If $\log b < 0$, what can you say about b?

78. Explain how the graph of $f(x) = \ln x$ can be used to obtain the graph of $g(x) = e^{x-2}$.

Skill Maintenance

Use synthetic division to find the function values.

79. $g(x) = x^3 - 6x^2 + 3x + 10$; find $f(-5)$

80. $f(x) = x^4 - 2x^3 + x - 6$; find $f(-1)$

Find a polynomial function of degree 3 with the given numbers as zeros.

81. $\sqrt{7}, -\sqrt{7}, 0$

82. $4i, -4i, 1$

Synthesis

Simplify.

83. $\dfrac{\log_5 8}{\log_5 2}$

84. $\dfrac{\log_3 64}{\log_3 16}$

Find the domain of the function.

85. $f(x) = \log_5 x^3$

86. $f(x) = \log_4 x^2$

87. $f(x) = \ln |x|$

88. $f(x) = \log (3x - 4)$

Solve.

89. $\log_2 (2x + 5) < 0$

90. $\log_2 (x - 3) \geq 4$

In Exercises 91–94, match the equation with one of figures (a)–(d), which follow.

a)

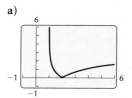

b)

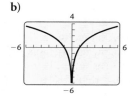

c)

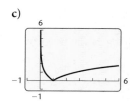

d)

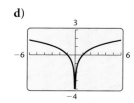

91. $f(x) = \ln |x|$

92. $f(x) = |\ln x|$

93. $f(x) = \ln x^2$

94. $g(x) = |\ln (x - 1)|$

4.4

Properties of Logarithmic Functions

- *Convert from logarithms of products, powers, and quotients to expressions in terms of individual logarithms, and conversely.*
- *Simplify expressions of the type $\log_a a^x$ and $a^{\log_a x}$.*

We now establish some properties of logarithmic functions. These properties are based on corresponding rules for exponents.

Logarithms of Products

The first property of logarithms corresponds to the product rule for exponents: $a^m \cdot a^n = a^{m+n}$.

The Product Rule

For any positive numbers M and N and any logarithmic base a,

$$\log_a MN = \log_a M + \log_a N.$$

(The logarithm of a product is the sum of the logarithms of the factors.)

EXAMPLE 1   Express as a sum of logarithms: $\log_3 (9 \cdot 27)$.

Solution   We have

$$\log_3 (9 \cdot 27) = \log_3 9 + \log_3 27. \qquad \text{Using the product rule}$$

As a check, note that

$$\log_3 (9 \cdot 27) = \log_3 243 = 5$$

and $\log_3 9 + \log_3 27 = 2 + 3 = 5.$

EXAMPLE 2   Express as a single logarithm: $\log_2 p^3 + \log_2 q$.

Solution   We have

$$\log_2 p^3 + \log_2 q = \log_2 (p^3 q).$$

A PROOF OF THE PRODUCT RULE: Let $\log_a M = x$ and $\log_a N = y$. Converting to exponential equations, we have $a^x = M$ and $a^y = N$. Then

$$MN = a^x \cdot a^y = a^{x+y}.$$

Converting back to a logarithmic equation, we get

$$\log_a MN = x + y.$$

Remembering what x and y represent, we know it follows that

$$\log_a MN = \log_a M + \log_a N.$$

Logarithms of Powers

The second property of logarithms corresponds to the power rule for exponents: $(a^m)^n = a^{mn}$.

The Power Rule

For any positive number M, any logarithmic base a, and any real number p,

$$\log_a M^p = p \log_a M.$$

(The logarithm of a power of M is the exponent times the logarithm of M.)

EXAMPLE 3   Express each of the following as a product.

a) $\log_a 11^{-3}$ **b)** $\log_a \sqrt[4]{7}$

Solution

a) $\log_a 11^{-3} = -3 \log_a 11$ Using the power rule

b) $\log_a \sqrt[4]{7} = \log_a 7^{1/4}$ Writing exponential notation

$\phantom{\log_a \sqrt[4]{7}} = \frac{1}{4} \log_a 7$ Using the power rule

A PROOF OF THE POWER RULE: Let $x = \log_a M$. The equivalent exponential equation is $a^x = M$. Raising both sides to the power p, we obtain

$$(a^x)^p = M^p, \quad \text{or} \quad a^{xp} = M^p.$$

Converting back to a logarithmic equation, we get

$$\log_a M^p = xp.$$

But $x = \log_a M$, so substituting gives us

$$\log_a M^p = (\log_a M)p$$

$$ = p \log_a M.$$

Logarithms of Quotients

The third property of logarithms corresponds to the quotient rule for exponents: $a^m/a^n = a^{m-n}$.

The Quotient Rule

For any positive numbers M and N, and any logarithmic base a,

$$\log_a \frac{M}{N} = \log_a M - \log_a N.$$

(The logarithm of a quotient is the logarithm of the numerator minus the logarithm of the denominator.)

EXAMPLE 4 Express as a difference of logarithms: $\log_t \dfrac{8}{w}$.

Solution

$$\log_t \frac{8}{w} = \log_t 8 - \log_t w \qquad \text{Using the quotient rule}$$

EXAMPLE 5 Express as a single logarithm: $\log_b 64 - \log_b 16$.

Solution

$$\log_b 64 - \log_b 16 = \log_b \frac{64}{16} = \log_b 4$$

A PROOF OF THE QUOTIENT RULE: The proof follows from both the product and the power rules:

$$\log_a \frac{M}{N} = \log_a MN^{-1}$$

$$= \log_a M + \log_a N^{-1} \qquad \text{Using the product rule}$$

$$= \log_a M + (-1) \log_a N \qquad \text{Using the power rule}$$

$$= \log_a M - \log_a N.$$

Note the following.

Common Errors

$\log_a MN \neq (\log_a M)(\log_a N)$	The logarithm of a product is *not* the product of the logarithms.
$\log_a (M + N) \neq \log_a M + \log_a N$	The logarithm of a sum is *not* the sum of the logarithms.
$\log_a \dfrac{M}{N} \neq \dfrac{\log_a M}{\log_a N}$	The logarithm of a quotient is *not* the quotient of the logarithms.
$(\log_a M)^P \neq P \log_a M$	The power of a logarithm is *not* the exponent times the logarithm.

Applying the Properties

EXAMPLE 6 Express each of the following in terms of sums and differences of logarithms.

a) $\log_a \dfrac{x^2 y^5}{z^4}$ **b)** $\log_a \sqrt[3]{\dfrac{a^2 b}{c^5}}$ **c)** $\log_b \dfrac{a y^5}{m^3 n^4}$

Solution

a) $\log_a \dfrac{x^2 y^5}{z^4} = \log_a (x^2 y^5) - \log_a z^4 \qquad \text{Using the quotient rule}$

$$= \log_a x^2 + \log_a y^5 - \log_a z^4 \qquad \text{Using the product rule}$$

$$= 2 \log_a x + 5 \log_a y - 4 \log_a z \qquad \text{Using the power rule}$$

b) $\log_a \sqrt[3]{\dfrac{a^2b}{c^5}} = \log_a \left(\dfrac{a^2b}{c^5}\right)^{1/3}$ Writing exponential notation

$= \dfrac{1}{3}\log_a \dfrac{a^2b}{c^5}$ Using the power rule

$= \dfrac{1}{3}(\log_a a^2b - \log_a c^5)$ Using the quotient rule. The parentheses are important.

$= \dfrac{1}{3}(2\log_a a + \log_a b - 5\log_a c)$ Using the product and power rules

$= \dfrac{1}{3}(2 + \log_a b - 5\log_a c)$ $\log_a a = 1$

$= \dfrac{2}{3} + \dfrac{1}{3}\log_a b - \dfrac{5}{3}\log_a c$ Multiplying to remove parentheses

c) $\log_b \dfrac{ay^5}{m^3n^4} = \log_b ay^5 - \log_b m^3n^4$ Using the quotient rule

$= (\log_b a + \log_b y^5) - (\log_b m^3 + \log_b n^4)$ Using the product rule

$= \log_b a + \log_b y^5 - \log_b m^3 - \log_b n^4$ Removing parentheses

$= \log_b a + 5\log_b y - 3\log_b m - 4\log_b n$ Using the power rule

EXAMPLE 7   Express as a single logarithm:

$$5\log_b x - \log_b y + \dfrac{1}{4}\log_b z.$$

Solution

$5\log_b x - \log_b y + \dfrac{1}{4}\log_b z = \log_b x^5 - \log_b y + \log_b z^{1/4}$

     Using the power rule

$= \log_b \dfrac{x^5}{y} + \log_b z^{1/4}$

     Using the quotient rule

$= \log_b \dfrac{x^5 z^{1/4}}{y}, \text{ or } \log_b \dfrac{x^5\sqrt[4]{z}}{y}$

     Using the product rule

EXAMPLE 8   Given that $\log_a 2 \approx 0.301$ and $\log_a 3 \approx 0.477$, find each of the following.

a) $\log_a 6$ **b)** $\log_a \dfrac{2}{3}$ **c)** $\log_a 81$

d) $\log_a \sqrt{a}$ **e)** $\log_a 5$ **f)** $\dfrac{\log_a 3}{\log_a 2}$

Solution

a) $\log_a 6 = \log_a (2 \cdot 3) = \log_a 2 + \log_a 3$ Using the product rule
$$\approx 0.301 + 0.477$$
$$\approx 0.778$$

b) $\log_a \frac{2}{3} = \log_a 2 - \log_a 3$ Using the quotient rule
$$\approx 0.301 - 0.477 \approx -0.176$$

c) $\log_a 81 = \log_a 3^4 = 4 \log_a 3$ Using the power rule
$$\approx 4(0.477) \approx 1.908$$

d) $\log_a \sqrt{a} = \log_a a^{1/2} = \frac{1}{2} \log_a a$ Using the power rule
$$= \frac{1}{2} \cdot 1 = \frac{1}{2}$$

e) $\log_a 5$ *cannot be found using these properties and the given information.*

$$(\log_a 5 \neq \log_a 2 + \log_a 3) \qquad \log_a 2 + \log_a 3 = \log_a 2 \cdot 3 = \log_a 6$$

f) $\dfrac{\log_a 3}{\log_a 2} \approx \dfrac{0.477}{0.301} \approx 1.585$ We simply divided, not using any of the properties.

In Example 8, a is actually 10 so we have common logarithms.

Simplifying Expressions of the Type $\log_a a^x$ and $a^{\log_a x}$

We have two final properties to consider. The first follows from the product rule: Since $\log_a a^x = x \log_a a = x \cdot 1 = x$, we have $\log_a a^x = x$. This property also follows from the definition of a logarithm: x is the power to which we raise a in order to get a^x.

The Logarithm of a Base to a Power

For any base a and any real number x,

$$\log_a a^x = x.$$

(The logarithm, base a, of a to a power is the power.)

EXAMPLE 9   Simplify each of the following.

a) $\log_a a^8$ **b)** $\ln e^{-t}$ **c)** $\log 10^{3k}$

Solution

a) $\log_a a^8 = 8$ 8 is the power to which we raise a in order to get a^8.

b) $\ln e^{-t} = \log_e e^{-t} = -t$ $\ln e^x = x$

c) $\log 10^{3k} = \log_{10} 10^{3k} = 3k$ —

Let $M = \log_a x$. Then $a^M = x$. Substituting $\log_a x$ for M, we obtain $a^{\log_a x} = x$. This also follows from the definition of a logarithm: $\log_a x$ is the power to which a is raised in order to get x.

A Base to a Logarithmic Power

For any base a and any positive real number x,

$$a^{\log_a x} = x.$$

(The number a raised to the power $\log_a x$ is x.)

EXAMPLE 10   Simplify each of the following.

a) $4^{\log_4 k}$ **b)** $e^{\ln 5}$ **c)** $10^{\log 7t}$

Solution

a) $4^{\log_4 k} = k$

b) $e^{\ln 5} = e^{\log_e 5} = 5$

c) $10^{\log 7t} = 10^{\log_{10} 7t} = 7t$ —

A PROOF OF THE CHANGE-OF-BASE FORMULA: We close this section by proving the change-of-base formula and summarizing the properties of logarithms considered thus far in this chapter. In Section 4.3, we used the change-of-base formula,

$$\log_b M = \frac{\log_a M}{\log_a b},$$

to make base conversions in order to find logarithmic values using a calculator. Let $x = \log_b M$. Then

$b^x = M$	Definition of logarithm
$\log_a b^x = \log_a M$	Taking the log on both sides
$x \log_a b = \log_a M$	Using the power rule
$x = \dfrac{\log_a M}{\log_a b},$	Dividing by $\log_a b$

so

$$x = \log_b M = \frac{\log_a M}{\log_a b}.$$

Following is a summary of the properties of logarithms.

> **Summary of the Properties of Logarithms**
>
> The Product Rule: $\qquad\qquad\qquad\log_a MN = \log_a M + \log_a N$
>
> The Power Rule: $\qquad\qquad\qquad\log_a M^p = p \log_a M$
>
> The Quotient Rule: $\qquad\qquad\log_a \dfrac{M}{N} = \log_a M - \log_a N$
>
> The Change-of-Base Formula: $\quad\log_b M = \dfrac{\log_a M}{\log_a b}$
>
> Other Properties: $\qquad\qquad\log_a a = 1, \qquad \log_a 1 = 0,$
> $\qquad\qquad\qquad\qquad\qquad\quad\log_a a^x = x, \qquad a^{\log_a x} = x$

Exercise Set

Express as a sum of logarithms.

1. $\log_3 (81 \cdot 27)$

2. $\log_2 (8 \cdot 64)$

3. $\log_5 (5 \cdot 125)$

4. $\log_4 (64 \cdot 32)$

5. $\log_t 8Y$

6. $\log_e Qx$

Express as a product.

7. $\log_b t^3$

8. $\log_a x^4$

9. $\log y^8$

10. $\ln y^5$

11. $\log_c K^{-6}$

12. $\log_b Q^{-8}$

Express as a difference of logarithms.

13. $\log_t \dfrac{M}{8}$

14. $\log_a \dfrac{76}{13}$

15. $\log_a \dfrac{x}{y}$

16. $\log_b \dfrac{3}{w}$

Express in terms of sums and differences of logarithms.

17. $\log_a 6xy^5z^4$

18. $\log_a x^3y^2z$

19. $\log_b \dfrac{p^2q^5}{m^4b^9}$

20. $\log_b \dfrac{x^2y}{b^3}$

21. $\log_a \sqrt{\dfrac{x^6}{p^5q^8}}$

22. $\log_c \sqrt[3]{\dfrac{y^3z^2}{x^4}}$

23. $\log_a \sqrt[4]{\dfrac{m^8n^{12}}{a^3b^5}}$

24. $\log_a \sqrt{\dfrac{a^6b^8}{a^2b^5}}$

Express as a single logarithm and, if possible, simplify.

25. $\log_a 75 + \log_a 2$

26. $\log 0.01 + \log 1000$

27. $\log 10{,}000 - \log 100$

28. $\ln 54 - \ln 6$

29. $\frac{1}{2} \log_a x + 4 \log_a y - 3 \log_a x$

30. $\frac{2}{5} \log_a x - \frac{1}{3} \log_a y$

31. $\ln x^2 - 2 \ln \sqrt{x}$

32. $\ln 2x + 3(\ln x - \ln y)$

33. $\ln (x^2 - 4) - \ln (x + 2)$

34. $\log_a \dfrac{a}{\sqrt{x}} - \log_a \sqrt{ax}$

35. $\ln x - 3[\ln (x - 5) + \ln (x + 5)]$

36. $\frac{2}{3}[\ln (x^2 - 9) - \ln (x + 3)] + \ln (x + y)$

37. $\frac{3}{2} \ln 4x^6 - \frac{4}{5} \ln 2y^{10}$

38. $120(\ln \sqrt[5]{x^3} + \ln \sqrt[3]{y^2} - \ln \sqrt[4]{16z^5})$

Given that $\log_b 3 = 1.0986$ and $\log_b 5 = 1.6094$, find each of the following.

39. $\log_b \frac{3}{5}$

40. $\log_b 15$

41. $\log_b \frac{1}{5}$

42. $\log_b \frac{5}{3}$

43. $\log_b \sqrt{b}$

44. $\log_b \sqrt{b^3}$

45. $\log_b 5b$

46. $\log_b 9$

47. $\log_b 75$

48. $\log_b \frac{1}{b}$

Simplify.

49. $\log_p p^3$

50. $\log_t t^{2713}$

51. $\log_e e^{|x-4|}$

52. $\log_q q^{\sqrt{3}}$

53. $3^{\log_3 4x}$

54. $5^{\log_5 (4x-3)}$

55. $10^{\log w}$

56. $e^{\ln x^3}$

57. $\ln e^{8t}$

58. $\log 10^{-k}$

Collaborative Discussion and Writing

59. Given that $f(x) = a^x$ and $g(x) = \log_a x$, find $(f \circ g)(x)$ and $(g \circ f)(x)$. These results are alternative proofs of what properties of logarithms already proven in this section? Explain.

60. Explain the errors, if any, in the following:
$$\log_a ab^3 = (\log_a a)(\log_a b^3) = 3 \log_a b.$$

Skill Maintenance

Simplify.

61. $(1 - 4i)(7 + 6i)$

62. $\dfrac{2 - i}{3 + i}$

Find the x-intercepts and the zeros of the function.

63. $f(x) = 2x^2 - 13x - 7$

64. $h(x) = x^3 - 3x^2 + 3x - 1$

Synthesis

Solve for x.

65. $5^{\log_5 8} = 2x$

66. $\ln e^{3x-5} = -8$

Express as a single logarithm and, if possible, simplify.

67. $\log_a (x^2 + xy + y^2) + \log_a (x - y)$

68. $\log_a (a^{10} - b^{10}) - \log_a (a + b)$

Express as a sum or a difference of logarithms.

69. $\log_a \dfrac{x - y}{\sqrt{x^2 - y^2}}$

70. $\log_a \sqrt{9 - x^2}$

71. Given that $\log_a x = 2$, $\log_a y = 3$, and $\log_a z = 4$, find
$$\log_a \frac{\sqrt[4]{y^2 z^5}}{\sqrt[4]{x^3 z^{-2}}}.$$

Determine whether each of the following is true. Assume that a, x, M, and N are positive.

72. $\log_a M + \log_a N = \log_a (M + N)$

73. $\log_a M - \log_a N = \log_a \dfrac{M}{N}$

74. $\dfrac{\log_a M}{\log_a N} = \log_a M - \log_a N$

75. $\dfrac{\log_a M}{x} = \log_a M^{1/x}$

76. $\log_a x^3 = 3 \log_a x$

77. $\log_a 8x = \log_a x + \log_a 8$

78. $\log_N (MN)^x = x \log_N M + x$

Suppose that $\log_a x = 2$. Find each of the following.

79. $\log_a \left(\dfrac{1}{x}\right)$

80. $\log_{1/a} x$

81. Simplify:
$$\log_{10} 11 \cdot \log_{11} 12 \cdot \log_{12} 13 \cdots \log_{998} 999 \cdot \log_{999} 1000.$$

Prove each of the following for any base a and any positive number x.

82. $\log_a \left(\dfrac{1}{x}\right) = -\log_a x = \log_{1/a} x$

83. $\log_a \left(\dfrac{x + \sqrt{x^2 - 5}}{5}\right) = -\log_a (x - \sqrt{x^2 - 5})$

4.5

Solving Exponential and Logarithmic Equations

• *Solve exponential and logarithmic equations.*

Solving Exponential Equations

Equations with variables in the exponents, such as
$$3^x = 20 \quad \text{and} \quad 2^{5x} = 64,$$

are called **exponential equations.** We now consider solving exponential equations.

Sometimes, as is the case with the equation $2^{5x} = 64$, we can write each side as a power of the same number:
$$2^{5x} = 2^6.$$

We can then set the exponents equal and solve:

$$5x = 6$$
$$x = \tfrac{6}{5}, \text{ or } 1.2.$$

We use the following property.

Base–Exponent Property

For any $a > 0$, $a \neq 1$,

$$a^x = a^y \longleftrightarrow x = y.$$

ONE–TO–ONE FUNCTIONS

REVIEW SECTION 4.1.

This property follows from the fact that for any $a > 0$, $a \neq 1$, $f(x) = a^x$ is a one-to-one function. If $a^x = a^y$, then $f(x) = f(y)$. Then since f is one-to-one, it follows that $x = y$. Conversely, if $x = y$, it follows that $a^x = a^y$, since we are raising a to the same power.

EXAMPLE 1   Solve: $2^{3x-7} = 32$.

Algebraic Solution

Note that $32 = 2^5$. Thus we can write each side as a power of the same number:

$$2^{3x-7} = 2^5.$$

Since the bases are the same number, 2, we can use the base–exponent property and set the exponents equal:

$$3x - 7 = 5$$
$$3x = 12$$
$$x = 4.$$

CHECK:
$$\frac{2^{3x-7} = 32}{}$$
$$2^{3(4)-7} \;?\; 32$$
$$2^{12-7}$$
$$2^5$$
$$32 \;\Big|\; 32 \quad \text{TRUE}$$

The solution is 4.

Visualizing the Solution

When we graph $y = 2^{3x-7}$ and $y = 32$, we find that the first coordinate of the point of intersection of the graphs is 4.

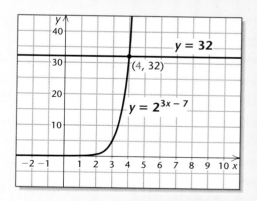

The solution of $2^{3x-7} = 32$ is 4.

When it does not seem possible to write each side as a power of the same base, we can take the common or natural logarithm on each side and use the power rule for logarithms.

EXAMPLE 2   Solve: $3^x = 20$.

Algebraic Solution

We have

$$3^x = 20$$

$$\log 3^x = \log 20 \qquad \text{Taking the common logarithm on both sides}$$

$$x \log 3 = \log 20 \qquad \text{Using the power rule}$$

$$x = \frac{\log 20}{\log 3}. \qquad \text{Dividing by } \log 3$$

This is an exact answer. We cannot simplify further, but we can approximate using a calculator:

$$x = \frac{\log 20}{\log 3} \approx 2.7268.$$

We can check this by finding $3^{2.7268}$:

$$3^{2.7268} \approx 20.$$

The solution is about 2.7268.

Visualizing the Solution

We graph $y = 3^x$ and $y = 20$. The first coordinate of the point of intersection of the graphs is the value of x for which $3^x = 20$ and is thus the solution of the equation.

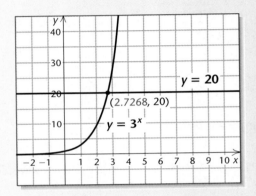

The solution is approximately 2.7268.

It will make our work easier if we take the natural logarithm when working with equations that have e as a base.

EXAMPLE 3 Solve: $e^{0.08t} = 2500$.

Algebraic Solution

We have

$$e^{0.08t} = 2500$$

$$\ln e^{0.08t} = \ln 2500 \quad \text{Taking the natural logarithm on both sides}$$

$$0.08t = \ln 2500 \quad \text{Finding the logarithm of a base to a power: } \log_a a^x = x$$

$$t = \frac{\ln 2500}{0.08} \quad \text{Dividing by 0.08}$$

$$\approx 97.8.$$

The solution is about 97.8.

Visualizing the Solution

The first coordinate of the point of intersection of the graphs of $y = e^{0.08t}$ and $y = 2500$ is about 97.8. This is the solution of the equation.

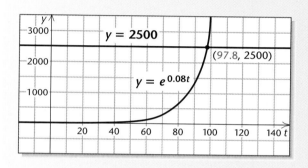

Technology Connection

We can solve the equations in Examples 1–3 using the Intersect method. In Example 3, for instance, we graph $y = e^{0.08x}$ and $y = 2500$ and use the INTERSECT feature to find the coordinates of the point of intersection.

The first coordinate of the point of intersection is the solution of the equation $e^{0.08x} = 2500$. The solution is about 97.8. We could also write the equation in the form $e^{0.08x} - 2500 = 0$ and use the Zero method.

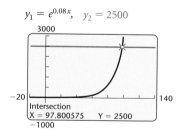

EXAMPLE 4   Solve: $e^x + e^{-x} - 6 = 0$.

Algebraic Solution

In this case, we have more than one term with x in the exponent:

$$e^x + e^{-x} - 6 = 0$$

$$e^x + \frac{1}{e^x} - 6 = 0 \qquad \text{Rewriting } e^{-x} \text{ with a positive exponent}$$

$$e^{2x} + 1 - 6e^x = 0. \qquad \text{Multiplying by } e^x \text{ on both sides}$$

This equation is reducible to quadratic with $u = e^x$:

$$u^2 - 6u + 1 = 0.$$

Using the quadratic formula, we have

$$u = \frac{-(-6) \pm \sqrt{(-6)^2 - 4 \cdot 1 \cdot 1}}{2 \cdot 1}$$

$$u = \frac{6 \pm \sqrt{32}}{2} = \frac{6 \pm 4\sqrt{2}}{2}$$

$$u = 3 \pm 2\sqrt{2}$$

$$e^x = 3 \pm 2\sqrt{2}. \qquad \text{Replacing } u \text{ with } e^x$$

We now take the natural logarithm on both sides:

$$\ln e^x = \ln (3 \pm 2\sqrt{2})$$

$$x = \ln (3 \pm 2\sqrt{2}). \qquad \text{Using } \ln e^x = x$$

Approximating each of the solutions, we obtain 1.76 and -1.76.

Visualizing the Solution

The solutions of the equation

$$e^x + e^{-x} - 6 = 0$$

are the zeros of the function

$$f(x) = e^x + e^{-x} - 6.$$

Note that the solutions are also the first coordinates of the x-intercepts of the graph of the function.

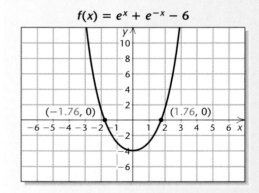

$$f(x) = e^x + e^{-x} - 6$$

The leftmost zero is about -1.76. The zero on the right is about 1.76. The solutions of the equation are approximately -1.76 and 1.76.

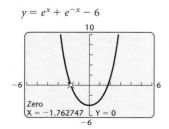

$$y = e^x + e^{-x} - 6$$

Zero
X = −1.762747   Y = 0

Technology Connection

We can use the Zero method in Example 4 to solve the equation $e^x + e^{-x} - 6 = 0$. We graph the function $y = e^x + e^{-x} - 6$ and use the ZERO feature to find the zeros.

The leftmost zero is about -1.76. Using the ZERO feature one more time, we find that the other zero is about 1.76.

Solving Logarithmic Equations

Equations containing variables in logarithmic expressions, such as $\log_2 x = 4$ and $\log x + \log (x + 3) = 1$, are called **logarithmic equations.**

To solve logarithmic equations algebraically, first try to obtain a single logarithmic expression on one side and then write an equivalent exponential equation.

EXAMPLE 5   Solve: $\log_3 x = -2$.

Algebraic Solution

We have

$$\log_3 x = -2$$

$$3^{-2} = x \qquad \text{Converting to an exponential equation}$$

$$\frac{1}{3^2} = x$$

$$\frac{1}{9} = x.$$

CHECK:
$$\log_3 x = -2$$

$$\log_3 \frac{1}{9} \; ? \; -2$$

$$\log_3 3^{-2} \;\Big|$$

$$-2 \;\Big|\; -2 \quad \text{TRUE}$$

The solution is $\frac{1}{9}$.

Visualizing the Solution

When we graph $y = \log_3 x$ and $y = -2$, we find that the first coordinate of the point of intersection of the graphs is $\frac{1}{9}$.

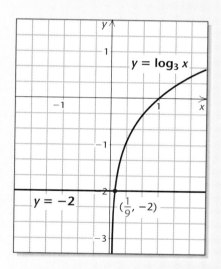

The solution of $\log_3 x = -2$ is $\frac{1}{9}$.

EXAMPLE 6 Solve: $\log x + \log (x + 3) = 1$.

Algebraic Solution

In this case, we have common logarithms. Writing the base of 10 will help us understand the problem:

$$\log_{10} x + \log_{10} (x + 3) = 1$$

$$\log_{10} [x(x + 3)] = 1 \qquad \text{Using the product rule to obtain a single logarithm}$$

$$x(x + 3) = 10^1 \qquad \text{Writing an equivalent exponential equation}$$

$$x^2 + 3x = 10$$

$$x^2 + 3x - 10 = 0$$

$$(x - 2)(x + 5) = 0 \qquad \text{Factoring}$$

$$x - 2 = 0 \quad or \quad x + 5 = 0$$

$$x = 2 \quad or \qquad x = -5.$$

CHECK: For 2:

$$\log x + \log (x + 3) = 1$$

$$\overline{\log 2 + \log (2 + 3) \; ? \; 1}$$

$$\log 2 + \log 5$$

$$\log (2 \cdot 5)$$

$$\log 10$$

$$1 \; \bigl| \; 1 \quad \text{TRUE}$$

For -5:

$$\log x + \log (x + 3) = 1$$

$$\overline{\log (-5) + \log (-5 + 3) \; ? \; 1} \quad \text{FALSE}$$

The number -5 is not a solution because negative numbers do not have real-number logarithms. The solution is 2.

Visualizing the Solution

The solution of the equation

$$\log x + \log (x + 3) = 1$$

is the zero of the function

$$f(x) = \log x + \log (x + 3) - 1.$$

The solution is also the first coordinate of the x-intercept of the graph of the function.

$$f(x) = \log x + \log (x + 3) - 1$$

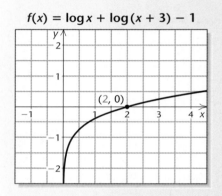

The solution of the equation is 2. From the graph, we can easily see that there is only one solution.

EXAMPLE 7   Solve: $\log_3 (2x - 1) - \log_3 (x - 4) = 2$.

Algebraic Solution

$\log_3 (2x - 1) - \log_3 (x - 4) = 2$

$$\log_3 \frac{2x - 1}{x - 4} = 2 \qquad \text{Using the}$$
quotient rule

$$\frac{2x - 1}{x - 4} = 3^2 \qquad \text{Writing an equivalent}$$
exponential equation

$$\frac{2x - 1}{x - 4} = 9$$

$$2x - 1 = 9(x - 4)$$
Multiplying by the LCD, $x - 4$

$$2x - 1 = 9x - 36$$

$$35 = 7x$$

$$5 = x.$$

CHECK: $\dfrac{\log_3 (2x - 1) - \log_3 (x - 4) = 2}{}$

$\log_3 (2 \cdot 5 - 1) - \log_3 (5 - 4) \ ? \ 2$

$\log_3 9 - \log_3 1 \ \bigg|$

$2 - 0 \ \bigg|$

$2 \ \bigg| \ 2 \quad$ TRUE

The solution is 5.

Visualizing the Solution

We see that the first coordinate of the point of intersection of the graphs of

$$y = \log_3 (2x - 1) - \log_3 (x - 4)$$

and

$$y = 2$$

is 5.

$y = \log_3 (2x - 1) - \log_3 (x - 4)$

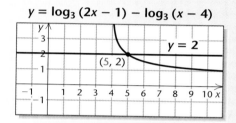

The solution is 5.

Technology Connection

In Example 6, we can graph the equations

$$y_1 = \log x + \log (x + 3)$$

and

$$y_2 = 1$$

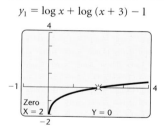

and use the Intersect method. The first coordinate of the point of intersection is the solution of the equation.

We could also graph the function

$$y = \log x + \log (x + 3) - 1$$

and use the Zero method. The zero of the function is the solution of the equation.

With either method, we see that the solution is 2. Note that the graphical solution gives only the one *true* solution.

Exercise Set 4.5

Solve the exponential equation.

1. $3^x = 81$

2. $2^x = 32$

3. $2^{2x} = 8$

4. $3^{7x} = 27$

5. $2^x = 33$

6. $2^x = 40$

7. $5^{4x-7} = 125$

8. $4^{3x-5} = 16$

9. $27 = 3^{5x} \cdot 9^{x^2}$

10. $3^{x^2+4x} = \frac{1}{27}$

11. $84^x = 70$

12. $28^x = 10^{-3x}$

13. $e^t = 1000$

14. $e^{-t} = 0.04$

15. $e^{-0.03t} = 0.08$

16. $1000e^{0.09t} = 5000$

17. $3^x = 2^{x-1}$

18. $5^{x+2} = 4^{1-x}$

19. $(3.9)^x = 48$

20. $250 - (1.87)^x = 0$

21. $e^x + e^{-x} = 5$

22. $e^x - 6e^{-x} = 1$

23. $\dfrac{e^x + e^{-x}}{e^x - e^{-x}} = 3$

24. $\dfrac{5^x - 5^{-x}}{5^x + 5^{-x}} = 8$

Solve the logarithmic equation.

25. $\log_5 x = 4$

26. $\log_2 x = -3$

27. $\log x = -4$

28. $\log x = 1$

29. $\ln x = 1$

30. $\ln x = -2$

31. $\log_2 (10 + 3x) = 5$

32. $\log_5 (8 - 7x) = 3$

33. $\log x + \log (x - 9) = 1$

34. $\log_2 (x + 1) + \log_2 (x - 1) = 3$

35. $\log_8 (x + 1) - \log_8 x = 2$

36. $\log x - \log (x + 3) = -1$

37. $\log_4 (x + 3) + \log_4 (x - 3) = 2$

38. $\ln (x + 1) - \ln x = \ln 4$

39. $\log (2x + 1) - \log (x - 2) = 1$

40. $\log_5 (x + 4) + \log_5 (x - 4) = 2$

Technology Connection

Find approximate solutions of the equation.

41. $e^{7.2x} = 14.009$

42. $0.082e^{0.05x} = 0.034$

43. $xe^{3x} - 1 = 3$

44. $5e^{5x} + 10 = 3x + 40$

45. $4 \ln (x + 3.4) = 2.5$

46. $\ln x^2 = -x^2$

47. $\log_8 x + \log_8 (x + 2) = 2$

48. $\log_3 x + 7 = 4 - \log_5 x$

49. $\log_5 (x + 7) - \log_5 (2x - 3) = 1$

Approximate the point(s) of intersection of the pair of equations.

50. $y = \ln 3x, \ y = 3x - 8$

51. $2.3x + 3.8y = 12.4, \ y = 1.1 \ln (x - 2.05)$

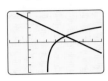

52. $y = 2.3 \ln (x + 10.7), \ y = 10e^{-0.07x^2}$

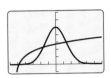

53. $y = 2.3 \ln (x + 10.7), \ y = 10e^{-0.007x^2}$

Collaborative Discussion and Writing

54. In Example 3, we took the natural logarithm on both sides of the equation. What would have happened had we taken the common logarithm? Explain which approach seems better to you and why.

55. Explain how Exercises 29 and 30 could be solved using the graph of $f(x) = \ln x$.

Skill Maintenance

In Exercises 56–59:

a) *Find the vertex.*

b) *Find the line of symmetry.*

c) *Determine whether there is a maximum or minimum value and find that value.*

56. $f(x) = -x^2 + 6x - 8$

57. $g(x) = x^2 - 6$

58. $H(x) = 3x^2 - 12x + 16$

59. $G(x) = -2x^2 - 4x - 7$

Synthesis

Solve using any method.

60. $\ln (\ln x) = 2$

61. $\ln (\log x) = 0$

62. $\ln \sqrt[4]{x} = \sqrt{\ln x}$

63. $\sqrt{\ln x} = \ln \sqrt{x}$

64. $\log_3 (\log_4 x) = 0$

65. $(\log_3 x)^2 - \log_3 x^2 = 3$

66. $(\log x)^2 - \log x^2 = 3$

67. $\ln x^2 = (\ln x)^2$

68. $e^{2x} - 9 \cdot e^x + 14 = 0$

69. $5^{2x} - 3 \cdot 5^x + 2 = 0$

70. $x \left(\ln \frac{1}{6} \right) = \ln 6$

71. $\log_3 |x| = 2$

72. $x^{\log x} = \dfrac{x^3}{100}$

73. $\ln x^{\ln x} = 4$

74. $\dfrac{(e^{3x+1})^2}{e^4} = e^{10x}$

75. $\dfrac{\sqrt{(e^{2x} \cdot e^{-5x})^{-4}}}{e^x \div e^{-x}} = e^7$

76. $e^x < \dfrac{4}{5}$

77. $|\log_5 x| + 3 \log_5 |x| = 4$

78. $|2^{x^2} - 8| = 3$

79. Given that $a = \log_8 225$ and $b = \log_2 15$, express a as a function of b.

80. Given that $a = (\log_{125} 5)^{\log_5 125}$, find the value of $\log_3 a$.

81. Given that

$$\begin{aligned} \log_2 [\log_3 (\log_4 x)] &= \log_3 [\log_2 (\log_4 y)] \\ &= \log_4 [\log_3 (\log_2 z)] \\ &= 0, \end{aligned}$$

find $x + y + z$.

82. Given that $f(x) = e^x - e^{-x}$, find $f^{-1}(x)$ if it exists.

4.6

Applications and Models: Growth and Decay

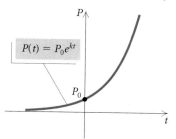

• *Solve applied problems involving exponential growth and decay.*

Exponential and logarithmic functions with base e are rich in applications to many fields such as business, science, psychology, and sociology.

Population Growth

The function

$$P(t) = P_0 e^{kt}, \quad k > 0$$

is a model of many kinds of population growth, whether it be a population of people, bacteria, cellular phones, or money. In this function, P_0 is the population at time 0, P is the population after time t, and k is called the **exponential growth rate.** The graph of such an equation is shown at left.

EXAMPLE 1 *Population Growth of India.* In 1998, the population of India was about 984 million and the exponential growth rate was 1.8% per year (*Source: Statistical Abstract of the United States*).

a) Find the exponential growth function.

b) What will the population be in 2005?

c) After how long will the population be double what it was in 1998?

Technology Connection

We can find function values using a graphing calculator. Below, we find $P(7)$ from Example 1(b) with the VALUE feature from the CALC menu. We see that $P(7) \approx 1116$.

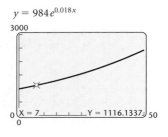

$y = 984e^{0.018x}$

Solution

a) At $t = 0$ (1998), the population was 984 million. We substitute 984 for P_0 and 1.8%, or 0.018, for k to obtain the exponential growth function

$$P(t) = 984e^{0.018t}.$$

b) In 2005, $t = 7$; that is, 7 yr have passed since 1998. To find the population in 2005, we substitute 7 for t:

$$P(7) = 984e^{0.018(7)} = 984e^{0.126} \approx 1116.$$

The population will be 1116 million, or 1,116,000,000, in 2005.

c) We are looking for the time T for which $P(T) = 2 \cdot 984$, or 1968. The number T is called the **doubling time.** To find T, we solve the equation

$$1968 = 984e^{0.018T}.$$

Algebraic Solution

We have

$1968 = 984e^{0.018T}$	Substituting 1968 for $P(T)$
$2 = e^{0.018T}$	Dividing by 984
$\ln 2 = \ln e^{0.018T}$	Taking the natural logarithm on both sides
$\ln 2 = 0.018T$	$\ln e^x = x$
$\dfrac{\ln 2}{0.018} = T$	Dividing by 0.018
$39 \approx T.$	

The population of India will be double what it was in 1998 about 39 yr after 1998.

Visualizing the Solution

From the graph of $y = 1968$ and $y = 984e^{0.018T}$, we see that the first coordinate of the point of intersection of the graphs is about 39.

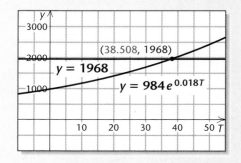

The solution of the equation is approximately 39.

$y_1 = 984e^{0.018x}$, $y_2 = 1968$

Technology Connection

Using the Intersect method in Example 1(c), we graph the equations

$$y_1 = 984e^{0.018x} \quad \text{and} \quad y_2 = 1968$$

and find the first coordinate of their point of intersection. It is about 39, so the population of India will be double that of 1998 about 39 yr after 1998.

Interest Compounded Continuously

Here we explore the mathematics behind the concept of **interest compounded continuously.** Suppose that an amount P_0 is invested in a savings account at interest rate k *compounded continuously.* The amount $P(t)$ in the account after t years is given by the exponential function

$$P(t) = P_0 e^{kt}.$$

EXAMPLE 2 *Interest Compounded Continuously.* Suppose that $2000 is invested at interest rate k, compounded continuously, and grows to $2983.65 in 5 yr.

a) What is the interest rate?

b) Find the exponential growth function.

c) What will the balance be after 10 yr?

d) After how long will the $2000 have doubled?

Solution

a) At $t = 0$, $P(0) = P_0 = 2000. Thus the exponential growth function is

$$P(t) = 2000e^{kt}.$$

We know that $P(5) = 2983.65. We substitute and solve for k:

$$2983.65 = 2000e^{k(5)} \qquad \text{Substituting 2983.65 for } P(t) \text{ and 5 for } t$$

$$2983.65 = 2000e^{5k}$$

$$\frac{2983.65}{2000} = e^{5k} \qquad \text{Dividing by 2000}$$

$$\ln \frac{2983.65}{2000} = \ln e^{5k} \qquad \text{Taking the natural logarithm}$$

$$\ln \frac{2983.65}{2000} = 5k \qquad \text{Using } \ln e^x = x$$

$$\frac{\ln \dfrac{2983.65}{2000}}{5} = k \qquad \text{Dividing by 5}$$

$$0.08 \approx k.$$

The interest rate is about 0.08, or 8%.

b) The exponential growth function is
$$P(t) = 2000e^{0.08t}.$$

c) The balance after 10 yr is
$$P(10) = 2000e^{0.08(10)}$$
$$= 2000e^{0.8}$$
$$\approx \$4451.08.$$

d) To find the doubling time T, we set $P(T) = 2 \cdot P_0 = \$4000$ and solve for T:

Algebraic Solution

We have

$$4000 = 2000e^{0.08T}$$

$2 = e^{0.08T}$	Dividing by 2000
$\ln 2 = \ln e^{0.08T}$	Taking the natural logarithm
$\ln 2 = 0.08T$	$\ln e^x = x$
$\dfrac{\ln 2}{0.08} = T$	Dividing by 0.08
$8.7 \approx T.$	

Thus the original investment of $2000 will double in about 8.7 yr.

Visualizing the Solution

The solution of the equation
$$4000 = 2000e^{0.08T},$$
or
$$2000e^{0.08T} - 4000 = 0,$$
is the zero of the function
$$y = 2000e^{0.08T} - 4000.$$

Let's observe the zero from the graph shown here.

$$y = 2000e^{0.08T} - 4000$$

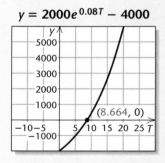

The zero is about 8.7. Thus the solution of the equation is approximately 8.7.

Technology Connection

The money in Example 2 will have doubled when $P(t) = 2 \cdot P_0 = 4000$, or when $2000e^{0.08t} = 4000$. We use the Zero method. We graph the equation

$$y = 2000e^{0.08x} - 4000$$

and find the zero of the function. The zero of the function is the solution of the equation. The zero is about 8.7, so the original investment of $2000 will double in about 8.7 yr.

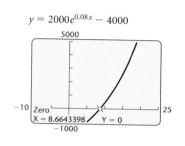

$y = 2000e^{0.08x} - 4000$

We can find a general expression relating the growth rate k and the doubling time T by solving the following equation:

$2P_0 = P_0 e^{kT}$	Substituting $2P_0$ for P and T for t
$2 = e^{kT}$	Dividing by P_0
$\ln 2 = \ln e^{kT}$	Taking the natural logarithm
$\ln 2 = kT$	Using $\ln e^x = x$
$\dfrac{\ln 2}{k} = T.$	

Growth Rate and Doubling Time

The **growth rate k** and the **doubling time T** are related by

$$kT = \ln 2, \quad \text{or} \quad k = \frac{\ln 2}{T}, \quad \text{or} \quad T = \frac{\ln 2}{k}.$$

Note that the relationship between k and T does not depend on P_0.

EXAMPLE 3 *World Population Growth.* The population of the world is now doubling every 54.6 yr. What is the exponential growth rate?

Solution We have

$$k = \frac{\ln 2}{T} = \frac{\ln 2}{54.6} \approx 1.3\%.$$

The growth rate of the world population is about 1.3% per year. ▬

Models of Limited Growth

The model $P(t) = P_0 e^{kt}$, $k > 0$, has many applications involving unlimited population growth. However, in some populations, there can be factors that prevent a population from exceeding some limiting value — perhaps a limitation on food, living space, or other natural resources. One model of such growth is

$$P(t) = \frac{a}{1 + be^{-kt}},$$

which is called a **logistic function.** This function increases toward a *limiting value a* as $t \to \infty$. Thus, $y = a$ is the horizontal asymptote of the graph of $P(t)$.

EXAMPLE 4 *Limited Population Growth.* A ship carrying 1000 passengers has the misfortune to be shipwrecked on a small island from which the passengers are never rescued. The natural resources of the island limit the population to 5780. The population gets closer and closer to this limiting value, but never reaches it. The population of the island after time t, in years, is given by the logistic equation

$$P(t) = \frac{5780}{1 + 4.78e^{-0.4t}}.$$

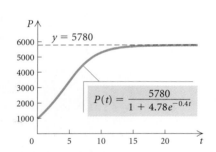

The graph of $P(t)$ is the S-shaped curve shown at left. Note that this function increases toward a limiting value of 5780. The graph has $y = 5780$ as a horizontal asymptote. Find the population after 0, 1, 2, 5, 10, and 20 yr.

Solution   Using a calculator, we compute the function values. We find that

$$P(0) = 1000, \qquad P(5) \approx 3509.6,$$
$$P(1) \approx 1374.8, \qquad P(10) \approx 5314.7,$$
$$P(2) \approx 1836.2, \qquad P(20) \approx 5770.7.$$

Thus the population will be about 1000 after 0 yr, 1375 after 1 yr, 1836 after 2 yr, 3510 after 5 yr, 5315 after 10 yr, and 5771 after 20 yr. ▬

Another model of limited growth is provided by the function

$$P(t) = L(1 - e^{-kt}), \quad k > 0,$$

which is shown graphed at right. This function also increases toward a limiting value L, as $x \to \infty$, so $y = L$ is the horizontal asymptote of the graph of $P(t)$.

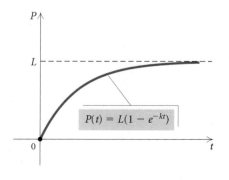

Exponential Decay

The function

$$P(t) = P_0 e^{-kt}, \quad k > 0$$

is an effective model of the decline, or decay, of a population. An example is the decay of a radioactive substance. In this case, P_0 is the amount of the substance at time $t = 0$, and $P(t)$ is the amount of the substance left after time t, where k is a positive constant that depends on the situation. The constant k is called the **decay rate.**

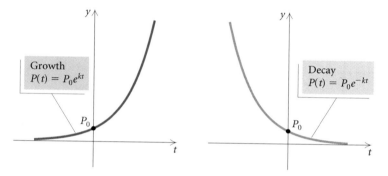

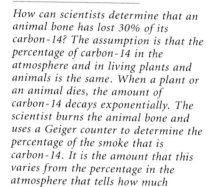

How can scientists determine that an animal bone has lost 30% of its carbon-14? The assumption is that the percentage of carbon-14 in the atmosphere and in living plants and animals is the same. When a plant or an animal dies, the amount of carbon-14 decays exponentially. The scientist burns the animal bone and uses a Geiger counter to determine the percentage of the smoke that is carbon-14. It is the amount that this varies from the percentage in the atmosphere that tells how much carbon-14 has been lost.

The process of carbon-14 dating was developed by the American chemist Willard E. Libby in 1952. It is known that the radioactivity in a living plant is 16 disintegrations per gram per minute. Since the half-life of carbon-14 is 5750 years, an object with an activity of 8 disintegrations per gram per minute is 5750 years old, one with an activity of 4 disintegrations per gram per minute is 11,500 years old, and so on. Carbon-14 dating can be used to measure the age of objects up to 40,000 years old. Beyond such an age, it is too difficult to measure the radioactivity and some other method would have to be used.

Carbon-14 was indeed used to find the age of the Dead Sea Scrolls. It was also used to refute the authenticity of the Shroud of Turin, presumed to have covered the body of Christ.

The **half-life** of bismuth is 5 days. This means that half of an amount of bismuth will cease to be radioactive in 5 days. The effect of half-life T is shown in the graph below for nonnegative inputs. The exponential function gets close to 0, but never reaches 0, as t gets very large. Thus, according to an exponential decay model, a radioactive substance never completely decays.

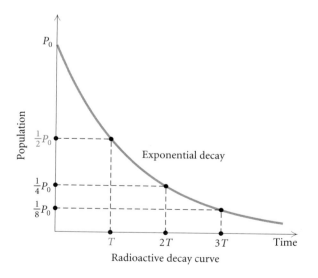

Radioactive decay curve

In 1947, a Bedouin youth looking for a stray goat climbed into a cave at Kirbet Qumran on the shores of the Dead Sea near Jericho and came upon earthenware jars containing an incalculable treasure of ancient manuscripts. Shown here are fragments of those so-called Dead Sea Scrolls, a portion of some 600 or so texts found so far and which concern the Jewish books of the Bible. Officials date them before 70 A.D., making them the oldest Biblical manuscripts by 1000 years.

EXAMPLE 5 *Carbon Dating.* The radioactive element carbon-14 has a half-life of 5750 yr. The percentage of carbon-14 present in the remains of organic matter can be used to determine the age of that organic matter. Archaeologists discovered that the linen wrapping from one of the Dead Sea Scrolls had lost 22.3% of its carbon-14 at the time it was found. How old was the linen wrapping?

Solution We first find k. When $t = 5750$ (the half-life), $P(t)$ will be half of P_0. We substitute $\frac{1}{2}P_0$ for $P(t)$ and 5750 for t and solve for k. Then

$$P(t) = P_0 e^{-kt}$$
$$\frac{1}{2}P_0 = P_0 e^{-k(5750)}$$
$$\frac{1}{2} = e^{-5750k} \qquad \text{Dividing by } P_0$$
$$\ln \frac{1}{2} = \ln e^{-5750k} \qquad \text{Taking the natural logarithm on both sides}$$
$$\ln 0.5 = -5750k.$$

Then

$$k = \frac{\ln 0.5}{-5750} \approx 0.00012.$$

Now we have the function

$$P(t) = P_0 e^{-0.00012t}.$$

(This function can be used for any subsequent carbon-dating problem.) If the linen wrapping has lost 22.3% of its carbon-14 from an initial amount P_0, then $77.7\%P_0$ is the amount present. To find the age t of the wrapping, we solve the following equation for t:

$$77.7\%P_0 = P_0 e^{-0.00012t} \qquad \text{Substituting } 77.7\%P_0 \text{ for } P$$
$$0.777 = e^{-0.00012t} \qquad \text{Dividing by } P_0 \text{ and writing } 77.7\% \text{ as } 0.777$$
$$\ln 0.777 = \ln e^{-0.00012t} \qquad \text{Taking the natural logarithm on both sides}$$
$$\ln 0.777 = -0.00012t \qquad \ln e^x = x$$
$$\frac{\ln 0.777}{-0.00012} = t \qquad \text{Dividing by } -0.00012$$
$$2103 \approx t.$$

Thus the linen wrapping on the Dead Sea Scrolls was about 2103 yr old when it was found.

YEAR, x	CREDIT CARD VOLUME, y (IN BILLIONS)
1988, 0	$261.0
1989, 1	296.3
1990, 2	338.4
1991, 3	361.0
1992, 4	403.1
1993, 5	476.7
1994, 6	584.8
1995, 7	701.2
1996, 8	798.3
1997, 9	885.2

Source: CardWeb Inc.'s CardData

```
ExpReg
y=a∗b^x
a=249.7267077
b=1.150881332
r²=.9877691696
r=.9938657704
```

Technology Connection

We now expand the regression procedure to include modeling data with an exponential function.

Credit Card Volume. The total credit card volume for Visa, MasterCard, American Express, and Discover has increased dramatically in recent years, as shown in the table at left.

a) Use a graphing calculator to fit an exponential function to the data.

b) Predict the total credit card volume for Visa, MasterCard, American Express, and Discover in 2003.

Solution

a) We will fit an equation of the type $y = a \cdot b^x$ to the data, where x is the number of years since 1988. Entering the data into the calculator and carrying out the regression procedure, we find that the equation is

$$y = 249.7267077(1.150881332)^x.$$

The correlation coefficient is very close to 1. This gives us a good indication that the exponential function fits the data well.

b) We evaluate the function found in part (a) for $x = 15$ ($2003 - 1988 = 15$) and estimate that the total credit card volume in 2003 will be about $2056 billion, or $2,056,000,000,000.

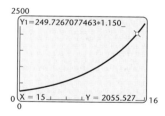

Exercise Set 4.6

1. *World Population Growth.* In 1999, the world population was 6.0 billion. The exponential growth rate was 1.3% per year.

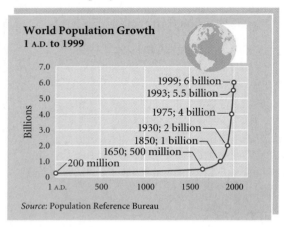

World Population Growth
1 A.D. to 1999

1999; 6 billion
1993; 5.5 billion
1975; 4 billion
1930; 2 billion
1850; 1 billion
1650; 500 million
200 million

Source: Population Reference Bureau

a) Find the exponential growth function.
b) Predict the population of the world in 2005 and 2010.
c) When will the world population be 8 billion?
d) Find the doubling time.

2. *Population Growth of Rabbits.* Under ideal conditions, a population of rabbits has an exponential growth rate of 11.7% per day. Consider an initial population of 100 rabbits.

a) Find the exponential growth function.
b) What will the population be after 7 days?
c) Find the doubling time.

3. *Population Growth.* Complete the following table.

POPULATION	GROWTH RATE, k	DOUBLING TIME, T
a) Mexico	1.9% per year	
b) Japan		346 yr
c) Mozambique	3.3% per year	
d) Norway	0.5% per year	
e) Syria		20.4 yr
f) Philippines		31.5 yr

4. *Female Olympic Athletes.* In 1985, the number of female athletes participating in Summer

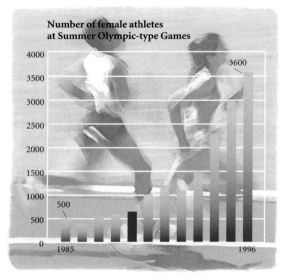

Number of female athletes at Summer Olympic-type Games

Olympic-type Games was 500. In 1996, about 3600 participated in the Summer Olympics in Atlanta. Assuming the exponential model applies:

a) Find the value of k ($P_0 = 500$), and write the function.
b) Estimate the number of female athletes in the Summer Olympics of 2000 and 2004.

5. *Population Growth of Israel.* The population of Israel has a growth rate of 2.6% per year. In 1998, the population was 5,644,000. The land area of Israel is 24,313,062,400 square yards. (*Source: Statistical Abstract of the United States*) Assuming this growth rate continues and is exponential, after how long will there be one person for every square yard of land?

6. *Value of Manhattan Island.* In 1626, Peter Minuit of the Dutch West India Company purchased Manhattan Island from the Indians for $24. Assuming an exponential rate of inflation of 8% per year, how much will Manhattan be worth in 2003?

7. *Interest Compounded Continuously.* Suppose that $10,000 is invested at an interest rate of 5.4% per year, compounded continuously.

a) Find the exponential function that describes the amount in the account after time t, in years.
b) What is the balance after 1 yr? 2 yr? 5 yr? 10 yr?
c) What is the doubling time?

8. *Interest Compounded Continuously.* Complete the following table.

INITIAL INVESTMENT AT $t = 0$, P_0	INTEREST RATE, k	DOUBLING TIME, T	AMOUNT AFTER 5 YR
a) $35,000	6.2%		
b) $5000			$ 7,130.90
c)	8.4%		$11,414.71
d)		11 yr	$17,539.32

9. *Carbon Dating.* A mummy discovered in the pyramid Khufu in Egypt has lost 46% of its carbon-14. Determine its age.

10. *Carbon Dating.* The statue of Zeus at Olympia in Greece is one of the Seven Wonders of the World. It is made of gold and ivory. The ivory was found to have lost 35% of its carbon-14. Determine the age of the statue.

11. *Radioactive Decay.* Complete the following table.

RADIOACTIVE SUBSTANCE	DECAY RATE, k	HALF-LIFE, T
a) Polonium		3 min
b) Lead		22 yr
c) Iodine-131	9.6% per day	
d) Krypton-85	6.3% per year	
e) Strontium-90		25 yr
f) Uranium-238		4560 yr
g) Plutonium		23,105 yr

12. *Number of Farms.* The number N of farms in the United States has declined continually since 1950. (See Example 5 in Section 3.1.) In 1950, there were 5,647,800 farms, and in 1995 that number had decreased to 2,071,520 (*Source:* U.S. Department of Agriculture). Assuming the number of farms decreased according to the exponential model:

a) Find the value of k, and write an exponential function that describes the number of farms after time t, in years, where t is the number of years since 1950.
b) Estimate the number of farms in 2000, 2005, and 2010.
c) At this decay rate, in what year will only 100,000 farms remain?

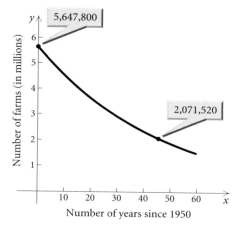

13. *Decline in Beef Consumption.* In 1985, the average annual consumption of beef B was about 80 lb per person. In 1996, it was about 67 lb per person. Assuming consumption is decreasing according to the exponential-decay model:

a) Find the value k, and write an equation that describes beef consumption after time t, in years.
b) Estimate the consumption of beef in 2002.
c) After how many years (according to the model from part a) will the average annual consumption of beef be 20 lb per person?

14. *The Value of Mark McGwire's Baseball Card.* The collecting of baseball cards has become a popular hobby. The card shown here is a photograph of Mark

McGwire when he was a member of the Olympic USA Baseball Team in 1984.

MARK McGWIRE   FIRST BASE

In 1987, the value of the card was $8 and in 1997, its value was $20 (*Source*: *SportsCards*, Joe Clemens, Price Guide Coordinator). Assuming the value V_0 of the card has grown exponentially:

a) Find the value k, and determine the exponential growth function, assuming $V_0 = 8$.

b) Using the function found in part (a), estimate the value of the card in 2000.

c) What is the doubling time for the value of the card?

d) After how long will the value of the card be $2000, assuming there is no change in the growth rate?

e) In the Baseball Price Guide in *SportsCards*, April 2000, this card was valued at $200. How does your estimate in part (b) compare with the $200 value? Explain why they differ.

15. *Spread of an Epidemic.*   In a town whose population is 2000, a disease creates an epidemic. The number of people N infected t days after the disease has begun is given by the function

$$N(t) = \frac{2000}{1 + 19.9e^{-0.6t}}.$$

a) How many are initially infected with the disease $(t = 0)$?

b) Find the number infected after 2 days, 5 days, 8 days, 12 days, and 16 days.

c) Using this model, can you say whether all 2000 people will ever be infected? Explain.

16. *Acceptance of Seat Belt Laws.*   In recent years, many states have passed mandatory seat belt laws. The total number of states N that have passed a seat belt law t years after 1984 is given by the function

$$N(t) = \frac{50}{1 + 22e^{-0.6t}}$$

(*Source*: National Highway Traffic Safety Administration).

a) How many states had passed the law in 1984? ($t = 0$ corresponds to 1984.)

b) Find the number of states that had passed the law by 1996 and 2002.

c) If the function were to continue to be appropriate, would all 50 states ever pass the law? Explain.

17. *Limited Population Growth in a Lake.*   A lake is stocked with 400 fish of a new variety. The size of the lake, the availability of food, and the number of other fish restrict growth in the lake to a *limiting value* of 2500. The population of fish in the lake after time t, in months, is given by the function

$$P(t) = \frac{2500}{1 + 5.25e^{-0.32t}}.$$

Find the population after 0, 1, 5, 10, 15, and 20 months.

Technology Connection

18. *Number of Physicians.*   The following table contains data regarding the number of physicians in the United States in selected years.

YEAR	TOTAL NUMBER OF PHYSICIANS
1950	219,997
1955	241,711
1960	260,484
1965	292,088
1970	334,028
1975	393,742
1980	467,679
1985	552,716
1990	615,421
1994	684,414
1995	720,325
1996	737,764

Source: American Medical Association

a) Create a scatterplot of these data. Determine whether an exponential function appears to fit the data.

b) Use a graphing calculator to fit the data with an exponential function and determine whether the function is a good fit.

c) Graph the function found in part (b) with a scatterplot of the data.

d) Estimate the number of physicians in 2000 and 2025.

19. *Projected Number of Alzheimer's Patients.* German psychiatrist Alois Alzheimer first described the disease, later called Alzheimer's disease, in 1906. Since life expectancy has significantly increased in the last century, the number of Alzheimer's patients has increased dramatically. The number of patients in the United States reached 4 million in 2000. The following table lists projected data regarding the number of Alzheimer's patients in years beyond 2000.

YEAR	PROJECTED NUMBER OF ALZHEIMER'S PATIENTS IN THE UNITED STATES (IN MILLIONS)
2000	4.0
2010	5.8
2020	6.8
2030	8.7
2040	11.8
2050	14.3

Source: "Alzheimer's, Unlocking the Mystery," by Geoffrey Cowley, *Newsweek*, January 31, 2000

a) Use a graphing calculator to fit the data with an exponential function and determine whether the function is a good fit.

b) Graph the function found in part (a) with a scatterplot of the data.

c) Estimate the number of Alzheimer's patients in 2005, 2025, and 2100.

20. *Forgetting.* In an art class, students were tested at the end of the course on a final exam. Then they were retested with an equivalent test at subsequent time intervals. Their scores after time *t*, in months, are given in the following table.

TIME, *t* (IN MONTHS)	SCORE, *y*
1	84.9%
2	84.6%
3	84.4%
4	84.2%
5	84.1%
6	83.9%

a) Use a graphing calculator to fit a logarithmic function $y = a + b \ln x$ to the data.

b) Use the function to predict test scores after 8, 10, 24, and 36 months.

c) After how long will the test scores fall below 82%?

21. *On-line Travel Revenue.* With the explosion of increased Internet use, more and more travelers are booking their travel reservations on-line. The following table lists the total on-line revenue for recent years. Most of the revenue is from airline tickets.

YEAR	ON-LINE TRAVEL REVENUE (IN MILLIONS)
1996	$ 276
1997	827
1998	1900
1999	3200
2000*	4700
2001*	6500
2002*	8900

*Projections
Source: *Travel and Interactive Technology 1999*; Travel Industry Association of America

a) Create a scatterplot of the data. Let *x* = the number of years since 1996.

b) Use a graphing calculator to fit the data with linear, quadratic, and exponential functions. Determine which function has the best fit.

c) Graph all three functions found in part (b) with the scatterplot in part (a).

d) Use the functions found in part (b) to estimate the on-line travel revenue in 2010. Which function provides the most realistic prediction?

22. *Effect of Advertising.* A company introduces a new software product on a trial run in a city. They advertised the product on television and found the following data relating the percent P of people who bought the product after x ads were run.

NUMBER OF ADS, x	PERCENTAGE WHO BOUGHT, P
0	0.2
10	0.7
20	2.7
30	9.2
40	27.0
50	57.6
60	83.3
70	94.8
80	98.5
90	99.6

a) Use a graphing calculator to fit a logistic function

$$P(x) = \frac{a}{1 + be^{-kx}}$$

to the data.
b) What percent will buy the product when 55 ads are run? 100 ads?
c) Find the horizontal asymptote for the graph. Interpret the asymptote in terms of the advertising situation.

Collaborative Discussion and Writing

23. Browse through some newspapers or magazines until you find some data and/or a graph that seem as though they can be fit to an exponential function. Make a case for why such a fit is appropriate. Then fit an exponential function to the data and make some predictions.

24. *Atmospheric Pressure.* Atmospheric pressure P at an altitude a is given by

$$P = P_0 e^{-0.00005a},$$

where P_0 is the pressure at sea level ≈ 14.7 lb/in^2 (pounds per square inch). Explain how a barometer, or some device for measuring atmospheric pressure, can be used to find the height of a skyscraper.

Skill Maintenance

Find the slope and the y-intercept of the line.

25. $y = 6$

26. $3x - 10y = 14$

27. $y = 2x - \frac{3}{13}$

28. $x = -4$

Synthesis

29. *Present Value.* Following the birth of a child, a parent wants to make an initial investment P_0 that will grow to $50,000 for the child's education at age 18. Interest is compounded continuously at 7%. What should the initial investment be? Such an amount is called the **present value** of $50,000 due 18 yr from now.

30. *Present Value.* Referring to Exercise 29:
a) Solve $P = P_0 e^{kt}$ for P_0.
b) Find the present value of $50,000 due 18 yr from now at interest rate 6.4%.

31. *Supply and Demand.* The supply and demand for the sale of a certain type of VCR are given by

$$S(p) = 480e^{-0.003p} \quad \text{and} \quad D(p) = 150e^{0.004p},$$

where $S(p)$ is the number of VCRs that the company is willing to sell at price p and $D(p)$ is the quantity that the public is willing to buy at price p. Find p, called the **equilibrium price,** such that $D(p) = S(p)$.

32. *Carbon Dating.* Recently, while digging in Chaco Canyon, New Mexico, archeologists found corn pollen that was 4000 yr old (*Source: American Anthropologist*). This was evidence that Native Americans had been cultivating crops in the Southwest centuries earlier than scientists had thought. What percent of the carbon-14 had been lost from the pollen?

33. *Newton's Law of Cooling.* Suppose that a body with temperature T_1 is placed in surroundings with temperature T_0 different from that of T_1. The body will either cool or warm to temperature $T(t)$ after time t, in minutes, where

$$T(t) = T_0 + |T_1 - T_0|e^{-kt}.$$

A cup of coffee with temperature 105°F is placed in a freezer with temperature 0°F. After 5 min, the temperature of the coffee is 70°F. What will its temperature be after 10 min?

34. *When Was the Murder Committed?* The police discover the body of a murder victim. Critical to solving the crime is determining when the murder was committed. The coroner arrives at the murder scene at 12:00 P.M. She immediately takes the temperature of the body and finds it to be 94.6°. She then takes the temperature 1 hr later and finds it to be 93.4°. The temperature of the room is 70°. When was the murder committed? (Use Newton's law of cooling in Exercise 33.)

35. *Electricity.* The formula

$$i = \frac{V}{R}[1 - e^{-(R/L)t}]$$

occurs in the theory of electricity. Solve for t.

36. *The Beer–Lambert Law.* A beam of light enters a medium such as water or smog with initial intensity I_0. Its intensity decreases depending on the thickness (or concentration) of the medium. The intensity I at a depth (or concentration) of x units is given by

$$I = I_0 e^{-\mu x}.$$

The constant μ (the Greek letter "mu") is called the **coefficient of absorption**, and it varies with the medium. For sea water, $\mu = 1.4$.

a) What percentage of light intensity I_0 remains at a depth of sea water that is 1 m? 3 m? 5 m? 50 m?

b) Plant life cannot exist below 10 m. What percentage of I_0 remains at 10 m?

37. Given that $y = ae^x$, take the natural logarithm on both sides. Let $Y = \ln y$. Consider Y as a function of x. What kind of function is Y?

38. Given that $y = ax^b$, take the natural logarithm on both sides. Let $Y = \ln y$ and $X = \ln x$. Consider Y as a function of X. What kind of function is Y?

4 Chapter Summary and Review

Important Properties and Formulas

The Composition of Two Functions:	$(f \circ g)(x) = f(g(x))$
One-to-One Function:	$f(a) = f(b) \rightarrow a = b$
Exponential Function:	$f(x) = a^x$
The Number e:	$e = 2.7182818284\ldots$
Logarithmic Function:	$f(x) = \log_a x$
A Logarithm Is an Exponent:	$\log_a x = y \longleftrightarrow x = a^y$
The Change-of-Base Formula:	$\log_b M = \dfrac{\log_a M}{\log_a b}$
The Product Rule:	$\log_a MN = \log_a M + \log_a N$
The Power Rule:	$\log_a M^p = p \log_a M$
The Quotient Rule:	$\log_a \dfrac{M}{N} = \log_a M - \log_a N$
Other Properties:	$\log_a a = 1, \quad \log_a 1 = 0,$
	$\log_a a^x = x, \quad a^{\log_a x} = x$
Base–Exponent Property:	$a^x = a^y \longleftrightarrow x = y$, for $a > 0, a \neq 1$
Exponential Growth Model:	$P(t) = P_0 e^{kt}, k > 0$
Exponential Decay Model:	$P(t) = P_0 e^{-kt}, k > 0$
Interest Compounded Continuously:	$P(t) = P_0 e^{kt}, k > 0$
Limited Growth:	$P(t) = \dfrac{a}{1 + be^{-kt}}, k > 0$

REVIEW EXERCISES

In Exercises 1 and 2, for the pair of functions:

a) *Find the domain of* $f \circ g$ *and* $g \circ f$.
b) *Find* $(f \circ g)(x)$ *and* $(g \circ f)(x)$.

1. $f(x) = \dfrac{4}{x^2}$; $g(x) = 3 - 2x$

2. $f(x) = 3x^2 + 4x$; $g(x) = 2x - 1$

Find $f(x)$ *and* $g(x)$ *such that* $h(x) = (f \circ g)x$.

3. $h(x) = \sqrt{5x + 2}$

4. $h(x) = 4(5x - 1)^2 + 9$

5. Find the inverse of the relation
$\{(1.3, -2.7), (8, -3), (-5, 3), (6, -3), (7, -5)\}$.

6. Find an equation of the inverse relation.

a) $y = 3x^2 + 2x - 1$
b) $0.8x^3 - 5.4y^2 = 3x$

In Exercises 7–10, given the function:

a) *Sketch the graph and determine whether the function is one-to-one.*
b) *If it is one-to-one, find a formula for the inverse.*

7. $f(x) = \sqrt{x - 6}$ **8.** $f(x) = x^3 - 8$

9. $f(x) = 3x^2 + 2x - 1$ **10.** $f(x) = e^x$

11. Find $f(f^{-1}(657))$: $f(x) = \dfrac{4x^5 - 16x^{37}}{119x}$, $x > 1$.

In Exercises 12–17, match the equation with one of figures (a)–(f), which follow.

a)

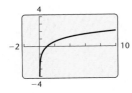

b)

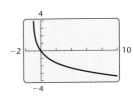

c)

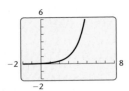

d)

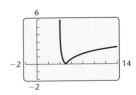

e)

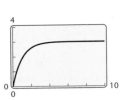

f)

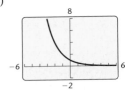

12. $f(x) = e^{x-3}$ **13.** $f(x) = \log_3 x$

14. $y = -\log_3(x + 1)$ **15.** $y = \left(\frac{1}{2}\right)^x$

16. $f(x) = 3(1 - e^{-x})$, $x \geq 0$

17. $f(x) = |\ln(x - 4)|$

18. Convert to an exponential equation:
$$\log_4 x = 2.$$

19. Convert to a logarithmic equation:
$$e^x = 80.$$

Solve.

20. $\log_4 x = 2$ **21.** $3^{1-x} = 9^{2x}$

22. $e^x = 80$ **23.** $4^{2x-1} - 3 = 61$

24. $\log_{16} 4 = x$ **25.** $\log_x 125 = 3$

26. $\log_2 x + \log_2(x - 2) = 3$

27. $\log(x^2 - 1) - \log(x - 1) = 1$

28. $\log x^2 = \log x$ **29.** $e^{-x} = 0.02$

Express as a single logarithm and, if possible, simplify.

30. $3 \log_b x - 4 \log_b y + \frac{1}{2} \log_b z$

31. $\ln(x^3 - 8) - \ln(x^2 + 2x + 4) + \ln(x + 2)$

Express in terms of sums and differences of logarithms.

32. $\ln \sqrt[4]{wr^2}$ **33.** $\log \sqrt[3]{\dfrac{M^2}{N}}$

Given that $\log_a 2 = 0.301$, $\log_a 5 = 0.699$, *and* $\log_a 6 = 0.778$, *find each of the following.*

34. $\log_a 3$ **35.** $\log_a 50$

36. $\log_a \frac{1}{5}$ **37.** $\log_a \sqrt[3]{5}$

Simplify.

38. $\ln e^{-5k}$ **39.** $\log_5 5^{-6t}$

40. How long will it take an investment to double itself if it is invested at 8.6%, compounded continuously?

41. The population of a city doubled in 30 yr. What was the exponential growth rate?

42. How old is a skeleton that has lost 27% of its carbon-14?

43. The hydrogen ion concentration of milk is 2.3×10^{-6}. What is the pH? (See Exercise 72 in Exercise Set 4.3.)

44. What is the loudness, in decibels, of a sound whose intensity is $1000I_0$? (See Exercise 75 in Exercise Set 4.3.)

45. *The Population of Zimbabwe.* The population of Zimbabwe was 11 million in 1998, and the exponential growth rate was 1.2% per year (*Source*: U.S. Bureau of the Census, World Population Profile).

a) Find the exponential growth function.
b) What will the population be in 2004? in 2020?
c) When will the population be 25 million?
d) What is the doubling time?

46. *Toll-free Numbers.* The use of toll-free numbers has grown exponentially. In 1967, there were 7 million such calls, and in 1991, there were 10.2 billion such calls (*Source*: Federal Communication Commission).

a) Find the exponential growth rate k.
b) Find the exponential growth function.
c) How many toll-free number calls will be placed in 2002? in 2005?
d) In what year will 1 trillion such calls be placed?

47. *Walking Speed.* The average walking speed w, in feet per second, of a person living in a city of population P, in thousands, is given by the function

$$w(P) = 0.37 \ln P + 0.05.$$

a) The population of Phoenix, Arizona, is 1,200,000. Find the average walking speed.
b) A city's population has an average walking speed of 3.4 ft/sec. Find the population.

Technology Connection

48. *Cholesterol Level and the Risk of Heart Attack.* The data in the following table show the relationship of cholesterol level in men to the risk of a heart attack.

CHOLESTEROL LEVEL, x	MEN, PER 10,000, WHO SUFFER A HEART ATTACK, y
100	30
200	65
250	100
275	130
300	180

Source: Nutrition Action Healthletter

a) Use a graphing calculator to fit an exponential function to the data.
b) Graph the function with a scatterplot of the data.
c) Predict the heart attack rate for men with cholesterol levels of 150, 350, and 400.
d) Compare your answers in part (c) with the answer to part (c) in Example 4 of Section 3.1. Which function, power or exponential, is a better fit to the given data?

49. Using only a graphing calculator, determine whether the following functions are inverses of each other:

$$f(x) = \frac{4 + 3x}{x - 2}, \qquad g(x) = \frac{x + 4}{x - 3}.$$

50. a) Use a graphing calculator to graph $f(x) = 5e^{-x} \ln x$ in the viewing window $[-1, 10, -5, 5]$.
b) Estimate the relative maximum and minimum values.

Collaborative Discussion and Writing

51. Suppose that you were trying to convince a fellow student that

$$\log_2 (x + 5) \neq \log_2 x + \log_2 5.$$

Give as many explanations as you can.

52. Describe the difference between $f^{-1}(x)$ and $[f(x)]^{-1}$.

Synthesis

Solve.

53. $|\log_4 x| = 3$

54. $\log x = \ln x$

55. $5^{\sqrt{x}} = 625$

56. Find the domain: $f(x) = \log_3 (\ln x)$.

Chapter Test

1. For $f(x) = x - 5$ and $g(x) = x^2 + 1$, find $(f \circ g)(x)$ and $(g \circ f)(x)$.

2. Find the inverse of the relation
$$\{(-2, 5), (4, 3), (0, -1), (-6, -3)\}.$$

3. Determine whether the function shown below is one-to-one. Answer yes or no.

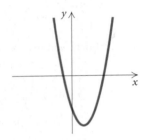

4. Find a formula for the inverse of the function $f(x) = x^3 + 1$.

Graph each of the following functions.

5. $f(x) = e^x - 3$ **6.** $f(x) = \ln (x + 2)$

7. Convert to an exponential equation: $\ln x = 4$.

8. Convert to a logarithmic equation: $3^x = 5.4$.

Solve.

9. $\log_{25} 5 = x$

10. $\log_3 x + \log_3 (x + 8) = 2$

11. $3^{4-x} = 27^x$

12. $e^x = 65$

13. Express $\ln \sqrt[5]{x^2 y}$ in terms of sums and differences of logarithms.

14. Given that $\log_a 2 = 0.328$ and $\log_a 8 = 0.984$, find $\log_a 4$.

15. Simplify: $\ln e^{-4t}$.

16. *Growth Rate.*   A country's population doubled in 45 yr. What was the exponential growth rate?

17. *Compound Interest.*   Suppose $1000 is invested at interest rate k, compounded continuously, and grows to $1144.54 in 3 yr.

   a) Find the interest rate.
   b) Find the exponential growth function.
   c) Find the balance after 8 yr.
   d) Find the doubling time.

Synthesis

18. Solve: $4^{\sqrt[3]{x}} = 8$.

The Trigonometric Functions 5

We now consider an important class of functions called *trigonometric,* or *circular, functions*. Historically, these functions arose from a study of triangles; hence the name trigonometric. We will begin our study with right triangles and degree measure and solve applied problems involving right triangles. Then we will consider trigonometric functions of angles or rotations of any size with both degree and radian measure. A circle of radius 1 (a unit circle) is then used to define the six basic trigonometric functions; hence the name circular functions. The domains and ranges of these functions consist of real numbers.

APPLICATION

Lance Armstrong won the 2000 Tour de France bicycle race. The wheel of his bicycle had a 63-cm diameter. His overall average linear speed during the race was 39.569 km/h. What was the angular speed of the wheel, in revolutions per hour? (*Source*: Wilcockson, John, with Charles Pelkey and Bryan Jew, *The 2000 Tour de France: Armstrong Encore*. Boulder, CO: VeloPress, 2000)

This problem appears as Exercise 56 in Exercise Set 5.4.

5.1

Trigonometric Functions of Acute Angles

- Determine the six trigonometric ratios for a given acute angle of a right triangle.
- Determine the trigonometric function values of 30°, 45°, and 60°.
- Using a calculator, find function values for any acute angle, and given a function value of an acute angle, find the angle.
- Given the function values of an acute angle, find the function values of its complement.

The Trigonometric Ratios

We begin our study of trigonometry by considering right triangles and acute angles measured in degrees. An **acute angle** is an angle with measure greater than 0° and less than 90°. Greek letters such as α (alpha), β (beta), γ (gamma), θ (theta), and ϕ (phi) are often used to denote an angle. Consider a right triangle with one of its acute angles labeled θ. The side opposite the right angle is called the **hypotenuse**. The other sides of the triangle are referenced by their position relative to the acute angle θ. One side is opposite θ and one is adjacent to θ.

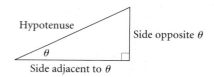

The *lengths* of the sides of the triangle are used to define the six trigonometric ratios:

sine (sin),	cosecant (csc),
cosine (cos),	secant (sec),
tangent (tan),	cotangent (cot).

The **sine of θ** is the *length* of the side opposite θ divided by the *length* of the hypotenuse:

$$\sin \theta = \frac{\text{length of side opposite } \theta}{\text{length of hypotenuse}}.$$

The ratio depends on the measure of angle θ and thus is a function of θ. The notation $\sin \theta$ actually means $\sin (\theta)$, where sin, or sine, is the name of the function.

The **cosine of θ** is the *length* of the side adjacent to θ divided by the *length* of the hypotenuse:

$$\cos \theta = \frac{\text{length of side adjacent to } \theta}{\text{length of hypotenuse}}.$$

The six trigonometric ratios, or trigonometric functions, are defined as follows.

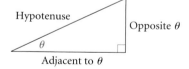

Hypotenuse — Opposite θ

θ

Adjacent to θ

Trigonometric Function Values of an Acute Angle θ

Let θ be an acute angle of a right triangle. Then the six trigonometric functions of θ are as follows:

$$\sin \theta = \frac{\text{side opposite } \theta}{\text{hypotenuse}}, \qquad \csc \theta = \frac{\text{hypotenuse}}{\text{side opposite } \theta},$$

$$\cos \theta = \frac{\text{side adjacent to } \theta}{\text{hypotenuse}}, \qquad \sec \theta = \frac{\text{hypotenuse}}{\text{side adjacent to } \theta},$$

$$\tan \theta = \frac{\text{side opposite } \theta}{\text{side adjacent to } \theta}, \qquad \cot \theta = \frac{\text{side adjacent to } \theta}{\text{side opposite } \theta}.$$

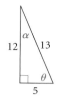

12 13

α

θ

5

EXAMPLE 1   In the right triangle shown at left, find the six trigonometric function values of (a) θ and (b) α.

Solution   We use the definitions.

a) $\sin \theta = \dfrac{\text{opp}}{\text{hyp}} = \dfrac{12}{13},$ $\csc \theta = \dfrac{\text{hyp}}{\text{opp}} = \dfrac{13}{12},$

$\cos \theta = \dfrac{\text{adj}}{\text{hyp}} = \dfrac{5}{13},$ $\sec \theta = \dfrac{\text{hyp}}{\text{adj}} = \dfrac{13}{5},$

$\tan \theta = \dfrac{\text{opp}}{\text{adj}} = \dfrac{12}{5},$ $\cot \theta = \dfrac{\text{adj}}{\text{opp}} = \dfrac{5}{12}$

The references to opposite, adjacent, and hypotenuse are relative to θ.

b) $\sin \alpha = \dfrac{\text{opp}}{\text{hyp}} = \dfrac{5}{13},$ $\csc \alpha = \dfrac{\text{hyp}}{\text{opp}} = \dfrac{13}{5},$

$\cos \alpha = \dfrac{\text{adj}}{\text{hyp}} = \dfrac{12}{13},$ $\sec \alpha = \dfrac{\text{hyp}}{\text{adj}} = \dfrac{13}{12},$

$\tan \alpha = \dfrac{\text{opp}}{\text{adj}} = \dfrac{5}{12},$ $\cot \alpha = \dfrac{\text{adj}}{\text{opp}} = \dfrac{12}{5}$

The references to opposite, adjacent, and hypotenuse are relative to α.

In Example 1(a), we note that the value of $\sin \theta$, $\frac{12}{13}$, is the reciprocal of $\frac{13}{12}$, the value of $\csc \theta$. Likewise, we see the same reciprocal relationship between the values of $\cos \theta$ and $\sec \theta$ and between the values of $\tan \theta$ and $\cot \theta$. For any angle, the cosecant, secant, and cotangent values are the reciprocals of the sine, cosine, and tangent function values, respectively.

Reciprocal Functions

$$\csc \theta = \frac{1}{\sin \theta}, \qquad \sec \theta = \frac{1}{\cos \theta}, \qquad \cot \theta = \frac{1}{\tan \theta}$$

If we know the values of the sine, cosine, and tangent functions of an angle, we can use these reciprocal relationships to find the values of the cosecant, secant, and cotangent functions of that angle.

EXAMPLE 2   Given that $\sin \phi = \frac{4}{5}$, $\cos \phi = \frac{3}{5}$, and $\tan \phi = \frac{4}{3}$, find $\csc \phi$, $\sec \phi$, and $\cot \phi$.

Solution   Using the reciprocal relationships, we have

$$\csc \phi = \frac{1}{\sin \phi} = \frac{1}{\dfrac{4}{5}} = \frac{5}{4}, \qquad \sec \phi = \frac{1}{\cos \phi} = \frac{1}{\dfrac{3}{5}} = \frac{5}{3},$$

and

$$\cot \phi = \frac{1}{\tan \phi} = \frac{1}{\dfrac{4}{3}} = \frac{3}{4}.$$

Triangles are said to be **similar** if their corresponding angles have the *same* measure. In similar triangles, the lengths of corresponding sides are in the same ratio. The right triangles shown below are similar. Note that the corresponding angles are equal and the length of each side of the second triangle is four times the length of the corresponding side of the first triangle.

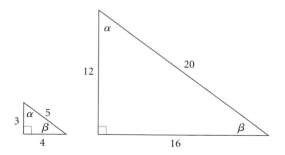

Let's observe the sine, cosine, and tangent values of β in each triangle. Can we expect corresponding function values to be the same?

FIRST TRIANGLE	SECOND TRIANGLE
$\sin \beta = \dfrac{3}{5}$	$\sin \beta = \dfrac{12}{20} = \dfrac{3}{5}$
$\cos \beta = \dfrac{4}{5}$	$\cos \beta = \dfrac{16}{20} = \dfrac{4}{5}$
$\tan \beta = \dfrac{3}{4}$	$\tan \beta = \dfrac{12}{16} = \dfrac{3}{4}$

For the two triangles, the corresponding values of $\sin \beta$, $\cos \beta$, and $\tan \beta$ are the same. The lengths of the sides are proportional—thus the *ratios* are the same. This must be the case because in order for the sine, cosine, and tangent to be functions, there must be only one output (the ratio) for each input (the angle β).

> The trigonometric function values of θ depend only on the measure of the angle, not on the size of the triangle.

The Six Functions Related

We can find the other five trigonometric function values of an acute angle when one of the function-value ratios is known.

EXAMPLE 3   If $\sin \beta = \frac{6}{7}$ and β is an acute angle, find the other five trigonometric function values of β.

Solution   We know from the definition of the sine function that the ratio

$$\frac{6}{7} \quad \text{is} \quad \frac{\text{opp}}{\text{hyp}}.$$

Using this information, let's consider a right triangle in which the hypotenuse has length 7 and the side opposite β has length 6. To find the length of the side adjacent to β, we recall the *Pythagorean theorem*:

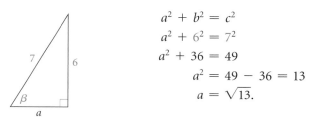

$$a^2 + b^2 = c^2$$
$$a^2 + 6^2 = 7^2$$
$$a^2 + 36 = 49$$
$$a^2 = 49 - 36 = 13$$
$$a = \sqrt{13}.$$

We now use the lengths of the three sides to find the other five ratios:

$$\sin \beta = \frac{6}{7}, \qquad\qquad \csc \beta = \frac{7}{6},$$

$$\cos \beta = \frac{\sqrt{13}}{7}, \qquad\qquad \sec \beta = \frac{7}{\sqrt{13}}, \quad \text{or} \quad \frac{7\sqrt{13}}{13},$$

$$\tan \beta = \frac{6}{\sqrt{13}}, \quad \text{or} \quad \frac{6\sqrt{13}}{13}, \qquad \cot \beta = \frac{\sqrt{13}}{6}.$$

Function Values of 30°, 45°, and 60°

In Examples 1 and 3, we found the trigonometric function values of an acute angle of a right triangle when the lengths of the three sides were known. In most situations, we are asked to find the function values when the measure of the acute angle is given. For certain special angles such as

30°, 45°, and 60°, which are frequently seen in applications, we can use geometry to determine the function values.

A right triangle with a 45° angle actually has two 45° angles. Thus the triangle is *isosceles*, and the legs are the same length. Let's consider such a triangle whose legs have length 1. Then we can find the length of its hypotenuse, c, using the Pythagorean theorem as follows:

$$1^2 + 1^2 = c^2, \quad \text{or} \quad c^2 = 2, \quad \text{or} \quad c = \sqrt{2}.$$

Such a triangle is shown below. From this diagram, we can easily determine the trigonometric function values of 45°.

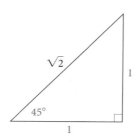

$$\sin 45° = \frac{\text{opp}}{\text{hyp}} = \frac{1}{\sqrt{2}} = \frac{\sqrt{2}}{2} \approx 0.7071,$$

$$\cos 45° = \frac{\text{adj}}{\text{hyp}} = \frac{1}{\sqrt{2}} = \frac{\sqrt{2}}{2} \approx 0.7071,$$

$$\tan 45° = \frac{\text{opp}}{\text{adj}} = \frac{1}{1} = 1$$

It is sufficient to find only the function values of the sine, cosine, and tangent, since the others are their reciprocals.

It is also possible to determine the function values of 30° and 60°. A right triangle with 30° and 60° acute angles is half of an equilateral triangle, as shown in the following figure. Thus if we choose an equilateral triangle whose sides have length 2 and take half of it, we obtain a right triangle that has a hypotenuse of length 2 and a leg of length 1. The other leg has length a, which can be found as follows:

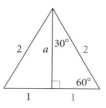

$$a^2 + 1^2 = 2^2$$
$$a^2 + 1 = 4$$
$$a^2 = 3$$
$$a = \sqrt{3}.$$

We can now determine the function values of 30° and 60°:

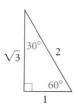

$$\sin 30° = \frac{1}{2} = 0.5, \qquad\qquad \sin 60° = \frac{\sqrt{3}}{2} \approx 0.8660,$$

$$\cos 30° = \frac{\sqrt{3}}{2} \approx 0.8660, \qquad\qquad \cos 60° = \frac{1}{2} = 0.5,$$

$$\tan 30° = \frac{1}{\sqrt{3}} = \frac{\sqrt{3}}{3} \approx 0.5774, \qquad \tan 60° = \frac{\sqrt{3}}{1} = \sqrt{3} \approx 1.7321.$$

Since we will often use the function values of 30°, 45°, and 60°, either the triangles that yield them or the values themselves should be memorized.

	30°	45°	60°
sin	$1/2$	$\sqrt{2}/2$	$\sqrt{3}/2$
cos	$\sqrt{3}/2$	$\sqrt{2}/2$	$1/2$
tan	$\sqrt{3}/3$	1	$\sqrt{3}$

Let's now use what we have learned about trigonometric functions of special angles to solve problems. We will consider such applications in greater detail in Section 5.2.

EXAMPLE 4 *Height of a Hot-air Balloon.* As a hot-air balloon began to rise, the ground crew drove 1.2 mi to an observation station. The initial observation from the station estimated the angle between the ground and the line of sight to the balloon to be 30°. Approximately how high was the balloon at that point? (We are assuming that the wind velocity was low and that the balloon rose vertically for the first few minutes.)

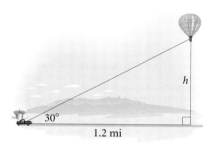

Solution We begin with a sketch of the situation. We know the measure of an acute angle and the length of its adjacent side. Since we want to determine the length of the opposite side, we can use the tangent ratio, or the cotangent ratio. Here we use the tangent ratio:

$$\tan 30° = \frac{\text{opp}}{\text{adj}} = \frac{h}{1.2}$$

$$1.2 \tan 30° = h$$

$$1.2\left(\frac{\sqrt{3}}{3}\right) = h \qquad \text{Substituting; } \tan 30° = \frac{\sqrt{3}}{3}$$

$$0.7 \approx h.$$

The balloon is approximately 0.7 mi, or 3696 ft, high.

Function Values of Any Acute Angle

Historically, the measure of an angle has been expressed in degrees, minutes, and seconds. One minute, denoted $1'$, is such that $60' = 1°$, or $1' = \frac{1}{60} \cdot (1°)$. One second, denoted $1''$, is such that $60'' = 1'$, or $1'' = \frac{1}{60} \cdot (1')$. Then 61 degrees, 27 minutes, 4 seconds could be written as $61°27'4''$. This D°M'S'' form was common before the widespread use of scientific calculators. Now the preferred notation is to express fractional parts of degrees in *decimal degree form*. Although the D°M'S'' notation is still widely used in navigation, we will most often use the decimal form in this text.

Most scientific calculators can convert D°M'S'' notation to decimal degree notation and vice versa. Procedures among calculators vary.

We can use a graphing calculator set in DEGREE mode to convert between D°M′S″ form and decimal degree form.

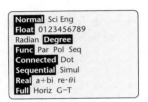

To convert D°M′S″ form to decimal degree form in Example 5, we enter 5°42′30″ using the ANGLE menu for the degree and minute symbols and $\boxed{\text{ALPHA}}$ $\boxed{+}$ for the second symbol. Pressing $\boxed{\text{ENTER}}$ gives us

$$5°42′30″ \approx 5.71°.$$

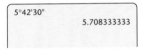

To convert decimal degree form to D°M′S″ form in Example 6, we enter 72.18 and access the ▶DMS feature in the ANGLE menu.

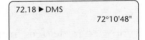

EXAMPLE 5 Convert 5°42′30″ to decimal degree notation.

Solution We enter 5°42′30″. The calculator gives us

$$5°42′30″ \approx 5.71°,$$

rounded to the nearest hundredth of a degree.

Without a calculator, we can convert as follows:

$$5°42′30″ = 5° + 42′ + 30″$$

$$= 5° + 42′ + \frac{30}{60}′ \qquad 1″ = \frac{1}{60}′ ; 30″ = \frac{30}{60}′$$

$$= 5° + 42.5′ \qquad \frac{30}{60}′ = 0.5′$$

$$= 5° + \frac{42.5}{60}° \qquad 1′ = \frac{1}{60}° ; 42.5′ = \frac{42.5}{60}°$$

$$\approx 5.71°. \qquad \frac{42.5}{60}° \approx 0.71°$$

EXAMPLE 6 Convert 72.18° to D°M′S″ notation.

Solution On a calculator, we enter 72.18. The result is

$$72.18° = 72°10′48″.$$

Without a calculator, we can convert as follows:

$$72.18° = 72° + 0.18 \times 1°$$

$$= 72° + 0.18 \times 60′ \qquad 1° = 60′$$

$$= 72° + 10.8′$$

$$= 72° + 10′ + 0.8 \times 1′$$

$$= 72° + 10′ + 0.8 \times 60″ \qquad 1′ = 60″$$

$$= 72° + 10′ + 48″$$

$$= 72°10′48″.$$

So far we have measured angles using degrees. Another useful unit for angle measure is the radian, which we will study in Section 5.4. Calculators work with either degrees or radians. Be sure to use whichever mode is appropriate. In this section, we use the degree mode.

Keep in mind the difference between an exact answer and an approximation. For example,

$$\sin 60° = \frac{\sqrt{3}}{2}. \qquad \textbf{This is exact!}$$

But using a calculator, you might get an answer like

$$\sin 60° \approx 0.8660254038. \qquad \textbf{This is an approximation!}$$

Calculators generally provide values only of the sine, cosine, and tangent functions. You can find values of the cosecant, secant, and cotangent by taking reciprocals of the sine, cosine, and tangent functions, respectively.

EXAMPLE 7 Find the trigonometric function value, rounded to four decimal places, of each of the following.

a) tan 29.7°   **b)** sec 48°   **c)** sin 84°10′39″

Solution

a) We check to be sure that the calculator is in DEGREE mode. The function value is

$$\tan 29.7° \approx 0.5703899297$$
$$\approx 0.5704. \qquad \text{Rounded to four decimal places}$$

b) The secant function value can be found by taking the reciprocal of the cosine function value:

$$\sec 48° = \frac{1}{\cos 48°} \approx 1.49447655 \approx 1.4945.$$

c) We enter sin 84°10′39″. The result is

$$\sin 84°10′39″ \approx 0.9948409474 \approx 0.9948. \qquad \rule{1em}{0.5pt}$$

We can use a calculator to find an angle for which we know a trigonometric function value.

EXAMPLE 8 Find the acute angle, to the nearest tenth of a degree, whose sine value is approximately 0.20113.

Solution The quickest way to find the angle with a calculator is to use an inverse function key. (We first studied inverse functions in Section 4.1 and will consider inverse *trigonometric* functions in Section 6.4.) First check to be sure that your calculator is in DEGREE mode. Usually two keys must be pressed in sequence. For this example, if we press

.20113 $\boxed{\text{SHIFT}}$ $\boxed{\text{SIN}}$, or
$\boxed{\text{2ND}}$ $\boxed{\text{SIN}}$ .20113 $\boxed{\text{ENTER}}$

we find that the acute angle whose sine is 0.20113 is approximately 11.60304613°, or 11.6°. $\rule{1em}{0.5pt}$

EXAMPLE 9 *Ladder Safety.* A paint crew has purchased new 30-ft extension ladders. The manufacturer states that the safest placement on a wall is to extend the ladder to 25 ft and to position the base 6.5 ft from the wall. (*Source*: R. D. Werner Co., Inc.) What angle does the ladder make with the ground in this position?

Solution We make a drawing and then use the most convenient trigonometric function. Because we know the length of the side adjacent to θ and the length of the hypotenuse, we choose the cosine function.

From the definition of the cosine function, we have

$$\cos \theta = \frac{\text{adj}}{\text{hyp}} = \frac{6.5 \text{ ft}}{25 \text{ ft}} = 0.26.$$

Using a calculator, we find that

$$\theta \approx 74.92993786°.$$

Thus when the ladder is in its safest position, it makes an angle of about 75° with the ground.

Cofunctions and Complements

We recall that two angles are **complementary** whenever the sum of their measures is 90°. Each is the complement of the other. In a right triangle, the acute angles are complementary, since the sum of all three angle measures is 180° and the right angle accounts for 90° of this total. Thus if one acute angle of a right triangle is θ, the other is $90° - \theta$.

The six trigonometric function values of each of the acute angles in the triangle below are listed at the right. Note that 53° and 37° are complementary angles since $53° + 37° = 90°$.

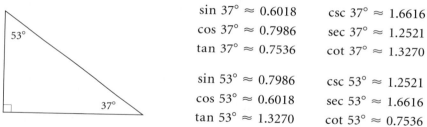

$\sin 37° \approx 0.6018$	$\csc 37° \approx 1.6616$
$\cos 37° \approx 0.7986$	$\sec 37° \approx 1.2521$
$\tan 37° \approx 0.7536$	$\cot 37° \approx 1.3270$
$\sin 53° \approx 0.7986$	$\csc 53° \approx 1.2521$
$\cos 53° \approx 0.6018$	$\sec 53° \approx 1.6616$
$\tan 53° \approx 1.3270$	$\cot 53° \approx 0.7536$

Try this with the acute, complementary angles 20.3° and 69.7° as well. What pattern do you observe? Look for this same pattern in Example 1 earlier in this section.

Note that the sine of an angle is also the cosine of the angle's complement. Similarly, the tangent of an angle is the cotangent of the angle's complement, and the secant of an angle is the cosecant of the angle's complement. These pairs of functions are called **cofunctions**. A list of cofunction identities follows.

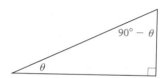

Cofunction Identities

$$\sin \theta = \cos (90° - \theta), \qquad \cos \theta = \sin (90° - \theta),$$
$$\tan \theta = \cot (90° - \theta), \qquad \cot \theta = \tan (90° - \theta),$$
$$\sec \theta = \csc (90° - \theta), \qquad \csc \theta = \sec (90° - \theta)$$

EXAMPLE 10 Given that $\sin 18° \approx 0.3090$, $\cos 18° \approx 0.9511$, and $\tan 18° \approx 0.3249$, find the six trigonometric function values of 72°.

Solution   Using reciprocal relationships, we know that

$$\csc 18° = \frac{1}{\sin 18°} \approx 3.2361,$$

$$\sec 18° = \frac{1}{\cos 18°} \approx 1.0515,$$

and $\cot 18° = \dfrac{1}{\tan 18°} \approx 3.0777.$

Since 72° and 18° are complementary, we have

$\sin 72° = \cos 18° \approx 0.9511,$	$\cos 72° = \sin 18° \approx 0.3090,$
$\tan 72° = \cot 18° \approx 3.0777,$	$\cot 72° = \tan 18° \approx 0.3249,$
$\sec 72° = \csc 18° \approx 3.2361,$	$\csc 72° = \sec 18° \approx 1.0515.$

Exercise Set 5.1

In Exercises 1–6, find the six trigonometric function values of the specified angle.

1.

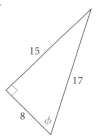

2.

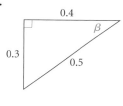

3.

4.

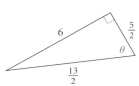

5.

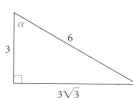

6.

Given a function value of an acute angle, find the other five trigonometric function values.

7. $\sin \theta = \frac{24}{25}$ **8.** $\cos \sigma = 0.7$

9. $\tan \phi = 2$ **10.** $\cot \theta = \frac{1}{3}$

11. $\csc \theta = 1.5$ **12.** $\sec \beta = \sqrt{17}$

Find the exact function value.

13. $\cos 45°$ **14.** $\tan 30°$

15. $\sec 60°$ **16.** $\sin 45°$

17. $\cot 60°$ **18.** $\csc 45°$

19. $\sin 30°$ **20.** $\cos 60°$

21. *Distance Between Bases.*   A baseball diamond is actually a square 90 ft on a side. If a line is drawn from third base to first base, then a right triangle *QPH* is formed, where ∠*QPH* is 45°. Using a trigonometric function, find the distance from third base to first base.

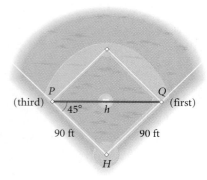

22. *Distance Across a River.* Find the distance *a* across the river.

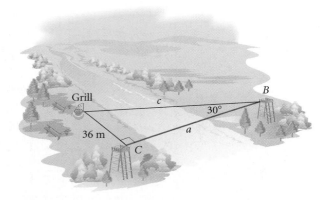

Convert to decimal degree notation. Round to two decimal places.

23. 9°43′

24. 52°15′

25. 35°50″

26. 64°53′

27. 3°2′

28. 19°47′23″

29. 49°38′46″

30. 76°11′34″

Convert to degrees, minutes, and seconds. Round to the nearest second.

31. 17.6°

32. 20.14°

33. 83.025°

34. 67.84°

35. 11.75°

36. 29.8°

37. 47.8268°

38. 0.253°

Find the function value. Round to four decimal places.

39. cos 51°

40. cot 17°

41. tan 4°13′

42. sin 26.1°

43. sec 38.43°

44. cos 74°10′40″

45. cos 40.35°

46. csc 45.2°

47. sin 69°

48. tan 63°48′

49. tan 85.4°

50. cos 4°

51. csc 89.5°

52. sec 35.28°

53. cot 30°25′6″

54. sin 59.2°

Find the acute angle θ, to the nearest tenth of a degree, for the given function value.

55. sin θ = 0.5125

56. tan θ = 2.032

57. tan θ = 0.2226

58. cos θ = 0.3842

59. sin θ = 0.9022

60. tan θ = 3.056

61. cos θ = 0.6879

62. sin θ = 0.4005

63. cot θ = 2.127

64. csc θ = 1.147

$$\left(\text{Hint: } \tan \theta = \frac{1}{\cot \theta}.\right)$$

65. sec θ = 1.279

66. cot θ = 1.351

Find the exact acute angle θ for the given function value.

67. $\sin \theta = \dfrac{\sqrt{2}}{2}$

68. $\cot \theta = \dfrac{\sqrt{3}}{3}$

69. $\cos \theta = \dfrac{1}{2}$

70. $\sin \theta = \dfrac{1}{2}$

71. tan θ = 1

72. $\cos \theta = \dfrac{\sqrt{3}}{2}$

Use the cofunction and reciprocal identities to complete each of the following.

73. $\cos 20° = \underline{\quad} 70° = \dfrac{1}{\underline{\quad} 20°}$

74. $\sin 64° = \underline{\quad} 26° = \dfrac{1}{\underline{\quad} 64°}$

75. $\tan 52° = \cot \underline{\quad} = \dfrac{1}{\underline{\quad} 52°}$

76. $\sec 13° = \csc \underline{\quad} = \dfrac{1}{\underline{\quad} 13°}$

77. Given that

sin 65° ≈ 0.9063,   cos 65° ≈ 0.4226,
tan 65° ≈ 2.145,   cot 65° ≈ 0.4663,
sec 65° ≈ 2.366,   csc 65° ≈ 1.103,

find the six function values of 25°.

78. Given that sin 38.7° ≈ 0.6252, cos 38.7° ≈ 0.7804, and tan 38.7° ≈ 0.8012, find the six function values of 51.3°.

79. Given that sin 82° = *p*, cos 82° = *q*, and tan 82° = *r*, find the six function values of 8° in terms of *p*, *q*, and *r*.

Technology Connection

80. Using the TABLE feature, scroll through a table of values to find the acute angle θ in each of Exercises 55–64, to the nearest tenth of a degree, for the given function value.

Collaborative Discussion and Writing

81. Explain why it is not necessary to memorize the function values for both 30° and 60°.

82. Explain the difference between reciprocal functions and cofunctions.

Skill Maintenance

Graph the function.

83. $f(x) = e^{x/2}$

84. $f(x) = 2^{-x}$

85. $h(x) = \ln\ x$

86. $g(x) = \log_2 x$

Synthesis

87. Given that $\sec \beta = 1.5304$, find $\sin (90° - \beta)$.

88. Find the six trigonometric function values of α.

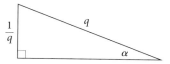

89. Show that the area of this right triangle is $\frac{1}{2} bc \sin A$.

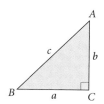

90. Show that the area of this triangle is $\frac{1}{2} ab \sin \theta$.

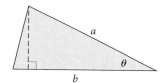

5.2

Applications of Right Triangles

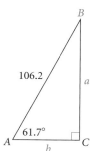

- *Solve right triangles.*
- *Solve applied problems involving right triangles and trigonometric functions.*

Solving Right Triangles

Now that we can find function values for any acute angle, it is possible to *solve* right triangles. To **solve** a triangle means to find the lengths of all sides and the measures of all angles.

EXAMPLE 1 In $\triangle ABC$ (shown at left), find a, b, and B, where a and b represent lengths of sides and B represents the measure of $\angle B$. Here we use standard lettering for naming the sides and angles of a right triangle; side a is opposite angle A, side b is opposite angle B, where a and b are the legs, and side c, the hypotenuse, is opposite angle C, the right angle.

Solution In △*ABC*, we know three of the measures:

$$A = 61.7°, \qquad a = ?,$$
$$B = ?, \qquad b = ?,$$
$$C = 90°, \qquad c = 106.2.$$

Since the sum of the angle measures of any triangle is 180° and $C = 90°$, the sum of A and B is 90°. Thus,

$$B = 90° - A = 90° - 61.7° = 28.3°.$$

We are given an acute angle and the hypotenuse. This suggests that we can use the sine and cosine ratios to find a and b, respectively:

$$\sin 61.7° = \frac{\text{opp}}{\text{hyp}} = \frac{a}{106.2} \quad \text{and} \quad \cos 61.7° = \frac{\text{adj}}{\text{hyp}} = \frac{b}{106.2}.$$

Solving for a and b, we get

$$a = 106.2 \sin 61.7° \quad \text{and} \quad b = 106.2 \cos 61.7°$$
$$a \approx 93.5 \qquad\qquad b \approx 50.3.$$

Thus,

$$A = 61.7°, \qquad a \approx 93.5,$$
$$B = 28.3°, \qquad b \approx 50.3,$$
$$C = 90°, \qquad c = 106.2.$$

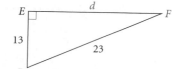

EXAMPLE 2   In △*DEF* (shown at left), find D and F. Then find d.

Solution In △*DEF*, we know three of the measures:

$$D = ?, \qquad d = ?,$$
$$E = 90°, \qquad e = 23,$$
$$F = ?, \qquad f = 13.$$

We know the side adjacent to D and the hypotenuse. This suggests the use of the cosine ratio:

$$\cos D = \frac{\text{adj}}{\text{hyp}} = \frac{13}{23}.$$

We now find the angle whose cosine is $\frac{13}{23}$. To the nearest hundredth of a degree,

$$D \approx 55.58°. \qquad \text{Pressing (13/23)} \boxed{\text{SHIFT}} \boxed{\text{cos}},$$
$$\text{or } \boxed{\text{2ND}} \boxed{\text{cos}} \text{ (13/23)} \boxed{\text{ENTER}}$$

Since the sum of D and F is 90°, we can find F by subtracting:

$$F = 90° - D \approx 90° - 55.58° \approx 34.42°.$$

We could use the Pythagorean theorem to find d, but we will use a trigonometric function here. We could use cos F, sin D, or the tangent or cotangent ratios for either D or F. Let's use $\tan D$:

$$\tan D = \frac{\text{opp}}{\text{adj}} = \frac{d}{13}, \quad \text{or} \quad \tan 55.58° \approx \frac{d}{13}.$$

Then

$$d \approx 13 \tan 55.58° \approx 19.$$

The six measures are

$$
\begin{aligned}
D &\approx 55.58°, & d &\approx 19, \\
E &= 90°, & e &= 23, \\
F &\approx 34.42°, & f &= 13.
\end{aligned}
$$

Applications

Right triangles can be used to model and solve many applied problems in the real world.

EXAMPLE 3 *Hiking at the Grand Canyon.* A backpacker hiking east along the North Rim of the Grand Canyon notices an unusual rock formation directly across the canyon. She decides to continue watching the landmark while hiking along the rim. In 2 hr, she has gone 10 km due east and the landmark is still visible but at approximately a 50° angle to the North Rim. (See the figure at left.)

a) How many kilometers is she from the rock formation?

b) How far is it across the canyon from her starting point?

Solution

a) We know the side adjacent to the 50° angle and want to find the hypotenuse. We can use the cosine function:

$$\cos 50° = \frac{10\text{ km}}{c}$$

$$c = \frac{10\text{ km}}{\cos 50°} \approx 15.6\text{ km}.$$

After hiking 10 km, she is approximately 15.6 km from the rock formation.

b) We know the side adjacent to the 50° angle and want to find the opposite side. We can use the tangent function:

$$\tan 50° = \frac{b}{10\text{ km}}$$

$$b = 10\text{ km} \cdot \tan 50° \approx 11.9\text{ km}.$$

Thus it is approximately 11.9 km across the canyon from her starting point.

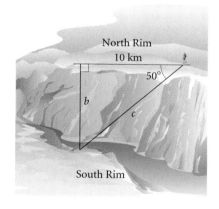

North Rim
10 km
50°
b
c
South Rim

Many applications with right triangles involve an *angle of elevation* or an *angle of depression*. The angle between the horizontal and a line of sight above the horizontal is called an **angle of elevation.** The angle between the horizontal and a line of sight below the horizontal is called an **angle of depression.** For example, suppose that you are looking straight ahead and then you move your eyes up to look at an approaching airplane. The angle that your eyes pass through is an angle of elevation. If the pilot of the plane is looking forward and then looks down, the pilot's eyes pass through an angle of depression.

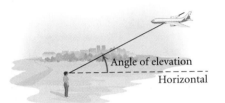

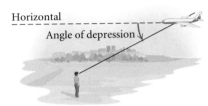

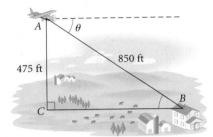

EXAMPLE 4   *Aerial Photography.* An aerial photographer who photographs farm properties for a real estate company has determined from experience that the best photo is taken at a height of approximately 475 ft and a distance of 850 ft from the farmhouse. What is the angle of depression from the plane to the house?

Solution   When parallel lines are cut by a transversal, alternate interior angles are equal. Thus the angle of depression from the plane to the house, θ, is equal to the angle of elevation from the house to the plane, so we can use the right triangle shown in the figure. Since we know the side opposite $\angle B$ and the hypotenuse, we can find θ by using the sine function:

$$\sin \theta = \sin B = \frac{475 \text{ ft}}{850 \text{ ft}}.$$

Using a calculator, we find that

$$\theta \approx 34°.$$

Thus the angle of depression is approximately 34°.   ▬

EXAMPLE 5   *Cloud Height.* To measure cloud height at night, a vertical beam of light is directed on a spot on the cloud. From a point 135 ft away from the light source, the angle of elevation to the spot is found to be 67.35°. Find the height of the cloud.

Solution   From the figure, we have

$$\tan 67.35° = \frac{h}{135 \text{ ft}}$$

$$h = 135 \text{ ft} \cdot \tan 67.35° \approx 324 \text{ ft}.$$

The height of the cloud is about 324 ft.   ▬

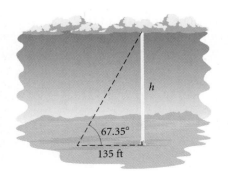

Some applications of trigonometry involve the concept of direction, or bearing. In this text we present two ways of giving direction, the first below and the second in Exercise Set 5.3.

BEARING: FIRST-TYPE   One method of giving direction, or bearing, involves reference to a north–south line using an acute angle. For example, N55°W means 55° west of north and S67°E means 67° east of south.

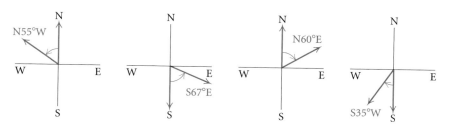

EXAMPLE 6   *Distance to a Forest Fire.* A forest ranger at point *A* sights a fire directly south. A second ranger at point *B*, 7.5 mi east, sights the same fire at a bearing of S27°23′W. How far from *A* is the fire?

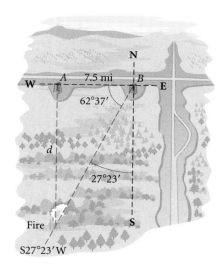

Solution   We first find the complement of 27°23′:

$$B = 90° - 27°23'$$ Angle *B* is opposite side *d* in the right triangle.

$$= 62°37'$$

$$\approx 62.62°.$$

From the figure shown above, we see that the desired distance *d* is part of a right triangle. We have

$$\frac{d}{7.5 \text{ mi}} \approx \tan 62.62°$$

$$d \approx 7.5 \text{ mi} \tan 62.62° \approx 14.5 \text{ mi.}$$

The forest ranger at *A* is about 14.5 mi from the fire.

EXAMPLE 7   *Comiskey Park.* In the new Comiskey Park, the home of the Chicago White Sox baseball team, the first row of seats in the upper deck is farther away from home plate than the last row of seats in the old Comiskey Park. Although there is no obstructed view in the new park, some of the fans still complain about the present distance from home plate to the upper deck of seats. (*Source: Chicago Tribune*, September 19, 1993) From a seat in the last row of the upper deck directly behind the batter, the angle of depression to home plate is 29.9° and the angle of depression to the pitcher's mound is 24.2°. Find (a) the viewing distance to home plate and (b) the viewing distance to the pitcher's mound.

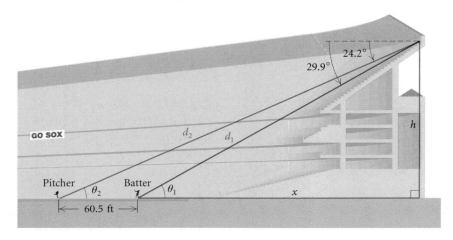

Solution   From geometry we know that $\theta_1 = 29.9°$ and $\theta_2 = 24.2°$. The standard distance from home plate to the pitcher's mound is 60.5 ft. In the drawing, we let d_1 be the viewing distance to home plate, d_2 the viewing distance to the pitcher's mound, h the elevation of the last row, and x the horizontal distance from the batter to a point directly below the seat in the last row of the upper deck.

We begin by determining the distance x. We use the tangent function with $\theta_1 = 29.9°$ and $\theta_2 = 24.2°$:

$$\tan 29.9° = \frac{h}{x} \qquad \text{and} \qquad \tan 24.2° = \frac{h}{x + 60.5}$$

or

$$h = x \tan 29.9° \qquad \text{and} \qquad h = (x + 60.5) \tan 24.2°.$$

Then substituting $x \tan 29.9°$ for h in the second equation, we obtain

$$x \tan 29.9° = (x + 60.5) \tan 24.2°.$$

Solving for x, we get

$$x \tan 29.9° = x \tan 24.2° + 60.5 \tan 24.2°$$

$$x \tan 29.9° - x \tan 24.2° = x \tan 24.2° + 60.5 \tan 24.2° - x \tan 24.2°$$

$$x(\tan 29.9° - \tan 24.2°) = 60.5 \tan 24.2°$$

$$x = \frac{60.5 \tan 24.2°}{\tan 29.9° - \tan 24.2°}$$

$$x \approx 216.5.$$

We can then find d_1 and d_2 using the cosine function:

$$\cos 29.9° = \frac{216.5}{d_1} \quad \text{and} \quad \cos 24.2° = \frac{216.5 + 60.5}{d_2}$$

or

$$d_1 = \frac{216.5}{\cos 29.9°} \quad \text{and} \quad d_2 = \frac{277}{\cos 24.2°}$$
$$d_1 \approx 249.7 \qquad\qquad\qquad d_2 \approx 303.7.$$

The distance to home plate is about 250 ft,* and the distance to the pitcher's mound is about 304 ft.

Exercise Set 5.2

In Exercises 1–6, solve the right triangle.

1.

2.

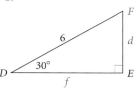

3.

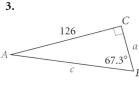

4.

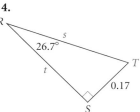

5.

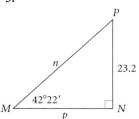

6.

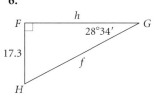

In Exercises 7-16, solve the right triangle. (Standard lettering has been used.)

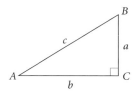

7. $A = 87°43'$, $a = 9.73$

8. $a = 12.5$, $b = 18.3$

9. $b = 100$, $c = 450$

10. $B = 56.5°$, $c = 0.0447$

11. $A = 47.58°$, $c = 48.3$

12. $B = 20.6°$, $a = 7.5$

13. $A = 35°$, $b = 40$

14. $B = 69.3°$, $b = 93.4$

15. $b = 1.86$, $c = 4.02$

16. $a = 10.2$, $c = 20.4$

17. *Safety Line to Raft.* Each spring Madison uses his vacation time to ready his lake property for the summer. He wants to run a new safety line from point B on the shore to the corner of the anchored diving raft. The current safety line, which runs perpendicular to the shore line to point A, is 40 ft

---

*In the old Comiskey Park, the distance to home plate was only 150 ft.

long. He estimates the angle from B to the corner of the raft to be 50°. Approximately how much rope does he need for the new safety line if he allows 5 ft of rope at each end to fasten the rope?

18. *Enclosing an Area.* Glynis is enclosing a triangular area in a corner of her fenced rectangular backyard for her Labrador retriever. In order for a certain tree to be included in this pen, one side needs to be 14.5 ft and make a 53° angle with the new side. How long is the new side?

19. *Easel Display.* A marketing group is designing an easel to display posters advertising their newest products. They want the easel to be 6 ft tall and the back of it to fit flush against a wall. For optimal eye contact, the best angle between the front and back legs of the easel is 23°. How far from the wall should the front legs be placed in order to obtain this angle?

20. *Height of a Tree.* A supervisor must train a new team of loggers to estimate the heights of trees. As an example, she walks off 40 ft from the base of a tree and estimates the angle of elevation to the tree's peak to be 70°. Approximately how tall is the tree?

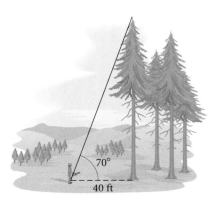

21. *Sand Dunes National Park.* While visiting the Sand Dunes National Park in Colorado, Ian approximated the angle of elevation to the top of a sand dune to be 20°. After walking 800 ft closer, he guessed that the angle of elevation had increased by 15°. Approximately how tall is the dune he was observing?

22. *Tee Shirt Design.* A new tee shirt design is to have a regular pentagon inscribed in a circle, as shown in the figure. Each side of the pentagon is to be 3.5 in. long. Find the radius of the circumscribed circle.

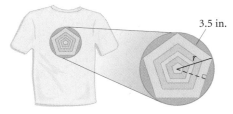

23. *Inscribed Octagon.* A regular octagon is inscribed in a circle of radius 15.8 cm. Find the perimeter of the octagon.

24. *Height of a Weather Balloon.* A weather balloon is directly west of two observing stations that are 10 mi apart. The angles of elevation of the balloon from the two stations are 17.6° and 78.2°. How high is the balloon?

25. *Height of a Kite.* For a science fair project, a group of students tested different materials used to construct kites. Their instructor provided an instrument that accurately measures the angle of elevation. In one of the tests, the angle of elevation was 63.4° with 670 ft of string out. Assuming the string was taut, how high was the kite?

26. *Height of a Building.* A window washer on a ladder looks at a nearby building 100 ft away, noting that the angle of elevation of the top of the building is 18.7° and the angle of depression of the bottom of the building is 6.5°. How tall is the nearby building?

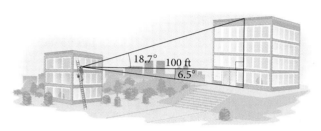

27. *Distance Between Towns.* From a hot-air balloon 2 km high, the angles of depression to two towns, in line with the balloon, are 81.2° and 13.5°. How far apart are the towns?

28. *Angle of Elevation.* What is the angle of elevation of the sun when a 35-ft mast casts a 20-ft shadow?

29. *Distance from a Lighthouse.* From the top of a lighthouse 55 ft above sea level, the angle of depression to a small boat is 11.3°. How far from the foot of the lighthouse is the boat?

30. *Lightning Detection.* In extremely large forests, it is not cost-effective to position forest rangers in towers or to use small aircraft to continually watch for fires. Since lightning is a frequent cause of fire, lightning detectors are now commonly used instead.

These devices not only give a bearing on the location but also measure the intensity of the lightning. A detector at point *Q* is situated 15 mi west of a central fire station at point *R*. The bearing from *Q* to where lightning hits due south of *R* is S37.6°E. How far is the hit from point *R*?

31. *Lobster Boat.* A lobster boat is situated due west of a lighthouse. A barge is 12 km south of the lobster boat. From the barge, the bearing to the lighthouse is N63°20′E. How far is the lobster boat from the lighthouse?

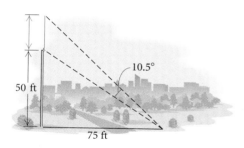

32. *Length of an Antenna.* A vertical antenna is mounted atop a 50-ft pole. From a point on the level ground 75 ft from the base of the pole, the antenna subtends an angle of 10.5°. Find the length of the antenna.

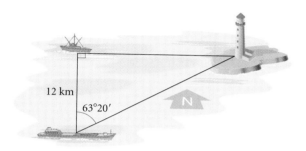

Collaborative Discussion and Writing

33. In this section, the trigonometric functions have been defined as functions of acute angles. Thus the set of angles whose measures are greater than 0° and less than 90° is the domain for each function. What appear to be the ranges for the sine, the cosine, and the tangent functions given this domain?

34. Explain in your own words five ways in which length c can be determined in this triangle. Which way seems the most efficient?

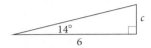

Skill Maintenance

Find the distance between the points.

35. $(8, -2)$ and $(-6, -4)$ **36.** $(-9, 3)$ and $(0, 0)$

37. Convert to an exponential equation:
$\log 0.001 = -3$.

38. Convert to a logarithmic equation: $e^4 = t$.

Synthesis

39. Find h, to the nearest tenth.

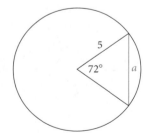

40. Find a, to the nearest tenth.

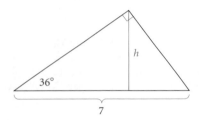

41. *Construction of Picnic Pavilions.* A construction company is mass-producing picnic pavilions for national parks, as shown in the figure. The rafter ends are to be sawed in such a way that they will be vertical when in place. The front wall is 8 ft high, the back wall is $6\frac{1}{2}$ ft high, and the distance between walls is 8 ft. At what angle should the rafters be cut?

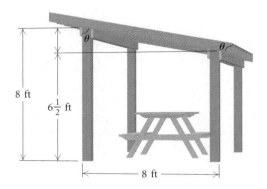

42. *Diameter of a Pipe.* A V-gauge is used to find the diameter of a pipe. The advantage of such a device is that it is rugged, it is accurate, and it has no moving parts to break down. In the figure, the measure of angle AVB is 54°. A pipe is placed in the V-shaped slot and the distance VP is used to estimate the diameter. The line VP is calibrated by listing as its units the corresponding diameters. This, in effect, establishes a function between VP and d.

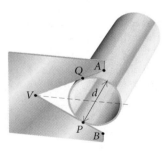

a) Suppose that the diameter of a pipe is 2 cm. What is the distance VP?

b) Suppose that the distance VP is 3.93 cm. What is the diameter of the pipe?

c) Find a formula for d in terms of VP.

d) Find a formula for VP in terms of d.

43. *Sound of an Airplane.* It is common experience to hear the sound of a low-flying airplane and look at the wrong place in the sky to see the plane. Suppose that a plane is traveling directly at you at a speed of 200 mph and an altitude of 3000 ft, and you hear the sound at what seems to be an angle of inclination of 20°. At what angle θ should you actually look in order to see the plane? Consider the speed of sound to be 1100 ft/sec.

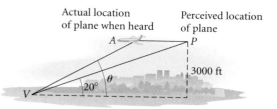

Actual location of plane when heard

Perceived location of plane

44. *Measuring the Radius of the Earth.* One way to measure the radius of the earth is to climb to the top of a mountain whose height above sea level is known and measure the angle between a vertical line to the center of the earth from the top of the mountain and a line drawn from the top of the mountain to the horizon, as shown in the figure. The height of Mt. Shasta in California is 14,162 ft.

From the top of Mt. Shasta, one can see the horizon on the Pacific Ocean. The angle formed between a line to the horizon and the vertical is found to be 87°53′. Use this information to estimate the radius of the earth, in miles.

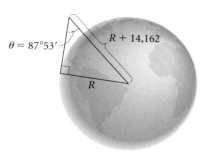

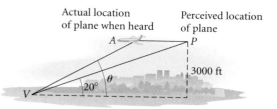

5.3

Trigonometric Functions of Any Angle

- *Find angles that are coterminal with a given angle and find the complement and the supplement of a given angle.*
- *Determine the six trigonometric function values for any angle in standard position when the coordinates of a point on the terminal side are given.*
- *Find the function values for any angle whose terminal side lies on an axis.*
- *Find the function values for an angle whose terminal side makes an angle of 30°, 45°, or 60° with the x-axis.*
- *Use a calculator to find function values and angles.*

Angles, Rotations, and Degree Measure

An *angle* is a familiar figure in the world around us.

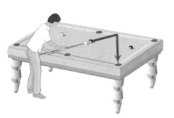

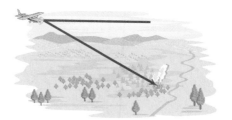

An **angle** is the union of two rays with a common endpoint called the **vertex**. In trigonometry, we often think of an angle as a **rotation**. To do so, think of locating a ray along the positive *x*-axis with its endpoint at the origin. This ray is called the **initial side** of the angle. Though we leave that ray fixed, think of making a copy of it and rotating it. A rotation *counterclockwise* is a **positive rotation,** and a rotation *clockwise* is a **negative rotation.** The ray at the end of the rotation is called the **terminal side** of the angle. The angle formed is said to be in **standard position.**

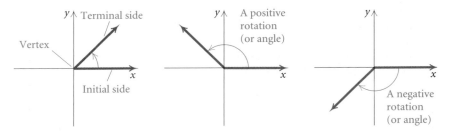

The measure of an angle or rotation may be given in degrees. The Babylonians developed the idea of dividing the circumference of a circle into 360 equal parts, or degrees. If we let the measure of one of these parts be 1°, then one complete positive revolution or rotation has a measure of 360°. One half of a revolution has a measure of 180°, one fourth of a revolution has a measure of 90°, and so on. We can also speak of an angle of measure 60°, 135°, 330°, or 420°. The terminal sides of these angles lie in quadrants I, II, IV, and I, respectively. The negative rotations −30°, −110°, and −225° represent angles with terminal sides in quadrants IV, III, and II, respectively.

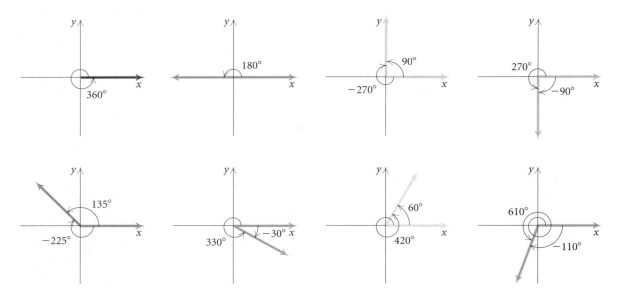

If two or more angles have the same terminal side, the angles are said to be **coterminal**. To find angles coterminal with a given angle, we add or subtract multiples of 360°. For example, 420°, shown above, has the same terminal side as 60°, since 420° = 360° + 60°. Thus we say that angles of measure 60° and 420° are coterminal. The negative rotation that measures −300° is also coterminal with 60° because 60° − 360° = −300°. Other examples of coterminal angles shown above are 90° and −270°, −90° and 270°, 135° and −225°, −30° and 330°, and −110° and 610°.

EXAMPLE 1 Find two positive and two negative angles that are coterminal with (a) 51° and (b) −7°.

Solution

a) We add and subtract multiples of 360°. Many answers are possible.

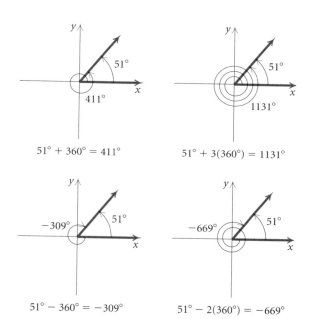

$51° + 360° = 411°$ $51° + 3(360°) = 1131°$

$51° − 360° = −309°$ $51° − 2(360°) = −669°$

Thus angles of measure 411°, 1131°, −309°, and −669° are coterminal with 51°.

b) We have the following:

$$-7° + 360° = 353°, \qquad -7° + 2(360°) = 713°,$$
$$-7° - 360° = -367°, \qquad -7° - 10(360°) = -3607°.$$

Thus angles of measure 353°, 713°, −367°, and −3607° are coterminal with −7°.

Angles can be classified by their measures, as seen in the following figure.

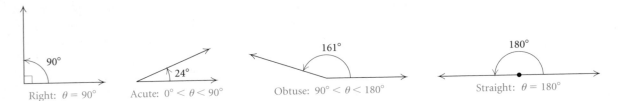

Right: $\theta = 90°$ Acute: $0° < \theta < 90°$ Obtuse: $90° < \theta < 180°$ Straight: $\theta = 180°$

Recall that two acute angles are **complementary** if their sum is 90°. For example, angles that measure 10° and 80° are complementary because $10° + 80° = 90°$. Two positive angles are **supplementary** if their sum is 180°. For example, angles that measure 45° and 135° are supplementary because $45° + 135° = 180°$.

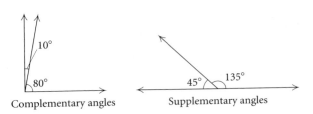

Complementary angles   Supplementary angles

EXAMPLE 2   Find the complement and the supplement of 71.46°.

Solution   We have

$$90° - 71.46° = 18.54°,$$
$$180° - 71.46° = 108.54°.$$

Thus the complement of 71.46° is 18.54° and the supplement is 108.54°.

Trigonometric Functions of Angles or Rotations

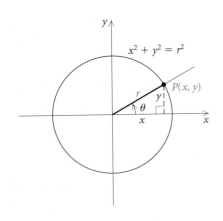

Many applied problems in trigonometry involve the use of angles that are not acute. Thus we need to extend the domains of the trigonometric functions defined in Section 5.1 to angles, or rotations, of any size. To do this, we first consider a right triangle with one vertex at the origin of a coordinate system and one vertex on the positive x-axis. (See the figure at left.) The other vertex is at P, a point on the circle whose center is at the origin and whose radius r is the length of the hypotenuse of the triangle. This triangle is a **reference triangle** for angle θ, which is in standard position. Note that y is the length of the side opposite θ and x is the length of the side adjacent to θ.

Recalling the definitions in Section 5.1, we note that three of the trigonometric functions of angle θ are defined as follows:

$$\sin\theta = \frac{\text{opp}}{\text{hyp}} = \frac{y}{r}, \qquad \cos\theta = \frac{\text{adj}}{\text{hyp}} = \frac{x}{r}, \qquad \tan\theta = \frac{\text{opp}}{\text{adj}} = \frac{y}{x}.$$

Since x and y are the coordinates of the point P and the length of the radius is the length of the hypotenuse, we can also define these functions as follows:

$$\sin \theta = \frac{y\text{-coordinate}}{\text{radius}}, \qquad \cos \theta = \frac{x\text{-coordinate}}{\text{radius}},$$

$$\tan \theta = \frac{y\text{-coordinate}}{x\text{-coordinate}}.$$

We will use these definitions for functions of angles of any measure. The following figures show angles whose terminal sides lie in quadrants II, III, and IV.

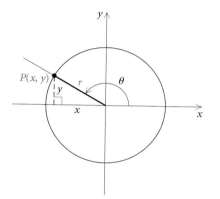
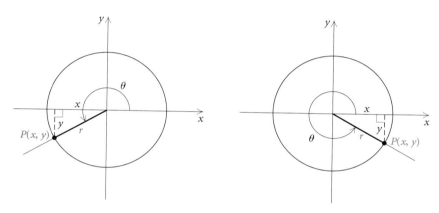

A reference triangle can be drawn for angles in any quadrant, as shown. Note that the angle is in standard position; that is, it is always measured from the positive half of the x-axis. The point $P(x, y)$ is a point, other than the vertex, on the terminal side of the angle. Each of its two coordinates may be positive, negative, or zero, depending on the location of the terminal side. *The length of the radius, which is also the length of the hypotenuse of the reference triangle, is always considered positive.* (Note that $x^2 + y^2 = r^2$, or $r = \sqrt{x^2 + y^2}$.) Regardless of the location of P, we have the following definitions.

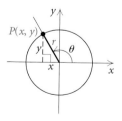

Trigonometric Functions of Any Angle θ

Suppose that $P(x, y)$ is any point other than the vertex on the terminal side of any angle θ in standard position, and r is the radius, or distance from the origin to $P(x, y)$. Then the trigonometric functions are defined as follows.

$$\sin \theta = \frac{y\text{-coordinate}}{\text{radius}} = \frac{y}{r} \qquad \csc \theta = \frac{\text{radius}}{y\text{-coordinate}} = \frac{r}{y}$$

$$\cos \theta = \frac{x\text{-coordinate}}{\text{radius}} = \frac{x}{r} \qquad \sec \theta = \frac{\text{radius}}{x\text{-coordinate}} = \frac{r}{x}$$

$$\tan \theta = \frac{y\text{-coordinate}}{x\text{-coordinate}} = \frac{y}{x} \qquad \cot \theta = \frac{x\text{-coordinate}}{y\text{-coordinate}} = \frac{x}{y}$$

Values of the trigonometric functions can be positive, negative, or zero, depending on where the terminal side of the angle lies. The length of the radius is always positive. Thus the signs of the function values depend only on the coordinates of the point P on the terminal side of the angle. In the first quadrant, all function values are positive because both coordinates are positive. In the second quadrant, first coordinates are negative and second coordinates are positive; thus only the sine and the cosecant values are positive. Similarly, we can determine the signs of the function values in the third and fourth quadrants. Because of the reciprocal relationships, we need to learn only the signs for the sine, cosine, and tangent functions.

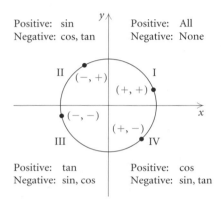

Positive: sin
Negative: cos, tan

Positive: All
Negative: None

Positive: tan
Negative: sin, cos

Positive: cos
Negative: sin, tan

EXAMPLE 3   Find the six trigonometric function values for each angle shown.

a)

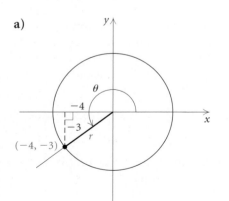

b)

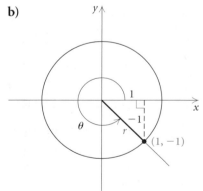

c)

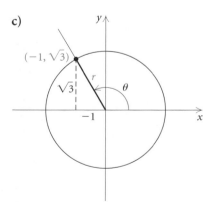

Solution

a) We first determine r, the distance from the origin $(0, 0)$ to the point $(-4, -3)$. The distance between $(0, 0)$ and any point (x, y) on the terminal side of the angle is

$$r = \sqrt{(x - 0)^2 + (y - 0)^2}$$
$$= \sqrt{x^2 + y^2}.$$

Substituting -4 for x and -3 for y, we find

$$r = \sqrt{(-4)^2 + (-3)^2}$$
$$= \sqrt{16 + 9} = \sqrt{25} = 5.$$

Using the definitions of the trigonometric functions, we can now find the function values of θ. We substitute -4 for x, -3 for y, and 5 for r:

$$\sin \theta = \frac{y}{r} = \frac{-3}{5} = -\frac{3}{5}, \qquad \csc \theta = \frac{r}{y} = \frac{5}{-3} = -\frac{5}{3},$$

$$\cos \theta = \frac{x}{r} = \frac{-4}{5} = -\frac{4}{5}, \qquad \sec \theta = \frac{r}{x} = \frac{5}{-4} = -\frac{5}{4},$$

$$\tan \theta = \frac{y}{x} = \frac{-3}{-4} = \frac{3}{4}, \qquad \cot \theta = \frac{x}{y} = \frac{-4}{-3} = \frac{4}{3}.$$

As expected, the tangent and the cotangent values are positive and the other four are negative. This is true for all angles in quadrant III.

b) We first determine r, the distance from the origin to the point $(1, -1)$:

$$r = \sqrt{1^2 + (-1)^2} = \sqrt{1 + 1} = \sqrt{2}.$$

Substituting 1 for x, -1 for y, and $\sqrt{2}$ for r, we find

$$\sin \theta = \frac{y}{r} = \frac{-1}{\sqrt{2}} = -\frac{\sqrt{2}}{2}, \qquad \csc \theta = \frac{r}{y} = \frac{\sqrt{2}}{-1} = -\sqrt{2},$$

$$\cos \theta = \frac{x}{r} = \frac{1}{\sqrt{2}} = \frac{\sqrt{2}}{2}, \qquad \sec \theta = \frac{r}{x} = \frac{\sqrt{2}}{1} = \sqrt{2},$$

$$\tan \theta = \frac{y}{x} = \frac{-1}{1} = -1, \qquad \cot \theta = \frac{x}{y} = \frac{1}{-1} = -1.$$

c) We determine r, the distance from the origin to the point $(-1, \sqrt{3})$:

$$r = \sqrt{(-1)^2 + (\sqrt{3})^2} = \sqrt{1 + 3} = \sqrt{4} = 2.$$

Substituting -1 for x, $\sqrt{3}$ for y, and 2 for r, we find the trigonometric function values of θ are

$$\sin \theta = \frac{\sqrt{3}}{2}, \qquad \csc \theta = \frac{2}{\sqrt{3}} = \frac{2\sqrt{3}}{3},$$

$$\cos \theta = \frac{-1}{2} = -\frac{1}{2}, \qquad \sec \theta = \frac{2}{-1} = -2,$$

$$\tan \theta = \frac{\sqrt{3}}{-1} = -\sqrt{3}, \qquad \cot \theta = \frac{-1}{\sqrt{3}} = -\frac{\sqrt{3}}{3}.$$

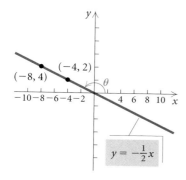

Any point other than the origin on the terminal side of an angle can be used to determine the trigonometric function values of that angle. The function values are the same regardless of which point is used. To illustrate this, let's consider an angle θ in standard position whose terminal side lies on the line $y = -\frac{1}{2}x$. We can determine two second-quadrant solutions of the equation, find the length r for each point, and then compare the sine, cosine, and tangent function values using each point.

If $x = -4$, then $y = -\frac{1}{2}(-4) = 2$.

If $x = -8$, then $y = -\frac{1}{2}(-8) = 4$.

For $(-4, 2)$, $r = \sqrt{(-4)^2 + 2^2} = \sqrt{20} = 2\sqrt{5}$.

For $(-8, 4)$, $r = \sqrt{(-8)^2 + 4^2} = \sqrt{80} = 4\sqrt{5}$.

Using $(-4, 2)$ and $r = 2\sqrt{5}$, we find that

$$\sin \theta = \frac{2}{2\sqrt{5}} = \frac{1}{\sqrt{5}} = \frac{\sqrt{5}}{5}, \qquad \cos \theta = \frac{-4}{2\sqrt{5}} = \frac{-2}{\sqrt{5}} = -\frac{2\sqrt{5}}{5},$$

and $\qquad \tan \theta = \frac{2}{-4} = -\frac{1}{2}$.

Using $(-8, 4)$ and $r = 4\sqrt{5}$, we find that

$$\sin \theta = \frac{4}{4\sqrt{5}} = \frac{1}{\sqrt{5}} = \frac{\sqrt{5}}{5}, \qquad \cos \theta = \frac{-8}{4\sqrt{5}} = \frac{-2}{\sqrt{5}} = -\frac{2\sqrt{5}}{5},$$

and $\qquad \tan \theta = \frac{4}{-8} = -\frac{1}{2}$.

We see that the function values are the same using either point. Any point other than the origin on the terminal side of an angle can be used to determine the trigonometric function values.

> The trigonometric function values of θ depend only on the angle, not on the choice of the point on the terminal side that is used to compute them.

The Six Functions Related

When we know one of the function values of an angle, we can find the other five if we know the quadrant in which the terminal side lies. The procedure is to sketch a reference triangle in the appropriate quadrant, use the Pythagorean theorem as needed to find the lengths of its sides, and then find the ratios of the sides.

EXAMPLE 4   Given that $\tan \theta = -\frac{2}{3}$ and θ is in the second quadrant, find the other function values.

Solution   We first sketch a second-quadrant angle. Since

$$\tan \theta = \frac{y}{x} = -\frac{2}{3} = \frac{2}{-3}, \qquad \text{Expressing } -\frac{2}{3} \text{ as } \frac{2}{-3} \text{ since } \theta \text{ is in quadrant II}$$

we make the legs of lengths 2 and 3. The hypotenuse must then have

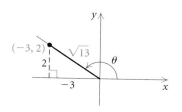

length $\sqrt{2^2 + 3^2}$, or $\sqrt{13}$. Now we can read off the appropriate ratios:

$$\sin \theta = \frac{2}{\sqrt{13}}, \quad \text{or} \quad \frac{2\sqrt{13}}{13}, \qquad \csc \theta = \frac{\sqrt{13}}{2},$$

$$\cos \theta = -\frac{3}{\sqrt{13}}, \quad \text{or} \quad -\frac{3\sqrt{13}}{13}, \qquad \sec \theta = -\frac{\sqrt{13}}{3},$$

$$\tan \theta = -\frac{2}{3}, \qquad\qquad\qquad \cot \theta = -\frac{3}{2}.$$

Terminal Side on an Axis

An angle whose terminal side falls on one of the axes is a **quadrantal angle.** Then one of the coordinates of any point on that side is 0. The definitions of the trigonometric functions still apply, but in some cases, function values will not be defined because a denominator will be 0.

EXAMPLE 5 Find the sine, cosine, and tangent values for 90°, 180°, 270°, and 360°.

Solution We first draw a sketch of each angle in standard position and label a point on the terminal side. Since the function values are the same for all points on the terminal side, we choose (0, 1), (−1, 0), (0, −1), and (1, 0) for convenience. Note also that $r = 1$ for each choice.

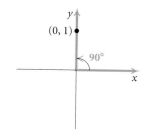 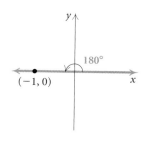 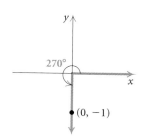 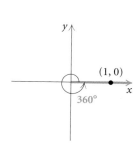

Then by the definitions we get

$$\sin 90° = \frac{1}{1} = 1, \qquad \sin 180° = \frac{0}{1} = 0, \qquad \sin 270° = \frac{-1}{1} = -1, \qquad \sin 360° = \frac{0}{1} = 0,$$

$$\cos 90° = \frac{0}{1} = 0, \qquad \cos 180° = \frac{-1}{1} = -1, \qquad \cos 270° = \frac{0}{1} = 0, \qquad \cos 360° = \frac{1}{1} = 1,$$

$$\tan 90° = \frac{1}{0}, \quad \text{Undefined} \qquad \tan 180° = \frac{0}{-1} = 0, \qquad \tan 270° = \frac{-1}{0}, \quad \text{Undefined} \qquad \tan 360° = \frac{0}{1} = 0.$$

In Example 5, all the values can be found using a calculator, but you will find that it is convenient to be able to compute them mentally. It is also helpful to note that coterminal angles have the same function values.

EXAMPLE 6   Find each of the following.

a) $\sin(-90°)$ **b)** $\csc 540°$

Solution

a) We note that $-90°$ is coterminal with $270°$. Thus,

$$\sin(-90°) = \sin 270° = \frac{-1}{1} = -1.$$

b) Since $540° = 180° + 360°$, $540°$ and $180°$ are coterminal. Thus,

$$\csc 540° = \csc 180° = \frac{1}{\sin 180°} = \frac{1}{0}, \quad \text{which is undefined.} \quad \rule{0.7em}{0.5em}$$

Reference Angles: 30°, 45°, and 60°

We can also mentally determine trigonometric function values whenever the terminal side makes a 30°, 45°, or 60° angle with the x-axis. Consider, for example, an angle of 150°. The terminal side makes a 30° angle with the x-axis, since $180° - 150° = 30°$.

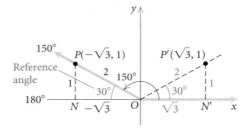

As the figure shows, $\triangle ONP$ is congruent to $\triangle ON'P'$; therefore, the ratios of the sides of the two triangles are the same. Thus the trigonometric function values are the same except perhaps for the sign. We could determine the function values directly from $\triangle ONP$, but this is not necessary. If we remember that in quadrant II, the sine is positive and the cosine and the tangent are negative, we can simply use the function values of 30° that we already know and prefix the appropriate sign. Thus,

$$\sin 150° = \sin 30° = \frac{1}{2}, \qquad \cos 150° = -\cos 30° = -\frac{\sqrt{3}}{2},$$

and $\tan 150° = -\tan 30° = -\dfrac{\sqrt{3}}{3}.$

Triangle ONP is the reference triangle and the acute angle $\angle NOP$ is called a *reference angle*.

Reference Angle

The **reference angle** for an angle is the acute angle formed by the terminal side of the angle and the x-axis.

EXAMPLE 7 Find the sine, cosine, and tangent function values for each of the following.

a) 225°   **b)** −780°

Solution

a) We draw a figure showing the terminal side of a 225° angle. The reference angle is 225° − 180°, or 45°.

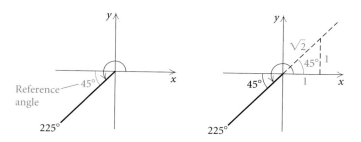

Recall from Section 5.1 that sin 45° = $\sqrt{2}/2$, cos 45° = $\sqrt{2}/2$, and tan 45° = 1. Also note that in the third quadrant, the sine and the cosine are negative and the tangent is positive. Thus we have

$$\sin 225° = -\frac{\sqrt{2}}{2}, \quad \cos 225° = -\frac{\sqrt{2}}{2}, \quad \text{and} \quad \tan 225° = 1.$$

b) We draw a figure showing the terminal side of a −780° angle. Since −780° + 2(360°) = −60°, we know that −780° and −60° are coterminal.

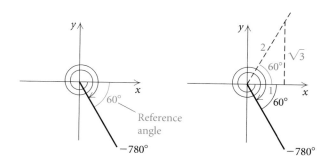

The reference angle for −60° is the acute angle formed by the terminal side of the angle and the *x*-axis. Thus the reference angle for −60° is 60°. We know that since −780° is a fourth-quadrant angle, the cosine is positive and the sine and the tangent are negative. Recalling that sin 60° = $\sqrt{3}/2$, cos 60° = 1/2, and tan 60° = $\sqrt{3}$, we have

$$\sin(-780°) = -\frac{\sqrt{3}}{2}, \quad \cos(-780°) = \frac{1}{2},$$

and tan (−780°) = $-\sqrt{3}$.

Function Values for Any Angle

When the terminal side of an angle falls on one of the axes or makes a 30°, 45°, or 60° angle with the *x*-axis, we can find exact function values without the use of a calculator. But this group is only a small subset of *all* angles. Using a calculator, we can approximate the trigonometric function values of *any* angle. In fact, we can approximate or find exact function values of all angles without using a reference angle.

EXAMPLE 8   Find each of the following function values using a calculator and round the answer to four decimal places, where appropriate.

a) $\cos 112°$

b) $\sec 500°$

c) $\tan (-83.4°)$

d) $\csc 351.75°$

e) $\cos 2400°$

f) $\sin 175°40'9''$

g) $\cot (-135°)$

Solution   Using a calculator set in DEGREE mode, we find the values.

a) $\cos 112° \approx -0.3746$

b) $\sec 500° = \dfrac{1}{\cos 500°} \approx -1.3054$

c) $\tan (-83.4°) \approx -8.6427$

d) $\csc 351.75° = \dfrac{1}{\sin 351.75°} \approx -6.9690$

e) $\cos 2400° = -0.5$

f) $\sin 175°40'9'' \approx 0.0755$

g) $\cot (-135°) = \dfrac{1}{\tan (-135°)} = 1$

In many applications, we have a trigonometric function value and want to find the measure of a corresponding angle. When only acute angles are considered, there is only one angle for each trigonometric function value. This is not the case when we extend the domain of the trigonometric functions to the set of *all* angles. For a given function value, there is an infinite number of angles that have that function value. There can be two such angles for each value in the range from 0° to 360°. To determine a unique answer in the interval (0°, 360°), the quadrant in which the terminal side lies must be specified.

The calculator gives the reference angle as an output for each function value that is entered as an input. Knowing the reference angle and the quadrant in which the terminal side lies, we can find the specified angle.

EXAMPLE 9   Given the function value and the quadrant restriction, find θ.

a) $\sin \theta = 0.2812, \ 90° < \theta < 180°$

b) $\cot \theta = -0.1611, \ 270° < \theta < 360°$

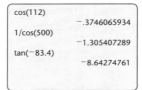
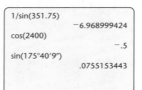

Solution

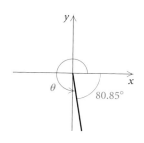

a) We first sketch the angle in the second quadrant. We use the calculator to find the acute angle (reference angle) whose sine is 0.2812. The reference angle is approximately 16.33°. We find the angle θ by subtracting 16.33° from 180°:

$$180° - 16.33° = 163.67°.$$

Thus, $\theta \approx 163.67°$.

b) We begin by sketching the angle in the fourth quadrant. Because the cotangent value is the reciprocal of the tangent value, we know that

$$\tan \theta \approx \frac{1}{-0.1611} \approx -6.2073.$$

We use the calculator to find the acute angle (reference angle) whose tangent is 6.2073, ignoring the fact that tan θ is negative. The reference angle is approximately 80.85°. We find angle θ by subtracting 80.85° from 360°:

$$360° - 80.85° = 279.15°.$$

Thus, $\theta \approx 279.15°$.

Exercise Set 5.3

For angles of the following measures, state in which quadrant the terminal side lies. It helps to sketch the angle in standard position.

1. 187°   **2.** −14.3°   **3.** 245°15′

4. −120°   **5.** 800°   **6.** 1075°

7. −460.5°   **8.** 315°   **9.** −912°

10. 13°15′60″   **11.** 537°   **12.** −345.14°

Find two positive angles and two negative angles that are coterminal with the given angle. Answers may vary.

13. 74°   **14.** −81°   **15.** 115.3°

16. 275°10′   **17.** −180°   **18.** −310°

Find the complement and the supplement.

19. 17.11°   **20.** 47°38′   **21.** 12°3′14″

22. 9.038°   **23.** 45.2°   **24.** 67.31°

Find the six trigonometric function values for the angle shown.

25.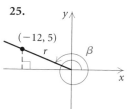

26.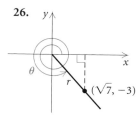

27.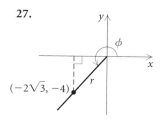

28.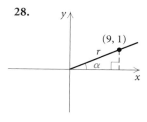

The terminal side of angle θ in standard position lies on the given line in the given quadrant. Find sin θ, cos θ, and tan θ.

29. $2x + 3y = 0$; quadrant IV

30. $4x + y = 0$; quadrant II

31. $5x - 4y = 0$; quadrant I

32. $y = 0.8x$; quadrant III

A function value and a quadrant are given. Find the other five function values. Give exact answers.

33. $\sin \theta = -\dfrac{1}{3}$, quadrant III

34. $\tan \beta = 5$, quadrant I

35. $\cot \theta = -2$, quadrant IV

36. $\cos \alpha = -\dfrac{4}{5}$, quadrant II

37. $\cos \phi = \dfrac{3}{5}$, quadrant IV

38. $\sin \theta = -\dfrac{5}{13}$, quadrant III

Find the reference angle and the exact function value if it exists.

39. $\cos 150°$	**40.** $\sec (-225°)$	**41.** $\tan (-135°)$
42. $\sin (-45°)$	**43.** $\sin 7560°$	**44.** $\tan 270°$
45. $\cos 495°$	**46.** $\tan 675°$	**47.** $\csc (-210°)$
48. $\sin 300°$	**49.** $\cot 570°$	**50.** $\cos (-120°)$
51. $\tan 330°$	**52.** $\cot 855°$	**53.** $\sec (-90°)$
54. $\sin 90°$	**55.** $\cos (-180°)$	**56.** $\csc 90°$
57. $\tan 240°$	**58.** $\cot (-180°)$	**59.** $\sin 495°$
60. $\sin 1050°$	**61.** $\csc 225°$	**62.** $\sin (-450°)$
63. $\cos 0°$	**64.** $\tan 480°$	**65.** $\cot (-90°)$
66. $\sec 315°$	**67.** $\cos 90°$	**68.** $\sin (-135°)$
69. $\cos 270°$	**70.** $\tan 0°$	

Find the signs of the six trigonometric function values for the given angles.

71. $319°$	**72.** $-57°$	**73.** $194°$
74. $-620°$	**75.** $-215°$	**76.** $290°$
77. $-272°$	**78.** $91°$	

Use a calculator in Exercises 79–82, but do not use the trigonometric function keys.

79. Given that

$$\sin 41° = 0.6561,$$
$$\cos 41° = 0.7547,$$
$$\tan 41° = 0.8693,$$

find the trigonometric function values for 319°.

80. Given that

$$\sin 27° = 0.4540,$$
$$\cos 27° = 0.8910,$$
$$\tan 27° = 0.5095,$$

find the trigonometric function values for 333°.

81. Given that

$$\sin 65° = 0.9063,$$
$$\cos 65° = 0.4226,$$
$$\tan 65° = 2.1445,$$

find the trigonometric function values for 115°.

82. Given that

$$\sin 35° = 0.5736,$$
$$\cos 35° = 0.8192,$$
$$\tan 35° = 0.7002,$$

find the trigonometric function values for 215°.

*Aerial Navigation.　In aerial navigation, directions are given in degrees clockwise from north. Thus, east is 90°, south is 180°, and west is 270°. Several aerial directions or **bearings** are given below.*

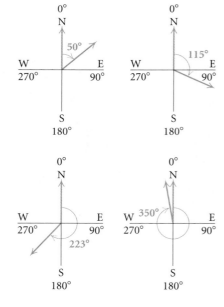

83. An airplane flies 150 km from an airport in a direction of 120°. How far east of the airport is the plane then? How far south?

84. An airplane leaves an airport and travels for 100 mi in a direction of 300°. How far north of the airport is the plane then? How far west?

85. An airplane travels at 150 km/h for 2 hr in a direction of 138° from Omaha. At the end of this time, how far south of Omaha is the plane?

86. An airplane travels at 120 km/h for 2 hr in a direction of 319° from Chicago. At the end of this time, how far north of Chicago is the plane?

Find the function value. Round to four decimal places.

87. tan 310.8°   **88.** cos 205.5°

89. cot 146.15°   **90.** sin (−16.4°)

91. sin 118°42′   **92.** cos 273°45′

93. cos (−295.8°)   **94.** tan 1086.2°

95. cos 5417°   **96.** sec 240°55′

97. csc 520°   **98.** sin 3824°

Given the function value and the quadrant restriction, find θ.

	FUNCTION VALUE	INTERVAL	θ
99.	$\sin \theta = -0.9956$	(270°, 360°)	
100.	$\tan \theta = 0.2460$	(180°, 270°)	
101.	$\cos \theta = -0.9388$	(180°, 270°)	
102.	$\sec \theta = -1.0485$	(90°, 180°)	
103.	$\tan \theta = -3.0545$	(270°, 360°)	
104.	$\sin \theta = -0.4313$	(180°, 270°)	
105.	$\csc \theta = 1.0480$	(0°, 90°)	
106.	$\cos \theta = -0.0990$	(90°, 180°)	

Collaborative Discussion and Writing

107. Why do the function values of θ depend only on the angle and not on the choice of a point on the terminal side?

108. Why is the domain of the tangent function different from the domains of the sine and the cosine functions?

Skill Maintenance

Graph the function. Sketch and label any vertical asymptotes.

109. $f(x) = \dfrac{1}{x^2 - 25}$ **110.** $g(x) = x^3 - 2x + 1$

Determine the domain and the range of the function.

111. $f(x) = \dfrac{x - 4}{x + 2}$

112. $g(x) = \dfrac{x^2 - 9}{2x^2 - 7x - 15}$

Synthesis

113. *Valve Cap on a Bicycle.* The valve cap on a bicycle wheel is 12.5 in. from the center of the wheel. From the position shown, the wheel starts to roll. After the wheel has turned 390°, how far above the ground is the valve cap? Assume that the outer radius of the tire is 13.375 in.

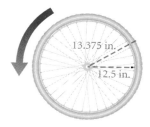

114. *Seats of a Ferris Wheel.* The seats of a ferris wheel are 35 ft from the center of the wheel. When you board the wheel, you are 5 ft above the ground. After you have rotated through an angle of 765°, how far above the ground are you?

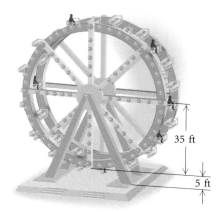

5.4

Radians, Arc Length, and Angular Speed

CIRCLES
REVIEW SECTION 1.1.

- Find points on the unit circle determined by real numbers.
- Convert between radian and degree measure; find coterminal, complementary, and supplementary angles.
- Find the length of an arc of a circle; find the measure of a central angle of a circle.
- Convert between linear speed and angular speed.

Another useful unit of angle measure is called a *radian*. To introduce radian measure, we use a circle centered at the origin with a radius of length 1. Such a circle is called a **unit circle.** Its equation is $x^2 + y^2 = 1$.

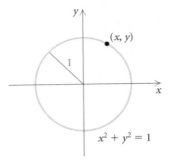

Distances on the Unit Circle

The circumference of a circle of radius r is $2\pi r$. Thus for the unit circle, where $r = 1$, the circumference is 2π. If a point starts at A and travels around the circle (Fig. 1), it will travel a distance of 2π. If it travels halfway around the circle (Fig. 2), it will travel a distance of $\frac{1}{2} \cdot 2\pi$, or π.

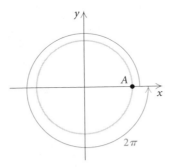

FIGURE 1

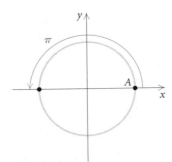

FIGURE 2

If a point C travels $\frac{1}{8}$ of the way around the circle (Fig. 3), it will travel a distance of $\frac{1}{8} \cdot 2\pi$, or $\pi/4$. Note that C is $\frac{1}{4}$ of the way from A to B. If a point D travels $\frac{1}{6}$ of the way around the circle (Fig. 4), it will travel a distance of $\frac{1}{6} \cdot 2\pi$, or $\pi/3$. Note that D is $\frac{1}{3}$ of the way from A to B.

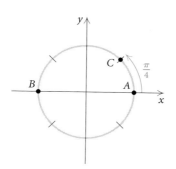

FIGURE 3

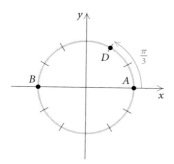

FIGURE 4

EXAMPLE 1 How far will a point travel if it goes (a) $\frac{1}{4}$, (b) $\frac{1}{12}$, (c) $\frac{3}{8}$, and (d) $\frac{5}{6}$ of the way around the unit circle?

Solution

a) $\frac{1}{4}$ of the total distance around the circle is $\frac{1}{4} \cdot 2\pi$, which is $\frac{1}{2} \cdot \pi$, or $\pi/2$.

b) The distance will be $\frac{1}{12} \cdot 2\pi$, which is $\frac{1}{6}\pi$, or $\pi/6$.

c) The distance will be $\frac{3}{8} \cdot 2\pi$, which is $\frac{3}{4}\pi$, or $3\pi/4$.

d) The distance will be $\frac{5}{6} \cdot 2\pi$, which is $\frac{5}{3}\pi$, or $5\pi/3$. Think of $5\pi/3$ as $\pi + \frac{2}{3}\pi$.

These distances are illustrated in the following figures.

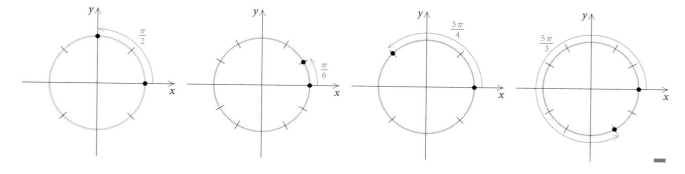

A point may travel completely around the circle and then continue. For example, if it goes around once and then continues $\frac{1}{4}$ of the way around, it will have traveled a distance of $2\pi + \frac{1}{4} \cdot 2\pi$, or $5\pi/2$ (Fig. 5). *Every* real number determines a point on the unit circle. For the positive number 10, for example, we start at A and travel counterclockwise a distance of 10. The point at which we stop is the point "determined" by the number 10 (Fig. 6).

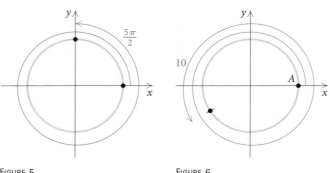

FIGURE 5   FIGURE 6

For a negative number, we move clockwise around the circle. Points for $-\pi/4$ and $-3\pi/2$ are shown in the figure below. The number 0 determines the point A.

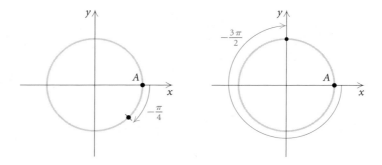

EXAMPLE 2   On the unit circle, mark the point determined by each of the following real numbers.

a) $\dfrac{9\pi}{4}$ **b)** $-\dfrac{7\pi}{6}$

Solution

a) Think of $9\pi/4$ as $2\pi + \frac{1}{4}\pi$ (see the figure on the left below). Since $9\pi/4 > 0$, the point moves counterclockwise. The point goes completely around once and then continues $\frac{1}{4}$ of the way from A to B.

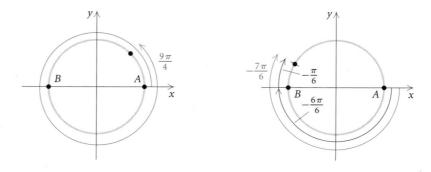

b) The number is negative, so the point moves clockwise. From A to B, the distance is π, or $\frac{6}{6}\pi$, so we need to go beyond B another distance of $\pi/6$, clockwise. (See the figure on the right above.)

Radian Measure

Degree measure is a common unit of angle measure in many everyday applications. But in many scientific fields and in mathematics (calculus, in particular), there is another commonly used unit of measure called the *radian*.

Consider the unit circle. Recall that this circle has radius 1. Suppose we measure, moving counterclockwise, an arc of length 1, and mark a point T on the circle.

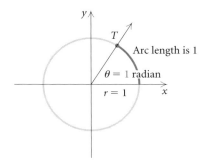

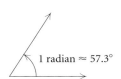

If we draw a ray from the origin through T, we have formed an angle. The measure of that angle is 1 **radian**. The word radian comes from the word *radius*. Thus measuring 1 "radius" along the circumference of the circle determines an angle whose measure is 1 *radian*. One radian is about 57.3°. Angles that measure 2 radians, 3 radians, and 6 radians are shown below.

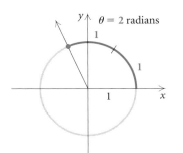

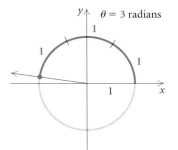

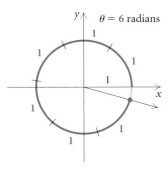

When we make a complete (counterclockwise) revolution, the terminal side coincides with the initial side on the positive x-axis. We then have an angle whose measure is 2π radians, or about 6.28 radians, which is the circumference of the circle:

$$2\pi r = 2\pi(1) = 2\pi.$$

Thus a rotation of 360° (1 revolution) has a measure of 2π radians. A half revolution is a rotation of 180°, or π radians. A quarter revolution is a rotation of 90°, or $\pi/2$ radians, and so on.

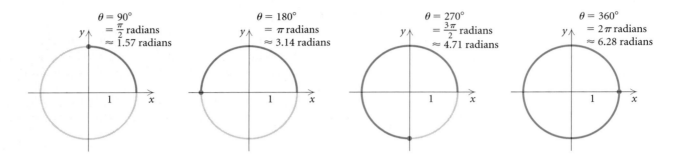

To convert between degrees and radians, we first note that

$$360° = 2\pi \text{ radians.}$$

It follows that

$$180° = \pi \text{ radians.}$$

To make conversions, we multiply by 1, noting that:

$$\frac{\pi \text{ radians}}{180°} = \frac{180°}{\pi \text{ radians}} = 1.$$

To convert from degree to radian measure, multiply by $\dfrac{\pi \text{ radians}}{180°}$.

To convert from radian to degree measure, multiply by $\dfrac{180°}{\pi \text{ radians}}$.

EXAMPLE 3   Convert each of the following to radians.

a) 120°

b) −297.25°

Solution

a) $120° = 120° \cdot \dfrac{\pi \text{ radians}}{180°}$ **Multiplying by 1**

$\qquad = \dfrac{120°}{180°} \pi \text{ radians}$

$\qquad = \dfrac{2\pi}{3} \text{ radians,}$ or about   2.09 radians

To convert degrees to radians, we set the calculator in RADIAN mode. Then we enter the angle measure followed by ^{r} (radians) from the ANGLE menu. Example 3 is shown here.

120°^{r}	
	2.094395102
−297.25°^{r}	
	−5.187991202

To convert radians to degrees, we set the calculator in DEGREE mode. Then we enter the angle measure followed by ^{r} (radians) from the ANGLE menu. Example 4 is shown here.

(3π/4)^r	
	135
8.5^r	
	487.0141259

b) $-297.25° = -297.25° \cdot \dfrac{\pi \text{ radians}}{180°}$

$\qquad = -\dfrac{297.25°}{180°} \pi \text{ radians}$

$\qquad = -\dfrac{297.25\pi}{180} \text{ radians}$

$\qquad \approx -5.19 \text{ radians}$

EXAMPLE 4 Convert each of the following to degrees.

a) $\dfrac{3\pi}{4}$ radians **b)** 8.5 radians

Solution

a) $\dfrac{3\pi}{4}$ radians $= \dfrac{3\pi}{4}$ radians $\cdot \dfrac{180°}{\pi \text{ radians}}$ **Multiplying by 1**

$\qquad\qquad\quad = \dfrac{3\pi}{4\pi} \cdot 180° = \dfrac{3}{4} \cdot 180° = 135°$

b) 8.5 radians $= 8.5$ radians $\cdot \dfrac{180°}{\pi \text{ radians}}$

$\qquad\qquad\quad = \dfrac{8.5(180°)}{\pi} \approx 487.01°$

The radian–degree equivalents of the most commonly used angle measures are illustrated in the following figures.

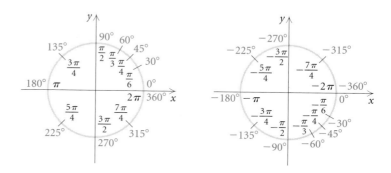

When a rotation is given in radians, the word "radians" is optional and is most often omitted. *Thus if no unit is given for a rotation, the rotation is understood to be in radians.*

We can also find coterminal, complementary, and supplementary angles in radian measure just as we did for degree measure in Section 5.3.

EXAMPLE 5 Find a positive angle and a negative angle that are coterminal with $2\pi/3$. Many answers are possible.

Solution   To find angles coterminal with a given angle, we add or subtract multiples of 2π:

$$\frac{2\pi}{3} + 2\pi = \frac{2\pi}{3} + \frac{6\pi}{3} = \frac{8\pi}{3},$$

$$\frac{2\pi}{3} - 3(2\pi) = \frac{2\pi}{3} - \frac{18\pi}{3} = -\frac{16\pi}{3}.$$

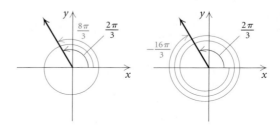

Thus, $8\pi/3$ and $-16\pi/3$ are two of the many angles coterminal with $2\pi/3$.

EXAMPLE 6   Find the complement and the supplement of $\pi/6$.

Solution   Since $90°$ equals $\pi/2$ radians, the complement of $\pi/6$ is

$$\frac{\pi}{2} - \frac{\pi}{6} = \frac{3\pi}{6} - \frac{\pi}{6} = \frac{2\pi}{6}, \quad \text{or} \quad \frac{\pi}{3}.$$

Since $180°$ equals π radians, the supplement of $\pi/6$ is

$$\pi - \frac{\pi}{6} = \frac{6\pi}{6} - \frac{\pi}{6} = \frac{5\pi}{6}.$$

Thus the complement of $\pi/6$ is $\pi/3$ and the supplement is $5\pi/6$.

Arc Length and Central Angles

Radian measure can be determined using a circle other than a unit circle. In the figure at left, a unit circle (with radius 1) is shown along with another circle (with radius r, $r \neq 1$). The angle shown is a **central angle** of both circles.

From geometry, we know that the arcs that the angle subtends have their lengths in the same ratio as the radii of the circles. The radii of the circles are r and 1. The corresponding arc lengths are s and s_1. Thus we have the proportion

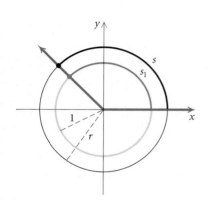

$$\frac{s}{s_1} = \frac{r}{1},$$

which can also be written as

$$\frac{s_1}{1} = \frac{s}{r}.$$

Now s_1 is the *radian measure* of the rotation in question. It is common to use a Greek letter, such as θ, for the measure of an angle or rotation and the letter s for arc length. Adopting this convention, we rewrite the proportion above as

$$\theta = \frac{s}{r}.$$

In any circle, the measure (in radians) of a central angle, the arc length the angle subtends, and the length of the radius are related in this fashion. Or, in general, the following is true.

Radian Measure

The **radian measure** θ of a rotation is the ratio of the distance s traveled by a point at a radius r from the center of rotation, to the length of the radius r:

$$\theta = \frac{s}{r}.$$

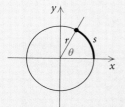

When using the formula $\theta = s/r$, you must make sure that θ is in radians and that s and r are expressed in the same unit.

EXAMPLE 7   Find the measure of a rotation in radians when a point 2 m from the center of rotation travels 4 m.

Solution

$$\theta = \frac{s}{r}$$

$$= \frac{4 \text{ m}}{2 \text{ m}} = 2 \qquad \text{The unit is understood to be radians.}$$

EXAMPLE 8   Find the length of an arc of a circle of radius 5 cm associated with an angle of $\pi/3$ radians.

Solution   We have

$$\theta = \frac{s}{r}, \quad \text{or} \quad s = r\theta.$$

Thus, $s = 5 \cdot \pi/3$ cm, or about 5.24 cm.

Linear Speed and Angular Speed

Linear speed is defined to be distance traveled per unit of time. If we use v for linear speed, s for distance, and t for time, then

$$v = \frac{s}{t}.$$

Similarly, **angular speed** is defined to be amount of rotation per unit of time. For example, we might speak of the angular speed of a bicycle wheel as 150 revolutions per minute or the angular speed of the earth as 2π radians per day. The Greek letter ω (omega) is generally used for angular speed. Thus angular speed is defined as

$$\omega = \frac{\theta}{t}.$$

As an example of how these definitions can be applied, let's consider the refurbished carousel at the Children's Museum in Indianapolis, Indiana. It consists of three circular rows of animals. All animals, regardless of the row, travel at the same angular speed. But the animals in the outer row travel at a greater linear speed than those in the inner rows. What is the relationship between the linear speed v and the angular speed ω?

To develop the relationship we seek, recall that, for rotations measured in radians, $\theta = s/r$. This is equivalent to

$$s = r\theta.$$

We divide by time, t, to obtain

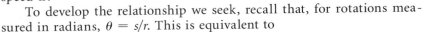
$$\frac{s}{t} = r\frac{\theta}{t}.$$

$$\downarrow \qquad \downarrow$$
$$v \qquad \omega$$

Now s/t is linear speed v and θ/t is angular speed ω. Thus we have the relationship we seek, $v = r\omega$.

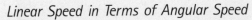

Linear Speed in Terms of Angular Speed

The **linear speed** v of a point a distance r from the center of rotation is given by

$$v = r\omega,$$

where ω is the **angular speed** in radians per unit of time.

For our new formula $v = r\omega$, the units of distance for v and r must be the same, ω must be in radians per unit of time, and the units of time for v and ω must be the same.

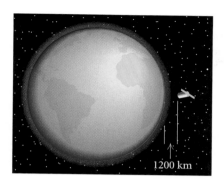

1200 km

EXAMPLE 9 *Linear Speed of an Earth Satellite.* An earth satellite in circular orbit 1200 km high makes one complete revolution every 90 min. What is its linear speed? Use 6400 km for the length of a radius of the earth.

Solution To use the formula $v = r\omega$, we need to know r and ω:

$$r = 6400 \text{ km} + 1200 \text{ km}$$
Radius of earth plus height of satellite

$$= 7600 \text{ km},$$

$$\omega = \frac{\theta}{t} = \frac{2\pi}{90 \text{ min}} = \frac{\pi}{45 \text{ min}}.$$
We have, as usual, omitted the word radians.

Now, using $v = r\omega$, we have

$$v = 7600 \text{ km} \cdot \frac{\pi}{45 \text{ min}} = \frac{7600\pi}{45} \cdot \frac{\text{km}}{\text{min}} \approx 531 \frac{\text{km}}{\text{min}}.$$

Thus the linear speed of the satellite is approximately 531 km/min. ▬

Capstan

1.8 yd

Chain

Anchor

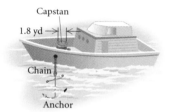

EXAMPLE 10 *Angular Speed of a Capstan.* An anchor is hoisted at a rate of 2 ft/sec as the chain is wound around a capstan with a 1.8-yd diameter. What is the angular speed of the capstan?

Solution We will use the formula $v = r\omega$ in the form $\omega = v/r$, taking care to use the proper units. Since v is given in feet per second, we need r in feet:

$$r = \frac{d}{2} = \frac{1.8}{2} \text{ yd} \cdot \frac{3 \text{ ft}}{1 \text{ yd}} = 2.7 \text{ ft}.$$

Then ω will be in radians per second:

$$\omega = \frac{v}{r} = \frac{2 \text{ ft/sec}}{2.7 \text{ ft}} = \frac{2 \text{ ft}}{\text{sec}} \cdot \frac{1}{2.7 \text{ ft}} \approx 0.741/\text{sec}.$$

Thus the angular speed is approximately 0.741 radian/sec. ▬

The formulas $\theta = \omega t$ and $v = r\omega$ can be used in combination to find distances and angles in various situations involving rotational motion.

EXAMPLE 11 *Angle of Revolution.* A Porsche 911 is traveling at a speed of 65 mph. Its tires have an outside diameter of 25.086 in. Find the angle through which a tire turns in 10 sec.

Solution   Recall that $\omega = \theta/t$, or $\theta = \omega t$. Thus we can find θ if we know ω and t. To find ω, we use the formula $v = r\omega$. The linear speed v of a point on the outside of the tire is the speed of the Porsche, 65 mph. For convenience, we first convert 65 mph to feet per second:

$$v = 65 \frac{\text{mi}}{\text{hr}} \cdot \frac{1 \text{ hr}}{60 \text{ min}} \cdot \frac{1 \text{ min}}{60 \text{ sec}} \cdot \frac{5280 \text{ ft}}{1 \text{ mi}}$$

$$\approx 95.333 \frac{\text{ft}}{\text{sec}}.$$

The radius of the tire is half the diameter. Now $r = d/2 = 25.086$ in./2 = 12.543 in. We will convert to feet, since v is in feet per second:

$$r = 12.543 \text{ in.} \cdot \frac{1 \text{ ft}}{12 \text{ in.}}$$

$$= \frac{12.543}{12} \text{ ft} \approx 1.045 \text{ ft}.$$

Using $v = r\omega$, we have

$$95.333 \frac{\text{ft}}{\text{sec}} = 1.045 \text{ ft} \cdot \omega,$$

so

$$\omega = \frac{95.333 \text{ ft/sec}}{1.045 \text{ ft}} \approx \frac{91.228}{\text{sec}}.$$

Then in 10 sec,

$$\theta = \omega t = \frac{91.228}{\text{sec}} \cdot 10 \text{ sec} \approx 912.$$

Thus the angle, in radians, through which a tire turns in 10 sec is 912.

Exercise Set

For each of Exercises 1–4, sketch a unit circle and mark the points determined by the given real numbers.

1. a) $\dfrac{\pi}{4}$ b) $\dfrac{3\pi}{2}$ c) $\dfrac{3\pi}{4}$

d) π e) $\dfrac{11\pi}{4}$ f) $\dfrac{17\pi}{4}$

2. a) $\dfrac{\pi}{2}$ b) $\dfrac{5\pi}{4}$ c) 2π

d) $\dfrac{9\pi}{4}$ e) $\dfrac{13\pi}{4}$ f) $\dfrac{23\pi}{4}$

3. a) $\dfrac{\pi}{6}$ b) $\dfrac{2\pi}{3}$ c) $\dfrac{7\pi}{6}$

d) $\dfrac{10\pi}{6}$ e) $\dfrac{14\pi}{6}$ f) $\dfrac{23\pi}{4}$

4. a) $-\dfrac{\pi}{2}$ b) $-\dfrac{3\pi}{4}$ c) $-\dfrac{5\pi}{6}$

d) $-\dfrac{5\pi}{2}$ e) $-\dfrac{17\pi}{6}$ f) $-\dfrac{9\pi}{4}$

Find two real numbers between -2π *and* 2π *that determine each of the points on the unit circle.*

5.

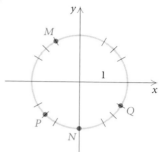

6.

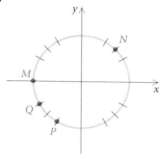

For Exercises 7 and 8, sketch a unit circle and mark the approximate location of the point determined by the given real number.

7. a) 2.4   **b)** 7.5
    c) 32   **d)** 320

8. a) 0.25   **b)** 1.8
    c) 47   **d)** 500

Find a positive angle and a negative angle that are coterminal with the given angle. Answers may vary.

9. $\dfrac{\pi}{4}$ **10.** $\dfrac{5\pi}{3}$

11. $\dfrac{7\pi}{6}$ **12.** π

Find the complement and the supplement.

13. $\dfrac{\pi}{3}$ **14.** $\dfrac{5\pi}{12}$

15. $\dfrac{3\pi}{8}$ **16.** $\dfrac{\pi}{4}$

Convert to radian measure. Leave the answer in terms of π.

17. 75°   **18.** 30°

19. 200°   **20.** −135°

21. −214.6°   **22.** 37.71°

23. −180°   **24.** 90°

Convert to radian measure. Round the answer to two decimal places.

25. 240°   **26.** 15°

27. −60°   **28.** 145°

29. 117.8°   **30.** −231.2°

31. 1.354°   **32.** 584°

Convert to degree measure. Round the answer to two decimal places.

33. $-\dfrac{3\pi}{4}$ **34.** $\dfrac{7\pi}{6}$

35. 8π **36.** $-\dfrac{\pi}{3}$

37. 1   **38.** −17.6

39. 2.347   **40.** 25

41. Certain positive angles are marked here in degrees. Find the corresponding radian measures.

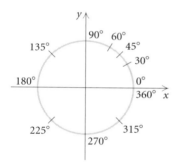

42. Certain negative angles are marked here in degrees. Find the corresponding radian measures.

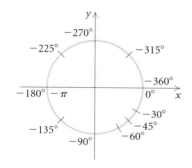

43. In a circle with a 120-cm radius, an arc 132 cm long subtends an angle of how many radians? how many degrees, to the nearest degree?

44. In a circle with a 10-ft diameter, an arc 20 ft long subtends an angle of how many radians? how many degrees, to the nearest degree?

45. In a circle with a 2-yd radius, how long is an arc associated with an angle of 1.6 radians?

46. In a circle with a 5-m radius, how long is an arc associated with an angle of 2.1 radians?

47. *Angle of Revolution.* Through how many radians does the minute hand of a clock rotate from 12:40 P.M. to 1:30 P.M.?

48. *Angle of Revolution.* A tire on a Dodge Neon has an outside diameter of 23.468 in. Through what angle (in radians) does the tire turn while traveling 1 mi?

23.468 in.

49. *Linear Speed.* A flywheel with a 15-cm diameter is rotating at a rate of 7 radians/sec. What is the linear speed of a point on its rim, in centimeters per minute?

50. *Linear Speed.* A wheel with a 30-cm radius is rotating at a rate of 3 radians/sec. What is the linear speed of a point on its rim, in meters per minute?

51. *Angular Speed on a Printing Press.* This text was printed on a four-color web heatset offset press. A cylinder on this press has a 13.37-in. diameter. The linear speed of a point on the cylinder's surface is 18.33 feet per second. What is the angular speed of the cylinder, in revolutions per hour? Printers often refer to the angular speed as impressions per hour

(IPH). (*Source:* Scott Coulter, Quebecor World, Taunton, MA)

52. *Linear Speeds on a Carousel.* When Alicia and Zoe ride the carousel described earlier in this section, Alicia always selects a horse on the outside row, whereas Zoe prefers the row closest to the center. These rows are 19 ft 3 in. and 13 ft 11 in. from the center, respectively. The angular speed of the carousel is 2.4 revolutions per minute. What is the difference, in miles per hour, in the linear speeds of Alicia and Zoe? (*Source:* The Children's Museum, Indianapolis, IN)

53. *Linear Speed at the Equator.* The earth has a 4000-mi radius and rotates one revolution every 24 hr. What is the linear speed of a point on the equator, in miles per hour?

54. *Linear Speed of the Earth.* The earth is 93,000,000 mi from the sun and traverses its orbit, which is nearly circular, every 365.25 days. What is the linear velocity of the earth in its orbit, in miles per hour?

55. *Determining the Speed of a River.* A water wheel has a 10-ft radius. To get a good approximation of the speed of the river, you count the revolutions of the wheel and find that it makes 14 revolutions per minute (rpm). What is the speed of the river, in miles per hour?

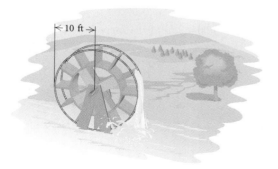

←10 ft→

56. *The Tour de France.* Lance Armstrong won the 2000 Tour de France bicycle race. The wheel of his bicycle had a 63-cm diameter. His overall average linear speed during the race was 39.569 km/h. What was the angular speed of the wheel, in revolutions per hour? (*Source:* Wilcockson, John, with Charles Pelkey and Bryan Jew, *The 2000 Tour de France: Armstrong Encore.* Boulder, CO: VeloPress, 2000)

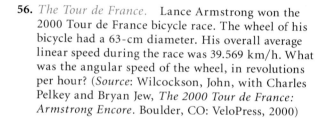

57. *John Deere Tractor.* A rear wheel on a John Deere 8300 farm tractor has a 23-in. radius. Find the angle (in radians) through which a wheel rotates in 12 sec if the tractor is traveling at a speed of 22 mph.

23 in.

Technology Connection

58. In each of Exercises 25–32, convert to radian measure using a graphing calculator.

59. In each of Exercises 33–40, convert to degree measure using a graphing calculator.

Collaborative Discussion and Writing

60. Explain in your own words why it is preferable to omit the word, or unit, *radians* in radian measures.

61. In circular motion with a fixed angular speed, the length of the radius is directly proportional to the linear speed. Explain why with an example.

62. Two new cars are each driven at an average speed of 60 mph for an extended highway test drive of 2000 mi. The diameter of the wheels of the two cars are 15 in. and 16 in., respectively. If the cars use tires of equal durability and profile, differing only by the diameter, which car will probably need new tires first? Explain your answer.

Skill Maintenance

Solve.

63. $5^x = 625$

64. $e^t = 10,000$

65. $\log_7 x = 3$

66. $\log (3x + 1) - \log (x - 1) = 2$

Synthesis

67. On the earth, one degree of latitude is how many kilometers? how many miles? (Assume that the radius of the earth is 6400 km, or 4000 mi, approximately.)

68. A point on the unit circle has y-coordinate $-\sqrt{21}/5$. What is its x-coordinate? Check using a calculator.

69. A **mil** is a unit of angle measure. A right angle has a measure of 1600 mils. Convert each of the following to degrees, minutes, and seconds.

   a) 100 mils   **b)** 350 mils

70. A **grad** is a unit of angle measure similar to a degree. A right angle has a measure of 100 grads. Convert each of the following to grads.

   a) 48°   **b)** $\dfrac{5\pi}{7}$

71. *Angular Speed of a Gear Wheel.* One gear wheel turns another, the teeth being on the rims. The wheels have 40-cm and 50-cm radii, and the smaller wheel rotates at 20 rpm. Find the angular speed of the larger wheel, in radians per second.

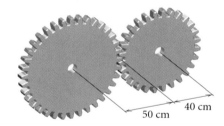

40 cm

50 cm

72. *Angular Speed of a Pulley.* Two pulleys, 50 cm and 30 cm in diameter, respectively, are connected by a belt. The larger pulley makes 12 revolutions per minute. Find the angular speed of the smaller pulley, in radians per second.

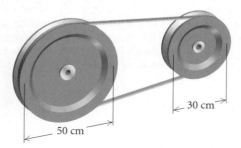

73. *Distance between Points on the Earth.* To find the distance between two points on the earth when their latitude and longitude are known, we can use a right triangle for an excellent approximation if the points are not too far apart. Point *A* is at latitude 38°27′30″ N, longitude 82°57′15″ W; and point *B* is at latitude 38°28′45″ N, longitude 82°56′30″ W. Find the distance from *A* to *B* in nautical miles (one minute of latitude is one nautical mile).

74. *Hands of a Clock.* At what times between noon and 1:00 P.M. are the hands of a clock perpendicular?

5.5

Circular Functions: Graphs and Properties

- *Given the coordinates of a point on the unit circle, find its reflections across the x-axis, the y-axis, and the origin.*
- *Determine the six trigonometric function values for a real number when the coordinates of the point on the unit circle determined by that real number are given.*
- *Find function values for any real number using a calculator.*
- *Graph the six circular functions and state their properties.*

The domains of the trigonometric functions, defined in Sections 5.1 and 5.3, have been sets of angles or rotations measured in a real number of degree units. We can also consider the domains to be sets of real numbers, or radians, introduced in Section 5.4. Many applications in calculus that use the trigonometric functions refer only to radians.

Let's again consider radian measure and the unit circle. We defined radian measure for θ as

$$\theta = \frac{s}{r}.$$

When $r = 1$,

$$\theta = \frac{s}{1}, \quad \text{or} \quad \theta = s.$$

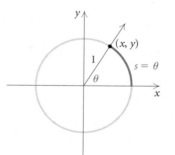

> The arc length s on the unit circle is the same as the radian measure of the angle θ.

In the figure above, the point (x, y) is the point where the terminal side of the angle with radian measure s intersects the unit circle. We can now extend our definitions of the trigonometric functions using domains composed of real numbers, or radians.

In the definitions, s can be considered the radian measure of an angle or the measure of an arc length on the unit circle. Either way, s is a real number. To each real number s, there corresponds an arc length s on the unit circle. Trigonometric functions with domains composed of real numbers are called **circular functions.**

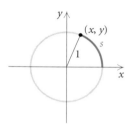

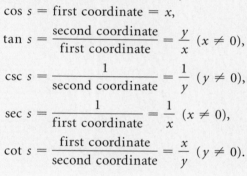

Basic Circular Functions

For a real number s that determines a point (x, y) on the unit circle:

$$\sin s = \text{second coordinate} = y,$$
$$\cos s = \text{first coordinate} = x,$$
$$\tan s = \frac{\text{second coordinate}}{\text{first coordinate}} = \frac{y}{x} \ (x \neq 0),$$
$$\csc s = \frac{1}{\text{second coordinate}} = \frac{1}{y} \ (y \neq 0),$$
$$\sec s = \frac{1}{\text{first coordinate}} = \frac{1}{x} \ (x \neq 0),$$
$$\cot s = \frac{\text{first coordinate}}{\text{second coordinate}} = \frac{x}{y} \ (y \neq 0).$$

We can consider the domains of trigonometric functions to be real numbers rather than angles. We can determine these values for a specific real number if we know the coordinates of the point on the unit circle determined by that number. As with degree measure, we can also find these function values directly using a calculator.

Reflections on the Unit Circle

Let's consider the unit circle and a few of its points. For any point (x, y) on the unit circle, $x^2 + y^2 = 1$, we know that $-1 \leq x \leq 1$ and $-1 \leq y \leq 1$. If we know the x- or y-coordinate of a point on the unit circle, we can find the other coordinate. If $x = \frac{3}{5}$, then

$$\left(\tfrac{3}{5}\right)^2 + y^2 = 1$$
$$y^2 = 1 - \tfrac{9}{25} = \tfrac{16}{25}$$
$$y = \pm \tfrac{4}{5}.$$

Thus, $\left(\frac{3}{5}, \frac{4}{5}\right)$ and $\left(\frac{3}{5}, -\frac{4}{5}\right)$ are points on the unit circle. There are two points with an x-coordinate of $\frac{3}{5}$.

Now let's consider the radian measure $\pi/3$ and determine the coordinates of the point on the unit circle determined by $\pi/3$. We construct a right triangle by dropping a perpendicular segment from the point to the x-axis.

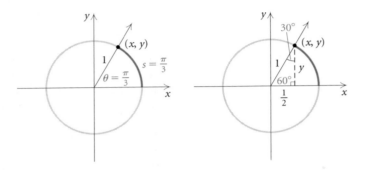

Since $\pi/3 = 60°$, we have a 30°–60° right triangle in which the side opposite the 30° angle is one half of the hypotenuse. The hypotenuse, or radius, is 1, so the side opposite the 30° angle is $\frac{1}{2}$. Using the Pythagorean theorem, we can find the other side:

$$\left(\frac{1}{2}\right)^2 + y^2 = 1$$

$$y^2 = 1 - \frac{1}{4} = \frac{3}{4}$$

$$y = \sqrt{\frac{3}{4}} = \frac{\sqrt{3}}{2}.$$

We know that y is positive since the point is in the first quadrant. Thus the coordinates of the point determined by $\pi/3$ are $x = 1/2$ and $y = \sqrt{3}/2$, or $(1/2, \sqrt{3}/2)$. We can always check to see if a point is on the unit circle by substituting into the equation $x^2 + y^2 = 1$.

Because a unit circle is symmetric with respect to the x-axis, the y-axis, and the origin, we can use the coordinates of one point on the unit circle to find coordinates of its reflections.

EXAMPLE 1   Each of the following points lies on the unit circle. Find their reflections across the x-axis, the y-axis, and the origin.

a) $\left(\dfrac{3}{5}, \dfrac{4}{5}\right)$ b) $\left(\dfrac{\sqrt{2}}{2}, \dfrac{\sqrt{2}}{2}\right)$

c) $\left(\dfrac{1}{2}, \dfrac{\sqrt{3}}{2}\right)$

Solution

a)

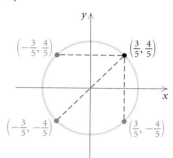

b)

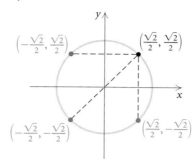

c)

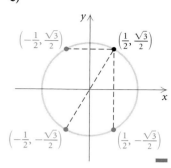

Finding Function Values

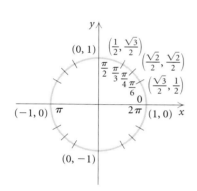

Knowing the coordinates of only a few points on the unit circle along with their reflections allows us to find trigonometric function values of the most frequently used real numbers, or radians.

EXAMPLE 2   Find each of the following function values.

a) $\tan \dfrac{\pi}{3}$

b) $\cos \dfrac{3\pi}{4}$

c) $\sin\left(-\dfrac{\pi}{6}\right)$

d) $\cos \dfrac{4\pi}{3}$

e) $\cot \pi$

f) $\csc\left(-\dfrac{7\pi}{2}\right)$

Solution   We locate the point on the unit circle determined by the rotation, and then find its coordinates using reflection if necessary.

a) The coordinates of the point determined by $\pi/3$ are $(1/2, \sqrt{3}/2)$.

b) The reflection of $(\sqrt{2}/2, \sqrt{2}/2)$ across the y-axis is $(-\sqrt{2}/2, \sqrt{2}/2)$.

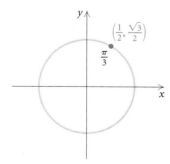

Thus, $\tan \dfrac{\pi}{3} = \dfrac{y}{x} = \dfrac{\sqrt{3}/2}{1/2} = \sqrt{3}$.

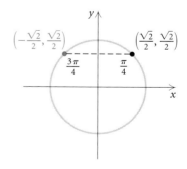

Thus, $\cos \dfrac{3\pi}{4} = x = -\dfrac{\sqrt{2}}{2}$.

c) The reflection of $(\sqrt{3}/2,\ 1/2)$ across the x-axis is $(\sqrt{3}/2,\ -1/2)$.

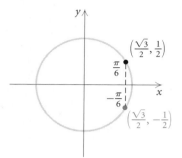

Thus, $\sin\left(-\dfrac{\pi}{6}\right) = y = -\dfrac{1}{2}.$

e) The coordinates of the point determined by π are $(-1,\ 0)$.

We can also think of $\cot \pi$ as the reciprocal of $\tan \pi$. Since $\tan \pi = y/x = 0/-1 = 0$ and the reciprocal of 0 is not defined, we know that $\cot \pi$ is undefined.

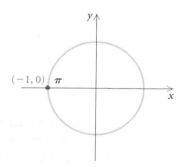

Thus, $\cot \pi = \dfrac{x}{y} = \dfrac{-1}{0},$ which is undefined.

d) The reflection of $(1/2,\ \sqrt{3}/2)$ across the origin is $(-1/2,\ -\sqrt{3}/2)$.

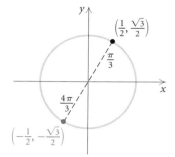

Thus, $\cos\dfrac{4\pi}{3} = x = -\dfrac{1}{2}.$

f) The coordinates of the point determined by $-7\pi/2$ are $(0,\ 1)$.

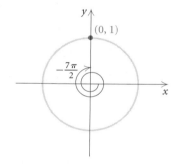

Thus, $\csc\left(-\dfrac{7\pi}{2}\right) = \dfrac{1}{y} = \dfrac{1}{1} = 1.$

Using a calculator, we can find trigonometric function values of any real number without knowing the coordinates of the point that it determines on the unit circle. Most calculators have both degree and radian modes. When finding function values of radian measures, or real numbers, we *must* set the calculator in RADIAN mode.

EXAMPLE 3 Find each of the following function values of radian measures using a calculator. Round the answers to four decimal places.

a) $\cos\dfrac{2\pi}{5}$ **b)** $\tan(-3)$ **c)** $\sin 24.9$ **d)** $\sec\dfrac{\pi}{7}$

Technology Connection

To find trigonometric function values of angles measured in radians, we set the calculator in RADIAN mode.

Parts (a)–(c) of Example 3 are shown in the window below.

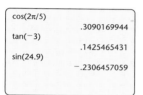

Solution   Using a calculator set in RADIAN mode, we find the values.

a) $\cos \dfrac{2\pi}{5} \approx 0.3090$

b) $\tan(-3) \approx 0.1425$

c) $\sin 24.9 \approx -0.2306$

d) $\sec \dfrac{\pi}{7} = \dfrac{1}{\cos \dfrac{\pi}{7}} \approx 1.1099$

Note in part (d) that the secant function value can be found by taking the reciprocal of the cosine value. Thus we can enter $\cos \pi/7$ and use the reciprocal key.

Technology Connection: *Exploration*

We can graph the unit circle using a graphing calculator. We use PARAMETRIC mode with the following window and let $X_{1T} = \cos T$ and $Y_{1T} = \sin T$. Here we use DEGREE mode.

WINDOW

Tmin = 0
Tmax = 360
Tstep = 15
Xmin = −1.5
Xmax = 1.5
Xscl = 1
Ymin = −1
Ymax = 1
Yscl = 1

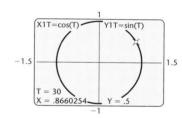

Using the trace key and an arrow key to move the cursor around the unit circle, we see the T, X, and Y values appear on the screen. What do they represent? Repeat this exercise in RADIAN mode. What do the T, X, and Y values represent? (For more on parametric equations, see Appendix A.)

From the definitions on p. 387, we can relabel any point (x, y) on the unit circle as $(\cos s, \sin s)$, where s is any real number.

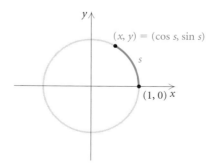

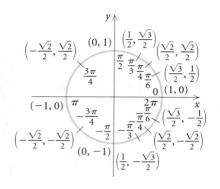

Graphs of the Sine and Cosine Functions

Properties of functions can be observed from their graphs. We begin by graphing the sine and cosine functions. We make a table of values, plot the points, and then connect those points with a smooth curve. It is helpful to first draw a unit circle and label a few points with coordinates. We can either use the coordinates as the function values or find approximate sine and cosine values directly with a calculator.

s	$\sin s$	$\cos s$
0	0	1
$\pi/6$	0.5	0.8660
$\pi/4$	0.7071	0.7071
$\pi/3$	0.8660	0.5
$\pi/2$	1	0
$3\pi/4$	0.7071	-0.7071
π	0	-1
$5\pi/4$	-0.7071	-0.7071
$3\pi/2$	-1	0
$7\pi/4$	-0.7071	0.7071
2π	0	1

s	$\sin s$	$\cos s$
0	0	1
$-\pi/6$	-0.5	0.8660
$-\pi/4$	-0.7071	0.7071
$-\pi/3$	-0.8660	0.5
$-\pi/2$	-1	0
$-3\pi/4$	-0.7071	-0.7071
$-\pi$	0	-1
$-5\pi/4$	0.7071	-0.7071
$-3\pi/2$	1	0
$-7\pi/4$	0.7071	0.7071
-2π	0	1

Technology Connection

The graphing calculator provides an efficient way to graph trigonometric functions. Here we use RADIAN mode to graph $y = \sin x$ and $y = \cos x$.

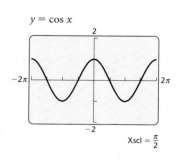

The graphs are as follows.

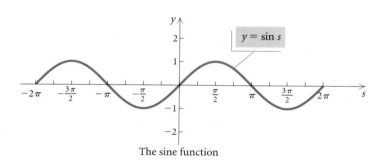

The sine function

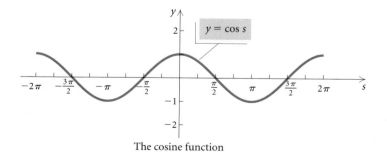

The cosine function

The sine and cosine functions are continuous functions. Note in the graph of the sine function that function values increase from 0 at $s = 0$ to 1 at $s = \pi/2$, then decrease to 0 at $s = \pi$, decrease further to -1 at $s = 3\pi/2$, and increase to 0 at 2π. The reverse pattern follows when s decreases from 0 to -2π. Note in the graph of the cosine function that function values start at 1 when $s = 0$, and decrease to 0 at $s = \pi/2$. They decrease further to -1 at $s = \pi$, then increase to 0 at $s = 3\pi/2$, and increase further to 1 at $s = 2\pi$. An identical pattern follows when s decreases from 0 to -2π.

From the unit circle and the graphs of the functions, we know that the domain of both the sine and cosine functions is the entire set of real numbers, $(-\infty, \infty)$. The range of each function is the set of all real numbers from -1 to 1, $[-1, 1]$.

> The *domain* of the sine and cosine functions is $(-\infty, \infty)$.
>
> The *range* of the sine and cosine functions is $[-1, 1]$.

Technology Connection

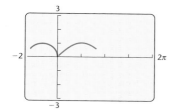

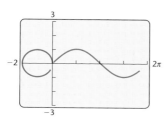

Another way to construct the sine and cosine graphs is by considering the unit circle and transferring vertical distances for the sine function and horizontal distances for the cosine function. Using a graphing calculator, we can visualize the transfer of these distances. We use the calculator set in PARAMETRIC and RADIAN modes and let $X_{1T} = \cos T - 1$ and $Y_{1T} = \sin T$ for the unit circle centered at $(-1, 0)$ and $X_{2T} = T$ and $Y_{2T} = \sin T$ for the sine curve. Use the following window settings.

Tmin = 0	Xmin = -2	Ymin = -3
Tmax = 2π	Xmax = 2π	Ymax = 3
Tstep = .1	Xscl = $\pi/2$	Yscl = 1

With the calculator set in SIMULTANEOUS mode, we can actually watch the sine function (in red) "unwind" from the unit circle (in blue). In the two screens at left, we partially illustrate this animated procedure.

Consult your calculator's instruction manual for specific keystrokes and graph both the sine curve and the cosine curve in this manner. (For more on parametric equations, see Appendix A.)

A function with a repeating pattern is called **periodic**. The sine and cosine functions are examples of periodic functions. The values of each function repeat themselves every 2π units. In other words, for any s, we have

$$\sin(s + 2\pi) = \sin s \quad \text{and} \quad \cos(s + 2\pi) = \cos s.$$

To see this another way, think of the part of the graph between 0 and 2π and note that the rest of the graph consists of copies of it. If we translate the graph of $y = \sin x$ or $y = \cos x$ to the left or right 2π units, we will obtain the original graph. We say that each of these functions has a period of 2π.

Periodic Function

A function f is said to be **periodic** if there exists a positive constant p such that

$$f(s + p) = f(s)$$

for all s in the domain of f. The smallest such positive number p is called the period of the function.

The period p can be thought of as the length of the shortest recurring interval.

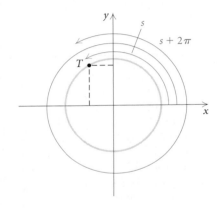

We can also use the unit circle to verify that the period of the sine and cosine functions is 2π. Consider any real number s and the point T that it determines on a unit circle, as shown at left. If we increase s by 2π, the point determined by $s + 2\pi$ is again the point T. Hence for any real number s,

$$\sin (s + 2\pi) = \sin s \quad \text{and} \quad \cos (s + 2\pi) = \cos s.$$

It is also true that $\sin (s + 4\pi) = \sin s$, $\sin (s + 6\pi) = \sin s$, and so on. In fact, for *any* integer k, the following equations are identities:

$$\sin [s + k(2\pi)] = \sin s \quad \text{and} \quad \cos [s + k(2\pi)] = \cos s,$$

or

$$\sin s = \sin (s + 2k\pi) \quad \text{and} \quad \cos s = \cos (s + 2k\pi).$$

The **amplitude** of a periodic function is defined as one half of the distance between its maximum and minimum function values. It is always positive. Both the graphs and the unit circle verify that the maximum value of the sine and cosine functions is 1, whereas the minimum value of each is -1. Thus,

the amplitude of the sine function $= \frac{1}{2}|1 - (-1)| = 1$

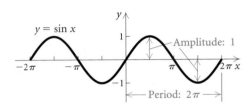

REVIEW SECTION 1.6.

Technology Connection: Exploration

Using the TABLE feature on a graphing calculator, compare the y-values for $y_1 = \sin x$ and $y_2 = \sin(-x)$ and for $y_3 = \cos x$ and $y_4 = \cos(-x)$. We set TblMin = 0 and $\triangle$Tbl = $\pi/12$.

X	Y1	Y2
0	0	0
.2618	.25882	−.2588
.5236	.5	−.5
.7854	.70711	−.7071
1.0472	.86603	−.866
1.309	.96593	−.9659
1.5708	1	−1

X = 0

X	Y3	Y4
0	1	1
.2618	.96593	.96593
.5236	.86603	.86603
.7854	.70711	.70711
1.0472	.5	.5
1.309	.25882	.25882
1.5708	0	0

X = 0

What appears to be the relationship between $\sin x$ and $\sin(-x)$ and between $\cos x$ and $\cos(-x)$?

EVEN AND ODD FUNCTIONS

and

the amplitude of the cosine function is $\frac{1}{2}|1 - (-1)| = 1$.

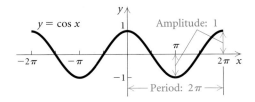

Consider any real number s and its opposite, $-s$. These numbers determine points T and T_1 on a unit circle that are symmetric with respect to the x-axis.

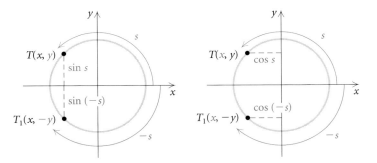

Because their second coordinates are opposites of each other, we know that for any number s,

$$\sin(-s) = -\sin s.$$

Because their first coordinates are the same, we know that for any number s,

$$\cos(-s) = \cos s.$$

Thus we have shown that the sine function is odd and the cosine function is even.

The following is a summary of the properties of the sine and cosine functions.

CONNECTING THE CONCEPTS

COMPARING THE SINE AND COSINE FUNCTIONS

SINE FUNCTION

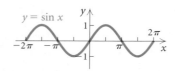

1. Continuous
2. Period: 2π
3. Domain: All real numbers
4. Range: $[-1, 1]$
5. Amplitude: 1
6. Odd: $\sin(-s) = -\sin s$

COSINE FUNCTION

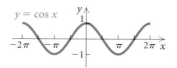

1. Continuous
2. Period: 2π
3. Domain: All real numbers
4. Range: $[-1, 1]$
5. Amplitude: 1
6. Even: $\cos(-s) = \cos s$

Graphs of the Tangent, Cotangent, Cosecant, and Secant Functions

To graph the tangent function, we could make a table of values using a calculator, but in this case it is easier to begin with the definition of tangent and the coordinates of a few points on the unit circle. We recall that

$$\tan s = \frac{y}{x} = \frac{\sin s}{\cos s}.$$

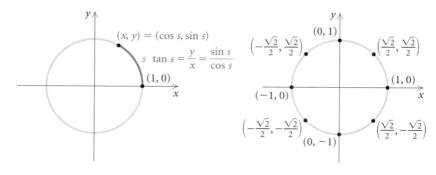

The tangent function is undefined when x, the first coordinate, is 0. That is, it is undefined for any number s whose cosine is 0:

$$s = \pm\frac{\pi}{2}, \ \pm\frac{3\pi}{2}, \ \pm\frac{5\pi}{2}, \ \ldots.$$

We draw vertical asymptotes at these locations (see Fig. 1 below).

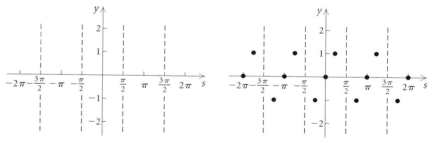

FIGURE 1   FIGURE 2

We also note that

$$\tan s = 0 \text{ at } s = 0, \pm\pi, \pm 2\pi, \pm 3\pi, \ldots,$$

$$\tan s = 1 \text{ at } s = \ldots -\frac{7\pi}{4}, -\frac{3\pi}{4}, \frac{\pi}{4}, \frac{5\pi}{4}, \frac{9\pi}{4}, \ldots,$$

$$\tan s = -1 \text{ at } s = \ldots -\frac{9\pi}{4}, -\frac{5\pi}{4}, -\frac{\pi}{4}, \frac{3\pi}{4}, \frac{7\pi}{4}, \ldots.$$

We can add these ordered pairs to the graph (see Fig. 2 above) and investigate the values in $(-\pi/2, \pi/2)$ using a calculator. Note that the function value is 0 when $s = 0$, and the values increase without bound as s increases toward $\pi/2$. The graph gets closer and closer to an asymptote as s gets closer to $\pi/2$, but it never touches the line. As s decreases from 0 to $-\pi/2$, the values decrease without bound. Again the graph gets closer and closer to an asymptote, but it never touches it. We now complete the graph.

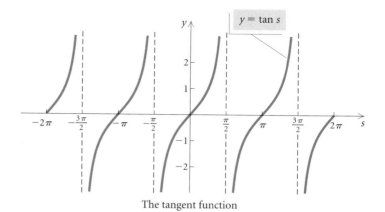

$y = \tan s$

The tangent function

From the graph, we see that the tangent function is continuous except where it is not defined. The period of the tangent function is π. Note that although there is a period, there is no amplitude because there are no maximum and minimum values. Tan s is not defined where

$\cos s = 0$. Thus the domain of the tangent function is the set of all real numbers except $(\pi/2) + k\pi$, where k is an integer. The range of the function is the set of all real numbers.

The cotangent function ($\cot s = \cos s/\sin s$) is undefined when y, the second coordinate, is 0—that is, it is undefined for any number s whose sine is 0. Thus the cotangent is undefined for $s = 0$, $\pm\pi$, $\pm2\pi$, $\pm3\pi$, ... The graph of the function is shown below.

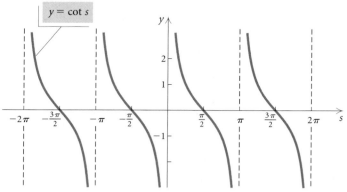

The cotangent function

The cosecant and sine functions are reciprocal functions, as are the secant and cosine functions. The graphs of the cosecant and secant functions can be constructed by finding the reciprocals of the values of the sine and cosine functions, respectively. Thus the functions will be positive together and negative together. The cosecant function is not defined for those numbers s whose sine is 0. The secant function is not defined for those numbers s whose cosine is 0. In the graphs below, the sine and cosine functions are shown by the gray curves for reference.

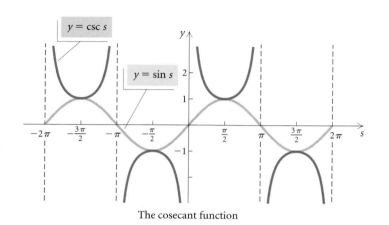

The cosecant function

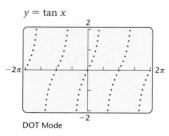
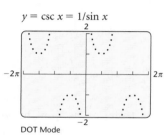

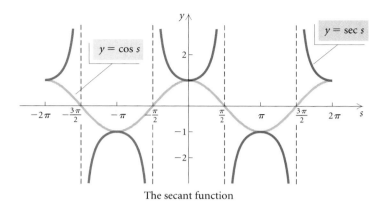

The secant function

The following is a summary of the basic properties of the tangent, cotangent, cosecant, and secant functions. These functions are continuous except where they are not defined.

CONNECTING THE CONCEPTS

COMPARING THE TANGENT, COTANGENT, COSECANT, AND SECANT FUNCTIONS

TANGENT FUNCTION

1. Period: π
2. Domain: All real numbers except $(\pi/2) + k\pi$, where k is an integer
3. Range: All real numbers

COTANGENT FUNCTION

1. Period: π
2. Domain: All real numbers except $k\pi$, where k is an integer
3. Range: All real numbers

COSECANT FUNCTION

1. Period: 2π
2. Domain: All real numbers except $k\pi$, where k is an integer
3. Range: $(-\infty, -1] \cup [1, \infty)$

SECANT FUNCTION

1. Period: 2π
2. Domain: All real numbers except $(\pi/2) + k\pi$, where k is an integer
3. Range: $(-\infty, -1] \cup [1, \infty)$

In this chapter, we have used the letter s for arc length and have avoided the letters x and y, which generally represent first and second coordinates. Nevertheless, we can represent the arc length on a unit circle by any variable, such as s, t, x, or θ. Each arc length determines a point that can be labeled with an ordered pair. The first coordinate of that ordered pair is the cosine of the arc length, and the second coordinate is the sine of the arc length. The identities we have developed hold no matter what symbols are used for variables—for example, $\cos{(-s)} = \cos{s}$, $\cos{(-x)} = \cos{x}$, $\cos{(-\theta)} = \cos{\theta}$, and $\cos{(-t)} = \cos{t}$.

Exercise Set **5.5**

The following points are on the unit circle. Find the coordinates of their reflections across (a) the x-axis, (b) the y-axis, and (c) the origin.

1. $\left(-\dfrac{3}{4}, \dfrac{\sqrt{7}}{4}\right)$ **2.** $\left(\dfrac{2}{3}, \dfrac{\sqrt{5}}{3}\right)$

3. $\left(\dfrac{2}{5}, -\dfrac{\sqrt{21}}{5}\right)$ **4.** $\left(-\dfrac{\sqrt{3}}{2}, -\dfrac{1}{2}\right)$

5. The number $\pi/4$ determines a point on the unit circle with coordinates $(\sqrt{2}/2, \sqrt{2}/2)$. What are the coordinates of the point determined by $-\pi/4$?

6. A number β determines a point on the unit circle with coordinates $(-2/3, \sqrt{5}/3)$. What are the coordinates of the point determined by $-\beta$?

Find the function value using coordinates of points on the unit circle. Give exact answers.

7. $\sin{\pi}$ **8.** $\cos{\left(-\dfrac{\pi}{3}\right)}$

9. $\cot{\dfrac{7\pi}{6}}$ **10.** $\tan{\dfrac{11\pi}{4}}$

11. $\sin{(-3\pi)}$ **12.** $\csc{\dfrac{3\pi}{4}}$

13. $\cos{\dfrac{5\pi}{6}}$ **14.** $\tan{\left(-\dfrac{\pi}{4}\right)}$

15. $\cos{10\pi}$ **16.** $\sec{\dfrac{\pi}{2}}$

17. $\cos{\dfrac{\pi}{6}}$ **18.** $\sin{\dfrac{2\pi}{3}}$

19. $\sin{\dfrac{5\pi}{4}}$ **20.** $\cos{\dfrac{11\pi}{6}}$

21. $\sin{(-5\pi)}$ **22.** $\tan{\dfrac{3\pi}{2}}$

23. $\cot{\dfrac{5\pi}{2}}$ **24.** $\tan{\dfrac{5\pi}{3}}$

Find the function value using a calculator. Round the answer to four decimal places, where appropriate.

25. $\tan{\dfrac{\pi}{7}}$ **26.** $\cos{\left(-\dfrac{2\pi}{5}\right)}$

27. $\sec{37}$ **28.** $\sin{11.7}$

29. $\cot{342}$ **30.** $\tan{1.3}$

31. $\cos{6\pi}$ **32.** $\sin{\dfrac{\pi}{10}}$

33. $\csc{4.16}$ **34.** $\sec{\dfrac{10\pi}{7}}$

35. $\tan{\dfrac{7\pi}{4}}$ **36.** $\cos{2000}$

37. $\sin{\left(-\dfrac{\pi}{4}\right)}$ **38.** $\cot{7\pi}$

39. $\sin{0}$ **40.** $\cos{(-29)}$

41. $\tan{\dfrac{2\pi}{9}}$ **42.** $\sin{\dfrac{8\pi}{3}}$

43. a) Sketch a graph of $y = \sin{x}$.
   b) By reflecting the graph in part (a), sketch a graph of $y = \sin{(-x)}$.
   c) By reflecting the graph in part (a), sketch a graph of $y = -\sin{x}$.
   d) How do the graphs in parts (b) and (c) compare?

44. a) Sketch a graph of $y = \cos{x}$.
   b) By reflecting the graph in part (a), sketch a graph of $y = \cos{(-x)}$.
   c) By reflecting the graph in part (a), sketch a graph of $y = -\cos{x}$.
   d) How do the graphs in parts (a) and (b) compare?

45. a) Sketch a graph of $y = \sin{x}$.
   b) By translating, sketch a graph of $y = \sin{(x + \pi)}$.
   c) By reflecting the graph of part (a), sketch a graph of $y = -\sin{x}$.
   d) How do the graphs of parts (b) and (c) compare?

46. a) Sketch a graph of $y = \sin x$.
  b) By translating, sketch a graph of
  $y = \sin (x - \pi)$.
  c) By reflecting the graph of part (a), sketch a graph
  of $y = -\sin x$.
  d) How do the graphs of parts (b) and (c) compare?

47. a) Sketch a graph of $y = \cos x$.
  b) By translating, sketch a graph of
  $y = \cos (x + \pi)$.
  c) By reflecting the graph of part (a), sketch a graph
  of $y = -\cos x$.
  d) How do the graphs of parts (b) and (c) compare?

48. a) Sketch a graph of $y = \cos x$.
  b) By translating, sketch a graph of
  $y = \cos (x - \pi)$.
  c) By reflecting the graph of part (a), sketch a graph
  of $y = -\cos x$.
  d) How do the graphs of parts (b) and (c) compare?

49. Of the six circular functions, which are even?
Which are odd?

50. Of the six circular functions, which have period π?
Which have period 2π?

*Consider the coordinates on the unit circle for
Exercises 51–54.*

51. In which quadrants is the tangent function positive?
negative?

52. In which quadrants is the sine function positive?
negative?

53. In which quadrants is the cosine function positive?
negative?

54. In which quadrants is the cosecant function
positive? negative?

Technology Connection

*Use a graphing calculator to determine the domain, the
range, the period, and the amplitude of the function.*

55. $y = (\sin x)^2$

56. $y = |\cos x| + 1$

57. Using a calculator, consider $(\sin x)/x$, where x is
between 0 and $\pi/2$. As x approaches 0, this function
approaches a limiting value. What is it?

58. Using graphs, determine all numbers x that satisfy

$$\sin x < \cos x.$$

Collaborative Discussion and Writing

59. Describe how the graphs of the sine and cosine
functions are related.

60. Explain why both the sine and cosine functions are
continuous, but the tangent function, defined as
sine/cosine, is not continuous.

Skill Maintenance

*Graph both functions on the same set of axes, and
describe how g is a transformation of f.*

61. $f(x) = x^2$, $g(x) = 2x^2 - 3$

62. $f(x) = x^2$, $g(x) = (x - 2)^2$

63. $f(x) = |x|$, $g(x) = \frac{1}{2}|x - 4| + 1$

64. $f(x) = x^3$, $g(x) = -x^3$

*Write an equation for a function that has a graph with
the given characteristics.*

65. The shape of $y = x^3$, but reflected across the x-axis,
shifted right 2 units, and shifted down 1 unit

66. The shape of $y = 1/x$, but shrunk vertically by a
factor of $\frac{1}{4}$ and shifted up 3 units

Synthesis

Complete. (For example, $\sin (x + 2\pi) = \sin x$.)

67. $\cos (-x) = $ _____

68. $\sin (-x) = $ _____

69. $\sin (x + 2k\pi)$, $k \in \mathbb{Z} = $ _____

70. $\cos (x + 2k\pi)$, $k \in \mathbb{Z} = $ _____

71. $\sin (\pi - x) = $ _____

72. $\cos (\pi - x) = $ _____

73. $\cos (x - \pi) = $ _____

74. $\cos (x + \pi) = $ _____

75. $\sin (x + \pi) = $ _____

76. $\sin (x - \pi) = $ _____

77. Find all numbers x that satisfy the following.
  a) $\sin x = 1$
  b) $\cos x = -1$
  c) $\sin x = 0$

78. Find $f \circ g$ and $g \circ f$, where $f(x) = x^2 + 2x$ and
$g(x) = \cos x$.

Determine the domain of the function.

79. $f(x) = \sqrt{\cos x}$

80. $g(x) = \dfrac{1}{\sin x}$

81. $f(x) = \dfrac{\sin x}{\cos x}$

82. $g(x) = \log (\sin x)$

Graph.

83. $y = 3 \sin x$

84. $y = \sin |x|$

85. $y = \sin x + \cos x$

86. $y = |\cos x|$

87. One of the motivations for developing trigonometry with a unit circle is that you can actually "see" $\sin \theta$ and $\cos \theta$ on the circle. Note in the figure at right that $AP = \sin \theta$ and $OA = \cos \theta$. It turns out that you can also "see" the other four trigonometric functions. Prove each of the following.

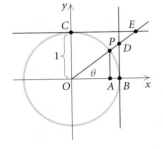

a) $BD = \tan \theta$

b) $OD = \sec \theta$

c) $OE = \csc \theta$

d) $CE = \cot \theta$

5.6

Graphs of Transformed Sine and Cosine Functions

- *Graph transformations of* $y = \sin x$ *and* $y = \cos x$ *in the form*

$$y = A \sin (Bx - C) + D$$

and

$$y = A \cos (Bx - C) + D$$

and determine the amplitude, the period, and the phase shift.

- *Graph sums of functions.*

Variations of Basic Graphs

In Section 5.5, we graphed all six trigonometric functions. In this section, we will consider variations of the graphs of the sine and cosine functions. For example, we will graph equations like the following:

$$y = 5 \sin \tfrac{1}{2}x, \qquad y = \cos (2x - \pi), \quad \text{and} \quad y = \tfrac{1}{2} \sin x - 3.$$

In particular, we are interested in graphs of functions in the form

$$y = A \sin (Bx - C) + D$$

and

$$y = A \cos (Bx - C) + D,$$

TRANSFORMATIONS OF FUNCTIONS
REVIEW SECTION 1.5.

where A, B, C, and D are constants. These constants have the effect of translating, reflecting, stretching, and shrinking the basic graphs. Let's first examine the effect of each constant individually. Then we will consider the combined effects of more than one constant.

Let's observe the effect of the constant D in the graphs below.

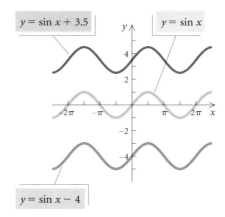

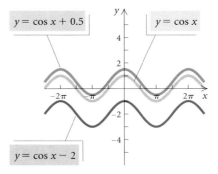

The constant D in

$$y = A \sin (Bx - C) + D \quad \text{and} \quad y = A \cos (Bx - C) + D$$

translates the graphs up D units if $D > 0$ or down $|D|$ units if $D < 0$.

EXAMPLE 1   Sketch a graph of $y = \sin x + 3$.

Solution   The graph of $y = \sin x + 3$ is a *vertical* translation of the graph of $y = \sin x$ up 3 units. One way to sketch the graph is to first consider $y = \sin x$ on an interval of length 2π, say, $[0, 2\pi]$. The zeros of the function and the maximum and minimum values can be considered key points. These are

$$(0, 0), \quad \left(\frac{\pi}{2}, 1\right), \quad (\pi, 0), \quad \left(\frac{3\pi}{2}, -1\right), \quad (2\pi, 0).$$

These key points are transformed up 3 units to obtain the key points of the graph of $y = \sin x + 3$. These are

$$(0, 3), \quad \left(\frac{\pi}{2}, 4\right), \quad (\pi, 3), \quad \left(\frac{3\pi}{2}, 2\right), \quad (2\pi, 3).$$

The graph of $y = \sin x + 3$ can be sketched on the interval $[0, 2\pi]$ and extended to obtain the rest of the graph by repeating the graph on intervals of length 2π.

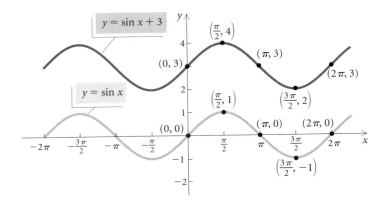

Next, we consider the effect of the constant A. What can we observe in the following graphs? What is the effect of the constant A on the graph of the basic function (a) when $0 < A < 1$? (b) when $A > 1$? (c) when $-1 < A < 0$? (d) when $A < -1$?

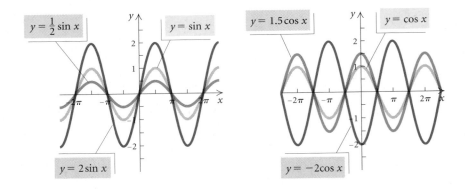

If $|A| > 1$, then there will be a vertical stretching. If $|A| < 1$, then there will be a vertical shrinking. If $A < 0$, the graph is also reflected across the x-axis.

Amplitude

The **amplitude** of the graphs of $y = A \sin (Bx - C) + D$ and $y = A \cos (Bx - C) + D$ is $|A|$.

EXAMPLE 2   Sketch a graph of $y = 2 \cos x$. What is the amplitude?

Solution   The constant 2 in $y = 2 \cos x$ has the effect of stretching the graph of $y = \cos x$ vertically by a factor of 2 units. Since the function values of $y = \cos x$ are such that $-1 \le \cos x \le 1$, the function values of $y = 2 \cos x$ are such that $-2 \le 2 \cos x \le 2$. The maximum value of $y = 2 \cos x$ is 2, and the minimum value is -2. Thus the *amplitude, A,* is $\frac{1}{2}|2 - (-2)|$, or 2.

We draw the graph of $y = \cos x$ and consider its key points,

$$(0, 1), \quad \left(\frac{\pi}{2}, 0\right), \quad (\pi, -1), \quad \left(\frac{3\pi}{2}, 0\right), \quad (2\pi, 1),$$

on the interval $[0, 2\pi]$.

We then multiply the second coordinates by 2 to obtain the key points of $y = 2 \cos x$. These are

$$(0, 2), \quad \left(\frac{\pi}{2}, 0\right), \quad (\pi, -2), \quad \left(\frac{3\pi}{2}, 0\right), \quad (2\pi, 2).$$

We plot these points and sketch the graph on the interval $[0, 2\pi]$. Then we repeat this part of the graph on adjacent intervals of length 2π.

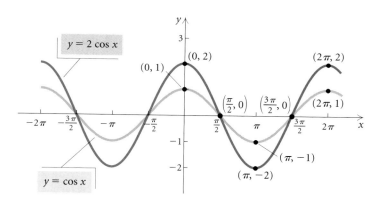

EXAMPLE 3   Sketch a graph of $y = -\frac{1}{2} \sin x$.

Solution   The amplitude of the graph is $\left|-\frac{1}{2}\right|$, or $\frac{1}{2}$. The graph of $y = -\frac{1}{2} \sin x$ is a vertical shrinking and a reflection of the graph of $y = \sin x$ across the x-axis. In graphing, the key points of $y = \sin x$,

$$(0, 0), \quad \left(\frac{\pi}{2}, 1\right), \quad (\pi, 0), \quad \left(\frac{3\pi}{2}, -1\right), \quad (2\pi, 0),$$

are transformed to

$$(0, 0), \quad \left(\frac{\pi}{2}, -\frac{1}{2}\right), \quad (\pi, 0), \quad \left(\frac{3\pi}{2}, \frac{1}{2}\right), \quad (2\pi, 0).$$

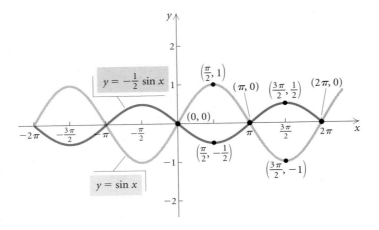

Now, we consider the effect of the constant B. Changes in the constants A and D do not change the period. But what effect, if any, does a change in B have on the period of the function? Let's observe the period of each of the following graphs.

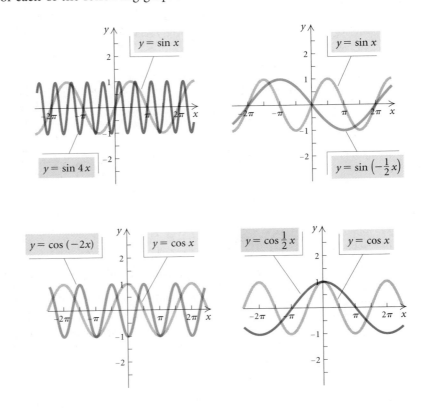

If $|B| < 1$, then there will be a horizontal stretching. If $|B| > 1$, then there will be a horizontal shrinking. If $B < 0$, the graph is also reflected across the y-axis.

Period

The **period** of the graphs of $y = A \sin (Bx - C) + D$ and $y = A \cos (Bx - C) + D$ is $\left| \dfrac{2\pi}{B} \right|$.

EXAMPLE 4   Sketch a graph of $y = \sin 4x$. What is the period?

Solution   The constant B has the effect of changing the period. The graph of $y = f(4x)$ is obtained from the graph of $y = f(x)$ by shrinking the graph horizontally. The new graph is obtained by dividing the first coordinate of each ordered-pair solution of $y = f(x)$ by 4. The key points of $y = \sin x$ are

$$(0, 0), \quad \left(\frac{\pi}{2}, 1\right), \quad (\pi, 0), \quad \left(\frac{3\pi}{2}, -1\right), \quad (2\pi, 0).$$

These are transformed to the key points of $y = \sin 4x$, which are

$$(0, 0), \quad \left(\frac{\pi}{8}, 1\right), \quad \left(\frac{\pi}{4}, 0\right), \quad \left(\frac{3\pi}{8}, -1\right), \quad \left(\frac{\pi}{2}, 0\right).$$

We plot these key points and sketch in the graph on the shortened interval $[0, \pi/2]$. Then we repeat the graph on other intervals of length $\pi/2$.

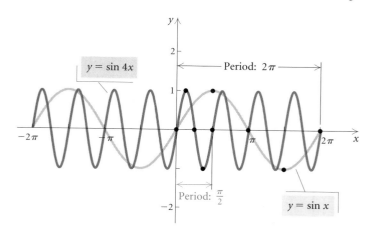

Next, we examine the effect of the constant C. The curve in each of the following graphs has an amplitude of 1 and a period of 2π, but there are six distinct graphs. What is the effect of the constant C?

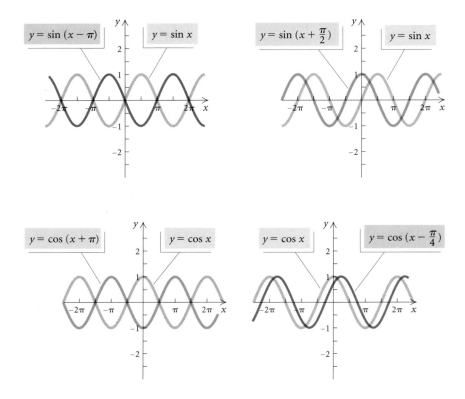

For each of the functions of the form

$$y = A \sin (Bx - C) + D \quad \text{and} \quad y = A \cos (Bx - C) + D$$

that are graphed above, the coefficient of x, which is B, is 1. In this case, the effect of the constant C on the graph of the basic function is a horizontal translation of $|C|$ units. In Example 5, which follows, $B = 1$. We will consider functions where $B \neq 1$ in Examples 6 and 7. When $B \neq 1$, the horizontal translation will be $|C/B|$.

EXAMPLE 5   Sketch a graph of $y = \sin \left(x - \dfrac{\pi}{2} \right)$.

Solution   The amplitude is 1, and the period is 2π. The graph of $y = f(x - c)$ is obtained from the graph of $y = f(x)$ by translating the graph horizontally—to the right c units if $c > 0$ and to the left $|c|$ units if $c < 0$. The graph of $y = \sin (x - \pi/2)$ is a translation of the graph of

$y = \sin x$ to the right $\pi/2$ units. The value $\pi/2$ is called the *phase shift*. The key points of $y = \sin x$,

$$(0, 0), \quad \left(\frac{\pi}{2}, 1\right), \quad (\pi, 0), \quad \left(\frac{3\pi}{2}, -1\right), \quad (2\pi, 0),$$

are transformed by adding $\pi/2$ to each of the first coordinates to obtain the following key points of $y = \sin (x - \pi/2)$:

$$\left(\frac{\pi}{2}, 0\right), \quad (\pi, 1), \quad \left(\frac{3\pi}{2}, 0\right), \quad (2\pi, -1), \quad \left(\frac{5\pi}{2}, 0\right).$$

We plot these key points and sketch the curve on the interval $[\pi/2, 5\pi/2]$. Then we repeat the graph on other intervals of length 2π.

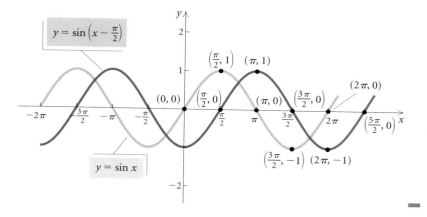

Now we consider combined transformations of graphs. It is helpful to rewrite

$$y = A \sin (Bx - C) + D \qquad \text{and} \quad y = A \cos (Bx - C) + D$$

as

$$y = A \sin \left[B\left(x - \frac{C}{B}\right)\right] + D \quad \text{and} \quad y = A \cos \left[B\left(x - \frac{C}{B}\right)\right] + D.$$

EXAMPLE 6 Sketch a graph of $y = \cos (2x - \pi)$.

Solution The graph of

$$y = \cos (2x - \pi)$$

is the same as the graph of

$$y = 1 \cdot \cos \left[2\left(x - \frac{\pi}{2}\right)\right] + 0.$$

The amplitude is 1. The factor 2 shrinks the period by half, making the period $|2\pi/2|$, or π. The $\pi/2$ translates the graph of $y = \cos 2x$ to the right $\pi/2$ units. Thus, to form the graph, we first graph $y = \cos x$, followed by $y = \cos 2x$ and then $y = \cos [2(x - \pi/2)]$.

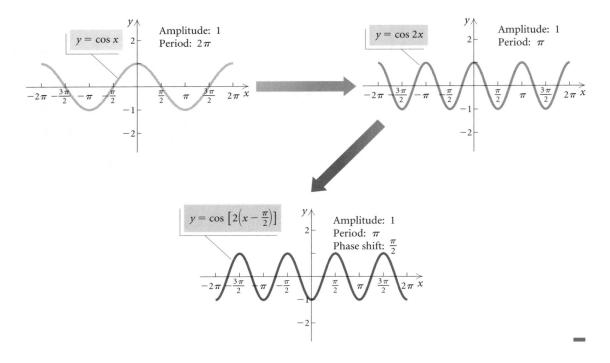

Phase Shift

The **phase shift** of the graphs

$$y = A \sin (Bx - C) + D = A \sin \left[B\left(x - \frac{C}{B} \right) \right] + D$$

and

$$y = A \cos (Bx - C) + D = A \cos \left[B\left(x - \frac{C}{B} \right) \right] + D$$

is the quantity $\dfrac{C}{B}$.

If $C/B > 0$, the graph is translated to the right C/B units. If $C/B < 0$, the graph is translated to the left $|C/B|$ units. Be sure that the horizontal stretching or shrinking based on the constant B is done before the translation based on the phase shift C/B.

Let's now summarize the effect of the constants. We carry out the procedures in the order listed.

Transformations of Sine and Cosine Functions

To graph

$$y = A \sin (Bx - C) + D = A \sin \left[B\left(x - \frac{C}{B} \right) \right] + D$$

and

$$y = A \cos (Bx - C) + D = A \cos \left[B\left(x - \frac{C}{B} \right) \right] + D,$$

follow the steps listed below in the order in which they are listed.

1. Stretch or shrink the graph horizontally according to B.

   $|B| < 1$ Stretch horizontally

   $|B| > 1$ Shrink horizontally

   $B < 0$ Reflect across the y-axis

   The *period* is $\left| \dfrac{2\pi}{B} \right|$.

2. Stretch or shrink the graph vertically according to A.

   $|A| < 1$ Shrink vertically

   $|A| > 1$ Stretch vertically

   $A < 0$ Reflect across the x-axis

   The *amplitude* is $|A|$.

3. Translate the graph horizontally according to C/B.

   $\dfrac{C}{B} < 0$ $\left| \dfrac{C}{B} \right|$ units to the left

   $\dfrac{C}{B} > 0$ $\dfrac{C}{B}$ units to the right

   The *phase shift* is $\dfrac{C}{B}$.

4. Translate the graph vertically according to D.

   $D < 0$ $|D|$ units down

   $D > 0$ D units up

EXAMPLE 7   Sketch a graph of $y = 3 \sin (2x + \pi/2) + 1$. Find the amplitude, the period, and the phase shift.

Solution   We first note that

$$y = 3 \sin \left(2x + \frac{\pi}{2} \right) + 1 = 3 \sin \left[2\left(x - \left(-\frac{\pi}{4} \right) \right) \right] + 1.$$

Then we have the following:

$$\text{Amplitude} = |A| = |3| = 3,$$

$$\text{Period} = \left|\frac{2\pi}{B}\right| = \left|\frac{2\pi}{2}\right| = \pi,$$

$$\text{Phase shift} = \frac{C}{B} = \frac{-\pi/2}{2} = -\frac{\pi}{4}.$$

To create the final graph, we begin with the basic sine curve, $y = \sin x$. Then we sketch graphs of each of the following equations in sequence.

1. $y = \sin 2x$ **2.** $y = 3 \sin 2x$

3. $y = 3 \sin \left[2\left(x - \left(-\frac{\pi}{4}\right)\right)\right]$ **4.** $y = 3 \sin \left[2\left(x - \left(-\frac{\pi}{4}\right)\right)\right] + 1$

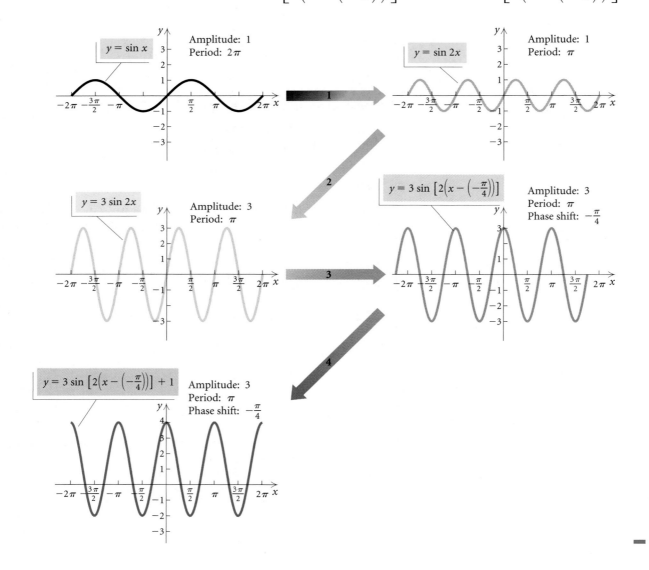

All the graphs in Examples 1–7 can be checked using a graphing calculator. Even though it is faster and more accurate to graph using a calculator, graphing by hand gives us a greater understanding of the effect of changing the constants A, B, C, and D.

Graphing calculators are especially convenient when a period or a phase shift is not a multiple of $\pi/4$.

EXAMPLE 8 Graph $y = 3 \cos 2\pi x - 1$. Find the amplitude, the period, and the phase shift.

Solution First we note the following:

$$\text{Amplitude} = |A| = |3| = 3,$$

$$\text{Period} = \left|\frac{2\pi}{B}\right| = \left|\frac{2\pi}{2\pi}\right| = |1| = 1,$$

$$\text{Phase shift} = \frac{C}{B} = \frac{0}{2\pi} = 0.$$

There is no phase shift in this case because the constant $C = 0$. The graph has a vertical translation of the graph of the cosine function down 1 unit, an amplitude of 3, and a period of 1, so we can use $[-4, 4, -5, 5]$ as the viewing window. ▬

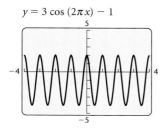

$y = 3 \cos (2\pi x) - 1$

The transformation techniques that we learned in this section for graphing the sine and cosine functions can also be applied in the same manner to the other trigonometric functions. Transformations of this type appear in the synthesis exercises in Exercise Set 5.6.

An **oscilloscope** is an electronic device that converts electrical signals into graphs like those in the preceding examples. These graphs are often called sine waves. By manipulating the controls, we can change the amplitude, the period, and the phase of sine waves. The oscilloscope has many applications, and the trigonometric functions play a major role in many of them.

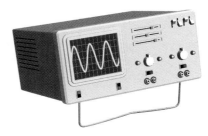

Graphs of Sums: Addition of Ordinates

The output of an electronic synthesizer used in the recording and playing of music can be converted into sine waves by an oscilloscope. The following graphs illustrate simple tones of different frequencies. The frequency of a simple tone is the number of vibrations in the signal of the tone per second. The loudness or intensity of the tone is reflected in the height of the graph (its amplitude). The three tones in the diagrams below all have the same intensity but different frequencies.

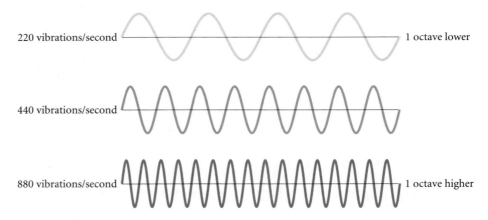

220 vibrations/second — 1 octave lower

440 vibrations/second —

880 vibrations/second — 1 octave higher

Musical instruments can generate extremely complex sine waves. On a single instrument, overtones can become superimposed on a simple tone. When multiple notes are played simultaneously, graphs become very complicated. This can happen when multiple notes are played on a single instrument or a group of instruments, or even when the same simple note is played on different instruments.

Combinations of simple tones produce interesting curves. Consider two tones whose graphs are $y_1 = 2 \sin x$ and $y_2 = \sin 2x$. The combination of the two tones produces a new sound whose graph is $y = 2 \sin x + \sin 2x$, shown in the following example.

EXAMPLE 9 Graph: $y = 2 \sin x + \sin 2x$.

Solution We graph $y = 2 \sin x$ and $y = \sin 2x$ using the same set of axes.

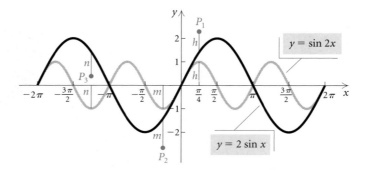

Now we graphically add some y-coordinates, or ordinates, to obtain points on the graph that we seek. At $x = \pi/4$, we transfer the distance h, which is the value of $\sin 2x$, up to add it to the value of $2 \sin x$. Point P_1 is on the graph that we seek. At $x = -\pi/4$, we use a similar procedure, but this time both ordinates are negative. Point P_2 is on the graph. At $x = -5\pi/4$, we add the negative ordinate of sin $2x$ to the positive ordinate of 2 sin x. Point P_3 is also on the graph. We continue to plot points in this fashion and then connect them to get the desired graph, shown below. This method is called *addition of ordinates*, because we add the y-values (ordinates) of $y = \sin 2x$ to the y-values (ordinates) of $y = 2 \sin x$.

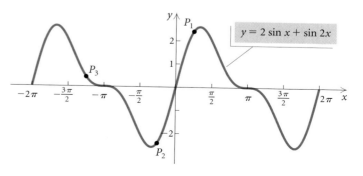

$y = 2 \sin x + \sin 2x$

Exercise Set 5.6

Determine the amplitude, the period, and the phase shift of the function, and, sketch the graph of the function.

1. $y = \sin x + 1$

2. $y = \frac{1}{4} \cos x$

3. $y = -3 \cos x$

4. $y = \sin(-2x)$

5. $y = 2 \sin\left(\frac{1}{2} x\right)$

6. $y = \cos\left(x - \frac{\pi}{2}\right)$

7. $y = \frac{1}{2} \sin\left(x + \frac{\pi}{2}\right)$

8. $y = \cos x - \frac{1}{2}$

9. $y = 3 \cos(x - \pi)$

10. $y = -\sin\left(\frac{1}{4} x\right) + 1$

11. $y = \frac{1}{3} \sin x - 4$

12. $y = \cos\left(\frac{1}{2} x + \frac{\pi}{2}\right)$

13. $y = -\cos(-x) + 2$

14. $y = \frac{1}{2} \sin\left(2x - \frac{\pi}{4}\right)$

Determine the amplitude, the period, and the phase shift of the function.

15. $y = 2 \cos\left(\frac{1}{2} x - \frac{\pi}{2}\right)$

16. $y = 4 \sin\left(\frac{1}{4} x + \frac{\pi}{8}\right)$

17. $y = -\frac{1}{2} \sin\left(2x + \frac{\pi}{2}\right)$

18. $y = -3 \cos(4x - \pi) + 2$

19. $y = 2 + 3 \cos(\pi x - 3)$

20. $y = 5 - 2 \cos\left(\frac{\pi}{2} x + \frac{\pi}{2}\right)$

21. $y = -\frac{1}{2} \cos(2\pi x) + 2$

22. $y = -2 \sin(-2x + \pi) - 2$

23. $y = -\sin\left(\frac{1}{2} x - \frac{\pi}{2}\right) + \frac{1}{2}$

24. $y = \frac{1}{3} \cos(-3x) + 1$

25. $y = \cos(-2\pi x) + 2$

26. $y = \frac{1}{2} \sin(2\pi x + \pi)$

27. $y = -\frac{1}{4} \cos(\pi x - 4)$

28. $y = 2 \sin(2\pi x + 1)$

In Exercises 29–36, match the function with one of graphs (a)–(h), which follow.

a)

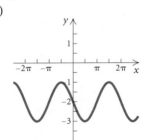

b)

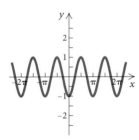

c)

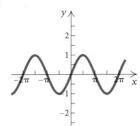

d)

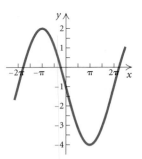

e)

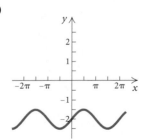

f)

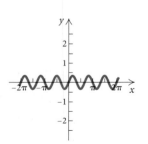

g)

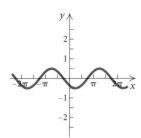

h)

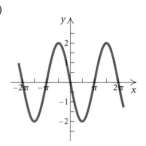

29. $y = -\cos 2x$

30. $y = \dfrac{1}{2} \sin x - 2$

31. $y = 2 \cos \left(x + \dfrac{\pi}{2} \right)$

32. $y = -3 \sin \dfrac{1}{2} x - 1$

33. $y = \sin (x - \pi) - 2$

34. $y = -\dfrac{1}{2} \cos \left(x - \dfrac{\pi}{4} \right)$

35. $y = \dfrac{1}{3} \sin 3x$

36. $y = \cos \left(x - \dfrac{\pi}{2} \right)$

In Exercises 37–40, determine the equation of the function that is graphed.

37.

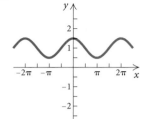

38.

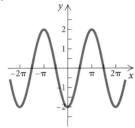

39.

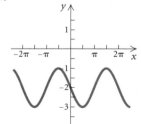

40.

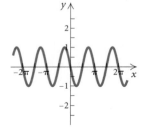

Graph using addition of ordinates.

41. $y = 2 \cos x + \cos 2x$ **42.** $y = 3 \cos x + \cos 3x$

43. $y = \sin x + \cos 2x$ **44.** $y = 2 \sin x + \cos 2x$

45. $y = \sin x - \cos x$ **46.** $y = 3 \cos x - \sin x$

47. $y = 3 \cos x + \sin 2x$ **48.** $y = 3 \sin x - \cos 2x$

Technology Connection

Use a graphing calculator to graph the function.

49. $y = x + \sin x$ **50.** $y = -x - \sin x$

51. $y = \cos x - x$ **52.** $y = -(\cos x - x)$

53. $y = \cos 2x + 2x$ **54.** $y = \cos 3x + \sin 3x$

55. $y = 4 \cos 2x - 2 \sin x$ **56.** $y = 7.5 \cos x + \sin 2x$

Use a graphing calculator to graph each of the following on the given interval and approximate the zeros.

57. $f(x) = \dfrac{\sin x}{x}$; $[-12, 12]$

58. $f(x) = \dfrac{\cos x - 1}{x}$; $[-12, 12]$

59. $f(x) = x^3 \sin x$; $[-5, 5]$

60. $f(x) = \dfrac{(\sin x)^2}{x}$; $[-4, 4]$

61. *Temperature During an Illness.* The temperature T of a patient during a 12-day illness is given by

$$T(t) = 101.6° + 3° \sin\left(\frac{\pi}{8}t\right).$$

a) Graph the function on the interval $[0, 12]$.
b) What are the maximum and the minimum temperatures during the illness?

62. *Periodic Sales.* A company in a northern climate has sales of skis as given by

$$S(t) = 10\left(1 - \cos\frac{\pi}{6}t\right),$$

where t is the time, in months ($t = 0$ corresponds to July 1), and $S(t)$ is in thousands of dollars.

a) Graph the function on a 12-month interval $[0, 12]$.
b) What is the period of the function?
c) What is the minimum amount of sales and when does it occur?
d) What is the maximum amount of sales and when does it occur?

Collaborative Discussion and Writing

63. In the equations $y = A \sin (Bx - C) + D$ and $y = A \cos (Bx - C) + D$, which constants translate the graphs and which constants stretch and shrink the graphs? Describe in your own words the effect of each constant.

64. In the transformation steps listed in this section, why must step (1) precede step (3)? Give an example that illustrates this.

Skill Maintenance

Find the zeros of the function.

65. $f(x) = 12 - x$ **66.** $g(x) = x^2 - x - 6$

Find the x-intercepts of the graph of the function.

67. $f(x) = 12 - x$ **68.** $g(x) = x^2 - x - 6$

Synthesis

The transformation techniques that we learned in this section for graphing the sine and cosine functions can also be applied to the other trigonometric functions. Sketch a graph of each of the following.

69. $y = -\tan x$

70. $y = \tan (-x)$

71. $y = -2 + \cot x$

72. $y = -\dfrac{3}{2} \csc x$

73. $y = 2 \tan \dfrac{1}{2} x$

74. $y = \cot 2x$

75. $y = 2 \sec (x - \pi)$

76. $y = 4 \tan \left(\dfrac{1}{4}x + \dfrac{\pi}{8}\right)$

77. $y = 2 \csc \left(\dfrac{1}{2}x - \dfrac{3\pi}{4}\right)$

78. $y = 4 \sec (2x - \pi)$

79. *Satellite Location.* A satellite circles the earth in such a way that it is y miles from the equator (north or south, height not considered) t minutes after its launch, where

$$y(t) = 3000\left[\cos\frac{\pi}{45}(t - 10)\right].$$

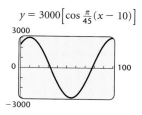

$$y = 3000\left[\cos\tfrac{\pi}{45}(x - 10)\right]$$

What are the amplitude, the period, and the phase shift?

80. *Water Wave.* The cross-section of a water wave is given by

$$y = 3 \sin\left(\frac{\pi}{4} x + \frac{\pi}{4}\right),$$

where y is the vertical height of the water wave and x is the distance from the origin to the wave.

$$y = 3 \sin\left(\frac{\pi}{4}x + \frac{\pi}{4}\right)$$

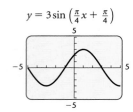

What are the amplitude, the period, and the phase shift?

81. *Damped Oscillations.* Suppose that the motion of a spring is given by

$$d(t) = 6e^{-0.8t} \cos(6\pi t) + 4,$$

where d is the distance, in inches, of a weight from the point at which the spring is attached to a ceiling, after t seconds. How far do you think the spring is from the ceiling when the spring stops bobbing?

82. *Rotating Beacon.* A police car is parked 10 ft from a wall. On top of the car is a beacon rotating in such a way that the light is at a distance $d(t)$ from point Q after t seconds, where

$$d(t) = 10 \tan(2\pi t).$$

When d is positive, as shown in the figure, the light is pointing north of Q, and when d is negative, the light is pointing south of Q.

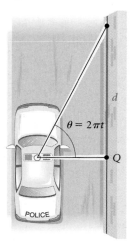

Explain the meaning of the values of t for which the function is undefined.

---

5 | Chapter Summary and Review

Important Properties and Formulas

Trigonometric Function Values of an Acute Angle θ

Let θ be an acute angle of a right triangle. The six trigonometric functions of θ are as follows:

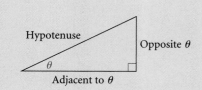

$$\sin\theta = \frac{\text{opp}}{\text{hyp}}, \quad \cos\theta = \frac{\text{adj}}{\text{hyp}}, \quad \tan\theta = \frac{\text{opp}}{\text{adj}},$$

$$\csc\theta = \frac{\text{hyp}}{\text{opp}}, \quad \sec\theta = \frac{\text{hyp}}{\text{adj}}, \quad \cot\theta = \frac{\text{adj}}{\text{opp}}.$$

Reciprocal Functions

$$\csc \theta = \frac{1}{\sin \theta}, \qquad \sec \theta = \frac{1}{\cos \theta}, \qquad \cot \theta = \frac{1}{\tan \theta}$$

Function Values of Special Angles

	0°	30°	45°	60°	90°
sin	0	1/2	$\sqrt{2}/2$	$\sqrt{3}/2$	1
cos	1	$\sqrt{3}/2$	$\sqrt{2}/2$	1/2	0
tan	0	$\sqrt{3}/3$	1	$\sqrt{3}$	Undefined

Cofunction Identities

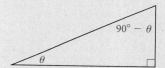

$$\sin \theta = \cos (90° - \theta), \qquad \cos \theta = \sin (90° - \theta),$$
$$\tan \theta = \cot (90° - \theta), \qquad \cot \theta = \tan (90° - \theta),$$
$$\sec \theta = \csc (90° - \theta), \qquad \csc \theta = \sec (90° - \theta)$$

Trigonometric Functions of Any Angle θ

If $P(x, y)$ is any point on the terminal side of any angle θ in standard position, and r is the distance from the origin to $P(x, y)$, where $r = \sqrt{x^2 + y^2}$, then

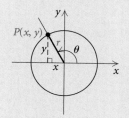

$$\sin \theta = \frac{y}{r}, \qquad \cos \theta = \frac{x}{r}, \qquad \tan \theta = \frac{y}{x}$$

(continued)

Signs of Function Values

The signs of the function values depend only on the coordinates of the point P on the terminal side of an angle.

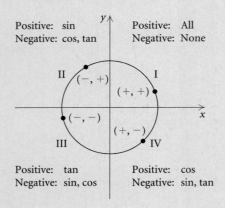

Positive: sin
Negative: cos, tan

Positive: All
Negative: None

Positive: tan
Negative: sin, cos

Positive: cos
Negative: sin, tan

Radian–Degree Equivalents

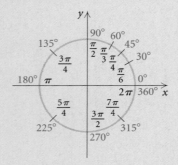

Linear Speed in Terms of Angular Speed

$$v = r\omega$$

Basic Circular Functions

For a real number s that determines a point (x, y) on the unit circle:

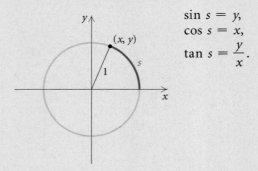

$$\sin s = y,$$
$$\cos s = x,$$
$$\tan s = \frac{y}{x}.$$

Sine (Odd Function): $\sin(-s) = -\sin s$
Cosine (Even Function): $\cos(-s) = \cos s$

Transformations of Sine and Cosine Functions

To graph $y = A \sin(Bx - C) + D$ and $y = A \cos(Bx - C) + D$:

1. Stretch or shrink the graph horizontally according to B.
2. Stretch or shrink the graph vertically according to A.
3. Translate the graph horizontally according to C/B.
4. Translate the graph vertically according to D.

REVIEW EXERCISES

1. Find the six trigonometric function values of the specified angle.

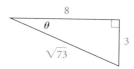

Find the exact function value, if it exists.

2. cos 45°

3. cot 60°

4. cos 495°

5. sin 150°

6. sec (−270°)

7. tan (−600°)

8. Convert 22.27° to degrees, minutes, and seconds. Round to the nearest second.

9. Convert 47°33′27″ to decimal degree notation. Round to two decimal places.

Find the function value. Round to four decimal places.

10. tan 2184°

11. sec 27.9°

12. cos 18°13′42″

13. sin 245°24′

14. cot (−33.2°)

15. sin 556.13°

Find θ in the interval indicated. Round the answer to the nearest tenth of a degree.

16. cos θ = −0.9041, (180°, 270°)

17. tan θ = 1.0799, (0°, 90°)

Find the exact acute angle θ, in degrees, given the function value.

18. sin θ = $\dfrac{\sqrt{3}}{2}$

19. tan θ = $\sqrt{3}$

20. Given that sin 59.1° ≈ 0.8581, cos 59.1° ≈ 0.5135, and tan 59.1° ≈ 1.6709, find the six function values for 30.9°.

Solve each of the following right triangles. Standard lettering has been used.

21. a = 7.3, c = 8.6

22. a = 30.5, B = 51.17°

23. One leg of a right triangle bears east. The hypotenuse is 734 m long and bears N57°23′E. Find the perimeter of the triangle.

24. An observer's eye is 6 ft above the floor. A mural is being viewed. The bottom of the mural is at floor level. The observer looks down 13° to see the bottom and up 17° to see the top. How tall is the mural?

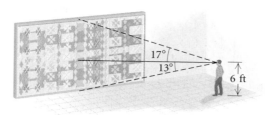

Find a positive angle and a negative angle that are coterminal with the given angle. Answers may vary.

25. 65°

26. $\dfrac{7\pi}{3}$

Find the complement and the supplement.

27. 13.4°

28. $\dfrac{\pi}{6}$

29. Find the six trigonometric function values for the angle θ shown.

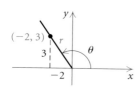

30. Given that tan θ = 2/√5 and that the terminal side is in quadrant III, find the other five function values.

31. An airplane travels at 530 mph for 3½ hr in a direction of 160° from Minneapolis, Minnesota. At the end of that time, how far south of Minneapolis is the airplane?

32. On a unit circle, mark and label the points determined by 7π/6, −3π/4, −π/3, and 9π/4.

For angles of the following measures, state in which quadrant the terminal side lies, convert to radian measure in terms of π, and convert to radian measure not in terms of π.

33. 145.2°

34. −30°

Convert to degree measure. Round the answer to two decimal places.

35. $\dfrac{3\pi}{2}$

36. 3

37. Find the length of an arc of a circle, given a central angle of $\pi/4$ and a radius of 7 cm.

38. An arc 18 m long on a circle of radius 8 m subtends an angle of how many radians? how many degrees, to the nearest degree?

39. Inside La Madeleine French Bakery and Cafe in Houston, Texas, there is one of the few remaining working watermills in the world. The 300-yr-old French-built waterwheel has a radius of 7 ft and makes one complete revolution in 70 sec. What is the linear speed in feet per minute of a point on the rim? (*Source:* La Madeleine French Bakery and Cafe, Houston, TX)

40. An automobile wheel has a diameter of 14 in. If the car travels at a speed of 55 mph, what is the angular velocity, in radians per hour, of a point on the edge of the wheel?

41. The point $\left(\frac{3}{5}, -\frac{4}{5}\right)$ is on a unit circle. Find the coordinates of its reflections across the x-axis, the y-axis, and the origin.

Find the exact function value, if it exists.

42. $\cos \pi$

43. $\tan \dfrac{5\pi}{4}$

44. $\sin \dfrac{5\pi}{3}$

45. $\sin \left(-\dfrac{7\pi}{6}\right)$

46. $\tan \dfrac{\pi}{6}$

47. $\cos (-13\pi)$

Find the function value. Round to four decimal places.

48. $\sin 24$

49. $\cos (-75)$

50. $\cot 16\pi$

51. $\tan \dfrac{3\pi}{7}$

52. $\sec 14.3$

53. $\cos \left(-\dfrac{\pi}{5}\right)$

54. Graph each of the six trigonometric functions from -2π to 2π.

55. What is the period of each of the six trigonometric functions?

56. Complete the following table.

FUNCTION	DOMAIN	RANGE
sine		
cosine		
tangent		

57. Complete the following table with the sign of the specified trigonometric function value in each of the four quadrants.

FUNCTION	I	II	III	IV
sine				
cosine				
tangent				

Determine the amplitude, the period, and the phase shift of the function, and sketch the graph of the function.

58. $y = \sin \left(x + \dfrac{\pi}{2}\right)$

59. $y = 3 + \dfrac{1}{2} \cos \left(2x - \dfrac{\pi}{2}\right)$

In Exercises 60–63, match the function with one of graphs (a)–(d), which follow.

a)

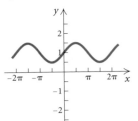

b)

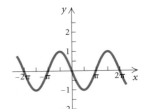

c)

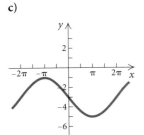

d)

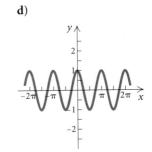

60. $y = \cos 2x$

61. $y = \dfrac{1}{2} \sin x + 1$

62. $y = -2 \sin \dfrac{1}{2} x - 3$

63. $y = -\cos \left(x - \dfrac{\pi}{2}\right)$

64. Sketch a graph of $y = 3 \cos x + \sin x$ for values of x between 0 and 2π.

Collaborative Discussion and Writing

65. Compare the terms radian and degree.

66. Describe the shape of the graph of the cosine function. How many maximum values are there of the cosine function? Where do they occur?

67. Does $5 \sin x = 7$ have a solution for x? Why or why not?

Synthesis

68. Graph $y = 3 \sin (x/2)$, and determine the domain, the range, and the period.

69. In the graph below, $y_1 = \sin x$ is shown and y_2 is shown in red. Express y_2 as a transformation of the graph of y_1.

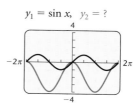

$$y_1 = \sin x, \quad y_2 = ?$$

70. Find the domain of $y = \log (\cos x)$.

71. Given that $\sin x = 0.6144$ and that the terminal side is in quadrant II, find the other basic circular function values.

---

5 Chapter Test

1. Find the six trigonometric function values of θ.

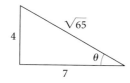

Find the exact function value, if it exists.

2. $\sin 120°$

3. $\tan (-45°)$

4. $\cos 3\pi$

5. $\sec \dfrac{5\pi}{4}$

6. Convert $38°27'56''$ to decimal degree notation. Round to two decimal places.

Find the function values. Round to four decimal places.

7. $\tan 526.4°$

8. $\sin (-12°)$

9. $\sec \dfrac{5\pi}{9}$

10. $\cos 76.07$

11. Find the exact acute angle θ, in degrees, for which $\sin \theta = \frac{1}{2}$.

12. Given that $\sin 28.4° \approx 0.4756$, $\cos 28.4° \approx 0.8796$, and $\tan 28.4° \approx 0.5407$, find the six trigonometric function values for $61.6°$.

13. Solve the right triangle with $b = 45.1$ and $A = 35.9°$. Standard lettering has been used.

14. Find a positive angle and a negative angle coterminal with a $112°$ angle.

15. Find the supplement of $\dfrac{5\pi}{6}$.

16. Given that $\sin \theta = -4/\sqrt{41}$ and that the terminal side is in quadrant IV, find the other five trigonometric function values.

17. Convert $210°$ to radian measure in terms of π.

18. Convert $\dfrac{3\pi}{4}$ to degree measure.

19. Find the length of an arc of a circle given a central angle of $\pi/3$ and a radius of 16 cm.

Consider the function $y = -\sin(x - \pi/2) + 1$ for Exercises 20–23.

20. Find the amplitude.

21. Find the period.

22. Find the phase shift.

23. Which is the graph of the function?

a)

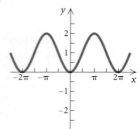

b)

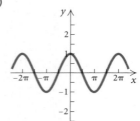

c)

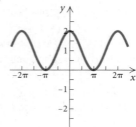

d)

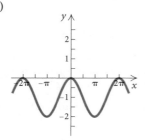

24. *Height of a Kite.*   The angle of elevation of a kite is 65° with 490 ft of string out. Assuming the string is taut, how high is the kite?

25. *Location.*   A pickup-truck camper travels at 50 mph for 6 hr in a direction of 115° from Buffalo, Wyoming. At the end of that time, how far east of Buffalo is the camper?

26. *Linear Speed.*   A ferris wheel has a radius of 6 m and revolves at 1.5 rpm. What is the linear speed, in meters per minute?

Synthesis

27. Determine the domain of $f(x) = \dfrac{-3}{\sqrt{\cos x}}$.

Trigonometric Identities, Inverse Functions, and Equations

6

*T*here are a number of relationships among the trigonometric functions, expressed as identities, that are important in algebraic and trigonometric manipulations. A large part of this chapter is devoted to those identities and their use in solving trigonometric equations. We also provide a detailed examination of inverses of the trigonometric functions. This chapter provides not only a basis for applications but also a foundation for more advanced work in mathematics.

APPLICATION

The number of daylight hours in Kajaani, Finland, varies from about 4.3 hr per day to 20.7 hr per day (*Source*: *The Astronomical Almanac*). The function

$$H(d) = 7.8787 \sin (0.0166d - 1.2723) + 12.1840$$

can be used to approximate the number of daylight hours H on a certain day of the year d in Kajaani. We will use this function to approximate the number of daylight hours in Kajaani for April 22, July 4, and December 15.

This problem appears as Exercise 50 in Exercise Set 6.5.

6.1

Identities: Pythagorean and Sum and Difference

- *State the Pythagorean identities.*
- *Simplify and manipulate expressions containing trigonometric expressions.*
- *Use the sum and difference identities to find function values.*

An **identity** is an equation that is true for all *possible* replacements of the variables. The following is a list of the identities studied in Chapter 5.

Basic Identities

$$\sin x = \frac{1}{\csc x}, \qquad \csc x = \frac{1}{\sin x}, \qquad \sin(-x) = -\sin x,$$
$$\cos(-x) = \cos x,$$

$$\cos x = \frac{1}{\sec x}, \qquad \sec x = \frac{1}{\cos x}, \qquad \tan(-x) = -\tan x,$$

$$\tan x = \frac{1}{\cot x}, \qquad \cot x = \frac{1}{\tan x}, \qquad \tan x = \frac{\sin x}{\cos x},$$

$$\cot x = \frac{\cos x}{\sin x}$$

In this section, we will develop some other important identities.

Pythagorean Identities

CIRCLES

REVIEW SECTION 1.1.

We now consider three other identities that are fundamental to a study of trigonometry. They are called the *Pythagorean identities*. Recall that the equation of a unit circle in the *xy*-plane is

$$x^2 + y^2 = 1.$$

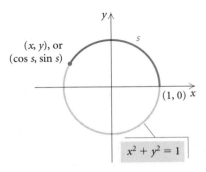

(x, y), or $(\cos s, \sin s)$

s

$(1, 0)$ x

$x^2 + y^2 = 1$

For any point on the unit circle, the coordinates *x* and *y* satisfy this equation. Suppose that a real number *s* determines a point on the unit circle with coordinates (x, y), or $(\cos s, \sin s)$. Then $x = \cos s$ and $y = \sin s$. Substituting $\cos s$ for x and $\sin s$ for y in the equation of the unit circle gives us the identity

$$(\cos s)^2 + (\sin s)^2 = 1,$$

which can be expressed as

$\sin^2 s + \cos^2 s = 1$.

It is conventional in trigonometry to use the notation $\sin^2 s$ rather than $(\sin s)^2$. Note that $\sin^2 s \neq \sin s^2$.

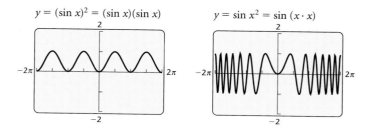

$y = (\sin x)^2 = (\sin x)(\sin x)$ $\qquad$ $y = \sin x^2 = \sin (x \cdot x)$

The identity $\sin^2 s + \cos^2 s = 1$ gives a relationship between the sine and the cosine of any real number s. It is an important **Pythagorean identity.**

We can divide by $\sin^2 s$ on both sides of the preceding identity:

$$\frac{\sin^2 s}{\sin^2 s} + \frac{\cos^2 s}{\sin^2 s} = \frac{1}{\sin^2 s}. \qquad \text{Dividing by } \sin^2 s$$

Simplifying gives us a second identity:

$1 + \cot^2 s = \csc^2 s$.

This equation is true for any replacement of s with a real number for which $\sin^2 s \neq 0$, since we divided by $\sin^2 s$. But the numbers for which $\sin^2 s = 0$ (or $\sin s = 0$) are exactly the ones for which the cotangent and cosecant functions are undefined. Hence our new equation holds for all real numbers s for which $\cot s$ and $\csc s$ are defined and is thus an identity.

The third Pythagorean identity, obtained by dividing by $\cos^2 s$ on both sides of the first Pythagorean identity, is

$\tan^2 s + 1 = \sec^2 s$.

The identities we have developed hold no matter what symbols are used for the variables. For example, we could write $\sin^2 s + \cos^2 s = 1$, $\sin^2 \theta + \cos^2 \theta = 1$, or $\sin^2 x + \cos^2 x = 1$.

> *Pythagorean Identities*
>
> $\sin^2 x + \cos^2 x = 1,$
>
> $1 + \cot^2 x = \csc^2 x,$
>
> $1 + \tan^2 x = \sec^2 x$

It is often helpful to express the Pythagorean identities in equivalent forms. For example, $\sin^2 x + \cos^2 x = 1$ is frequently seen as $\sin^2 x = 1 - \cos^2 x$ or $\cos^2 x = 1 - \sin^2 x$.

Simplifying Trigonometric Expressions

We can factor, simplify, and manipulate trigonometric expressions in the same way that we manipulate strictly algebraic expressions.

EXAMPLE 1   Multiply and simplify: $\cos x (\tan x - \sec x)$.

Solution

$\cos x (\tan x - \sec x)$

$\qquad = \cos x \tan x - \cos x \sec x$ Multiplying

$\qquad = \cos x \dfrac{\sin x}{\cos x} - \cos x \dfrac{1}{\cos x}$ Recalling the identities $\tan x = \dfrac{\sin x}{\cos x}$

$\qquad\qquad\qquad\qquad\qquad\qquad$ and $\sec x = \dfrac{1}{\cos x}$ and substituting

$\qquad = \sin x - 1$ Simplifying   ▬

There is no general procedure for manipulating trigonometric expressions, but it is often helpful to write everything in terms of sines and cosines, as we did in Example 1. We also look for the Pythagorean identiy, $\sin^2 x + \cos^2 x = 1$, within a trigonometric expression.

EXAMPLE 2   Factor and simplify: $\sin^2 x \cos^2 x + \cos^4 x$.

Solution

$\qquad \sin^2 x \cos^2 x + \cos^4 x$

$\qquad\quad = \cos^2 x (\sin^2 x + \cos^2 x)$ Removing a common factor

$\qquad\quad = \cos^2 x$ Using $\sin^2 x + \cos^2 x = 1$ ▬

Technology Connection

A graphing calculator can be used to perform a partial check of an identity. First, we graph the expression on the left side of the equals sign. Then we graph the expression on the right side using the same screen. If the two graphs are indistinguishable, then we have a partial verification that the equation is an identity. Of course, we can never see the entire graph, so there can always be some doubt. Also, the graphs may not overlap precisely, but you may not be able to tell because the difference between the graphs may be less than the width of a pixel. However, if the graphs are obviously different, we know that a mistake has been made.

For example, consider the identity in Example 1:

$$\cos x\,(\tan x - \sec x) = \sin x - 1.$$

Recalling that $\sec x = 1/\cos x$, we enter

$$y_1 = \cos x\,[\tan x - (1/\cos x)] \quad \text{and} \quad y_2 = \sin x - 1.$$

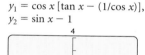

$y_1 = \cos x\,[\tan x - (1/\cos x)],$
$y_2 = \sin x - 1$

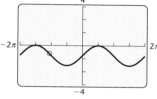

To graph, we first select SEQUENTIAL mode. Then we select the "line"-graph style for y_1 and the "path"-graph style, denoted by $-\bigcirc$, for y_2. The calculator will graph y_1 first. Then it will graph y_2 as the circular cursor traces the leading edge of the graph, allowing us to determine whether the graphs coincide. As you can see in the first screen on the left, the graphs appear to be identical. Thus, $\cos x\,(\tan x - \sec x) = \sin x - 1$ is most likely an identity.

The TABLE feature can also be used to check identities. Note in the table at left that the function values are the same except for those values of x for which $\cos x = 0$. The domain of y_1 excludes these values. The domain of y_2 is the set of all real numbers. Thus all real numbers except $\pm\pi/2$, $\pm 3\pi/2$, $\pm 5\pi/2,\ldots$ are possible replacements for x in the identity. Recall that an identity is an equation that is true for all *possible* replacements.

X	Y1	Y2
−6.283	−1	−1
−5.498	−.2929	−.2929
−4.712	ERROR	0
−3.927	−.2929	−.2929
−3.142	−1	−1
−2.356	−1.707	−1.707
−1.571	ERROR	−2

X = −6.28318530718

TblStart = −2π
ΔTbl = π/4

EXAMPLE 3   Simplify each of the following trigonometric expressions.

a) $\dfrac{\cot(-\theta)}{\csc(-\theta)}$

b) $\dfrac{2\sin^2 t + \sin t - 3}{1 - \cos^2 t - \sin t}$

Solution

a) $\dfrac{\cot(-\theta)}{\csc(-\theta)} = \dfrac{\dfrac{\cos(-\theta)}{\sin(-\theta)}}{\dfrac{1}{\sin(-\theta)}}$ Rewriting in terms of sines and cosines

$\quad\quad\quad\quad = \dfrac{\cos(-\theta)}{\sin(-\theta)} \cdot \sin(-\theta)$ Multiplying by the reciprocal

$\quad\quad\quad\quad = \cos(-\theta) = \cos\theta$ The cosine function is even.

b) $\dfrac{2\sin^2 t + \sin t - 3}{1 - \cos^2 t - \sin t}$

$\quad\quad = \dfrac{2\sin^2 t + \sin t - 3}{\sin^2 t - \sin t}$ Substituting $\sin^2 t$ for $1 - \cos^2 t$

$\quad\quad = \dfrac{(2\sin t + 3)(\sin t - 1)}{\sin t\,(\sin t - 1)}$ Factoring in both numerator and denominator

$\quad\quad = \dfrac{2\sin t + 3}{\sin t}$ Simplifying

$\quad\quad = \dfrac{2\sin t}{\sin t} + \dfrac{3}{\sin t}$

$\quad\quad = 2 + \dfrac{3}{\sin t}, \quad\text{or}\quad 2 + 3\csc t$ ▬

We can add and subtract trigonometric rational expressions in the same way that we do algebraic expressions.

EXAMPLE 4 Add and simplify: $\dfrac{\cos x}{1 + \sin x} + \tan x$.

Solution

$\dfrac{\cos x}{1 + \sin x} + \tan x = \dfrac{\cos x}{1 + \sin x} + \dfrac{\sin x}{\cos x}$ Using $\tan x = \dfrac{\sin x}{\cos x}$

$\quad\quad = \dfrac{\cos x}{1 + \sin x} \cdot \dfrac{\cos x}{\cos x} + \dfrac{\sin x}{\cos x} \cdot \dfrac{1 + \sin x}{1 + \sin x}$ Multiplying by forms of 1

$\quad\quad = \dfrac{\cos^2 x + \sin x + \sin^2 x}{\cos x\,(1 + \sin x)}$ Adding

$\quad\quad = \dfrac{1 + \sin x}{\cos x\,(1 + \sin x)}$ Using $\sin^2 x + \cos^2 x = 1$

$\quad\quad = \dfrac{1}{\cos x}, \quad\text{or}\quad \sec x$ Simplifying   ▬

When radicals occur, the use of absolute value is sometimes necessary, but it can be difficult to determine when to use it. In Examples 5 and 6, we will assume that all radicands are nonnegative. This means that the identities are meant to be confined to certain quadrants.

EXAMPLE 5   Multiply and simplify: $\sqrt{\sin^3 x \cos x} \cdot \sqrt{\cos x}$.

Solution

$$\sqrt{\sin^3 x \cos x} \cdot \sqrt{\cos x} = \sqrt{\sin^3 x \cos^2 x}$$
$$= \sqrt{\sin^2 x \cos^2 x \sin x}$$
$$= \sin x \cos x \sqrt{\sin x}$$

EXAMPLE 6   Rationalize the denominator: $\sqrt{\dfrac{2}{\tan x}}$.

Solution

$$\sqrt{\frac{2}{\tan x}} = \sqrt{\frac{2}{\tan x} \cdot \frac{\tan x}{\tan x}}$$
$$= \sqrt{\frac{2 \tan x}{\tan^2 x}}$$
$$= \frac{\sqrt{2 \tan x}}{\tan x}$$

Often in calculus, a substitution is a useful manipulation, as we show in the following example.

EXAMPLE 7   Express $\sqrt{9 + x^2}$ as a trigonometric function of θ without using radicals by letting $x = 3 \tan \theta$. Assume that $0 < \theta < \pi/2$. Then find $\sin \theta$ and $\cos \theta$.

Solution   We have

$$\sqrt{9 + x^2} = \sqrt{9 + (3 \tan \theta)^2} \qquad \text{Substituting 3 tan } \theta \text{ for } x$$
$$= \sqrt{9 + 9 \tan^2 \theta}$$
$$= \sqrt{9(1 + \tan^2 \theta)} \qquad \text{Factoring}$$
$$= \sqrt{9 \sec^2 \theta} \qquad \text{Using } 1 + \tan^2 x = \sec^2 x$$
$$= 3|\sec \theta| = 3 \sec \theta. \qquad \text{For } 0 < \theta < \pi/2, \sec \theta > 0.$$

We can express $\sqrt{9 + x^2} = 3 \sec \theta$ as

$$\sec \theta = \frac{\sqrt{9 + x^2}}{3}.$$

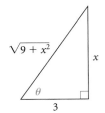

In a right triangle, we know that $\sec \theta$ is hypotenuse/adjacent, when θ is one of the acute angles. Using the Pythagorean theorem, we can determine that the side opposite θ is x. Then from the right triangle, we see that

$$\sin \theta = \frac{x}{\sqrt{9 + x^2}} \quad \text{and} \quad \cos \theta = \frac{3}{\sqrt{9 + x^2}}.$$

Sum and Difference Identities

We now develop some important identities involving sums or differences of two numbers (or angles), beginning with an identity for the cosine of the difference of two numbers. We use the letters u and v for these numbers.

Let's consider a real number u in the interval $[\pi/2, \pi]$ and a real number v in the interval $[0, \pi/2]$. These determine points A and B on the unit circle, as shown below. The arc length s is $u - v$, and we know that $0 \le s \le \pi$. Recall that the coordinates of A are $(\cos u, \sin u)$, and the coordinates of B are $(\cos v, \sin v)$.

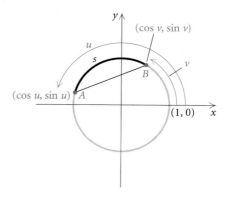

Using the distance formula, we can write an expression for the distance AB:

$$AB = \sqrt{(\cos u - \cos v)^2 + (\sin u - \sin v)^2}.$$

This can be simplified as follows:

$$
\begin{aligned}
AB &= \sqrt{\cos^2 u - 2\cos u \cos v + \cos^2 v + \sin^2 u - 2\sin u \sin v + \sin^2 v} \\
&= \sqrt{(\sin^2 u + \cos^2 u) + (\sin^2 v + \cos^2 v) - 2(\cos u \cos v + \sin u \sin v)} \\
&= \sqrt{2 - 2(\cos u \cos v + \sin u \sin v)}.
\end{aligned}
$$

Now let's imagine rotating the circle above so that point B is at $(1, 0)$. Although the coordinates of point A are now $(\cos s, \sin s)$, the distance AB has not changed.

Again we use the distance formula to write an expression for the distance AB:

$$AB = \sqrt{(\cos s - 1)^2 + (\sin s - 0)^2}.$$

This can be simplified as follows:

$$
\begin{aligned}
AB &= \sqrt{\cos^2 s - 2\cos s + 1 + \sin^2 s} \\
&= \sqrt{(\sin^2 s + \cos^2 s) + 1 - 2\cos s} \\
&= \sqrt{2 - 2\cos s}.
\end{aligned}
$$

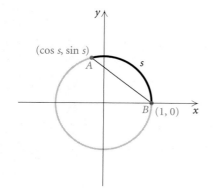

Equating our two expressions for AB, we obtain

$$\sqrt{2 - 2(\cos u \cos v + \sin u \sin v)} = \sqrt{2 - 2\cos s}.$$

Solving this equation for cos *s* gives

$$\cos s = \cos u \cos v + \sin u \sin v. \tag{1}$$

But *s* = *u* − *v*, so we have the equation

$$\cos (u - v) = \cos u \cos v + \sin u \sin v. \tag{2}$$

Formula (1) above holds when *s* is the length of the shortest arc from *A* to *B*. Given any real numbers *u* and *v*, the length of the shortest arc from *A* to *B* is not always *u* − *v*. In fact, it could be *v* − *u*. However, since cos (−*x*) = cos *x*, we know that cos (*v* − *u*) = cos (*u* − *v*). Thus, cos *s* is always equal to cos (*u* − *v*). Formula (2) holds for all real numbers *u* and *v*. That formula is thus the identity we sought:

cos (*u* − *v*) = cos *u* cos *v* + sin *u* sin *v*.

The cosine sum formula follows easily from the one we have just derived. Let's consider cos (*u* + *v*). This is equal to cos [*u* − (−*v*)], and by the identity above, we have

$$\cos (u + v) = \cos [u - (-v)]$$
$$= \cos u \cos (-v) + \sin u \sin (-v).$$

But cos (−*v*) = cos *v* and sin (−*v*) = −sin *v*, so the identity we seek is the following:

cos (*u* + *v*) = cos *u* cos *v* − sin *u* sin *v*.

EXAMPLE 8   Find cos (5π/12) exactly.

Solution   We can express 5π/12 as a difference of two numbers whose sine and cosine values are known:

$$\frac{5\pi}{12} = \frac{9\pi}{12} - \frac{4\pi}{12}, \quad \text{or} \quad \frac{3\pi}{4} - \frac{\pi}{3}.$$

Then, using cos (*u* − *v*) = cos *u* cos *v* + sin *u* sin *v*, we have

$$\cos \frac{5\pi}{12} = \cos \left(\frac{3\pi}{4} - \frac{\pi}{3} \right) = \cos \frac{3\pi}{4} \cos \frac{\pi}{3} + \sin \frac{3\pi}{4} \sin \frac{\pi}{3}$$
$$= -\frac{\sqrt{2}}{2} \cdot \frac{1}{2} + \frac{\sqrt{2}}{2} \cdot \frac{\sqrt{3}}{2}$$
$$= -\frac{\sqrt{2}}{4} + \frac{\sqrt{6}}{4} = \frac{\sqrt{6} - \sqrt{2}}{4}.$$

Technology Connection

We can check the result of Example 8 using a graphing calculator set in RADIAN mode.

cos(5π/12)	
	.2588190451
(√(6)−√(2))/4	
	.2588190451

Consider cos (π/2 − θ). We can use the identity for the cosine of a difference to simplify as follows:

$$\cos \left(\frac{\pi}{2} - \theta \right) = \cos \frac{\pi}{2} \cos \theta + \sin \frac{\pi}{2} \sin \theta$$
$$= 0 \cdot \cos \theta + 1 \cdot \sin \theta = \sin \theta.$$

Thus we have developed the identity

$$\sin \theta = \cos \left(\frac{\pi}{2} - \theta \right). \qquad \text{This cofunction identity first appeared in Section 5.1.} \qquad (3)$$

This identity holds for any real number θ. From it we can obtain an identity for the sine function. We first let α be any real number. Then we replace θ in $\sin \theta = \cos (\pi/2 - \theta)$ with $\pi/2 - \alpha$. This gives us

$$\sin \left(\frac{\pi}{2} - \alpha \right) = \cos \left[\frac{\pi}{2} - \left(\frac{\pi}{2} - \alpha \right) \right] = \cos \alpha,$$

which yields the identity

$$\cos \alpha = \sin \left(\frac{\pi}{2} - \alpha \right). \qquad (4)$$

Using identities (3) and (4) and the identity for the cosine of a difference, we can obtain an identity for the sine of a sum. We start with identity (3) and substitute $u + v$ for θ:

$$\sin \theta = \cos \left(\frac{\pi}{2} - \theta \right) \qquad \text{Identity (3)}$$

$$\sin (u + v) = \cos \left[\frac{\pi}{2} - (u + v) \right] \qquad \text{Substituting } u + v \text{ for } \theta$$

$$= \cos \left[\left(\frac{\pi}{2} - u \right) - v \right]$$

$$= \cos \left(\frac{\pi}{2} - u \right) \cos v + \sin \left(\frac{\pi}{2} - u \right) \sin v$$

Using the identity for the cosine of a difference

$$= \sin u \cos v + \cos u \sin v. \qquad \text{Using identities (3) and (4)}$$

Thus the identity we seek is

$$\sin (u + v) = \sin u \cos v + \cos u \sin v.$$

To find a formula for the sine of a difference, we can use the identity just derived, substituting $-v$ for v:

$$\sin (u + (-v)) = \sin u \cos (-v) + \cos u \sin (-v).$$

Simplifying gives us

$$\sin (u - v) = \sin u \cos v - \cos u \sin v.$$

EXAMPLE 9 Find $\sin 105°$ exactly.

Solution We express $105°$ as the sum of two measures:

$$105° = 45° + 60°.$$

Then

$$\begin{aligned}
\sin 105° &= \sin (45° + 60°) \\
&= \sin 45° \cos 60° + \cos 45° \sin 60°
\end{aligned}$$

Using $\sin (u + v) = \sin u \cos v + \cos u \sin v$

$$\begin{aligned}
&= \frac{\sqrt{2}}{2} \cdot \frac{1}{2} + \frac{\sqrt{2}}{2} \cdot \frac{\sqrt{3}}{2} \\
&= \frac{\sqrt{2} + \sqrt{6}}{4}.
\end{aligned}$$

Formulas for the tangent of a sum or a difference can be derived using identities already established. A summary of the sum and difference identities follows.*

Sum and Difference Identities

$$\sin (u \pm v) = \sin u \cos v \pm \cos u \sin v,$$

$$\cos (u \pm v) = \cos u \cos v \mp \sin u \sin v,$$

$$\tan (u \pm v) = \frac{\tan u \pm \tan v}{1 \mp \tan u \tan v}$$

EXAMPLE 10   Find tan 15° exactly.

Solution   We rewrite 15° as 45° − 30° and use the identity for the tangent of a difference:

$$\begin{aligned}
\tan 15° = \tan (45° - 30°) &= \frac{\tan 45° - \tan 30°}{1 + \tan 45° \tan 30°} \\
&= \frac{1 - \sqrt{3}/3}{1 + 1 \cdot \sqrt{3}/3} = \frac{3 - \sqrt{3}}{3 + \sqrt{3}}.
\end{aligned}$$

EXAMPLE 11   Assume that $\sin \alpha = \frac{2}{3}$ and $\sin \beta = \frac{1}{3}$ and that α and β are between 0 and $\pi/2$. Then evaluate $\sin (\alpha + \beta)$.

Solution   Using the identity for the sine of a sum, we have

$$\begin{aligned}
\sin (\alpha + \beta) &= \sin \alpha \cos \beta + \cos \alpha \sin \beta \\
&= \tfrac{2}{3} \cos \beta + \tfrac{1}{3} \cos \alpha.
\end{aligned}$$

---

*There are six identities here, half of them obtained by using the signs shown in color.

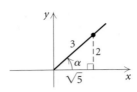

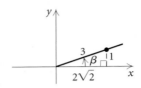

To finish, we need to know the values of cos β and cos α. Using reference triangles and the Pythagorean theorem, we can determine these values from the diagrams:

$$\cos \alpha = \frac{\sqrt{5}}{3} \quad \text{and} \quad \cos \beta = \frac{2\sqrt{2}}{3}.$$

Cosine values are positive in the first quadrant.

Substituting these values gives us

$$\sin (\alpha + \beta) = \frac{2}{3} \cdot \frac{2\sqrt{2}}{3} + \frac{1}{3} \cdot \frac{\sqrt{5}}{3}$$

$$= \frac{4}{9}\sqrt{2} + \frac{1}{9}\sqrt{5}, \quad \text{or} \quad \frac{4\sqrt{2} + \sqrt{5}}{9}.$$

Exercise Set

Multiply and simplify.

1. $(\sin x - \cos x)(\sin x + \cos x)$

2. $\tan x \,(\cos x - \csc x)$

3. $\cos y \sin y \,(\sec y + \csc y)$

4. $(\sin x + \cos x)(\sec x + \csc x)$

5. $(\sin \phi - \cos \phi)^2$

6. $(1 + \tan x)^2$

7. $(\sin x + \csc x)(\sin^2 x + \csc^2 x - 1)$

8. $(1 - \sin t)(1 + \sin t)$

Factor and simplify.

9. $\sin x \cos x + \cos^2 x$

10. $\tan^2 \theta - \cot^2 \theta$

11. $\sin^4 x - \cos^4 x$

12. $4 \sin^2 y + 8 \sin y + 4$

13. $2 \cos^2 x + \cos x - 3$

14. $3 \cot^2 \beta + 6 \cot \beta + 3$

15. $\sin^3 x + 27$

16. $1 - 125 \tan^3 s$

Simplify.

17. $\dfrac{\sin^2 x \cos x}{\cos^2 x \sin x}$

18. $\dfrac{30 \sin^3 x \cos x}{6 \cos^2 x \sin x}$

19. $\dfrac{\sin^2 x + 2 \sin x + 1}{\sin x + 1}$

20. $\dfrac{\cos^2 \alpha - 1}{\cos \alpha + 1}$

21. $\dfrac{4 \tan t \sec t + 2 \sec t}{6 \tan t \sec t + 2 \sec t}$

22. $\dfrac{\csc (-x)}{\cot (-x)}$

23. $\dfrac{\sin^4 x - \cos^4 x}{\sin^2 x - \cos^2 x}$

24. $\dfrac{4 \cos^3 x}{\sin^2 x} \cdot \left(\dfrac{\sin x}{4 \cos x}\right)^2$

25. $\dfrac{5 \cos \phi}{\sin^2 \phi} \cdot \dfrac{\sin^2 \phi - \sin \phi \cos \phi}{\sin^2 \phi - \cos^2 \phi}$

26. $\dfrac{\tan^2 y}{\sec y} \div \dfrac{3 \tan^3 y}{\sec y}$

27. $\dfrac{1}{\sin^2 s - \cos^2 s} - \dfrac{2}{\cos s - \sin s}$

28. $\left(\dfrac{\sin x}{\cos x}\right)^2 - \dfrac{1}{\cos^2 x}$

29. $\dfrac{\sin^2 \theta - 9}{2 \cos \theta + 1} \cdot \dfrac{10 \cos \theta + 5}{3 \sin \theta + 9}$

30. $\dfrac{9 \cos^2 \alpha - 25}{2 \cos \alpha - 2} \cdot \dfrac{\cos^2 \alpha - 1}{6 \cos \alpha - 10}$

Simplify. Assume that all radicands are nonnegative.

31. $\sqrt{\sin^2 x \cos x} \cdot \sqrt{\cos x}$

32. $\sqrt{\cos^2 x \sin x} \cdot \sqrt{\sin x}$

33. $\sqrt{\cos \alpha \sin^2 \alpha} - \sqrt{\cos^3 \alpha}$

34. $\sqrt{\tan^2 x - 2 \tan x \sin x + \sin^2 x}$

35. $(1 - \sqrt{\sin y})(\sqrt{\sin y} + 1)$

36. $\sqrt{\cos \theta}(\sqrt{2 \cos \theta} + \sqrt{\sin \theta \cos \theta})$

Rationalize the denominator.

37. $\sqrt{\dfrac{\sin x}{\cos x}}$

38. $\sqrt{\dfrac{\cos x}{\tan x}}$

39. $\sqrt{\dfrac{\cos^2 y}{2 \sin^2 y}}$

40. $\sqrt{\dfrac{1 - \cos \beta}{1 + \cos \beta}}$

Rationalize the numerator.

41. $\sqrt{\dfrac{\cos x}{\sin x}}$

42. $\sqrt{\dfrac{\sin x}{\cot x}}$

43. $\sqrt{\dfrac{1 + \sin y}{1 - \sin y}}$

44. $\sqrt{\dfrac{\cos^2 x}{2 \sin^2 x}}$

Use the given substitution to express the given radical expression as a trigonometric function without radicals. Assume that $a > 0$ and $0 < \theta < \pi/2$. Then find expressions for the indicated trigonometric functions.

45. Let $x = a \sin \theta$ in $\sqrt{a^2 - x^2}$. Then find $\cos \theta$ and $\tan \theta$.

46. Let $x = 2 \tan \theta$ in $\sqrt{4 + x^2}$. Then find $\sin \theta$ and $\cos \theta$.

47. Let $x = 3 \sec \theta$ in $\sqrt{x^2 - 9}$. Then find $\sin \theta$ and $\cos \theta$.

48. Let $x = a \sec \theta$ in $\sqrt{x^2 - a^2}$. Then find $\sin \theta$ and $\cos \theta$.

Use the given substitution to express the given radical expression as a trigonometric function without radicals. Assume that $0 < \theta < \pi/2$.

49. Let $x = \sin \theta$ in $\dfrac{x^2}{\sqrt{1 - x^2}}$.

50. Let $x = 4 \sec \theta$ in $\dfrac{\sqrt{x^2 - 16}}{x^2}$.

Use the sum and difference identities to evaluate exactly.

51. $\sin \dfrac{\pi}{12}$

52. $\cos 75°$

53. $\tan 105°$

54. $\tan \dfrac{5\pi}{12}$

55. $\cos 15°$

56. $\sin \dfrac{7\pi}{12}$

First write each of the following as a trigonometric function of a single angle; then evaluate.

57. $\sin 37° \cos 22° + \cos 37° \sin 22°$

58. $\cos 83° \cos 53° + \sin 83° \sin 53°$

59. $\dfrac{\tan 20° + \tan 32°}{1 - \tan 20° \tan 32°}$

60. $\dfrac{\tan 35° - \tan 12°}{1 + \tan 35° \tan 12°}$

61. Derive the formula for the tangent of a sum.

62. Derive the formula for the tangent of a difference.

Assuming that $\sin u = \frac{3}{5}$ and $\sin v = \frac{4}{5}$ and that u and v are between 0 and $\pi/2$, evaluate each of the following exactly.

63. $\cos (u + v)$

64. $\tan (u - v)$

65. $\sin (u - v)$

Assuming that $\sin \theta = 0.6249$ and $\cos \phi = 0.1102$ and that both θ and ϕ are first-quadrant angles, evaluate each of the following.

66. $\sin (\theta - \phi)$

67. $\tan (\theta + \phi)$

68. $\cos (\theta + \phi)$

Simplify.

69. $\sin (\alpha + \beta) + \sin (\alpha - \beta)$

70. $\cos (\alpha + \beta) - \cos (\alpha - \beta)$

71. $\cos (u + v) \cos v + \sin (u + v) \sin v$

72. $\sin (u - v) \cos v + \cos (u - v) \sin v$

Technology Connection

73. Check your answers to each of Exercises 17–30 by graphing the original expression and the simplified result in the same window.

74. Check your solutions to each of Exercises 51–56 with a calculator.

Collaborative Discussion and Writing

75. What is the difference between a trigonometric equation that is an identity and a trigonometric equation that is not an identity? Give an example of each.

76. Why is it possible to use a graph to *disprove* that an equation is an identity but not to *prove* that one is?

Skill Maintenance

Solve.

77. $2x - 3 = 2\left(x - \frac{3}{2}\right)$ **78.** $x - 7 = x + 3.4$

Given that $\sin 31° = 0.5150$ and $\cos 31° = 0.8572$, find the specified function value.

79. $\sec 59°$ **80.** $\tan 59°$

Synthesis

Angles Between Lines.   One of the identities gives an easy way to find an angle formed by two lines. Consider two lines with equations $l_1: y = m_1x + b_1$ and $l_2: y = m_2x + b_2$.

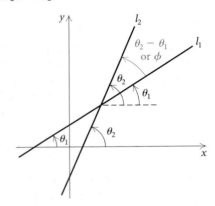

The slopes m_1 and m_2 are the tangents of the angles θ_1 and θ_2 that the lines form with the positive direction of the x-axis. Thus we have $m_1 = \tan \theta_1$ and $m_2 = \tan \theta_2$. To find the measure of $\theta_2 - \theta_1$, or ϕ, we proceed as follows:

$$\tan \phi = \tan (\theta_2 - \theta_1)$$
$$= \frac{\tan \theta_2 - \tan \theta_1}{1 + \tan \theta_2 \tan \theta_1}$$
$$= \frac{m_2 - m_1}{1 + m_2 m_1}.$$

This formula also holds when the lines are taken in the reverse order. When ϕ is acute, $\tan \phi$ will be positive. When ϕ is obtuse, $\tan \phi$ will be negative.

Find the measure of the angle from l_1 to l_2.

81. $l_1: 2x = 3 - 2y$,
$\quad l_2: x + y = 5$

82. $l_1: 3y = \sqrt{3}x + 3$,
$\quad l_2: y = \sqrt{3}x + 2$

83. $l_1: y = 3$,
$\quad l_2: x + y = 5$

84. $l_1: 2x + y - 4 = 0$,
$\quad l_2: y - 2x + 5 = 0$

85. *Circus Guy Wire.*   In a circus, a guy wire A is attached to the top of a 30-ft pole. Wire B is used for performers to walk up to the tight wire, 10 ft above the ground. Find the angle ϕ between the wires if they are attached to the ground 40 ft from the pole.

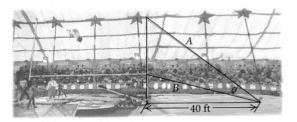

86. Given that $f(x) = \sin x$, show that

$$\frac{f(x + h) - f(x)}{h} = \sin x \left(\frac{\cos h - 1}{h}\right) + \cos x \left(\frac{\sin h}{h}\right).$$

87. Given that $f(x) = \cos x$, show that

$$\frac{f(x + h) - f(x)}{h} = \cos x \left(\frac{\cos h - 1}{h}\right) - \sin x \left(\frac{\sin h}{h}\right).$$

Show that each of the following is not an identity by finding a replacement or replacements for which the sides of the equation do not name the same number.

88. $\sqrt{\sin^2 \theta} = \sin \theta$

89. $\dfrac{\sin 5x}{x} = \sin 5$

90. $\sin (-x) = \sin x$

91. $\cos (2\alpha) = 2 \cos \alpha$

92. $\tan^2 \theta + \cot^2 \theta = 1$

93. $\dfrac{\cos 6x}{\cos x} = 6$

Find the slope of line l_1, where m_2 is the slope of line l_2 and ϕ is the smallest positive angle from l_1 to l_2.

94. $m_2 = \frac{4}{3}$, $\phi = 45°$

95. $m_2 = \frac{2}{3}$, $\phi = 30°$

96. Line l_1 contains the points $(-2, 4)$ and $(5, -1)$. Find the slope of line l_2 such that the angle from l_1 to l_2 is $45°$.

97. Line l_1 contains the points $(-3, 7)$ and $(-3, -2)$. Line l_2 contains $(0, -4)$ and $(2, 6)$. Find the smallest positive angle from l_1 to l_2.

98. Find an identity for $\sin 2\theta$. (*Hint:* $2\theta = \theta + \theta$.)

99. Find an identity for $\cos 2\theta$. (*Hint:* $2\theta = \theta + \theta$.)

Derive the identity.

100. $\sin\left(x - \dfrac{3\pi}{2}\right) = \cos x$

101. $\tan\left(x + \dfrac{\pi}{4}\right) = \dfrac{1 + \tan x}{1 - \tan x}$

102. $\dfrac{\sin(\alpha + \beta)}{\cos(\alpha - \beta)} = \dfrac{\tan \alpha + \tan \beta}{1 + \tan \alpha \tan \beta}$

103. $\sin(\alpha + \beta) + \sin(\alpha - \beta) = 2 \sin \alpha \cos \beta$

6.2

Identities: Cofunction, Double-Angle, and Half-Angle

- *Use cofunction identities to derive other identities.*
- *Use the double-angle identities to find function values of twice an angle when one function value is known for that angle.*
- *Use the half-angle identities to find function values of half an angle when one function value is known for that angle.*
- *Simplify trigonometric expressions using the double-angle and half-angle identities.*

Cofunction Identities

Each of the identities listed below yields a conversion to a *cofunction*. For this reason, we call them cofunction identities.

Cofunction Identities

$$\sin\left(\frac{\pi}{2} - x\right) = \cos x, \qquad \cos\left(\frac{\pi}{2} - x\right) = \sin x,$$

$$\tan\left(\frac{\pi}{2} - x\right) = \cot x, \qquad \cot\left(\frac{\pi}{2} - x\right) = \tan x,$$

$$\sec\left(\frac{\pi}{2} - x\right) = \csc x, \qquad \csc\left(\frac{\pi}{2} - x\right) = \sec x$$

We verified the first two of these identities in Section 6.1. The other four can be proved using the first two and the definitions of the trigonometric functions. These identities hold for all real numbers, and thus, for all angle measures, but if we restrict θ to values such that $0° < \theta < 90°$, or $0 < \theta < \pi/2$, then we have a special application to the acute angles of a right triangle.

Comparing graphs can lead to possible identities. On the left below, we see that the graph of $y = \sin(x + \pi/2)$ is a translation of the graph of $y = \sin x$ to the left $\pi/2$ units. On the right, we see the graph of $y = \cos x$.

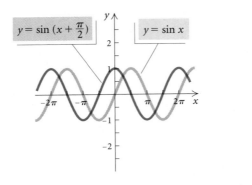
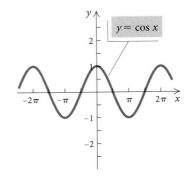

Comparing the graphs, we observe a possible identity:

$$\sin\left(x + \frac{\pi}{2}\right) = \cos x.$$

The identity can be proved using the identity for the sine of a sum developed in Section 6.1.

EXAMPLE 1 Prove the identity $\sin(x + \pi/2) = \cos x$.

Solution

$$\sin\left(x + \frac{\pi}{2}\right) = \sin x \cos \frac{\pi}{2} + \cos x \sin \frac{\pi}{2} \qquad \text{Using } \sin(u + v) = \\ \sin u \cos v + \\ \cos u \sin v$$

$$= \sin x \cdot 0 + \cos x \cdot 1$$

$$= \cos x$$

We now state four more cofunction identities. These new identities that involve the sine and cosine functions can be verified using previously established identities as seen in Example 1.

Cofunction Identities for the Sine and Cosine

$$\sin\left(x \pm \frac{\pi}{2}\right) = \pm\cos x, \qquad \cos\left(x \pm \frac{\pi}{2}\right) = \mp\sin x$$

EXAMPLE 2 Find an identity for each of the following.

a) $\tan\left(x + \dfrac{\pi}{2}\right)$

b) $\sec(x - 90°)$

Solution

a) We have

$$\tan\left(x + \frac{\pi}{2}\right) = \frac{\sin\left(x + \dfrac{\pi}{2}\right)}{\cos\left(x + \dfrac{\pi}{2}\right)} \qquad \text{Using } \tan x = \frac{\sin x}{\cos x}$$

$$= \frac{\cos x}{-\sin x} \qquad \text{Using cofunction identities}$$

$$= -\cot x.$$

Thus the identity we seek is

$$\tan\left(x + \frac{\pi}{2}\right) = -\cot x.$$

b) We have

$$\sec\,(x - 90°) = \frac{1}{\cos\,(x - 90°)} = \frac{1}{\sin x} = \csc x.$$

Thus, $\sec\,(x - 90°) = \csc x.$

Double-Angle Identities

If we double an angle of measure x, the new angle will have measure $2x$. **Double-angle identities** give trigonometric function values of $2x$ in terms of function values of x. To develop these identities, we will use the sum formulas from the preceding section. We first develop a formula for $\sin 2x$. Recall that

$$\sin\,(u + v) = \sin u \cos v + \cos u \sin v.$$

We will consider a number x and substitute it for both u and v in this identity. Doing so gives us

$$\sin\,(x + x) = \sin 2x$$
$$= \sin x \cos x + \cos x \sin x$$
$$= 2 \sin x \cos x.$$

Our first double-angle identity is thus

$$\mathbf{\sin 2x = 2 \sin x \cos x.}$$

Double-angle identities for the cosine and tangent functions can be derived in much the same way as the identity above:

$$\mathbf{\cos 2x = \cos^2 x - \sin^2 x,} \qquad \mathbf{\tan 2x = \frac{2 \tan x}{1 - \tan^2 x}.}$$

Technology Connection

Graphing calculators provide visual partial checks of identities. We can graph

$$y_1 = \sin 2x, \quad \text{and}$$
$$y_2 = 2 \sin x \cos x$$

using the "line"-graph style for y_1 and the "path"-graph style for y_2 and see that they appear to have the same graph. We can also use the TABLE feature.

$y_1 = \sin 2x, \quad y_2 = 2 \sin x \cos x$

X	Y₁	Y₂
−6.283	2E−13	0
−5.498	1	1
−4.712	0	0
−3.927	−1	−1
−3.142	0	0
−2.356	1	1
−1.571	0	0

X = −1.57079632679

EXAMPLE 3   Given that $\tan \theta = -\frac{3}{4}$ and θ is in quadrant II, find each of the following.

a) $\sin 2\theta$ **b)** $\cos 2\theta$

c) $\tan 2\theta$ **d)** The quadrant in which 2θ lies

Solution   By drawing a diagram as shown, we find that

$$\sin \theta = \frac{3}{5}$$

and

$$\cos \theta = -\frac{4}{5}.$$

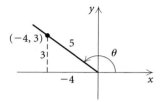

Thus we have the following.

a) $\sin 2\theta = 2 \sin \theta \cos \theta = 2 \cdot \frac{3}{5} \cdot \left(-\frac{4}{5}\right) = -\frac{24}{25}$

b) $\cos 2\theta = \cos^2 \theta - \sin^2 \theta = \left(-\frac{4}{5}\right)^2 - \left(\frac{3}{5}\right)^2 = \frac{16}{25} - \frac{9}{25} = \frac{7}{25}$

c) $\tan 2\theta = \dfrac{2 \tan \theta}{1 - \tan^2 \theta} = \dfrac{2 \cdot \left(-\frac{3}{4}\right)}{1 - \left(-\frac{3}{4}\right)^2} = \dfrac{-\frac{3}{2}}{1 - \frac{9}{16}} = -\frac{3}{2} \cdot \frac{16}{7} = -\frac{24}{7}$

Note that $\tan 2\theta$ could have been found more easily in this case by simply dividing:

$$\tan 2\theta = \frac{\sin 2\theta}{\cos 2\theta} = \frac{-\frac{24}{25}}{\frac{7}{25}} = -\frac{24}{7}.$$

d) Since $\sin 2\theta$ is negative and $\cos 2\theta$ is positive, we know that 2θ is in quadrant IV.

Two other useful identities for $\cos 2x$ can be derived easily, as follows.

$$\cos 2x = \cos^2 x - \sin^2 x \qquad\qquad \cos 2x = \cos^2 x - \sin^2 x$$
$$= (1 - \sin^2 x) - \sin^2 x \qquad\qquad = \cos^2 x - (1 - \cos^2 x)$$
$$= 1 - 2 \sin^2 x \qquad\qquad = 2 \cos^2 x - 1$$

Double-Angle Identities

$$\sin 2x = 2 \sin x \cos x, \qquad\qquad \cos 2x = \cos^2 x - \sin^2 x$$
$$\qquad\qquad = 1 - 2 \sin^2 x$$
$$\tan 2x = \frac{2 \tan x}{1 - \tan^2 x} \qquad\qquad = 2 \cos^2 x - 1$$

Solving the last two cosine double-angle identities for $\sin^2 x$ and $\cos^2 x$, respectively, we obtain two more identities:

$$\sin^2 x = \frac{1 - \cos 2x}{2} \quad \text{and} \quad \cos^2 x = \frac{1 + \cos 2x}{2}.$$

Using division and these two identities gives us the following useful identity:

$$\tan^2 x = \frac{1 - \cos 2x}{1 + \cos 2x}.$$

EXAMPLE 4 Find an equivalent expression for each of the following.

a) $\sin 3\theta$ in terms of function values of θ

b) $\cos^3 x$ in terms of function values of x or $2x$, raised only to the first power

Solution

a) $\sin 3\theta = \sin (2\theta + \theta)$

$\qquad = \sin 2\theta \cos \theta + \cos 2\theta \sin \theta$

$\qquad = (2 \sin \theta \cos \theta) \cos \theta + (2 \cos^2 \theta - 1) \sin \theta$

$\qquad\qquad$ Using $\sin 2\theta = 2 \sin \theta \cos \theta$ and $\cos 2\theta = 2 \cos^2 \theta - 1$

$\qquad = 2 \sin \theta \cos^2 \theta + 2 \sin \theta \cos^2 \theta - \sin \theta$

$\qquad = 4 \sin \theta \cos^2 \theta - \sin \theta$

We could also substitute $\cos^2 \theta - \sin^2 \theta$ or $1 - 2 \sin^2 \theta$ for $\cos 2\theta$. Each substitution leads to a different result, but all results are equivalent.

b) $\cos^3 x = \cos^2 x \cos x$

$$= \frac{1 + \cos 2x}{2} \cos x = \frac{\cos x + \cos x \cos 2x}{2}$$

Half-Angle Identities

If we take half of an angle of measure x, the new angle will have measure $x/2$. **Half-angle identities** give trigonometric function values of $x/2$ in terms of function values of x. To develop these identities, we take square roots and replace x with $x/2$. For example,

$$\sin^2 x = \frac{1 - \cos 2x}{2} \rightarrow \left| \sin \frac{x}{2} \right| = \sqrt{\frac{1 - \cos x}{2}}.$$

The formula on the right above is called a *half-angle formula*. We can eliminate the absolute-value sign by using $\pm$ signs, with the understanding that our use of $+$ and $-$ depends on the quadrant in which the angle $x/2$ lies. Half-angle identities for the cosine and tangent functions can be derived in a similar manner. Two additional formulas for the half-angle tangent identity are listed at the top of the next page.

Half-Angle Identities

$$\sin \frac{x}{2} = \pm \sqrt{\frac{1 - \cos x}{2}},$$

$$\cos \frac{x}{2} = \pm \sqrt{\frac{1 + \cos x}{2}},$$

$$\tan \frac{x}{2} = \pm \sqrt{\frac{1 - \cos x}{1 + \cos x}}$$

$$= \frac{\sin x}{1 + \cos x} = \frac{1 - \cos x}{\sin x}$$

EXAMPLE 5 Find $\tan (\pi/8)$ exactly.

Solution

$$\tan \frac{\pi}{8} = \tan \frac{\frac{\pi}{4}}{2} = \frac{\sin \frac{\pi}{4}}{1 + \cos \frac{\pi}{4}} = \frac{\frac{\sqrt{2}}{2}}{1 + \frac{\sqrt{2}}{2}} = \frac{\frac{\sqrt{2}}{2}}{\frac{2 + \sqrt{2}}{2}}$$

$$= \frac{\sqrt{2}}{2 + \sqrt{2}} = \frac{\sqrt{2}}{2 + \sqrt{2}} \cdot \frac{2 - \sqrt{2}}{2 - \sqrt{2}} = \sqrt{2} - 1$$

The identities that we have developed are also useful for simplifying trigonometric expressions.

EXAMPLE 6 Simplify each of the following.

a) $\dfrac{\sin x \cos x}{\frac{1}{2} \cos 2x}$ **b)** $2 \sin^2 \dfrac{x}{2} + \cos x$

Solution

a) We can obtain $2 \sin x \cos x$ in the numerator by multiplying the expression by $\frac{2}{2}$:

$$\frac{\sin x \cos x}{\frac{1}{2} \cos 2x} = \frac{2}{2} \cdot \frac{\sin x \cos x}{\frac{1}{2} \cos 2x} = \frac{2 \sin x \cos x}{\cos 2x}$$

$$= \frac{\sin 2x}{\cos 2x} \qquad \text{Using } \sin 2x = 2 \sin x \cos x$$

$$= \tan 2x.$$

Technology Connection

Here we show a partial check of Example 6(b) using a graph and a table.

$$y_1 = 2 \sin^2 \frac{x}{2} + \cos x, \quad y_2 = 1$$

X	Y1	Y2
−6.283	1	1
−5.498	1	1
−4.712	1	1
−3.927	1	1
−3.142	1	1
−2.356	1	1
−1.571	1	1

X = −6.28318530718
ΔTbl = π/4

b) We have

$$2 \sin^2 \frac{x}{2} + \cos x = 2\left(\frac{1 - \cos x}{2}\right) + \cos x$$

$$\text{Using } \sin \frac{x}{2} = \pm\sqrt{\frac{1 - \cos x}{2}}, \text{ or } \sin^2 \frac{x}{2} = \frac{1 - \cos x}{2}$$

$$= 1 - \cos x + \cos x = 1.$$

Exercise Set

1. Given that $\sin (3\pi/10) \approx 0.8090$ and $\cos (3\pi/10) \approx 0.5878$, find each of the following.

a) The other four function values for $3\pi/10$
b) The six function values for $\pi/5$

2. Given that

$$\sin \frac{\pi}{12} = \frac{\sqrt{2 - \sqrt{3}}}{2} \quad \text{and} \quad \cos \frac{\pi}{12} = \frac{\sqrt{2 + \sqrt{3}}}{2},$$

find exact answers for each of the following.

a) The other four function values for $\pi/12$
b) The six function values for $5\pi/12$

3. Given that $\sin \theta = \frac{1}{3}$ and that the terminal side is in quadrant II, find exact answers for each of the following.

a) The other function values for θ
b) The six function values for $\pi/2 - \theta$
c) The six function values for $\theta - \pi/2$

4. Given that $\cos \phi = \frac{4}{5}$ and that the terminal side is in quadrant IV, find exact answers for each of the following.

a) The other function values for ϕ
b) The six function values for $\pi/2 - \phi$
c) The six function values for $\phi + \pi/2$

Find an equivalent expression for each of the following.

5. $\sec \left(x + \frac{\pi}{2}\right)$

6. $\cot \left(x - \frac{\pi}{2}\right)$

7. $\tan \left(x - \frac{\pi}{2}\right)$

8. $\csc \left(x + \frac{\pi}{2}\right)$

Find $\sin 2\theta$, $\cos 2\theta$, $\tan 2\theta$, and the quadrant in which 2θ lies.

9. $\sin \theta = \frac{4}{5}$, θ in quadrant I

10. $\cos \theta = \frac{5}{13}$, θ in quadrant I

11. $\cos \theta = -\frac{3}{5}$, θ in quadrant III

12. $\tan \theta = -\frac{15}{8}$, θ in quadrant II

13. $\tan \theta = -\frac{5}{12}$, θ in quadrant II

14. $\sin \theta = -\frac{\sqrt{10}}{10}$, θ in quadrant IV

15. Find an equivalent expression for $\cos 4x$ in terms of function values of x.

16. Find an equivalent expression for $\sin^4 \theta$ in terms of function values of θ, 2θ, or 4θ, raised only to the first power.

Use the half-angle identities to evaluate exactly.

17. $\cos 15°$

18. $\tan 67.5°$

19. $\sin 112.5°$

20. $\cos \frac{\pi}{8}$

21. $\tan 75°$

22. $\sin \frac{5\pi}{12}$

Given that $\sin\theta = 0.3416$ *and* θ *is in quadrant I, find each of the following using identities.*

23. $\sin 2\theta$

24. $\cos \dfrac{\theta}{2}$

25. $\sin \dfrac{\theta}{2}$

26. $\sin 4\theta$

Simplify.

27. $2\cos^2 \dfrac{x}{2} - 1$

28. $\cos^4 x - \sin^4 x$

29. $(\sin x - \cos x)^2 + \sin 2x$

30. $(\sin x + \cos x)^2$

31. $\dfrac{2 - \sec^2 x}{\sec^2 x}$

32. $\dfrac{1 + \sin 2x + \cos 2x}{1 + \sin 2x - \cos 2x}$

33. $(-4\cos x \sin x + 2\cos 2x)^2 +$
$(2\cos 2x + 4\sin x \cos x)^2$

34. $2\sin x \cos^3 x - 2\sin^3 x \cos x$

Technology Connection

In Exercises 35–38, use a graphing calculator to determine which of the following expressions asserts an identity. Then prove the identity algebraically.

35. $\dfrac{\cos 2x}{\cos x - \sin x} = \cdots$

a) $1 + \cos x$

b) $\cos x - \sin x$

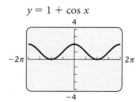
$y = 1 + \cos x$

$y = \cos x - \sin x$

c) $-\cot x$

d) $\sin x\,(\cot x + 1)$

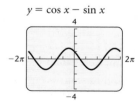
$y = -\cot x$

$y = \sin x\,(\cot x + 1)$

36. $2\cos^2 \dfrac{x}{2} = \cdots$

a) $\sin x\,(\csc x + \tan x)$ **b)** $\sin x - 2\cos x$
c) $2(\cos^2 x - \sin^2 x)$ **d)** $1 + \cos x$

37. $\dfrac{\sin 2x}{2\cos x} = \cdots$

a) $\cos x$ **b)** $\tan x$
c) $\cos x + \sin x$ **d)** $\sin x$

38. $2\sin \dfrac{\theta}{2}\cos \dfrac{\theta}{2} = \cdots$

a) $\cos^2 \theta$ **b)** $\sin \dfrac{\theta}{2}$

c) $\sin \theta$ **d)** $\sin \theta - \cos \theta$

Collaborative Discussion and Writing

39. Discuss and compare the graphs of $y = \sin x$, $y = \sin 2x$, and $y = \sin (x/2)$.

40. Find all errors in the following:
$$2\sin^2 2x + \cos 4x$$
$$= 2(2\sin x \cos x)^2 + 2\cos 2x$$
$$= 8\sin^2 x \cos^2 x + 2(\cos^2 x + \sin^2 x)$$
$$= 8\sin^2 x \cos^2 x + 2.$$

Skill Maintenance

Complete the identity.

41. $1 - \cos^2 x =$

42. $\sec^2 x - \tan^2 x =$

43. $\sin^2 x - 1 =$

44. $1 + \cot^2 x =$

Synthesis

45. Given that $\cos 51° \approx 0.6293$, find the six function values for $141°$.

Simplify.

46. $\sin\left(\dfrac{\pi}{2} - x\right)[\sec x - \cos x]$

47. $\cos(\pi - x) + \cot x \sin\left(x - \dfrac{\pi}{2}\right)$

48. $\dfrac{\cos x - \sin\left(\dfrac{\pi}{2} - x\right)\sin x}{\cos x - \cos(\pi - x)\tan x}$

49. $\dfrac{\cos^2 y \sin\left(y + \dfrac{\pi}{2}\right)}{\sin^2 y \sin\left(\dfrac{\pi}{2} - y\right)}$

Find sin θ, cos θ, and tan θ under the given conditions.

50. $\cos 2\theta = \dfrac{7}{12}, \ \dfrac{3\pi}{2} \le 2\theta \le 2\pi$

51. $\tan \dfrac{\theta}{2} = -\dfrac{5}{3}, \ \pi < \theta \le \dfrac{3\pi}{2}$

52. *Nautical Mile.* *Latitude* is used to measure north–south location on the earth between the equator and the poles. For example, Chicago has latitude 42°N. (See the figure.) In Great Britain, the *nautical mile* is defined as the length of a minute of arc of the earth's radius. Since the earth is flattened slightly at the poles, a British nautical mile varies with latitude. In fact, it is given, in feet, by the function

$$N(\phi) = 6066 - 31 \cos 2\phi,$$

where ϕ is the latitude in degrees.

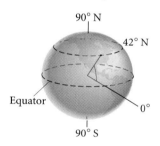

a) What is the length of a British nautical mile at Chicago?

b) What is the length of a British nautical mile at the North Pole?
c) Express $N(\phi)$ in terms of $\cos \phi$ only. That is, do not use the double angle.

53. *Acceleration Due to Gravity.* The acceleration due to gravity is often denoted by g in a formula such as $S = \frac{1}{2}gt^2$, where S is the distance that an object falls in time t. The number g relates to motion near the earth's surface and is usually considered constant. In fact, however, g is not constant, but varies slightly with latitude. If ϕ stands for latitude, in degrees, g is given with good approximation by the formula

$$g = 9.78049(1 + 0.005288 \sin^2 \phi - 0.000006 \sin^2 2\phi),$$

where g is measured in meters per second per second at sea level.

a) Chicago has latitude 42°N. Find g.
b) Philadelphia has latitude 40°N. Find g.
c) Express g in terms of $\sin \phi$ only. That is, eliminate the double angle.

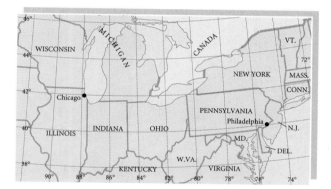

6.3

Proving Trigonometric Identities

• *Prove identities using other identities.*

The Logic of Proving Identities

We outline two algebraic methods for proving identities.

Method 1. Start with either the left or the right side of the equation and obtain the other side. For example, suppose you are trying to prove that the equation $P = Q$ is an identity. You might try to produce a string of statements (R_1, R_2, ... or T_1, T_2, ...) like the following, which start with P and end with Q or start with Q and end with P:

$$
\begin{aligned}
P &= R_1 \quad \text{or} \quad Q = T_1 \\
&= R_2 \qquad\qquad\; = T_2 \\
&\;\;\vdots \qquad\qquad\qquad \vdots \\
&= Q \qquad\qquad\;\;\, = P.
\end{aligned}
$$

Method 2. Work with each side separately until you obtain the same expression. For example, suppose you are trying to prove that $P = Q$ is an identity. You might be able to produce two strings of statements like the following, each ending with the same statement S.

$$
\begin{aligned}
P &= R_1 \qquad Q = T_1 \\
&= R_2 \qquad\quad\; = T_2 \\
&\;\;\vdots \qquad\qquad\;\; \vdots \\
&= S \qquad\quad\;\; = S.
\end{aligned}
$$

The number of steps in each string might be different, but in each case the result is S.

A first step in learning to prove identities is to have at hand a list of the identities that you have already learned. Such a list is on the inside back cover of this text. Ask your instructor which ones you are expected to memorize. The more identities you prove, the easier it will be to prove new ones. A list of helpful hints follows.

Hints for Proving Identities

1. Use method 1 or 2 above.
2. Work with the more complex side first.
3. Carry out any algebraic manipulations, such as adding, subtracting, multiplying, or factoring.
4. Multiplying by 1 can be helpful when rational expressions are involved.
5. Converting all expressions to sines and cosines is often helpful.
6. Try something! Put your pencil to work and get involved. You will be amazed at how often this leads to success.

Proving Identities

In what follows, method 1 is used in Examples 1–3 and method 2 is used in Examples 4 and 5.

EXAMPLE 1 Prove the identity $1 + \sin 2\theta = (\sin \theta + \cos \theta)^2$.

Solution Let's use method 1. We begin with the right side and obtain the left side:

$$(\sin \theta + \cos \theta)^2 = \sin^2 \theta + 2 \sin \theta \cos \theta + \cos^2 \theta \qquad \text{Squaring}$$
$$= 1 + 2 \sin \theta \cos \theta \qquad \text{Recalling the identity } \sin^2 x + \cos^2 x = 1 \text{ and substituting}$$
$$= 1 + \sin 2\theta. \qquad \text{Using } \sin 2x = 2 \sin x \cos x$$

We could also begin with the left side and obtain the right side:

$$1 + \sin 2\theta = 1 + 2 \sin \theta \cos \theta \qquad \text{Using } \sin 2x = 2 \sin x \cos x$$
$$= \sin^2 \theta + 2 \sin \theta \cos \theta + \cos^2 \theta \qquad \text{Replacing 1 with } \sin^2 \theta + \cos^2 \theta$$
$$= (\sin \theta + \cos \theta)^2. \qquad \text{Factoring} \qquad \blacksquare$$

EXAMPLE 2 Prove the identity

$$\frac{\sec t - 1}{t \sec t} = \frac{1 - \cos t}{t}.$$

Solution We use method 1, starting with the left side. Note that the left side involves sec t, whereas the right side involves cos t, so it might be wise to make use of a basic identity that involves these two expressions: $\sec t = 1/\cos t$.

$$\frac{\sec t - 1}{t \sec t} = \frac{\dfrac{1}{\cos t} - 1}{t \dfrac{1}{\cos t}} \qquad \text{Substituting } 1/\cos t \text{ for sec } t$$
$$= \left(\frac{1}{\cos t} - 1 \right) \cdot \frac{\cos t}{t}$$
$$= \frac{1}{t} - \frac{\cos t}{t} \qquad \text{Multiplying}$$
$$= \frac{1 - \cos t}{t}$$

We started with the left side and obtained the right side, so the proof is complete. $\blacksquare$

EXAMPLE 3 Prove the identity

$$\frac{\sin 2x}{\sin x} - \frac{\cos 2x}{\cos x} = \sec x.$$

Solution

$$\frac{\sin 2x}{\sin x} - \frac{\cos 2x}{\cos x} = \frac{2 \sin x \cos x}{\sin x} - \frac{\cos^2 x - \sin^2 x}{\cos x} \qquad \text{Using double-angle identities}$$

$$= 2 \cos x - \frac{\cos^2 x - \sin^2 x}{\cos x} \qquad \text{Simplifying}$$

$$= \frac{2 \cos^2 x}{\cos x} - \frac{\cos^2 x - \sin^2 x}{\cos x} \qquad \text{Multiplying 2 cos } x \text{ by 1, or cos } x/\cos x$$

$$= \frac{2 \cos^2 x - \cos^2 x + \sin^2 x}{\cos x} \qquad \text{Subtracting}$$

$$= \frac{\cos^2 x + \sin^2 x}{\cos x}$$

$$= \frac{1}{\cos x} \qquad \text{Using a Pythagorean identity}$$

$$= \sec x \qquad \text{Recalling a basic identity}$$

EXAMPLE 4   Prove the identity

$$\sin^2 x \tan^2 x = \tan^2 x - \sin^2 x.$$

Solution   For this proof, we are going to work with each side separately using method 2. We try to obtain the same expression on each side. In actual practice, you might work on one side for awhile, then work on the other side, and then go back to the first side. In other words, you work back and forth until you arrive at the same expression. Let's start with the right side:

$$\tan^2 x - \sin^2 x = \frac{\sin^2 x}{\cos^2 x} - \sin^2 x \qquad \text{Recalling the identity } \tan x = \frac{\sin x}{\cos x} \text{ and substituting}$$

$$= \frac{\sin^2 x}{\cos^2 x} - \sin^2 x \cdot \frac{\cos^2 x}{\cos^2 x} \qquad \text{Multiplying by 1 in order to subtract}$$

$$= \frac{\sin^2 x - \sin^2 x \cos^2 x}{\cos^2 x} \qquad \text{Carrying out the subtraction}$$

$$= \frac{\sin^2 x \, (1 - \cos^2 x)}{\cos^2 x} \qquad \text{Factoring}$$

$$= \frac{\sin^2 x \sin^2 x}{\cos^2 x} \qquad \text{Recalling the identity } 1 - \cos^2 x = \sin^2 x \text{ and substituting}$$

$$= \frac{\sin^4 x}{\cos^2 x}.$$

At this point, we stop and work with the left side, $\sin^2 x \tan^2 x$, of the original identity and try to end with the same expression that we ended with on the right side:

$$\sin^2 x \tan^2 x = \sin^2 x \frac{\sin^2 x}{\cos^2 x}$$

Recalling the identity $\tan x = \dfrac{\sin x}{\cos x}$ and substituting

$$= \frac{\sin^4 x}{\cos^2 x}.$$

We have obtained the same expression from each side, so the proof is complete.

EXAMPLE 5 Prove the identity

$$\cot \phi + \csc \phi = \frac{\sin \phi}{1 - \cos \phi}.$$

Solution We are again using method 2, beginning with the left side:

$$\cot \phi + \csc \phi = \frac{\cos \phi}{\sin \phi} + \frac{1}{\sin \phi}$$

Using basic identities

$$= \frac{1 + \cos \phi}{\sin \phi}.$$

Adding

At this point, we stop and work with the right side of the original identity:

$$\frac{\sin \phi}{1 - \cos \phi} = \frac{\sin \phi}{1 - \cos \phi} \cdot \frac{1 + \cos \phi}{1 + \cos \phi}$$

Multiplying by 1

$$= \frac{\sin \phi (1 + \cos \phi)}{1 - \cos^2 \phi}$$

$$= \frac{\sin \phi (1 + \cos \phi)}{\sin^2 \phi}$$

Using $\sin^2 x = 1 - \cos^2 x$

$$= \frac{1 + \cos \phi}{\sin \phi}.$$

Simplifying

The proof is complete since we obtained the same expression from each side.

Exercise Set 6.3

Prove each of the following identities.

1. $\sec x - \sin x \tan x = \cos x$

$$y_1 = \sec x - \sin x \tan x,$$
$$y_2 = \cos x$$

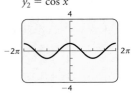

2. $\dfrac{1 + \cos \theta}{\sin \theta} + \dfrac{\sin \theta}{\cos \theta} = \dfrac{\cos \theta + 1}{\sin \theta \cos \theta}$

$$y_1 = \dfrac{1 + \cos x}{\sin x} + \dfrac{\sin x}{\cos x},$$
$$y_2 = \dfrac{\cos x + 1}{\sin x \cos x}$$

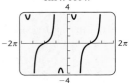

3. $\dfrac{1 - \cos x}{\sin x} = \dfrac{\sin x}{1 + \cos x}$

4. $\dfrac{1 + \tan y}{1 + \cot y} = \dfrac{\sec y}{\csc y}$

5. $\dfrac{1 + \tan \theta}{1 - \tan \theta} + \dfrac{1 + \cot \theta}{1 - \cot \theta} = 0$

6. $\dfrac{\sin x + \cos x}{\sec x + \csc x} = \dfrac{\sin x}{\sec x}$

7. $\dfrac{\cos^2 \alpha + \cot \alpha}{\cos^2 \alpha - \cot \alpha} = \dfrac{\cos^2 \alpha \tan \alpha + 1}{\cos^2 \alpha \tan \alpha - 1}$

8. $\sec 2\theta = \dfrac{\sec^2 \theta}{2 - \sec^2 \theta}$

9. $\dfrac{2 \tan \theta}{1 + \tan^2 \theta} = \sin 2\theta$

10. $\dfrac{\cos (u - v)}{\cos u \sin v} = \tan u + \cot v$

11. $1 - \cos 5\theta \cos 3\theta - \sin 5\theta \sin 3\theta = 2 \sin^2 \theta$

12. $\cos^4 x - \sin^4 x = \cos 2x$

13. $2 \sin \theta \cos^3 \theta + 2 \sin^3 \theta \cos \theta = \sin 2\theta$

14. $\dfrac{\tan 3t - \tan t}{1 + \tan 3t \tan t} = \dfrac{2 \tan t}{1 - \tan^2 t}$

15. $\dfrac{\tan x - \sin x}{2 \tan x} = \sin^2 \dfrac{x}{2}$

16. $\dfrac{\cos^3 \beta - \sin^3 \beta}{\cos \beta - \sin \beta} = \dfrac{2 + \sin 2\beta}{2}$

17. $\sin (\alpha + \beta) \sin (\alpha - \beta) = \sin^2 \alpha - \sin^2 \beta$

18. $\cos^2 x (1 - \sec^2 x) = -\sin^2 x$

19. $\tan \theta (\tan \theta + \cot \theta) = \sec^2 \theta$

20. $\dfrac{\cos \theta + \sin \theta}{\cos \theta} = 1 + \tan \theta$

21. $\dfrac{1 + \cos^2 x}{\sin^2 x} = 2 \csc^2 x - 1$

22. $\dfrac{\tan y + \cot y}{\csc y} = \sec y$

23. $\dfrac{1 + \sin x}{1 - \sin x} + \dfrac{\sin x - 1}{1 + \sin x} = 4 \sec x \tan x$

24. $\tan \theta - \cot \theta = (\sec \theta - \csc \theta)(\sin \theta + \cos \theta)$

25. $\cos^2 \alpha \cot^2 \alpha = \cot^2 \alpha - \cos^2 \alpha$

26. $\dfrac{\tan x + \cot x}{\sec x + \csc x} = \dfrac{1}{\cos x + \sin x}$

27. $2 \sin^2 \theta \cos^2 \theta + \cos^4 \theta = 1 - \sin^4 \theta$

28. $\dfrac{\cot \theta}{\csc \theta - 1} = \dfrac{\csc \theta + 1}{\cot \theta}$

29. $\dfrac{1 + \sin x}{1 - \sin x} = (\sec x + \tan x)^2$

30. $\sec^4 s - \tan^2 s = \tan^4 s + \sec^2 s$

Technology Connection

In Exercises 31–36, use a graphing calculator to determine which expression (A)–(F) on the right can be used to complete the identity. Then try to prove that identity algebraically.

31. $\dfrac{\cos x + \cot x}{1 + \csc x}$

A. $\dfrac{\sin^3 x - \cos^3 x}{\sin x - \cos x}$

32. $\cot x + \csc x$

B. $\cos x$

33. $\sin x \cos x + 1$

C. $\tan x + \cot x$

34. $2 \cos^2 x - 1$

D. $\cos^3 x + \sin^3 x$

35. $\dfrac{1}{\cot x \sin^2 x}$

E. $\dfrac{\sin x}{1 - \cos x}$

36. $(\cos x + \sin x)(1 - \sin x \cos x)$ **F.** $\cos^4 x - \sin^4 x$

Collaborative Discussion and Writing

37. What restrictions must be placed on the variable in each of the following identities? Why?

a) $\sin 2x = \dfrac{2 \tan x}{1 + \tan^2 x}$

b) $\dfrac{1 - \cos x}{\sin x} = \dfrac{\sin x}{1 + \cos x}$

c) $2 \sin x \cos^3 x + 2 \sin^3 x \cos x = \sin 2x$

38. Explain why $\tan (x + 450°)$ cannot be simplified using the tangent sum formula, but can be simplified using the sine and cosine sum formulas.

Skill Maintenance

For each function:

a) *Graph the function.*
b) *Determine whether the function is one-to-one.*
c) *If the function is one-to-one, find an equation for its inverse.*
d) *Graph the inverse of the function.*

39. $f(x) = 3x - 2$

40. $f(x) = x^3 + 1$

41. $f(x) = x^2 - 4, \ x \geq 0$

42. $f(x) = \sqrt{x + 2}$

Synthesis

Prove the identity.

43. $\ln |\tan x| = -\ln |\cot x|$

44. $\ln |\sec \theta + \tan \theta| = -\ln |\sec \theta - \tan \theta|$

45. Prove the identity

$\log (\cos x - \sin x) + \log (\cos x + \sin x) = \log \cos 2x.$

46. *Mechanics.* The following equation occurs in the study of mechanics:

$$\sin \theta = \frac{I_1 \cos \phi}{\sqrt{(I_1 \cos \phi)^2 + (I_2 \sin \phi)^2}}.$$

It can happen that $I_1 = I_2$. Assuming that this happens, simplify the equation.

47. *Alternating Current.* In the theory of alternating current, the following equation occurs:

$$R = \frac{1}{\omega C (\tan \theta + \tan \phi)}.$$

Show that this equation is equivalent to

$$R = \frac{\cos \theta \cos \phi}{\omega C \sin (\theta + \phi)}.$$

48. *Electrical Theory.* In electrical theory, the following equations occur:

$$E_1 = \sqrt{2} E_t \cos \left(\theta + \frac{\pi}{P} \right)$$

and

$$E_2 = \sqrt{2} E_t \cos \left(\theta - \frac{\pi}{P} \right).$$

Assuming that these equations hold, show that

$$\frac{E_1 + E_2}{2} = \sqrt{2} E_t \cos \theta \cos \frac{\pi}{P}$$

and

$$\frac{E_1 - E_2}{2} = -\sqrt{2} E_t \sin \theta \sin \frac{\pi}{P}.$$

6.4

Inverses of the Trigonometric Functions

INVERSE FUNCTIONS
REVIEW SECTION 4.1.

- *Find values of the inverse trigonometric functions.*
- *Simplify expressions such as sin (sin⁻¹ x) and sin⁻¹ (sin x).*
- *Simplify expressions involving compositions such as sin $\left(\cos^{-1} \frac{1}{2}\right)$ without using a calculator.*
- *Simplify expressions such as sin arctan (a/b) by making a drawing and reading off appropriate ratios.*

In this section, we develop inverse trigonometric functions. The graphs of the sine, cosine, and tangent functions follow. Do these functions have inverses that are functions? They do have inverses if they are one-to-one, which means that they pass the horizontal-line test.

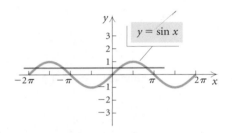
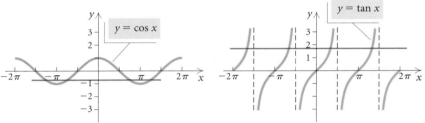

Note that for each function, a horizontal line (shown in red) crosses the graph more than once. Therefore, none of them has an inverse that is a function.

The graphs of an equation and its inverse are reflections of each other across the line $y = x$. Let's examine the graphs of the inverses of each of the three functions graphed above.

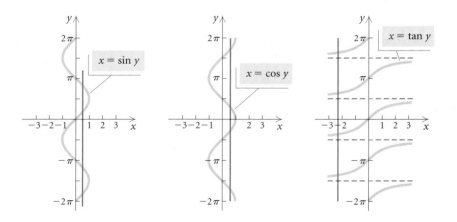

We can check again to see whether these are graphs of functions by using the vertical-line test. In each case, there is a vertical line (shown in red) that crosses the graph more than once, so each *fails* to be a function.

Restricting Ranges to Define Inverse Functions

Recall that a function like $f(x) = x^2$ does not have an inverse that is a function, but by restricting the domain of f to nonnegative numbers, we have a new squaring function, $f(x) = x^2$, $x \geq 0$, that has an inverse, $f^{-1}(x) = \sqrt{x}$. This is equivalent to restricting the range of the inverse relation to exclude ordered pairs that contain negative numbers.

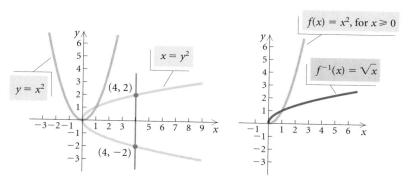

In a similar manner, we can define new trigonometric functions whose inverses are functions. We can do this by restricting either the domains of the basic trigonometric functions or the ranges of their inverse relations. This can be done in many ways, but the restrictions illustrated below with solid red curves are fairly standard in mathematics.

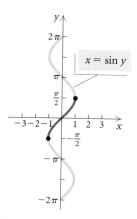

FIGURE 1

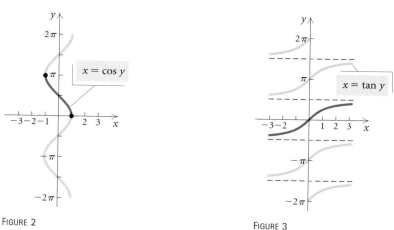

FIGURE 2

FIGURE 3

For the inverse sine function, we choose a range close to the origin that allows all inputs on the interval $[-1, 1]$ to have function values. Thus we choose the interval $[-\pi/2, \pi/2]$ for the range (Fig. 1). For the inverse cosine function, we choose a range close to the origin that allows all inputs on the interval $[-1, 1]$ to have function values. We choose the interval $[0, \pi]$ (Fig. 2). For the inverse tangent function, we choose a range close to the origin that allows all real numbers to have function values. The interval $(-\pi/2, \pi/2)$ satisfies this requirement (Fig. 3).

Inverse Trigonometric Functions

FUNCTION	DOMAIN	RANGE
$y = \sin^{-1} x$ $= \arcsin x$, where $x = \sin y$	$[-1, 1]$	$[-\pi/2, \pi/2]$
$y = \cos^{-1} x$ $= \arccos x$, where $x = \cos y$	$[-1, 1]$	$[0, \pi]$
$y = \tan^{-1} x$ $= \arctan x$, where $x = \tan y$	$(-\infty, \infty)$	$(-\pi/2, \pi/2)$

The notation arcsin x arises because the function value, y, is the length of an arc on the unit circle for which the sine is x. Either of the two kinds of notation above can be read "the inverse sine of x" or "the arc sine of x" or "the number (or angle) whose sine is x."

The notation $\sin^{-1} x$ is not exponential notation.

It does *not* mean $\dfrac{1}{\sin\ x}$!

Technology Connection:
Exploration

Inverse trigonometric functions can be graphed using a graphing calculator. Graph $y = \sin^{-1} x$ using the viewing window $[-3, 3, -\pi, \pi]$, with Xscl = 1 and Yscl = $\pi/2$. Now try graphing $y = \cos^{-1} x$ and $y = \tan^{-1} x$. Then use the graphs to confirm the domain and the range of each inverse.

The graphs of the inverse trigonometric functions are as follows.

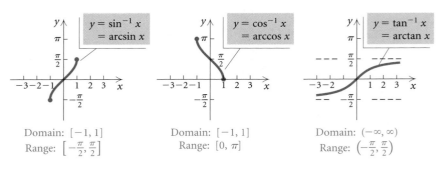

Domain: $[-1, 1]$
Range: $\left[-\frac{\pi}{2}, \frac{\pi}{2}\right]$

Domain: $[-1, 1]$
Range: $[0, \pi]$

Domain: $(-\infty, \infty)$
Range: $\left(-\frac{\pi}{2}, \frac{\pi}{2}\right)$

The following diagrams show the restricted ranges for the inverse trigonometric functions on a unit circle. Compare these graphs with the graphs above. The ranges of these functions should be memorized. The missing endpoints in the graph of the arctangent function indicate inputs that are not in the domain of the original function.

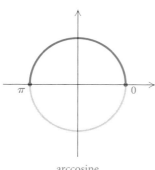

arcsine
Range $\left[-\frac{\pi}{2}, \frac{\pi}{2}\right]$

arccosine
Range $[0, \pi]$

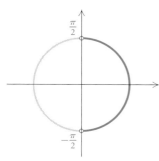

arctangent
Range $\left(-\frac{\pi}{2}, \frac{\pi}{2}\right)$

EXAMPLE 1 Find each of the following function values.

a) arcsin $\dfrac{\sqrt{2}}{2}$ **b)** $\cos^{-1}\left(-\dfrac{1}{2}\right)$ **c)** $\tan^{-1}\left(-\dfrac{\sqrt{3}}{3}\right)$

Solution

a) Another way to state "find arcsin $\sqrt{2}/2$" is to say "find β such that $\sin \beta = \sqrt{2}/2$." In the restricted range $[-\pi/2, \pi/2]$, the only number with a sine of $\sqrt{2}/2$ is $\pi/4$. Thus, arcsin $(\sqrt{2}/2) = \pi/4$, or 45°. (See Fig. 4 below.)

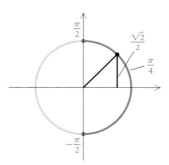

FIGURE 4

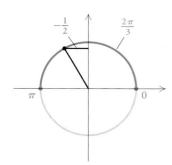

FIGURE 5

b) The only number with a cosine of $-\frac{1}{2}$ in the restricted range $[0, \pi]$ is $2\pi/3$. Thus, $\cos^{-1}\left(-\frac{1}{2}\right) = 2\pi/3$, or 120°. (See Fig. 5 above.)

c) The only number in the restricted range $(-\pi/2, \pi/2)$ with a tangent of $-\sqrt{3}/3$ is $-\pi/6$. Thus, $\tan^{-1}(-\sqrt{3}/3)$ is $-\pi/6$, or $-30°$. (See Fig. 6 at left.)

We can also use a calculator to find inverse trigonometric function values. On most graphing calculators, we can find inverse function values in either radians or degrees simply by selecting the appropriate mode. The key strokes involved in finding inverse function values vary with the calculator. Be sure to read the instructions for the particular calculator that you are using.

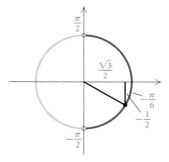

FIGURE 6

EXAMPLE 2 Approximate each of the following function values in both radians and degrees. Round radian measure to four decimal places and degree measure to the nearest tenth of a degree.

a) $\cos^{-1}(-0.2689)$

b) $\arctan(-0.2623)$

c) $\sin^{-1} 0.20345$

d) $\arccos 1.318$

e) $\csc^{-1} 8.205$

Solution

Function Value	Mode	Readout	Rounded
a) $\cos^{-1}(-0.2689)$	Radian	1.843047111	1.8430
	Degree	105.5988209	105.6°
b) $\arctan(-0.2623)$	Radian	−.256521214	−0.2565
	Degree	−14.69758292	−14.7°
c) $\sin^{-1} 0.20345$	Radian	.2048803359	0.2049
	Degree	11.73877855	11.7°
d) $\arccos 1.318$	Radian	ERR:DOMAIN	
	Degree	ERR:DOMAIN	

The value 1.318 is not in $[-1, 1]$, the domain of the arccosine function.

e) The cosecant function is the reciprocal of the sine function:
$$\csc^{-1} 8.205 =$$

$\sin^{-1}(1/8.205)$	Radian	.1221806653	0.1222
	Degree	7.000436462	7.0°

CONNECTING THE CONCEPTS

DOMAINS AND RANGES

The following is a summary of the domains and ranges of the trigonometric functions together with a summary of the domains and ranges of the inverse trigonometric functions. For completeness, we have included the arccosecant, the arcsecant, and the arccotangent, though there is a lack of uniformity in their definitions in mathematical literature.

FUNCTION	DOMAIN	RANGE
sin	All reals, $(-\infty, \infty)$	$[-1, 1]$
cos	All reals, $(-\infty, \infty)$	$[-1, 1]$
tan	All reals except $k\pi/2$, k odd	All reals, $(-\infty, \infty)$
csc	All reals except $k\pi$	$(-\infty, -1] \cup [1, \infty)$
sec	All reals except $k\pi/2$, k odd	$(-\infty, -1] \cup [1, \infty)$
cot	All reals except $k\pi$	All reals, $(-\infty, \infty)$

INVERSE FUNCTION	DOMAIN	RANGE
$\sin^{-1}$	$[-1, 1]$	$\left[-\dfrac{\pi}{2}, \dfrac{\pi}{2}\right]$
$\cos^{-1}$	$[-1, 1]$	$[0, \pi]$
$\tan^{-1}$	All reals, $(-\infty, \infty)$	$\left(-\dfrac{\pi}{2}, \dfrac{\pi}{2}\right)$
$\csc^{-1}$	$(-\infty, -1] \cup [1, \infty)$	$\left[-\dfrac{\pi}{2}, 0\right) \cup \left(0, \dfrac{\pi}{2}\right]$
$\sec^{-1}$	$(-\infty, -1] \cup [1, \infty)$	$\left[0, \dfrac{\pi}{2}\right) \cup \left(\dfrac{\pi}{2}, \pi\right]$
$\cot^{-1}$	All reals, $(-\infty, \infty)$	$\left[-\dfrac{\pi}{2}, 0\right) \cup \left(0, \dfrac{\pi}{2}\right]$

Composition of Trigonometric Functions and Their Inverses

Various compositions of trigonometric functions and their inverses often occur in practice. For example, we might want to try to simplify an expression such as

$$\sin\,(\sin^{-1} x) \quad \text{or} \quad \sin\left(\text{arccot}\,\frac{x}{2}\right).$$

COMPOSITION OF FUNCTIONS

REVIEW SECTION 4.1.

In the expression on the left, we are finding "the sine of a number whose sine is x." Recall from Section 4.1 that if a function f has an inverse that is also a function, then

$$f(f^{-1}(x)) = x, \quad \text{for all } x \text{ in the domain of } f^{-1},$$

and

$$f^{-1}(f(x)) = x, \quad \text{for all } x \text{ in the domain of } f.$$

Thus, if $f(x) = \sin x$ and $f^{-1}(x) = \sin^{-1} x$, then

$$\mathbf{\sin\,(\sin^{-1} x) = x, \quad \text{for all } x \text{ in the domain of } \sin^{-1},}$$

which is any number on the interval $[-1, 1]$. Similar results hold for the other trigonometric functions.

> ### Composition of Trigonometric Functions
>
> $\sin(\sin^{-1} x) = x,$ for all x in the domain of $\sin^{-1}$.
>
> $\cos(\cos^{-1} x) = x,$ for all x in the domain of $\cos^{-1}$.
>
> $\tan(\tan^{-1} x) = x,$ for all x in the domain of $\tan^{-1}$.

EXAMPLE 3   Simplify each of the following.

a) $\cos\left(\cos^{-1}\dfrac{\sqrt{3}}{2}\right)$ **b)** $\sin(\sin^{-1} 1.8)$

Solution

a) Since $\sqrt{3}/2$ is in $[-1, 1]$, the domain of $\cos^{-1}$, it follows that

$$\cos\left(\cos^{-1}\frac{\sqrt{3}}{2}\right) = \frac{\sqrt{3}}{2}.$$

b) Since 1.8 is not in $[-1, 1]$, the domain of $\sin^{-1}$, we cannot evaluate this expression. We know that there is no number with a sine of 1.8. Since we cannot find $\sin^{-1} 1.8$, we state that $\sin(\sin^{-1} 1.8)$ does not exist.

Now let's consider an expression like $\sin^{-1}(\sin x)$. We might also suspect that this is equal to x for any x in the domain of $\sin x$, but this is not true unless x is in the range of the $\sin^{-1}$ function. Note that in order to define $\sin^{-1}$, we had to restrict the domain of the sine function. In doing so, we restricted the range of the inverse sine function. Thus,

$$\sin^{-1}(\sin x) = x, \quad \textbf{for all } x \textbf{ in the range of } \sin^{-1}.$$

Similar results hold for the other trigonometric functions.

> ### Special Cases
>
> $\sin^{-1}(\sin x) = x,$ for all x in the range of $\sin^{-1}$.
>
> $\cos^{-1}(\cos x) = x,$ for all x in the range of $\cos^{-1}$.
>
> $\tan^{-1}(\tan x) = x,$ for all x in the range of $\tan^{-1}$.

EXAMPLE 4   Simplify each of the following.

a) $\tan^{-1}\left(\tan\dfrac{\pi}{6}\right)$ **b)** $\sin^{-1}\left(\sin\dfrac{3\pi}{4}\right)$

Solution

a) Since $\pi/6$ is in $(-\pi/2, \pi/2)$, the range of the $\tan^{-1}$ function, we can use $\tan^{-1}(\tan x) = x$. Thus,

$$\tan^{-1}\left(\tan \frac{\pi}{6}\right) = \frac{\pi}{6}.$$

b) Note that $3\pi/4$ is not in $[-\pi/2, \pi/2]$, the range of the $\sin^{-1}$ function. Thus we *cannot* apply $\sin^{-1}(\sin x) = x$. Instead we first find $\sin(3\pi/4)$, which is $\sqrt{2}/2$, and substitute:

$$\sin^{-1}\left(\sin \frac{3\pi}{4}\right) = \sin^{-1}\left(\frac{\sqrt{2}}{2}\right) = \frac{\pi}{4}.$$

Now we find some other function compositions.

EXAMPLE 5 Simplify each of the following.

a) $\sin[\arctan(-1)]$ **b)** $\cos^{-1}\left(\sin \frac{\pi}{2}\right)$

Solution

a) Arctan (-1) is the number (or angle) θ in $(-\pi/2, \pi/2)$ whose tangent is -1. That is, $\tan \theta = -1$. Thus, $\theta = -\pi/4$ and

$$\sin[\arctan(-1)] = \sin\left[-\frac{\pi}{4}\right] = -\frac{\sqrt{2}}{2}.$$

b) $\cos^{-1}\left(\sin \frac{\pi}{2}\right) = \cos^{-1}(1) = 0 \qquad \sin \frac{\pi}{2} = 1$

Next, let's consider

$$\cos\left(\arcsin \frac{3}{5}\right).$$

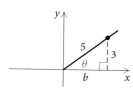

Without using a calculator, we cannot find $\arcsin \frac{3}{5}$. However, we can still evaluate the entire expression by sketching a reference triangle. We are looking for angle θ such that $\arcsin \frac{3}{5} = \theta$, or $\sin \theta = \frac{3}{5}$. Since arcsin is defined in $[-\pi/2, \pi/2]$ and $\frac{3}{5} > 0$, we know that θ is in quadrant I. We sketch a reference right triangle, as shown at left. The angle θ in this triangle is an angle whose sine is $\frac{3}{5}$. We wish to find the cosine of this angle. Since the triangle is a right triangle, we can find the length of the base, b. It is 4. Thus we know that $\cos \theta = b/5$, or $\frac{4}{5}$. Therefore,

$$\cos\left(\arcsin \frac{3}{5}\right) = \frac{4}{5}.$$

EXAMPLE 6 Find $\sin\left(\operatorname{arccot} \frac{x}{2}\right)$.

Solution We draw a right triangle whose legs have lengths x and 2, so that $\cot \theta = x/2$. Since the range of arccot is $[-\pi/2, 0) \cup (0, \pi/2]$, we consider only positive values of *x*.

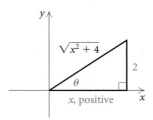

We find the length of the hypotenuse and then read off the sine ratio. We get

$$\sin\left(\text{arccot}\,\frac{x}{2}\right) = \frac{2}{\sqrt{x^2 + 4}}.$$

In the following example, we use a sum identity to evaluate an expression.

EXAMPLE 7   Evaluate:

$$\sin\left(\sin^{-1}\frac{1}{2} + \cos^{-1}\frac{5}{13}\right).$$

Solution   Since $\sin^{-1}\frac{1}{2}$ and $\cos^{-1}\frac{5}{13}$ are both angles, the expression is the sine of a sum of two angles, so we use the identity

$$\sin(u + v) = \sin u \cos v + \cos u \sin v.$$

Thus,

$$\sin\left(\sin^{-1}\frac{1}{2} + \cos^{-1}\frac{5}{13}\right)$$

$$= \sin\left(\sin^{-1}\frac{1}{2}\right)\cdot\cos\left(\cos^{-1}\frac{5}{13}\right) + \cos\left(\sin^{-1}\frac{1}{2}\right)\cdot\sin\left(\cos^{-1}\frac{5}{13}\right)$$

$$= \frac{1}{2}\cdot\frac{5}{13} + \cos\left(\sin^{-1}\frac{1}{2}\right)\cdot\sin\left(\cos^{-1}\frac{5}{13}\right)\qquad\text{\textbf{Using composition identities}}$$

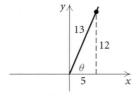

Now since $\sin^{-1}\frac{1}{2} = \pi/6$, $\cos\left(\sin^{-1}\frac{1}{2}\right)$ simplifies to $\cos\pi/6$, or $\sqrt{3}/2$. To find $\sin\left(\cos^{-1}\frac{5}{13}\right)$, we use a reference triangle in quadrant I and determine that the sine of the angle whose cosine is $\frac{5}{13}$ is $\frac{12}{13}$. Our expression now simplifies to

$$\frac{1}{2}\cdot\frac{5}{13} + \frac{\sqrt{3}}{2}\cdot\frac{12}{13}, \quad\text{or}\quad \frac{5 + 12\sqrt{3}}{26}.$$

Thus,

$$\sin\left(\sin^{-1}\frac{1}{2} + \cos^{-1}\frac{5}{13}\right) = \frac{5 + 12\sqrt{3}}{26}.$$

Exercise Set 6.4

Find each of the following exactly in radians and degrees.

1. $\sin^{-1}\left(-\dfrac{\sqrt{3}}{2}\right)$

2. $\cos^{-1}\dfrac{1}{2}$

3. $\arctan 1$

4. $\arcsin 0$

5. $\arccos \dfrac{\sqrt{2}}{2}$

6. $\sec^{-1}\sqrt{2}$

7. $\tan^{-1} 0$

8. $\arctan \dfrac{\sqrt{3}}{3}$

9. $\cos^{-1}\dfrac{\sqrt{3}}{2}$

10. $\cot^{-1}\left(-\dfrac{\sqrt{3}}{3}\right)$

11. $\csc^{-1} 2$

12. $\sin^{-1}\dfrac{1}{2}$

13. $\text{arccot}\,(-\sqrt{3})$

14. $\tan^{-1}(-1)$

15. $\arcsin\left(-\dfrac{1}{2}\right)$

16. $\arccos\left(-\dfrac{\sqrt{2}}{2}\right)$

17. $\cos^{-1} 0$

18. $\sin^{-1}\dfrac{\sqrt{3}}{2}$

19. $\text{arcsec}\,2$

20. $\text{arccsc}\,(-1)$

Use a calculator to find each of the following in radians, rounded to four decimal places, and in degrees, rounded to the nearest tenth of a degree.

21. $\arctan 0.3673$

22. $\cos^{-1}(-0.2935)$

23. $\sin^{-1} 0.9613$

24. $\arcsin(-0.6199)$

25. $\cos^{-1}(-0.9810)$

26. $\tan^{-1} 158$

27. $\csc^{-1}(-6.2774)$

28. $\sec^{-1} 1.1677$

29. $\tan^{-1}(1.091)$

30. $\cot^{-1} 1.265$

31. $\arcsin(-0.8192)$

32. $\arccos(-0.2716)$

33. State the domains of the inverse sine, inverse cosine, and inverse tangent functions.

34. State the ranges of the inverse sine, inverse cosine, and inverse tangent functions.

35. *Angle of Depression.* An airplane is flying at an altitude of 2000 ft toward an island. The straight-line distance from the airplane to the island is d feet. Express θ, the angle of depression, as an inverse sine function of d.

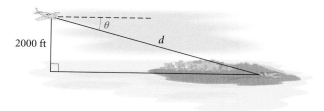

36. *Angle of Inclination.* A guy wire is attached to the top of a 50-ft pole and stretched to a point that is d feet from the bottom of the pole. Express β, the angle of inclination, as a function of d.

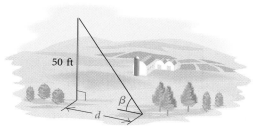

Evaluate.

37. $\sin(\arcsin 0.3)$

38. $\tan[\tan^{-1}(-4.2)]$

39. $\cos^{-1}\left[\cos\left(-\dfrac{\pi}{4}\right)\right]$

40. $\arcsin\left(\sin\dfrac{2\pi}{3}\right)$

41. $\sin^{-1}\left(\sin\dfrac{\pi}{5}\right)$

42. $\cot^{-1}\left(\cot\dfrac{2\pi}{3}\right)$

43. $\tan^{-1}\left(\tan\dfrac{2\pi}{3}\right)$

44. $\cos^{-1}\left(\cos\dfrac{\pi}{7}\right)$

45. $\sin\left(\arctan\dfrac{\sqrt{3}}{3}\right)$

46. $\cos\left(\arcsin\dfrac{\sqrt{3}}{2}\right)$

47. $\tan\left(\cos^{-1}\dfrac{\sqrt{2}}{2}\right)$

48. $\cos^{-1}(\sin\pi)$

49. $\arcsin\left(\cos\dfrac{\pi}{6}\right)$

50. $\sin^{-1}\left[\tan\left(-\dfrac{\pi}{4}\right)\right]$

51. $\tan(\arcsin 0.1)$

52. $\cos\left(\tan^{-1}\dfrac{\sqrt{3}}{4}\right)$

53. $\sin^{-1}\left(\sin\dfrac{7\pi}{6}\right)$

54. $\tan^{-1}\left(\tan -\dfrac{3\pi}{4}\right)$

Find.

55. $\sin\left(\arctan\dfrac{a}{3}\right)$

56. $\tan\left(\cos^{-1}\dfrac{3}{x}\right)$

57. $\cot\left(\sin^{-1}\dfrac{p}{q}\right)$

58. $\sin\left(\cos^{-1}x\right)$

59. $\tan\left(\arcsin\dfrac{p}{\sqrt{p^2+9}}\right)$

60. $\tan\left(\dfrac{1}{2}\arcsin\dfrac{1}{2}\right)$

61. $\cos\left(\dfrac{1}{2}\arcsin\dfrac{\sqrt{3}}{2}\right)$

62. $\sin\left(2\cos^{-1}\dfrac{3}{5}\right)$

Evaluate.

63. $\cos\left(\sin^{-1}\dfrac{\sqrt{2}}{2}+\cos^{-1}\dfrac{3}{5}\right)$

64. $\sin\left(\sin^{-1}\dfrac{1}{2}+\cos^{-1}\dfrac{3}{5}\right)$

65. $\sin\left(\sin^{-1}x+\cos^{-1}y\right)$

66. $\cos\left(\sin^{-1}x-\cos^{-1}y\right)$

67. $\sin\left(\sin^{-1}0.6032+\cos^{-1}0.4621\right)$

68. $\cos\left(\sin^{-1}0.7325-\cos^{-1}0.4838\right)$

Collaborative Discussion and Writing

69. Explain in your own words why the ranges of the inverse trigonometric functions are restricted.

70. How does the graph of $y=\sin^{-1}x$ differ from the graph of $y=\sin x$?

71. Why is it that

$$\sin\dfrac{5\pi}{6}=\dfrac{1}{2},$$

but

$$\sin^{-1}\left(\dfrac{1}{2}\right)\neq\dfrac{5\pi}{6}?$$

Skill Maintenance

Solve.

72. $3x^2+5x-10=18$

73. $2x^2=5x$

74. $x^2-10x+1=0$

75. $x^4+5x^2-36=0$

76. $x=\sqrt{x+7}+5$

77. $\sqrt{x-2}=5$

Synthesis

78. Use a calculator to approximate the following expression:

$$16\tan^{-1}\dfrac{1}{5}-4\tan^{-1}\dfrac{1}{239}.$$

What number does this expression seem to approximate?

Prove the identity.

79. $\sin^{-1}x+\cos^{-1}x=\dfrac{\pi}{2}$

80. $\tan^{-1}x+\cot^{-1}x=\dfrac{\pi}{2}$

81. $\sin^{-1}x=\tan^{-1}\dfrac{x}{\sqrt{1-x^2}}$

82. $\tan^{-1}x=\sin^{-1}\dfrac{x}{\sqrt{x^2+1}}$

83. $\arcsin x=\arccos\sqrt{1-x^2}$,   for $x\geq0$

84. $\arccos x=\arctan\dfrac{\sqrt{1-x^2}}{x}$,   for $x>0$

85. *Height of a Mural.*   An art student's eye is at a point A, looking at a mural of height h, with the bottom of the mural y feet above the eye. The eye is x feet from the wall. Write an expression for θ in terms of x, y, and h. Then evaluate the expression when $x=20$ ft, $y=7$ ft, and $h=25$ ft.

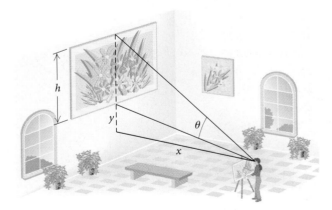

6.5

Solving Trigonometric Equations

- *Solve trigonometric equations.*

When an equation contains a trigonometric expression with a variable, such as cos x, it is called a trigonometric equation. Some trigonometric equations are identities, such as $\sin^2 x + \cos^2 x = 1$. Now we consider equations, such as $2 \cos x = -1$, that are usually not identities. As we have done for other types of equations, we will solve such equations by finding all values for x that make the equation true.

EXAMPLE 1 Solve: $2 \cos x = -1$.

Algebraic Solution

We first solve for cos x:

$$2 \cos x = -1$$

$$\cos x = -\frac{1}{2}.$$

The solutions are numbers that have a cosine of $-\frac{1}{2}$. To find them, we use the unit circle (see Section 5.5).

There are just two points on the unit circle for which the cosine is $-\frac{1}{2}$, as shown in the following figure.

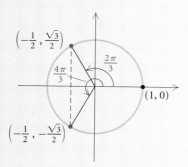

They are the points corresponding to $2\pi/3$ and $4\pi/3$. These numbers, plus any multiple of 2π, are the solutions:

$$\frac{2\pi}{3} + 2k\pi \quad \text{and} \quad \frac{4\pi}{3} + 2k\pi,$$

where k is any integer. In degrees, the solutions are

$$120° + k \cdot 360° \quad \text{and} \quad 240° + k \cdot 360°,$$

where k is any integer.

Visualizing the Solution

We graph $y = 2 \cos x$ and $y = -1$. The first coordinates of the points of intersection of the graphs are the values of x for which $2 \cos x = -1$. If we use $\pi/3$ as the x-scale, the solutions can be seen easily.

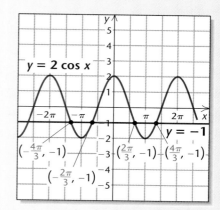

The only solutions in $[-2\pi, 2\pi]$ are

$$-\frac{4\pi}{3}, \quad -\frac{2\pi}{3}, \quad \frac{2\pi}{3}, \quad \text{and} \quad \frac{4\pi}{3}.$$

Since the cosine is periodic, there is an infinite number of solutions. Thus the entire set of solutions is

$$\frac{2\pi}{3} + 2k\pi \quad \text{and} \quad \frac{4\pi}{3} + 2k\pi,$$

where k is an integer.

EXAMPLE 2 Solve: $4 \sin^2 x = 1$.

Algebraic Solution

We begin by solving for $\sin x$:

$$4 \sin^2 x = 1$$

$$\sin^2 x = \frac{1}{4}$$

$$\sin x = \pm \frac{1}{2}.$$

Again, we use the unit circle to find those numbers having a sine of $\frac{1}{2}$ or $-\frac{1}{2}$.

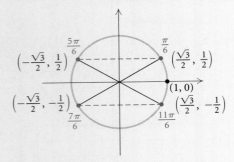

The solutions are

$$\frac{\pi}{6} + 2k\pi, \quad \frac{5\pi}{6} + 2k\pi, \quad \frac{7\pi}{6} + 2k\pi,$$

and

$$\frac{11\pi}{6} + 2k\pi,$$

where k is any integer. In degrees, the solutions are

$$30° + k \cdot 360°, \quad 150° + k \cdot 360°,$$
$$210° + k \cdot 360°, \quad \text{and} \quad 330° + k \cdot 360°,$$

where k is any integer.

The general solutions listed above could be condensed using odd as well as even multiples of π:

$$\frac{\pi}{6} + k\pi \quad \text{and} \quad \frac{5\pi}{6} + k\pi,$$

or, in degrees,

$$30° + k \cdot 180° \quad \text{and} \quad 150° + k \cdot 180°,$$

where k is any integer.

Visualizing the Solution

From the graph shown here, we see that the first coordinates of the points of intersection of the graphs of

$$y = 4 \sin^2 x \quad \text{and} \quad y = 1$$

in $[0, 2\pi)$ are

$$\frac{\pi}{6}, \quad \frac{5\pi}{6}, \quad \frac{7\pi}{6}, \quad \text{and} \quad \frac{11\pi}{6}.$$

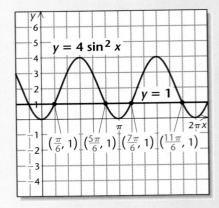

Thus, since the sine function is periodic, the general solutions are

$$\frac{\pi}{6} + k\pi \quad \text{and} \quad \frac{5\pi}{6} + k\pi,$$

or, in degrees,

$$30° + k \cdot 180° \quad \text{and} \quad 150° + k \cdot 180°,$$

where k is an integer.

> In most applications, it is sufficient to find just the solutions from 0 to 2π or from 0° to 360°. We then remember that any multiple of 2π, or 360°, can be added to obtain the rest of the solutions.

We must be careful to find all solutions in $[0, 2\pi)$ when solving trigonometric equations involving double angles.

EXAMPLE 3 Solve $3 \tan 2x = -3$ in the interval $[0, 2\pi)$.

Solution We first solve for $\tan 2x$:

$$3 \tan 2x = -3$$
$$\tan 2x = -1.$$

We are looking for solutions x to the equation for which

$$0 \leq x < 2\pi.$$

Multiplying by 2, we get

$$0 \leq 2x < 4\pi,$$

which is the interval we use when solving $\tan 2x = -1$.

Using the unit circle, we find points $2x$ in $[0, 4\pi)$ for which $\tan 2x = -1$. These values of $2x$ are as follows:

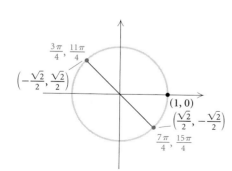

$$2x = \frac{3\pi}{4}, \quad \frac{7\pi}{4}, \quad \frac{11\pi}{4}, \quad \text{and} \quad \frac{15\pi}{4}.$$

Thus the desired values of x in $[0, 2\pi)$ are each of these values divided by 2. Therefore,

$$x = \frac{3\pi}{8}, \quad \frac{7\pi}{8}, \quad \frac{11\pi}{8}, \quad \text{and} \quad \frac{15\pi}{8}.$$

Calculators are needed to solve some trigonometric equations. Answers can be found in radians or degrees, depending on the mode setting.

EXAMPLE 4 Solve $\dfrac{1}{2} \cos \phi + 1 = 1.2108$ in $[0, 360°)$.

Solution We have

$$\frac{1}{2} \cos \phi + 1 = 1.2108$$

$$\frac{1}{2} \cos \phi = 0.2108$$

$$\cos \phi = 0.4216.$$

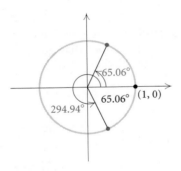

Using a calculator set in DEGREE mode, we find that the reference angle, arccos 0.4216, is

$$\phi \approx 65.06°.$$

Since cos ϕ is positive, the solutions are in quadrants I and IV. The solutions in $[0, 360°)$ are

$$65.06° \quad \text{and} \quad 360° - 65.06° = 294.94°.$$

EXAMPLE 5   Solve $2\cos^2 u = 1 - \cos u$ in $[0°, 360°)$.

Algebraic Solution

We use the principle of zero products:

$$2\cos^2 u = 1 - \cos u$$
$$2\cos^2 u + \cos u - 1 = 0$$
$$(2\cos u - 1)(\cos u + 1) = 0$$
$$2\cos u - 1 = 0 \quad \text{or} \quad \cos u + 1 = 0$$
$$2\cos u = 1 \quad \text{or} \quad \cos u = -1$$
$$\cos u = \frac{1}{2} \quad \text{or} \quad \cos u = -1.$$

Thus,

$$u = 60°, 300° \quad \text{or} \quad u = 180°.$$

The solutions in $[0°, 360°)$ are 60°, 180°, and 300°.

Visualizing the Solution

The solutions of the equation are the zeros of the function

$$y = 2\cos^2 u + \cos u - 1.$$

Note that they are also the first coordinates of the x-intercepts of the graph.

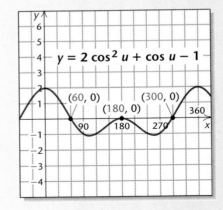

The zeros in $[0°, 360°)$ are 60°, 180°, and 300°. Thus the solutions of the equation in $[0°, 360°)$ are 60°, 180°, and 300°.

Technology Connection

We can use either the Intersect method or the Zero method to solve trigonometric equations. Here we illustrate by solving the equation in Example 5 using both methods.

Intersect Method. We graph the equations

$$y_1 = 2 \cos^2 x \quad \text{and} \quad y_2 = 1 - \cos x$$

and use the INTERSECT feature to find the first coordinates of the points of intersection.

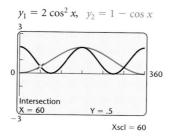

$y_1 = 2 \cos^2 x, \quad y_2 = 1 - \cos x$

The leftmost solution is 60°. Using the INTERSECT feature two more times, we find the other solutions, 180° and 300°.

Zero Method. We write the equation in the form

$$2 \cos^2 u + \cos u - 1 = 0.$$

Then we graph

$$y = 2 \cos^2 x + \cos x - 1$$

and use the ZERO feature to determine the zeros of the function.

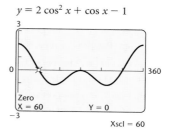

$y = 2 \cos^2 x + \cos x - 1$

The leftmost zero is 60°. Using the ZERO feature two more times, we find the other zeros, 180° and 300°. The solutions in $[0°, 360°)$ are 60°, 180°, and 300°.

EXAMPLE 6   Solve $\sin^2 \beta - \sin \beta = 0$ in $[0, 2\pi)$.

Algebraic Solution

We have

$$\sin^2 \beta - \sin \beta = 0$$
$$\sin \beta (\sin \beta - 1) = 0 \qquad \text{Factoring}$$
$$\sin \beta = 0 \quad or \quad \sin \beta - 1 = 0$$
$$\sin \beta = 0 \quad or \qquad \sin \beta = 1$$
$$\beta = 0, \pi \quad or \qquad \beta = \frac{\pi}{2}.$$

The solutions in $[0, 2\pi)$ are 0, $\pi/2$, and π.

Visualizing the Solution

The solutions of the equation

$$\sin^2 \beta - \sin \beta = 0$$

are the zeros of the function

$$f(\beta) = \sin^2 \beta - \sin \beta.$$

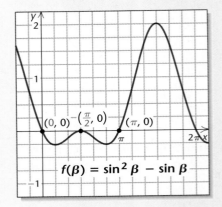

The zeros in $[0, 2\pi)$ are 0, $\pi/2$, and π. Thus the solutions of $\sin^2 \beta - \sin \beta = 0$ are 0, $\pi/2$, and π.

If a trigonometric equation is quadratic but difficult or impossible to factor, we use the quadratic formula.

EXAMPLE 7   Solve $10 \sin^2 x - 12 \sin x - 7 = 0$ in $[0°, 360°)$.

Solution   This equation is quadratic in $\sin x$ with $a = 10$, $b = -12$, and $c = -7$. Substituting into the quadratic formula, we get

$$\sin x = \frac{12 \pm \sqrt{144 + 280}}{20} \qquad \text{Using the quadratic formula}$$

$$= \frac{12 \pm \sqrt{424}}{20}$$

$$\approx \frac{12 \pm 20.5913}{20}$$

$$\sin x \approx 1.6296 \quad or \quad \sin x \approx -0.4296.$$

Since sine values are never greater than 1, the first of the equations has no solution. Using the other equation, we find the reference angle to be 25.44°. Since $\sin x$ is negative, the solutions are in quadrants III and IV.

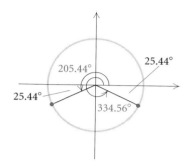

Thus the solutions in $[0°, 360°)$ are

$$180° + 25.44° = 205.44° \quad \text{and} \quad 360° - 25.44° = 334.56°. \quad \blacksquare$$

Trigonometric equations can involve more than one function.

EXAMPLE 8   Solve $2 \cos^2 x \tan x = \tan x$ in $[0, 2\pi)$.

Solution   We have

$$2 \cos^2 x \tan x = \tan x$$
$$2 \cos^2 x \tan x - \tan x = 0$$
$$\tan x \, (2 \cos^2 x - 1) = 0$$
$$\tan x = 0 \qquad or \quad 2 \cos^2 x - 1 = 0$$
$$\cos^2 x = \frac{1}{2}$$
$$\cos x = \pm \frac{\sqrt{2}}{2}$$
$$x = 0, \pi \quad or \qquad\qquad x = \frac{\pi}{4}, \frac{3\pi}{4}, \frac{5\pi}{4}, \frac{7\pi}{4}.$$

Thus, $x = 0, \pi/4, 3\pi/4, \pi, 5\pi/4,$ and $7\pi/4$. $\blacksquare$

When a trigonometric equation involves more than one function, it is sometimes helpful to use identities to rewrite the equation in terms of a single function.

Technology Connection

In Example 9, we can graph the left side and then the right side of the equation as seen in the first window below. Then we look for points of intersection. We could also rewrite the equation as $\sin x + \cos x - 1 = 0$, graph the left side, and look for the zeros of the function, as illustrated in the second window below. In each window, we see the solutions in $[0, 2\pi)$ as 0 and $\pi/2$.

This example illustrates a valuable advantage of the calculator—that is, with a graphing calculator, extraneous solutions do not appear.

$y_1 = \sin x + \cos x, \ y_2 = 1$

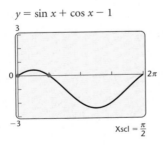

$y = \sin x + \cos x - 1$

EXAMPLE 9   Solve $\sin x + \cos x = 1$ in $[0, 2\pi)$.

Solution   We have

$$\sin x + \cos x = 1$$
$$(\sin x + \cos x)^2 = 1^2 \qquad \text{Squaring both sides}$$
$$\sin^2 x + 2 \sin x \cos x + \cos^2 x = 1$$
$$2 \sin x \cos x + 1 = 1 \qquad \text{Using } \sin^2 x + \cos^2 x = 1$$
$$2 \sin x \cos x = 0$$
$$\sin 2x = 0. \qquad \text{Using } 2 \sin x \cos x = \sin 2x$$

We are looking for solutions x to the equation for which $0 \leq x < 2\pi$. Multiplying by 2, we get $0 \leq 2x < 4\pi$, which is the interval we consider to solve $\sin 2x = 0$. These values of $2x$ are 0, π, 2π, and 3π. Thus the desired values of x in $[0, 2\pi)$ satisfying this equation are 0, $\pi/2$, π, and $3\pi/2$. Now we check these in the original equation:

$$\sin 0 + \cos 0 = 0 + 1 = 1,$$
$$\sin \frac{\pi}{2} + \cos \frac{\pi}{2} = 1 + 0 = 1,$$
$$\sin \pi + \cos \pi = 0 + (-1) = -1,$$
$$\sin \frac{3\pi}{2} + \cos \frac{3\pi}{2} = (-1) + 0 = -1.$$

We find that π and $3\pi/2$ do not check, but the other values do. Thus the solutions in $[0, 2\pi)$ are

$$0 \quad \text{and} \quad \frac{\pi}{2}.$$

When the solution process involves squaring both sides, values are sometimes obtained that are not solutions of the original equation. As we saw in this example, it is important to check the possible solutions. ▬

EXAMPLE 10   Solve $\cos 2x + \sin x = 1$ in $[0, 2\pi)$.

Algebraic Solution

We have

$$\cos 2x + \sin x = 1$$
$$1 - 2\sin^2 x + \sin x = 1 \qquad \text{Using the identity}$$
$$\cos 2x = 1 - 2\sin^2 x$$
$$-2\sin^2 x + \sin x = 0$$
$$\sin x\,(-2\sin x + 1) = 0 \qquad \text{Factoring}$$
$$\sin x = 0 \qquad or \quad -2\sin x + 1 = 0 \qquad \text{Principle of zero products}$$
$$\sin x = 0 \qquad or \qquad \sin x = \frac{1}{2}$$
$$x = 0,\, \pi \quad or \qquad x = \frac{\pi}{6},\, \frac{5\pi}{6}.$$

All four values check. The solutions in $[0, 2\pi)$ are 0, $\pi/6$, $5\pi/6$, and π.

Visualizing the Solution

We graph the function
$y = \cos 2x + \sin x - 1$ and look for the zeros of the function.

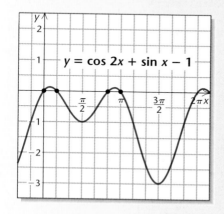

The zeros, or solutions, in $[0, 2\pi)$ are 0, $\pi/6$, $5\pi/6$, and π.

EXAMPLE 11   Solve $\tan^2 x + \sec x - 1 = 0$ in $[0, 2\pi)$.

Solution   We have

$$\tan^2 x + \sec x - 1 = 0$$
$$\sec^2 x - 1 + \sec x - 1 = 0 \qquad \text{Using the identity}$$
$$1 + \tan^2 x = \sec^2 x$$
$$\sec^2 x + \sec x - 2 = 0$$
$$(\sec x + 2)(\sec x - 1) = 0 \qquad \text{Factoring}$$
$$\sec x = -2 \qquad or \quad \sec x = 1 \qquad \text{Principle of zero products}$$
$$x = \frac{2\pi}{3},\, \frac{4\pi}{3} \quad or \qquad x = 0$$

All these values check. The solutions in $[0, 2\pi)$ are 0, $2\pi/3$, and $4\pi/3$.

Exercise Set 6.5

Solve, finding all solutions. Express the solutions in both radians and degrees.

1. $\cos x = \dfrac{\sqrt{3}}{2}$ **2.** $\sin x = -\dfrac{\sqrt{2}}{2}$

3. $\tan x = -\sqrt{3}$ **4.** $\cos x = -\dfrac{1}{2}$

Solve, finding all solutions in $[0, 2\pi)$ or $[0°, 360°)$.

5. $2\cos x - 1 = -1.2814$

6. $\sin x + 3 = 2.0816$

7. $2\sin x + \sqrt{3} = 0$ **8.** $2\tan x - 4 = 1$

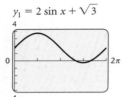

$y_1 = 2\sin x + \sqrt{3}$

$y_1 = 2\tan x - 4,$
$y_2 = 1$

9. $2\cos^2 x = 1$ **10.** $\csc^2 x - 4 = 0$

11. $2\sin^2 x + \sin x = 1$ **12.** $\cos^2 x + 2\cos x = 3$

13. $2\cos^2 x - \sqrt{3}\cos x = 0$

14. $2\sin^2 \theta + 7\sin \theta = 4$

15. $6\cos^2 \phi + 5\cos \phi + 1 = 0$

16. $2\sin t \cos t + 2\sin t - \cos t - 1 = 0$

17. $\sin 2x \cos x - \sin x = 0$

18. $5\sin^2 x - 8\sin x = 3$

19. $\cos^2 x + 6\cos x + 4 = 0$

20. $2\tan^2 x = 3\tan x + 7$

21. $7 = \cot^2 x + 4\cot x$

22. $3\sin^2 x = 3\sin x + 2$

Solve, finding all solutions in $[0, 2\pi)$.

23. $\cos 2x - \sin x = 1$

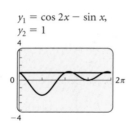

$y_1 = \cos 2x - \sin x,$
$y_2 = 1$

24. $2\sin x \cos x + \sin x = 0$

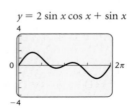

$y = 2\sin x \cos x + \sin x$

25. $\sin 4x - 2\sin 2x = 0$

26. $\tan x \sin x - \tan x = 0$

27. $\sin 2x \cos x + \sin x = 0$

28. $\cos 2x \sin x + \sin x = 0$

29. $2\sec x \tan x + 2\sec x + \tan x + 1 = 0$

30. $\sin 2x \sin x - \cos 2x \cos x = -\cos x$

31. $\sin 2x + \sin x + 2\cos x + 1 = 0$

32. $\tan^2 x + 4 = 2\sec^2 x + \tan x$

33. $\sec^2 x - 2\tan^2 x = 0$

34. $\cot x = \tan(2x - 3\pi)$

35. $2\cos x + 2\sin x = \sqrt{6}$

36. $\sqrt{3}\cos x - \sin x = 1$

37. $\sec^2 x + 2\tan x = 6$

38. $5\cos 2x + \sin x = 4$

39. $\cos(\pi - x) + \sin\left(x - \dfrac{\pi}{2}\right) = 1$

40. $\dfrac{\sin^2 x - 1}{\cos\left(\dfrac{\pi}{2} - x\right) + 1} = \dfrac{\sqrt{2}}{2} - 1$

Technology Connection

41. Check the solutions of Exercises 10, 15, 16, and 17 using the Zero method.

42. Check the solutions of Exercises 11, 14, 20, and 21 using the Intersect method.

Solve using a calculator, finding all solutions in $[0, 2\pi)$.

43. $x\sin x = 1$ **44.** $x^2 + 2 = \sin x$

45. $2\cos^2 x = x + 1$ **46.** $x\cos x - 2 = 0$

47. $\cos x - 2 = x^2 - 3x$ **48.** $\sin x = \tan\dfrac{x}{2}$

Some graphing calculators can use regression to fit a trigonometric function to a set of data.

49. *Sales.* Sales of certain products fluctuate in cycles. The data in the following table show the total sales of skis per month for a business in a northern climate.

MONTH, *x*		TOTAL SALES, *y* (IN THOUSANDS)
August,	8	0
November,	11	7
February,	2	$14
May,	5	7
August,	8	0

a) Using the SINE REGRESSION feature on a graphing calculator, fit a sine function of the form $y = A \sin(Bx - C) + D$ to this set of data.

b) Approximate the total sales for December and for July.

50. *Daylight Hours.* The data in the following table give the number of daylight hours for certain days in Kajaani, Finland.

DAY, *x*		NUMBER OF DAYLIGHT HOURS, *y*
January 10,	10	5.0
February 19,	50	9.1
March 3,	62	10.4
April 28,	118	16.4
May 14,	134	18.2
June 11,	162	20.7
July 17,	198	19.5
August 22,	234	15.7
September 19,	262	12.7
October 1,	274	11.4
November 14,	318	6.7
December 28,	362	4.3

Source: The Astronomical Almanac, 1995, Washington: U.S. Government Printing Office.

a) Using the SINE REGRESSION feature on a graphing calculator, model these data with an equation of the form $y = A \sin(Bx - C) + D$.

b) Approximate the number of daylight hours in Kajaani for April 22 ($x = 112$), July 4 ($x = 185$), and December 15 ($x = 349$).

Collaborative Discussion and Writing

51. Jan lists her answer to a problem as $\pi/6 + k\pi$, for any integer k, while Jacob lists his answer as $\pi/6 + 2k\pi$ and $7\pi/6 + 2\pi k$, for any integer k. Are their answers equivalent? Why or why not?

52. An identity is an equation that is true for all possible replacements of the variables. Explain the meaning of "possible" in this definition.

Skill Maintenance

Solve the right triangle.

53.

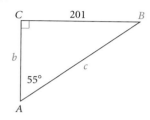

54.

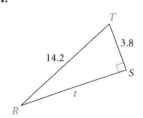

Solve.

55. $\dfrac{x}{27} = \dfrac{4}{3}$

56. $\dfrac{0.01}{0.7} = \dfrac{0.2}{h}$

Synthesis

Solve in $[0, 2\pi)$.

57. $|\sin x| = \dfrac{\sqrt{3}}{2}$

58. $|\cos x| = \dfrac{1}{2}$

59. $\sqrt{\tan x} = \sqrt[4]{3}$

60. $12 \sin x - 7\sqrt{\sin x} + 1 = 0$

61. $\ln(\cos x) = 0$

62. $e^{\sin x} = 1$

63. $\sin(\ln x) = -1$

64. $e^{\ln(\sin x)} = 1$

65. *Temperature During an Illness.* The temperature T, in degrees Fahrenheit, of a patient t days into a 12-day illness is given by

$$T(t) = 101.6° + 3° \sin\left(\frac{\pi}{8}t\right).$$

Find the times t during the illness at which the patient's temperature was 103°.

66. *Satellite Location.* A satellite circles the earth in such a manner that it is y miles from the equator (north or south, height from the surface not considered) t minutes after its launch, where

$$y = 5000\left[\cos\frac{\pi}{45}(t - 10)\right].$$

At what times t in the interval [0, 240], the first 4 hr, is the satellite 3000 mi north of the equator?

67. *Nautical Mile.* (See Exercise 52 in Exercise Set 6.2.) In Great Britain, the *nautical mile* is defined as the length of a minute of arc of the earth's radius. Since the earth is flattened at the poles, a British nautical mile varies with latitude. In fact, it is given, in feet, by the function

$$N(\phi) = 6066 - 31 \cos 2\phi,$$

where ϕ is the latitude in degrees. At what latitude north is the length of a British nautical mile found to be 6040 ft?

68. *Acceleration Due to Gravity.* (See Exercise 53 in Exercise Set 6.2.) The acceleration due to gravity is often denoted by g in a formula such as $S = \frac{1}{2}gt^2$, where S is the distance that an object falls in t seconds. The number g is generally considered constant, but in fact it varies slightly with latitude. If ϕ stands for latitude, in degrees, an excellent approximation of g is given by the formula

$$g = 9.78049(1 + 0.005288 \sin^2 \phi - 0.000006 \sin^2 2\phi),$$

where g is measured in meters per second per second at sea level. At what latitude north does $g = 9.8$?

Solve.

69. $\arccos x = \arccos \frac{3}{5} - \arcsin \frac{4}{5}$

70. $\sin^{-1} x = \tan^{-1} \frac{1}{3} + \tan^{-1} \frac{1}{2}$

71. Suppose that $\sin x = 5 \cos x$. Find $\sin x \cos x$.

6 Chapter Summary and Review

Important Properties and Formulas

Basic Identities

$$\sin x = \frac{1}{\csc x}, \qquad \tan x = \frac{\sin x}{\cos x},$$

$$\cos x = \frac{1}{\sec x}, \qquad \cot x = \frac{\cos x}{\sin x}$$

$$\tan x = \frac{1}{\cot x}$$

$$\sin (-x) = -\sin x,$$
$$\cos (-x) = \cos x,$$
$$\tan (-x) = -\tan x$$

Pythagorean Identities

$$\sin^2 x + \cos^2 x = 1,$$
$$1 + \cot^2 x = \csc^2 x,$$
$$1 + \tan^2 x = \sec^2 x$$

Sum and Difference Identities

$$\sin (u \pm v) = \sin u \cos v \pm \cos u \sin v,$$
$$\cos (u \pm v) = \cos u \cos v \mp \sin u \sin v,$$
$$\tan (u \pm v) = \frac{\tan u \pm \tan v}{1 \mp \tan u \tan v}$$

Cofunction Identities

$$\sin\left(\frac{\pi}{2} - x\right) = \cos x,$$

$$\tan\left(\frac{\pi}{2} - x\right) = \cot x,$$

$$\sec\left(\frac{\pi}{2} - x\right) = \csc x,$$

$$\sin\left(x \pm \frac{\pi}{2}\right) = \pm\cos x,$$

$$\cos\left(x \pm \frac{\pi}{2}\right) = \mp \sin x$$

Double-Angle Identities

$$\sin 2x = 2 \sin x \cos x,$$

$$\cos 2x = \cos^2 x - \sin^2 x$$
$$= 1 - 2 \sin^2 x$$
$$= 2 \cos^2 x - 1,$$

$$\tan 2x = \frac{2 \tan x}{1 - \tan^2 x}$$

Half-Angle Identities

$$\sin \frac{x}{2} = \pm \sqrt{\frac{1 - \cos x}{2}},$$

$$\cos \frac{x}{2} = \pm \sqrt{\frac{1 + \cos x}{2}},$$

$$\tan \frac{x}{2} = \pm \sqrt{\frac{1 - \cos x}{1 + \cos x}}$$

$$= \frac{\sin x}{1 + \cos x}$$

$$= \frac{1 - \cos x}{\sin x}$$

Inverse Trigonometric Functions

FUNCTION	DOMAIN	RANGE
$y = \sin^{-1} x$	$[-1, 1]$	$\left[-\dfrac{\pi}{2}, \dfrac{\pi}{2}\right]$
$y = \cos^{-1} x$	$[-1, 1]$	$[0, \pi]$
$y = \tan^{-1} x$	$(-\infty, \infty)$	$\left(-\dfrac{\pi}{2}, \dfrac{\pi}{2}\right)$

Composition of Trigonometric Functions

The following are true for any x in the domain of the inverse function:

$$\sin (\sin^{-1} x) = x,$$
$$\cos (\cos^{-1} x) = x,$$
$$\tan (\tan^{-1} x) = x.$$

The following are true for any x in the range of the inverse function:

$$\sin^{-1} (\sin x) = x,$$
$$\cos^{-1} (\cos x) = x,$$
$$\tan^{-1} (\tan x) = x.$$

REVIEW EXERCISES

Complete the Pythagorean identity.

1. $1 + \cot^2 x =$

2. $\sin^2 x + \cos^2 x =$

Multiply and simplify.

3. $(\tan y - \cot y)(\tan y + \cot y)$

4. $(\cos x + \sec x)^2$

Factor and simplify.

5. $\sec x \csc x - \csc^2 x$

6. $3 \sin^2 y - 7 \sin y - 20$

7. $1000 - \cos^3 u$

Simplify.

8. $\dfrac{\sec^4 x - \tan^4 x}{\sec^2 x + \tan^2 x}$

9. $\dfrac{2 \sin^2 x}{\cos^3 x} \cdot \left(\dfrac{\cos x}{2 \sin x}\right)^2$

10. $\dfrac{3 \sin x}{\cos^2 x} \cdot \dfrac{\cos^2 x + \cos x \sin x}{\sin^2 x - \cos^2 x}$

11. $\dfrac{3}{\cos y - \sin y} - \dfrac{2}{\sin^2 y - \cos^2 y}$

12. $\left(\dfrac{\cot x}{\csc x}\right)^2 + \dfrac{1}{\csc^2 x}$

13. $\dfrac{4 \sin x \cos^2 x}{16 \sin^2 x \cos x}$

14. Simplify. Assume the radicand is nonnegative.

$$\sqrt{\sin^2 x + 2 \cos x \sin x + \cos^2 x}$$

15. Rationalize the denominator: $\sqrt{\dfrac{1 + \sin x}{1 - \sin x}}$.

16. Rationalize the numerator: $\sqrt{\dfrac{\cos x}{\tan x}}$.

17. Given that $x = 3 \tan \theta$, express $\sqrt{9 + x^2}$ as a trigonometric function without radicals. Assume that $0 < \theta < \pi/2$.

Use the sum and difference formulas to write equivalent expressions. You need not simplify.

18. $\cos\left(x + \dfrac{3\pi}{2}\right)$

19. $\tan (45° - 30°)$

20. Simplify: $\cos 27° \cos 16° + \sin 27° \sin 16°$.

21. Find $\cos 165°$ exactly.

22. Given that $\tan \alpha = \sqrt{3}$ and $\sin \beta = \sqrt{2}/2$ and that α and β are between 0 and $\pi/2$, evaluate $\tan (\alpha - \beta)$ exactly.

23. Assume that $\sin \theta = 0.5812$ and $\cos \phi = 0.2341$ and that both θ and ϕ are first-quadrant angles. Evaluate $\cos (\theta + \phi)$.

Complete the cofunction identity.

24. $\cos\left(x + \dfrac{\pi}{2}\right) =$

25. $\cos\left(\dfrac{\pi}{2} - x\right) =$

26. $\sin\left(x - \dfrac{\pi}{2}\right) =$

27. Given that $\cos \alpha = -\frac{3}{5}$ and that the terminal side is in quadrant III:

a) Find the other function values for α.
b) Find the six function values for $\pi/2 - \alpha$.
c) Find the six function values for $\alpha + \pi/2$.

28. Find an equivalent expression for $\csc\left(x - \dfrac{\pi}{2}\right)$.

29. Find $\tan 2\theta$, $\cos 2\theta$, and $\sin 2\theta$ and the quadrant in which 2θ lies, where $\cos \theta = -\frac{4}{5}$ and θ is in quadrant III.

30. Find $\sin \dfrac{\pi}{8}$ exactly.

31. Given that $\sin \beta = 0.2183$ and β is in quadrant I, find $\sin 2\beta$, $\cos \dfrac{\beta}{2}$, and $\cos 4\beta$.

Simplify.

32. $1 - 2 \sin^2 \dfrac{x}{2}$

33. $(\sin x + \cos x)^2 - \sin 2x$

34. $2 \sin x \cos^3 x + 2 \sin^3 x \cos x$

35. $\dfrac{2 \cot x}{\cot^2 x - 1}$

Prove the identity.

36. $\dfrac{1 - \sin x}{\cos x} = \dfrac{\cos x}{1 + \sin x}$

37. $\dfrac{1 + \cos 2\theta}{\sin 2\theta} = \cot \theta$

38. $\dfrac{\tan y + \sin y}{2 \tan \theta} = \cos^2 \dfrac{y}{2}$

39. $\dfrac{\sin x - \cos x}{\cos^2 x} = \dfrac{\tan^2 x - 1}{\sin x + \cos x}$

Find each of the following exactly in both radians and degrees.

40. $\sin^{-1}\left(-\dfrac{1}{2}\right)$

41. $\cos^{-1} \dfrac{\sqrt{3}}{2}$

42. $\arctan 1$

43. $\arcsin 0$

Use a calculator to find each of the following in radians, rounded to four decimal places, and in degrees, rounded to the nearest tenth of a degree.

44. $\arccos (-0.2194)$

45. $\cot^{-1} 2.381$

Evaluate.

46. $\cos\left(\cos^{-1} \dfrac{1}{2}\right)$

47. $\tan^{-1}\left(\tan \dfrac{\sqrt{3}}{3}\right)$

48. $\sin^{-1}\left(\sin \dfrac{\pi}{7}\right)$

49. $\cos\left(\arcsin \dfrac{\sqrt{2}}{2}\right)$

Find.

50. $\cos\left(\arctan \dfrac{b}{3}\right)$

51. $\cos\left(2 \sin^{-1} \dfrac{4}{5}\right)$

Solve, finding all solutions. Express the solutions in both radians and degrees.

52. $\cos x = -\dfrac{\sqrt{2}}{2}$

53. $\tan x = \sqrt{3}$

Solve, finding all solutions in $[0, 2\pi)$.

54. $4 \sin^2 x = 1$

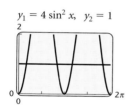

$y_1 = 4 \sin^2 x, \quad y_2 = 1$

55. $\sin 2x \sin x - \cos x = 0$

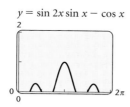

$y = \sin 2x \sin x - \cos x$

56. $2 \cos^2 x + 3 \cos x = -1$

57. $\sin^2 x - 7 \sin x = 0$

58. $\csc^2 x - 2 \cot^2 x = 0$

59. $\sin 4x + 2 \sin 2x = 0$

60. $2 \cos x + 2 \sin x = \sqrt{2}$

61. $6 \tan^2 x = 5 \tan x + \sec^2 x$

Technology Connection

In Exercises 62–65, use a graphing calculator to determine which expression (A)–(D) on the right can be used to complete the identity. Then prove the identity algebraically.

62. $\csc x - \cos x \cot x$ **A.** $\dfrac{\csc x}{\sec x}$

63. $\dfrac{1}{\sin x \cos x} - \dfrac{\cos x}{\sin x}$ **B.** $\sin x$

64. $\dfrac{\cot x - 1}{1 - \tan x}$ **C.** $\dfrac{2}{\sin x}$

65. $\dfrac{\cos x + 1}{\sin x} + \dfrac{\sin x}{\cos x + 1}$ **D.** $\dfrac{\sin x \cos x}{1 - \sin^2 x}$

Solve using a graphing calculator, finding all solutions in $[0, 2\pi)$.

66. $x \cos x = 1$

67. $2 \sin^2 x = x + 1$

Collaborative Discussion and Writing

68. Prove the identity $2 \cos^2 x - 1 = \cos^4 x - \sin^4 x$ in three ways:

   a) Start with the left side and deduce the right (method 1).

   b) Start with the right side and deduce the left (method 1).

   c) Work with each side separately until you deduce the same expression (method 2).

   Then determine the most efficient method and explain why you chose that method.

69. Why are the ranges of the inverse trigonometric functions restricted?

Synthesis

70. Find the measure of the angle from l_1 to l_2:

$$l_1: x + y = 3 \qquad l_2: 2x - y = 5.$$

71. Find an identity for $\cos (u + v)$ involving only cosines.

72. Simplify: $\cos \left(\dfrac{\pi}{2} - x \right) [\csc x - \sin x]$.

73. Find $\sin \theta$, $\cos \theta$, and $\tan \theta$ under the given conditions:

$$\sin 2\theta = \frac{1}{5}, \quad \frac{\pi}{2} \le 2\theta < \pi.$$

74. Prove the following equation to be an identity:

$$\ln e^{\sin t} = \sin t.$$

75. Graph: $y = \sec^{-1} x$.

76. Show that

$$\tan^{-1} x = \frac{\sin^{-1} x}{\cos^{-1} x}$$

   is *not* an identity.

77. Solve $e^{\cos x} = 1$ in $[0, 2\pi)$.

Chapter Test

1. Simplify: $\dfrac{2\cos^2 x - \cos x - 1}{\cos x - 1}$.

2. Given that $x = 2\sin\theta$, express $\sqrt{4 - x^2}$ as a trigonometric function without radicals. Assume $0 < \theta < \pi/2$.

3. Use a sum or difference identity to find $\sin 75°$ exactly.

4. Given that $\cos\theta = -\frac{2}{3}$ and that the terminal side is in quadrant II, find $\cos(\pi/2 - \theta)$.

5. Given that $\sin\theta = -\frac{4}{5}$ and θ is in quadrant III, find $\sin 2\theta$.

6. Use a half-angle identity to evaluate $\cos\dfrac{\pi}{12}$ exactly.

Prove each of the following identities.

7. $\csc x - \cos x \cot x = \sin x$

8. $(\sin x + \cos x)^2 = 1 + \sin 2x$

9. Find $\sin^{-1}\left(-\dfrac{\sqrt{2}}{2}\right)$ exactly in degrees.

10. Find $\arctan\sqrt{3}$ exactly in radians.

11. Use a calculator to find $\cos^{-1}(-0.6716)$ in radians, rounded to four decimal places.

12. Evaluate $\cos\left(\sin^{-1}\dfrac{1}{2}\right)$.

Solve, finding all solutions in $[0, 2\pi)$.

13. $4\cos^2 x = 3$

14. $2\sin^2 x = \sqrt{2}\sin x$

Synthesis

15. Find $\cos\theta$, given that $\cos 2\theta = \dfrac{5}{6}$, $\dfrac{3\pi}{2} < \theta < 2\pi$.

Applications of Trigonometry 7

Triangle trigonometry is important in applications such as large-scale construction, navigation, and surveying. In this chapter, we continue the study of triangle trigonometry that we began in Chapter 5. We will find that the trigonometric functions can be used to solve triangles that are not right triangles.

The study of complex numbers begun in Chapter 2 is also continued in this chapter. Complex numbers have applications in fields such as electronics and engineering. In addition, we introduce the polar coordinate system and graphs of polar equations.

The study of triangles also leads to the study of vectors. A vector is a quantity that has a direction. Vectors have many practical applications in the physical sciences.

APPLICATION

Rangers in fire detection lookout towers can use an Osborne fire finder to determine the location of a forest fire. (*Source*: National Interagency Fire Center, Boise, Idaho) If a ranger in fire tower A spots a fire at a direction of 295° and a ranger in fire tower B, located 45 mi at a direction of 045° from tower A, spots the same fire at a direction of 255°, how far from tower A is the fire? from tower B? We can use the law of sines to determine these distances.

This problem appears as Exercise 29 in Exercise Set 7.1.

7.1

The Law of Sines

- Use the law of sines to solve triangles.
- Find the area of any triangle given the lengths of two sides and the measure of the included angle.

To **solve a triangle** means to find the lengths of all its sides and the measures of all its angles. We solved right triangles in Section 5.2. For review, let's solve the right triangle shown below. We begin by listing the known measures.

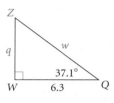

$$Q = 37.1° \qquad q = ?$$
$$W = 90° \qquad w = ?$$
$$Z = ? \qquad z = 6.3$$

Since the sum of the three angle measures of any triangle is 180°, we can immediately find the measure of the third angle:

$$Z = 180° - (90° + 37.1°)$$
$$= 52.9°.$$

Then using the tangent and cosine ratios, respectively, we can find q and w:

$$\tan 37.1° = \frac{q}{6.3}, \quad \text{or}$$
$$q = 6.3 \tan 37.1° \approx 4.8,$$

and

$$\cos 37.1° = \frac{6.3}{w}, \quad \text{or}$$
$$w = \frac{6.3}{\cos 37.1°} \approx 7.9.$$

Now all six measures are known and we have solved triangle QWZ.

$$Q = 37.1° \qquad q \approx 4.8$$
$$W = 90° \qquad w \approx 7.9$$
$$Z = 52.9° \qquad z = 6.3$$

Solving Oblique Triangles

The trigonometric functions can also be used to solve triangles that are not right triangles. Such triangles are called **oblique**. Any triangle, right or oblique, can be solved *if at least one side and any other two measures are known*. The five possible situations are illustrated on the next page.

1. AAS: Two angles of a triangle and a side opposite one of them are known.

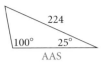

AAS

2. ASA: Two angles of a triangle and the included side are known.

ASA

3. SSA: Two sides of a triangle and an angle opposite one of them are known. (In this case, there may be no solution, one solution, or two solutions. The latter is known as the ambiguous case.)

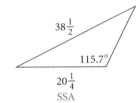

SSA

4. SAS: Two sides of a triangle and the included angle are known.

SAS

5. SSS: All three sides of the triangle are known.

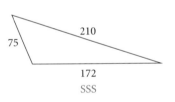

SSS

The list above does not include the situation in which only the three angle measures are given. The reason for this lies in the fact that the angle measures determine *only the shape* of the triangle and *not the size,* as shown with the following triangles. Thus we cannot solve a triangle when only the three angle measures are given.

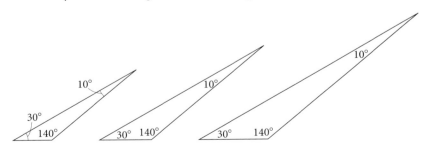

In order to solve oblique triangles, we need to derive the *law of sines* and the *law of cosines*. The law of sines applies to the first three situations listed above. The law of cosines, which we develop in Section 7.2, applies to the last two situations.

The Law of Sines

We consider any oblique triangle. It may or may not have an obtuse angle. Although we look at only the acute-triangle case, the derivation of the obtuse-triangle case is essentially the same.

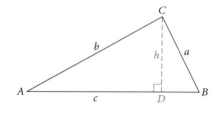

In acute $\triangle ABC$ at left, we have drawn an altitude from vertex C. It has length h. From $\triangle ADC$, we have

$$\sin A = \frac{h}{b}, \quad \text{or} \quad h = b \sin A.$$

From $\triangle BDC$, we have

$$\sin B = \frac{h}{a}, \quad \text{or} \quad h = a \sin B.$$

With $h = b \sin A$ and $h = a \sin B$, we now have

$$a \sin B = b \sin A$$

$$\frac{a \sin B}{\sin A \sin B} = \frac{b \sin A}{\sin A \sin B} \qquad \text{Dividing by } \sin A \sin B$$

$$\frac{a}{\sin A} = \frac{b}{\sin B}. \qquad \text{Simplifying}$$

There is no danger of dividing by 0 here because we are dealing with triangles whose angles are never 0° or 180°. Thus the sine value will never be 0.

If we were to consider altitudes from vertex A and vertex B in the triangle shown above, the same argument would give us

$$\frac{b}{\sin B} = \frac{c}{\sin C} \quad \text{and} \quad \frac{a}{\sin A} = \frac{c}{\sin C}.$$

We combine these results to obtain the law of sines.

The Law of Sines
In any triangle ABC,

$$\frac{a}{\sin A} = \frac{b}{\sin B} = \frac{c}{\sin C}.$$

In any triangle, the sides are proportional to the sines of the opposite angles.

Solving Triangles (AAS and ASA)

When two angles and a side of any triangle are known, the law of sines can be used to solve the triangle.

EXAMPLE 1   In $\triangle EFG$, $e = 4.56$, $E = 43°$, and $G = 57°$. Solve the triangle.

Solution   We first make a drawing. We know three of the six measures.

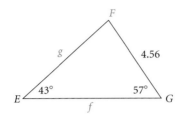

$$E = 43° \qquad e = 4.56$$
$$F = ? \qquad f = ?$$
$$G = 57° \qquad g = ?$$

From the figure, we see that we have the AAS situation. We begin by finding F:

$$F = 180° - (43° + 57°) = 80°.$$

We can now find the other two sides, using the law of sines:

$$\frac{f}{\sin F} = \frac{e}{\sin E}$$

$$\frac{f}{\sin 80°} = \frac{4.56}{\sin 43°} \qquad \text{Substituting}$$

$$f = \frac{4.56 \sin 80°}{\sin 43°} \qquad \text{Solving for } f$$

$$f \approx 6.58;$$

$$\frac{g}{\sin G} = \frac{e}{\sin E}$$

$$\frac{g}{\sin 57°} = \frac{4.56}{\sin 43°} \qquad \text{Substituting}$$

$$g = \frac{4.56 \sin 57°}{\sin 43°} \qquad \text{Solving for } g$$

$$g \approx 5.61.$$

Thus, we have solved the triangle:

$$E = 43°, \qquad e = 4.56,$$
$$F = 80°, \qquad f \approx 6.58,$$
$$G = 57°, \qquad g \approx 5.61.$$

The law of sines is frequently used in determining distances.

EXAMPLE 2 *Rescue Mission.* During a rescue mission, a Marine fighter pilot receives data on an unidentified aircraft from an AWACS plane and is instructed to intercept the aircraft. The diagram shown below appears on the screen, but before the distance to the point of interception appears on the screen, communications are jammed. Fortunately, the pilot remembers the law of sines. How far must the pilot fly?

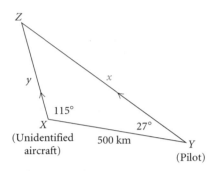

Solution We let x represent the distance that the pilot must fly in order to intercept the aircraft and Z represent the point of interception. We first find angle Z:

$$Z = 180° - (115° + 27°)$$
$$= 38°.$$

Because this application involves the ASA situation, we use the law of sines to determine x:

$$\frac{x}{\sin X} = \frac{z}{\sin Z}$$

$$\frac{x}{\sin 115°} = \frac{500}{\sin 38°} \qquad \text{Substituting}$$

$$x = \frac{500 \sin 115°}{\sin 38°} \qquad \text{Solving for } x$$

$$x \approx 736.$$

Thus the pilot must fly approximately 736 km in order to intercept the unidentified aircraft. ▬

Solving Triangles (SSA)

When two sides of a triangle and an angle opposite one of them are known, the law of sines can be used to solve the triangle.

Suppose for $\triangle ABC$ that b, c, and B are given. The various possibilities are as shown in the eight cases below: 5 cases when B is acute and 3 cases when B is obtuse. Note that $b < c$ in cases 1, 2, 3, and 6; $b = c$ in cases 4 and 7; and $b > c$ in cases 5 and 8.

Angle B Is Acute

Case 1: No solution
$b < c$; side b is too short to
reach the base. No triangle is
formed.

Case 2: One solution
$b < c$; side b just reaches the
base and is perpendicular to it.

Case 3: Two solutions
$b < c$; an arc of radius b
meets the base at two points.
(This case is called the
ambiguous case.)

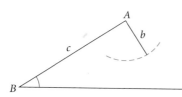

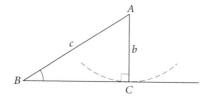

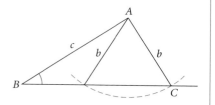

Case 4: One solution
$b = c$; an arc of radius b
meets the base at just one
point, other than B.

Case 5: One solution
$b > c$; an arc of radius b meets
the base at just one point.

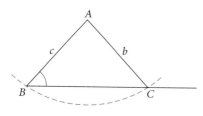

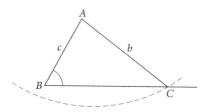

Angle B Is Obtuse

Case 6: No solution
$b < c$; side b is too short to
reach the base. No triangle is
formed.

Case 7: No solution
$b = c$; an arc of radius b meets
the base only at point B. No
triangle is formed.

Case 8: One solution
$b > c$; an arc of radius b meets
the base at just one point.

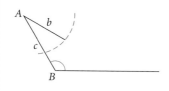

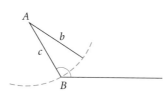

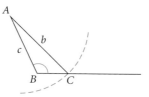

The eight cases above lead us to three possibilities in the SSA situation: *no* solution, *one* solution, or *two* solutions. Let's investigate these possibilities further, looking for ways to recognize the number of solutions.

EXAMPLE 3 *No solution.* In $\triangle QRS$, $q = 15$, $r = 28$, and $Q = 43.6°$. Solve the triangle.

Solution We make a drawing and list the known measures.

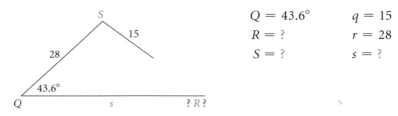

$$Q = 43.6° \qquad q = 15$$
$$R = ? \qquad r = 28$$
$$S = ? \qquad s = ?$$

We observe the SSA situation and use the law of sines to find R:

$$\frac{q}{\sin Q} = \frac{r}{\sin R}$$

$$\frac{15}{\sin 43.6°} = \frac{28}{\sin R} \qquad \text{Substituting}$$

$$\sin R = \frac{28 \sin 43.6°}{15} \qquad \text{Solving for } \sin R$$

$$\sin R \approx 1.2873.$$

Since there is no angle with a sine greater than 1, there is *no solution.*

EXAMPLE 4 *One solution.* In $\triangle XYZ$, $x = 23.5$, $y = 9.8$, and $X = 39.7°$. Solve the triangle.

Solution We make a drawing and organize the given information.

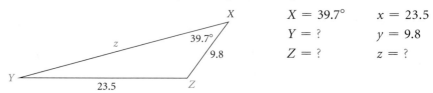

$$X = 39.7° \qquad x = 23.5$$
$$Y = ? \qquad y = 9.8$$
$$Z = ? \qquad z = ?$$

We see the SSA situation and begin by finding Y with the law of sines:

$$\frac{x}{\sin X} = \frac{y}{\sin Y}$$

$$\frac{23.5}{\sin 39.7°} = \frac{9.8}{\sin Y} \qquad \text{Substituting}$$

$$\sin Y = \frac{9.8 \sin 39.7°}{23.5} \qquad \text{Solving for } \sin Y$$

$$\sin Y \approx 0.2664.$$

Then $Y = 15.4°$ or $Y = 164.6°$, to the nearest tenth of a degree. An angle of $164.6°$ cannot be an angle of this triangle because it already has an angle of $39.7°$ and these two angles would total more than $180°$. Thus,

15.4° is the only possibility for Y. Therefore,

$$Z \approx 180° - (39.7° + 15.4°) \approx 124.9°.$$

We now find z:

$$\frac{z}{\sin Z} = \frac{x}{\sin X}$$

$$\frac{z}{\sin 124.9°} = \frac{23.5}{\sin 39.7°} \qquad \text{Substituting}$$

$$z = \frac{23.5 \sin 124.9°}{\sin 39.7°} \qquad \text{Solving for } z$$

$$z \approx 30.2.$$

We now have solved the triangle:

$$X = 39.7°, \qquad x = 23.5,$$
$$Y \approx 15.4°, \qquad y = 9.8,$$
$$Z \approx 124.9°, \qquad z \approx 30.2.$$

The next example illustrates the ambiguous case in which there are two possible solutions.

EXAMPLE 5 *Two solutions.* In $\triangle ABC$, $b = 15$, $c = 20$, and $B = 29°$. Solve the triangle.

Solution We make a drawing, list the known measures, and see that we again have the SSA (ambiguous) situation.

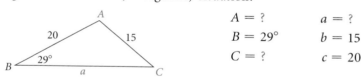

$$A = ? \qquad a = ?$$
$$B = 29° \qquad b = 15$$
$$C = ? \qquad c = 20$$

We first find C:

$$\frac{b}{\sin B} = \frac{c}{\sin C}$$

$$\frac{15}{\sin 29°} = \frac{20}{\sin C} \qquad \text{Substituting}$$

$$\sin C = \frac{20 \sin 29°}{15} \approx 0.6464. \qquad \text{Solving for } \sin C$$

There are two angles less than 180° with a sine of 0.6464. They are 40° and 140°, to the nearest degree. This gives us two possible solutions.

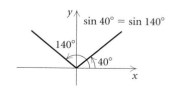

Possible Solution I.

If $C = 40°$, then

$$A = 180° - (29° + 40°) = 111°.$$

Then we find a:

$$\frac{a}{\sin A} = \frac{b}{\sin B}$$

$$\frac{a}{\sin 111°} = \frac{15}{\sin 29°}$$

$$a = \frac{15 \sin 111°}{\sin 29°} \approx 29.$$

These measures make a triangle as shown below; thus we have a solution.

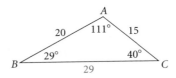

Possible Solution II.

If $C = 140°$, then

$$A = 180° - (29° + 140°) = 11°.$$

Then we find a:

$$\frac{a}{\sin A} = \frac{b}{\sin B}$$

$$\frac{a}{\sin 11°} = \frac{15}{\sin 29°}$$

$$a = \frac{15 \sin 11°}{\sin 29°} \approx 6.$$

These measures make a triangle as shown below; thus we have a second solution.

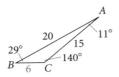

Examples 3–5 illustrate the SSA situation. Note that we need not memorize the eight cases or the procedures in finding no solution, one solution, or two solutions. When we are using the law of sines, the sine value leads us directly to the correct solution or solutions.

The Area of a Triangle

The familiar formula for the area of a triangle, $A = \frac{1}{2}bh$, can be used only when h is known. However, we can use the method used to derive the law of sines to derive an area formula that does not involve the height.

Consider a general triangle $\triangle ABC$, with area K, as shown below.

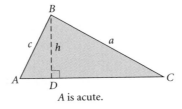
A is acute.

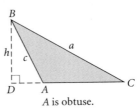
A is obtuse.

In each $\triangle ADB$,

$$\sin A = \frac{h}{c}, \quad \text{or} \quad h = c \sin A.$$

In the second triangle, $\sin A = h/c$ because $\sin A = \sin (180° - A)$. Substituting into the formula $K = \frac{1}{2}bh$, we get

$$K = \frac{1}{2}bc \sin A.$$

Any pair of sides and the included angle could have been used. Thus we also have

$$K = \frac{1}{2} ab \sin C \quad \text{and} \quad K = \frac{1}{2} ac \sin B.$$

The Area of a Triangle

The area K of any $\triangle ABC$ is one half the product of the lengths of two sides and the sine of the included angle:

$$K = \frac{1}{2} bc \sin A = \frac{1}{2} ab \sin C = \frac{1}{2} ac \sin B.$$

EXAMPLE 6 *Area of the Peace Monument.* Through the Mentoring in the City Program sponsored by Marian College, in Indianapolis, Indiana, children have turned a vacant downtown lot into a monument for peace.[*] This community project brought together neighborhood volunteers, businesses, and government in hopes of showing children how to develop positive, nonviolent ways of dealing with conflict. A landscape architect[†] used the children's drawings and ideas to design a triangular-shaped peace garden. Two sides of the property, formed by Indiana Avenue and Senate Avenue, measure 182 ft and 230 ft, respectively, and together form a 44.7° angle. The third side of the garden, formed by an apartment building, measures 163 ft. What is the area of this property?

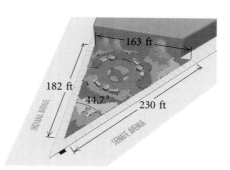

Solution Since we do not know a height of the triangle, we use the area formula:

$$K = \frac{1}{2} bc \sin A$$
$$K = \frac{1}{2} \cdot 182 \text{ ft} \cdot 230 \text{ ft} \cdot \sin 44.7°$$
$$K \approx 14{,}722 \text{ ft}^2.$$

The area of the property is approximately 14,722 ft².

---

[*]*The Indianapolis Star*, August 6, 1995, p. J8.
[†]Alan Day, a landscape architect with Browning Day Mullins Dierdorf, Inc., donated his time to this project.

Exercise Set

Solve the triangle, if possible.

1. $B = 38°$, $C = 21°$, $b = 24$
2. $A = 131°$, $C = 23°$, $b = 10$
3. $A = 36.5°$, $a = 24$, $b = 34$
4. $B = 118.3°$, $C = 45.6°$, $b = 42.1$
5. $C = 61°10'$, $c = 30.3$, $b = 24.2$
6. $A = 126.5°$, $a = 17.2$, $c = 13.5$
7. $c = 3$ mi, $B = 37.48°$, $C = 32.16°$
8. $a = 2345$ mi, $b = 2345$ mi, $A = 124.67°$
9. $b = 56.78$ yd, $c = 56.78$ yd, $C = 83.78°$
10. $A = 129°32'$, $C = 18°28'$, $b = 1204$ in.
11. $a = 20.01$ cm, $b = 10.07$ cm, $A = 30.3°$
12. $b = 4.157$ km, $c = 3.446$ km, $C = 51°48'$
13. $A = 89°$, $a = 15.6$ in., $b = 18.4$ in.
14. $C = 46°32'$, $a = 56.2$ m, $c = 22.1$ m
15. $a = 200$ m, $A = 32.76°$, $C = 21.97°$
16. $B = 115°$, $c = 45.6$ yd, $b = 23.8$ yd

Find the area of the triangle.

17. $B = 42°$, $a = 7.2$ ft, $c = 3.4$ ft
18. $A = 17°12'$, $b = 10$ in., $c = 13$ in.
19. $C = 82°54'$, $a = 4$ yd, $b = 6$ yd
20. $C = 75.16°$, $a = 1.5$ m, $b = 2.1$ m
21. $B = 135.2°$, $a = 46.12$ ft, $c = 36.74$ ft
22. $A = 113°$, $b = 18.2$ cm, $c = 23.7$ cm

Solve.

23. *Area of Back Yard.* A new homeowner has a triangular-shaped back yard. Two of the three sides measure 53 ft and 42 ft and form an included angle of 135°. To determine the amount of fertilizer and grass seed to be purchased, the owner has to know, or at least approximate, the area of the yard. Find the area of the yard to the nearest square foot.

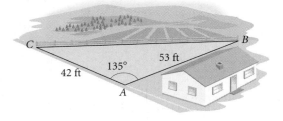

24. *Boarding Stable.* A rancher operates a boarding stable and temporarily needs to make an extra pen. He has a piece of rope 38 ft long and plans to tie the rope to one end of the barn (S) and run the rope around a tree (T) and back to the barn (Q). The tree is 21 ft from where the rope is first tied, and the rope from the barn to the tree makes an angle of 35° with the barn. Does the rancher have enough rope if he allows $4\frac{1}{2}$ ft at each end to fasten the rope?

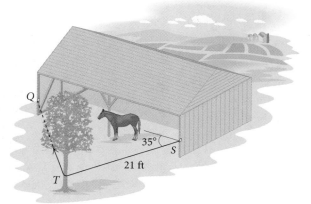

25. *Rock Concert.* In preparation for an outdoor rock concert, a stage crew must determine how far apart to place the two large speaker columns on stage. What generally works best is to place them at 50° angles to the center of the front row. The distance from the center of the front row to each of the speakers is 10 ft. How far apart does the crew need to place the speakers on stage?

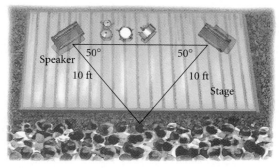

26. *Lunar Crater.* Points A and B are on opposite sides of a lunar crater. Point C is 50 m from A. The measure of $\angle BAC$ is determined to be 112° and the

measure of ∠*ACB* is determined to be 42°. What is the width of the crater?

27. *Length of Pole.* A pole leans away from the sun at an angle of 7° to the vertical. When the angle of elevation of the sun is 51°, the pole casts a shadow 47 ft long on level ground. How long is the pole?

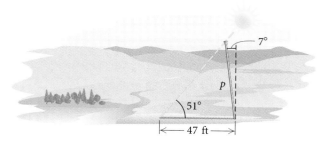

In Exercises 28–31, keep in mind the two types of bearing considered in Sections 5.2 and 5.3.

28. *Reconnaissance Airplane.* A reconnaissance airplane leaves its airport on the east coast of the United States and flies in a direction of 085°. Because of bad weather, it returns to another airport 230 km to the north of its home base. For the return trip, it flies in a direction of 283°. What is the total distance that the airplane flew?

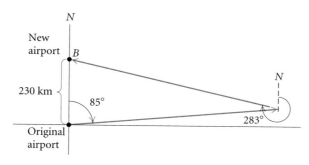

29. *Fire Tower.* A ranger in fire tower A spots a fire at a direction of 295°. A ranger in fire tower B, located 45 mi at a direction of 045° from tower A, spots the same fire at a direction of 255°. How far from tower A is the fire? from tower B?

30. *Lighthouse.* A boat leaves a lighthouse A and sails 5.1 km. At this time it is sighted from lighthouse B, 7.2 km west of A. The bearing of the boat from B is N65°10′E. How far is the boat from B?

31. *Mackinac Island.* Mackinac Island is located 35 mi N65°20′W of Cheboygan, Michigan, where the Coast Guard cutter Mackinaw is stationed. A freighter in distress radios the Coast Guard cutter

for help. It radios its position as N25°40′E of Mackinac Island and N10°10′W of Cheboygan. How far is the freighter from Cheboygan?

32. *Gears.* Three gears are arranged as shown in the figure below. Find the angle *ϕ*.

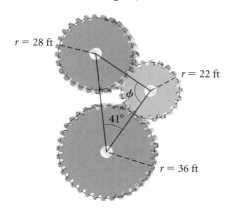

Collaborative Discussion and Writing

33. Explain why the law of sines cannot be used to find the first angle when solving a triangle given three sides.

34. We considered eight cases of solving triangles given two sides and an angle opposite one of them. Describe the relationship between side *b* and the height *h* in each.

Skill Maintenance

Find the acute angle, A, in both radians and degrees, for the given function value.

35. cos *A* = 0.2213

36. cos *A* = 1.5612

Convert to decimal degree notation.

37. 18°14′20″

38. 125°3′42″

Synthesis

39. Prove the following area formulas for a general triangle *ABC* with area represented by *K*.

$$K = \frac{a^2 \sin B \sin C}{2 \sin A}$$

$$K = \frac{c^2 \sin A \sin B}{2 \sin C}$$

$$K = \frac{b^2 \sin C \sin A}{2 \sin B}$$

40. *Area of a Parallelogram.* Prove that the area of a parallelogram is the product of two sides and the sine of the included angle.

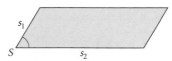

41. *Area of a Quadrilateral.* Prove that the area of a quadrilateral is one half the product of the lengths of its diagonals and the sine of the angle between the diagonals.

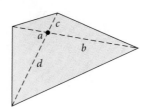

42. Find *d*.

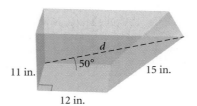

7.2

The Law of Cosines

- *Use the law of cosines to solve triangles.*
- *Determine whether the law of sines or the law of cosines should be applied to solve a triangle.*

The law of sines is used to solve triangles given a side and two angles (AAS and ASA) or given two sides and an angle opposite one of them (SSA). A second law, called the *law of cosines,* is needed to solve triangles given two sides and the included angle (SAS) or given three sides (SSS).

The Law of Cosines

To derive this property, we consider any △*ABC* placed on a coordinate system. We position the origin at one of the vertices—say, *C*—and the positive half of the *x*-axis along one of the sides—say, *CB*. Let (x, y) be the coordinates of vertex *A*. Point *B* has coordinates $(a, 0)$ and point *C* has coordinates $(0, 0)$.

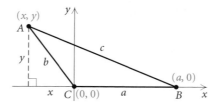

Then

$$\cos C = \frac{x}{b}, \quad \text{so} \quad x = b \cos C$$

and

$$\sin C = \frac{y}{b}, \quad \text{so} \quad y = b \sin C.$$

Thus point A has coordinates

$$(b \cos C, \; b \sin C).$$

Next, we use the distance formula to determine c^2:

$$c^2 = (x - a)^2 + (y - 0)^2,$$

or $\quad c^2 = (b \cos C - a)^2 + (b \sin C - 0)^2.$

Now we multiply and simplify:

$$\begin{aligned}
c^2 &= b^2 \cos^2 C - 2ab \cos C + a^2 + b^2 \sin^2 C \\
&= a^2 + b^2(\sin^2 C + \cos^2 C) - 2ab \cos C \\
&= a^2 + b^2 - 2ab \cos C.
\end{aligned}$$

Had we placed the origin at one of the other vertices, we would have obtained

$$a^2 = b^2 + c^2 - 2bc \cos A$$

or $\quad b^2 = a^2 + c^2 - 2ac \cos B.$

The Law of Cosines

In any triangle ABC,

$$a^2 = b^2 + c^2 - 2bc \cos A,$$
$$b^2 = a^2 + c^2 - 2ac \cos B,$$

or $\quad c^2 = a^2 + b^2 - 2ab \cos C.$

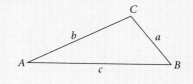

Thus, in any triangle, the square of a side is the sum of the squares of the other two sides, minus twice the product of those sides and the cosine of the included angle. When the included angle is 90°, the law of cosines reduces to the Pythagorean theorem.

Solving Triangles (SAS)

When two sides of a triangle and the included angle are known, we can use the law of cosines to find the third side. The law of cosines or the law of sines can then be used to finish solving the triangle.

EXAMPLE 1   Solve $\triangle ABC$ if $a = 32$, $c = 48$, and $B = 125.2°$.

Solution   We first label a triangle with the known and unknown measures.

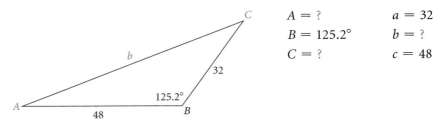

$A = ?$	$a = 32$
$B = 125.2°$	$b = ?$
$C = ?$	$c = 48$

We can find the third side using the law of cosines, as follows:

$$b^2 = a^2 + c^2 - 2ac \cos B$$
$$b^2 = 32^2 + 48^2 - 2 \cdot 32 \cdot 48 \cos 125.2° \qquad \text{Substituting}$$
$$b^2 \approx 5098.8$$
$$b \approx 71.$$

We now have $a = 32$, $b \approx 71$, and $c = 48$, and we need to find the other two angle measures. At this point, we can find them in two ways. One way uses the law of sines. The ambiguous case may arise, however, and we would have to be alert to this possibility. The advantage of using the law of cosines again is that if we solve for the cosine and find that its value is *negative*, then we know that the angle is obtuse. If the value of the cosine is *positive*, then the angle is acute. Thus we use the law of cosines to find a second angle.

Let's find angle A. We select the formula from the law of cosines that contains $\cos A$ and substitute:

$$a^2 = b^2 + c^2 - 2bc \cos A$$
$$32^2 = 71^2 + 48^2 - 2 \cdot 71 \cdot 48 \cos A \qquad \text{Substituting}$$
$$1024 = 5041 + 2304 - 6816 \cos A$$
$$-6321 = -6816 \cos A$$
$$\cos A \approx 0.9273768$$
$$A \approx 22.0°.$$

The third angle is now easy to find:

$$C \approx 180° - (125.2° + 22.0°)$$
$$\approx 32.8°.$$

Thus,

$A \approx 22.0°,$	$a = 32,$
$B = 125.2°,$	$b \approx 71,$
$C \approx 32.8°,$	$c = 48.$

Due to errors created by rounding, answers may vary depending on the order in which they are found. Had we found the measure of angle C first in Example 1, the angle measures would have been $C \approx 34.1°$ and $A \approx 20.7°$. Variances in rounding also change the answers. Had we used 71.4 for b in Example 1, the angle measures would have been $A \approx 21.5°$ and $C \approx 33.3°$.

Suppose we used the law of sines at the outset in Example 1 to find b. We were given only three measures: $a = 32$, $c = 48$, and $B = 125.2°$. When substituting these measures into the proportions, we see that there is not enough information to use the law of sines:

$$\frac{a}{\sin A} = \frac{b}{\sin B} \rightarrow \frac{32}{\sin A} = \frac{b}{\sin 125.2°},$$

$$\frac{b}{\sin B} = \frac{c}{\sin C} \rightarrow \frac{b}{\sin 125.2°} = \frac{48}{\sin C},$$

$$\frac{a}{\sin A} = \frac{c}{\sin C} \rightarrow \frac{32}{\sin A} = \frac{48}{\sin C}.$$

In all three situations, the resulting equation, after the substitutions, still has two unknowns. Thus we cannot easily use the law of sines to find b.

Solving Triangles (SSS)

When all three sides of a triangle are known, the law of cosines can be used to solve the triangle.

EXAMPLE 2 Solve $\triangle RST$ if $r = 3.5$, $s = 4.7$, and $t = 2.8$.

Solution We sketch a triangle and label it with the given measures.

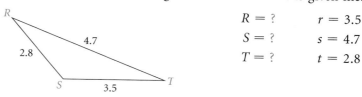

$$R = ? \qquad r = 3.5$$
$$S = ? \qquad s = 4.7$$
$$T = ? \qquad t = 2.8$$

Since we do not know any of the angle measures, we cannot use the law of sines. We begin instead by finding an angle with the law of cosines. We choose to find S first and select the formula that contains $\cos S$:

$$s^2 = r^2 + t^2 - 2rt \cos S$$
$$(4.7)^2 = (3.5)^2 + (2.8)^2 - 2(3.5)(2.8) \cos S \qquad \text{Substituting}$$
$$\cos S = \frac{(3.5)^2 + (2.8)^2 - (4.7)^2}{2(3.5)(2.8)}$$
$$\cos S \approx -0.1020408$$
$$S \approx 95.86°.$$

Similarly, we find angle R:

$$r^2 = s^2 + t^2 - 2st \cos R$$

$$(3.5)^2 = (4.7)^2 + (2.8)^2 - 2(4.7)(2.8) \cos R$$

$$\cos R = \frac{(4.7)^2 + (2.8)^2 - (3.5)^2}{2(4.7)(2.8)}$$

$$\cos R \approx 0.6717325$$

$$R \approx 47.80°.$$

Then

$$T \approx 180° - (95.86° + 47.80°) \approx 36.34°.$$

Thus,

$$R \approx 47.80°, \qquad r = 3.5,$$
$$S \approx 95.86°, \qquad s = 4.7,$$
$$T \approx 36.34°, \qquad t = 2.8.$$

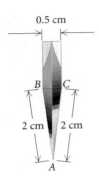

0.5 cm

B C

2 cm   2 cm

A

EXAMPLE 3 *Knife Bevel.* Knifemakers know that the *bevel* of the blade (the angle formed at the cutting edge of the blade) determines the cutting characteristics of the knife. A small bevel like that of a straight razor makes for a keen edge, but is impractical for heavy-duty cutting because the edge dulls quickly and is prone to chipping. A large bevel is suitable for heavy-duty work like chopping wood. Survival knives, being universal in application, are a compromise between small and large bevels. The diagram at left illustrates the blade of a hand-made Randall Model 18 survival knife. What is its bevel? (*Source*: Randall Made Knives, P.O. Box 1988, Orlando, FL 32802)

Solution   We know three sides of a triangle. We can use the law of cosines to find the bevel, angle A.

$$a^2 = b^2 + c^2 - 2bc \cos A$$

$$(0.5)^2 = 2^2 + 2^2 - 2 \cdot 2 \cdot 2 \cdot \cos A$$

$$0.25 = 4 + 4 - 8 \cos A$$

$$\cos A = \frac{4 + 4 - 0.25}{8}$$

$$\cos A = 0.96875$$

$$A \approx 14.36°.$$

Thus the bevel is approximately $14.36°$.

CONNECTING THE CONCEPTS

CHOOSING THE APPROPRIATE LAW

The following summarizes the situations in which to use the law of sines and the law of cosines.

To solve an oblique triangle:

Use the *law of sines* for:	Use the *law of cosines* for:
AAS	SAS
ASA	SSS
SSA	

The law of cosines can also be used for the SSA situation, but since the process involves solving a quadratic equation, we do not include that option in the list above.

EXAMPLE 4 In $\triangle ABC$, three measures are given. Determine which law to use when solving the triangle. You need not solve the triangle.

a) $a = 14$, $b = 23$, $c = 10$
b) $a = 207$, $B = 43.8°$, $C = 57.6°$
c) $A = 112°$, $C = 37°$, $a = 84.7$
d) $B = 101°$, $a = 960$, $c = 1042$
e) $b = 17.26$, $a = 27.29$, $A = 39°$
f) $A = 61°$, $B = 39°$, $C = 80°$

Solution It is helpful to make a drawing of a triangle with the given information. The triangle need not be drawn to scale. The given parts are shown in color.

FIGURE	SITUATION	LAW TO USE
a) 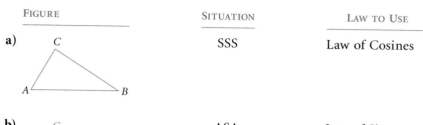	SSS	Law of Cosines
b) 	ASA	Law of Sines

FIGURE	SITUATION	LAW TO USE
c) 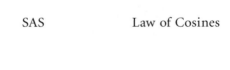	AAS	Law of Sines
d) 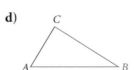	SAS	Law of Cosines
e) 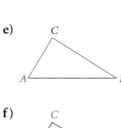	SSA	Law of Sines
f) 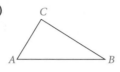	AAA	Cannot be solved

STUDY TIP

The InterAct Math Tutorial software that accompanies this text provides practice exercises that correlate at the objective level to the odd-numbered exercises in the text. Each practice exercise is accompanied by an example and guided solution designed to involve students in the solution process. This software is available in your campus lab or on CD-ROM.

Exercise Set 7.2

Solve the triangle, if possible.

1. $A = 30°$, $b = 12$, $c = 24$

2. $B = 133°$, $a = 12$, $c = 15$

3. $a = 12$, $b = 14$, $c = 20$

4. $a = 22.3$, $b = 22.3$, $c = 36.1$

5. $B = 72°40'$, $c = 16$ m, $a = 78$ m

6. $C = 22.28°$, $a = 25.4$ cm, $b = 73.8$ cm

7. $a = 16$ m, $b = 20$ m, $c = 32$ m

8. $B = 72.66°$, $a = 23.78$ km, $c = 25.74$ km

9. $a = 2$ ft, $b = 3$ ft, $c = 8$ ft

10. $A = 96°13'$, $b = 15.8$ yd, $c = 18.4$ yd

11. $a = 26.12$ km, $b = 21.34$ km, $c = 19.25$ km

12. $C = 28°43'$, $a = 6$ mm, $b = 9$ mm

13. $a = 60.12$ mi, $b = 40.23$ mi, $C = 48.7°$

14. $a = 11.2$ cm, $b = 5.4$ cm, $c = 7$ cm

15. $b = 10.2$ in., $c = 17.3$ in., $A = 53.456°$

16. $a = 17$ yd, $b = 15.4$ yd, $c = 1.5$ yd

Determine which law applies. Then solve the triangle.

17. $A = 70°$, $B = 12°$, $b = 21.4$

18. $a = 15$, $c = 7$, $B = 62°$

19. $a = 3.3$, $b = 2.7$, $c = 2.8$

20. $a = 1.5$, $b = 2.5$, $A = 58°$

21. $A = 40.2°$, $B = 39.8°$, $C = 100°$

22. $a = 60$, $b = 40$, $C = 47°$

23. $a = 3.6$, $b = 6.2$, $c = 4.1$

24. $B = 110°30'$, $C = 8°10'$, $c = 0.912$

Solve.

25. *Poachers.* A park ranger establishes an observation post from which to watch for poachers. Despite losing her map, the ranger does have a compass and a rangefinder. She observes some poachers, and the rangefinder indicates that they are 500 ft from her position. They are headed toward big game that she knows to be 375 ft from her position. Using her compass, she finds that the poachers' azimuth (the direction measured as an angle from north) is 355° and that of the big game is 42°. What is the distance between the poachers and the game?

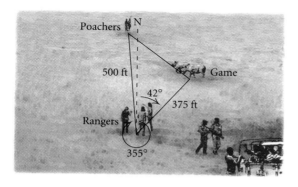

26. *Circus Highwire Act.* A circus highwire act walks up an approach wire to reach a highwire. The approach wire is 122 ft long and is currently anchored so that it forms the maximum allowable angle of 35° with the ground. A greater approach angle causes the aerialists to slip. However, the aerialists find that there is enough room to anchor the approach wire 30 ft back in order to make the approach angle less severe. When this is done, how much farther will they have to walk up the approach wire, and what will the new approach angle be?

27. *In-line Skater.* An in-line skater skates on a fitness trail along the Pacific Ocean from point *A* to point *B*. As shown below, two streets intersecting at point *C* also intersect the trail at *A* and *B*. In his

car, the skater found the lengths of *AC* and *BC* to be approximately 0.5 mi and 1.3 mi, respectively. From a map, he estimates the included angle at *C* to be 110°. How far did he skate from *A* to *B*?

28. *Baseball Bunt.* A batter in a baseball game drops a bunt down the first-base line. It rolls 34 ft at an angle of 25° with the base path. The pitcher's mound is 60.5 ft from home plate. How far must the pitcher travel to pick up the ball? (*Hint*: A baseball diamond is a square.)

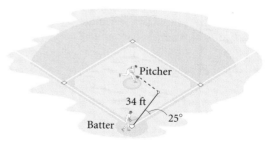

29. *Ships.* Two ships leave harbor at the same time. The first sails N15°W at 25 knots (a knot is one nautical mile per hour). The second sails N32°E at 20 knots. After 2 hr, how far apart are the ships?

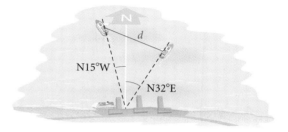

30. *Survival Trip.* A group of college students is learning to navigate for an upcoming survival trip. On a map, they have been given three points at which they are to check in. The map also shows the distances between the points. However, to navigate they need to know the angle measurements. Calculate the angles for them.

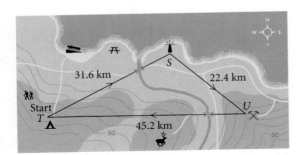

31. *Airplanes.*   Two airplanes leave an airport at the same time. The first flies 150 km/h in a direction of 320°. The second flies 200 km/h in a direction of 200°. After 3 hr, how far apart are the planes?

32. *Slow-pitch Softball.*   A slow-pitch softball diamond is a square 65 ft on a side. The pitcher's mound is 46 ft from home. How far is it from the pitcher's mound to first base?

33. *Isosceles Trapezoid.*   The longer base of an isosceles trapezoid measures 14 ft. The nonparallel sides measure 10 ft, and the base angles measure 80°.

  a) Find the length of a diagonal.
  b) Find the area.

34. *Area of Sail.*   A sail that is in the shape of an isosceles triangle has a vertex angle of 38°. The angle is included by two sides, each measuring 20 ft. Find the area of the sail.

35. Three circles are arranged as shown in the figure below. Find the length *PQ*.

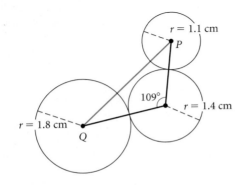

36. *Swimming Pool.*   A triangular swimming pool measures 44 ft on one side and 32.8 ft on another side. These sides form an angle that measures 40.8°. How long is the other side?

Collaborative Discussion and Writing

37. Try to solve this triangle using the law of cosines. Then explain why it is easier to solve it using the law of sines.

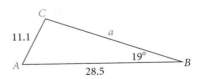

38. Explain why we cannot solve a triangle given SAS with the law of sines.

Skill Maintenance

39. Find the absolute value: $|-5|$.

Find the values.

40. $\cos \dfrac{\pi}{6}$

41. $\sin 45°$

42. $\sin 300°$

43. $\cos \left(-\dfrac{2\pi}{3}\right)$

44. Multiply: $(1 - i)(1 + i)$.

Synthesis

45. *Canyon Depth.*   A bridge is being built across a canyon. The length of the bridge is 5045 ft. From the deepest point in the canyon, the angles of elevation of the ends of the bridge are 78° and 72°. How deep is the canyon?

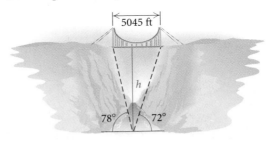

46. *Heron's Formula.*   If *a*, *b*, and *c* are the lengths of the sides of a triangle, then the area *K* of the triangle is given by

$$K = \sqrt{s(s - a)(s - b)(s - c)},$$

where $s = \frac{1}{2}(a + b + c)$. The number *s* is called the *semiperimeter*. Prove Heron's formula. (*Hint*: Use the area formula $K = \frac{1}{2}bc \sin A$ developed in Section 7.1.) Then use Heron's formula to find the area of the triangular swimming pool described in Exercise 36.

47. *Area of Isosceles Triangle.*   Find a formula for the area of an isosceles triangle in terms of the congruent sides and their included angle. Under what conditions will the area of a triangle with fixed congruent sides be maximum?

48. *Reconnaissance Plane.*   A reconnaissance plane patrolling at 5000 ft sights a submarine at bearing 35° and at an angle of depression of 25°. A carrier is at bearing 105° and at an angle of depression of 60°. How far is the submarine from the carrier?

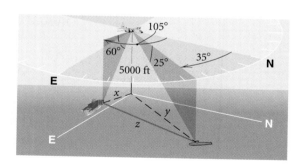

7.3

Complex Numbers: Trigonometric Form

- *Graph complex numbers.*
- *Given a complex number in standard form, find trigonometric, or polar, notation; and given a complex number in trigonometric form, find standard notation.*
- *Use trigonometric notation to multiply and divide complex numbers.*
- *Use DeMoivre's theorem to raise complex numbers to powers.*
- *Find the nth roots of a complex number.*

Graphical Representation

Just as real numbers can be graphed on a line, complex numbers can be graphed on a plane. We graph a complex number $a + bi$ in the same way that we graph an ordered pair of real numbers (a, b). However, in place of an x-axis, we have a real axis, and in place of a y-axis, we have an imaginary axis. Horizontal distances correspond to the real part of a number. Vertical distances correspond to the imaginary part.

COMPLEX NUMBERS
REVIEW SECTION 2.2.

EXAMPLE 1   Graph each of the following complex numbers.

a) $3 + 2i$ **b)** $-4 - 5i$ **c)** $-3i$

d) $-1 + 3i$ **e)** 2

Solution

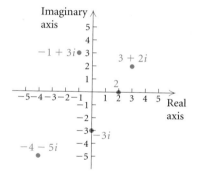

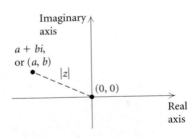

We recall that the absolute value of a real number is its distance from 0 on the number line. The absolute value of a complex number is its distance from the origin in the complex plane. For example, if $z = a + bi$, then using the distance formula, we have

$$|z| = |a + bi| = \sqrt{(a - 0)^2 + (b - 0)^2} = \sqrt{a^2 + b^2}.$$

Absolute Value of a Complex Number

The **absolute value of a complex number** $a + bi$ is

$$|a + bi| = \sqrt{a^2 + b^2}.$$

EXAMPLE 2   Find the absolute value of each of the following.

a) $3 + 4i$ **b)** $-2 - i$ **c)** $\dfrac{4}{5}i$

Solution

a) $|3 + 4i| = \sqrt{3^2 + 4^2} = \sqrt{9 + 16} = \sqrt{25} = 5$

b) $|-2 - i| = \sqrt{(-2)^2 + (-1)^2} = \sqrt{5}$

c) $\left|\dfrac{4}{5}i\right| = \left|0 + \dfrac{4}{5}i\right| = \sqrt{0^2 + \left(\dfrac{4}{5}\right)^2} = \dfrac{4}{5}$

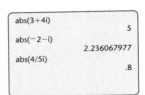
Trigonometric Notation for Complex Numbers

Now let's consider a nonzero complex number $a + bi$. Suppose that its absolute value is r. If we let θ be an angle in standard position whose terminal side passes through the point (a, b), as shown in the figure, then

$$\cos \theta = \frac{a}{r}, \quad \text{or} \quad a = r \cos \theta$$

and

$$\sin \theta = \frac{b}{r}, \quad \text{or} \quad b = r \sin \theta.$$

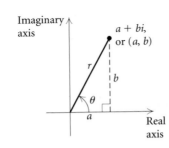

Substituting these values for a and b into the $(a + bi)$ notation, we get

$$a + bi = r \cos \theta + (r \sin \theta)i$$
$$= r(\cos \theta + i \sin \theta).$$

This is **trigonometric notation** for a complex number $a + bi$. The number r is called the **absolute value** of $a + bi$, and θ is called the **argument** of $a + bi$. Trigonometric notation for a complex number is also called **polar notation**.

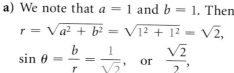

Trigonometric Notation for Complex Numbers

$$a + bi = r(\cos \theta + i \sin \theta)$$

To find trigonometric notation for a complex number given in **standard notation**, $a + bi$, we must find r and determine the angle θ for which $\sin \theta = b/r$ and $\cos \theta = a/r$.

EXAMPLE 3 Find trigonometric notation for each of the following complex numbers.

a) $1 + i$ **b)** $\sqrt{3} - i$

Solution

a) We note that $a = 1$ and $b = 1$. Then

$$r = \sqrt{a^2 + b^2} = \sqrt{1^2 + 1^2} = \sqrt{2},$$

$$\sin \theta = \frac{b}{r} = \frac{1}{\sqrt{2}}, \quad \text{or} \quad \frac{\sqrt{2}}{2},$$

and

$$\cos \theta = \frac{1}{\sqrt{2}}, \quad \text{or} \quad \frac{\sqrt{2}}{2}.$$

Since θ is in quadrant I, $\theta = \pi/4$, or 45°, and we have

$$1 + i = \sqrt{2}\left(\cos \frac{\pi}{4} + i \sin \frac{\pi}{4}\right),$$

or

$$1 + i = \sqrt{2}(\cos 45° + i \sin 45°).$$

b) We see that $a = \sqrt{3}$ and $b = -1$. Then

$$r = \sqrt{(\sqrt{3})^2 + (-1)^2} = 2,$$

$$\sin \theta = \frac{-1}{2} = -\frac{1}{2},$$

and

$$\cos \theta = \frac{\sqrt{3}}{2}.$$

Technology Connection

When finding trigonometric notation for complex numbers, as in Example 3, we can use a graphing calculator to determine angle values in degrees.

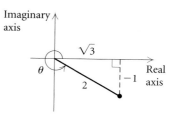

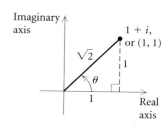

Since θ is in quadrant IV, $\theta = 11\pi/6$, or $330°$, and we have

$$\sqrt{3} - i = 2\left(\cos\frac{11\pi}{6} + i\sin\frac{11\pi}{6}\right),$$

or

$$\sqrt{3} - i = 2(\cos 330° + i\sin 330°).$$

    In changing to trigonometric notation, note that there are many angles satisfying the given conditions. We ordinarily choose the *smallest positive* angle.

    To change from trigonometric notation to standard notation, $a + bi$, we recall that $a = r\cos\theta$ and $b = r\sin\theta$.

EXAMPLE 4 Find standard notation, $a + bi$, for each of the following complex numbers.

a) $2(\cos 120° + i\sin 120°)$ **b)** $\sqrt{8}\left(\cos\frac{7\pi}{4} + i\sin\frac{7\pi}{4}\right)$

Solution

a) Rewriting, we have

$$2(\cos 120° + i\sin 120°) = 2\cos 120° + (2\sin 120°)i.$$

Thus,

$$a = 2\cos 120° = 2\cdot\left(-\frac{1}{2}\right) = -1$$

and

$$b = 2\sin 120° = 2\cdot\frac{\sqrt{3}}{2} = \sqrt{3},$$

so

$$2(\cos 120° + i\sin 120°) = -1 + \sqrt{3}i.$$

b) Rewriting, we have

$$\sqrt{8}\left(\cos\frac{7\pi}{4} + i\sin\frac{7\pi}{4}\right) = \sqrt{8}\cos\frac{7\pi}{4} + \left(\sqrt{8}\sin\frac{7\pi}{4}\right)i.$$

Thus,

$$a = \sqrt{8}\cos\frac{7\pi}{4} = \sqrt{8}\cdot\frac{\sqrt{2}}{2} = 2$$

and

$$b = \sqrt{8}\sin\frac{7\pi}{4} = \sqrt{8}\cdot\left(-\frac{\sqrt{2}}{2}\right) = -2,$$

Technology Connection

We can perform the computations in Example 4 on a graphing calculator.

Degree Mode

```
2(cos(120)+isin(120))
           −1+1.732050808i
```

Radian Mode

```
√(8)(cos(7π/4)+isin(7π/4))
                    2−2i
```

so

$$\sqrt{8}\left(\cos\frac{7\pi}{4} + i\sin\frac{7\pi}{4}\right) = 2 - 2i.$$

Multiplication and Division with Trigonometric Notation

Multiplication of complex numbers is easier to manage with trigonometric notation than with standard notation. We simply multiply the absolute values and add the arguments. Let's state this in a more formal manner.

Complex Numbers: Multiplication

For any complex numbers $r_1(\cos\theta_1 + i\sin\theta_1)$ and $r_2(\cos\theta_2 + i\sin\theta_2)$,

$$r_1(\cos\theta_1 + i\sin\theta_1) \cdot r_2(\cos\theta_2 + i\sin\theta_2)$$
$$= r_1r_2[\cos(\theta_1 + \theta_2) + i\sin(\theta_1 + \theta_2)].$$

PROOF

$$r_1(\cos\theta_1 + i\sin\theta_1) \cdot r_2(\cos\theta_2 + i\sin\theta_2) =$$
$$r_1r_2(\cos\theta_1\cos\theta_2 - \sin\theta_1\sin\theta_2) + r_1r_2(\sin\theta_1\cos\theta_2 + \cos\theta_1\sin\theta_2)i$$

Now, using identities for sums of angles, we simplify, obtaining

$$r_1r_2\cos(\theta_1 + \theta_2) + r_1r_2\sin(\theta_1 + \theta_2)i,$$

or

$$r_1r_2[\cos(\theta_1 + \theta_2) + i\sin(\theta_1 + \theta_2)],$$

which was to be shown.

EXAMPLE 5 Multiply and express the answer to each of the following in standard notation.

a) $3(\cos 40° + i\sin 40°)$ and $4(\cos 20° + i\sin 20°)$

b) $2(\cos\pi + i\sin\pi)$ and $3\left[\cos\left(-\frac{\pi}{2}\right) + i\sin\left(-\frac{\pi}{2}\right)\right]$

Solution

a) $3(\cos 40° + i\sin 40°) \cdot 4(\cos 20° + i\sin 20°)$

$$= 3\cdot 4\cdot[\cos(40° + 20°) + i\sin(40° + 20°)]$$
$$= 12(\cos 60° + i\sin 60°)$$
$$= 12\left(\frac{1}{2} + \frac{\sqrt{3}}{2}i\right)$$
$$= 6 + 6\sqrt{3}i$$

Technology Connection

We can multiply complex numbers on a graphing calculator. The products in Example 5 are shown below.

Degree Mode

```
3(cos(40)+isin(40))*4(cos(20)+
isin(20))
                  6+10.39230485i
```

Radian Mode

```
2(cos(π)+isin(π))*3(cos(−π/2)+
isin(−π/2))
                            6i
```

b) $2(\cos \pi + i \sin \pi) \cdot 3\left[\cos\left(-\frac{\pi}{2}\right) + i \sin\left(-\frac{\pi}{2}\right)\right]$

$$= 2 \cdot 3 \cdot \left[\cos\left(\pi + \left(-\frac{\pi}{2}\right)\right) + i \sin\left(\pi + \left(-\frac{\pi}{2}\right)\right)\right]$$

$$= 6\left(\cos\frac{\pi}{2} + i \sin\frac{\pi}{2}\right)$$

$$= 6(0 + i \cdot 1)$$

$$= 6i$$

EXAMPLE 6   Convert to trigonometric notation and multiply:

$$(1 + i)(\sqrt{3} - i).$$

Solution   We first find trigonometric notation:

$$1 + i = \sqrt{2}(\cos 45° + i \sin 45°), \qquad \text{See Example 3(a).}$$
$$\sqrt{3} - i = 2(\cos 330° + i \sin 330°). \qquad \text{See Example 3(b).}$$

Then we multiply:

$\sqrt{2}(\cos 45° + i \sin 45°) \cdot 2(\cos 330° + i \sin 330°)$

$$= 2\sqrt{2}[\cos(45° + 330°) + i \sin(45° + 330°)]$$
$$= 2\sqrt{2}(\cos 375° + i \sin 375°)$$
$$= 2\sqrt{2}(\cos 15° + i \sin 15°). \qquad \text{375° has the same terminal side as 15°.}$$

To divide complex numbers, we divide the absolute values and subtract the arguments. We state this fact below, but omit the proof.

Complex Numbers: Division

For any complex numbers $r_1(\cos \theta_1 + i \sin \theta_1)$ and $r_2(\cos \theta_2 + i \sin \theta_2)$, $r_2 \neq 0$,

$$\frac{r_1(\cos \theta_1 + i \sin \theta_1)}{r_2(\cos \theta_2 + i \sin \theta_2)} = \frac{r_1}{r_2}[\cos(\theta_1 - \theta_2) + i \sin(\theta_1 - \theta_2)].$$

EXAMPLE 7   Divide

$$2\left(\cos\frac{3\pi}{2} + i \sin\frac{3\pi}{2}\right) \quad \text{by} \quad 4\left(\cos\frac{\pi}{2} + i \sin\frac{\pi}{2}\right)$$

and express the solution in standard notation.

Solution   We have

$$2\left(\cos \frac{3\pi}{2} + i \sin \frac{3\pi}{2}\right) \over 4\left(\cos \frac{\pi}{2} + i \sin \frac{\pi}{2}\right)$$

$$= \frac{2}{4}\left[\cos\left(\frac{3\pi}{2} - \frac{\pi}{2}\right) + i \sin\left(\frac{3\pi}{2} - \frac{\pi}{2}\right)\right]$$

$$= \frac{1}{2}(\cos \pi + i \sin \pi)$$

$$= \frac{1}{2}(-1 + i \cdot 0)$$

$$= -\frac{1}{2}.$$

EXAMPLE 8   Convert to trigonometric notation and divide:

$$\frac{1 + i}{1 - i}.$$

Solution   We first convert to trigonometric notation:

$$1 + i = \sqrt{2}(\cos 45° + i \sin 45°), \qquad \text{See Example 3(a).}$$
$$1 - i = \sqrt{2}(\cos 315° + i \sin 315°).$$

We now divide:

$$\frac{\sqrt{2}(\cos 45° + i \sin 45°)}{\sqrt{2}(\cos 315° + i \sin 315°)}$$

$$= 1[\cos (45° - 315°) + i \sin (45° - 315°)]$$
$$= \cos (-270°) + i \sin (-270°)$$
$$= 0 + i \cdot 1 = i.$$

Powers of Complex Numbers

An important theorem about powers and roots of complex numbers is named for the French mathematician Abraham DeMoivre (1667–1754). Let's consider the square of a complex number $r(\cos \theta + i \sin \theta)$:

$$[r(\cos \theta + i \sin \theta)]^2 = [r(\cos \theta + i \sin \theta)] \cdot [r(\cos \theta + i \sin \theta)]$$
$$= r \cdot r \cdot [\cos (\theta + \theta) + i \sin (\theta + \theta)]$$
$$= r^2(\cos 2\theta + i \sin 2\theta).$$

Similarly, we see that

$$[r(\cos \theta + i \sin \theta)]^3$$
$$= r \cdot r \cdot r \cdot [\cos (\theta + \theta + \theta) + i \sin (\theta + \theta + \theta)]$$
$$= r^3(\cos 3\theta + i \sin 3\theta).$$

DeMoivre's theorem is the generalization of these results.

DeMoivre's Theorem

For any complex number $r(\cos \theta + i \sin \theta)$ and any natural number n,

$$[r(\cos \theta + i \sin \theta)]^n = r^n(\cos n\theta + i \sin n\theta).$$

EXAMPLE 9 Find each of the following.

a) $(1 + i)^9$ 　　　　　　　　　　　　**b)** $(\sqrt{3} - i)^{10}$

Solution

a) We first find trigonometric notation:

$$1 + i = \sqrt{2}(\cos 45° + i \sin 45°).$$

Then

$$(1 + i)^9 = [\sqrt{2}(\cos 45° + i \sin 45°)]^9$$
$$= (\sqrt{2})^9[\cos (9 \cdot 45°) + i \sin (9 \cdot 45°)] \qquad \text{DeMoivre's theorem}$$
$$= 2^{9/2}(\cos 405° + i \sin 405°)$$
$$= 16\sqrt{2}(\cos 45° + i \sin 45°) \qquad \text{405° has the same terminal side as 45°.}$$
$$= 16\sqrt{2}\left(\frac{\sqrt{2}}{2} + i\frac{\sqrt{2}}{2}\right)$$
$$= 16 + 16i.$$

b) We first convert to trigonometric notation:

$$\sqrt{3} - i = 2(\cos 330° + i \sin 330°).$$

Then

$$(\sqrt{3} - i)^{10} = [2(\cos 330° + i \sin 330°)]^{10}$$
$$= 2^{10}(\cos 3300° + i \sin 3300°)$$
$$= 1024(\cos 60° + i \sin 60°) \qquad \text{3300° has the same terminal side as 60°.}$$
$$= 1024\left(\frac{1}{2} + i\frac{\sqrt{3}}{2}\right)$$
$$= 512 + 512\sqrt{3}i.$$

Technology Connection

We can find powers of complex numbers, like those in Example 9, on a graphing calculator.

```
(1+i)^9
                    16+16i
(√(3)−i)^10
          512+886.8100135i
```

Roots of Complex Numbers

As we will see, every nonzero complex number has two square roots. A nonzero complex number has three cube roots, four fourth roots, and so on. In general, a nonzero complex number has n different nth roots. They can be found using the formula that we now state but do not prove.

Roots of Complex Numbers

The nth roots of a complex number $r(\cos \theta + i \sin \theta)$, $r \neq 0$, are given by

$$r^{1/n}\left[\cos\left(\frac{\theta}{n} + k \cdot \frac{360°}{n}\right) + i \sin\left(\frac{\theta}{n} + k \cdot \frac{360°}{n}\right)\right],$$

where $k = 0, 1, 2, \ldots, n - 1$.

EXAMPLE 10 Find the square roots of $2 + 2\sqrt{3}i$.

Solution We first find trigonometric notation:

$$2 + 2\sqrt{3}i = 4(\cos 60° + i \sin 60°).$$

Then

$$[4(\cos 60° + i \sin 60°)]^{1/2}$$

$$= 4^{1/2}\left[\cos\left(\frac{60°}{2} + k \cdot \frac{360°}{2}\right) + i \sin\left(\frac{60°}{2} + k \cdot \frac{360°}{2}\right)\right], \quad k = 0, 1$$

$$= 2[\cos(30° + k \cdot 180°) + i \sin(30° + k \cdot 180°), \quad k = 0, 1.$$

Thus the roots are

$$2(\cos 30° + i \sin 30°) \text{ for } k = 0$$

and

$$2(\cos 210° + i \sin 210°) \text{ for } k = 1,$$

or

$$\sqrt{3} + i \quad \text{and} \quad -\sqrt{3} - i.$$

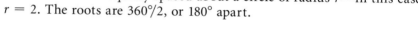

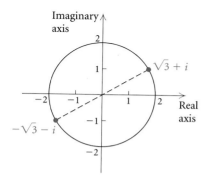

In Example 10, we see that the two square roots of the number are opposites of each other. We can illustrate this graphically. We also note that the roots are equally spaced about a circle of radius r—in this case, $r = 2$. The roots are $360°/2$, or $180°$ apart.

EXAMPLE 11 Find the cube roots of 1. Then locate them on a graph.

Solution We begin by finding trigonometric notation:

$$1 = 1(\cos 0° + i \sin 0°).$$

Then

$$[1(\cos 0° + i \sin 0°)]^{1/3}$$

$$= 1^{1/3} \left[\cos \left(\frac{0°}{3} + k \cdot \frac{360°}{3} \right) + i \sin \left(\frac{0°}{3} + k \cdot \frac{360°}{3} \right) \right], \quad k = 0, 1, 2.$$

The roots are

$$1(\cos 0° + i \sin 0°), \quad 1(\cos 120° + i \sin 120°),$$

and

$$1(\cos 240° + i \sin 240°),$$

or

$$1, \quad -\frac{1}{2} + \frac{\sqrt{3}}{2} i, \quad \text{and} \quad -\frac{1}{2} - \frac{\sqrt{3}}{2} i.$$

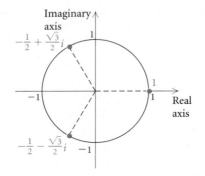

The graphs of the cube roots lie equally spaced about a circle of radius 1. The roots are 360°/3, or 120° apart. ▬

The nth roots of 1 are often referred to as the **nth roots of unity.** In Example 11, we found the cube roots of unity.

Technology Connection

Using a graphing calculator set in PARAMETRIC mode, we can approximate the nth roots of a number p. We use the following window and let

$$X_{1T} = (p\wedge(1/n)) \cos T \quad \text{and} \quad Y_{1T} = (p\wedge(1/n)) \sin T.$$

WINDOW

Tmin = 0

Tmax = 360, if in degree mode, or
        = 2π, if in radian mode

Tstep = 360/n, or $2\pi/n$

Xmin = -3, Xmax = 3, Xscl = 1

Ymin = -2, Ymax = 2, Yscl = 1

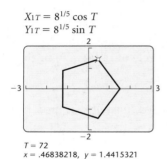

$X_{1T} = 8^{1/5} \cos T$
$Y_{1T} = 8^{1/5} \sin T$

$T = 72$
$x = .46838218, \; y = 1.4415321$

To find the fifth roots of 8, enter $X_{1T} = (8\wedge(1/5)) \cos T$ and $Y_{1T} = (8\wedge(1/5)) \sin T$. In this case, use DEGREE mode. After the graph has been generated, use the TRACE feature to locate the fifth roots. The T, X, and Y values appear on the screen. What do they represent?

Three of the fifth roots of 8 are approximately

$$1.5157, \quad 0.46838 + 1.44153i, \quad \text{and} \quad -1.22624 + 0.89092i.$$

Find the other two. Then use a calculator to approximate the cube roots of unity that were found in Example 11. Also approximate the fourth roots of 5 and the tenth roots of unity.

Exercise Set 7.3

Graph the complex number and find its absolute value.

1. $4 + 3i$ **2.** $-2 - 3i$

3. i **4.** $-5 - 2i$

5. $4 - i$ **6.** $6 + 3i$

7. 3 **8.** $-2i$

Express the indicated number in both standard notation and trigonometric notation.

9.

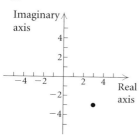

10.

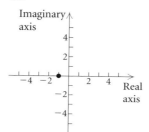

11.

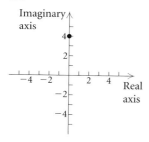

12.

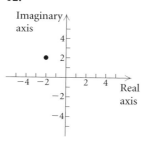

Find trigonometric notation.

13. $1 - i$ **14.** $-10\sqrt{3} + 10i$

15. $-3i$ **16.** $-5 + 5i$

17. $\sqrt{3} + i$ **18.** 4

19. $\dfrac{2}{5}$ **20.** $7.5i$

Find standard notation, $a + bi$.

21. $3(\cos 30° + i \sin 30°)$

22. $6(\cos 120° + i \sin 120°)$

23. $10(\cos 270° + i \sin 270°)$

24. $3(\cos 0° + i \sin 0°)$

25. $\sqrt{8}\left(\cos \dfrac{\pi}{4} + i \sin \dfrac{\pi}{4}\right)$

26. $5\left(\cos \dfrac{\pi}{3} + i \sin \dfrac{\pi}{3}\right)$

27. $2\left(\cos \dfrac{\pi}{2} + i \sin \dfrac{\pi}{2}\right)$

28. $3\left[\cos \left(-\dfrac{3\pi}{4}\right) + i \sin \left(-\dfrac{3\pi}{4}\right)\right]$

Multiply or divide and leave the answer in trigonometric notation.

29. $\dfrac{12(\cos 48° + i \sin 48°)}{3(\cos 6° + i \sin 6°)}$

30. $5\left(\cos \dfrac{\pi}{3} + i \sin \dfrac{\pi}{3}\right) \cdot 2\left(\cos \dfrac{\pi}{4} + i \sin \dfrac{\pi}{4}\right)$

31. $2.5(\cos 35° + i \sin 35°) \cdot 4.5(\cos 21° + i \sin 21°)$

32. $\dfrac{\dfrac{1}{2}\left(\cos \dfrac{2\pi}{3} + i \sin \dfrac{2\pi}{3}\right)}{\dfrac{3}{8}\left(\cos \dfrac{\pi}{6} + i \sin \dfrac{\pi}{6}\right)}$

Convert to trigonometric notation and then multiply or divide.

33. $(1 - i)(2 + 2i)$ **34.** $(1 + i\sqrt{3})(1 + i)$

35. $\dfrac{1 - i}{1 + i}$ **36.** $\dfrac{1 - i}{\sqrt{3} - i}$

37. $(3\sqrt{3} - 3i)(2i)$ **38.** $(2\sqrt{3} + 2i)(2i)$

39. $\dfrac{2\sqrt{3} - 2i}{1 + \sqrt{3}i}$ **40.** $\dfrac{3 - 3\sqrt{3}i}{\sqrt{3} - i}$

Raise the number to the given power and write trigonometric notation for the answer.

41. $\left[2\left(\cos \dfrac{\pi}{3} + i \sin \dfrac{\pi}{3}\right)\right]^3$

42. $[2(\cos 120° + i \sin 120°)]^4$

43. $(1 + i)^6$

44. $(-\sqrt{3} + i)^5$

Raise the number to the given power and write standard notation for the answer.

45. $[3(\cos 20° + i \sin 20°)]^3$

46. $[2(\cos 10° + i \sin 10°)]^9$

47. $(1 - i)^5$ **48.** $(2 + 2i)^4$

49. $\left(\dfrac{1}{\sqrt{2}} - \dfrac{1}{\sqrt{2}}i\right)^{12}$ **50.** $\left(\dfrac{\sqrt{3}}{2} + \dfrac{1}{2}i\right)^{10}$

Find the square roots of the number.

51. $-i$

52. $1 + i$

53. $2\sqrt{2} - 2\sqrt{2}i$

54. $-\sqrt{3} - i$

Find the cube roots of the number.

55. i

56. $-64i$

57. $2\sqrt{3} - 2i$

58. $1 - \sqrt{3}i$

59. Find and graph the fourth roots of 16.

60. Find and graph the fourth roots of i.

61. Find and graph the fifth roots of -1.

62. Find and graph the sixth roots of 1.

63. Find the tenth roots of 8.

64. Find the ninth roots of -4.

65. Find the sixth roots of -1.

66. Find the fourth roots of 12.

Find all the complex solutions of the equation.

67. $x^3 = 1$

68. $x^5 - 1 = 0$

69. $x^4 + i = 0$

70. $x^4 + 81 = 0$

71. $x^6 + 64 = 0$

72. $x^5 + \sqrt{3} + i = 0$

Technology Connection

73. Using a graphing calculator, check your work in each of the odd-numbered Exercises 13–49.

74. Using a graphing calculator, check your work in each of the even-numbered Exercises 14–50.

Collaborative Discussion and Writing

75. Find and graph the square roots of $1 - i$. Explain geometrically why they are the opposites of each other.

76. Explain why trigonometric notation for a complex number is not unique, but rectangular, or standard, notation is unique.

Skill Maintenance

Convert to degree measure.

77. $\dfrac{\pi}{12}$

78. 3π

Convert to radian measure.

79. $330°$

80. $-225°$

81. Find r.

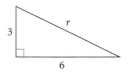

82. Graph these points in the rectangular coordinate system: $(2, -1)$, $(0, 3)$, and $\left(-\frac{1}{2}, -4\right)$.

Synthesis

Solve.

83. $x^2 + (1 - i)x + i = 0$

84. $3x^2 + (1 + 2i)x + 1 - i = 0$

85. Find polar notation for $(\cos\theta + i\sin\theta)^{-1}$.

86. Show that for any complex number z,
$$|z| = |-z|.$$

87. Show that for any complex number z,
$$|z| = |\bar{z}|.$$
(*Hint*: Let $z = a + bi$ and $\bar{z} = a - bi$.)

88. Show that for any complex number z,
$$|z\bar{z}| = |z^2|.$$
(*Hint*: Let $z = a + bi$ and $\bar{z} = a - bi$.)

89. Show that for any complex number z,
$$|z^2| = |z|^2.$$

90. Show that for any complex numbers z and w,
$$|z \cdot w| = |z| \cdot |w|.$$
(*Hint*: Let $z = r_1(\cos\theta_1 + i\sin\theta_1)$ and $w = r_2(\cos\theta_2 + i\sin\theta_2)$.)

91. Show that for any complex number z and any nonzero, complex number w,
$$\left|\frac{z}{w}\right| = \frac{|z|}{|w|}. \quad \text{(Use the hint for Exercise 90.)}$$

92. On a complex plane, graph $|z| = 1$.

93. On a complex plane, graph $z + \bar{z} = 3$.

7.4

Polar Coordinates and Graphs

- *Graph points given their polar coordinates.*
- *Convert from rectangular to polar coordinates and from polar to rectangular coordinates.*
- *Convert from rectangular to polar equations and from polar to rectangular equations.*
- *Graph polar equations.*

Polar Coordinates

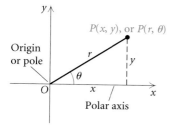

All graphing throughout this text has been done with rectangular coordinates, (x, y), in the Cartesian coordinate system. We now introduce the polar coordinate system. As shown in the diagram at left, any point P has rectangular coordinates (x, y) and polar coordinates (r, θ). On a polar graph, the origin is called the **pole** and the positive half of the x-axis is called the **polar axis.** The point P can be plotted given the directed angle θ from the polar axis to the ray OP and the directed distance r from the pole to the point. The angle θ can be expressed in degrees or radians.

To plot points on a polar graph:

1. Locate the directed angle θ.

2. Move a directed distance r from the pole. If $r > 0$, move along ray OP. If $r < 0$, move in the opposite direction of ray OP.

Polar graph paper, shown below, facilitates plotting. Points B and G illustrate that θ may be in radians. Points E and F illustrate that the polar coordinates of a point are not unique.

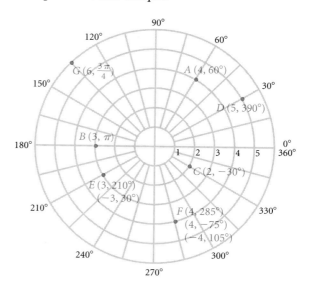

EXAMPLE 1 Graph each of the following points.

a) $A(3, 60°)$

b) $B(0, 10°)$

c) $C(-5, 120°)$

d) $D(1, -60°)$

e) $E\left(2, \dfrac{3\pi}{2}\right)$

f) $F\left(-4, \dfrac{\pi}{3}\right)$

Solution

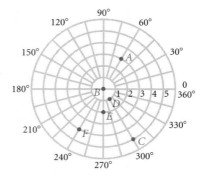

To convert from rectangular to polar coordinates and from polar to rectangular coordinates, we need to recall the following relationships.

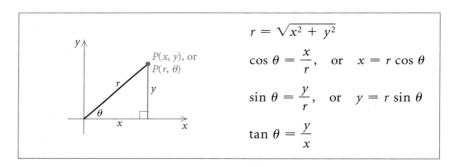

$$r = \sqrt{x^2 + y^2}$$

$$\cos \theta = \frac{x}{r}, \quad \text{or} \quad x = r \cos \theta$$

$$\sin \theta = \frac{y}{r}, \quad \text{or} \quad y = r \sin \theta$$

$$\tan \theta = \frac{y}{x}$$

EXAMPLE 2 Convert each of the following to polar coordinates.

a) $(3, 3)$

b) $(2\sqrt{3}, -2)$

Solution

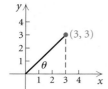

a) We first find r:

$$r = \sqrt{3^2 + 3^2} = \sqrt{18} = 3\sqrt{2}.$$

Then we determine θ:

$$\tan \theta = \frac{3}{3} = 1; \quad \text{therefore,} \quad \theta = 45°, \text{ or } \frac{\pi}{4}.$$

We know that $\theta = \pi/4$ and not $5\pi/4$ since $(3, 3)$ is in quadrant I. Thus, $(r, \theta) = (3\sqrt{2}, 45°)$, or $(3\sqrt{2}, \pi/4)$. Other possibilities for polar coordinates include $(3\sqrt{2}, -315°)$ and $(-3\sqrt{2}, 5\pi/4)$.

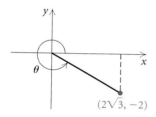

(2√3, −2)

b) We first find r:

$$r = \sqrt{(2\sqrt{3})^2 + (-2)^2} = \sqrt{12 + 4} = \sqrt{16} = 4.$$

Then we determine θ:

$$\tan \theta = \frac{-2}{2\sqrt{3}} = -\frac{1}{\sqrt{3}}; \quad \text{therefore,} \quad \theta = 330°, \text{ or } \frac{11\pi}{6}.$$

Thus, $(r, \theta) = (4, 330°)$, or $(4, 11\pi/6)$. Other possibilities for polar coordinates for this point include $(4, -\pi/6)$ and $(-4, 150°)$. ▬

It is easier to convert from polar to rectangular coordinates than from rectangular to polar coordinates.

EXAMPLE 3 Convert each of the following to rectangular coordinates.

a) $\left(10, \dfrac{\pi}{3}\right)$ **b)** $(-5, 135°)$

Solution

a) The ordered pair $(10, \pi/3)$ gives us $r = 10$ and $\theta = \pi/3$. We now find x and y:

$$x = r \cos \theta = 10 \cos \frac{\pi}{3} = 10 \cdot \frac{1}{2} = 5$$

and

$$y = r \sin \theta = 10 \sin \frac{\pi}{3} = 10 \cdot \frac{\sqrt{3}}{2} = 5\sqrt{3}.$$

Thus, $(x, y) = (5, 5\sqrt{3})$.

b) From the ordered pair $(-5, 135°)$, we know that $r = -5$ and $\theta = 135°$. We now find x and y:

$$x = -5 \cos 135° = -5 \cdot \left(-\frac{\sqrt{2}}{2}\right) = \frac{5\sqrt{2}}{2}$$

and

$$y = -5 \sin 135° = -5 \cdot \left(\frac{\sqrt{2}}{2}\right) = -\frac{5\sqrt{2}}{2}.$$

Thus, $(x, y) = (5\sqrt{2}/2, -5\sqrt{2}/2)$. ▬

Polar and Rectangular Equations

Some curves have simpler equations in polar coordinates than in rectangular coordinates. For others, the reverse is true.

EXAMPLE 4   Convert each of the following to a polar equation.

a) $x^2 + y^2 = 25$

b) $2x - y = 5$

Solution

a) We have

$$x^2 + y^2 = 25$$
$$(r \cos \theta)^2 + (r \sin \theta)^2 = 25 \qquad \text{Substituting for } x \text{ and } y$$
$$r^2 \cos^2 \theta + r^2 \sin^2 \theta = 25$$
$$r^2(\cos^2 \theta + \sin^2 \theta) = 25$$
$$r^2 = 25 \qquad \cos^2 \theta + \sin^2 \theta = 1$$
$$r = 5.$$

This example illustrates that the polar equation of a circle centered at the origin is much simpler than the rectangular equation.

b) We have

$$2x - y = 5$$
$$2(r \cos \theta) - (r \sin \theta) = 5$$
$$r(2 \cos \theta - \sin \theta) = 5.$$

In this example, we see that the rectangular equation is simpler than the polar equation.   ▬

EXAMPLE 5   Convert each of the following to a rectangular equation.

a) $r = 4$

b) $r \cos \theta = 6$

c) $r = 2 \cos \theta + 3 \sin \theta$

Solution

a) We have

$$r = 4$$
$$\sqrt{x^2 + y^2} = 4 \qquad \text{Substituting for } r$$
$$x^2 + y^2 = 16. \qquad \text{Squaring}$$

In squaring, we must be careful not to introduce solutions of the equation that are not already present. In this case, we did not, because the graph of either equation is a circle of radius 4 centered at the origin.

b) We have

$$r \cos \theta = 6$$
$$x = 6. \qquad x = r \cos \theta$$

The graph of $r \cos \theta = 6$, or $x = 6$, is a vertical line.

c) We have

$$r = 2 \cos \theta + 3 \sin \theta$$
$$r^2 = 2r \cos \theta + 3r \sin \theta \qquad \text{Multiplying by } r \text{ on both sides}$$
$$x^2 + y^2 = 2x + 3y. \qquad \text{Substituting } x^2 + y^2 \text{ for } r^2, x \text{ for } r \cos \theta, \text{ and } y \text{ for } r \sin \theta$$

Graphing Polar Equations

To graph a polar equation, we can make a table of values, choosing values of θ and calculating corresponding values of r. We plot the points and complete the graph, as we do when graphing a rectangular equation. A difference occurs in the case of a polar equation however, because as θ increases sufficiently, points may begin to repeat and the curve will be traced again and again. When this happens, the curve is complete.

EXAMPLE 6 Graph $r = 1 - \sin \theta$.

Solution We first make a table of values. Note that the points begin to repeat at $\theta = 360°$. We plot these points and draw the curve, as shown below.

θ	r
0°	1
15°	0.74118
30°	0.5
45°	0.29289
60°	0.13397
75°	0.03407
90°	0
105°	0.03407
120°	0.13397
135°	0.29289
150°	0.5
165°	0.74118
180°	1

θ	r
195°	1.2588
210°	1.5
225°	1.7071
240°	1.866
255°	1.9659
270°	2
285°	1.9659
300°	1.866
315°	1.7071
330°	1.5
345°	1.2588
360°	1
375°	0.74118
390°	0.5

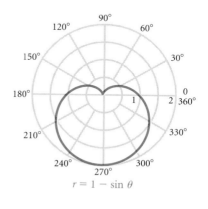

$r = 1 - \sin \theta$

Because of its heart shape, this curve is called a *cardioid*.

Technology Connection

We can graph polar equations using a graphing calculator. The equation usually must be written first in the form $r = f(\theta)$. It is necessary to decide on not only the best window dimensions but also the range of values for θ. Typically, we begin with a range of 0 to 2π for θ in radians and 0° to 360° for θ in degrees. Because most polar graphs are curved, it is important to square the window to minimize distortion.

Graph $r = 4 \sin 3\theta$. Begin by setting the calculator in POLAR mode, and use either of the following windows:

WINDOW
(Radians)
  θmin = 0
  θmax = 2π
  θstep = $\pi/24$
  Xmin = −9
  Xmax = 9
  Xscl = 1
  Ymin = −6
  Ymax = 6
  Yscl = 1

WINDOW
(Degrees)
  θmin = 0
  θmax = 360
  θstep = 1
  Xmin = −9
  Xmax = 9
  Xscl = 1
  Ymin = −6
  Ymax = 6
  Yscl = 1

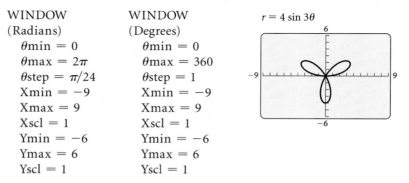

We observe the same graph in both windows. The calculator allows us to view the curve as it is formed.

Now graph each of the following equations and observe the effect of changing the coefficient of sin 3θ and the coefficient of θ:

$$r = 2 \sin 3\theta, \qquad r = 6 \sin 3\theta, \qquad r = 4 \sin \theta,$$
$$r = 4 \sin 5\theta, \qquad r = 4 \sin 2\theta, \qquad r = 4 \sin 4\theta.$$

Polar equations of the form $r = a \cos n\theta$ and $r = a \sin n\theta$ have rose-shaped curves. The number a determines the length of the petals, and the number n determines the number of petals. If n is odd, there are n petals. If n is even, there are $2n$ petals.

EXAMPLE 7 Graph each of the following polar equations. Try to visualize the shape of the curve before graphing it.

a) $r = 3$ **b)** $r = 5 \sin \theta$ **c)** $r = 2 \csc \theta$

Solution

For each graph, we can begin with a table of values. Then we plot points and complete the graph. With some polar equations, it is easier to first convert the equation to the equivalent rectangular equation.

a) $r = 3$

For all values of θ, r is 3. Thus the graph of $r = 3$ is a circle of radius 3 centered at the origin.

θ	r
0°	3
60°	3
135°	3
210°	3
300°	3
360°	3

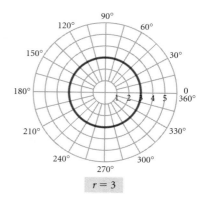

$r = 3$

We can verify our graph by converting to the equivalent rectangular equation. For $r = 3$, we substitute $\sqrt{x^2 + y^2}$ for r and square. The resulting equation,

$$x^2 + y^2 = 3^2,$$

is the equation of a circle with radius 3 centered at the origin.

b) $r = 5 \sin \theta$

θ	r
0°	0
15°	1.2941
30°	2.5
45°	3.5355
60°	4.3301
75°	4.8296
90°	5
105°	4.8296
120°	4.3301
135°	3.5355
150°	2.5
165°	1.2941
180°	0

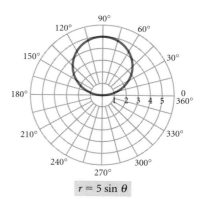

$r = 5 \sin \theta$

c) $r = 2 \csc \theta$

We can rewrite $r = 2 \csc \theta$ as $r = 2/\sin \theta$.

θ	r
0°	Undefined
15°	7.7274
30°	4
45°	2.8284
60°	2.3094
75°	2.0706
90°	2
105°	2.0706
120°	2.3094
135°	2.8284
150°	4
165°	7.7274
180°	Undefined

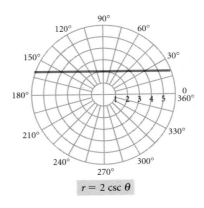

$r = 2 \csc \theta$

We can check our graph in Example 7(c) by converting the polar equation to the equivalent rectangular equation:

$$r = 2 \csc \theta$$

$$r = \frac{2}{\sin \theta}$$

$$r \sin \theta = 2$$

$$y = 2. \qquad \text{Substituting } y \text{ for } r \sin \theta$$

The graph of $y = 2$ is a horizontal line passing through (0, 2) on a rectangular grid.

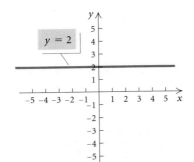

Exercise Set **7.4**

Graph the point on a polar grid.

1. (2, 45°)

2. (4, π)

3. (3.5, 210°)

4. (−3, 135°)

5. $\left(1, \dfrac{\pi}{6}\right)$

6. (2.75, 150°)

7. $\left(-5, \dfrac{\pi}{2}\right)$

8. (0, 15°)

9. (3, −315°)

10. $\left(1.2, -\dfrac{2\pi}{3}\right)$

11. (4.3, −60°)

12. (3, 405°)

Find polar coordinates of points A, B, C, and D. Give three answers for each point.

13.

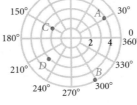

14.

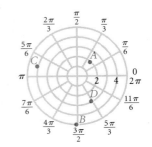

Find the polar coordinates of the point. Express the angle in degrees and then in radians, using the smallest possible positive angle.

15. $(0, -3)$ **16.** $(-4, 4)$

17. $(3, -3\sqrt{3})$ **18.** $(-\sqrt{3}, 1)$

19. $(4\sqrt{3}, -4)$ **20.** $(2\sqrt{3}, 2)$

21. $(-\sqrt{2}, -\sqrt{2})$ **22.** $(-3, 3\sqrt{3})$

Find the rectangular coordinates of the point.

23. $(5, 60°)$ **24.** $(0, -23°)$

25. $(-3, 45°)$ **26.** $(6, 30°)$

27. $(3, -120°)$ **28.** $\left(7, \dfrac{\pi}{6}\right)$

29. $\left(-2, \dfrac{5\pi}{3}\right)$ **30.** $(1.4, 225°)$

Convert to a polar equation.

31. $3x + 4y = 5$ **32.** $5x + 3y = 4$

33. $x = 5$ **34.** $y = 4$

35. $x^2 + y^2 = 36$ **36.** $x^2 - 4y^2 = 4$

37. $x^2 = 25y$ **38.** $2x - 9y + 3 = 0$

39. $y^2 - 5x - 25 = 0$ **40.** $x^2 + y^2 = 8y$

Convert to a rectangular equation.

41. $r = 5$ **42.** $\theta = \dfrac{3\pi}{4}$

43. $r \sin \theta = 2$ **44.** $r = -3 \sin \theta$

45. $r + r \cos \theta = 3$ **46.** $r = \dfrac{2}{1 - \sin \theta}$

47. $r - 9 \cos \theta = 7 \sin \theta$ **48.** $r + 5 \sin \theta = 7 \cos \theta$

49. $r = 5 \sec \theta$ **50.** $r = 3 \cos \theta$

Graph the equation.

51. $r = \sin \theta$ **52.** $r = 1 - \cos \theta$

53. $r = 4 \cos 2\theta$ **54.** $r = 1 - 2 \sin \theta$

55. $r = \cos \theta$ **56.** $r = 2 \sec \theta$

Technology Connection

Use a graphing calculator to convert from rectangular to polar coordinates. Express the answer in both degrees and radians, using the smallest possible positive angle.

57. $(3, 7)$ **58.** $(-2, -\sqrt{5})$

59. $(-\sqrt{10}, 3.4)$ **60.** $(0.9, -6)$

Use a graphing calculator to convert from polar to rectangular coordinates. Round the coordinates to the nearest hundredth.

61. $(3, -43°)$ **62.** $\left(-5, \dfrac{\pi}{7}\right)$

63. $\left(-4.2, \dfrac{3\pi}{5}\right)$ **64.** $(2.8, 166°)$

In Exercises 65–76, use a graphing calculator to match the equation with one of figures (a)–(l), which follow. Try matching the graphs mentally before using a calculator.

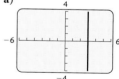

65. $r = 3 \sin 2\theta$ **66.** $r = 4 \cos \theta$

67. $r = \theta$ **68.** $r^2 = \sin 2\theta$

69. $r = \dfrac{5}{1 + \cos \theta}$ **70.** $r = 1 + 2 \sin \theta$

71. $r = 3 \cos 2\theta$ **72.** $r = 3 \sec \theta$

73. $r = 3 \sin \theta$ **74.** $r = 4 \cos 5\theta$

75. $r = 2 \sin 3\theta$ **76.** $r \sin \theta = 6$

Graph.

77. $r = \sin \theta \tan \theta$ (Cissoid)

78. $r = 3\theta$ (Spiral of Archimedes)

79. $r = e^{\theta/10}$ (Logarithmic spiral)

80. $r = 10^{2\theta}$ (Logarithmic spiral)

81. $r = \cos 2\theta \sec \theta$ (Strophoid)

82. $r = \cos 2\theta - 2$ (Peanut)

83. $r = \frac{1}{4} \tan^2 \theta \sec \theta$ (Semicubical parabola)

84. $r = \sin 2\theta + \cos \theta$ (Twisted sister)

Collaborative Discussion and Writing

85. Explain why the rectangular coordinates of a point are unique and the polar coordinates of a point are not unique.

86. Give an example of an equation that is easier to graph in polar notation than in rectangular notation and explain why.

Skill Maintenance

Solve.

87. $2x - 4 = x + 8$

88. $4 - 5y = 3$

Graph.

89. $y = 2x - 5$

90. $4x - y = 6$

91. $x = -3$

92. $y = 0$

Synthesis

93. Convert to a rectangular equation:

$$r = \sec^2 \frac{\theta}{2}.$$

94. The center of a regular hexagon is at the origin, and one vertex is the point $(4, 0°)$. Find the coordinates of the other vertices.

7.5

Vectors and Applications

- *Determine whether two vectors are equivalent.*
- *Find the sum, or resultant, of two vectors.*
- *Resolve a vector into its horizontal and vertical components.*
- *Solve applied problems involving vectors.*

We measure some quantities using only their magnitudes. For example, we describe time, length, and mass using units like seconds, feet, and kilograms, respectively. However, to measure quantities like **displacement**, **velocity**, or **force**, we need to describe a *magnitude* and a *direction*. Together magnitude and direction describe a **vector**. The following are some examples.

DISPLACEMENT An object moves a certain distance in a certain direction.

A surveyor steps 20 yd to the northeast.

A hiker follows a trail 5 mi to the west.

A batter hits a ball 100 m along the left-field line.

VELOCITY An object travels at a certain speed in a certain direction.

A breeze is blowing 15 mph from the northwest.

An airplane is traveling 450 km/h in a direction of 243°.

FORCE A push or pull is exerted on an object in a certain direction.

A force of 200 lb is required to pull a cart up a 30° incline.

A 25-lb force is required to lift a box upward.

A force of 15 newtons is exerted downward on the handle of a jack. (A newton, abbreviated N, is a unit of force used in physics, and 1 N ≈ 0.22 lb.)

Vectors

Vectors can be graphically represented by directed line segments. The length is chosen, according to some scale, to represent the **magnitude of the vector,** and the direction of the directed line segment represents the **direction of the vector.** For example, if we let 1 cm represent 5 km/h, then a 15-km/h wind from the northwest would be represented by a directed line segment 3 cm long, as shown in the figure at left.

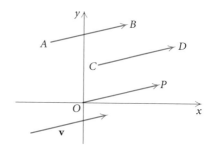

Vector

A **vector** in the plane is a directed line segment. Two vectors are **equivalent** if they have the same *magnitude* and *direction.*

Consider a vector drawn from point A to point B. Point A is called the **initial point** of the vector, and point B is called the **terminal point.** Symbolic notation for this vector is $\overrightarrow{AB}$ (read "vector AB"). Vectors are also denoted by boldface letters such as **u**, **v**, and **w**. The four vectors in the figure at left have the *same* length and direction. Thus they represent **equivalent** vectors; that is,

$$\overrightarrow{AB} = \overrightarrow{CD} = \overrightarrow{OP} = \mathbf{v}.$$

In the context of vectors, we use = to mean equivalent.

The length, or **magnitude**, of $\overrightarrow{AB}$ is expressed as $|\overrightarrow{AB}|$. In order to determine whether vectors are equivalent, we find their magnitudes and directions.

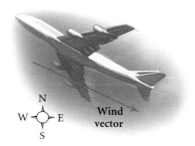

EXAMPLE 1 The vectors **u**, $\overrightarrow{OR}$, and **w** are shown in the figure below. Show that $\mathbf{u} = \overrightarrow{OR} = \mathbf{w}$.

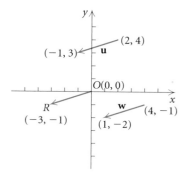

u ≠ v (not equivalent)
*Different magnitudes;
different directions*

u ≠ v
*Same magnitude;
different directions*

u ≠ v
*Different magnitudes;
same direction*

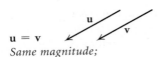

u = v
*Same magnitude;
same direction*

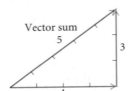

Vector sum

Solution   We first find the length of each vector using the distance formula.

$$|\mathbf{u}| = \sqrt{[2-(-1)]^2 + (4-3)^2} = \sqrt{9+1} = \sqrt{10},$$
$$|\overrightarrow{OR}| = \sqrt{[0-(-3)]^2 + [0-(-1)]^2} = \sqrt{9+1} = \sqrt{10},$$
$$|\mathbf{w}| = \sqrt{(4-1)^2 + [-1-(-2)]^2} = \sqrt{9+1} = \sqrt{10}.$$

Thus

$$|\mathbf{u}| = |\overrightarrow{OR}| = |\mathbf{w}|.$$

The vectors **u**, $\overrightarrow{OR}$, and **w** appear to go in the same direction so we check their slopes. If the lines that they are on all have the same slope, the vectors have the same direction. We calculate the slopes:

$$\text{Slope} = \underset{\mathbf{u}}{\frac{4-3}{2-(-1)}} = \underset{\overrightarrow{OR}}{\frac{0-(-1)}{0-(-3)}} = \underset{\mathbf{w}}{\frac{-1-(-2)}{4-1}} = \frac{1}{3}.$$

Since **u**, $\overrightarrow{OR}$, and **w** have the *same* magnitude and the *same* direction,

$$\mathbf{u} = \overrightarrow{OR} = \mathbf{w}. \qquad \blacksquare$$

Keep in mind that the equivalence of vectors requires only the same magnitude and the same direction—not the same location. In the illustrations at left, each of the first three pairs of vectors are not equivalent. The fourth set of vectors is an example of equivalence.

Vector Addition

Suppose a person takes 4 steps east and then 3 steps north. He or she will then be 5 steps from the starting point in the direction shown at left. A vector 4 units long and pointing to the right represents 4 steps east and a vector 3 units long and pointing up represents 3 steps north. The **sum** of the two vectors is the vector 5 steps in magnitude and in the direction shown. The sum is also called the **resultant** of the two vectors.

In general, two nonzero vectors **u** and **v** can be added geometrically by placing the initial point of **v** at the terminal point of **u** and then finding the vector that has the same initial point as **u** and the same terminal point as **v**, as shown in the following figure.

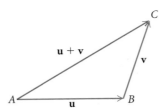

The sum **u** + **v** is the vector represented by the directed line segment from the initial point *A* of **u** to the terminal point *C* of **v**. That is, if

$$\mathbf{u} = \overrightarrow{AB} \quad \text{and} \quad \mathbf{v} = \overrightarrow{BC},$$

then

$$\mathbf{u} + \mathbf{v} = \overrightarrow{AB} + \overrightarrow{BC} = \overrightarrow{AC}.$$

We can also describe vector addition by placing the initial points of the vectors together, completing a parallelogram, and finding the diagonal of the parallelogram. (See the figure on the left below.) This description of addition is sometimes called the **parallelogram law** of vector addition. Vector addition is **commutative**. As shown in the figure on the right below, both $\mathbf{u} + \mathbf{v}$ and $\mathbf{v} + \mathbf{u}$ are represented by the same directed line segment.

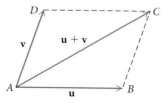
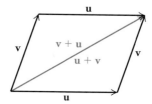

Applications

If two forces F_1 and F_2 act on an object, the *combined* effect is the sum, or resultant, $F_1 + F_2$ of the separate forces.

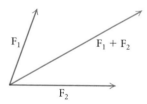

EXAMPLE 2   Forces of 15 newtons and 25 newtons act on an object at right angles to each other. Find their sum, or resultant, giving the magnitude of the resultant and the angle that it makes with the larger force.

Solution   We make a drawing—this time, a rectangle—using $\mathbf{v}$ or $\overrightarrow{OB}$ to represent the resultant. To find the magnitude, we use the Pythagorean theorem:

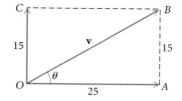

$$|\mathbf{v}|^2 = 15^2 + 25^2 \quad \text{Here } |\mathbf{v}| \text{ denotes the length, or magnitude, of v.}$$
$$|\mathbf{v}| = \sqrt{15^2 + 25^2}$$
$$|\mathbf{v}| \approx 29.2.$$

To find the direction, we note that since OAB is a right triangle,

$$\tan \theta = \tfrac{15}{25} = 0.6.$$

Using a calculator, we find θ, the angle that the resultant makes with the larger force:

$$\theta = \tan^{-1}(0.6) \approx 31°.$$

The resultant $\overrightarrow{OB}$ has a magnitude of 29.2 and makes an angle of 31° with the larger force.

AERIAL BEARINGS

REVIEW SECTION 5.3.

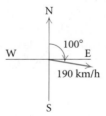

Airplane airspeed

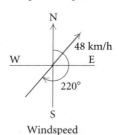

Windspeed

Pilots must adjust the direction of their flight when there is a crosswind. Both the wind and the aircraft velocities can be described by vectors.

EXAMPLE 3 *Airplane Speed and Direction.* An airplane travels on a bearing of 100° at an airspeed of 190 km/h while a wind is blowing 48 km/h from 220°. Find the ground speed of the airplane and the direction of its track, or course, over the ground.

Solution We first make a drawing. The wind is represented by $\overrightarrow{OC}$ and the velocity vector of the airplane by $\overrightarrow{OA}$. The resultant velocity vector is **v**, the sum of the two vectors. The angle θ between **v** and $\overrightarrow{OA}$ is called a **drift angle.**

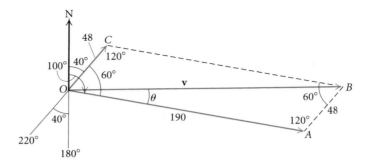

Note that the measure of $\angle COA = 100° - 40° = 60°$. Thus the measure of $\angle CBA$ is also 60° (opposite angles of a parallelogram are equal). Since the sum of all the angles of the parallelogram is 360° and $\angle OCB$ and $\angle OAB$ have the same measure, each must be 120°. By the *law of cosines* in $\triangle OAB$, we have

$$|\mathbf{v}|^2 = 48^2 + 190^2 - 2 \cdot 48 \cdot 190 \cos 120°$$
$$|\mathbf{v}|^2 = 47{,}524$$
$$|\mathbf{v}| = 218.$$

Thus, $|\mathbf{v}|$ is 218 km/h. By the *law of sines* in the same triangle,

$$\frac{48}{\sin \theta} = \frac{218}{\sin 120°},$$

or

$$\sin \theta = \frac{48 \sin 120°}{218} \approx 0.1907$$
$$\theta \approx 11°.$$

Thus, $\theta = 11°$, to the nearest degree. The ground speed of the airplane is 218 km/h, and its track is in the direction of $100° - 11°$, or 89°. ▬

Components

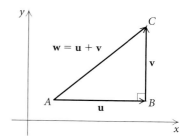

Given a vector **w**, we may want to find two other vectors **u** and **v** whose sum is **w**. The vectors **u** and **v** are called **components** of **w** and the process of finding them is called **resolving**, or **representing**, a vector into its vector components.

When we resolve a vector, we generally look for perpendicular components. Most often, one component will be parallel to the x-axis and the other will be parallel to the y-axis. For this reason, they are often called the **horizontal** and **vertical** components of a vector. In the figure at left, the vector $\mathbf{w} = \overrightarrow{AC}$ is resolved as the sum of $\mathbf{u} = \overrightarrow{AB}$ and $\mathbf{v} = \overrightarrow{BC}$. The horizontal component of **w** is **u** and the vertical component is **v**.

EXAMPLE 4   A vector **w** has a magnitude of 130 and is inclined 40° with the horizontal. Resolve the vector into horizontal and vertical components.

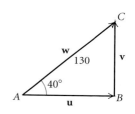

Solution   We first make a drawing showing horizontal and vertical vectors **u** and **v** whose sum is **w**.

From $\triangle ABC$, we find $|\mathbf{u}|$ and $|\mathbf{v}|$ using the definitions of the cosine and sine functions:

$$\cos 40° = \frac{|\mathbf{u}|}{130}, \quad \text{or} \quad |\mathbf{u}| = 130 \cos 40° \approx 100,$$

$$\sin 40° = \frac{|\mathbf{v}|}{130}, \quad \text{or} \quad |\mathbf{v}| = 130 \sin 40° \approx 84.$$

Thus the horizontal component of **w** is 100 right, and the vertical component of **w** is 84 up.

EXAMPLE 5   *Shipping Crate.* A wooden shipping crate that weighs 816 lb is placed on a loading ramp that makes an angle of 25° with the horizontal. To keep the crate from sliding, a chain is hooked to the crate and to a pole at the top of the ramp. Find the magnitude of the components of the crate's weight (disregarding friction) perpendicular and parallel to the incline.

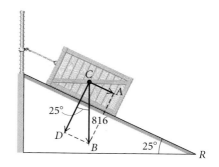

Solution   We first make a drawing illustrating the forces with a rectangle. We let

$|\overrightarrow{CB}|$ = the weight of the crate = 816 lb (force of gravity),

$|\overrightarrow{CD}|$ = the magnitude of the component of the crate's weight perpendicular to the incline (force against the ramp), and

$|\overrightarrow{CA}|$ = the magnitude of the component of the crate's weight parallel to the incline (force that pulls the crate down the ramp).

The angle at R is given to be $25°$ and $\angle BCD = \angle R = 25°$ because the sides of these angles are, respectively, perpendicular. Using the cosine and sine functions, we find that

$$\cos 25° = \frac{|\overrightarrow{CD}|}{816}, \quad \text{or} \quad |\overrightarrow{CD}| = 816 \cos 25° \approx 740 \text{ lb}, \quad \text{and}$$

$$\sin 25° = \frac{|\overrightarrow{CA}|}{816}, \quad \text{or} \quad |\overrightarrow{CA}| = 816 \sin 25° \approx 345 \text{ lb}.$$

Exercise Set 7.5

Sketch the pair of vectors and determine whether they are equivalent. Use the following ordered pairs for the initial and terminal points.

$A(-2, 2)$ $E(-4, 1)$ $I(-6, -3)$
$B(3, 4)$ $F(2, 1)$ $J(3, 1)$
$C(-2, 5)$ $G(-4, 4)$ $K(-3, -3)$
$D(-1, -1)$ $H(1, 2)$ $O(0, 0)$

1. $\overrightarrow{GE}, \overrightarrow{BJ}$ **2.** $\overrightarrow{DJ}, \overrightarrow{OF}$

3. $\overrightarrow{DJ}, \overrightarrow{AB}$ **4.** $\overrightarrow{CG}, \overrightarrow{FO}$

5. $\overrightarrow{DK}, \overrightarrow{BH}$ **6.** $\overrightarrow{BA}, \overrightarrow{DI}$

7. $\overrightarrow{EG}, \overrightarrow{BJ}$ **8.** $\overrightarrow{GC}, \overrightarrow{FO}$

9. $\overrightarrow{GA}, \overrightarrow{BH}$ **10.** $\overrightarrow{JD}, \overrightarrow{CG}$

11. $\overrightarrow{AB}, \overrightarrow{ID}$ **12.** $\overrightarrow{OF}, \overrightarrow{HB}$

13. Two forces of 32 N (newtons) and 45 N act on an object at right angles. Find the magnitude of the resultant and the angle that it makes with the smaller force.

14. Two forces of 50 N and 60 N act on an object at right angles. Find the magnitude of the resultant and the angle that it makes with the larger force.

15. Two forces of 410 N and 600 N act on an object. The angle between the forces is $47°$. Find the magnitude of the resultant and the angle that it makes with the larger force.

16. Two forces of 255 N and 325 N act on an object. The angle between the forces is $64°$. Find the magnitude of the resultant and the angle that it makes with the smaller force.

In Exercises 17–24, magnitudes of vectors **u** and **v** and the angle θ between the vectors are given. Find the sum of **u** + **v**. Give the magnitude to the nearest tenth and give the direction by specifying to the nearest degree the angle that the resultant makes with **u**.

17. $|\mathbf{u}| = 45,\ |\mathbf{v}| = 35,\ \theta = 90°$

18. $|\mathbf{u}| = 54,\ |\mathbf{v}| = 43,\ \theta = 150°$

19. $|\mathbf{u}| = 10,\ |\mathbf{v}| = 12,\ \theta = 67°$

20. $|\mathbf{u}| = 25,\ |\mathbf{v}| = 30,\ \theta = 75°$

21. $|\mathbf{u}| = 20,\ |\mathbf{v}| = 20,\ \theta = 117°$

22. $|\mathbf{u}| = 30,\ |\mathbf{v}| = 30,\ \theta = 123°$

23. $|\mathbf{u}| = 23,\ |\mathbf{v}| = 47,\ \theta = 27°$

24. $|\mathbf{u}| = 32,\ |\mathbf{v}| = 74,\ \theta = 72°$

25. *Hot-air Balloon.* A hot-air balloon is rising vertically 10 ft/sec while the wind is blowing horizontally 5 ft/sec. Find the speed of the balloon and the angle that it makes with the horizontal.

26. *Boat.* A boat heads $35°$, propelled by a force of 750 lb. A wind from $320°$ exerts a force of 150 lb on the boat. How large is the resultant force, and in what direction is the boat moving?

27. *Ship.* A ship sails first N80°E for 120 nautical mi, and then S20°W for 200 nautical mi. How far is the ship, then, from the starting point, and in what direction?

28. *Airplane.* An airplane flies $032°$ for 210 km, and then $280°$ for 170 km. How far is the airplane, then, from the starting point, and in what direction?

29. *Airplane.* An airplane has an airspeed of 150 km/h. It is to make a flight in a direction of 070° while there is a 25-km/h wind from 340°. What will the airplane's actual heading be?

30. *Wind.* A wind has an easterly component (*from* the east) of 10 km/h and a southerly component (*from* the south) of 16 km/h. Find the magnitude and the direction of the wind.

31. A vector **w** of magnitude 100 points southeast. Resolve the vector into easterly and southerly components.

32. A vector **u** with a magnitude of 150 lb is inclined to the right and upward 52° from the horizontal. Resolve the vector into components.

33. *Airplane.* An airplane takes off at a speed **S** of 225 mph at an angle of 17° with the horizontal. Resolve the vector **S** into components.

34. *Wheelbarrow.* A wheelbarrow is pushed by applying a 97-lb force **F** that makes a 38° angle with the horizontal. Resolve **F** into its horizontal and vertical components. (The horizontal component is the effective force in the direction of motion and the vertical component adds weight to the wheelbarrow.)

35. *Luggage Wagon.* A luggage wagon is being pulled with vector force **V**, which has a magnitude of 780 lb at an angle of elevation of 60°. Resolve the vector **V** into components.

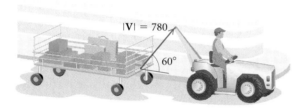

36. *Hot-air Balloon.* A hot-air balloon exerts a 1200-lb pull on a tether line at a 45° angle with the horizontal. Resolve the vector **B** into components.

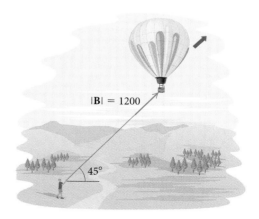

37. *Airplane.* An airplane is flying at 200 km/h in a direction of 305°. Find the westerly and northerly components of its velocity.

38. *Baseball.* A baseball player throws a baseball with a speed **S** of 72 mph at an angle of 45° with the horizontal. Resolve the vector **S** into components.

39. A block weighing 100 lb rests on a 25° incline. Find the magnitude of the components of the block's weight perpendicular and parallel to the incline.

40. A shipping crate that weighs 450 kg is placed on a loading ramp that makes an angle of 30° with the horizontal. Find the magnitude of the components of the crate's weight perpendicular and parallel to the incline.

41. An 80-lb block of ice rests on a 37° incline. What force parallel to the incline is necessary in order to keep the ice from sliding down?

42. What force is necessary to pull a 3500-lb truck up a 9° incline?

Collaborative Discussion and Writing

43. Describe the concept of a vector as though you were explaining it to a classmate. Use an arrow shot from a bow in the explanation.

44. Explain why vectors $\overrightarrow{QR}$ and $\overrightarrow{RQ}$ are not equivalent.

Skill Maintenance

Find the function value using coordinates of points on the unit circle.

45. $\sin \dfrac{2\pi}{3}$

46. $\cos \dfrac{\pi}{6}$

47. $\cos \dfrac{\pi}{4}$

48. $\sin \dfrac{5\pi}{6}$

Synthesis

49. *Eagle's Flight.*   An eagle flies from its nest 7 mi in the direction northeast, where it stops to rest on a cliff. It then flies 8 mi in the direction S30°W to land on top of a tree. Place an *xy*-coordinate system so that the origin is the bird's nest, the *x*-axis points east, and the *y*-axis points north.

a) At what point is the cliff located?
b) At what point is the tree?

7.6

Vector Operations

- *Perform calculations with vectors in component form.*
- *Express a vector as a linear combination of unit vectors.*
- *Express a vector in terms of its magnitude and its direction.*
- *Find the angle between two vectors using the dot product.*
- *Solve applied problems involving forces in equilibrium.*

Position Vectors

Let's consider a vector **v** whose initial point is the *origin* in an *xy*-coordinate system and whose terminal point is (a, b). We say that the vector is in **standard position** and refer to it as a position vector. Note that the ordered pair (a, b) defines the vector uniquely. Thus we can use (a, b) to denote the vector. To emphasize that we are thinking of a vector and to avoid the confusion of notation with ordered-pair and interval notation, we generally write

$$\mathbf{v} = \langle a, b \rangle.$$

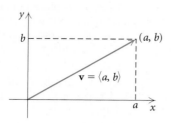

The coordinate a is the *scalar* **horizontal component** of the vector, and the coordinate b is the *scalar* **vertical component** of the vector. By **scalar**, we mean a *numerical* quantity rather than a *vector* quantity. Thus, $\langle a, b \rangle$ is considered to be the *component form* of **v**. Note that a and b are *not* vectors and should not be confused with the vector component definition given in Section 7.5.

Now consider $\overrightarrow{AC}$ with $A = (x_1, y_1)$ and $C = (x_2, y_2)$. Let's see how to find the position vector equivalent to $\overrightarrow{AC}$. As you can see in the figure below, the initial point A is relocated to the origin $(0, 0)$. The coordinates of P are found by subtracting the coordinates of A from the coordinates of C. Thus, $P = (x_2 - x_1, y_2 - y_1)$ and the position vector is $\overrightarrow{OP}$.

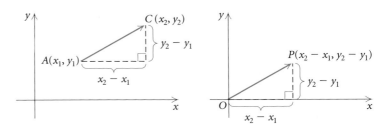

It can be shown that $\overrightarrow{OP}$ and $\overrightarrow{AC}$ have the same magnitude and direction and are therefore equivalent. Thus, $\overrightarrow{AC} = \overrightarrow{OP} = \langle x_2 - x_1, y_2 - y_1 \rangle$.

Component Form of a Vector

The **component form** of $\overrightarrow{AC}$ with $A = (x_1, y_1)$ and $C = (x_2, y_2)$ is

$$\overrightarrow{AC} = \langle x_2 - x_1, y_2 - y_1 \rangle.$$

EXAMPLE 1 Find the component form of $\overrightarrow{CF}$ if $C = (-4, -3)$ and $F = (1, 5)$.

Solution We have

$$\overrightarrow{CF} = \langle 1 - (-4), 5 - (-3) \rangle = \langle 5, 8 \rangle.$$

Note that vector $\overrightarrow{CF}$ is equivalent to *position vector* $\overrightarrow{OP}$ with $P = (5, 8)$ as shown in the figure at left. ■

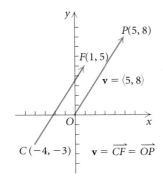

Now that we know how to write vectors in component form, let's restate some definitions that we first considered in Section 7.5.

The length of a vector $\mathbf{v}$ is easy to determine when the components of the vector are known. For $\mathbf{v} = \langle v_1, v_2 \rangle$, we have

$$|\mathbf{v}|^2 = v_1^2 + v_2^2 \qquad \text{Using the Pythagorean theorem}$$
$$|\mathbf{v}| = \sqrt{v_1^2 + v_2^2}.$$

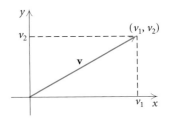

Length of a Vector

The **length**, or **magnitude**, of a vector $\mathbf{v} = \langle v_1, v_2 \rangle$ is given by

$$|\mathbf{v}| = \sqrt{v_1^2 + v_2^2}.$$

Two vectors are **equivalent** if they have the *same* magnitude and the *same* direction.

Equivalent Vectors

Let $\mathbf{u} = \langle u_1, u_2 \rangle$ and $\mathbf{v} = \langle v_1, v_2 \rangle$. Then

$$\langle u_1, u_2 \rangle = \langle v_1, v_2 \rangle \quad \text{if and only if} \quad u_1 = v_1 \quad \text{and} \quad u_2 = v_2.$$

Operations on Vectors

To multiply a vector $\mathbf{v}$ by a positive real number, we multiply its length by the number. Its direction stays the same. When a vector $\mathbf{v}$ is multiplied by 2 for instance, its length is doubled and its direction is not changed. When a vector is multiplied by 1.6, its length is increased by 60% and its direction stays the same. To multiply a vector $\mathbf{v}$ by a negative real number, we multiply its length by the number and reverse its direction. When a vector is multiplied by -2, its length is doubled and its direction is reversed. Since real numbers work like scaling factors in vector multiplication, we call them **scalars** and the products $k\mathbf{v}$ are called **scalar multiples** of $\mathbf{v}$.

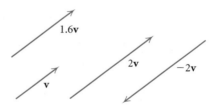

Scalar multiples of $\mathbf{v}$

Scalar Multiplication

For a real number k and a vector $\mathbf{v} = \langle v_1, v_2 \rangle$, the **scalar product** of k and $\mathbf{v}$ is

$$k\mathbf{v} = k\langle v_1, v_2 \rangle = \langle kv_1, kv_2 \rangle.$$

The vector $k\mathbf{v}$ is a **scalar multiple** of the vector $\mathbf{v}$.

EXAMPLE 2   Let $\mathbf{u} = \langle -5, \ 4\rangle$ and $\mathbf{w} = \langle 1, \ -1\rangle$. Find $-7\mathbf{w}$, $3\mathbf{u}$, and $-1\mathbf{w}$.

Solution

$$-7\mathbf{w} = -7\langle 1, \ -1\rangle = \langle -7, \ 7\rangle,$$
$$3\mathbf{u} = 3\langle -5, \ 4\rangle = \langle -15, \ 12\rangle,$$
$$-1\mathbf{w} = -1\langle 1, \ -1\rangle = \langle -1, \ 1\rangle$$

In Section 7.5, we used the parallelogram law to add two vectors, but now we can add two vectors using components. To add two vectors given in component form, we add the corresponding components. Let $\mathbf{u} = \langle u_1, \ u_2\rangle$ and $\mathbf{v} = \langle v_1, \ v_2\rangle$. Then

$$\mathbf{u} + \mathbf{v} = \langle u_1 + v_1, \ u_2 + v_2\rangle.$$

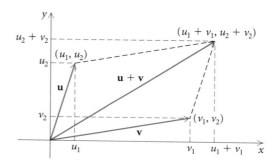

For example, if $\mathbf{v} = \langle -3, \ 2\rangle$ and $\mathbf{w} = \langle 5, \ -9\rangle$, then

$$\mathbf{v} + \mathbf{w} = \langle -3 + 5, \ 2 + (-9)\rangle = \langle 2, \ -7\rangle.$$

Vector Addition

If $\mathbf{u} = \langle u_1, \ u_2\rangle$ and $\mathbf{v} = \langle v_1, \ v_2\rangle$, then

$$\mathbf{u} + \mathbf{v} = \langle u_1 + v_1, \ u_2 + v_2\rangle.$$

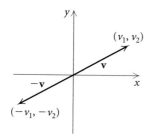

Before we define vector subtraction, we need to define $-\mathbf{v}$. The opposite of $\mathbf{v} = \langle v_1, \ v_2\rangle$, shown at left, is

$$-\mathbf{v} = (-1)\mathbf{v} = (-1)\langle v_1, \ v_2\rangle = \langle -v_1, \ -v_2\rangle.$$

Vector subtraction such as $\mathbf{u} - \mathbf{v}$ involves subtracting corresponding components. We show this by rewriting $\mathbf{u} - \mathbf{v}$ as $\mathbf{u} + (-\mathbf{v})$. If $\mathbf{u} = \langle u_1, \ u_2\rangle$ and $\mathbf{v} = \langle v_1, \ v_2\rangle$, then

$$\mathbf{u} - \mathbf{v} = \mathbf{u} + (-\mathbf{v}) = \langle u_1, \ u_2\rangle + \langle -v_1, \ -v_2\rangle$$
$$= \langle u_1 + (-v_1), \ u_2 + (-v_2)\rangle$$
$$= \langle u_1 - v_1, \ u_2 - v_2\rangle.$$

We can illustrate vector subtraction with parallelograms, just as we did vector addition.

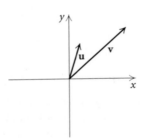

Sketch **u** and **v**.

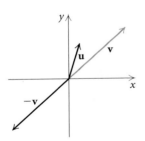

Sketch −**v**.

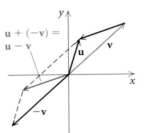

Sketch **u** + (−**v**), or **u** − **v**, using the parallelogram law.

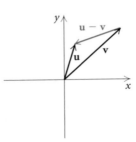

u − **v** is the vector from the terminal point of **v** to the terminal point of **u**.

Vector Subtraction

If **u** = $\langle u_1, u_2 \rangle$ and **v** = $\langle v_1, v_2 \rangle$, then

$$\mathbf{u} - \mathbf{v} = \langle u_1 - v_1, u_2 - v_2 \rangle.$$

It is interesting to compare the sum of two vectors with the difference of the same two vectors in the same parallelogram. The vectors **u** + **v** and **u** − **v** are the diagonals of the parallelogram.

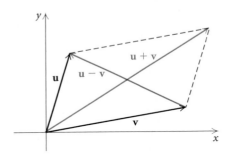

EXAMPLE 3 Do the following calculations, where **u** = $\langle 7, 2 \rangle$ and **v** = $\langle -3, 5 \rangle$.

a) **u** + **v**   b) **u** − 6**v**

c) 3**u** + 4**v**   d) $|5\mathbf{v} - 2\mathbf{u}|$

Solution

a) **u** + **v** = $\langle 7, 2 \rangle + \langle -3, 5 \rangle = \langle 7 + (-3), 2 + 5 \rangle = \langle 4, 7 \rangle$

b) **u** − 6**v** = $\langle 7, 2 \rangle - 6\langle -3, 5 \rangle = \langle 7, 2 \rangle - \langle -18, 30 \rangle = \langle 25, -28 \rangle$

c) 3**u** + 4**v** = $3\langle 7, 2 \rangle + 4\langle -3, 5 \rangle = \langle 21, 6 \rangle + \langle -12, 20 \rangle = \langle 9, 26 \rangle$

d) $|5\mathbf{v} - 2\mathbf{u}| = |5\langle -3, 5\rangle - 2\langle 7, 2\rangle| = |\langle -15, 25\rangle - \langle 14, 4\rangle|$
$$= |\langle -29, 21\rangle|$$
$$= \sqrt{(-29)^2 + 21^2}$$
$$= \sqrt{1282}$$
$$\approx 35.8$$

Before we state the properties of vector addition and scalar multiplication, we need to define another special vector—the zero vector. The vector whose initial and terminal points are both $(0, 0)$ is the **zero vector,** denoted by $\mathbf{O}$, or $\langle 0, 0\rangle$. Its magnitude is 0. In vector addition, the zero vector is the additive identity vector:

$$\mathbf{v} + \mathbf{O} = \mathbf{v}. \qquad \langle v_1, v_2\rangle + \langle 0, 0\rangle = \langle v_1, v_2\rangle$$

Operations on vectors share many of the same properties as operations on real numbers.

Properties of Vector Addition and Scalar Multiplication
For all vectors $\mathbf{u}$, $\mathbf{v}$, and $\mathbf{w}$, and for all scalars b and c:

1. $\mathbf{u} + \mathbf{v} = \mathbf{v} + \mathbf{u}$.
2. $\mathbf{u} + (\mathbf{v} + \mathbf{w}) = (\mathbf{u} + \mathbf{v}) + \mathbf{w}$.
3. $\mathbf{v} + \mathbf{O} = \mathbf{v}$.
4. $1\mathbf{v} = \mathbf{v}; \quad 0\mathbf{v} = \mathbf{O}$.
5. $\mathbf{v} + (-\mathbf{v}) = \mathbf{O}$.
6. $b(c\mathbf{v}) = (bc)\mathbf{v}$.
7. $(b + c)\mathbf{v} = b\mathbf{v} + c\mathbf{v}$.
8. $b(\mathbf{u} + \mathbf{v}) = b\mathbf{u} + b\mathbf{v}$.

Unit Vectors

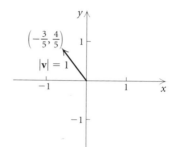

A vector of magnitude, or length, 1 is called a **unit vector.** The vector $\mathbf{v} = \left\langle -\frac{3}{5}, \frac{4}{5}\right\rangle$ is a unit vector because

$$|\mathbf{v}| = \left|\left\langle -\tfrac{3}{5}, \tfrac{4}{5}\right\rangle\right| = \sqrt{\left(-\tfrac{3}{5}\right)^2 + \left(\tfrac{4}{5}\right)^2}$$

$$= \sqrt{\tfrac{9}{25} + \tfrac{16}{25}}$$

$$= \sqrt{\tfrac{25}{25}}$$

$$= \sqrt{1} = 1.$$

EXAMPLE 4 Find a unit vector that has the same direction as the vector $\mathbf{w} = \langle -3, 5 \rangle$.

Solution We first find the length of $\mathbf{w}$:

$$|\mathbf{w}| = \sqrt{(-3)^2 + 5^2} = \sqrt{34}.$$

Thus we want a vector whose length is $1/\sqrt{34}$ of $\mathbf{w}$ and whose direction is the same as vector $\mathbf{w}$. That vector is

$$\mathbf{u} = \frac{1}{\sqrt{34}}\mathbf{w} = \frac{1}{\sqrt{34}}\langle -3, 5 \rangle = \left\langle \frac{-3}{\sqrt{34}}, \frac{5}{\sqrt{34}} \right\rangle.$$

The vector $\mathbf{u}$ is a *unit vector* because

$$|\mathbf{u}| = \left| \frac{1}{\sqrt{34}}\mathbf{w} \right| = \sqrt{\left(\frac{-3}{\sqrt{34}}\right)^2 + \left(\frac{5}{\sqrt{34}}\right)^2} = \sqrt{\frac{9}{34} + \frac{25}{34}}$$

$$= \sqrt{\frac{34}{34}} = \sqrt{1} = 1.$$

Unit Vector

If $\mathbf{v}$ is a vector and $\mathbf{v} \neq \mathbf{O}$, then

$$\frac{1}{|\mathbf{v}|} \cdot \mathbf{v}, \quad \text{or} \quad \frac{\mathbf{v}}{|\mathbf{v}|},$$

is a **unit vector** in the direction of $\mathbf{v}$.

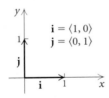

Although unit vectors can have any direction, the unit vectors parallel to the x- and y-axes are particularly useful. They are defined as

$$\mathbf{i} = \langle 1, 0 \rangle \quad \text{and} \quad \mathbf{j} = \langle 0, 1 \rangle.$$

Any vector can be expressed as a **linear combination** of unit vectors $\mathbf{i}$ and $\mathbf{j}$. For example, let $\mathbf{v} = \langle v_1, v_2 \rangle$. Then

$$\mathbf{v} = \langle v_1, v_2 \rangle = \langle v_1, 0 \rangle + \langle 0, v_2 \rangle$$
$$= v_1\langle 1, 0 \rangle + v_2\langle 0, 1 \rangle = v_1\mathbf{i} + v_2\mathbf{j}.$$

EXAMPLE 5 Express the vector $\mathbf{r} = \langle 2, -6 \rangle$ as a linear combination of $\mathbf{i}$ and $\mathbf{j}$.

Solution

$$\mathbf{r} = \langle 2, -6 \rangle = 2\mathbf{i} + (-6)\mathbf{j} = 2\mathbf{i} - 6\mathbf{j}$$

EXAMPLE 6 Write the vector $\mathbf{q} = -\mathbf{i} + 7\mathbf{j}$ in component form.

Solution

$$\mathbf{q} = -\mathbf{i} + 7\mathbf{j} = -1\mathbf{i} + 7\mathbf{j} = \langle -1, 7 \rangle$$

Vector operations can also be performed when vectors are written as linear combinations of **i** and **j**.

EXAMPLE 7 If $\mathbf{a} = 5\mathbf{i} - 2\mathbf{j}$ and $\mathbf{b} = -\mathbf{i} + 8\mathbf{j}$, find $3\mathbf{a} - \mathbf{b}$.

Solution

$$
\begin{aligned}
3\mathbf{a} - \mathbf{b} &= 3(5\mathbf{i} - 2\mathbf{j}) - (-\mathbf{i} + 8\mathbf{j}) \\
&= 15\mathbf{i} - 6\mathbf{j} + \mathbf{i} - 8\mathbf{j} \\
&= 16\mathbf{i} - 14\mathbf{j}
\end{aligned}
$$

Direction Angles

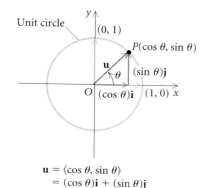

Unit circle

$$\mathbf{u} = \langle \cos\theta, \sin\theta \rangle$$
$$= (\cos\theta)\mathbf{i} + (\sin\theta)\mathbf{j}$$

UNIT CIRCLE

REVIEW SECTION 5.5.

The terminal point P of a unit vector in standard position is a point on the unit circle denoted by $(\cos\theta, \sin\theta)$. Thus the unit vector can be expressed in component form,

$$\mathbf{u} = \langle \cos\theta, \sin\theta \rangle,$$

or as a linear combination of the unit vectors **i** and **j**,

$$\mathbf{u} = (\cos\theta)\mathbf{i} + (\sin\theta)\mathbf{j},$$

where the components of **u** are functions of the **direction angle** θ measured counterclockwise from the x-axis to the vector. As θ varies from 0 to 2π, the point P traces the circle $x^2 + y^2 = 1$. This takes in all possible directions for unit vectors so the equation $\mathbf{u} = (\cos\theta)\mathbf{i} + (\sin\theta)\mathbf{j}$ describes every possible unit vector in the plane.

EXAMPLE 8 Calculate and sketch the unit vector $\mathbf{u} = (\cos\theta)\mathbf{i} + (\sin\theta)\mathbf{j}$ for $\theta = 2\pi/3$. Include the unit circle in your sketch.

Solution

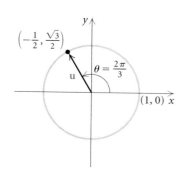

$$
\begin{aligned}
\mathbf{u} &= \left(\cos\frac{2\pi}{3}\right)\mathbf{i} + \left(\sin\frac{2\pi}{3}\right)\mathbf{j} \\
&= \left(-\frac{1}{2}\right)\mathbf{i} + \left(\frac{\sqrt{3}}{2}\right)\mathbf{j}
\end{aligned}
$$

Let $\mathbf{v} = \langle v_1, v_2 \rangle$ with direction angle θ. Using the definition of the tangent function, we can determine the direction angle from the components of $\mathbf{v}$:

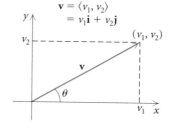

$$\tan\theta = \frac{v_2}{v_1}$$

$$\theta = \tan^{-1}\frac{v_2}{v_1}.$$

EXAMPLE 9   Determine the direction angle θ of the vector $\mathbf{w} = -4\mathbf{i} - 3\mathbf{j}$.

Solution   We know that

$$\mathbf{w} = -4\mathbf{i} - 3\mathbf{j} = \langle -4, -3 \rangle.$$

Thus we have

$$\tan \theta = \frac{-3}{-4} = \frac{3}{4} \quad \text{and} \quad \theta = \tan^{-1} \frac{3}{4}.$$

Since $\mathbf{w}$ is in the third quadrant, we know that θ is a third-quadrant angle. The reference angle is

$$\tan^{-1} \frac{3}{4} \approx 37°, \text{ and } \theta \approx 180° + 37°, \text{ or } 217°.$$

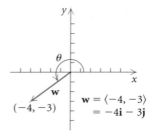

It is convenient for work with applied problems and in subsequent courses, such as calculus, to have a way to express a vector so that both its magnitude and its direction can be determined, or read, easily. Let $\mathbf{v}$ be a vector. Then $\mathbf{v}/|\mathbf{v}|$ is a unit vector in the same direction as $\mathbf{v}$. Thus we have

$$\frac{\mathbf{v}}{|\mathbf{v}|} = (\cos \theta)\mathbf{i} + (\sin \theta)\mathbf{j}$$

$$\mathbf{v} = |\mathbf{v}|[(\cos \theta)\mathbf{i} + (\sin \theta)\mathbf{j}] \qquad \text{Multiplying by } |v|$$

$$\mathbf{v} = |\mathbf{v}|(\cos \theta)\mathbf{i} + |\mathbf{v}|(\sin \theta)\mathbf{j}.$$

Let's revisit the applied problem in Example 3 of Section 7.5 and use this new notation.

EXAMPLE 10   *Airplane Speed and Direction.* An airplane travels on a bearing of 100° at an airspeed of 190 km/h while a wind is blowing 48 km/h from 220°. Find the ground speed of the airplane and the direction of its track, or course, over the ground.

Solution   We first make a drawing. The wind is represented by $\overrightarrow{OC}$ and the velocity vector of the airplane by $\overrightarrow{OA}$. The resultant velocity vector is $\mathbf{v}$, the sum of the two vectors:

$$\mathbf{v} = \overrightarrow{OC} + \overrightarrow{OA}.$$

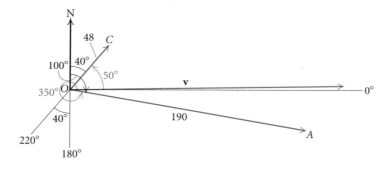

The bearing (measured from north) of the airspeed vector $\overrightarrow{OA}$ is 100°. Its *direction angle* (measured counterclockwise from the positive *x*-axis) is 350°. The bearing (measured from north) of the wind vector $\overrightarrow{OC}$ is 220°. Its direction angle (measured counterclockwise from the positive *x*-axis) is 50°. The magnitudes of $\overrightarrow{OA}$ and $\overrightarrow{OC}$ are 190 and 48, respectively. We have

$$\overrightarrow{OA} = 190(\cos 350°)\mathbf{i} + 190(\sin 350°)\mathbf{j}, \quad \text{and}$$
$$\overrightarrow{OC} = 48(\cos 50°)\mathbf{i} + 48(\sin 50°)\mathbf{j}.$$

Thus,

$$\mathbf{v} = \overrightarrow{OA} + \overrightarrow{OC}$$
$$= [190(\cos 350°)\mathbf{i} + 190(\sin 350°)\mathbf{j}] + [48(\cos 50°)\mathbf{i} + 48(\sin 50°)\mathbf{j}]$$
$$= [190(\cos 350°) + 48(\cos 50°)]\mathbf{i} + [190(\sin 350°) + 48(\sin 50°)]\mathbf{j}$$
$$\approx 217.97\mathbf{i} + 3.78\mathbf{j}.$$

From this form, we can determine the ground speed and the course:

$$\text{Ground speed} \approx \sqrt{(217.97)^2 + (3.78)^2}$$
$$\approx 218 \text{ km/h}.$$

We let α be the direction angle of $\mathbf{v}$. Then

$$\tan \alpha = \frac{3.78}{217.97}$$
$$\alpha = \tan^{-1} \frac{3.78}{217.97} \approx 1°.$$

Thus the course of the airplane (the direction from north) is $90° - 1°$, or 89°.

Angle Between Vectors

When a vector is multiplied by a scalar, the result is a vector. When two vectors are added, the result is also a vector. Thus we might expect the product of two vectors to be a vector as well, but it is not. The *dot product* of two vectors is a real number, or scalar. This product is useful in finding the angle between two vectors and in determining whether two vectors are perpendicular.

Dot Product

The **dot product** of two vectors $\mathbf{u} = \langle u_1, u_2 \rangle$ and $\mathbf{v} = \langle v_1, v_2 \rangle$ is

$$\mathbf{u} \cdot \mathbf{v} = u_1 v_1 + u_2 v_2.$$

(Note that $u_1 v_1 + u_2 v_2$ is a *scalar*, not a vector.)

EXAMPLE 11 Find the indicated dot product when

$$\mathbf{u} = \langle 2, -5 \rangle, \quad \mathbf{v} = \langle 0, 4 \rangle, \quad \text{and} \quad \mathbf{w} = \langle -3, 1 \rangle.$$

a) $\mathbf{u} \cdot \mathbf{w}$

b) $\mathbf{w} \cdot \mathbf{v}$

Solution

a) $\mathbf{u} \cdot \mathbf{w} = 2(-3) + (-5)1 = -6 - 5 = -11$

b) $\mathbf{w} \cdot \mathbf{v} = -3(0) + 1(4) = 0 + 4 = 4$

    The dot product can be used to find the angle between two vectors. The angle *between* two vectors is the smallest positive angle formed by the two directed line segments. Thus the angle θ between $\mathbf{u}$ and $\mathbf{v}$ is the same angle as between $\mathbf{v}$ and $\mathbf{u}$, and $0 \leq \theta \leq \pi$.

> ### Angle Between Two Vectors
> If θ is the angle between two *nonzero* vectors $\mathbf{u}$ and $\mathbf{v}$, then
>
> $$\cos \theta = \frac{\mathbf{u} \cdot \mathbf{v}}{|\mathbf{u}| \, |\mathbf{v}|}.$$

EXAMPLE 12 Find the angle between $\mathbf{u} = \langle 3, 7 \rangle$ and $\mathbf{v} = \langle -4, 2 \rangle$.

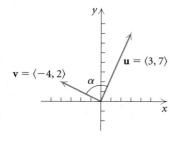

Solution We begin by finding $\mathbf{u} \cdot \mathbf{v}$, $|\mathbf{u}|$, and $|\mathbf{v}|$:

$$\mathbf{u} \cdot \mathbf{v} = 3(-4) + 7(2) = 2,$$
$$|\mathbf{u}| = \sqrt{3^2 + 7^2} = \sqrt{58}, \quad \text{and}$$
$$|\mathbf{v}| = \sqrt{(-4)^2 + 2^2} = \sqrt{20}.$$

Then

$$\cos \alpha = \frac{\mathbf{u} \cdot \mathbf{v}}{|\mathbf{u}| \, |\mathbf{v}|} = \frac{2}{\sqrt{58} \, \sqrt{20}}$$

$$\alpha = \cos^{-1} \frac{2}{\sqrt{58} \, \sqrt{20}}$$

$$\alpha \approx 86.6°.$$

Forces in Equilibrium

When several forces act through the same point on an object, their vector sum must be $\mathbf{O}$ in order for a balance to occur. When a balance occurs, then the object is either stationary or moving in a straight line without acceleration. The fact that the vector sum must be $\mathbf{O}$ for a balance, and vice versa, allows us to solve many applied problems involving forces.

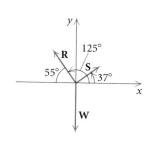

EXAMPLE 13 *Suspended Block.* A 350-lb block is suspended by two cables, as shown at left. At point *A*, there are three forces acting: **W**, the block pulling down, and **R** and **S**, the two cables pulling upward and outward. Find the tension in each cable.

Solution We draw a force diagram with the initial points of each vector at the origin. For there to be a balance, the vector sum must be the vector **O**:

$$\mathbf{R} + \mathbf{S} + \mathbf{W} = \mathbf{O}.$$

We can express each vector in terms of its magnitude and its direction angle:

$$\mathbf{R} = |\mathbf{R}|[(\cos 125°)\mathbf{i} + (\sin 125°)\mathbf{j}],$$
$$\mathbf{S} = |\mathbf{S}|[(\cos 37°)\mathbf{i} + (\sin 37°)\mathbf{j}], \quad \text{and}$$
$$\mathbf{W} = |\mathbf{W}|[(\cos 270°)\mathbf{i} + (\sin 270°)\mathbf{j}]$$
$$= 350(\cos 270°)\mathbf{i} + 350(\sin 270°)\mathbf{j}$$
$$= -350\mathbf{j}. \qquad \cos 270° = 0; \sin 270° = -1$$

Substituting for **R**, **S**, and **W** in **R** + **S** + **W** = **O**, we have

$$[|\mathbf{R}|(\cos 125°) + |\mathbf{S}|(\cos 37°)]\mathbf{i} + [|\mathbf{R}|(\sin 125°) + |\mathbf{S}|(\sin 37°) - 350]\mathbf{j}$$
$$= 0\mathbf{i} + 0\mathbf{j}.$$

This gives us a system of equations:

$$|\mathbf{R}|(\cos 125°) + |\mathbf{S}|(\cos 37°) = 0,$$
$$|\mathbf{R}|(\sin 125°) + |\mathbf{S}|(\sin 37°) - 350 = 0.$$

Solving this system, we get

$$|\mathbf{R}| \approx 280 \quad \text{and} \quad |\mathbf{S}| \approx 201.$$

The tensions in the cables are 280 lb and 201 lb.

Exercise Set **7.6**

Find the component form of the vector given the initial and terminal points. Then find the length of the vector.

1. $\overrightarrow{MN}$; $M(6, -7)$, $N(-3, -2)$
2. $\overrightarrow{CD}$; $C(1, 5)$, $D(5, 7)$
3. $\overrightarrow{FE}$; $E(8, 4)$, $F(11, -2)$
4. $\overrightarrow{BA}$; $A(9, 0)$, $B(9, 7)$
5. $\overrightarrow{KL}$; $K(4, -3)$, $L(8, -3)$
6. $\overrightarrow{GH}$; $G(-6, 10)$, $H(-3, 2)$

7. Find the magnitude of vector **u** if $\mathbf{u} = \langle -1, 6 \rangle$.
8. Find the magnitude of vector $\overrightarrow{ST}$ if $\overrightarrow{ST} = \langle -12, 5 \rangle$.

Do the indicated calculations in Exercises 9–26 for the vectors

$$\mathbf{u} = \langle 5, -2 \rangle, \quad \mathbf{v} = \langle -4, 7 \rangle, \quad \text{and} \quad \mathbf{w} = \langle -1, -3 \rangle.$$

9. **u** + **w**
10. **w** + **u**
11. $|3\mathbf{w} - \mathbf{v}|$
12. 6**v** + 5**u**
13. **v** − **u**
14. $|2\mathbf{w}|$

15. $5\mathbf{u} - 4\mathbf{v}$

16. $-5\mathbf{v}$

17. $|3\mathbf{u}| - |\mathbf{v}|$

18. $|\mathbf{v}| + |\mathbf{u}|$

19. $\mathbf{v} + \mathbf{u} + 2\mathbf{w}$

20. $\mathbf{w} - (\mathbf{u} + 4\mathbf{v})$

21. $2\mathbf{v} + \mathbf{O}$

22. $10|7\mathbf{w} - 3\mathbf{u}|$

23. $\mathbf{u} \cdot \mathbf{w}$

24. $\mathbf{w} \cdot \mathbf{u}$

25. $\mathbf{u} \cdot \mathbf{v}$

26. $\mathbf{v} \cdot \mathbf{w}$

*The vectors **u**, **v**, and **w** are drawn below. Copy them on a sheet of paper. Then sketch each of the vectors in Exercises 27–30.*

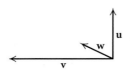

27. $\mathbf{u} + \mathbf{v}$

28. $\mathbf{u} - 2\mathbf{v}$

29. $\mathbf{u} + \mathbf{v} + \mathbf{w}$

30. $\frac{1}{2}\mathbf{u} - \mathbf{w}$

31. Vectors **u**, **v**, and **w** are determined by the sides of $\triangle ABC$ below.

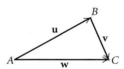

    a) Find an expression for **w** in terms of **u** and **v**.
    b) Find an expression for **v** in terms of **u** and **w**.

32. In $\triangle ABC$, vectors **u** and **w** are determined by the sides shown, where P is the midpoint of side BC. Find an expression for **v** in terms of **u** and **w**.

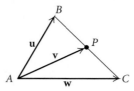

Find a unit vector that has the same direction as the given vector.

33. $\mathbf{v} = \langle -5, 12 \rangle$

34. $\mathbf{u} = \langle 3, 4 \rangle$

35. $\mathbf{w} = \langle 1, -10 \rangle$

36. $\mathbf{a} = \langle 6, -7 \rangle$

37. $\mathbf{r} = \langle -2, -8 \rangle$

38. $\mathbf{t} = \langle -3, -3 \rangle$

*Express the vector as a linear combination of the unit vectors **i** and **j**.*

39. $\mathbf{w} = \langle -4, 6 \rangle$

40. $\mathbf{r} = \langle -15, 9 \rangle$

41. $\mathbf{s} = \langle 2, 5 \rangle$

42. $\mathbf{u} = \langle 2, -1 \rangle$

*Express the vector as a linear combination of **i** and **j**.*

43.

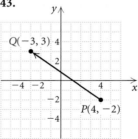

44.

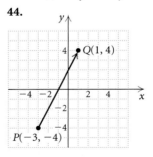

For Exercises 45–48, use the vectors

$$\mathbf{u} = 2\mathbf{i} + \mathbf{j}, \qquad \mathbf{v} = -3\mathbf{i} - 10\mathbf{j}, \quad and \quad \mathbf{w} = \mathbf{i} - 5\mathbf{j}.$$

*Perform the indicated vector operations and state the answer in two forms: (a) as a linear combination of **i** and **j** and (b) in component form.*

45. $4\mathbf{u} - 5\mathbf{w}$

46. $\mathbf{v} + 3\mathbf{w}$

47. $\mathbf{u} - (\mathbf{v} + \mathbf{w})$

48. $(\mathbf{u} - \mathbf{v}) + \mathbf{w}$

Sketch (include the unit circle) and calculate the unit vector $\mathbf{u} = (\cos \theta)\mathbf{i} + (\sin \theta)\mathbf{j}$ for the given direction angle.

49. $\theta = \dfrac{\pi}{2}$

50. $\theta = \dfrac{\pi}{3}$

51. $\theta = \dfrac{4\pi}{3}$

52. $\theta = \dfrac{3\pi}{2}$

Determine the direction angle θ of the vector, to the nearest degree.

53. $\mathbf{u} = \langle -2, -5 \rangle$

54. $\mathbf{w} = \langle 4, -3 \rangle$

55. $\mathbf{q} = \mathbf{i} + 2\mathbf{j}$

56. $\mathbf{w} = 5\mathbf{i} - \mathbf{j}$

57. $\mathbf{t} = \langle 5, 6 \rangle$

58. $\mathbf{b} = \langle -8, -4 \rangle$

Find the magnitude and the direction angle θ of the vector.

59. $\mathbf{u} = 3[(\cos 45°)\mathbf{i} + (\sin 45°)\mathbf{j}]$

60. $\mathbf{w} = 6[(\cos 150°)\mathbf{i} + (\sin 150°)\mathbf{j}]$

61. $\mathbf{v} = \left\langle -\dfrac{1}{2}, \dfrac{\sqrt{3}}{2} \right\rangle$

62. $\mathbf{u} = -\mathbf{i} - \mathbf{j}$

Find the angle between the given vectors, to the nearest tenth of a degree.

63. $\mathbf{u} = \langle 2, -5 \rangle$, $\mathbf{v} = \langle 1, 4 \rangle$

64. $\mathbf{a} = \langle -3, -3 \rangle$, $\mathbf{b} = \langle -5, 2 \rangle$

65. $\mathbf{w} = \langle 3, 5 \rangle$, $\mathbf{r} = \langle 5, 5 \rangle$

66. $\mathbf{v} = \langle -4, 2 \rangle$, $\mathbf{t} = \langle 1, -4 \rangle$

67. $a = i + j$, $b = 2i - 3j$

68. $u = 3i + 2j$, $v = -i + 4j$

Express each vector in Exercises 69–72 in the form
$ai + bj$ and sketch each in the coordinate plane.

69. The unit vectors $u = (\cos \theta)i + (\sin \theta)j$ for
$\theta = \pi/6$ and $\theta = 3\pi/4$. Include the unit circle
$x^2 + y^2 = 1$ in your sketch.

70. The unit vectors $u = (\cos \theta)i + (\sin \theta)j$ for
$\theta = -\pi/4$ and $\theta = -3\pi/4$. Include the unit circle
$x^2 + y^2 = 1$ in your sketch.

71. The unit vector obtained by rotating j
counterclockwise $3\pi/4$ radians about the origin

72. The unit vector obtained by rotating j clockwise
$2\pi/3$ radians about the origin

For the vectors in Exercises 73 and 74, find the unit
vectors $u = (\cos \theta)i + (\sin \theta)j$ in the same direction.

73. $-i + 3j$ **74.** $6i - 8j$

For the vectors in Exercises 75 and 76, express each
vector in terms of its magnitude and its direction.

75. $2i - 3j$ **76.** $5i + 12j$

77. Use a sketch to show that
$$v = 3i - 6j \quad \text{and} \quad u = -i + 2j$$
have opposite directions.

78. Use a sketch to show that
$$v = 3i - 6j \quad \text{and} \quad u = \tfrac{1}{2}i - j$$
have the same direction.

Exercises 79–82 appeared first in Exercise Set 7.5, where
we used the law of cosines and the law of sines to solve
the applied problems. For this exercise set, solve the
problem using the vector form
$$v = |v|[(\cos \theta)i + (\sin \theta)j].$$

79. *Ship.* A ship sails first N80°E for 120 nautical mi,
and then S20°W for 200 nautical mi. How far is the
ship, then, from the starting point, and in what
direction?

80. *Boat.* A boat heads 35°, propelled by a force of
750 lb. A wind from 320° exerts a force of 150 lb on
the boat. How large is the resultant force, and in
what direction is the boat moving?

81. *Airplane.* An airplane has an airspeed of
150 km/h. It is to make a flight in a direction of
070° while there is a 25-km/h wind from 340°.
What will the airplane's actual heading be?

82. *Airplane.* An airplane flies 032° for 210 mi, and
then 280° for 170 mi. How far is the airplane, then,
from the starting point, and in what direction?

83. Two cables support a 1000-lb weight, as shown.
Find the tension in each cable.

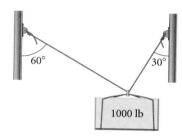

84. A 2500-kg block is suspended by two ropes, as
shown. Find the tension in each rope.

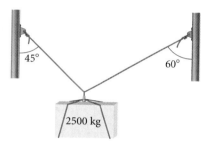

85. A 150-lb sign is hanging from the end of a hinged
boom, supported by a cable inclined 42° with the
horizontal. Find the tension in the cable and the
compression in the boom.

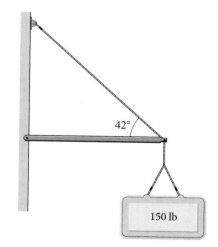

86. A weight of 200 lb is supported by a frame made of two rods and hinged at points A, B, and C. Find the forces exerted by the two rods.

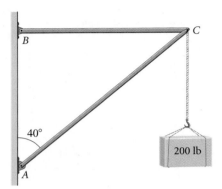

Let $\mathbf{u} = \langle u_1, u_2 \rangle$ *and* $\mathbf{v} = \langle v_1, v_2 \rangle$. *Prove each of the following properties.*

87. $\mathbf{u} + \mathbf{v} = \mathbf{v} + \mathbf{u}$

88. $\mathbf{u} \cdot \mathbf{v} = \mathbf{v} \cdot \mathbf{u}$

Collaborative Discussion and Writing

89. Explain how unit vectors are related to the unit circle.

90. Write a vector sum problem for a classmate for which the answer is $\mathbf{v} = 5\mathbf{i} - 8\mathbf{j}$.

Skill Maintenance

Find the slope and the y-intercept of the line with the given equation.

91. $-\frac{1}{5}x - y = 15$ **92.** $y = 7$

Find the zeros of the function.

93. $x^3 - 4x^2 = 0$ **94.** $6x^2 + 7x = 55$

Synthesis

95. If the dot product of two nonzero vectors $\mathbf{u}$ and $\mathbf{v}$ is 0, then the vectors are perpendicular (**orthogonal**). Let $\mathbf{u} = \langle u_1, u_2 \rangle$ and $\mathbf{v} = \langle v_1, v_2 \rangle$.

  a) Prove that if $\mathbf{u} \cdot \mathbf{v} = 0$, then $\mathbf{u}$ and $\mathbf{v}$ are perpendicular.

  b) Give an example of two perpendicular vectors and show that their dot product is 0.

96. If $\overrightarrow{PQ}$ is any vector, what is $\overrightarrow{PQ} + \overrightarrow{QP}$?

97. Find all the unit vectors that are parallel to the vector $\langle 3, -4 \rangle$.

98. Find a vector of length 2 whose direction is the opposite of the direction of the vector $\mathbf{v} = -\mathbf{i} + 2\mathbf{j}$. How many such vectors are there?

99. Given the vector $\overrightarrow{AB} = 3\mathbf{i} - \mathbf{j}$ and A is the point $(2, 9)$, find the point B.

100. Find vector $\mathbf{v}$ from point A to the origin, where $\overrightarrow{AB} = 4\mathbf{i} - 2\mathbf{j}$ and B is the point $(-2, 5)$.

7   Chapter Summary and Review

Important Properties and Formulas

The Law of Sines

$$\frac{a}{\sin A} = \frac{b}{\sin B} = \frac{c}{\sin C}$$

The Area of a Triangle

$$K = \frac{1}{2} bc \sin A = \frac{1}{2} ab \sin C = \frac{1}{2} ac \sin B$$

The Law of Cosines

$$a^2 = b^2 + c^2 - 2bc \cos A,$$
$$b^2 = a^2 + c^2 - 2ac \cos B,$$
$$c^2 = a^2 + b^2 - 2ab \cos C$$

Complex Numbers

Absolute Value: $|a + bi| = \sqrt{a^2 + b^2}$

Trigonometric Notation: $a + bi = r(\cos \theta + i \sin \theta)$

Multiplication: $r_1(\cos \theta_1 + i \sin \theta_1) \cdot r_2(\cos \theta_2 + i \sin \theta_2)$
$$= r_1 r_2 [\cos (\theta_1 + \theta_2) + i \sin (\theta_1 + \theta_2)]$$

Division: $\dfrac{r_1(\cos \theta_1 + i \sin \theta_1)}{r_2(\cos \theta_2 + i \sin \theta_2)} = \dfrac{r_1}{r_2} [\cos (\theta_1 - \theta_2) + i \sin (\theta_1 - \theta_2)], \quad r_2 \neq 0$

DeMoivre's Theorem

$$[r(\cos \theta + i \sin \theta)]^n = r^n(\cos n\theta + i \sin n\theta)$$

Roots of Complex Numbers

The nth roots of $r(\cos \theta + i \sin \theta)$ are

$$r^{1/n} \left[\cos \left(\frac{\theta}{n} + k \cdot \frac{360°}{n} \right) + i \sin \left(\frac{\theta}{n} + k \cdot \frac{360°}{n} \right) \right], \quad r \neq 0, k = 0, 1, 2, \ldots, n - 1.$$

Vectors

If $\mathbf{u} = \langle u_1, u_2 \rangle$ and $\mathbf{v} = \langle v_1, v_2 \rangle$ and k is a scalar, then:

Length: $|\mathbf{v}| = \sqrt{v_1^2 + v_2^2}$

Addition: $\mathbf{u} + \mathbf{v} = \langle u_1 + v_1, u_2 + v_2 \rangle$

Subtraction: $\mathbf{u} - \mathbf{v} = \langle u_1 - v_1, u_2 - v_2 \rangle$

Scalar Multiplication: $k\mathbf{v} = \langle kv_1, kv_2 \rangle$

Dot Product: $\mathbf{u} \cdot \mathbf{v} = u_1 v_1 + u_2 v_2$

Angle Between Two Vectors: $\cos \theta = \dfrac{\mathbf{u} \cdot \mathbf{v}}{|\mathbf{u}| |\mathbf{v}|}$

REVIEW EXERCISES

Solve $\triangle ABC$, if possible.

1. $a = 23.4$ ft, $b = 15.7$ ft, $c = 8.3$ ft

2. $B = 27°$, $C = 35°$, $b = 19$ in.

3. $A = 133°28'$, $C = 31°42'$, $b = 890$ m

4. $B = 37°$, $b = 4$ yd, $c = 8$ yd

5. Find the area of $\triangle ABC$ if $b = 9.8$ m, $c = 7.3$ m, and $A = 67.3°$.

6. A parallelogram has sides of lengths 3.21 ft and 7.85 ft. One of its angles measures 147°. Find the area of the parallelogram.

7. *Sandbox.* A child-care center has a triangular-shaped sandbox. Two of the three sides measure 15 ft and 12.5 ft and form an included angle of 42°.

To determine the amount of sand that is needed to fill the box, the director must determine the area of the floor of the box. Find the area of the floor of the box to the nearest square foot.

8. *Flower Garden.* A triangular flower garden has sides of lengths 11 m, 9 m, and 6 m. Find the angles of the garden to the nearest degree.

9. In an isosceles triangle, the base angles each measure 52.3° and the base is 513 ft long. Find the lengths of the other two sides to the nearest foot.

10. *Airplanes.* Two airplanes leave an airport at the same time. The first flies 175 km/h in a direction of 305.6°. The second flies 220 km/h in a direction of 195.5°. After 2 hr, how far apart are the planes?

Graph the complex number and find its absolute value.

11. $2 - 5i$

12. 4

13. $2i$

14. $-3 + i$

Find trigonometric notation.

15. $1 + i$

16. $-4i$

17. $-5\sqrt{3} + 5i$

18. $\dfrac{3}{4}$

Find standard notation, $a + bi$.

19. $4(\cos 60° + i \sin 60°)$

20. $7(\cos 0° + i \sin 0°)$

21. $5\left(\cos \dfrac{2\pi}{3} + i \sin \dfrac{2\pi}{3}\right)$

22. $2\left[\cos\left(-\dfrac{\pi}{3}\right) + i \sin\left(-\dfrac{\pi}{3}\right)\right]$

Convert to trigonometric notation and then multiply or divide, expressing the answer in standard notation.

23. $(1 + i\sqrt{3})(1 - i)$

24. $\dfrac{2 - 2i}{2 + 2i}$

25. $\dfrac{2 + 2\sqrt{3}i}{\sqrt{3} - i}$

26. $i(3 - 3\sqrt{3}i)$

Raise the number to the given power and write trigonometric notation for the answer.

27. $[2(\cos 60° + i \sin 60°)]^3$

28. $(1 - i)^4$

Raise the number to the given power and write standard notation for the answer.

29. $(1 + i)^6$

30. $\left(\dfrac{1}{2} + \dfrac{\sqrt{3}}{2}i\right)^{10}$

31. Find the square roots of $-1 + i$.

32. Find the cube roots of $3\sqrt{3} - 3i$.

33. Find and graph the fourth roots of 81.

34. Find and graph the fifth roots of 1.

Find all the complex solutions of the equation.

35. $x^4 - i = 0$

36. $x^3 + 1 = 0$

37. Find the polar coordinates of each of these points. Give three answers for each point.

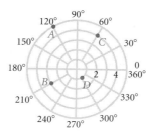

Find the polar coordinates of the point. Express the answer in degrees and then in radians.

38. $(-4\sqrt{2}, 4\sqrt{2})$

39. $(0, -5)$

Find the rectangular coordinates of the point.

40. $\left(3, \dfrac{\pi}{4}\right)$

41. $(-6, -120°)$

Convert to a polar equation.

42. $5x - 2y = 6$

43. $y = 3$

44. $x^2 + y^2 = 9$

45. $y^2 - 4x - 16 = 0$

Convert to a rectangular equation.

46. $r = 6$

47. $r + r \sin \theta = 1$

48. $r = \dfrac{3}{1 - \cos \theta}$

49. $r - 2 \cos \theta = 3 \sin \theta$

In Exercises 50–53, match the equation with one of figures (a)–(d), which follow.

a)

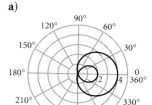

b)

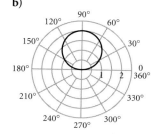

c)

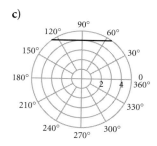

d)

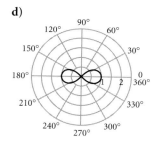

50. $r = 2 \sin \theta$

51. $r^2 = \cos 2\theta$

52. $r = 1 + 3 \cos \theta$

53. $r \sin \theta = 4$

*Magnitudes of vectors **u** and **v** and the angle θ between the vectors are given. Find the magnitude of the sum, **u** + **v**, to the nearest tenth and give the direction by specifying to the nearest degree the angle that it makes with the vector **u**.*

54. $|\mathbf{u}| = 12$, $|\mathbf{v}| = 15$, $\theta = 120°$

55. $|\mathbf{u}| = 41$, $|\mathbf{v}| = 60$, $\theta = 25°$

*The vectors **u**, **v**, and **w** are drawn below. Copy them on a sheet of paper. Then sketch each of the vectors in Exercises 56 and 57.*

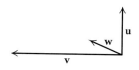

56. $\mathbf{u} - \mathbf{v}$

57. $\mathbf{u} + \frac{1}{2}\mathbf{w}$

58. Forces of 230 N and 500 N act on an object. The angle between the forces is 52°. Find the resultant, giving the angle that it makes with the smaller force.

59. *Wind.* A wind has an easterly component of 15 km/h and a southerly component of 25 km/h. Find the magnitude and the direction of the wind.

60. *Ship.* A ship sails N75°E for 90 nautical mi, and then S10°W for 100 nautical mi. How far is the ship, then, from the starting point, and in what direction?

Find the component form of the vector given the initial and terminal points.

61. $\overrightarrow{AB}$; $A(2, -8)$, $B(-2, -5)$

62. $\overrightarrow{TR}$; $R(0, 7)$, $T(-2, 13)$

63. Find the magnitude of vector **u** if $\mathbf{u} = \langle 5, -6 \rangle$.

Do the calculations in Exercises 64–67 for the vectors $\mathbf{u} = \langle 3, -4 \rangle$, $\mathbf{v} = \langle -3, 9 \rangle$, and $\mathbf{w} = \langle -2, -5 \rangle$.

64. $4\mathbf{u} + \mathbf{w}$

65. $2\mathbf{w} - 6\mathbf{v}$

66. $|\mathbf{u}| + |2\mathbf{w}|$

67. $\mathbf{u} \cdot \mathbf{w}$

68. Find a unit vector that has the same direction as $\mathbf{v} = \langle -6, -2 \rangle$.

69. Express the vector $\mathbf{t} = \langle -9, 4 \rangle$ as a linear combination of the unit vectors **i** and **j**.

70. Determine the direction angle θ of the vector $\mathbf{w} = \langle -4, -1 \rangle$ to the nearest degree.

71. Find the magnitude and the direction angle θ of $\mathbf{u} = -5\mathbf{i} - 3\mathbf{j}$.

72. Find the angle between $\mathbf{u} = \langle 3, -7 \rangle$ and $\mathbf{v} = \langle 2, 2 \rangle$ to the nearest tenth of a degree.

73. *Airplane.* An airplane has an airspeed of 160 mph. It is to make a flight in a direction of 080° while there is a 20-mph wind from 310°. What will the airplane's actual heading be?

Do the calculations in Exercises 74–77 for the vectors $\mathbf{u} = 2\mathbf{i} + 5\mathbf{j}$, $\mathbf{v} = -3\mathbf{i} + 10\mathbf{j}$, and $\mathbf{w} = 4\mathbf{i} + 7\mathbf{j}$.

74. $5\mathbf{u} - 8\mathbf{v}$

75. $\mathbf{u} - (\mathbf{v} + \mathbf{w})$

76. $|\mathbf{u} - \mathbf{v}|$

77. $3|\mathbf{w}| + |\mathbf{v}|$

78. Express the vector $\overrightarrow{PQ}$ in the form $a\mathbf{i} + b\mathbf{j}$, if P is the point $(1, -3)$ and Q is the point $(-4, 2)$.

Express each vector in Exercises 79 and 80 in the form $a\mathbf{i} + b\mathbf{j}$ *and sketch each in the coordinate plane.*

79. The unit vectors $\mathbf{u} = (\cos \theta)\mathbf{i} + (\sin \theta)\mathbf{j}$ for $\theta = \pi/4$ and $\theta = 5\pi/4$. Include the unit circle $x^2 + y^2 = 1$ in your sketch.

80. The unit vector obtained by rotating $\mathbf{j}$ counterclockwise $2\pi/3$ radians about the origin.

81. Express the vector $3\mathbf{i} - \mathbf{j}$ as a product of its magnitude and its direction.

Collaborative Discussion and Writing

82. Explain why these statements are not contradictory:

The number 1 has one real cube root.
The number 1 has three complex cube roots.

83. Summarize how you can tell algebraically when solving triangles whether there is no solution, one solution, or two solutions.

84. *Golf: Distance versus Accuracy.* It is often argued in golf that the farther you hit the ball, the more accurate it must be to stay safe. (Safe means not in the woods, water, or some other hazard.) In his book *Golf and the Spirit* (p. 54), M. Scott Peck asserts "Deviate 5° from your aiming point on a 150-yd shot, and your ball will land approximately 20 yd to the side of where you wanted it to be. Do the same on a 300-yd shot, and it will be 40 yd off target. Twenty yards may well be in the range of safety; 40 yards probably won't. This principle not infrequently allows a mediocre, short-hitting golfer like myself to score better than the long hitter." Check the accuracy of the mathematics in this statement, and comment on Peck's assertion.

Synthesis

85. Let $\mathbf{u} = 12\mathbf{i} + 5\mathbf{j}$. Find a vector that has the same direction as $\mathbf{u}$ but has length 3.

86. A parallelogram has sides of lengths 3.42 and 6.97. Its area is 18.4. Find the sizes of its angles.

7 Chapter Test

Solve $\triangle ABC$, *if possible.*

1. $a = 18$ ft, $b = 54°$, $c = 43°$

2. $b = 8$ m, $c = 5$ m, $C = 36°$

3. $a = 16.1$ in., $b = 9.8$ in., $c = 11.2$ in.

4. Find the area of $\triangle ABC$ if $C = 106.4°$, $a = 7$ cm, and $b = 13$ cm.

5. *Distance Across a Lake.* Points A and B are on opposite sides of a lake. Point C is 52 m from A. The measure of $\angle BAC$ is determined to be 108°, and the measure of $\angle ACB$ is determined to be 44°. What is the distance from A to B?

6. *Location of Airplanes.* Two airplanes leave an airport at the same time. The first flies 210 km/h in a direction of 290°. The second flies 180 km/h in a direction of 185°. After 3 hr, how far apart are the planes?

7. Graph: $-4 + i$.

8. Find the absolute value of $2 - 3i$.

9. Find trigonometric notation for $3 - 3i$.

10. Divide and express the result in standard notation $a + bi$:

$$\frac{2\left(\cos \dfrac{2\pi}{3} + i \sin \dfrac{2\pi}{3}\right)}{8\left(\cos \dfrac{\pi}{6} + i \sin \dfrac{\pi}{6}\right)}.$$

11. Find $(1 - i)^8$ and write standard notation for the answer.

12. Find the polar coordinates of $(-1, \sqrt{3})$. Express the angle in degrees using the smallest possible positive angle.

13. Convert $\left(-1, \dfrac{2\pi}{3}\right)$ to rectangular coordinates.

14. Convert to a polar equation: $x^2 + y^2 = 10$.

15. Graph: $r = 1 - \cos \theta$.

16. Which of the following is the graph of $r = 3 \cos \theta$?

a)

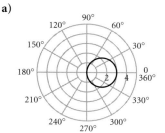

b)

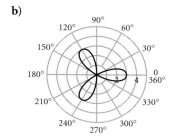

c)

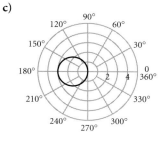

d)

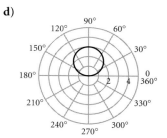

17. For vectors $\mathbf{u}$ and $\mathbf{v}$, $|\mathbf{u}| = 8$, $|\mathbf{v}| = 5$, and the angle between the vectors is $63°$. Find $\mathbf{u} + \mathbf{v}$. Give the magnitude to the nearest tenth, and give the direction by specifying the angle that the resultant makes with $\mathbf{u}$, to the nearest degree.

18. For $\mathbf{u} = 2\mathbf{i} - 7\mathbf{j}$ and $\mathbf{v} = 5\mathbf{i} + \mathbf{j}$, find $2\mathbf{u} - 3\mathbf{v}$.

19. Find a unit vector in the same direction as $-4\mathbf{i} + 3\mathbf{j}$.

Synthesis

20. A parallelogram has sides of length 15.4 and 9.8. Its area is 72.9. Find the measures of the angles.

Systems of Equations and Matrices

8

*I*t is often desirable or even necessary to use two or more variables to model a situation in a field such as business, science, psychology, engineering, education, or sociology. When this is the case, we write and solve a *system of equations* in order to answer questions about that situation. In this chapter, we study systems of equations and several methods for solving them, including the use of matrices and determinants. We also use *systems of inequalities,* along with *linear programming,* to model applications and to find the maximum and minimum values of functions subject to a set of restrictions, or constraints.

APPLICATION

The following table shows per capita coffee consumption, in gallons, in the United States, represented as years since 1992.

YEAR, x	COFFEE CONSUMPTION (IN GALLONS)
1992, 0	26
1994, 2	21
1996, 4	23

We can use a system of equations to fit a quadratic function to these data. Then we can use this function to predict coffee consumption in future years.

This problem appears as Exercise 28 in Section 8.2.

8.1

Systems of Equations in Two Variables

- Solve a system of two linear equations in two variables by graphing.
- Solve a system of two linear equations in two variables using the substitution and the elimination methods.
- Use systems of two linear equations to solve applied problems.

A **system of equations** is composed of two or more equations considered simultaneously. For example,

$$x - y = 5,$$
$$2x + y = 1$$

is a system of two linear equations in two variables. The solution set of this system consists of all ordered pairs that make *both* equations true. The ordered pair $(2, -3)$ is a solution of the system of equations above. We can verify this by substituting 2 for x and -3 for y in *each* equation.

$x - y = 5$		$2x + y = 1$	
$2 - (-3)$? 5		$2 \cdot 2 + (-3)$? 1	
$2 + 3$		$4 - 3$	
5	5 TRUE	1	1 TRUE

Solving Systems of Equations Graphically

Recall that the graph of a linear equation is a line that contains all the ordered pairs in the solution set of the equation. When we graph a system of linear equations, each point at which the equations intersect is a solution of *both* equations and therefore a solution of the system of equations.

EXAMPLE 1   Solve the following system of equations graphically.

$$x - y = 5,$$
$$2x + y = 1$$

Solution   We graph the equations on the same set of axes, as shown below.

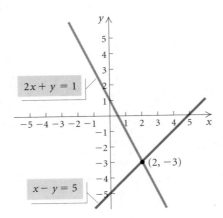

We see that the graphs intersect at a single point, $(2, -3)$, so $(2, -3)$ is the solution of the system of equations. To check this solution, we substitute 2 for x and -3 for y in both equations, as we did above. ▬

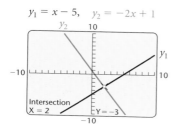
The graphs of most of the systems of equations that we use to model applications intersect at a single point, like the system above. However, it is possible that the graphs will have no points in common or infinitely many points in common. Each of these possibilities is illustrated below.

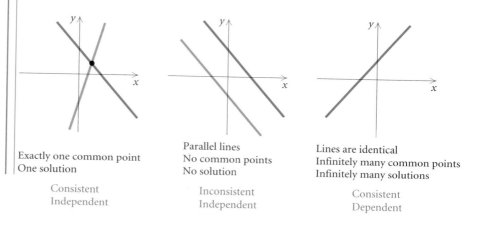

Exactly one common point
One solution

Consistent
Independent

Parallel lines
No common points
No solution

Inconsistent
Independent

Lines are identical
Infinitely many common points
Infinitely many solutions

Consistent
Dependent

If a system of equations has at least one solution, it is **consistent**. If the system has no solutions, it is **inconsistent**. In addition, if a system of two linear equations in two variables has an infinite number of solutions, the equations are **dependent**. Otherwise, they are **independent**.

The Substitution Method

Solving a system of equations graphically is not always accurate when the solutions are not integers. A pair like $\left(\frac{43}{27}, -\frac{19}{27}\right)$, for instance, will be difficult to determine from a hand-drawn graph.

Algebraic methods for solving systems of equations, when used correctly, always give accurate results. One such technique is the **substitution method.** It is used most often when a variable is alone on one side of an equation or when it is easy to solve for a variable. To apply the substitution method, we begin by using one of the equations to express one variable in terms of the other; then we substitute that expression in the other equation of the system.

EXAMPLE 2   Use the substitution method to solve the system

$$x - y = 5, \quad (1)$$
$$2x + y = 1. \quad (2)$$

Solution   First, we solve equation (1) for x. (We could have solved for y instead.) We have

$$x - y = 5, \quad (1)$$
$$x = y + 5. \quad \text{Solving for } x$$

Then we substitute $y + 5$ for x in equation (2). This gives an equation in one variable, which we know how to solve:

$$2x + y = 1 \quad (2)$$
$$2(y + 5) + y = 1 \quad \text{The parentheses are necessary.}$$
$$2y + 10 + y = 1 \quad \text{Removing parentheses}$$
$$3y + 10 = 1$$
$$3y = -9$$
$$y = -3.$$

Now we substitute -3 for y in either of the original equations (this is called **back-substitution**) and solve for y. We choose equation (1):

$$x - y = 5, \quad (1)$$
$$x - (-3) = 5 \quad \text{Substituting } -3 \text{ for } y$$
$$x + 3 = 5$$
$$x = 2.$$

We have previously checked the pair $(2, -3)$ in both equations. The solution of the system of equations is $(2, -3)$.   ▬

The Elimination Method

Another algebraic technique for solving systems of equations is the **elimination method.** With this method, we eliminate a variable by adding two equations. Before we add, it might be necessary to multiply one or both equations by suitable constants in order to find two equations in which coefficients are opposites.

EXAMPLE 3   Solve each of the following systems using the elimination method.

a) $2x + y = 2, \quad (1)$
$\quad x - y = 7 \quad (2)$

b) $4x + 3y = 11, \quad (1)$
$\quad -5x + 2y = 15 \quad (2)$

Solution

a) Since the y-coefficients, 1 and -1, are opposites, we can eliminate y by adding the equations:

$$2x + y = 2 \quad (1)$$
$$\underline{x - y = 7} \quad (2)$$
$$3x \quad\quad = 9 \quad \text{Adding}$$
$$x = 3.$$

We then back-substitute 3 for x in either equation and solve for y. We choose equation (1):

$$2x + y = 2 \qquad (1)$$
$$2 \cdot 3 + y = 2 \qquad \text{Substituting 3 for } x$$
$$6 + y = 2$$
$$y = -4.$$

We check the solution by substituting the pair $(3, -4)$ in both equations.

$2x + y = 2$		
$2 \cdot 3 + (-4)$? 2		
$6 - 4$		
2	2	TRUE

$x - y = 7$		
$3 - (-4)$? 7		
$3 + 4$		
7	7	TRUE

The solution is $(3, -4)$.

b) We can obtain x-coefficients that are opposites by multiplying the first equation by 5 and the second equation by 4:

$$20x + 15y = 55 \qquad \text{Multiplying equation (1) by 5}$$
$$\underline{-20x + 8y = 60} \qquad \text{Multiplying equation (2) by 4}$$
$$23y = 115 \qquad \text{Adding}$$
$$y = 5.$$

We then back-substitute 5 for y in either equation (1) or (2) and solve for x. We choose equation (1):

$$4x + 3y = 11 \qquad (1)$$
$$4x + 3 \cdot 5 = 11 \qquad \text{Substituting 5 for } y$$
$$4x + 15 = 11$$
$$4x = -4$$
$$x = -1.$$

We can check the pair $(-1, 5)$ by substituting in both equations. The solution is $(-1, 5)$.

In Example 3(b), the two systems

$$\begin{array}{l} 4x + 3y = 11, \\ -5x + 2y = 15 \end{array} \quad \text{and} \quad \begin{array}{l} 20x + 15y = 55, \\ -20x + 8y = 60 \end{array}$$

are **equivalent** because they have exactly the same solutions. When we use the elimination method, we often multiply one or both equations by constants to find equivalent equations that allow us to eliminate a variable by adding.

EXAMPLE 4 Solve each of the following systems using the elimination method.

a) $\quad x - 3y = 1,$ $\quad$ (1)
$\quad -2x + 6y = 5$ $\quad$ (2)

b) $2x + 3y = 6,$ $\quad$ (1)
$\quad 4x + 6y = 12$ $\quad$ (2)

Solution

a) We multiply equation (1) by 2 and add:

$$\begin{array}{ll} 2x - 6y = 2 & \text{Multiplying equation (1) by 2} \\ -2x + 6y = 5 & \text{(2)} \\ \hline 0 = 7. & \text{Adding} \end{array}$$

There are no values of x and y for which $0 = 7$ is true, so the system has no solution. The solution set is $\varnothing$. The system of equations is inconsistent. The graphs of the equations are parallel lines.

b) We multiply equation (1) by -2 and add:

$$\begin{array}{ll} -4x - 6y = -12 & \text{Multiplying equation (1) by } -2 \\ 4x + 6y = 12 & \text{(2)} \\ \hline 0 = 0. & \text{Adding} \end{array}$$

We obtain the equation $0 = 0$, which is true for all values of x and y. This tells us that the equations are dependent, so there are infinitely many solutions. That is, any solution of one equation of the system is also a solution of the other. The graphs of the equations are identical.

Using either equation, we can write $y = -\frac{2}{3}x + 2$, so we can write the solutions of the system as ordered pairs of the form $\left(x, -\frac{2}{3}x + 2\right)$. Any real value that we choose for x then gives us a value for y and thus an ordered pair in the solution set. For example,

if $x = -3$, then $-\frac{2}{3}x + 2 = -\frac{2}{3}(-3) + 2 = 4,$

if $x = 0$, then $-\frac{2}{3}x + 2 = -\frac{2}{3} \cdot 0 + 2 = 2,$ and

if $x = 6$, then $-\frac{2}{3}x + 2 = -\frac{2}{3} \cdot 6 + 2 = -2.$

Thus some of the solutions are $(-3, 4)$, $(0, 2)$, and $(6, -2)$. Similarly, we can write $x = -\frac{3}{2}y + 3$, so the solutions can also be expressed as $\left(-\frac{3}{2}y + 3, y\right)$.

Applications

Frequently the most challenging and time-consuming step in the problem-solving process is translating a situation to mathematical language. However, in many cases, this task is facilitated if we translate to more than one equation in more than one variable.

EXAMPLE 5 *Snack Mixtures.* At Sunda's Snacks, caramel corn worth $2.50 per pound is mixed with honey roasted mixed nuts worth $7.50 per pound in order to get 20 lb of a mixture worth $4.50 per pound. How much of each snack is used?

Sunda's Snacks

CARAMEL CORN
$2.50 per lb

HONEY ROASTED
MIXED NUTS
$7.50 per lb

Solution   We use the five-step problem-solving process.

1. Familiarize. Let's begin by making a guess. Suppose 16 lb of caramel corn and 4 lb of nuts were used. Then the total weight of the mixture would be 16 lb + 4 lb, or 20 lb, the desired weight. The total values of these amounts of ingredients are found by multiplying the price per pound by the number of pounds used:

Caramel corn: $2.50(16) = \$40$

Nuts: $7.50(4) = \underline{\$30}$

Total value: $70.

The desired value of the mixture is $4.50 per pound, so the value of 20 lb would be $4.50(20), or $90. Thus we see that our guess, which led to a total of $70, is incorrect. Nevertheless, these calculations will help us to translate.

We organize the information in a table. We let x = the number of pounds of caramel corn in the mixture and y = the number of pounds of nuts.

	CARAMEL CORN	NUTS	MIXTURE	
PRICE PER POUND	$2.50	$7.50	$4.50	
NUMBER OF POUNDS	x	y	20	$\longrightarrow x + y = 20$
VALUE OF MIXTURE	$2.50x$	$7.50y$	20(4.50), or 90	$\longrightarrow 2.50x + 7.50y = 90$

2. Translate. From the second row of the table, we get one equation:

$x + y = 20.$

The last row of the table yields a second equation:

$2.50x + 7.50y = 90.$

We can multiply by 10 on both sides of the second equation to clear the decimals. This gives us the following system of equations:

$x + y = 20,$ (1)

$25x + 75y = 900.$ (2)

3. Carry out. We carry out the solution as follows.

Algebraic Solution

Using the elimination method, we multiply equation (1) by -25 and add it to equation (2):

$$-25x - 25y = -500$$
$$\underline{25x + 75y = 900}$$
$$50y = 400$$
$$y = 8.$$

Then we back-substitute to find x:

$$x + y = 20 \qquad (1)$$
$$x + 8 = 20 \qquad \text{Substituting 8 for } y$$
$$x = 12.$$

Visualizing the Solution

The solution of the system of equations is the point of intersection of the graphs of the equations.

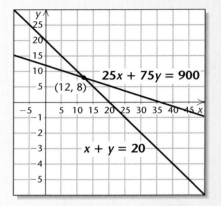

4. **Check.** If 12 lb of caramel corn and 8 lb of nuts are used, the mixture weighs $12 + 8$, or 20 lb. The value of the mixture is $2.50(12) + 7.50(8)$, or $30 + 60$, or $90. Since the possible solution yields the desired weight and value of the mixture, our result checks.

5. **State.** The mixture should consist of 12 lb of caramel corn and 8 lb of honey roasted nuts. ▬

EXAMPLE 6 *Supply and Demand.* Suppose that the price and the supply of the Great Tunes portable CD player are related by the equation

$$y = 90 + 30x,$$

where y is the price, in dollars, at which the seller is willing to supply x thousand units. Also suppose that the price and the demand for the same model of CD player are related by the equation

$$y = 200 - 25x,$$

where y is the price, in dollars, at which the consumer is willing to buy x thousand units.

The **equilibrium point** for this product is the pair (x, y) that is a solution of both equations. The **equilibrium price** is the price at which the amount of the product that the seller is willing to supply is the same as the amount demanded by the consumer. Find the equilibrium point for this product.

Solution

1., 2. Familiarize and **Translate.** We are given a system of equations in the statement of the problem, so no further translation is necessary.

$$y = 90 + 30x, \qquad (1)$$
$$y = 200 - 25x. \qquad (2)$$

We substitute some values for x in each equation to get an idea of the corresponding prices. When $x = 1$,

$$y = 90 + 30 \cdot 1 = 120, \qquad \text{Substituting in equation (1)}$$
$$y = 200 - 25 \cdot 1 = 175. \qquad \text{Substituting in equation (2)}$$

This indicates that the price when 1 thousand units are supplied is lower than the price when 1 thousand units are demanded.
When $x = 4$,

$$y = 90 + 30 \cdot 4 = 210, \qquad \text{Substituting in equation (1)}$$
$$y = 200 - 25 \cdot 4 = 100. \qquad \text{Substituting in equation (2)}$$

In this case, the price related to supply is higher than the price related to demand. It would appear that the x-value we are looking for is between 1 and 4.

3. Carry out. We use the substitution method:

$$200 - 25x = 90 + 30x \qquad \begin{array}{l} \text{Substituting } 200 - 25x \text{ for } y \\ \text{in equation (1)} \end{array}$$
$$110 = 55x \qquad \text{Adding } 25x \text{ and subtracting 90 on both sides}$$
$$2 = x. \qquad \text{Dividing by 55 on both sides}$$

We now back-substitute 2 for x in either equation and find y:

$$y = 200 - 25x \qquad (2)$$
$$y = 200 - 25 \cdot 2 \qquad \text{Substituting 2 for } x$$
$$y = 150.$$

4. Check. We can check by substituting 2 for x and 150 for y in both equations. Also note that 2 is between 1 and 4 as expected from the Familiarize and Translate step.

5. State. The equilibrium point is (2, \$150). That is, the equilibrium supply is 2 thousand units and the equilibrium price is \$150.

Exercise Set

In Exercises 1–6, match the system of equations with one of the graphs (a)–(f), which follow.

a)
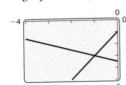

b)
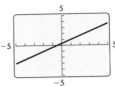

c)
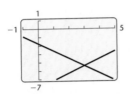

d)
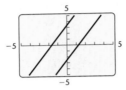

e)
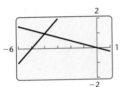

f)
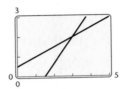

1. $x + y = -2,$
$y = x - 8$

2. $x - y = -5,$
$x = -4y$

3. $x - 2y = -1,$
$4x - 3y = 6$

4. $2x - y = 1,$
$x + 2y = -7$

5. $2x - 3y = -1,$
$-4x + 6y = 2$

6. $4x - 2y = 5,$
$6x - 3y = -10$

Solve graphically.

7. $x + y = 2,$
$3x + y = 0$

8. $x + y = 1,$
$3x + y = 7$

9. $x + 2y = 1,$
$x + 4y = 3$

10. $3x + 4y = 5,$
$x - 2y = 5$

11. $y + 1 = 2x,$
$y - 1 = 2x$

12. $2x - y = 1,$
$3y = 6x - 3$

Solve using the substitution method.

13. $x + y = 9,$
$2x - 3y = -2$

14. $3x - y = 5,$
$x + y = \frac{1}{2}$

15. $x - 2y = 7,$
$x = y + 4$

16. $x + 4y = 6,$
$x = -3y + 3$

17. $y = 2x - 6,$
$5x - 3y = 16$

18. $3x + 5y = 2,$
$2x - y = -3$

19. $x - 5y = 4,$
$y = 7 - 2x$

20. $5x + 3y = -1,$
$x + y = 1$

21. $2x - 3y = 5,$
$5x + 4y = 1$

22. $3x + 4y = 6,$
$2x + 3y = 5$

Solve using the elimination method. Also determine whether each system is consistent or inconsistent and dependent or independent.

23. $x + 2y = 7,$
$x - 2y = -5$

24. $3x + 4y = -2,$
$-3x - 5y = 1$

25. $x - 3y = 2,$
$6x + 5y = -34$

26. $x + 3y = 0,$
$20x - 15y = 75$

27. $0.3x - 0.2y = -0.9,$
$0.2x - 0.3y = -0.6$

28. $0.2x - 0.3y = 0.3,$
$0.4x + 0.6y = -0.2$

29. $3x - 12y = 6,$
$2x - 8y = 4$

30. $2x + 6y = 7,$
$3x + 9y = 10$

31. $\frac{1}{5}x + \frac{1}{2}y = 6,$
$\frac{3}{5}x - \frac{1}{2}y = 2$

32. $\frac{2}{3}x + \frac{3}{5}y = -17,$
$\frac{1}{2}x - \frac{1}{3}y = -1$

33. $2x = 5 - 3y,$
$4x = 11 - 7y$

34. $7(x - y) = 14,$
$2x = y + 5$

35. *Sales Promotions.* During a one-month promotional campaign, Video Village gave either a free video rental or a 12-serving box of microwave popcorn to new members. It cost the store $1 for each free rental and $2 for each box of popcorn. In all, 48 new members were signed up and the store's cost for the incentives was $86. How many of each incentive were given away?

36. *Concert Ticket Prices.* One evening 1500 concert tickets were sold for the Fairmont Summer Jazz Festival. Tickets cost $25 for covered pavilion seats and $15 for lawn seats. Total receipts were $28,500. How many of each type of ticket were sold?

37. *Museum Admission Prices.* Admission to the Indianapolis Children's Museum costs twice as much for adults as for children. Admission to the museum for 4 adults and 16 children from the Tiny Tots Day Care Center cost $72. Find the cost of each adult's admission and each child's admission.

38. *Mail-Order Business.* A mail-order lacrosse equipment business shipped 120 packages one day. Customers are charged $3.50 for each standard-delivery package and $7.50 for each express-delivery package. Total shipping charges for the day were $596. How many of each kind of package were shipped?

39. *Supply and Demand.* The supply and demand for a particular model of electronic organizer are related to price by the equations

$$y = 70 + 2x,$$
$$y = 175 - 5x,$$

respectively, where y is the price, in dollars, and x is the number of units, in thousands. Find the equilibrium point for this product.

40. *Supply and Demand.* The supply and demand for a particular model of treadmill are related to price by the equations

$$y = 240 + 40x,$$
$$y = 500 - 25x,$$

respectively, where y is the price, in dollars, and x is the number of units, in thousands. Find the equilibrium point for this product.

The point at which a company's costs equal its revenues is the break-even point. In Exercises 41–44, C represents the production cost, in dollars, of x units of a product and R represents the revenue, in dollars, from the sale of x units. Find the number of units that must be produced and sold in order to break even. That is, find the value of x for which C = R.

41. $C = 14x + 350,$
$R = 16.5x$

42. $C = 8.5x + 75,$
$R = 10x$

43. $C = 15x + 12,000,$
$R = 18x - 6000$

44. $C = 3x + 400,$
$R = 7x - 600$

45. *Nutrition.* A one-cup serving of spaghetti with meatballs contains 260 Cal (calories) and 32 g of carbohydrates. A one-cup serving of chopped iceberg lettuce contains 5 Cal and 1 g of carbohydrates. (*Source: Home and Garden Bulletin No. 72*, U.S. Government Printing Office, Washington, D.C. 20402) How many servings of each would be required to obtain 400 Cal and 50 g of carbohydrates?

46. *Nutrition.* One serving of tomato soup contains 100 Cal and 18 g of carbohydrates. One slice of whole wheat bread contains 70 Cal and 13 g of carbohydrates. (*Source: Home and Garden Bulletin*

No. 72, U.S. Government Printing Office, Washington, D.C. 20402) How many servings of each would be required to obtain 230 Cal and 42 g of carbohydrates?

47. *Motion.* A Leisure Time Cruises riverboat travels 46 km downstream in 2 hr. It travels 51 km upstream in 3 hr. Find the speed of the boat and the speed of the stream.

48. *Motion.* A DC10 travels 3000 km with a tail wind in 3 hr. It travels 3000 km with a head wind in 4 hr. Find the speed of the plane and the speed of the wind.

49. *Investment.* Bernadette inherited $15,000 and invested it in two municipal bonds, which pay 7% and 9% simple interest. The annual interest is $1230. Find the amount invested at each rate.

50. *Tee Shirt Sales.* Mack's Tee Shirt Shack sold 36 shirts one day. All short-sleeved tee shirts cost $12 and all long-sleeved tee shirts cost $18. Total receipts for the day were $522. How many of each kind of shirt were sold?

51. *Coffee Mixtures.* The owner of The Daily Grind coffee shop mixes French roast coffee worth $9.00 per pound with Kenyan coffee worth $7.50 per pound in order to get 10 lb of a mixture worth $8.40 per pound. How much of each type of coffee was used?

52. *Motion.* A Boeing 747 flies the 3000-mi distance from Los Angeles to New York, with a tail wind, in 5 hr. The return trip, against the wind, takes 6 hr. Find the speed of the plane and the speed of the wind.

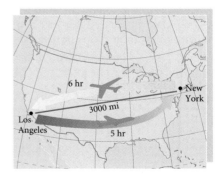

53. *Commissions.* Jackson Manufacturing offers its sales representatives a choice between being paid a commission of 8% of sales or being paid a monthly salary of $1500 plus a commission of 1% of sales. For what monthly sales do the two plans pay the same amount?

54. *Motion.* Two private airplanes travel toward each other from cities that are 780 km apart at speeds of 190 km/h and 200 km/h. They left at the same time. In how many hours will they meet?

Technology Connection

55. Use a graphing calculator to solve the systems of equations in Exercises 7, 13, and 23.

56. Use a graphing calculator to solve the systems of equations in Exercises 8, 14, and 24.

57. *Take-out Meals.* The number of take-out meals purchased per person has surpassed the number of restaurant meals purchased per person in recent years, as shown by the data in the following table.

Year	Number of Take-out Meals per Person	Number of Restaurant Meals per Person
1984	43	69
1988	53	68
1992	57	63
1997	66	64

Source: NPD Group's CREST service

a) Find linear regression functions $t(x)$ and $r(x)$ that represent the number of take-out meals purchased per person and the number of restaurant meals purchased per person, respectively, x years after 1984.

b) Use the functions found in part (a) to estimate how many years after 1984 the number of take-out meals purchased was equal to the number of restaurant meals purchased.

58. *U.S. Labor Force.* The percentages of men and women in the civilian labor force in recent years are shown in the table below.

Year	Percentage of Men in the Labor Force	Percentage of Women in the Labor Force
1985	76.3	54.5
1990	76.4	57.5
1995	75.0	58.9
1997	75.0	59.8

Source: U.S. Bureau of Labor Statistics

a) Find linear regression functions $m(x)$ and $w(x)$ that represent the percentages of men and women in the civilian labor force, respectively, x years after 1985.

b) Use the functions found in part (a) to predict when the percentages of men and women in the labor force will be equal.

Collaborative Discussion and Writing

59. Explain in your own words when the elimination method for solving a system of equations is preferable to the substitution method.

60. Cassidy solves the equation $2x + 5 = 3x - 7$ by finding the point of intersection of the graphs of $y_1 = 2x + 5$ and $y_2 = 3x - 7$. She finds the same point when she solves the system of equations

$$y = 2x + 5,$$
$$y = 3x - 7.$$

Explain the difference between the solutions.

Skill Maintenance

Find the zero(s) of the function. Give exact answers.

61. $f(x) = 2x - 5$

62. $f(x) = 6 - 3x$

63. $f(x) = x^2 - 4x + 3$

64. $f(x) = 3x^2 - x - 5$

Synthesis

65. *Motion.* Nancy jogs and walks to campus each day. She averages 4 km/h walking and 8 km/h jogging. The distance from home to the campus is 6 km and she makes the trip in 1 hr. How far does she jog on each trip?

66. *e-Commerce.* shirts.com advertises a limited-time sale, offering 1 turtleneck for $15 and 2 turtlenecks for $25. A total of 1250 turtlenecks are sold and $16,750 is taken in. How many customers ordered 2 turtlenecks?

67. *Motion.* A train leaves Union Station for Central Station, 216 km away, at 9 A.M. One hour later, a train leaves Central Station for Union Station. They meet at noon. If the second train had started at 9 A.M. and the first train at 10:30 A.M., they would still have met at noon. Find the speed of each train.

68. *Antifreeze Mixtures.* An automobile radiator contains 16 L of antifreeze and water. This mixture is 30% antifreeze. How much of this mixture should be drained and replaced with pure antifreeze so that the final mixture will be 50% antifreeze?

69. Two solutions of the equation $Ax + By = 1$ are $(3, -1)$ and $(-4, -2)$. Find A and B.

70. *Ticket Line.* You are in line at a ticket window. There are 2 more people ahead of you in line than there are behind you. In the entire line, there are three times as many people as there are behind you. How many people are ahead of you?

71. *Gas Mileage.* A 6-cylinder Oldsmobile Aurora gets 19 miles per gallon (mpg) in city driving and 28 mpg in highway driving. The car is driven 405 mi on a full tank of 18 gal of gasoline. How many miles were driven in the city and how many were driven on the highway?

72. *Motion.* Heather is out hiking and is standing on a railroad bridge, as shown in the figure below. A train is approaching from the direction shown by the arrow. If Heather runs at a speed of 10 mph toward the train, she will reach point P on the bridge at the same moment that the train does. If she runs to point Q at the other end of the bridge at a speed of 10 mph, she will reach point Q also at the same moment that the train does. How fast, in miles per hour, is the train traveling?

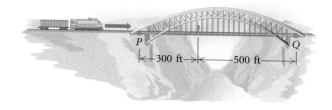

8.2
Systems of Equations in Three Variables

- Solve systems of linear equations in three variables.
- Use systems of three equations to solve applied problems.
- Model a situation using a quadratic function.

A **linear equation in three variables** is an equation equivalent to one of the form $Ax + By + Cz = D$, where A, B, C, and D are real numbers and A, B, and C are not all 0. A **solution of a system of three equations in three variables** is an ordered triple that makes all three equations true. For example, the triple $(2, -1, 0)$ is a solution of the system of equations

$$4x + 2y + 5z = 6,$$
$$2x - y + z = 5,$$
$$3x + 2y - z = 4.$$

We can verify this by substituting 2 for x, -1 for y, and 0 for z in each equation.

Solving Systems of Equations in Three Variables

We will solve systems of equations in three variables using an algebraic method called **Gaussian elimination,** named for the German mathematician Karl Friedrich Gauss (1777–1855). Our goal is to transform the original system to an equivalent one of the form

$$Ax + By + Cz = D,$$
$$Ey + Fz = G,$$
$$Hz = K.$$

Then we solve the third equation for z and back-substitute to find y and then x.

Each of the following operations can be used to transform the original system to an equivalent system in the desired form.

---

1. Interchange any two equations.

2. Multiply both sides of one of the equations by a nonzero constant.

3. Add a nonzero multiple of one equation to another equation.

EXAMPLE 1 Solve the following system:

$$x - 2y + 3z = 11, \qquad (1)$$
$$4x + 2y - 3z = 4, \qquad (2)$$
$$3x + 3y - z = 4. \qquad (3)$$

Solution We multiply equation (1) by -4 and add it to equation (2). We also multiply equation (1) by -3 and add it to equation (3). These operations produce an equivalent system of equations in which the x-terms have been eliminated from both the second and the third equations.

$$-4x + 8y - 12z = -44 \longleftarrow$$
$$\underline{4x + 2y - 3z = 4} \quad \text{Eq. (2)}$$
$$10y - 15z = -40 \quad \text{Eq. (4)}$$

$$x - 2y + 3z = 11, \longrightarrow$$
$$10y - 15z = -40,$$
$$9y - 10z = -29 \longleftarrow$$

$$-3x + 6y - 9z = -33$$
$$\underline{3x + 3y - z = 4} \quad \text{Eq. (3)}$$
$$9y - 10z = -29 \quad \text{Eq. (5)}$$

Now we multiply the last equation by 10 to make the y-coefficient a multiple of the y-coefficient in the equation above it:

$$x - 2y + 3z = 11, \qquad (1)$$
$$10y - 15z = -40, \qquad (4)$$
$$90y - 100z = -290. \qquad (5)$$

Next, we multiply equation (4) by -9 and add it to equation (5).

$$-90y + 135z = 360 \longleftarrow$$
$$\underline{90y - 100z = -290} \quad \text{Eq. (5)}$$
$$35z = 70 \longrightarrow$$

$$x - 2y + 3z = 11, \qquad (1)$$
$$10y - 15z = -40, \qquad (4)$$
$$35z = 70. \qquad (6)$$

Now we solve equation (6) for z:

$$35z = 70$$
$$z = 2.$$

Then we back-substitute 2 for z in equation (4) and solve for y:

$$10y - 15 \cdot 2 = -40$$
$$10y - 30 = -40$$
$$10y = -10$$
$$y = -1.$$

Finally, we back-substitute -1 for y and 2 for z in equation (1) and solve for x:

$$x - 2(-1) + 3 \cdot 2 = 11$$
$$x + 2 + 6 = 11$$
$$x = 3.$$

We can check the triple $(3, -1, 2)$ in each of the three original equations. Since it makes all three equations true, the solution is $(3, -1, 2)$. ▬

EXAMPLE 2   Solve the following system:

$$x + y + z = 7, \qquad (1)$$
$$3x - 2y + z = 3, \qquad (2)$$
$$x + 6y + 3z = 25. \qquad (3)$$

Solution   We multiply equation (1) by -3 and add it to equation (2). We also multiply equation (1) by -1 and add it to equation (3).

$$x + y + z = 7, \qquad (1)$$
$$-5y - 2z = -18, \qquad (4)$$
$$5y + 2z = 18 \qquad (5)$$

Next, we add equation (4) to equation (5):

$$x + y + z = 7, \qquad (1)$$
$$-5y - 2z = -18, \qquad (4)$$
$$0 = 0.$$

The equation $0 = 0$ tells us that equation (3) of the original system is dependent on the first two equations. Thus the original system is equivalent to

$$x + y + z = 7, \qquad (1)$$
$$3x - 2y + z = 3. \qquad (2)$$

In this particular case, the original system has infinitely many solutions. (In some cases, the system composed of equations (1) and (2) could be inconsistent.) To find an expression for these solutions, we first solve equation (4) for either y or z. We choose to solve for y:

$$-5y - 2z = -18 \qquad (4)$$
$$-5y = 2z - 18$$
$$y = -\tfrac{2}{5}z + \tfrac{18}{5}.$$

Then we back-substitute in equation (1) to find an expression for x in terms of z:

$$x - \tfrac{2}{5}z + \tfrac{18}{5} + z = 7$$
$$x + \tfrac{3}{5}z + \tfrac{18}{5} = 7$$
$$x + \tfrac{3}{5}z = \tfrac{17}{5}$$
$$x = -\tfrac{3}{5}z + \tfrac{17}{5}.$$

The solutions of the system of equations are ordered triples of the form $\left(-\tfrac{3}{5}z + \tfrac{17}{5}, -\tfrac{2}{5}z + \tfrac{18}{5}, z\right)$, where z can be any real number. Any real number that we use for z then gives us values for x and y and thus an ordered triple in the solution set. For example, if we choose $z = 0$, we have the solution $\left(\tfrac{17}{5}, \tfrac{18}{5}, 0\right)$. If we choose $z = -1$, we have $(4, 4, -1)$. ▬

If we get a false equation, such as $0 = -5$, at some stage of the elimination process, we conclude that the original system is *inconsistent*. That is, it has no solutions.

Although systems of three linear equations in three variables do not lend themselves well to graphical solutions, it is of interest to picture some possible solutions. The graph of a linear equation in three variables is a plane. Thus the solution set of such a system is the intersection of three planes. Some possibilities are shown below.

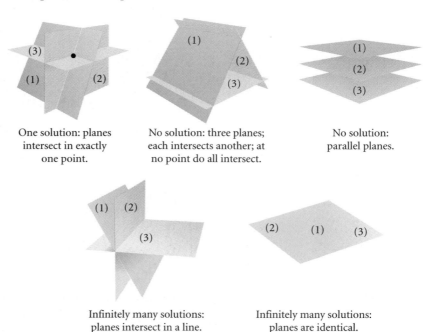

One solution: planes intersect in exactly one point.

No solution: three planes; each intersects another; at no point do all intersect.

No solution: parallel planes.

Infinitely many solutions: planes intersect in a line.

Infinitely many solutions: planes are identical.

Applications

Systems of equations in three or more variables allow us to solve many problems in fields such as business, the social and natural sciences, and engineering.

EXAMPLE 3 *Investment.* Moira inherited $15,000 and invested part of it in a money market account, part in municipal bonds, and part in a mutual fund. After 1 yr, she received a total of $730 in simple interest from the three investments. The money market account paid 4% annually, the bonds paid 5% annually, and the mutual fund paid 6% annually. There was $2000 more invested in the mutual fund than in bonds. Find the amount that Moira invested in each category.

Solution

1. **Familiarize.** We let x, y, and z represent the amounts invested in the money market account, the bonds, and the mutual fund, respectively. Then the amounts of income produced annually by each investment are given by 4%x, 5%y, and 6%z, or $0.04x$, $0.05y$, and $0.06z$.

2. **Translate.** The fact that a total of $15,000 is invested gives us one equation:

$$x + y + z = 15,000.$$

Since the total interest is $730, we have a second equation:

$$0.04x + 0.05y + 0.06z = 730.$$

Another statement in the problem gives us a third equation.

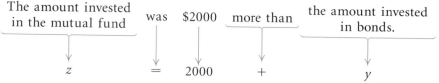

We now have a system of three equations:

$$x + y + z = 15,000, \qquad\qquad x + y + z = 15,000,$$
$$0.04x + 0.05y + 0.06z = 730, \quad\text{or}\quad 4x + 5y + 6z = 73,000,$$
$$z = 2000 + y; \qquad\qquad\qquad -y + z = 2000.$$

3. **Carry out.** Solving the system of equations, we get

$$(7000, 3000, 5000).$$

4. **Check.** The sum of the numbers is 15,000. The income produced is

$$0.04(7000) + 0.05(3000) + 0.06(5000) = 280 + 150 + 300, \quad\text{or}\quad \$730.$$

Also the amount invested in the mutual fund, $5000, is $2000 more than the amount invested in bonds, $3000. Our solution checks in the original problem.

5. **State.** Moira invested $7000 in a money market account, $3000 in municipal bonds, and $5000 in a mutual fund.

Mathematical Models and Applications

In a situation in which a quadratic function will serve as a mathematical model, we may wish to find an equation, or formula, for the function. For a linear model, we can find an equation if we know two data points. For a quadratic function, we need three data points.

EXAMPLE 4 *The Cost of Operating an Automobile at Various Speeds.* Under certain conditions, it is found that the cost of operating an automobile as a function of speed is approximated by a quadratic function. Use the data shown below to find an equation of the function. Then use the equation to determine the cost of operating the automobile at 60 mph and at 80 mph.

SPEED (IN MILES PER HOUR)	OPERATING COST PER MILE (IN CENTS)
10	22
20	20
50	20

Solution Letting x = the speed and $f(x)$ = the cost, we use the three data points $(10, 22)$, $(20, 20)$, and $(50, 20)$ to find a, b, and c in the equation $f(x) = ax^2 + bx + c$. First we substitute:

$$f(x) = ax^2 + bx + c$$

For $(10, 22)$: $\quad 22 = a \cdot 10^2 + b \cdot 10 + c;$

For $(20, 20)$: $\quad 20 = a \cdot 20^2 + b \cdot 20 + c;$

For $(50, 20)$: $\quad 20 = a \cdot 50^2 + b \cdot 50 + c.$

We now have a system of equations in the variables a, b, and c:

$$100a + 10b + c = 22,$$
$$400a + 20b + c = 20,$$
$$2500a + 50b + c = 20.$$

Solving this system of equations, we obtain $(0.005, -0.35, 25)$. Thus,

$$f(x) = 0.005x^2 - 0.35x + 25.$$

To determine the cost of operating an automobile at 60 mph, we find $f(60)$:

$$f(60) = 0.005(60)^2 - 0.35(60) + 25$$
$$= 22¢ \text{ per mile.}$$

To find the cost of operating an automobile at 80 mph, we find $f(80)$:

$$f(80) = 0.005(80)^2 - 0.35(80) + 25$$
$$= 29¢ \text{ per mile.}$$

Technology Connection

The function in Example 4 can also be found using the QUADRATIC REGRESSION feature on a graphing calculator. Note that the method of Example 4 works when we have exactly three data points, whereas the QUADRATIC REGRESSION feature on a graphing calculator can be used for three or more points.

Exercise Set 8.2

Solve each of the following systems.

1. $x + y + z = 2,$
$6x - 4y + 5z = 31,$
$5x + 2y + 2z = 13$

2. $x + 6y + 3z = 4,$
$2x + y + 2z = 3,$
$3x - 2y + z = 0$

3. $x - y + 2z = -3,$
$x + 2y + 3z = 4,$
$2x + y + z = -3$

4. $x + y + z = 6,$
$2x - y - z = -3,$
$x - 2y + 3z = 6$

5. $x + 2y - z = 5,$
$2x - 4y + z = 0,$
$3x + 2y + 2z = 3$

6. $2x + 3y - z = 1,$
$x + 2y + 5z = 4,$
$3x - y - 8z = -7$

7. $x + 2y - z = -8,$
$2x - y + z = 4,$
$8x + y + z = 2$

8. $x + 2y - z = 4,$
$4x - 3y + z = 8,$
$5x - y = 12$

9. $2x + y - 3z = 1,$
$x - 4y + z = 6,$
$4x - 7y - z = 13$

10. $x + 3y + 4z = 1,$
$3x + 4y + 5z = 3,$
$x + 8y + 11z = 2$

11. $4a + 9b = 8,$
$8a + 6c = -1,$
$6b + 6c = -1$

12. $3p + 2r = 11,$
$q - 7r = 4,$
$p - 6q = 1$

13. $w + x + y + z = 2,$
$w + 2x + 2y + 4z = 1,$
$-w + x - y - z = -6,$
$-w + 3x + y - z = -2$

14. $w + x - y + z = 0,$
$-w + 2x + 2y + z = 5,$
$-w + 3x + y - z = -4,$
$-2w + x + y - 3z = -7$

15. *e-Commerce.* computerwarehouse.com charges $3 for shipping orders up to 10 lb, $5 for orders from 10 lb up to 15 lb, and $7.50 for orders of 15 lb or more. One day shipping charges for 150 orders totaled $680. The number of orders under 10 lb was three times the number of orders weighing 15 lb

or more. Find the number of packages shipped at each rate.

16. *Mail-Order Business.* Natural Fibers Clothing charges $4 for shipping orders of $30 or less, $6 for orders from $30 to $70, and $7 for orders over $70. One week shipping charges for 600 orders totaled $3340. Eighty more orders for $30 or less were shipped than orders for more than $70. Find the number of orders shipped at each rate.

17. *Nutrition.* A hospital dietician must plan a lunch menu that provides 485 Cal, 41.5 g of carbohydrates, and 35 mg of calcium. A 3-oz serving of broiled ground beef contains 245 Cal, 0 g of carbohydrates, and 9 mg of calcium. One baked potato contains 145 Cal, 34 g of carbohydrates, and 8 mg of calcium. A one-cup serving of strawberries contains 45 Cal, 10 g of carbohydrates, and 21 mg of calcium. (*Source: Home and Garden Bulletin No. 72*, U.S. Government Printing Office, Washington, D.C. 20402) How many servings of each are required to provide the desired nutritional values?

18. *Nutrition.* A diabetic patient wishes to prepare a meal consisting of roasted chicken breast, mashed potatoes, and peas. A 3-oz serving of roasted, skinless chicken breast contains 140 Cal, 27 g of protein, and 64 mg of sodium. A one-cup serving of mashed potatoes contains 160 Cal, 4 g of protein, and 636 mg of sodium, and a one-cup serving of peas contains 125 Cal, 8 g of protein, and 139 mg of sodium. (*Source: Home and Garden Bulletin No. 72*, U.S. Government Printing Office, Washington, D.C. 20402) How many servings of each should be used if the meal is to contain 415 Cal, 50.5 g of protein, and 553 mg of sodium?

19. *Investment.* Jamal earns a year-end bonus of $5000 and puts it in 3 one-year investments that pay $302 in simple interest. Part is invested at 4%, part at 6%, and part at 7%. There is $1500 more invested at 7% than at 4%. Find the amount invested at each rate.

20. *Investment.* Casey receives $336 per year in simple interest from three investments. Part is invested at 8%, part at 9%, and part at 10%. There is $500 more invested at 9% than at 8%. The amount invested at 10% is three times the amount invested at 9%. Find the amount invested at each rate.

21. *Price Increases.* Orange juice, a raisin bagel, and a cup of coffee from Kelly's Koffee Kart cost a total of

$3. Kelly posts a notice announcing that, effective next week, the price of orange juice will increase 50% and the price of bagels will increase 20%. After the increase, the same purchase will cost a total of $3.75, and orange juice will cost twice as much as coffee. Find the price of each item before the increase.

22. *Cost of Snack Food.*   Martin and Eva pool their loose change to buy snacks on their coffee break. One day, they spent $1.85 on 1 carton of milk, 2 donuts, and 1 cup of coffee. The next day, they spent $2.30 on 3 donuts and 2 cups of coffee. The third day, they bought 1 carton of milk, 1 donut, and 2 cups of coffee and spent $1.75. On the fourth day, they have a total of $1.80 left. Is this enough to buy 2 cartons of milk and 2 donuts?

23. *Passenger Transportation.*   The total volume of passenger traffic on domestic airways, by bus (excluding school buses and urban transit buses), and by railroads in the United States in a recent year was 405 billion passenger-miles. (One passenger-mile is the transportation of one passenger the distance of one mile.) The volume of bus traffic was 10 billion passenger-miles more than the volume of railroad traffic. The total volume of railroad traffic was 329 billion passenger-miles less than the volume of traffic on domestic airways. (*Source: Transportation in America*, Eno Transportation Foundation, Inc., Landsdowne, VA) What was the volume of each type of passenger traffic?

24. *Cheese Consumption.*   The total per capita consumption (the average amount consumed per person) of cheddar, mozzarella, and Swiss cheese in the United States in a recent year was 18.1 lb. The total consumption of mozzarella and Swiss cheese was 0.3 lb less than that of cheddar cheese. The consumption of mozzarella cheese was 6.5 lb more than that of Swiss cheese. (*Source:* U.S. Department of Agriculture, Economic Research Service, *Food Consumption, Prices, and Expenditures*) What was the per capita consumption of each type of cheese?

25. *Golf.*   On an 18-hole golf course, there are par-3 holes, par-4 holes, and par-5 holes. A golfer who shoots par on every hole has a score of 72. The sum of the number of par-3 holes and the number of par-5 holes is 8. How many of each type of hole are there on the golf course?

26. *Golf.*   On an 18-hole golf course, there are par-3 holes, par-4 holes, and par-5 holes. A golfer who shoots par on every hole has a score of 70. There are twice as many par-4 holes as there are par-5 holes.

How many of each type of hole are there on the golf course?

27. *Number of Marriages.*   The following table shows the number of marriages, in thousands, in California, represented as years since 1980.

YEAR, x	NUMBER OF MARRIAGES (IN THOUSANDS)
1980, 0	211
1990, 10	237
1996, 16	203

Source: U.S. National Center for Health Statistics, *Vital Statistics of the United States,* annual, Monthly Vital Statistics Reports

a) Fit a quadratic function $f(x) = ax^2 + bx + c$ to the data.

b) Use the function to estimate the number of marriages in California in 2005.

28. *Coffee Consumption.*   The following table shows per capita coffee consumption, in gallons, in the United States, represented as years since 1992.

YEAR, x	COFFEE CONSUMPTION (IN GALLONS)
1992, 0	26
1994, 2	21
1996, 4	23

a) Fit a quadratic function $f(x) = ax^2 + bx + c$ to the data.

b) Use the function to estimate the per capita consumption of coffee in 2003.

29. *Milk Consumption.*   The following table shows per capita milk consumption, in pounds, in the United States, represented as years since 1985.

YEAR, x	MILK CONSUMPTION (IN POUNDS)
1985, 0	594
1990, 5	567
1996, 11	576

Source: U.S. Department of Agriculture, Economic Research Service, *Food Consumption, Prices, and Expenditures*

a) Fit a quadratic function $f(x) = ax^2 + bx + c$ to the data.

b) Use the function to estimate the per capita consumption of milk in 2005.

30. *Crude Steel Production.* The following table shows the world production of crude steel, in millions of metric tons, represented as years since 1990.

YEAR, x	CRUDE STEEL PRODUCTION (IN MILLIONS OF METRIC TONS)
1990, 0	771
1995, 5	755
1997, 7	773

Source: U.S. Geological Survey

a) Fit a quadratic function $f(x) = ax^2 + bx + c$ to the data.

b) Use the function to estimate the world production of crude steel in 2007.

Technology Connection

31. *Morning Newspapers.* The number of morning newspapers in the United States in various years is shown in the following table.

YEAR	NUMBER OF MORNING NEWSPAPERS
1920	437
1940	380
1960	312
1980	387
1990	559
1997	705

Source: Editor & Publisher

a) Use a graphing calculator to fit a quadratic function $f(x)$ to the data, where x is the number of years after 1920.

b) Use the function found in part (a) to estimate the number of morning newspapers in 2004 and 2007.

32. *Home Education.* The number of children who were home-educated in the United States in various years is shown in the following table.

YEAR	NUMBER OF HOME-EDUCATED CHILDREN (IN THOUSANDS)
1983	92.5
1988	225
1992	703
1995	1060
1997	1347

Source: National Home Education Research Institute

a) Use a graphing calculator to fit a quadratic function $f(x)$ to the data, where x is the number of years after 1983.

b) Use the function found in part (a) to estimate the number of children who will be home-educated in 2005 and 2010.

Collaborative Discussion and Writing

33. Given two linear equations in three variables, $Ax + By + Cz = D$ and $Ex + Fy + Gz = H$, explain how you would find a third equation such that the system contains dependent equations.

34. Write a problem for a classmate to solve that can be translated to a system of three equations in three variables.

Skill Maintenance

Simplify. Write answers in the form $a + bi$, where a and b are real numbers.

35. $(3 - 4i) - (-2 - i)$

36. $(5 + 2i) + (1 - 4i)$

37. $(1 - 2i)(6 + 2i)$

38. $\dfrac{3 + i}{4 - 3i}$

Synthesis

In Exercises 39 and 40, let u represent 1/x, v represent 1/y, and w represent 1/z. Solve first for u, v, and w. Then solve the system.

39. $\dfrac{2}{x} - \dfrac{1}{y} - \dfrac{3}{z} = -1,$

$\dfrac{2}{x} - \dfrac{1}{y} + \dfrac{1}{z} = -9,$

$\dfrac{1}{x} + \dfrac{2}{y} - \dfrac{4}{z} = 17$

40. $\dfrac{2}{x} + \dfrac{2}{y} - \dfrac{3}{z} = 3,$

$\dfrac{1}{x} - \dfrac{2}{y} - \dfrac{3}{z} = 9,$

$\dfrac{7}{x} - \dfrac{2}{y} + \dfrac{9}{z} = -39$

41. Find the sum of the angle measures at the tips of the star.

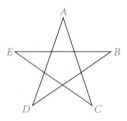

42. *Transcontinental Railroad.*   Use the following facts to find the year in which the first U.S. transcontinental railroad was completed. The sum of the digits in the year is 24. The units digit is 1 more than the hundreds digit. Both the tens and the units digits are multiples of three.

In Exercises 43 and 44, three solutions of an equation are given. Use a system of three equations in three variables to find the constants and write the equation.

43. $Ax + By + Cz = 12;$

$\left(1, \frac{3}{4}, 3\right), \left(\frac{4}{3}, 1, 2\right),$ and $(2, 1, 1)$

44. $y = B - Mx - Nz;$

$(1, 1, 2), (3, 2, -6),$ and $\left(\frac{3}{2}, 1, 1\right)$

In Exercises 45 and 46, four solutions of the equation $y = ax^3 + bx^2 + cx + d$ are given. Use a system of four equations in four variables to find the constants and write the equation.

45. $(-2, 59), (-1, 13), (1, -1),$ and $(2, -17)$

46. $(-2, -39), (-1, -12), (1, -6),$ and $(3, 16)$

47. *Theater Attendance.*   A performance at the Bingham Performing Arts Center was attended by 100 people. The audience consisted of adults, students, and children. The ticket prices were $10 for adults, $3 for students, and 50 cents for children. The total amount of money taken in was $100. How many adults, students, and children were in attendance? Does there seem to be some information missing? Do some careful reasoning.

8.3

Matrices and Systems of Equations

•   *Solve systems of equations using matrices.*

Matrices and Row-Equivalent Operations

In this section, we consider additional techniques for solving systems of equations. You have probably observed that when we solve a system of equations, we perform computations with the coefficients and the constants and continually rewrite the variables. We can streamline the solution process by omitting the variables until a solution is found. For example, the system

$$2x - 3y = 7,$$
$$x + 4y = -2$$

can be written more simply as

$$\left[\begin{array}{rr|r} 2 & -3 & 7 \\ 1 & 4 & -2 \end{array}\right].$$

The vertical line replaces the equals signs.

A rectangular array of numbers like the one above is called a **matrix** (pl., **matrices**). The matrix above is called an **augmented matrix** for the given system of equations, because it contains not only the coefficients but also the constant terms. The matrix

$$\left[\begin{array}{rr} 2 & -3 \\ 1 & 4 \end{array}\right]$$

is called the **coefficient matrix** of the system.

The **rows** of a matrix are horizontal, and the **columns** are vertical. The augmented matrix above has 2 rows and 3 columns, and the coefficient matrix has 2 rows and 2 columns. A matrix with m rows and n columns is said to be of **order** $m \times n$. Thus the order of the augmented matrix above is 2×3, and the order of the coefficient matrix is 2×2. When $m = n$, a matrix is said to be **square**. The coefficient matrix above is a square matrix. The numbers 2 and 4 lie on the **main diagonal** of the coefficient matrix. The numbers in a matrix are called **entries**.

Gaussian Elimination with Matrices

In Section 8.2, we described a series of operations that can be used to transform a system of equations to an equivalent system. Each of these operations corresponds to one that can be used to produce *row-equivalent matrices*.

Row-Equivalent Operations

1. Interchange any two rows.

2. Multiply each entry in a row by the same nonzero constant.

3. Add a nonzero multiple of one row to another row.

We can use these operations on the augmented matrix of a system of equations to solve the system.

EXAMPLE 1   Solve the following system:

$$\begin{array}{rcrcrcr} 2x & - & y & + & 4z & = & -3, \\ x & - & 2y & - & 10z & = & -6, \\ 3x & & & + & 4z & = & 7. \end{array}$$

Solution   First, we write the augmented matrix, writing 0 for the missing *y*-term in the last equation:

$$\left[\begin{array}{ccc|c} 2 & -1 & 4 & -3 \\ 1 & -2 & -10 & -6 \\ 3 & 0 & 4 & 7 \end{array}\right].$$

Our goal is to find a row-equivalent matrix of the form

$$\left[\begin{array}{ccc|c} 1 & a & b & c \\ 0 & 1 & d & e \\ 0 & 0 & 1 & f \end{array}\right].$$

The variables can then be reinserted to form equations from which we can complete the solution. This is done by working from the bottom equation to the top and using back-substitution.

The first step is to multiply and/or interchange rows so that each number in the first column below the first number is a multiple of that number. In this case, we interchange the first and second rows to obtain a 1 in the upper left-hand corner.

$$\left[\begin{array}{ccc|c} 1 & -2 & -10 & -6 \\ 2 & -1 & 4 & -3 \\ 3 & 0 & 4 & 7 \end{array}\right] \qquad \begin{array}{l} \text{New row 1 = row 2} \\ \text{New row 2 = row 1} \end{array}$$

Next, we multiply the first row by -2 and add it to the second row. We also multiply the first row by -3 and add it to the third row.

$$\left[\begin{array}{ccc|c} 1 & -2 & -10 & -6 \\ 0 & 3 & 24 & 9 \\ 0 & 6 & 34 & 25 \end{array}\right] \qquad \begin{array}{l} \text{New row 2 = } -2(\text{row 1}) + \text{row 2} \\ \text{New row 3 = } -3(\text{row 1}) + \text{row 3} \end{array}$$

Now we multiply the second row by $\frac{1}{3}$ to get a 1 in the second row, second column.

$$\left[\begin{array}{ccc|c} 1 & -2 & -10 & -6 \\ 0 & 1 & 8 & 3 \\ 0 & 6 & 34 & 25 \end{array}\right] \qquad \text{New row 2 = } \tfrac{1}{3}(\text{row 2})$$

Then we multiply the second row by -6 and add it to the third row.

$$\left[\begin{array}{ccc|c} 1 & -2 & -10 & -6 \\ 0 & 1 & 8 & 3 \\ 0 & 0 & -14 & 7 \end{array}\right] \qquad \text{New row 3 = } -6(\text{row 2}) + \text{row 3}$$

Technology Connection

Row-equivalent operations can be performed on a graphing calculator. For example, to interchange the first and second rows of the augmented matrix, as we did in the first step in Example 1, we enter the matrix as matrix **A** and select "rowSwap" from the MATRIX MATH menu. Some graphing calculators will not automatically store the matrix produced using a row-equivalent operation, so when several operations are to be performed in succession, it is helpful to store the result of each operation as it is produced. In the window below, we see both the matrix produced by the rowSwap operation and the indication that this matrix is stored as matrix **B**.

```
rowSwap([A],1,2)→[B]
[[1  −2 −10 −6]
 [2  −1  4    −3]
 [3   0  4     7 ]]
```

STUDY TIP

The *Graphing Calculator Manual* that accompanies this text contains the keystrokes for performing all the row-equivalent operations in Example 1.

Finally, we multiply the third row by $-\frac{1}{14}$ to get a 1 in the third row, third column.

$$\begin{bmatrix} 1 & -2 & -10 & \bigm| & -6 \\ 0 & 1 & 8 & \bigm| & 3 \\ 0 & 0 & 1 & \bigm| & -\frac{1}{2} \end{bmatrix} \qquad \text{New row } 3 = -\tfrac{1}{14}(\text{row 3})$$

Now we can write the system of equations that corresponds to the last matrix above:

$$\begin{aligned} x - 2y - 10z &= -6, & (1) \\ y + 8z &= 3, & (2) \\ z &= -\tfrac{1}{2}. & (3) \end{aligned}$$

We back-substitute $-\frac{1}{2}$ for z in equation (2) and solve for y:

$$y + 8\left(-\tfrac{1}{2}\right) = 3$$
$$y - 4 = 3$$
$$y = 7.$$

Next, we back-substitute 7 for y and $-\frac{1}{2}$ for z in equation (1) and solve for x:

$$x - 2 \cdot 7 - 10\left(-\tfrac{1}{2}\right) = -6$$
$$x - 14 + 5 = -6$$
$$x - 9 = -6$$
$$x = 3.$$

The triple $\left(3, 7, -\frac{1}{2}\right)$ checks in the original system of equations, so it is the solution.

The procedure followed in Example 1 is called **Gaussian elimination with matrices.** The matrix at the top of this page is in **row-echelon form.** To be in this form, a matrix must have the following properties.

Row-Echelon Form

1. If a row does not consist entirely of 0's, then the first nonzero element in the row is a 1 (called a **leading 1**).

2. For any two successive nonzero rows, the leading 1 in the lower row is farther to the right than the leading 1 in the higher row.

3. All the rows consisting entirely of 0's are at the bottom of the matrix.

If a fourth property is also satisfied, a matrix is said to be in **reduced row-echelon form:**

4. Each column that contains a leading 1 has 0's everywhere else.

EXAMPLE 2   Which of the following matrices are in row-echelon form? Which, if any, are in reduced row-echelon form?

a) $\begin{bmatrix} 1 & -3 & 5 & | & -2 \\ 0 & 1 & -4 & | & 3 \\ 0 & 0 & 1 & | & 10 \end{bmatrix}$ b) $\begin{bmatrix} 0 & -1 & | & 2 \\ 0 & 1 & | & 5 \end{bmatrix}$ c) $\begin{bmatrix} 1 & -2 & -6 & 4 & | & 7 \\ 0 & 3 & 5 & -8 & | & -1 \\ 0 & 0 & 1 & 9 & | & 2 \end{bmatrix}$

d) $\begin{bmatrix} 1 & 0 & 0 & | & -2.4 \\ 0 & 1 & 0 & | & 0.8 \\ 0 & 0 & 1 & | & 5.6 \end{bmatrix}$ e) $\begin{bmatrix} 1 & 0 & 0 & 0 & | & \frac{2}{3} \\ 0 & 1 & 0 & 0 & | & -\frac{1}{4} \\ 0 & 0 & 1 & 0 & | & \frac{6}{7} \\ 0 & 0 & 0 & 0 & | & 0 \end{bmatrix}$ f) $\begin{bmatrix} 1 & -4 & 2 & | & 5 \\ 0 & 0 & 0 & | & 0 \\ 0 & 1 & -3 & | & -8 \end{bmatrix}$

Solution   The matrices in (a), (d), and (e) satisfy the row-echelon criteria and, thus, are in row-echelon form. In (b) and (c), the first nonzero elements of the first and second rows, respectively, are not 1. In (f), the row consisting entirely of 0's is not at the bottom of the matrix. Thus the matrices in (b), (c), and (f) are not in row-echelon form. In (d) and (e), not only are the row-echelon criteria met but each column that contains a leading 1 also has 0's elsewhere, so these matrices are in reduced row-echelon form.

Gauss–Jordan Elimination

We have seen that with Gaussian elimination we perform row-equivalent operations on a matrix to obtain a row-equivalent matrix in row-echelon form. When we continue to apply these operations until we have a matrix in *reduced* row-echelon form, we are using **Gauss–Jordan elimination.** This method is named for Karl Friedrich Gauss and Wilhelm Jordan (1842–1899).

EXAMPLE 3   Use Gauss–Jordan elimination to solve the system of equations in Example 1.

Solution   Using Gaussian elimination in Example 1, we obtained the matrix

$$\begin{bmatrix} 1 & -2 & -10 & | & -6 \\ 0 & 1 & 8 & | & 3 \\ 0 & 0 & 1 & | & -\frac{1}{2} \end{bmatrix}.$$

We continue to perform row-equivalent operations until we have a matrix in reduced row-echelon form. We multiply the third row by 10 and add it to the first row. We also multiply the third row by -8 and add it to the second row.

$$\begin{bmatrix} 1 & -2 & 0 & | & -11 \\ 0 & 1 & 0 & | & 7 \\ 0 & 0 & 1 & | & -\frac{1}{2} \end{bmatrix}$$ New row 1 = 10(row 3) + row 1
New row 2 = −8(row 3) + row 2

Next, we multiply the second row by 2 and add it to the first row.

$$\left[\begin{array}{ccc|c} 1 & 0 & 0 & 3 \\ 0 & 1 & 0 & 7 \\ 0 & 0 & 1 & -\frac{1}{2} \end{array}\right]$$ New row 1 = 2(row 2) + row 1

Writing the system of equations that corresponds to this matrix, we have

$$
\begin{aligned}
x & = 3, \\
y & = 7, \\
z & = -\tfrac{1}{2}.
\end{aligned}
$$

We can actually read the solution, $\left(3, 7, -\frac{1}{2}\right)$, directly from the last column of the reduced row-echelon matrix.   ▬

EXAMPLE 4   Solve the following system:

$$
\begin{aligned}
3x - 4y - z &= 6, \\
2x - y + z &= -1, \\
4x - 7y - 3z &= 13.
\end{aligned}
$$

Solution We write the augmented matrix and use Gauss–Jordan elimination.

$$\left[\begin{array}{ccc|c} 3 & -4 & -1 & 6 \\ 2 & -1 & 1 & -1 \\ 4 & -7 & -3 & 13 \end{array}\right]$$

We begin by multiplying the second and third rows by 3 so that each number in the first column below the first number, 3, is a multiple of that number.

$$\left[\begin{array}{ccc|c} 3 & -4 & -1 & 6 \\ 6 & -3 & 3 & -3 \\ 12 & -21 & -9 & 39 \end{array}\right]$$ New row 2 = 3(row 2)
New row 3 = 3(row 3)

Next, we multiply the first row by −2 and add it to the second row. We also multiply the first row by −4 and add it to the third row.

$$\left[\begin{array}{ccc|c} 3 & -4 & -1 & 6 \\ 0 & 5 & 5 & -15 \\ 0 & -5 & -5 & 15 \end{array}\right]$$ New row 2 = −2(row 1) + row 2
New row 3 = −4(row 1) + row 3

Now we add the second row to the third row.

$$\left[\begin{array}{ccc|c} 3 & -4 & -1 & 6 \\ 0 & 5 & 5 & -15 \\ 0 & 0 & 0 & 0 \end{array}\right]$$ New row 3 = row 2 + row 3

We can stop at this stage because we have a row consisting entirely of 0's. The last row of the matrix corresponds to the equation $0 = 0$, which is true for all values of x, y, and z. Consequently, the equations are dependent and the system is equivalent to

$$3x - 4y - z = 6,$$
$$5y + 5z = -15.$$

This particular system has infinitely many solutions. (A system containing dependent equations could be inconsistent.)
Solving the second equation for y gives us

$$y = -z - 3.$$

Substituting for y in the first equation, we get

$$3x - 4(-z - 3) - z = 6$$
$$x = -z - 2.$$

Then the solutions of this system are of the form

$$(-z - 2, -z - 3, z),$$

where z can be any real number.

Similarly, if we obtain a row whose only nonzero entry occurs in the last column, we have an inconsistent system of equations. For example, in the matrix

$$\begin{bmatrix} 1 & 0 & 3 & -2 \\ 0 & 1 & 5 & 4 \\ 0 & 0 & 0 & 6 \end{bmatrix},$$

the last row corresponds to the false equation $0 = 6$, so we know the original system of equations has no solution.

Exercise Set

Determine the order of the matrix.

1. $\begin{bmatrix} 1 & -6 \\ -3 & 2 \\ 0 & 5 \end{bmatrix}$

2. $\begin{bmatrix} 7 \\ -5 \\ -1 \\ 3 \end{bmatrix}$

3. $[2 \ -4 \ 0 \ 9]$

4. $[-8]$

5. $\begin{bmatrix} 1 & -5 & -8 \\ 6 & 4 & -2 \\ -3 & 0 & 7 \end{bmatrix}$

6. $\begin{bmatrix} 13 & 2 & -6 & 4 \\ -1 & 18 & 5 & -12 \end{bmatrix}$

Write the augmented matrix for the system of equations.

7. $2x - y = 7,$
$\quad x + 4y = -5$

8. $3x + 2y = 8,$
$\quad 2x - 3y = 15$

9. $x - 2y + 3z = 12,$
$\quad 2x \qquad - 4z = 8,$
$\qquad 3y + z = 7$

10. $\quad x + y - z = 7,$
$\qquad 3y + 2z = 1,$
$\quad -2x - 5y \qquad = 6$

Write the system of equations that corresponds to the augmented matrix.

11. $\begin{bmatrix} 3 & -5 & | & 1 \\ 1 & 4 & | & -2 \end{bmatrix}$

12. $\begin{bmatrix} 1 & 2 & | & -6 \\ 4 & 1 & | & -3 \end{bmatrix}$

13. $\begin{bmatrix} 2 & 1 & -4 & | & 12 \\ 3 & 0 & 5 & | & -1 \\ 1 & -1 & 1 & | & 2 \end{bmatrix}$

14. $\begin{bmatrix} -1 & -2 & 3 & | & 6 \\ 0 & 4 & 1 & | & 2 \\ 2 & -1 & 0 & | & 9 \end{bmatrix}$

Solve the system of equations using Gaussian elimination or Gauss–Jordan elimination.

15. $4x + 2y = 11,$
$\quad 3x - y = 2$

16. $2x + y = 1,$
$\quad 3x + 2y = -2$

17. $5x - 2y = -3,$
$\quad 2x + 5y = -24$

18. $2x + y = 1,$
$\quad 3x - 6y = 4$

19. $\quad 3x + 4y = 7,$
$\quad -5x + 2y = 10$

20. $5x - 3y = -2,$
$\quad 4x + 2y = 5$

21. $3x + 2y = 6,$
$\quad 2x - 3y = -9$

22. $\quad x - 4y = 9,$
$\quad 2x + 5y = 5$

23. $\quad x - 3y = 8,$
$\quad -2x + 6y = 3$

24. $4x - 8y = 12,$
$\quad -x + 2y = -3$

25. $-2x + 6y = 4,$
$\quad 3x - 9y = -6$

26. $\quad 6x + 2y = -10,$
$\quad -3x - y = 6$

27. $x + 2y - 3z = 9,$
$\quad 2x - y + 2z = -8,$
$\quad 3x - y - 4z = 3$

28. $\quad x - y + 2z = 0,$
$\quad x - 2y + 3z = -1,$
$\quad 2x - 2y + z = -3$

29. $4x - y - 3z = 1,$
$\quad 8x + y - z = 5,$
$\quad 2x + y + 2z = 5$

30. $3x + 2y + 2z = 3,$
$\quad x + 2y - z = 5,$
$\quad 2x - 4y + z = 0$

31. $\quad x - 2y + 3z = -4,$
$\quad 3x + y - z = 0,$
$\quad 2x + 3y - 5z = 1$

32. $\quad 2x - 3y + 2z = 2,$
$\quad x + 4y - z = 9,$
$\quad -3x + y - 5z = 5$

33. $2x - 4y - 3z = 3,$
$\quad x + 3y + z = -1,$
$\quad 5x + y - 2z = 2$

34. $\quad x + y - 3z = 4,$
$\quad 4x + 5y + z = 1,$
$\quad 2x + 3y + 7z = -7$

35. $p + q + r = 1,$
$\quad p + 2q + 3r = 4,$
$\quad 4p + 5q + 6r = 7$

36. $m + n + t = 9,$
$\quad m - n - t = -15,$
$\quad 3m + n + t = 2$

37. $a + b - c = 7,$
$\quad a - b + c = 5,$
$\quad 3a + b - c = -1$

38. $a - b + c = 3,$
$\quad 2a + b - 3c = 5,$
$\quad 4a + b - c = 11$

39. $-2w + 2x + 2y - 2z = -10,$
$\quad w + x + y + z = -5,$
$\quad 3w + x - y + 4z = -2,$
$\quad w + 3x - 2y + 2z = -6$

40. $-w + 2x - 3y + z = -8,$
$\quad -w + x + y - z = -4,$
$\quad w + x + y + z = 22,$
$\quad -w + x - y - z = -14$

Use Gaussian elimination or Gauss–Jordan elimination in Exercises 41–44.

41. *Time of Return.* The Houlihans pay their babysitter $5 per hour before 11 P.M. and $7.50 per hour after 11 P.M. One evening they went out for 5 hr and paid the sitter $30. What time did they come home?

42. *Advertising Expense.* eAuction.com spent a total of $11 million on advertising in fiscal years 1998, 1999, and 2000. The amount spent in 2000 was three times the amount spent in 1998. The amount spent in 1999 was $3 million less than the amount spent in 2000. How much was spent on advertising each year?

43. *Borrowing.* Gonzalez Manufacturing borrowed $30,000 to buy a new piece of equipment. Part of the money was borrowed at 8%, part at 10%, and part at 12%. The annual interest was $3040, and the total amount borrowed at 8% and 10% was twice the amount borrowed at 12%. How much was borrowed at each rate?

44. *Stamp Purchase.* Ricardo spent $18.45 on 34¢ and 21¢ stamps. He bought a total of 60 stamps. How many of each type did he buy?

Collaborative Discussion and Writing

45. Solve the following system of equations using matrices and Gaussian elimination. Then solve it again using Gauss–Jordan elimination. Do you prefer one method over the other? Why or why not?

$$3x + 4y + 2z = 0,$$
$$x - y - z = 10,$$
$$2x + 3y + 3z = -10$$

46. Explain in your own words why the augmented matrix below represents a system of dependent equations.

$$\begin{bmatrix} 1 & -3 & 2 & | & -5 \\ 0 & 1 & -4 & | & 8 \\ 0 & 0 & 0 & | & 0 \end{bmatrix}$$

Skill Maintenance

Solve.

47. $2x^2 + x = 7$

48. $\dfrac{1}{x+1} - \dfrac{6}{x-1} = 1$

49. $\sqrt{2x+1} - 1 = \sqrt{2x-4}$

50. $x - \sqrt{x} - 6 = 0$

Synthesis

In Exercises 51 and 52, three solutions of the equation $y = ax^2 + bx + c$ are given. Use a system of three equations in three variables and Gaussian elimination or Gauss–Jordan elimination to find the constants and write the equation.

51. $(-3, 12)$, $(-1, -7)$, and $(1, -2)$

52. $(-1, 0)$, $(1, -3)$, and $(3, -22)$

53. Find two different row-echelon forms of

$$\begin{bmatrix} 1 & 5 \\ 3 & 2 \end{bmatrix}.$$

54. Consider the system of equations

$$\begin{aligned} x - y + 3z &= -8, \\ 2x + 3y - z &= 5, \\ 3x + 2y + 2kz &= -3k. \end{aligned}$$

For what value(s) of k, if any, will the system have

a) no solution?
b) exactly one solution?
c) infinitely many solutions?

Solve using matrices.

55. $\begin{aligned} y &= x + z, \\ 3y + 5z &= 4, \\ x + 4 &= y + 3z \end{aligned}$

56. $\begin{aligned} x + y &= 2z, \\ 2x - 5z &= 4, \\ x - z &= y + 8 \end{aligned}$

57. $\begin{aligned} x - 4y + 2z &= 7, \\ 3x + y + 3z &= -5 \end{aligned}$

58. $\begin{aligned} x - y - 3z &= 3, \\ -x + 3y + z &= -7 \end{aligned}$

59. $\begin{aligned} 4x + 5y &= 3, \\ -2x + y &= 9, \\ 3x - 2y &= -15 \end{aligned}$

60. $\begin{aligned} 2x - 3y &= -1, \\ -x + 2y &= -2, \\ 3x - 5y &= 1 \end{aligned}$

8.4

Matrix Operations

• *Add, subtract, and multiply matrices when possible.*
• *Write a matrix equation equivalent to a system of equations.*

In Section 8.3, we used matrices to solve systems of equations. Matrices are useful in many other types of applications as well. In this section, we study matrices and some of their properties.

A capital letter is generally used to name a matrix, and lower-case letters with double subscripts generally denote its entries. For example, a_{47}, read "a sub four seven," indicates the entry in the fourth row and the seventh column. A general term is represented by a_{ij}. The notation a_{ij} indicates the entry in row i and column j. In general, we can write a matrix as

$$\mathbf{A} = [a_{ij}] = \begin{bmatrix} a_{11} & a_{12} & a_{13} & \cdots & a_{1n} \\ a_{21} & a_{22} & a_{23} & \cdots & a_{2n} \\ a_{31} & a_{32} & a_{33} & \cdots & a_{3n} \\ \vdots & \vdots & \vdots & & \vdots \\ a_{m1} & a_{m2} & a_{m3} & \cdots & a_{mn} \end{bmatrix}.$$

The matrix above has m rows and n columns. That is, its order is $m \times n$.

Two matrices are **equal** if they have the same order and corresponding entries are equal.

Matrix Addition and Subtraction

To add or subtract matrices, we add or subtract their corresponding entries. The matrices must have the same order for this to be possible.

Addition and Subtraction of Matrices

Given two $m \times n$ matrices $\mathbf{A} = [a_{ij}]$ and $\mathbf{B} = [b_{ij}]$, their sum is

$$\mathbf{A} + \mathbf{B} = [a_{ij} + b_{ij}]$$

and their difference is

$$\mathbf{A} - \mathbf{B} = [a_{ij} - b_{ij}].$$

Addition of matrices is both commutative and associative.

EXAMPLE 1 Find $\mathbf{A} + \mathbf{B}$ for each of the following.

a) $\mathbf{A} = \begin{bmatrix} -5 & 0 \\ 4 & \frac{1}{2} \end{bmatrix}$, $\mathbf{B} = \begin{bmatrix} 6 & -3 \\ 2 & 3 \end{bmatrix}$

b) $\mathbf{A} = \begin{bmatrix} 1 & 3 \\ -1 & 5 \\ 6 & 0 \end{bmatrix}$, $\mathbf{B} = \begin{bmatrix} -1 & -2 \\ 1 & -2 \\ -3 & 1 \end{bmatrix}$

Solution We have a pair of 2×2 matrices in part (a) and a pair of 3×2 matrices in part (b). Since each pair of matrices has the same order, we can add the corresponding entries.

a) $\mathbf{A} + \mathbf{B} = \begin{bmatrix} -5 & 0 \\ 4 & \frac{1}{2} \end{bmatrix} + \begin{bmatrix} 6 & -3 \\ 2 & 3 \end{bmatrix}$

$\qquad = \begin{bmatrix} -5 + 6 & 0 + (-3) \\ 4 + 2 & \frac{1}{2} + 3 \end{bmatrix} = \begin{bmatrix} 1 & -3 \\ 6 & 3\frac{1}{2} \end{bmatrix}$

b) $A + B = \begin{bmatrix} 1 & 3 \\ -1 & 5 \\ 6 & 0 \end{bmatrix} + \begin{bmatrix} -1 & -2 \\ 1 & -2 \\ -3 & 1 \end{bmatrix}$

$= \begin{bmatrix} 1 + (-1) & 3 + (-2) \\ -1 + 1 & 5 + (-2) \\ 6 + (-3) & 0 + 1 \end{bmatrix} = \begin{bmatrix} 0 & 1 \\ 0 & 3 \\ 3 & 1 \end{bmatrix}$

EXAMPLE 2 Find $C - D$ for each of the following.

a) $C = \begin{bmatrix} 1 & 2 \\ -2 & 0 \\ -3 & -1 \end{bmatrix}, \quad D = \begin{bmatrix} 1 & -1 \\ 1 & 3 \\ 2 & 3 \end{bmatrix}$

b) $C = \begin{bmatrix} 5 & -6 \\ -3 & 4 \end{bmatrix}, \quad D = \begin{bmatrix} -4 \\ 1 \end{bmatrix}$

Solution

a) We subtract corresponding entries:

$C - D = \begin{bmatrix} 1 & 2 \\ -2 & 0 \\ -3 & -1 \end{bmatrix} - \begin{bmatrix} 1 & -1 \\ 1 & 3 \\ 2 & 3 \end{bmatrix}$

$= \begin{bmatrix} 1 - 1 & 2 - (-1) \\ -2 - 1 & 0 - 3 \\ -3 - 2 & -1 - 3 \end{bmatrix} = \begin{bmatrix} 0 & 3 \\ -3 & -3 \\ -5 & -4 \end{bmatrix}.$

b) C is a 2×2 matrix and D is a 2×1 matrix. Since the matrices do not have the same order, we cannot subtract.

The **opposite**, or **additive inverse,** of a matrix is obtained by replacing each entry with its opposite.

EXAMPLE 3 Find $-A$ and $A + (-A)$ for

$A = \begin{bmatrix} 1 & 0 & 2 \\ 3 & -1 & 5 \end{bmatrix}.$

Solution To find $-A$, we replace each entry of A with its opposite.

$-A = \begin{bmatrix} -1 & 0 & -2 \\ -3 & 1 & -5 \end{bmatrix},$

$A + (-A) = \begin{bmatrix} 1 & 0 & 2 \\ 3 & -1 & 5 \end{bmatrix} + \begin{bmatrix} -1 & 0 & -2 \\ -3 & 1 & -5 \end{bmatrix}$

$= \begin{bmatrix} 0 & 0 & 0 \\ 0 & 0 & 0 \end{bmatrix}$

A matrix having 0's for all its entries is called a **zero matrix.** When a zero matrix is added to a second matrix of the same order, the second matrix is unchanged. Thus a zero matrix is an **additive identity.** For example,

$$\begin{bmatrix} 0 & 0 & 0 \\ 0 & 0 & 0 \end{bmatrix}$$

is the additive identity for any 2×3 matrix.

Scalar Multiplication

When we find the product of a number and a matrix, we obtain a **scalar product.**

> ### Scalar Product
> The **scalar product** of a number k and a matrix **A** is the matrix denoted $k\mathbf{A}$, obtained by multiplying each entry of **A** by the number k. The number k is called a **scalar.**

EXAMPLE 4 Find $3\mathbf{A}$ and $(-1)\mathbf{A}$ for

$$\mathbf{A} = \begin{bmatrix} -3 & 0 \\ 4 & 5 \end{bmatrix}.$$

Solution We have

$$3\mathbf{A} = 3\begin{bmatrix} -3 & 0 \\ 4 & 5 \end{bmatrix} = \begin{bmatrix} 3(-3) & 3 \cdot 0 \\ 3 \cdot 4 & 3 \cdot 5 \end{bmatrix} = \begin{bmatrix} -9 & 0 \\ 12 & 15 \end{bmatrix},$$

$$(-1)\mathbf{A} = -1\begin{bmatrix} -3 & 0 \\ 4 & 5 \end{bmatrix} = \begin{bmatrix} -1(-3) & -1 \cdot 0 \\ -1 \cdot 4 & -1 \cdot 5 \end{bmatrix} = \begin{bmatrix} 3 & 0 \\ -4 & -5 \end{bmatrix}.$$

EXAMPLE 5 *Production.* Mitchell Fabricators, Inc., manufactures three styles of bicycle frames in its two plants. The following table shows the number of each style produced at each plant in April.

	MOUNTAIN BIKE	RACING BIKE	TOURING BIKE
NORTH PLANT	150	120	100
SOUTH PLANT	180	90	130

a) Write a 2×3 matrix **A** that represents the information in the table.

b) The manufacturer increased production by 20% in May. Find a matrix **M** that represents the increased production figures.

c) Find the matrix $\mathbf{A} + \mathbf{M}$ and tell what it represents.

Technology Connection

Scalar products, like those in Example 4, can be found using a graphing calculator.

```
3[A]
            [[-9  0]
             [ 12 15]]
(-1)[A]
            [[3   0 ]
             [-4  -5]]
```

Solution

a) Write the entries in the table in a 2×3 matrix $\mathbf{A}$.

$$\mathbf{A} = \begin{bmatrix} 150 & 120 & 100 \\ 180 & 90 & 130 \end{bmatrix}$$

b) The production in May will be represented by $\mathbf{A} + 20\%\mathbf{A}$, or $\mathbf{A} + 0.2\mathbf{A}$, or $1.2\mathbf{A}$. Thus,

$$\mathbf{M} = (1.2) \begin{bmatrix} 150 & 120 & 100 \\ 180 & 90 & 130 \end{bmatrix} = \begin{bmatrix} 180 & 144 & 120 \\ 216 & 108 & 156 \end{bmatrix}.$$

c) $\mathbf{A} + \mathbf{M} = \begin{bmatrix} 150 & 120 & 100 \\ 180 & 90 & 130 \end{bmatrix} + \begin{bmatrix} 180 & 144 & 120 \\ 216 & 108 & 156 \end{bmatrix}$

$$= \begin{bmatrix} 330 & 264 & 220 \\ 396 & 198 & 286 \end{bmatrix}$$

The matrix $\mathbf{A} + \mathbf{M}$ represents the total production of each of the three types of frames at each plant in April and May. ▬

The properties of matrix addition and scalar multiplication are similar to the properties of addition and multiplication of real numbers.

Properties of Matrix Addition and Scalar Multiplication

For any $m \times n$ matrices $\mathbf{A}$, $\mathbf{B}$, and $\mathbf{C}$ and any scalars k and l:

$\mathbf{A} + \mathbf{B} = \mathbf{B} + \mathbf{A}$.	*Commutative Property of Addition*
$\mathbf{A} + (\mathbf{B} + \mathbf{C}) = (\mathbf{A} + \mathbf{B}) + \mathbf{C}$.	*Associative Property of Addition*
$(kl)\mathbf{A} = k(l\mathbf{A})$.	*Associative Property of Scalar Multiplication*
$k(\mathbf{A} + \mathbf{B}) = k\mathbf{A} + k\mathbf{B}$.	*Distributive Property*
$(k + l)\mathbf{A} = k\mathbf{A} + l\mathbf{A}$.	*Distributive Property*

There exists a unique matrix $\mathbf{0}$ such that:

$\mathbf{A} + \mathbf{0} = \mathbf{0} + \mathbf{A} = \mathbf{A}$.	*Additive Identity Property*

There exists a unique matrix $-\mathbf{A}$ such that:

$\mathbf{A} + (-\mathbf{A}) = -\mathbf{A} + \mathbf{A} = \mathbf{0}$.	*Additive Inverse Property*

Products of Matrices

Matrix multiplication is defined in such a way that it can be used in solving systems of equations and in many applications.

Matrix Multiplication

For an $m \times n$ matrix $\mathbf{A} = [a_{ij}]$ and an $n \times p$ matrix $\mathbf{B} = [b_{ij}]$, the **product** $\mathbf{AB} = [c_{ij}]$ is an $m \times p$ matrix, where

$$c_{ij} = a_{i1} \cdot b_{1j} + a_{i2} \cdot b_{2j} + a_{i3} \cdot b_{3j} + \cdots + a_{in} \cdot b_{nj}.$$

In other words, the entry c_{ij} in $\mathbf{AB}$ is obtained by multiplying the entries in row i of $\mathbf{A}$ by the corresponding entries in column j of $\mathbf{B}$ and adding the results.

Note that we can multiply two matrices only when the number of columns in the first matrix is equal to the number of rows in the second matrix.

Technology Connection

Matrix multiplication can be performed on a graphing calculator. The products in Examples 6(a) and 6(b) are shown below.

```
[A][B]
            [[8   9 ]
             [-4 24]]
[B][A]
          [[15  1  17]
           [-1  3 -18]
           [2  -2  14]]
```

EXAMPLE 6   For

$$\mathbf{A} = \begin{bmatrix} 3 & 1 & -1 \\ 2 & 0 & 3 \end{bmatrix}, \qquad \mathbf{B} = \begin{bmatrix} 1 & 6 \\ 3 & -5 \\ -2 & 4 \end{bmatrix}, \quad \text{and} \quad \mathbf{C} = \begin{bmatrix} 4 & -6 \\ 1 & 2 \end{bmatrix},$$

find each of the following.

a) $\mathbf{AB}$ b) $\mathbf{BA}$

c) $\mathbf{BC}$ d) $\mathbf{AC}$

Solution

a) $\mathbf{A}$ is a 2×3 matrix and $\mathbf{B}$ is a 3×2 matrix, so $\mathbf{AB}$ will be a 2×2 matrix.

$$\mathbf{AB} = \begin{bmatrix} 3 & 1 & -1 \\ 2 & 0 & 3 \end{bmatrix} \begin{bmatrix} 1 & 6 \\ 3 & -5 \\ -2 & 4 \end{bmatrix}$$

$$= \begin{bmatrix} 3 \cdot 1 + 1 \cdot 3 + (-1)(-2) & 3 \cdot 6 + 1(-5) + (-1)(4) \\ 2 \cdot 1 + 0 \cdot 3 + 3(-2) & 2 \cdot 6 + 0(-5) + 3 \cdot 4 \end{bmatrix} = \begin{bmatrix} 8 & 9 \\ -4 & 24 \end{bmatrix}$$

b) $\mathbf{B}$ is a 3×2 matrix and $\mathbf{A}$ is a 2×3 matrix, so $\mathbf{BA}$ will be a 3×3 matrix.

$$\mathbf{BA} = \begin{bmatrix} 1 & 6 \\ 3 & -5 \\ -2 & 4 \end{bmatrix} \begin{bmatrix} 3 & 1 & -1 \\ 2 & 0 & 3 \end{bmatrix}$$

$$= \begin{bmatrix} 1 \cdot 3 + 6 \cdot 2 & 1 \cdot 1 + 6 \cdot 0 & 1(-1) + 6 \cdot 3 \\ 3 \cdot 3 + (-5)(2) & 3 \cdot 1 + (-5)(0) & 3(-1) + (-5)(3) \\ -2 \cdot 3 + 4 \cdot 2 & -2 \cdot 1 + 4 \cdot 0 & -2(-1) + 4 \cdot 3 \end{bmatrix} = \begin{bmatrix} 15 & 1 & 17 \\ -1 & 3 & -18 \\ 2 & -2 & 14 \end{bmatrix}$$

Note in parts (a) and (b) that **AB** ≠ **BA**. Multiplication of matrices is generally not commutative.

c) **B** is a 3 × 2 matrix and **C** is a 2 × 2 matrix, so **BC** will be a 3 × 2 matrix.

$$\mathbf{BC} = \begin{bmatrix} 1 & 6 \\ 3 & -5 \\ -2 & 4 \end{bmatrix} \begin{bmatrix} 4 & -6 \\ 1 & 2 \end{bmatrix}$$

$$= \begin{bmatrix} 1 \cdot 4 + 6 \cdot 1 & 1(-6) + 6 \cdot 2 \\ 3 \cdot 4 + (-5)(1) & 3(-6) + (-5)(2) \\ -2 \cdot 4 + 4 \cdot 1 & -2(-6) + 4 \cdot 2 \end{bmatrix}$$

$$= \begin{bmatrix} 10 & 6 \\ 7 & -28 \\ -4 & 20 \end{bmatrix}$$

d) The product **AC** is not defined because the number of columns of **A**, 3, is not equal to the number of rows of **C**, 2. ▬

Technology Connection

When the product **AC** in Example 6(d) is entered on a graphing calculator, an ERROR message is returned, indicating that the dimensions of the matrices are mismatched.

[A] [C]	ERR:DIM MISMATCH
	1: Quit
	2: Goto

EXAMPLE 7 *Dairy Profit.* Dalton's Dairy produces no-fat ice cream and frozen yogurt. The following table shows the number of gallons of each product that are sold at the dairy's three retail outlets one week.

	STORE 1	STORE 2	STORE 3
NO-FAT ICE CREAM	100	80	120
FROZEN YOGURT	160	120	100

On each gallon of no-fat ice cream, the dairy's profit is $4, and on each gallon of frozen yogurt, it is $3. Use matrices to find the total profit on these items at each store for the given week.

Solution We can write the table showing the distribution of the products as a 2 × 3 matrix:

$$\mathbf{D} = \begin{bmatrix} 100 & 80 & 120 \\ 160 & 120 & 100 \end{bmatrix}.$$

The profit per gallon for each product can also be written as a matrix:

$$\mathbf{P} = [4 \quad 3].$$

Then the total profit at each store is given by the matrix product **PD**:

$$\begin{aligned}
\mathbf{PD} &= [4 \quad 3]\begin{bmatrix} 100 & 80 & 120 \\ 160 & 120 & 100 \end{bmatrix} \\
&= [4 \cdot 100 + 3 \cdot 160 \quad 4 \cdot 80 + 3 \cdot 120 \quad 4 \cdot 120 + 3 \cdot 100] \\
&= [880 \quad 680 \quad 780].
\end{aligned}$$

The total profit on no-fat ice cream and frozen yogurt for the given week was $880 at store 1, $680 at store 2, and $780 at store 3. ▬

A matrix that consists of a single row, like **P** in Example 7, is called a **row matrix.** Similarly, a matrix that consists of a single column, like

$$\begin{bmatrix} 8 \\ -3 \\ 5 \end{bmatrix},$$

is called a **column matrix.**

We have already seen that matrix multiplication is generally not commutative. Nevertheless, matrix multiplication does have some properties that are similar to those for multiplication of real numbers.

Properties of Matrix Multiplication

For matrices **A**, **B**, and **C**, assuming that the indicated operation is possible:

$\mathbf{A(BC)} = \mathbf{(AB)C}.$	*Associative Property of Multiplication*
$\mathbf{A(B + C)} = \mathbf{AB} + \mathbf{AC}.$	*Distributive Property*
$\mathbf{(B + C)A} = \mathbf{BA} + \mathbf{CA}.$	*Distributive Property*

Matrix Equations

We can write a matrix equation equivalent to a system of equations.

EXAMPLE 8 Write a matrix equation equivalent to the following system of equations:

$$
\begin{aligned}
4x + 2y - z &= 3, \\
9x \quad\; + z &= 5, \\
4x + 5y - 2z &= 1.
\end{aligned}
$$

Solution We write the coefficients on the left in a matrix. We then write the product of that matrix and the column matrix containing the variables, and set the result equal to the column matrix containing the constants on the right:

$$
\begin{bmatrix} 4 & 2 & -1 \\ 9 & 0 & 1 \\ 4 & 5 & -2 \end{bmatrix}
\begin{bmatrix} x \\ y \\ z \end{bmatrix}
=
\begin{bmatrix} 3 \\ 5 \\ 1 \end{bmatrix}.
$$

If we let

$$
A = \begin{bmatrix} 4 & 2 & -1 \\ 9 & 0 & 1 \\ 4 & 5 & -2 \end{bmatrix}, \quad
X = \begin{bmatrix} x \\ y \\ z \end{bmatrix}, \quad \text{and} \quad
B = \begin{bmatrix} 3 \\ 5 \\ 1 \end{bmatrix},
$$

we can write this matrix equation as $AX = B$.

In the next section, we will solve systems of equations using a matrix equation like the one in Example 8.

Exercise Set 8.4

Find x, y, and z.

1. $[5 \quad x] = [y \quad -3]$

2. $\begin{bmatrix} 6x \\ 25 \end{bmatrix} = \begin{bmatrix} -9 \\ 5y \end{bmatrix}$

3. $\begin{bmatrix} 3 & 2x \\ y & -8 \end{bmatrix} = \begin{bmatrix} 3 & -2 \\ 1 & -8 \end{bmatrix}$

4. $\begin{bmatrix} x-1 & 4 \\ y+3 & -7 \end{bmatrix} = \begin{bmatrix} 0 & 4 \\ -2 & -7 \end{bmatrix}$

For Exercises 5–20, let

$$
A = \begin{bmatrix} 1 & 2 \\ 4 & 3 \end{bmatrix}, \quad
B = \begin{bmatrix} -3 & 5 \\ 2 & -1 \end{bmatrix},
$$

$$
C = \begin{bmatrix} 1 & -1 \\ -1 & 1 \end{bmatrix}, \quad
D = \begin{bmatrix} 1 & 1 \\ 1 & 1 \end{bmatrix},
$$

$$
E = \begin{bmatrix} 1 & 3 \\ 2 & 6 \end{bmatrix}, \quad
F = \begin{bmatrix} 3 & 3 \\ -1 & -1 \end{bmatrix},
$$

$$
O = \begin{bmatrix} 0 & 0 \\ 0 & 0 \end{bmatrix}, \quad
I = \begin{bmatrix} 1 & 0 \\ 0 & 1 \end{bmatrix}.
$$

Find each of the following.

5. $A + B$ **6.** $B + A$ **7.** $E + O$ **8.** $2A$

9. $3F$ **10.** $(-1)D$ **11.** $3F + 2A$ **12.** $A - B$

13. $B - A$ **14.** AB **15.** BA **16.** OF

17. CD **18.** EF **19.** AI **20.** IA

Find each product, if possible.

21. $\begin{bmatrix} -1 & 0 & 7 \\ 3 & -5 & 2 \end{bmatrix} \begin{bmatrix} 6 \\ -4 \\ 1 \end{bmatrix}$

22. $\begin{bmatrix} 6 & -1 & 2 \end{bmatrix} \begin{bmatrix} 1 & 4 \\ -2 & 0 \\ 5 & -3 \end{bmatrix}$

23. $\begin{bmatrix} -2 & 4 \\ 5 & 1 \\ -1 & -3 \end{bmatrix} \begin{bmatrix} 3 & -6 \\ -1 & 4 \end{bmatrix}$

24. $\begin{bmatrix} 2 & -1 & 0 \\ 0 & 5 & 4 \end{bmatrix} \begin{bmatrix} -3 & 1 & 0 \\ 0 & 2 & -1 \\ 5 & 0 & 4 \end{bmatrix}$

25. $\begin{bmatrix} 1 \\ -5 \\ 3 \end{bmatrix} \begin{bmatrix} -6 & 5 & 8 \\ 0 & 4 & -1 \end{bmatrix}$

26. $\begin{bmatrix} 2 & 0 & 0 \\ 0 & -1 & 0 \\ 0 & 0 & 3 \end{bmatrix} \begin{bmatrix} 0 & -4 & 3 \\ 2 & 1 & 0 \\ -1 & 0 & 6 \end{bmatrix}$

27. $\begin{bmatrix} 1 & -4 & 3 \\ 0 & 8 & 0 \\ -2 & -1 & 5 \end{bmatrix} \begin{bmatrix} 3 & 0 & 0 \\ 0 & -4 & 0 \\ 0 & 0 & 1 \end{bmatrix}$

28. $\begin{bmatrix} 4 \\ -5 \end{bmatrix} \begin{bmatrix} 2 & 0 \\ 6 & -7 \\ 0 & -3 \end{bmatrix}$

29. *Budget.* For the month of June, Nelia budgets $150 for food, $80 for clothes, and $40 for entertainment.

  a) Write a 1×3 matrix **B** that represents the amounts budgeted for these items.

  b) After receiving a raise, Nelia increases the amount budgeted for each item in July by 5%. Find a matrix **R** that represents the new amounts.

  c) Find **B** + **R** and tell what the entries represent.

30. *Produce.* The produce manager at Dugan's Market orders 40 lb of tomatoes, 20 lb of zucchini, and 30 lb of onions from a local farmer one week.

  a) Write a 1×3 matrix **A** that represents the amount of each item ordered.

  b) The following week the produce manager increases her order by 10%. Find a matrix **B** that represents this order.

  c) Find **A** + **B** and tell what the entries represent.

31. *Nutrition.* A 3-oz serving of roasted, skinless chicken breast contains 140 Cal, 27 g of protein, 3 g of fat, 13 mg of calcium, and 64 mg of sodium.

One-half cup of potato salad contains 180 Cal, 4 g of protein, 11 g of fat, 24 mg of calcium, and 662 mg of sodium. One broccoli spear contains 50 Cal, 5 g of protein, 1 g of fat, 82 mg of calcium, and 20 mg of sodium. (*Source: Home and Garden Bulletin No. 72*, U.S. Government Printing Office, Washington, D.C. 20402)

  a) Write 1×5 matrices **C**, **P**, and **B** that represent the nutritional values of each food.

  b) Find **C** + 2**P** + 3**B** and tell what the entries represent.

32. *Nutrition.* One slice of cheese pizza contains 290 Cal, 15 g of protein, 9 g of fat, and 39 g of carbohydrates. One-half cup of gelatin dessert contains 70 Cal, 2 g of protein, 0 g of fat, and 17 g of carbohydrates. One cup of whole milk contains 150 Cal, 8 g of protein, 8 g of fat, and 11 g of carbohydrates. (*Source: Home and Garden Bulletin No. 72*, U.S. Government Printing Office, Washington, D.C. 20402)

  a) Write 1×4 matrices **P**, **G**, and **M** that represent the nutritional values of each food.

  b) Find 3**P** + 2**G** + 2**M** and tell what the entries represent.

33. *Food Service Management.* The food service manager at a large hospital is concerned about maintaining reasonable food costs. The table below shows the cost per serving, in cents, for items on four menus.

MENU	MEAT	POTATO	VEGETABLE	SALAD	DESSERT
1	45.29	6.63	10.94	7.42	8.01
2	53.78	4.95	9.83	6.16	12.56
3	47.13	8.47	12.66	8.29	9.43
4	51.64	7.12	11.57	9.35	10.72

On a particular day, a dietician orders 65 meals from menu 1, 48 from menu 2, 93 from menu 3, and 57 from menu 4.

  a) Write the information in the table as a 4×5 matrix **M**.

  b) Write a row matrix **N** that represents the number of each menu ordered.

  c) Find the product **NM**.

  d) State what the entries of **NM** represent.

34. *Food Service Management.* A college food service manager uses a table like the one below to show the number of units of ingredients, by weight, required for various menu items.

	WHITE CAKE	BREAD	COFFEE CAKE	SUGAR COOKIES
FLOUR	1	2.5	0.75	0.5
MILK	0	0.5	0.25	0
EGGS	0.75	0.25	0.5	0.5
BUTTER	0.5	0	0.5	1

The cost per unit of each ingredient is 15 cents for flour, 28 cents for milk, 54 cents for eggs, and 83 cents for butter.

a) Write the information in the table as a 4×4 matrix **M**.
b) Write a row matrix **C** that represents the cost per unit of each ingredient.
c) Find the product **CM**.
d) State what the entries of **CM** represent.

35. *Production Cost.* Karin supplies two small campus coffee shops with homemade chocolate chip cookies, oatmeal cookies, and peanut butter cookies. The table below shows the number of each type of cookie, in dozens, that Karin sold in one week.

	MUGSY'S COFFEE SHOP	THE COFFEE CLUB
CHOCOLATE CHIP	8	15
OATMEAL	6	10
PEANUT BUTTER	4	3

Karin spends $3 for the ingredients for one dozen chocolate chip cookies, $1.50 for the ingredients for one dozen oatmeal cookies, and $2 for the ingredients for one dozen peanut butter cookies.

a) Write the information in the table as a 3×2 matrix **S**.
b) Write a row matrix **C** that represents the cost, per dozen, of the ingredients for each type of cookie.
c) Find the product **CS**.
d) State what the entries of **CS** represent.

36. *Profit.* A manufacturer produces exterior plywood, interior plywood, and fiberboard, which are shipped to two distributors. The table below shows the number of units of each type of product that are shipped to each warehouse.

	DISTRIBUTOR 1	DISTRIBUTOR 2
EXTERIOR PLYWOOD	900	500
INTERIOR PLYWOOD	450	1000
FIBERBOARD	600	700

The profits from each unit of exterior plywood, interior plywood, and fiberboard are $5, $8, and $4, respectively.

a) Write the information in the table as a 3×2 matrix **M**.
b) Write a row matrix **P** that represents the profit, per unit, of each type of product.
c) Find the product **PM**.
d) State what the entries of **PM** represent.

37. *Profit.* In Exercise 35, suppose that Karin's profits on one dozen chocolate chip, oatmeal, and peanut butter cookies are $6, $4.50, and $5.20, respectively.

a) Write a row matrix **P** that represents this information.
b) Use the matrices **S** and **P** to find Karin's total profit from each coffee shop.

38. *Production Cost.* In Exercise 36, suppose that the manufacturer's production costs for each unit of exterior plywood, interior plywood, and fiberboard are $20, $25, and $15, respectively.

a) Write a row matrix **C** that represents this information.
b) Use the matrices **M** and **C** to find the total production cost for the products shipped to each distributor.

Write a matrix equation equivalent to the system of equations.

39. $2x - 3y = 7,$
$\quad\ x + 5y = -6$

40. $-x + \ y = 3,$
$\quad\ 5x - 4y = 16$

41. $x + y - 2z = 6,$
$\quad 3x - y + z = 7,$
$\quad 2x + 5y - 3z = 8$

42. $3x - y + z = 1,$
$\quad x + 2y - z = 3,$
$\quad 4x + 3y - 2z = 11$

43. $3x - 2y + 4z = 17,$
$\quad 2x + y - 5z = 13$

44. $3x + 2y + 5z = 9,$
$\quad 4x - 3y + 2z = 10$

45. $-4w + x - y + 2z = 12,$
$\quad w + 2x - y - z = 0,$
$\quad -w + x + 4y - 3z = 1,$
$\quad 2w + 3x + 5y - 7z = 9$

46. $12w + 2x + 4y - 5z = 2,$
$\quad -w + 4x - y + 12z = 5,$
$\quad 2w - x + 4y = 13,$
$\quad 2x + 10y + z = 5$

Technology Connection

47. Use a graphing calculator to perform the operations in Exercises 5, 9, 13, and 15.

48. Use a graphing calculator to perform the operations in Exercises 6, 8, 12, and 18.

Collaborative Discussion and Writing

49. Is it true that if $AB = 0$, for matrices A and B, then $A = 0$ or $B = 0$? Why or why not?

50. Explain how Karin could use the matrix products found in Exercises 35 and 37 in making business decisions.

Skill Maintenance

In Exercises 51–54:

a) *Find the vertex.*
b) *Find the line of symmetry.*
c) *Determine whether there is a maximum or minimum value and find that value.*

51. $f(x) = x^2 - 3x - 10$

52. $f(x) = 2x^2 - 5x - 3$

53. $f(x) = -x^2 - 3x + 5$

54. $f(x) = -3x^2 + 4x + 1$

Synthesis

For Exercises 55–58, let

$$A = \begin{bmatrix} -1 & 0 \\ 2 & 1 \end{bmatrix} \quad and \quad B = \begin{bmatrix} 1 & -1 \\ 0 & 2 \end{bmatrix}.$$

55. Show that
$$(A + B)(A - B) \neq A^2 - B^2,$$
where
$$A^2 = AA \quad and \quad B^2 = BB.$$

56. Show that
$$(A + B)(A + B) \neq A^2 + 2AB + B^2.$$

57. Show that
$$(A + B)(A - B) = A^2 + BA - AB - B^2.$$

58. Show that
$$(A + B)(A + B) = A^2 + BA + AB + B^2.$$

In Exercises 59–63, let

$$A = \begin{bmatrix} a_{11} & a_{12} & a_{13} & \cdots & a_{1n} \\ a_{21} & a_{22} & a_{23} & \cdots & a_{2n} \\ a_{31} & a_{32} & a_{33} & \cdots & a_{3n} \\ \vdots & \vdots & \vdots & & \vdots \\ a_{m1} & a_{m2} & a_{m3} & \cdots & a_{mn} \end{bmatrix},$$

$$B = \begin{bmatrix} b_{11} & b_{12} & b_{13} & \cdots & b_{1n} \\ b_{21} & b_{22} & b_{23} & \cdots & b_{2n} \\ b_{31} & b_{32} & b_{33} & \cdots & b_{3n} \\ \vdots & \vdots & \vdots & & \vdots \\ b_{m1} & b_{m2} & b_{m3} & \cdots & b_{mn} \end{bmatrix},$$

$$and\ C = \begin{bmatrix} c_{11} & c_{12} & c_{13} & \cdots & c_{1n} \\ c_{21} & c_{22} & c_{23} & \cdots & c_{2n} \\ c_{31} & c_{32} & c_{33} & \cdots & c_{3n} \\ \vdots & \vdots & \vdots & & \vdots \\ c_{m1} & c_{m2} & c_{m3} & \cdots & c_{mn} \end{bmatrix},$$

and let k and l be any scalars.

59. Prove that $A + B = B + A$.

60. Prove that $A + (B + C) = (A + B) + C$.

61. Prove that $(kl)A = k(lA)$.

62. Prove that $k(A + B) = kA + kB$.

63. Prove that $(k + l)A = kA + lA$.

8.5

Inverses of Matrices

- *Find the inverse of a square matrix, if it exists.*
- *Use inverses of matrices to solve systems of equations.*

In this section, we continue our study of matrix algebra, finding the **multiplicative inverse,** or simply **inverse**, of a square matrix, if it exists. Then we use such inverses to solve systems of equations.

The Identity Matrix

Recall that, for real numbers, $a \cdot 1 = 1 \cdot a = a$; 1 is the multiplicative identity. A multiplicative identity matrix is very similar to the number 1.

Identity Matrix

For any positive integer n, the $n \times n$ **identity matrix** is an $n \times n$ matrix with 1's on the main diagonal and 0's elsewhere and is denoted by

$$I = \begin{bmatrix} 1 & 0 & 0 & \cdots & 0 \\ 0 & 1 & 0 & \cdots & 0 \\ 0 & 0 & 1 & \cdots & 0 \\ \vdots & \vdots & \vdots & & \vdots \\ 0 & 0 & 0 & \cdots & 1 \end{bmatrix}.$$

Then $\mathbf{AI} = \mathbf{IA} = \mathbf{A}$, for any $n \times n$ matrix $\mathbf{A}$.

EXAMPLE 1 For

$$\mathbf{A} = \begin{bmatrix} 4 & -7 \\ -3 & 2 \end{bmatrix} \quad \text{and} \quad \mathbf{I} = \begin{bmatrix} 1 & 0 \\ 0 & 1 \end{bmatrix},$$

find each of the following.

a) $\mathbf{AI}$ **b)** $\mathbf{IA}$

Solution

a) $\mathbf{AI} = \begin{bmatrix} 4 & -7 \\ -3 & 2 \end{bmatrix} \begin{bmatrix} 1 & 0 \\ 0 & 1 \end{bmatrix}$

$$= \begin{bmatrix} 4 \cdot 1 - 7 \cdot 0 & 4 \cdot 0 - 7 \cdot 1 \\ -3 \cdot 1 + 2 \cdot 0 & -3 \cdot 0 + 2 \cdot 1 \end{bmatrix} = \begin{bmatrix} 4 & -7 \\ -3 & 2 \end{bmatrix} = \mathbf{A}$$

b) $\mathbf{IA} = \begin{bmatrix} 1 & 0 \\ 0 & 1 \end{bmatrix} \begin{bmatrix} 4 & -7 \\ -3 & 2 \end{bmatrix}$

$$= \begin{bmatrix} 1 \cdot 4 + 0(-3) & 1(-7) + 0 \cdot 2 \\ 0 \cdot 4 + 1(-3) & 0(-7) + 1 \cdot 2 \end{bmatrix} = \begin{bmatrix} 4 & -7 \\ -3 & 2 \end{bmatrix} = \mathbf{A}$$

The Inverse of a Matrix

Recall that for every nonzero real number a, there is a multiplicative inverse $1/a$ such that $a(1/a) = (1/a)a = 1$. The multiplicative inverse of a matrix behaves in a similar manner.

Inverse of a Matrix

For an $n \times n$ matrix $\mathbf{A}$, if there is a matrix $\mathbf{A}^{-1}$ for which $\mathbf{A}^{-1} \cdot \mathbf{A} = \mathbf{I} = \mathbf{A} \cdot \mathbf{A}^{-1}$, then $\mathbf{A}^{-1}$ is the **inverse** of $\mathbf{A}$.

Note that $\mathbf{A}^{-1}$ is read "$\mathbf{A}$ inverse." Also note that not every matrix has an inverse.

EXAMPLE 2   Verify that

$$\mathbf{B} = \begin{bmatrix} 4 & -3 \\ 3 & -2 \end{bmatrix} \quad \text{is the inverse of} \quad \mathbf{A} = \begin{bmatrix} -2 & 3 \\ -3 & 4 \end{bmatrix}.$$

Solution   We show that $\mathbf{BA} = \mathbf{I} = \mathbf{AB}$.

$$\mathbf{BA} = \begin{bmatrix} 4 & -3 \\ 3 & -2 \end{bmatrix} \begin{bmatrix} -2 & 3 \\ -3 & 4 \end{bmatrix} = \begin{bmatrix} 1 & 0 \\ 0 & 1 \end{bmatrix}$$

$$\mathbf{AB} = \begin{bmatrix} -2 & 3 \\ -3 & 4 \end{bmatrix} \begin{bmatrix} 4 & -3 \\ 3 & -2 \end{bmatrix} = \begin{bmatrix} 1 & 0 \\ 0 & 1 \end{bmatrix}$$

We can find the inverse of a square matrix, if it exists, by using row-equivalent operations as in the Gauss–Jordan elimination method. For example, consider the matrix

$$\mathbf{A} = \begin{bmatrix} -2 & 3 \\ -3 & 4 \end{bmatrix}.$$

To find its inverse, we first form an **augmented matrix** consisting of $\mathbf{A}$ on the left side and the 2×2 identity matrix on the right side:

$$\left[\begin{array}{cc|cc} -2 & 3 & 1 & 0 \\ -3 & 4 & 0 & 1 \end{array} \right].$$

The 2×2 The 2×2
matrix A   identity matrix

Then we attempt to transform the augmented matrix to one of the form

$$\left[\begin{array}{cc|cc} 1 & 0 & a & b \\ 0 & 1 & c & d \end{array} \right].$$

The 2×2 The matrix $\mathbf{A}^{-1}$
identity matrix

If we can do this, the matrix on the right, $\begin{bmatrix} a & b \\ c & d \end{bmatrix}$, is $\mathbf{A}^{-1}$.

EXAMPLE 3 Find $\mathbf{A}^{-1}$, where

$$\mathbf{A} = \begin{bmatrix} -2 & 3 \\ -3 & 4 \end{bmatrix}.$$

Solution First, we write the augmented matrix. Then we transform it to the desired form.

$$\begin{bmatrix} -2 & 3 & | & 1 & 0 \\ -3 & 4 & | & 0 & 1 \end{bmatrix}$$

$$\begin{bmatrix} 1 & -\frac{3}{2} & | & -\frac{1}{2} & 0 \\ -3 & 4 & | & 0 & 1 \end{bmatrix} \qquad \text{New row } 1 = -\frac{1}{2}(\text{row } 1)$$

$$\begin{bmatrix} 1 & -\frac{3}{2} & | & -\frac{1}{2} & 0 \\ 0 & -\frac{1}{2} & | & -\frac{3}{2} & 1 \end{bmatrix} \qquad \text{New row } 2 = 3(\text{row } 1) + \text{row } 2$$

$$\begin{bmatrix} 1 & -\frac{3}{2} & | & -\frac{1}{2} & 0 \\ 0 & 1 & | & 3 & -2 \end{bmatrix} \qquad \text{New row } 2 = -2(\text{row } 2)$$

$$\begin{bmatrix} 1 & 0 & | & 4 & -3 \\ 0 & 1 & | & 3 & -2 \end{bmatrix} \qquad \text{New row } 1 = \frac{3}{2}(\text{row } 2) + \text{row } 1$$

Thus,

$$\mathbf{A}^{-1} = \begin{bmatrix} 4 & -3 \\ 3 & -2 \end{bmatrix},$$

which we verified in Example 2.

Technology Connection

The $\boxed{x^{-1}}$ key on a graphing calculator can be used to find the inverse of a matrix like the one in Example 3.

```
[A]⁻¹
              [[4  -3]
               [3  -2]]
```

EXAMPLE 4 Find $\mathbf{A}^{-1}$, where

$$\mathbf{A} = \begin{bmatrix} 1 & 2 & -1 \\ 3 & 5 & 3 \\ 2 & 4 & 3 \end{bmatrix}.$$

Solution First, we write the augmented matrix. Then we transform it to the desired form.

$$\begin{bmatrix} 1 & 2 & -1 & | & 1 & 0 & 0 \\ 3 & 5 & 3 & | & 0 & 1 & 0 \\ 2 & 4 & 3 & | & 0 & 0 & 1 \end{bmatrix}$$

$$\begin{bmatrix} 1 & 2 & -1 & | & 1 & 0 & 0 \\ 0 & -1 & 6 & | & -3 & 1 & 0 \\ 0 & 0 & 5 & | & -2 & 0 & 1 \end{bmatrix} \qquad \begin{array}{l} \text{New row } 2 = -3(\text{row } 1) + \text{row } 2 \\ \text{New row } 3 = -2(\text{row } 1) + \text{row } 3 \end{array}$$

$$\begin{bmatrix} 1 & 2 & -1 & | & 1 & 0 & 0 \\ 0 & -1 & 6 & | & -3 & 1 & 0 \\ 0 & 0 & 1 & | & -\frac{2}{5} & 0 & \frac{1}{5} \end{bmatrix} \qquad \text{New row } 3 = \frac{1}{5}(\text{row } 3)$$

$$\left[\begin{array}{ccc|ccc} 1 & 2 & 0 & \frac{3}{5} & 0 & \frac{1}{5} \\ 0 & -1 & 0 & -\frac{3}{5} & 1 & -\frac{6}{5} \\ 0 & 0 & 1 & -\frac{2}{5} & 0 & \frac{1}{5} \end{array}\right]$$

New row 1 = row 3 + row 1
New row 2 = −6(row 3) + row 2

$$\left[\begin{array}{ccc|ccc} 1 & 0 & 0 & -\frac{3}{5} & 2 & -\frac{11}{5} \\ 0 & -1 & 0 & -\frac{3}{5} & 1 & -\frac{6}{5} \\ 0 & 0 & 1 & -\frac{2}{5} & 0 & \frac{1}{5} \end{array}\right]$$

New row 1 = 2(row 2) + row 1

$$\left[\begin{array}{ccc|ccc} 1 & 0 & 0 & -\frac{3}{5} & 2 & -\frac{11}{5} \\ 0 & 1 & 0 & \frac{3}{5} & -1 & \frac{6}{5} \\ 0 & 0 & 1 & -\frac{2}{5} & 0 & \frac{1}{5} \end{array}\right]$$

New row 2 = −1(row 2)

Thus,

$$\mathbf{A}^{-1} = \left[\begin{array}{ccc} -\frac{3}{5} & 2 & -\frac{11}{5} \\ \frac{3}{5} & -1 & \frac{6}{5} \\ -\frac{2}{5} & 0 & \frac{1}{5} \end{array}\right].$$

Technology Connection

When we try to find the inverse of a noninvertible, or singular, matrix using a graphing calculator, the calculator returns an error message similar to ERR: SINGULAR MATRIX.

MATRIX EQUATIONS

REVIEW SECTION 8.4.

If a matrix has an inverse, we say that it is **invertible**, or **nonsingular**. When we cannot obtain the identity matrix on the left using the Gauss–Jordan method, then no inverse exists. This occurs when we obtain a row consisting entirely of 0's in either of the two matrices in the augmented matrix. In this case, we say that **A** is a **singular matrix.**

Solving Systems of Equations

We can write a system of n linear equations in n variables as a matrix equation $\mathbf{AX} = \mathbf{B}$. If $\mathbf{A}$ has an inverse, then the system of equations has a unique solution that can be found by solving for $\mathbf{X}$, as follows:

$$\mathbf{AX} = \mathbf{B}$$
$$\mathbf{A}^{-1}(\mathbf{AX}) = \mathbf{A}^{-1}\mathbf{B} \qquad \text{Multiplying by } \mathbf{A}^{-1} \text{ on the left on both sides}$$
$$(\mathbf{A}^{-1}\mathbf{A})\mathbf{X} = \mathbf{A}^{-1}\mathbf{B} \qquad \text{Using the associative property of matrix multiplication}$$
$$\mathbf{IX} = \mathbf{A}^{-1}\mathbf{B} \qquad \mathbf{A}^{-1}\mathbf{A} = \mathbf{I}$$
$$\mathbf{X} = \mathbf{A}^{-1}\mathbf{B}. \qquad \mathbf{IX} = \mathbf{X}$$

Matrix Solutions of Systems of Equations

For a system of n linear equations in n variables, $\mathbf{AX} = \mathbf{B}$, if $\mathbf{A}$ is an invertible matrix, then the unique solution of the system is given by

$$\mathbf{X} = \mathbf{A}^{-1}\mathbf{B}.$$

Since matrix multiplication is not commutative in general, care must be taken to multiply *on the left* by $\mathbf{A}^{-1}$.

EXAMPLE 5 Use an inverse matrix to solve the following system of equations:

$$-2x + 3y = 4,$$
$$-3x + 4y = 5.$$

Solution We write an equivalent matrix equation, $\mathbf{AX = B}$:

$$\begin{bmatrix} -2 & 3 \\ -3 & 4 \end{bmatrix} \begin{bmatrix} x \\ y \end{bmatrix} \quad \begin{bmatrix} 4 \\ 5 \end{bmatrix}$$

$$A \qquad \cdot \quad X \ = \ B$$

In Example 3, we found that

$$\mathbf{A}^{-1} = \begin{bmatrix} 4 & -3 \\ 3 & -2 \end{bmatrix}.$$

We also verified this in Example 2. Now we have

$$\mathbf{X = A^{-1}B}$$

$$\begin{bmatrix} x \\ y \end{bmatrix} = \begin{bmatrix} 4 & -3 \\ 3 & -2 \end{bmatrix} \begin{bmatrix} 4 \\ 5 \end{bmatrix} = \begin{bmatrix} 1 \\ 2 \end{bmatrix}.$$

The solution of the system of equations is (1, 2).

Technology Connection

To use a graphing calculator to solve the system of equations in Example 5, we enter **A** and **B** and then enter the notation $\mathbf{A}^{-1}\mathbf{B}$ on the home screen.

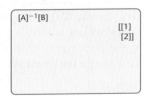

```
[A]⁻¹[B]
                [[1]
                 [2]]
```

Exercise Set 8.5

*Determine whether **B** is the inverse of **A**.*

1. $\mathbf{A} = \begin{bmatrix} 1 & -3 \\ -2 & 7 \end{bmatrix}$, $\mathbf{B} = \begin{bmatrix} 7 & 3 \\ 2 & 1 \end{bmatrix}$

2. $\mathbf{A} = \begin{bmatrix} 3 & 2 \\ 4 & 3 \end{bmatrix}$, $\mathbf{B} = \begin{bmatrix} 3 & -2 \\ -4 & 3 \end{bmatrix}$

3. $\mathbf{A} = \begin{bmatrix} -1 & -1 & 6 \\ 1 & 0 & -2 \\ 1 & 0 & -3 \end{bmatrix}$, $\mathbf{B} = \begin{bmatrix} 2 & 3 & 2 \\ 3 & 3 & 4 \\ 1 & 1 & 1 \end{bmatrix}$

4. $\mathbf{A} = \begin{bmatrix} -2 & 0 & -3 \\ 5 & 1 & 7 \\ -3 & 0 & 4 \end{bmatrix}$, $\mathbf{B} = \begin{bmatrix} 4 & 0 & -3 \\ 1 & 1 & 1 \\ -3 & 0 & 2 \end{bmatrix}$

Use the Gauss–Jordan method to find $\mathbf{A}^{-1}$, if it exists. Check your answers by finding $\mathbf{A}^{-1}\mathbf{A}$ and $\mathbf{AA}^{-1}$.

5. $\mathbf{A} = \begin{bmatrix} 3 & 2 \\ 5 & 3 \end{bmatrix}$

6. $\mathbf{A} = \begin{bmatrix} 3 & 5 \\ 1 & 2 \end{bmatrix}$

7. $\mathbf{A} = \begin{bmatrix} 6 & 9 \\ 4 & 6 \end{bmatrix}$

8. $\mathbf{A} = \begin{bmatrix} -4 & -6 \\ 2 & 3 \end{bmatrix}$

9. $\mathbf{A} = \begin{bmatrix} 4 & -3 \\ 1 & -2 \end{bmatrix}$

10. $\mathbf{A} = \begin{bmatrix} 0 & -1 \\ 1 & 0 \end{bmatrix}$

11. $\mathbf{A} = \begin{bmatrix} 3 & 1 & 0 \\ 1 & 1 & 1 \\ 1 & -1 & 2 \end{bmatrix}$

12. $\mathbf{A} = \begin{bmatrix} 1 & 0 & 1 \\ 2 & 1 & 0 \\ 1 & -1 & 1 \end{bmatrix}$

13. $A = \begin{bmatrix} 1 & -4 & 8 \\ 1 & -3 & 2 \\ 2 & -7 & 10 \end{bmatrix}$

14. $A = \begin{bmatrix} -2 & 5 & 3 \\ 4 & -1 & 3 \\ 7 & -2 & 5 \end{bmatrix}$

15. $A = \begin{bmatrix} 2 & 3 & 2 \\ 3 & 3 & 4 \\ -1 & -1 & -1 \end{bmatrix}$

16. $A = \begin{bmatrix} 1 & 2 & 3 \\ 2 & -1 & -2 \\ -1 & 3 & 3 \end{bmatrix}$

17. $A = \begin{bmatrix} 1 & 2 & -1 \\ -2 & 0 & 1 \\ 1 & -1 & 0 \end{bmatrix}$

18. $A = \begin{bmatrix} 7 & -1 & -9 \\ 2 & 0 & -4 \\ -4 & 0 & 6 \end{bmatrix}$

19. $A = \begin{bmatrix} 1 & 3 & -1 \\ 0 & 2 & -1 \\ 1 & 1 & 0 \end{bmatrix}$

20. $A = \begin{bmatrix} -1 & 0 & -1 \\ -1 & 1 & 0 \\ 0 & 1 & 1 \end{bmatrix}$

21. $A = \begin{bmatrix} 1 & 2 & 3 & 4 \\ 0 & 1 & 3 & -5 \\ 0 & 0 & 1 & -2 \\ 0 & 0 & 0 & -1 \end{bmatrix}$

22. $A = \begin{bmatrix} -2 & -3 & 4 & 1 \\ 0 & 1 & 1 & 0 \\ 0 & 4 & -6 & 1 \\ -2 & -2 & 5 & 1 \end{bmatrix}$

23. $A = \begin{bmatrix} 1 & -14 & 7 & 38 \\ -1 & 2 & 1 & -2 \\ 1 & 2 & -1 & -6 \\ 1 & -2 & 3 & 6 \end{bmatrix}$

24. $A = \begin{bmatrix} 10 & 20 & -30 & 15 \\ 3 & -7 & 14 & -8 \\ -7 & -2 & -1 & 2 \\ 4 & 4 & -3 & 1 \end{bmatrix}$

In Exercises 25–28, a system of equations is given, together with the inverse of the coefficient matrix. Use the inverse of the coefficient matrix to solve the system of equations.

25. $11x + 3y = -4,$
$7x + 2y = 5;$ $\quad A^{-1} = \begin{bmatrix} 2 & -3 \\ -7 & 11 \end{bmatrix}$

26. $8x + 5y = -6,$
$5x + 3y = 2;$ $\quad A^{-1} = \begin{bmatrix} -3 & 5 \\ 5 & -8 \end{bmatrix}$

27. $3x + y \quad\quad = 2,$
$2x - y + 2z = -5,$ $\quad A^{-1} = \dfrac{1}{9}\begin{bmatrix} 3 & 1 & -2 \\ 0 & -3 & 6 \\ -3 & 2 & 5 \end{bmatrix}$
$x + y + z = 5;$

28. $\quad\quad y - z = -4,$
$4x + y \quad\quad = -3,$ $\quad A^{-1} = \dfrac{1}{5}\begin{bmatrix} -3 & 2 & -1 \\ 12 & -3 & 4 \\ 7 & -3 & 4 \end{bmatrix}$
$3x - y + 3z = 1;$

Solve the system of equations using the inverse of the coefficient matrix of the equivalent matrix equation.

29. $4x + 3y = 2,$
$x - 2y = 6$

30. $2x - 3y = 7,$
$4x + y = -7$

31. $5x + y = 2,$
$3x - 2y = -4$

32. $x - 6y = 5,$
$-x + 4y = -5$

33. $x \quad\quad + z = 1,$
$2x + y \quad\quad = 3,$
$x - y + z = 4$

34. $x + 2y + 3z = -1,$
$2x - 3y + 4z = 2,$
$-3x + 5y - 6z = 4$

35. $2x + 3y + 4z = 2,$
$x - 4y + 3z = 2,$
$5x + y + z = -4$

36. $x + y \quad\quad = 2,$
$3x \quad\quad + 2z = 5,$
$2x + 3y - 3z = 9$

37. $2w - 3x + 4y - 5z = 0,$
$3w - 2x + 7y - 3z = 2,$
$w + x - y + z = 1,$
$-w - 3x - 6y + 4z = 6$

38. $5w - 4x + 3y - 2z = -6,$
$w + 4x - 2y + 3z = -5,$
$2w - 3x + 6y - 9z = 14,$
$3w - 5x + 2y - 4z = -3$

39. *Sales.* Stefan sold a total of 145 Italian sausages and hot dogs from his curbside pushcart and collected $242.05. He sold 45 more hot dogs than sausages. How many of each did he sell?

40. *Price of School Supplies.* Miranda bought 4 lab record books and 3 highlighters for $13.93. Victor bought 3 lab record books and 2 highlighters for $10.25. Find the price of each item.

41. *Cost.* Evergreen Landscaping bought 4 tons of topsoil, 3 tons of mulch, and 6 tons of pea gravel for $2825. The next week the firm bought 5 tons of topsoil, 2 tons of mulch, and 5 tons of pea gravel for $2663. Pea gravel costs $17 less per ton than topsoil. Find the price per ton for each item.

42. *Investment.* Selena receives $537.75 per year in simple interest from three investments totaling $8500. Part is invested at 5.15%, part at 6.05%, and the rest at 7.2%. There is $1500 more invested at 7.2% than at 5.15%. Find the amount invested at each rate.

Technology Connection

Use a graphing calculator to do the following.

43. Exercises 5, 13, and 21

44. Exercises 6, 14, and 24

45. Exercises 29, 35, and 37

46. Exercises 30, 34, and 38

Collaborative Discussion and Writing

47. For square matrices A and B, is it true, in general, that $(A + B)^{-1} = A^{-1} + B^{-1}$? Explain.

48. For square matrices A and B, is it true, in general, that $(AB)^{-1} = A^{-1}B^{-1}$? Explain.

Skill Maintenance

Use synthetic division to find the function values.

49. $f(x) = x^3 - 6x^2 + 4x - 8$; find $f(-2)$

50. $f(x) = 2x^4 - x^3 + 5x^2 + 6x - 4$; find $f(3)$

Factor the polynomial $f(x)$.

51. $f(x) = x^3 - 3x^2 - 6x + 8$

52. $f(x) = x^4 + 2x^3 - 16x^2 - 2x + 15$

Synthesis

State the conditions under which A^{-1} exists. Then find a formula for A^{-1}.

53. $A = [x]$

54. $A = \begin{bmatrix} x & 0 \\ 0 & y \end{bmatrix}$

55. $A = \begin{bmatrix} 0 & 0 & x \\ 0 & y & 0 \\ z & 0 & 0 \end{bmatrix}$

56. $A = \begin{bmatrix} x & 1 & 1 & 1 \\ 0 & y & 0 & 0 \\ 0 & 0 & z & 0 \\ 0 & 0 & 0 & w \end{bmatrix}$

8.6

Determinants and Cramer's Rule

- Evaluate determinants of square matrices.
- Use Cramer's rule to solve systems of equations.

Determinants of Square Matrices

With every square matrix, we associate a number called its *determinant*.

Determinant of a 2 × 2 Matrix

The **determinant** of the matrix $\begin{bmatrix} a & c \\ b & d \end{bmatrix}$ is denoted $\begin{vmatrix} a & c \\ b & d \end{vmatrix}$ and is defined as

$$\begin{vmatrix} a & c \\ b & d \end{vmatrix} = ad - bc.$$

EXAMPLE 1 Evaluate: $\begin{vmatrix} \sqrt{2} & -3 \\ -4 & -\sqrt{2} \end{vmatrix}$.

Solution

$$\begin{vmatrix} \sqrt{2} & -3 \\ -4 & -\sqrt{2} \end{vmatrix} \qquad \text{The arrows indicate the products involved.}$$

$$= \sqrt{2}(-\sqrt{2}) - (-4)(-3)$$
$$= -2 - 12 = -14$$

We now consider a way to evaluate determinants of square matrices of order 3×3 or higher.

Evaluating Determinants Using Cofactors

We first find minors and cofactors of matrices in order to evaluate determinants of many matrices.

Minor

For a square matrix $\mathbf{A} = [a_{ij}]$, the **minor** M_{ij} of an element a_{ij} is the determinant of the matrix formed by deleting the ith row and the jth column of $\mathbf{A}$.

EXAMPLE 2 For the matrix

$$[a_{ij}] = \begin{bmatrix} -8 & 0 & 6 \\ 4 & -6 & 7 \\ -1 & -3 & 5 \end{bmatrix},$$

find each of the following.

a) M_{11} **b)** M_{23}

Solution

a) We delete the first row and the first column and find the determinant of the 2×2 matrix formed by the remaining elements.

$$\begin{bmatrix} -8 & 0 & 6 \\ 4 & -6 & 7 \\ -1 & -3 & 5 \end{bmatrix} \qquad \begin{aligned} M_{11} &= \begin{vmatrix} -6 & 7 \\ -3 & 5 \end{vmatrix} \\ &= (-6) \cdot 5 - (-3) \cdot 7 \\ &= -30 - (-21) \\ &= -30 + 21 = -9 \end{aligned}$$

b) We delete the second row and the third column and find the determinant of the 2×2 matrix formed by the remaining elements.

$$\begin{bmatrix} -8 & 0 & 6 \\ 4 & -6 & 7 \\ -1 & -3 & 5 \end{bmatrix} \qquad \begin{aligned} M_{23} &= \begin{vmatrix} -8 & 0 \\ -1 & -3 \end{vmatrix} \\ &= -8(-3) - (-1)0 = 24 \end{aligned}$$

Cofactor

For a square matrix $\mathbf{A} = [a_{ij}]$, the cof[...] A_{ij} of an element a_{ij} is given by

$$A_{ij} = (-1)^{i+j}M_{ij},$$

where M_{ij} is the minor of a_{ij}.

EXAMPLE 3   For the matrix given in Example 2[...] of the following.

a) A_{11} **b)** A_{23}

Solution

a) In Example 2, we found that $M_{11} = -9$. Then

$$A_{11} = (-1)^{1+1}(-9) = (1)(-9) = -9.$$

b) In Example 2, we found that $M_{23} = 24$. Then

$$A_{23} = (-1)^{2+3}(24) = (-1)(24) = -24.$$

Consider the matrix $\mathbf{A}$ given by

$$\mathbf{A} = \begin{bmatrix} a_{11} & a_{12} & a_{13} \\ a_{21} & a_{22} & a_{23} \\ a_{31} & a_{32} & a_{33} \end{bmatrix}.$$

The determinant of the matrix, denoted $|\mathbf{A}|$, can be found by multiplying each element of the first column by its cofactor and adding:

$$|\mathbf{A}| = a_{11}A_{11} + a_{21}A_{21} + a_{31}A_{31}.$$

Because

$$A_{11} = (-1)^{1+1}M_{11} = M_{11},$$
$$A_{21} = (-1)^{2+1}M_{21} = -M_{21},$$

and $A_{31} = (-1)^{3+1}M_{31} = M_{31},$

we can write

$$|\mathbf{A}| = a_{11} \cdot \begin{vmatrix} a_{22} & a_{23} \\ a_{32} & a_{33} \end{vmatrix} - a_{21} \cdot \begin{vmatrix} a_{12} & a_{13} \\ a_{32} & a_{33} \end{vmatrix} + a_{31} \cdot \begin{vmatrix} a_{12} & a_{13} \\ a_{22} & a_{23} \end{vmatrix}.$$

It can be shown that we can determine $|\mathbf{A}|$ by picking *any* row or column, multiplying each element in that row or column by its cofactor, and adding. This is called *expanding* across a row or down a column. We just expanded down the first column. We now define the determinant of a square matrix of any order.

Determinant of Any Square Matrix

For any square matrix **A** of order $n \times n$ ($n > 1$), we define the **determinant** of **A**, denoted $|\mathbf{A}|$, as follows. Choose any row or column. Multiply each element in that row or column by its cofactor and add the results. The determinant of a 1×1 matrix is simply the element of the matrix. The value of a determinant will be the same no matter how it is evaluated.

EXAMPLE 4   Evaluate $|\mathbf{A}|$ by expanding across the third row.

$$\mathbf{A} = \begin{bmatrix} -8 & 0 & 6 \\ 4 & -6 & 7 \\ -1 & -3 & 5 \end{bmatrix}$$

Solution   We have

$$|\mathbf{A}| = (-1)A_{31} + (-3)A_{32} + 5A_{33}$$

$$= (-1)(-1)^{3+1} \cdot \begin{vmatrix} 0 & 6 \\ -6 & 7 \end{vmatrix} + (-3)(-1)^{3+2} \cdot \begin{vmatrix} -8 & 6 \\ 4 & 7 \end{vmatrix}$$

$$+ 5(-1)^{3+3} \cdot \begin{vmatrix} -8 & 0 \\ 4 & -6 \end{vmatrix}$$

$$= (-1) \cdot 1 \cdot [0 \cdot 7 - (-6)6] + (-3)(-1)[-8 \cdot 7 - 4 \cdot 6]$$

$$+ 5 \cdot 1 \cdot [-8(-6) - 4 \cdot 0]$$

$$= -[36] + 3[-80] + 5[48] = -36 - 240 + 240 = -36.$$

The value of this determinant is -36 no matter which row or column we expand upon.

Cramer's Rule

Determinants can be used to solve systems of linear equations. Consider a system of two linear equations:

$$a_1x + b_1y = c_1,$$
$$a_2x + b_2y = c_2.$$

Solving this system using the elimination method, we obtain

$$x = \frac{c_1b_2 - c_2b_1}{a_1b_2 - a_2b_1}$$

and

$$y = \frac{a_1c_2 - a_2c_1}{a_1b_2 - a_2b_1}.$$

Technology Connection

Determinants can be evaluated on a graphing calculator. After entering a matrix, we select the determinant operation from the MATRIX MATH menu and enter the name of the matrix. The calculator will return the value of the determinant of the matrix. For example, for

$$\mathbf{A} = \begin{bmatrix} 1 & 6 & -1 \\ -3 & -5 & 3 \\ 0 & 4 & 2 \end{bmatrix},$$

we have

```
det ([A])
                    26
```
.

The numerators and denominators of these expressions can be written as determinants:

$$x = \frac{\begin{vmatrix} c_1 & b_1 \\ c_2 & b_2 \end{vmatrix}}{\begin{vmatrix} a_1 & b_1 \\ a_2 & b_2 \end{vmatrix}} \quad \text{and} \quad y = \frac{\begin{vmatrix} a_1 & c_1 \\ a_2 & c_2 \end{vmatrix}}{\begin{vmatrix} a_1 & b_1 \\ a_2 & b_2 \end{vmatrix}}.$$

If we let

$$D = \begin{vmatrix} a_1 & b_1 \\ a_2 & b_2 \end{vmatrix}, \quad D_x = \begin{vmatrix} c_1 & b_1 \\ c_2 & b_2 \end{vmatrix}, \quad \text{and} \quad D_y = \begin{vmatrix} a_1 & c_1 \\ a_2 & c_2 \end{vmatrix},$$

we have

$$x = \frac{D_x}{D} \quad \text{and} \quad y = \frac{D_y}{D}.$$

This procedure for solving systems of equations is known as *Cramer's rule*.

Cramer's Rule for 2 × 2 Systems

The solution of the system of equations

$$a_1 x + b_1 y = c_1,$$
$$a_2 x + b_2 y = c_2$$

is given by

$$x = \frac{D_x}{D}, \quad y = \frac{D_y}{D},$$

where

$$D = \begin{vmatrix} a_1 & b_1 \\ a_2 & b_2 \end{vmatrix}, \quad D_x = \begin{vmatrix} c_1 & b_1 \\ c_2 & b_2 \end{vmatrix},$$

$$D_y = \begin{vmatrix} a_1 & c_1 \\ a_2 & c_2 \end{vmatrix}, \quad \text{and} \quad D \neq 0.$$

Note that the denominator D contains the coefficients of x and y, in the same position as in the original equations. For x, the numerator is obtained by replacing the x-coefficients in D (the a's) with the c's. For y, the numerator is obtained by replacing the y-coefficients in D (the b's) with the c's.

EXAMPLE 5 Solve using Cramer's rule:

$$2x + 5y = 7,$$
$$5x - 2y = -3.$$

Solution We have

$$x = \frac{\begin{vmatrix} 7 & 5 \\ -3 & -2 \end{vmatrix}}{\begin{vmatrix} 2 & 5 \\ 5 & -2 \end{vmatrix}} = \frac{7(-2) - (-3)5}{2(-2) - 5 \cdot 5} = \frac{1}{-29} = -\frac{1}{29},$$

$$y = \frac{\begin{vmatrix} 2 & 7 \\ 5 & -3 \end{vmatrix}}{\begin{vmatrix} 2 & 5 \\ 5 & -2 \end{vmatrix}} = \frac{2(-3) - 5 \cdot 7}{-29} = \frac{-41}{-29} = \frac{41}{29}.$$

The solution is $\left(-\frac{1}{29}, \frac{41}{29}\right)$.

Technology Connection

To use Cramer's rule to solve the system of equations in Example 5 on a graphing calculator, we first enter the matrices corresponding to D, D_x, and D_y. We enter

$$\mathbf{A} = \begin{bmatrix} 2 & 5 \\ 5 & -2 \end{bmatrix}, \quad \mathbf{B} = \begin{bmatrix} 7 & 5 \\ -3 & -2 \end{bmatrix}, \quad \text{and} \quad \mathbf{C} = \begin{bmatrix} 2 & 7 \\ 5 & -3 \end{bmatrix}.$$

Then

$$x = \frac{\det(\mathbf{B})}{\det(\mathbf{A})} \quad \text{and} \quad y = \frac{\det(\mathbf{C})}{\det(\mathbf{A})}.$$

```
det ([B])/det([A]
)▶Frac
                    -1/29
det ([C])/det([A]
)▶Frac
                    41/29
```

Cramer's rule works only when a system of equations has a unique solution. This occurs when $D \neq 0$. If $D = 0$, $D_x = 0$, and $D_y = 0$, then the equations are dependent. If $D = 0$ and D_x and/or D_y is not 0, then the system is inconsistent.

Cramer's rule can be extended to a system of n linear equations in n variables. We consider a 3×3 system.

Cramer's Rule for 3 × 3 Systems

The solution of the system of equations

$$a_1x + b_1y + c_1z = d_1,$$
$$a_2x + b_2y + c_2z = d_2,$$
$$a_3x + b_3y + c_3z = d_3$$

is given by

$$x = \frac{D_x}{D}, \qquad y = \frac{D_y}{D}, \qquad z = \frac{D_z}{D},$$

where

$$D = \begin{vmatrix} a_1 & b_1 & c_1 \\ a_2 & b_2 & c_2 \\ a_3 & b_3 & c_3 \end{vmatrix}, \qquad D_x = \begin{vmatrix} d_1 & b_1 & c_1 \\ d_2 & b_2 & c_2 \\ d_3 & b_3 & c_3 \end{vmatrix},$$

$$D_y = \begin{vmatrix} a_1 & d_1 & c_1 \\ a_2 & d_2 & c_2 \\ a_3 & d_3 & c_3 \end{vmatrix}, \qquad D_z = \begin{vmatrix} a_1 & b_1 & d_1 \\ a_2 & b_2 & d_2 \\ a_3 & b_3 & d_3 \end{vmatrix}, \quad \text{and } D \neq 0.$$

Note that the determinant D_x is obtained from D by replacing the x-coefficients with d_1, d_2, and d_3. A similar thing happens with D_y and D_z. When $D = 0$, Cramer's rule cannot be used. If $D = 0$ and D_x, D_y, and D_z are 0, the equations are dependent. If $D = 0$ and one of D_x, D_y, or D_z is not 0, then the system is inconsistent.

EXAMPLE 6 Solve using Cramer's rule:

$$x - 3y + 7z = 13,$$
$$x + y + z = 1,$$
$$x - 2y + 3z = 4.$$

Solution We have

$$D = \begin{vmatrix} 1 & -3 & 7 \\ 1 & 1 & 1 \\ 1 & -2 & 3 \end{vmatrix} = -10, \qquad D_x = \begin{vmatrix} 13 & -3 & 7 \\ 1 & 1 & 1 \\ 4 & -2 & 3 \end{vmatrix} = 20,$$

$$D_y = \begin{vmatrix} 1 & 13 & 7 \\ 1 & 1 & 1 \\ 1 & 4 & 3 \end{vmatrix} = -6, \qquad D_z = \begin{vmatrix} 1 & -3 & 13 \\ 1 & 1 & 1 \\ 1 & -2 & 4 \end{vmatrix} = -24.$$

Then

$$x = \frac{D_x}{D} = \frac{20}{-10} = -2,$$

$$y = \frac{D_y}{D} = \frac{-6}{-10} = \frac{3}{5},$$

$$z = \frac{D_z}{D} = \frac{-24}{-10} = \frac{12}{5}.$$

The solution is $\left(-2, \frac{3}{5}, \frac{12}{5}\right)$. In practice, it is not necessary to evaluate D_z. When we have found values for x and y, we can substitute them into one of the equations to find z. ▬

Exercise Set 8.6

Evaluate the determinant.

1. $\begin{vmatrix} -2 & -\sqrt{5} \\ -\sqrt{5} & 3 \end{vmatrix}$

2. $\begin{vmatrix} \sqrt{5} & -3 \\ 4 & 2 \end{vmatrix}$

3. $\begin{vmatrix} x & 4 \\ x & x^2 \end{vmatrix}$

4. $\begin{vmatrix} y^2 & -2 \\ y & 3 \end{vmatrix}$

5. $\begin{vmatrix} 3 & 1 & 2 \\ -2 & 3 & 1 \\ 3 & 4 & -6 \end{vmatrix}$

6. $\begin{vmatrix} 3 & -2 & 1 \\ 2 & 4 & 3 \\ -1 & 5 & 1 \end{vmatrix}$

7. $\begin{vmatrix} x & 0 & -1 \\ 2 & x & x^2 \\ -3 & x & 1 \end{vmatrix}$

8. $\begin{vmatrix} x & 1 & -1 \\ x^2 & x & x \\ 0 & x & 1 \end{vmatrix}$

Use the following matrix for Exercises 9–16:

$$\mathbf{A} = \begin{bmatrix} 7 & -4 & -6 \\ 2 & 0 & -3 \\ 1 & 2 & -5 \end{bmatrix}.$$

9. Find M_{11}, M_{32}, and M_{22}.

10. Find M_{13}, M_{31}, and M_{23}.

11. Find A_{11}, A_{32}, and A_{22}.

12. Find A_{13}, A_{31}, and A_{23}.

13. Evaluate $|\mathbf{A}|$ by expanding across the second row.

14. Evaluate $|\mathbf{A}|$ by expanding down the second column.

15. Evaluate $|\mathbf{A}|$ by expanding down the third column.

16. Evaluate $|\mathbf{A}|$ by expanding across the first row.

Use the following matrix for Exercises 17–22:

$$\mathbf{A} = \begin{bmatrix} 1 & 0 & 0 & -2 \\ 4 & 1 & 0 & 0 \\ 5 & 6 & 7 & 8 \\ -2 & -3 & -1 & 0 \end{bmatrix}$$

17. Find M_{41} and M_{33}.

18. Find M_{12} and M_{44}.

19. Find A_{24} and A_{43}.

20. Find A_{22} and A_{34}.

21. Evaluate $|\mathbf{A}|$ by expanding across the first row.

22. Evaluate $|\mathbf{A}|$ by expanding down the third column.

Solve using Cramer's rule.

23. $-2x + 4y = 3,$
      $3x - 7y = 1$

24. $5x - 4y = -3,$
      $7x + 2y = 6$

25. $2x - y = 5,$
      $x - 2y = 1$

26. $3x + 4y = -2,$
      $5x - 7y = 1$

27. $2x + 9y = -2,$
      $4x - 3y = 3$

28. $2x + 3y = -1,$
      $3x + 6y = -0.5$

29. $2x + 5y = 7,$
      $3x - 2y = 1$

30. $3x + 2y = 7,$
      $2x + 3y = -2$

31. $3x + 2y - z = 4,$
      $3x - 2y + z = 5,$
      $4x - 5y - z = -1$

32.
$$3x - y + 2z = 1,$$
$$x - y + 2z = 3,$$
$$-2x + 3y + z = 1$$

33.
$$3x + 5y - z = -2,$$
$$x - 4y + 2z = 13,$$
$$2x + 4y + 3z = 1$$

34.
$$3x + 2y + 2z = 1,$$
$$5x - y - 6z = 3,$$
$$2x + 3y + 3z = 4$$

35.
$$x - 3y - 7z = 6,$$
$$2x + 3y + z = 9,$$
$$4x + y = 7$$

36.
$$x - 2y - 3z = 4,$$
$$3x - 2z = 8,$$
$$2x + y + 4z = 13$$

37.
$$6y + 6z = -1,$$
$$8x + 6z = -1,$$
$$4x + 9y = 8$$

38.
$$3x + 5y = 2,$$
$$2x - 3z = 7,$$
$$4y + 2z = -1$$

Technology Connection

Use a graphing calculator to do the following.

39. Exercises 1 and 5   **40.** Exercises 2 and 6

*For each matrix **A**, evaluate $|\mathbf{A}|$ on a graphing calculator.*

41. $\mathbf{A} = \begin{bmatrix} 1 & 0 & 0 & -2 \\ 4 & 1 & 0 & 0 \\ 5 & 6 & 7 & 8 \\ -2 & -3 & -1 & 0 \end{bmatrix}$

42. $|\mathbf{A}| = \begin{vmatrix} 5 & -4 & 2 & -2 \\ 3 & -3 & -4 & 7 \\ -2 & 3 & 2 & 4 \\ -8 & 9 & 5 & -5 \end{vmatrix}.$

Use a graphing calculator to do the following.

43. Exercises 25 and 33   **44.** Exercises 26 and 34

Collaborative Discussion and Writing

45. Explain why the system of equations
$$a_1x + b_1y = c_1,$$
$$a_2x + b_2y = c_2$$
is either dependent or inconsistent when
$$\begin{vmatrix} a_1 & b_1 \\ a_2 & b_2 \end{vmatrix} = 0.$$

46. If the lines $a_1x + b_1y = c_1$ and $a_2x + b_2y = c_2$ are parallel, what can you say about the values of

$$\begin{vmatrix} a_1 & b_1 \\ a_2 & b_2 \end{vmatrix}, \quad \begin{vmatrix} c_1 & b_1 \\ c_2 & b_2 \end{vmatrix}, \quad \text{and} \quad \begin{vmatrix} a_1 & c_1 \\ a_2 & c_2 \end{vmatrix}?$$

Skill Maintenance

Determine whether the function is one-to-one, and if it is, find a formula for $f^{-1}(x)$.

47. $f(x) = 3x + 2$ **48.** $f(x) = x^2 - 4$

49. $f(x) = |x| + 3$ **50.** $f(x) = \sqrt[3]{x} + 1$

Synthesis

Solve.

51. $\begin{vmatrix} x & 5 \\ -4 & x \end{vmatrix} = 24$

52. $\begin{vmatrix} y & 2 \\ 3 & y \end{vmatrix} = y$

53. $\begin{vmatrix} x & -3 \\ -1 & x \end{vmatrix} \geq 0$

54. $\begin{vmatrix} y & -5 \\ -2 & y \end{vmatrix} < 0$

55. $\begin{vmatrix} x + 3 & 4 \\ x - 3 & 5 \end{vmatrix} = -7$

56. $\begin{vmatrix} m + 2 & -3 \\ m + 5 & -4 \end{vmatrix} = 3m - 5$

57. $\begin{vmatrix} 2 & x & 1 \\ 1 & 2 & -1 \\ 3 & 4 & -2 \end{vmatrix} = -6$

58. $\begin{vmatrix} x & 2 & x \\ 3 & -1 & 1 \\ 1 & -2 & 2 \end{vmatrix} = -10$

Rewrite the expression using a determinant. Answers may vary.

59. $2L + 2W$ **60.** $\pi r + \pi h$

61. $a^2 + b^2$ **62.** $\frac{1}{2}h(a + b)$

63. $2\pi r^2 + 2\pi rh$ **64.** $x^2y^2 - Q^2$

8.7

Systems of Inequalities and Linear Programming

- *Graph linear inequalities.*
- *Graph systems of linear inequalities.*
- *Solve linear programming problems.*

A graph of an inequality is a drawing that represents its solutions. We have already seen that an inequality in one variable can be graphed on a number line. An inequality in two variables can be graphed on a coordinate plane.

Graphs of Linear Inequalities

A statement like $5x - 4y < 20$ is a linear inequality in two variables.

Linear Inequality in Two Variables

A **linear inequality in two variables** is an inequality that can be written in the form

$$Ax + By < C,$$

where A, B, and C are real numbers and A and B are not both zero. The symbol $<$ may be replaced with $\leq$, $>$, or $\geq$.

A solution of a linear inequality in two variables is an ordered pair (x, y) for which the inequality is true. For example, $(1, 3)$ is a solution of $5x - 4y < 20$ because $5 \cdot 1 - 4 \cdot 3 < 20$, or $-7 < 20$, is true. On the other hand, $(2, -6)$ is not a solution of $5x - 4y < 20$ because $5 \cdot 2 - 4 \cdot (-6) \not< 20$, or $34 \not< 20$.

The **solution set** of an inequality is the set of all the ordered pairs that make it true. The **graph of an inequality** represents its solution set.

EXAMPLE 1   Graph: $y < x + 3$.

Solution   We begin by graphing the **related equation** $y = x + 3$. We use a dashed line because the inequality symbol is $<$. This indicates that the line itself is not in the solution set of the inequality.

Note that the line divides the coordinate plane into two regions called **half-planes,** one of which satisfies the inequality. Either *all* points in a half-plane are in the solution set of the inequality or *none* is.

To determine which half-plane satisfies the inequality, we try a test point in either region. The point $(0, 0)$ is usually a convenient choice so long as it does not lie on the line.

$$\frac{y < x + 3}{}$$

$$0 \ ?\ 0 + 3$$

$$0 \ \bigl|\ 3 \qquad \text{TRUE}$$

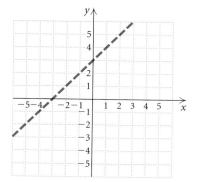

Since (0, 0) satisfies the inequality, so do all points in the half-plane that contains (0, 0). We shade this region to show the solution set of the inequality.

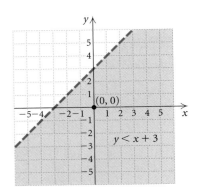

In general, we use the following procedure to graph linear inequalities in two variables.

To graph a linear inequality in two variables:

1. Replace the inequality symbol with an equals sign and graph this related equation. If the inequality symbol is $<$ or $>$, draw the line dashed. If the inequality symbol is $\leq$ or $\geq$, draw the line solid.

2. The graph consists of a half-plane on one side of the line and, if the line is solid, the line as well. To determine which half-plane to shade, test a point not on the line in the original inequality. If that point is a solution, shade the half-plane containing that point. If not, shade the opposite half-plane.

EXAMPLE 2 Graph: $3x + 4y \geq 12$.

Solution

1. First, we graph the related equation $3x + 4y = 12$. We use a solid line because the inequality symbol is $\geq$. This indicates that the line is included in the solution set.

2. To determine which half-plane to shade, we test a point in either region. We choose (0, 0).

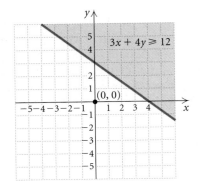

$$3x + 4y \geq 12$$

$$3 \cdot 0 + 4 \cdot 0 \ ? \ 12$$

$$0 \ | \ 12 \quad \text{FALSE}$$

Because (0, 0) is *not* a solution, all the points in the half-plane that does *not* contain (0, 0) are solutions. We shade that region, as shown in the figure at left.

Technology Connection

To graph the inequality in Example 2 on a graphing calculator, we first enter the related equation in the form $y = \dfrac{-3x + 12}{4}$. Then we select the "shade above" graph style.

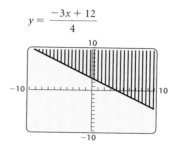

$$y = \frac{-3x + 12}{4}$$

EXAMPLE 3 Graph $x > -3$ on a plane.

Solution

1. First, we graph the related equation $x = -3$. We use a dashed line because the inequality symbol is $>$. This indicates that the line is not included in the solution set.

2. The inequality tells us that all points (x, y) for which $x > -3$ are solutions. These are the points to the right of the line. We can also use a test point to determine the solutions. We choose $(5, 1)$.

$$\frac{x > -3}{5 \ ? \ -3} \quad \text{TRUE}$$

Because $(5, 1)$ is a solution, we shade the region containing that point—that is, the region to the right of the dashed line.

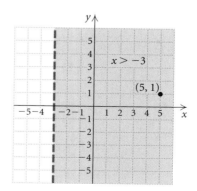

EXAMPLE 4   Graph $y \leq 4$ on a plane.

Solution

1. First, we graph the related equation $y = 4$. We use a solid line because the inequality symbol is $\leq$.

2. The inequality tells us that all points (x, y) for which $y \leq 4$ are solutions of the inequality. These are the points on or below the line. We can also use a test point to determine the solutions. We choose $(-2, 5)$.

$$\frac{y \leq 4}{5 \ ? \ 4 \quad \text{FALSE}}$$

Because $(-2, 5)$ is not a solution, we shade the half-plane that does not contain that point.

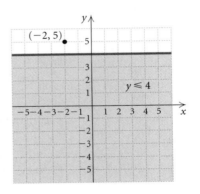

Technology Connection

We can graph the inequality $y \leq 4$ by first graphing $y = 4$ and then using the "shade below" graph style.

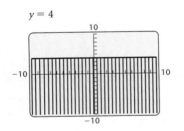

Systems of Linear Inequalities

A system of inequalities in two variables consists of two or more inequalities in two variables considered simultaneously. For example,

$$x + y \leq 4,$$
$$x - y \geq 2$$

is a system of two *linear* inequalities in two variables.

A solution of a system of inequalities is an ordered pair that is a solution of each inequality in the system. To graph a system of linear inequalities, we graph each inequality and determine the region that is common to all the solution sets.

EXAMPLE 5   Graph the solution set of the system

$$x + y \leq 4,$$
$$x - y \geq 2.$$

Solution We graph $x + y \leq 4$ by first graphing the equation $x + y = 4$ using a solid line. Next, we choose $(0, 0)$ as a test point and find that it is a solution of $x + y \leq 4$, so we shade the half-plane containing $(0, 0)$ using red. Next, we graph $x - y = 2$ using a solid line. We find that $(0, 0)$ is not a solution of $x - y \geq 2$, so we shade the half-plane that does not contain $(0, 0)$ using green. The arrows at the ends of each line help to indicate the half-plane that contains each solution set.

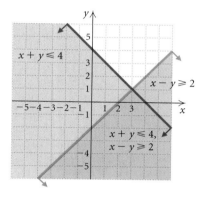

The solution set of the system of equations is the region shaded both red and green, or brown, including parts of the lines $x + y = 4$ and $x - y = 2$.

A system of inequalities may have a graph that consists of a polygon and its interior. As we will see later in this section, it is important in many applications to be able to find the vertices of such a polygon.

EXAMPLE 6 Graph the following system of inequalities and find the coordinates of any vertices formed:

$$3x - y \leq 6, \quad (1)$$
$$y - 3 \leq 0, \quad (2)$$
$$x + y \geq 0. \quad (3)$$

Solution We graph the related equations $3x - y = 6$, $y - 3 = 0$, and $x + y = 0$ using solid lines. The half-plane containing the solution set for each inequality is indicated by the arrows at the ends of each line. We shade the region common to all three solution sets.

To find the vertices, we solve three systems of equations. The system of equations from inequalities (1) and (2) is

$$3x - y = 6,$$
$$y - 3 = 0.$$

Solving, we obtain the vertex $(3, 3)$.

The system of equations from inequalities (1) and (3) is

$$3x - y = 6,$$
$$x + y = 0.$$

Technology Connection

We can use different shading patterns on a graphing calculator to graph the system of inequalities in Example 5. The solution set is the region shaded using both patterns.

$y_1 = 4 - x, \quad y_2 = x - 2$

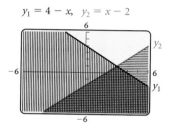

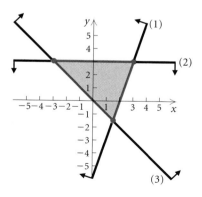

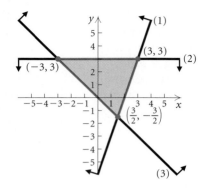

Solving, we obtain the vertex $\left(\frac{3}{2}, -\frac{3}{2}\right)$.

The system of equations from inequalities (2) and (3) is

$$y - 3 = 0,$$
$$x + y = 0.$$

Solving, we obtain the vertex $(-3, 3)$.

Applications: Linear Programming

In many applications, we want to find a maximum or minimum value. In business, for example, we might want to maximize profit and minimize cost. **Linear programming** can tell us how to do this.

In our study of linear programming, we will consider linear functions of two variables that are to be maximized or minimized subject to several conditions, or **constraints**. These constraints are expressed as inequalities. The solution set of the system of inequalities made up of the constraints contains all the **feasible solutions** of a linear programming problem. The function that we want to maximize or minimize is called the **objective function.**

It can be shown that the maximum and minimum values of the objective function occur at a vertex of the region of feasible solutions. Thus we have the following procedure.

Linear Programming Procedure

To find the maximum or minimum value of a linear objective function subject to a set of constraints:

1. Graph the region of feasible solutions.
2. Determine the coordinates of the vertices of the region.
3. Evaluate the objective function at each vertex. The largest and smallest of those values are the maximum and minimum values of the function, respectively.

EXAMPLE 7 *Maximizing Profit.* Dovetail Carpentry Shop makes bookcases and desks. Each bookcase requires 5 hr of woodworking and 4 hr of finishing. Each desk requires 10 hr of woodworking and 3 hr of finishing. Each month the shop has 600 hr of labor available for woodworking and 240 hr for finishing. The profit on each bookcase is $40 and on each desk is $75. How many of each product should be made each month in order to maximize profit?

Solution   We let $x =$ the number of bookcases to be produced and $y =$ the number of desks. Then the profit P is given by the function

$$P = 40x + 75y.$$ To emphasize that P is a function of two variables, we sometimes write $P(x, y) = 40x + 75y.$

We know that x bookcases require $5x$ hr of woodworking and y desks require $10y$ hr of woodworking. Since there is no more than 600 hr of labor available for woodworking, we have one constraint:

$$5x + 10y \leq 600.$$

Similarly, the bookcases and desks require $4x$ hr and $3y$ hr of finishing, respectively. There is no more than 240 hr of labor available for finishing, so we have a second constraint:

$$4x + 3y \leq 240.$$

We also know that $x \geq 0$ and $y \geq 0$ because the carpentry shop cannot make a negative number of either product.

Thus we want to maximize the objective function

$$P = 40x + 75y$$

subject to the constraints

$$5x + 10y \leq 600,$$
$$4x + 3y \leq 240,$$
$$x \geq 0,$$
$$y \geq 0.$$

We graph the system of inequalities and determine the vertices. Next, we evaluate the objective function P at each vertex.

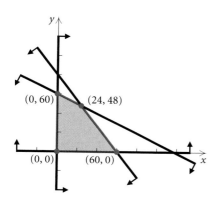

VERTICES (x, y)	PROFIT $P = 40x + 75y$	
$(0, 0)$	$P = 40 \cdot 0 + 75 \cdot 0 = 0$	
$(60, 0)$	$P = 40 \cdot 60 + 75 \cdot 0 = 2400$	
$(24, 48)$	$P = 40 \cdot 24 + 75 \cdot 48 = 4560$	←——— Maximum
$(0, 60)$	$P = 40 \cdot 0 + 75 \cdot 60 = 4500$	

The carpentry shop will make a maximum profit of $4560 when 24 bookcases and 48 desks are produced.

Exercise Set 8.7

In Exercises 1–8, match the inequality with one of the graphs (a)–(h), which follow.

a)

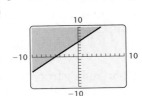

b)

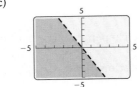

c)

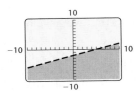

d)

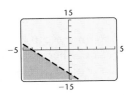

e)

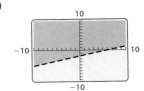

f)

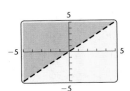

g)

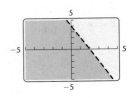

h)

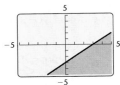

1. $y > x$

2. $y < -2x$

3. $y \le x - 3$

4. $y \ge x + 5$

5. $2x + y < 4$

6. $3x + y < -12$

7. $2x - 5y > 10$

8. $3x - 9y < 18$

Graph.

9. $y > 2x$

10. $2y < x$

11. $y + x \ge 0$

12. $y - x < 0$

13. $y > x - 3$

14. $y \le x + 4$

15. $x + y < 4$

16. $x - y \ge 5$

17. $3x - 2y \le 6$

18. $2x - 5y < 10$

19. $3y + 2x \ge 6$

20. $2y + x \le 4$

21. $3x - 2 \le 5x + y$

22. $2x - 6y \ge 8 + 2y$

23. $x < -4$

24. $y \ge 5$

25. $y > -3$

26. $x \le 5$

27. $-4 < y < -1$
   (*Hint:* Think of this as $-4 < y$ and $y < -1$.)

28. $-3 \le x \le 3$
   (*Hint:* Think of this as $-3 \le x$ and $x \le 3$.)

29. $y \ge |x|$

30. $y \le |x + 2|$

In Exercises 31–36, match the system of inequalities with one of the graphs (a)–(f), which follow.

a)

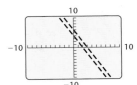

b)

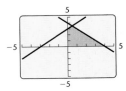

c)

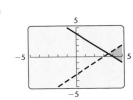

d)

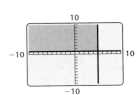

e)

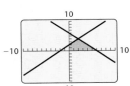

f)

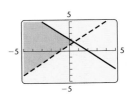

31. $y > x + 1,$
   $y \le 2 - x$

32. $y < x - 3,$
   $y \ge 4 - x$

33. $2x + y < 4,$
   $4x + 2y > 12$

34. $x \le 5,$
   $y \ge 1$

35. $x + y \le 4,$
   $x - y \ge -3,$
   $x \ge 0,$
   $y \ge 0$

36. $x - y \ge -2,$
   $x + y \le 6,$
   $x \ge 0,$
   $y \ge 0$

Graph the system of inequalities. Then find the coordinates of the vertices.

37. $y \le x,$
   $y \ge 3 - x$

38. $y \ge x,$
   $y \le x - 5$

39. $y \ge x,$
   $y \le x - 4$

40. $y \ge x,$
   $y \le 2 - x$

41. $y \geq -3,$
$\quad x \geq 1$

42. $y \leq -2,$
$\quad x \geq 2$

43. $x \leq 3,$
$\quad y \geq 2 - 3x$

44. $x \geq -2,$
$\quad y \leq 3 - 2x$

45. $x + y \leq 1,$
$\quad x - y \leq 2$

46. $y + 3x \geq 0,$
$\quad y + 3x \leq 2$

47. $2y - x \leq 2,$
$\quad y + 3x \geq -1$

48. $\quad y \leq 2x + 1,$
$\quad y \geq -2x + 1,$
$\quad x - 2 \leq 0$

49. $\quad x - y \leq 2,$
$\quad x + 2y \geq 8,$
$\quad y - 4 \leq 0$

50. $x + 2y \leq 12,$
$\quad 2x + y \leq 12,$
$\quad x \geq 0,$
$\quad y \geq 0$

51. $4y - 3x \geq -12,$
$\quad 4y + 3x \geq -36,$
$\quad y \leq 0,$
$\quad x \leq 0$

52. $8x + 5y \leq 40,$
$\quad x + 2y \leq 8,$
$\quad x \geq 0,$
$\quad y \geq 0$

53. $3x + 4y \geq 12,$
$\quad 5x + 6y \leq 30,$
$\quad 1 \leq x \leq 3$

54. $y - x \geq 1,$
$\quad y - x \leq 3,$
$\quad 2 \leq x \leq 5$

Find the maximum and the minimum values of the function and the values of x and y for which they occur.

55. $P = 17x - 3y + 60,$ subject to

$$6x + 8y \leq 48,$$
$$0 \leq y \leq 4,$$
$$0 \leq x \leq 7.$$

56. $Q = 28x - 4y + 72,$ subject to

$$5x + 4y \geq 20,$$
$$0 \leq y \leq 4,$$
$$0 \leq x \leq 3.$$

57. $F = 5x + 36y,$ subject to

$$5x + 3y \leq 34,$$
$$3x + 5y \leq 30,$$
$$x \geq 0,$$
$$y \geq 0.$$

58. $G = 16x + 14y,$ subject to

$$3x + 2y \leq 12,$$
$$7x + 5y \leq 29,$$
$$x \geq 0,$$
$$y \geq 0.$$

59. *Maximizing Income.* Golden Harvest Foods makes jumbo biscuits and regular biscuits. The oven can cook at most 200 biscuits per day. Each jumbo biscuit requires 2 oz of flour, each regular biscuit requires 1 oz of flour, and there is 300 oz of flour available. The income from each jumbo biscuit is $0.10 and from each regular biscuit is $0.08. How many of each size biscuit should be made in order to maximize income? What is the maximum income?

60. *Maximizing Mileage.* Omar owns a car and a moped. He can afford 12 gal of gasoline to be split between the car and the moped. Omar's car gets 20 mpg and holds at most 10 gal of gas. His moped gets 100 mpg and holds at most 3 gal of gas. How many gallons of gasoline should each vehicle use if Omar wants to travel as far as possible? What is the maximum number of miles that he can travel?

61. *Maximizing Profit.* Norris Mill can convert logs into lumber and plywood. In a given week, the mill can turn out 400 units of production, of which 100 units of lumber and 150 units of plywood are required by regular customers. The profit is $20 per unit of lumber and $30 per unit of plywood. How many units of each should the mill produce in order to maximize the profit?

62. *Maximizing Profit.* Sunnydale Farm includes 240 acres of cropland. The farm owner wishes to plant this acreage in corn and oats. The profit per acre in corn production is $40 and in oats is $30. A total of 320 hr of labor is available. Each acre of corn requires 2 hr of labor, whereas each acre of oats requires 1 hr of labor. How should the land be divided between corn and oats in order to yield the maximum profit? What is the maximum profit?

63. *Minimizing Cost.* An animal feed to be mixed from soybean meal and oats must contain at least 120 lb of protein, 24 lb of fat, and 10 lb of mineral ash. Each 100-lb sack of soybean meal costs $15 and contains 50 lb of protein, 8 lb of fat, and 5 lb of mineral ash. Each 100-lb sack of oats costs $5 and contains 15 lb of protein, 5 lb of fat, and 1 lb of mineral ash. How many sacks of each should be used to satisfy the minimum requirements at minimum cost?

64. *Minimizing Cost.* Suppose that in the preceding problem the oats were replaced by alfalfa, which costs $8 per 100 lb and contains 20 lb of protein, 6 lb of fat, and 8 lb of mineral ash. How much of each is now required in order to minimize the cost?

65. *Maximizing Income.* Clayton is planning to invest up to $40,000 in corporate and municipal bonds. The least he is allowed to invest in corporate bonds is $6000, and he does not want to invest more than $22,000 in corporate bonds. He also does not want to invest more than $30,000 in municipal bonds.

The interest is 8% on corporate bonds and $7\frac{1}{2}$% on municipal bonds. This is simple interest for one year. How much should he invest in each type of bond in order to maximize his income? What is the maximum income?

66. *Maximizing Income.* Margaret is planning to invest up to $22,000 in certificates of deposit at City Bank and People's Bank. She wants to invest at least $2000 but no more than $14,000 at City Bank. People's Bank does not insure more than a $15,000 investment, so she will invest no more than that in People's Bank. The interest is 6% at City Bank and $6\frac{1}{2}$% at People's Bank. This is simple interest for one year. How much should she invest in each bank in order to maximize her income? What is the maximum income?

67. *Minimizing Transportation Cost.* An airline with two types of airplanes, P_1 and P_2, has contracted with a tour group to provide transportation for a minimum of 2000 first-class, 1500 tourist-class, and 2400 economy-class passengers. For a certain trip, airplane P_1 costs $12 thousand to operate and can accommodate 40 first-class, 40 tourist-class, and 120 economy-class passengers, whereas airplane P_2 costs $10 thousand to operate and can accommodate 80 first-class, 30 tourist-class, and 40 economy-class passengers. How many of each type of airplane should be used in order to minimize the operating cost?

68. *Minimizing Transportation Cost.* Suppose that in the preceding problem a new airplane P_3 becomes available, having an operating cost for the same trip of $15 thousand and accommodating 40 first-class, 40 tourist-class, and 80 economy-class passengers. If airplane P_1 were replaced by airplane P_3, how many of P_2 and P_3 should be used in order to minimize the operating cost?

69. *Maximizing Profit.* It takes Just Sew 2 hr of cutting and 4 hr of sewing to make a knit suit. It

takes 4 hr of cutting and 2 hr of sewing to make a worsted suit. At most 20 hr per day are available for cutting and at most 16 hr per day are available for sewing. The profit is $34 on a knit suit and $31 on a worsted suit. How many of each kind of suit should be made each day in order to maximize profit? What is the maximum profit?

70. *Maximizing Profit.* Cambridge Metal Works manufactures two sizes of gears. The smaller gear requires 4 hr of machining and 1 hr of polishing and yields a profit of $25. The larger gear requires 1 hr of machining and 1 hr of polishing and yields a profit of $10. The firm has available at most 24 hr per day for machining and 9 hr per day for polishing. How many of each type of gear should be produced each day in order to maximize profit? What is the maximum profit?

71. *Minimizing Nutrition Cost.* Suppose that it takes 12 units of carbohydrates and 6 units of protein to satisfy Jacob's minimum weekly requirements. A particular type of meat contains 2 units of carbohydrates and 2 units of protein per pound. A particular cheese contains 3 units of carbohydrates and 1 unit of protein per pound. The meat costs $3.50 per pound and the cheese costs $4.60 per pound. How many pounds of each are needed in order to minimize the cost and still meet the minimum requirements?

72. *Minimizing Salary Cost.* The Spring Hill school board is analyzing education costs for Hill Top School. It wants to hire teachers and teacher's aides to make up a faculty that satisfies its needs at minimum cost. The average annual salary for a teacher is $35,000 and for a teacher's aide is $18,000. The school building can accommodate a faculty of no more than 50 but needs at least 20 faculty members to function properly. The school must have at least 12 aides, but the number of teachers must be at least twice the number of aides in order to accommodate the expectations of the community. How many teachers and teacher's aides should be hired in order to minimize salary costs?

73. *Maximizing Animal Support in a Forest.* A certain area of forest is populated by two species of animals, which scientists refer to as A and B for simplicity. The forest supplies two kinds of food, referred to as F_1 and F_2. For one year, species A requires 1 unit of F_1 and 0.5 unit of F_2. Species B requires 0.2 unit of F_1 and 1 unit of F_2. The forest can normally supply at most 600 units of F_1 and 525 units of F_2 per year. What is the maximum total number of these animals that the forest can support?

74. *Maximizing Animal Support in a Forest.* Refer to Exercise 73. If there is a wet spring, then supplies of food increase to 1080 units of F_1 and 810 units of F_2. In this case, what is the maximum total number of these animals that the forest can support?

Technology Connection

Use a graphing calculator to do the following.

75. Exercises 9, 13, and 25

76. Exercises 10, 14, and 24

Collaborative Discussion and Writing

77. Write an applied linear programming problem for a classmate to solve. Devise your problem so that the answer is "The bakery will make a maximum profit when 5 dozen pies and 12 dozen cookies are baked."

78. Describe how the graph of a linear inequality differs from the graph of a linear equation.

Skill Maintenance

Solve.

79. $-5 \le x + 2 < 4$

80. $|x - 3| \ge 2$

81. $x^2 - 2x \le 3$

82. $\dfrac{x - 1}{x + 2} > 4$

Synthesis

Graph the system of inequalities.

83. $y \ge x^2 - 2,$
$\quad y \le 2 - x^2$

84. $y < x + 1,$
$\quad y \ge x^2$

Graph the inequality.

85. $|x + y| \le 1$

86. $|x| + |y| \le 1$

87. $|x| > |y|$

88. $|x - y| > 0$

89. *Allocation of Resources.* Significant Sounds manufactures two types of speaker assemblies. The less expensive assembly, which sells for $350, consists of one midrange speaker and one tweeter. The more expensive speaker assembly, which sells for $600, consists of one woofer, one midrange speaker, and two tweeters. The manufacturer has in stock 44 woofers, 60 midrange speakers, and 90 tweeters. How many of each type of speaker assembly should be made in order to maximize income? What is the maximum income?

90. *Allocation of Resources.* Sitting Pretty Furniture produces chairs and sofas. The chairs require 20 ft of wood, 1 lb of foam rubber, and 2 yd^2 of fabric. The sofas require 100 ft of wood, 50 lb of foam rubber, and 20 yd^2 of fabric. The manufacturer has in stock 1900 ft of wood, 500 lb of foam rubber, and 240 yd^2 of fabric. The chairs can be sold for $80 each and the sofas for $300 each. How many of each should be produced in order to maximize income? What is the maximum income?

8.8

Partial Fractions

• *Decompose rational expressions into partial fractions.*

There are situations in calculus in which it is useful to write a rational expression as a sum of two or more simpler rational expressions. For example, in the equation

$$\frac{4x - 13}{2x^2 + x - 6} = \frac{3}{x + 2} + \frac{-2}{2x - 3},$$

each fraction on the right side is called a **partial fraction**. The expression on the right side is the **partial fraction decomposition** of the rational expression on the left side. In this section, we learn how such decompositions are created.

Partial Fraction Decompositions

The procedure for finding the partial fraction decomposition of a rational expression involves factoring its denominator into linear and quadratic factors.

Procedure for Decomposing a Rational Expression into Partial Fractions

Consider any rational expression $P(x)/Q(x)$ such that $P(x)$ and $Q(x)$ have no common factor other than 1 or -1.

1. If the degree of $P(x)$ is greater than or equal to the degree of $Q(x)$, divide to express $P(x)/Q(x)$ as a quotient $+$ remainder/$Q(x)$ and follow steps (2)–(5) to decompose the resulting rational expression.

2. If the degree of $P(x)$ is less than the degree of $Q(x)$, factor $Q(x)$ into linear factors of the form $(px + q)^n$ and/or quadratic factors of the form $(ax^2 + bx + c)^m$. Any quadratic factor $ax^2 + bx + c$ must be *irreducible*, meaning that it cannot be factored into linear factors with rational coefficients.

3. Assign to each linear factor $(px + q)^n$ the sum of n partial fractions:

$$\frac{A_1}{px + q} + \frac{A_2}{(px + q)^2} + \cdots + \frac{A_n}{(px + q)^n}.$$

4. Assign to each quadratic factor $(ax^2 + bx + c)^m$ the sum of m partial fractions:

$$\frac{B_1 x + C_1}{ax^2 + bx + c} + \frac{B_2 x + C_2}{(ax^2 + bx + c)^2} + \cdots + \frac{B_m x + C_m}{(ax^2 + bx + c)^m}.$$

5. Apply algebraic methods, as illustrated in the following examples, to find the constants in the numerators of the partial fractions.

EXAMPLE 1   Decompose into partial fractions:

$$\frac{4x - 13}{2x^2 + x - 6}.$$

Solution   The degree of the numerator is less than the degree of the denominator. We begin by factoring the denominator: $(x + 2)(2x - 3)$. We know that there are constants A and B such that

$$\frac{4x - 13}{(x + 2)(2x - 3)} = \frac{A}{x + 2} + \frac{B}{2x - 3}.$$

We can use the TABLE feature on a graphing calculator to check a partial fraction decomposition. To check the decomposition in Example 1, we compare values of

$$y_1 = \frac{4x - 13}{2x^2 + x - 6}$$

and

$$y_2 = \frac{3}{x + 2} - \frac{2}{2x - 3}$$

for the same values of x. Since $y_1 = y_2$ for the given values of x as we scroll through the table, the decomposition appears to be correct.

X	Y1	Y2
−1	3.4	3.4
0	2.1667	2.1667
1	3	3
2	−1.25	−1.25
3	−.0667	−.0667
4	.1	.1
5	.14286	.14286

X = −1

To determine A and B, we add the expressions on the right:

$$\frac{4x - 13}{(x + 2)(2x - 3)} = \frac{A(2x - 3) + B(x + 2)}{(x + 2)(2x - 3)}.$$

Next, we equate the numerators:

$$4x - 13 = A(2x - 3) + B(x + 2).$$

Since the last equation containing A and B is true for all x, we can substitute any value of x and still have a true equation. In order to have $2x - 3 = 0$, we choose $x = \frac{3}{2}$. This gives us

$$4\left(\tfrac{3}{2}\right) - 13 = A\left(2 \cdot \tfrac{3}{2} - 3\right) + B\left(\tfrac{3}{2} + 2\right)$$
$$-7 = 0 + \tfrac{7}{2}B.$$

Solving, we obtain $B = -2$.

In order to have $x + 2 = 0$, we choose $x = -2$, which gives us

$$4(-2) - 13 = A[2(-2) - 3] + B(-2 + 2).$$

Solving, we obtain $A = 3$.

The decomposition is as follows:

$$\frac{4x - 13}{2x^2 + x - 6} = \frac{3}{x + 2} + \frac{-2}{2x - 3}, \quad \text{or} \quad \frac{3}{x + 2} - \frac{2}{2x - 3}.$$

To check, we can add to see if we get the expression on the left. ▬

EXAMPLE 2 Decompose into partial fractions:

$$\frac{7x^2 - 29x + 24}{(2x - 1)(x - 2)^2}.$$

Solution The degree of the numerator is 2 and the degree of the denominator is 3, so the degree of the numerator is less than the degree of the denominator. The decomposition has the following form:

$$\frac{7x^2 - 29x + 24}{(2x - 1)(x - 2)^2} = \frac{A}{2x - 1} + \frac{B}{x - 2} + \frac{C}{(x - 2)^2}.$$

As in Example 1, we add the expressions on the right:

$$\frac{7x^2 - 29x + 24}{(2x - 1)(x - 2)^2} = \frac{A(x - 2)^2 + B(2x - 1)(x - 2) + C(2x - 1)}{(2x - 1)(x - 2)^2}.$$

Then we equate the numerators. This gives us

$$7x^2 - 29x + 24 = A(x - 2)^2 + B(2x - 1)(x - 2) + C(2x - 1).$$

Since the equation containing A, B, and C is true for all x, we can substitute any value of x and still have a true equation. In order to have $2x - 1 = 0$, we let $x = \frac{1}{2}$. This gives us

$$7\left(\tfrac{1}{2}\right)^2 - 29 \cdot \tfrac{1}{2} + 24 = A\left(\tfrac{1}{2} - 2\right)^2 + 0.$$

Solving, we obtain $A = 5$.

In order to have $x - 2 = 0$, we let $x = 2$. Substituting gives us

$$7(2)^2 - 29(2) + 24 = 0 + C(2 \cdot 2 - 1).$$

Solving, we obtain $C = -2$.

To find B, we choose any value for x except $\frac{1}{2}$ or 2 and replace A with 5 and C with -2. We let $x = 1$:

$$7 \cdot 1^2 - 29 \cdot 1 + 24 = 5(1 - 2)^2 + B(2 \cdot 1 - 1)(1 - 2)$$
$$+ (-2)(2 \cdot 1 - 1)$$
$$2 = 5 - B - 2$$
$$B = 1.$$

The decomposition is as follows:

$$\frac{7x^2 - 29x + 24}{(2x - 1)(x - 2)^2} = \frac{5}{2x - 1} + \frac{1}{x - 2} - \frac{2}{(x - 2)^2}.$$

EXAMPLE 3   Decompose into partial fractions:

$$\frac{6x^3 + 5x^2 - 7}{3x^2 - 2x - 1}.$$

Solution   The degree of the numerator is greater than that of the denominator. Thus we divide and find an equivalent expression:

$$
\begin{array}{r}
2x + 3 \\
3x^2 - 2x - 1 \overline{)6x^3 + 5x^2 - 7} \\
\underline{6x^3 - 4x^2 - 2x} \\
9x^2 + 2x - 7 \\
\underline{9x^2 - 6x - 3} \\
8x - 4
\end{array}
$$

The original expression is thus equivalent to

$$2x + 3 + \frac{8x - 4}{3x^2 - 2x - 1}.$$

We decompose the fraction to get

$$\frac{8x - 4}{(3x + 1)(x - 1)} = \frac{5}{3x + 1} + \frac{1}{x - 1}.$$

The final result is

$$2x + 3 + \frac{5}{3x + 1} + \frac{1}{x - 1}.$$

Systems of equations can be used to decompose rational expressions. Let's reconsider Example 2.

EXAMPLE 4 Decompose into partial fractions:

$$\frac{7x^2 - 29x + 24}{(2x - 1)(x - 2)^2}.$$

Solution   The decomposition has the following form:

$$\frac{A}{2x - 1} + \frac{B}{x - 2} + \frac{C}{(x - 2)^2}.$$

We first add as in Example 2:

$$\frac{7x^2 - 29x + 24}{(2x - 1)(x - 2)^2} = \frac{A}{2x - 1} + \frac{B}{x - 2} + \frac{C}{(x - 2)^2}$$

$$= \frac{A(x - 2)^2 + B(2x - 1)(x - 2) + C(2x - 1)}{(2x - 1)(x - 2)^2}.$$

Then we equate numerators:

$$7x^2 - 29x + 24$$
$$= A(x - 2)^2 + B(2x - 1)(x - 2) + C(2x - 1)$$
$$= A(x^2 - 4x + 4) + B(2x^2 - 5x + 2) + C(2x - 1)$$
$$= Ax^2 - 4Ax + 4A + 2Bx^2 - 5Bx + 2B + 2Cx - C,$$

or

$$7x^2 - 29x + 24$$
$$= (A + 2B)x^2 + (-4A - 5B + 2C)x + (4A + 2B - C).$$

Next, we equate corresponding coefficients:

$7 = A + 2B,$	The coefficients of the x^2-terms must be the same.
$-29 = -4A - 5B + 2C,$	The coefficients of the x-terms must be the same.
$24 = 4A + 2B - C.$	The constant terms must be the same.

We now have a system of three equations. You should confirm that the solution of the system is

$$A = 5, \qquad B = 1, \quad \text{and} \quad C = -2.$$

The decomposition is as follows:

$$\frac{7x^2 - 29x + 24}{(2x - 1)(x - 2)^2} = \frac{5}{2x - 1} + \frac{1}{x - 2} - \frac{2}{(x - 2)^2}.$$

STUDY TIP

The review icons in the text margins provide references to earlier sections where you can find content related to the concept at hand. Reviewing this earlier content will add to your understanding of the current concept.

SYSTEMS OF EQUATIONS IN THREE VARIABLES

REVIEW SECTION 8.2 or 8.5.

EXAMPLE 5   Decompose into partial fractions:

$$\frac{11x^2 - 8x - 7}{(2x^2 - 1)(x - 3)}.$$

Solution   The decomposition has the following form:

$$\frac{11x^2 - 8x - 7}{(2x^2 - 1)(x - 3)} = \frac{Ax + B}{2x^2 - 1} + \frac{C}{x - 3}.$$

Adding and equating numerators, we get

$$11x^2 - 8x - 7 = (Ax + B)(x - 3) + C(2x^2 - 1)$$
$$= Ax^2 - 3Ax + Bx - 3B + 2Cx^2 - C,$$

or $11x^2 - 8x - 7 = (A + 2C)x^2 + (-3A + B)x + (-3B - C).$

We then equate corresponding coefficients:

$$\begin{aligned} 11 &= A + 2C, &&\text{The coefficients of the } x^2\text{-terms} \\ -8 &= -3A + B, &&\text{The coefficients of the } x\text{-terms} \\ -7 &= -3B - C. &&\text{The constant terms} \end{aligned}$$

We solve this system of three equations and obtain

$$A = 3, \quad B = 1, \quad \text{and} \quad C = 4.$$

The decomposition is as follows:

$$\frac{11x^2 - 8x - 7}{(2x^2 - 1)(x - 3)} = \frac{3x + 1}{2x^2 - 1} + \frac{4}{x - 3}.$$

Exercise Set

Decompose into partial fractions.

1. $\dfrac{x + 7}{(x - 3)(x + 2)}$

2. $\dfrac{2x}{(x + 1)(x - 1)}$

3. $\dfrac{7x - 1}{6x^2 - 5x + 1}$

4. $\dfrac{13x + 46}{12x^2 - 11x - 15}$

5. $\dfrac{3x^2 - 11x - 26}{(x^2 - 4)(x + 1)}$

6. $\dfrac{5x^2 + 9x - 56}{(x - 4)(x - 2)(x + 1)}$

7. $\dfrac{9}{(x + 2)^2(x - 1)}$

8. $\dfrac{x^2 - x - 4}{(x - 2)^3}$

9. $\dfrac{2x^2 + 3x + 1}{(x^2 - 1)(2x - 1)}$

10. $\dfrac{x^2 - 10x + 13}{(x^2 - 5x + 6)(x - 1)}$

11. $\dfrac{x^4 - 3x^3 - 3x^2 + 10}{(x + 1)^2(x - 3)}$

12. $\dfrac{10x^3 - 15x^2 - 35x}{x^2 - x - 6}$

13. $\dfrac{-x^2 + 2x - 13}{(x^2 + 2)(x - 1)}$

14. $\dfrac{26x^2 + 208x}{(x^2 + 1)(x + 5)}$

15. $\dfrac{6 + 26x - x^2}{(2x - 1)(x + 2)^2}$

16. $\dfrac{5x^3 + 6x^2 + 5x}{(x^2 - 1)(x + 1)^3}$

17. $\dfrac{6x^3 + 5x^2 + 6x - 2}{2x^2 + x - 1}$

18. $\dfrac{2x^3 + 3x^2 - 11x - 10}{x^2 + 2x - 3}$

19. $\dfrac{2x^2 - 11x + 5}{(x - 3)(x^2 + 2x - 5)}$

20. $\dfrac{3x^2 - 3x - 8}{(x - 5)(x^2 + x - 4)}$

21. $\dfrac{-4x^2 - 2x + 10}{(3x + 5)(x + 1)^2}$

22. $\dfrac{26x^2 - 36x + 22}{(x - 4)(2x - 1)^2}$

23. $\dfrac{36x + 1}{12x^2 - 7x - 10}$

24. $\dfrac{-17x + 61}{6x^2 + 39x - 21}$

25. $\dfrac{-4x^2 - 9x + 8}{(3x^2 + 1)(x - 2)}$

26. $\dfrac{11x^2 - 39x + 16}{(x^2 + 4)(x - 8)}$

Collaborative Discussion and Writing

27. Describe the two methods used to find the constants in a partial fraction decomposition.

28. What would you say to a classmate who tells you that the partial fraction decomposition of

$$\frac{3x^2 - 8x + 9}{(x + 3)(x^2 - 5x + 6)}$$

is

$$\frac{2}{x + 3} + \frac{x - 1}{x^2 - 5x + 6}?$$

Explain.

29. Explain the error in the following process.

$$\frac{x^2 + 4}{(x + 2)(x + 1)} = \frac{A}{x + 2} + \frac{B}{x + 1}$$
$$= \frac{A(x + 1) + B(x + 2)}{(x + 2)(x + 1)}$$

Then

$$x^2 + 4 = A(x + 1) + B(x + 2).$$

When $x = -1$:

$$(-1)^2 + 4 = A(-1 + 1) + B(-1 + 2)$$
$$5 = B.$$

When $x = -2$:

$$(-2)^2 + 4 = A(-2 + 1) + B(-2 + 2)$$
$$8 = -A$$
$$-8 = A.$$

Thus,

$$\frac{x^2 + 4}{(x + 2)(x + 1)} = \frac{-8}{x + 2} + \frac{5}{x + 1}.$$

Skill Maintenance

Find the zeros of the polynomial function.

30. $f(x) = x^3 - 3x^2 + x - 3$

31. $f(x) = x^3 + x^2 - 3x - 2$

32. $f(x) = x^4 - x^3 - 5x^2 - x - 6$

33. $f(x) = x^3 + 5x^2 + 5x - 3$

Synthesis

Decompose into partial fractions.

34. $\dfrac{9x^3 - 24x^2 + 48x}{(x - 2)^4(x + 1)}$

[*Hint:* Let the expression equal

$$\frac{A}{x + 1} + \frac{P(x)}{(x - 2)^4}$$

and find $P(x)$.]

35. $\dfrac{x}{x^4 - a^4}$

36. $\dfrac{1}{e^{-x} + 3 + 2e^x}$

37. $\dfrac{1 + \ln x^2}{(\ln x + 2)(\ln x - 3)^2}$

Chapter Summary and Review

Important Properties and Formulas

Row-Equivalent Operations

1. Interchange any two rows.

2. Multiply each entry in a row by the same nonzero constant.

3. Add a nonzero multiple of one row to another row.

Row-Echelon Form

1. If a row does not consist entirely of 0's, then the first nonzero element in the row is a 1 (called a leading 1).

2. For any two successive nonzero rows, the leading 1 in the lower row is farther to the right than the leading 1 in the higher row.

(continued)

3. All the rows consisting entirely of 0's are at the bottom of the matrix.

If a fourth property is also satisfied, a matrix is said to be in reduced row-echelon form:

4. Each column that contains a leading 1 has 0's everywhere else.

Properties of Matrix Addition and Scalar Multiplication

For any $m \times n$ matrices **A**, **B**, and **C** and any scalars k and l:

Commutative Property of Addition:
  $\mathbf{A} + \mathbf{B} = \mathbf{B} + \mathbf{A}$.

Associative Property of Addition:
  $\mathbf{A} + (\mathbf{B} + \mathbf{C}) = (\mathbf{A} + \mathbf{B}) + \mathbf{C}$.

Associative Property of Scalar Multiplication:
  $(kl)\mathbf{A} = k(l\mathbf{A})$.

Additive Identity Property:
There exists a unique matrix **0** such that
$\mathbf{A} + \mathbf{0} = \mathbf{0} + \mathbf{A} = \mathbf{A}$.

Additive Inverse Property:
There exists a unique matrix $-\mathbf{A}$ such that
$\mathbf{A} + (-\mathbf{A}) = -\mathbf{A} + \mathbf{A} = \mathbf{0}$.

Distributive Properties:
  $k(\mathbf{A} + \mathbf{B}) = k\mathbf{A} + k\mathbf{B}$,
  $(k + l)\mathbf{A} = k\mathbf{A} + l\mathbf{A}$.

Properties of Matrix Multiplication

For matrices **A**, **B**, and **C**, assuming that the indicated operation is possible:

Associative Property of Multiplication:
  $\mathbf{A}(\mathbf{BC}) = (\mathbf{AB})\mathbf{C}$.

Distributive Properties:
  $\mathbf{A}(\mathbf{B} + \mathbf{C}) = \mathbf{AB} + \mathbf{AC}$,
  $(\mathbf{B} + \mathbf{C})\mathbf{A} = \mathbf{BA} + \mathbf{CA}$.

Matrix Solutions of Systems of Equations

For a system of n linear equations in n variables, $\mathbf{AX} = \mathbf{B}$, if **A** is an invertible matrix, then the unique solution of the system is given by $\mathbf{X} = \mathbf{A}^{-1}\mathbf{B}$.

Determinant of a 2 × 2 Matrix

The **determinant** of the matrix $\begin{bmatrix} a & c \\ b & d \end{bmatrix}$ is

denoted $\begin{vmatrix} a & c \\ b & d \end{vmatrix}$ and is defined as

$$\begin{vmatrix} a & c \\ b & d \end{vmatrix} = ad - bc.$$

Determinant of Any Square Matrix

For any square matrix **A** of order $n \times n$ $(n > 1)$, we define the **determinant** of **A**, denoted $|\mathbf{A}|$, as follows. Choose any row or column. Multiply each element in that row or column by its cofactor and add the results. The determinant of a 1×1 matrix is simply the element of the matrix. The value of a determinant will be the same, no matter how it is evaluated.

Cramer's Rule for 2 × 2 Systems

The solution of the system of equations

$$a_1 x + b_1 y = c_1,$$
$$a_2 x + b_2 y = c_2$$

is given by

$$x = \frac{D_x}{D}, \qquad y = \frac{D_y}{D},$$

where

$$D = \begin{vmatrix} a_1 & b_1 \\ a_2 & b_2 \end{vmatrix}, \qquad D_x = \begin{vmatrix} c_1 & b_1 \\ c_2 & b_2 \end{vmatrix},$$

$$D_y = \begin{vmatrix} a_1 & c_1 \\ a_2 & c_2 \end{vmatrix}, \quad \text{and} \quad D \neq 0.$$

Cramer's Rule for 3 × 3 Systems

The solution of the system of equations

$$a_1 x + b_1 y + c_1 z = d_1,$$
$$a_2 x + b_2 y + c_2 z = d_2,$$
$$a_3 x + b_3 y + c_3 z = d_3$$

is given by

$$x = \frac{D_x}{D}, \qquad y = \frac{D_y}{D}, \qquad z = \frac{D_z}{D},$$

where

$$D = \begin{vmatrix} a_1 & b_1 & c_1 \\ a_2 & b_2 & c_2 \\ a_3 & b_3 & c_3 \end{vmatrix}, \qquad D_x = \begin{vmatrix} d_1 & b_1 & c_1 \\ d_2 & b_2 & c_2 \\ d_3 & b_3 & c_3 \end{vmatrix},$$

$$D_y = \begin{vmatrix} a_1 & d_1 & c_1 \\ a_2 & d_2 & c_2 \\ a_3 & d_3 & c_3 \end{vmatrix}, \qquad D_z = \begin{vmatrix} a_1 & b_1 & d_1 \\ a_2 & b_2 & d_2 \\ a_3 & b_3 & d_3 \end{vmatrix},$$

and $D \neq 0$.

To Graph a Linear Inequality in Two Variables:

1. Replace the inequality symbol with an equals sign and graph this related equation. If the inequality symbol is $<$ or $>$, draw the line dashed. If the inequality symbol is $\leq$ or $\geq$, draw the line solid.

2. The graph consists of a half-plane on one side of the line and, if the line is solid, the line as well. To determine which half-plane to shade, test a point not on the line in the original inequality. If that point is a solution, shade the half-plane containing that point. If not, shade the opposite half-plane.

Linear Programming Procedure

To find the maximum or minimum value of a linear objective function subject to a set of constraints:

1. Graph the region of feasible solutions.

2. Determine the coordinates of the vertices of the region.

3. Evaluate the objective function at each vertex. The largest and smallest of those values are the maximum and minimum values of the function, respectively.

Procedure for Decomposing a Rational Expression into Partial Fractions

Consider any rational expression $P(x)/Q(x)$ such that $P(x)$ and $Q(x)$ have no common factor other than 1 or -1.

1. If the degree of $P(x)$ is greater than or equal to the degree of $Q(x)$, divide to express $P(x)/Q(x)$ as a quotient $+$ remainder/$Q(x)$ and follow steps (2)–(5) to decompose the resulting rational expression.

2. If the degree of $P(x)$ is less than the degree of $Q(x)$, factor $Q(x)$ into linear factors of the form $(px + q)^n$ and/or quadratic factors of the form $(ax^2 + bx + c)^m$. Any quadratic factor $ax^2 + bx + c$ must be irreducible, meaning that it cannot be factored into linear factors with rational coefficients.

3. Assign to each linear factor $(px + q)^n$ the sum of n partial fractions:

$$\frac{A_1}{px + q} + \frac{A_2}{(px + q)^2} + \cdots + \frac{A_n}{(px + q)^n}.$$

4. Assign to each quadratic factor $(ax^2 + bx + c)^m$ the sum of m partial fractions:

$$\frac{B_1 x + C_1}{ax^2 + bx + c} + \frac{B_2 x + C_2}{(ax^2 + bx + c)^2} + \cdots + \frac{B_m x + C_m}{(ax^2 + bx + c)^m}.$$

5. Apply algebraic methods to find the constants in the numerators of the partial fractions.

REVIEW EXERCISES

In Exercises 1–8, match the equations or inequalities with one of the graphs (a)–(h), which follow.

a)

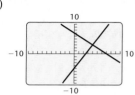

b)

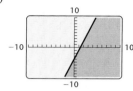

c)

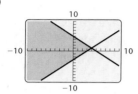

d)

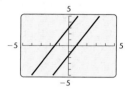

e)

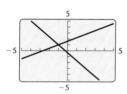

f)

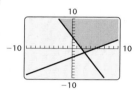

g)

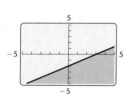

h)

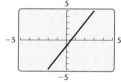

1. $x + y = 7,$
$2x - y = 5$

2. $3x - 5y = -8,$
$4x + 3y = -1$

3. $y = 2x - 1,$
$4x - 2y = 2$

4. $6x - 3y = 5,$
$y = 2x + 3$

5. $y \leq 3x - 4$

6. $2x - 3y \geq 6$

7. $x - y \leq 3,$
$x + y \leq 5$

8. $2x + y \geq 4,$
$3x - 5y \leq 15$

Solve.

9. $5x - 3y = -4,$
$3x - y = -4$

10. $2x + 3y = 2,$
$5x - y = -29$

11. $x + 5y = 12,$
$5x + 25y = 12$

12. $x - y = -2,$
$-3x + 3y = 6$

13. $2x - 4y + 3z = -3,$
$-5x + 2y - z = 7,$
$3x + 2y - 2z = 4$

14. $x + 5y + 3z = 0,$
$3x - 2y + 4z = 0,$
$2x + 3y - z = 0$

15. $x - y = 5,$
$y - z = 6,$
$z - w = 7,$
$x + w = 8$

16. Classify each of the systems in Exercises 9–15 as consistent or inconsistent.

17. Classify each of the systems in Exercises 9–15 as dependent or independent.

Solve the system of equations using Gaussian elimination or Gauss–Jordan elimination.

18. $x + 2y = 5,$
$2x - 5y = -8$

19. $3x + 4y + 2z = 3,$
$5x - 2y - 13z = 3,$
$4x + 3y - 3z = 6$

20. $3x + 5y + z = 0,$
$2x - 4y - 3z = 0,$
$x + 3y + z = 0$

21. $w + x + y + z = -2,$
$-3w - 2x + 3y + 2z = 10,$
$2w + 3x + 2y - z = -12,$
$2w + 4x - y + z = 1$

22. *Coins.* The value of 75 coins, consisting of nickels and dimes, is $5.95. How many of each kind are there?

23. *Investment.* The Mendez family invested $5000, part at 10% and the remainder at 10.5%. The annual income from both investments is $517. What is the amount invested at each rate?

24. *Nutrition.* A dietician must plan a breakfast menu that provides 460 Cal, 9 g of fat, and 55 mg of calcium. One plain bagel contains 200 Cal, 2 g of fat, and 29 mg of calcium. A one-tablespoon serving of cream cheese contains 100 Cal, 10 g of fat, and 24 mg of calcium. One banana contains 105 Cal, 1 g of fat, and 7 mg of calcium. (*Source: Home and Garden Bulletin No. 72*, U.S. Government Printing Office, Washington, D.C. 20402) How many servings of each are required to provide the desired nutritional values?

25. *Test Scores.* A student has a total of 225 on three tests. The sum of the scores on the first and second tests exceeds the score on the third test by 61. The first score exceeds the second by 6. Find the three scores.

26. *Ice Milk Consumption.* The table below shows the per capita ice milk consumption, in pounds, in the United States, represented as years since 1990.

YEAR, x	ICE MILK CONSUMPTION (IN POUNDS)
1990, 0	7.7
1993, 3	6.9
1996, 6	7.6

Source: U.S. Department of Agriculture, Economic Research Service, *Food Consumption, Prices, and Expenditures,* annual

a) Use a system of equations to fit a quadratic function $f(x) = ax^2 + bx + c$ to the data.

b) Use the function to estimate the per capita ice milk consumption in 2005.

For Exercises 27–34, let

$$A = \begin{bmatrix} 1 & -1 & 0 \\ 2 & 3 & -2 \\ -2 & 0 & 1 \end{bmatrix},$$

$$B = \begin{bmatrix} -1 & 0 & 6 \\ 1 & -2 & 0 \\ 0 & 1 & -3 \end{bmatrix},$$

and

$$C = \begin{bmatrix} -2 & 0 \\ 1 & 3 \end{bmatrix}.$$

Find each of the following, if possible.

27. $A + B$ **28.** $-3A$

29. $-A$ **30.** AB

31. $B + C$ **32.** $A - B$

33. $2A - B$ **34.** $A + 3B$

35. *Food Service Management.* The table below shows the cost per serving, in cents, for items on four menus that are served at an elder-care facility.

MENU	MEAT	POTATO	VEGETABLE	SALAD	DESSERT
1	46.1	5.9	10.1	8.5	11.4
2	54.6	4.6	9.6	7.6	10.6
3	48.9	5.5	12.7	9.4	9.3
4	51.3	4.8	11.3	6.9	12.7

On a particular day, a dietician orders 32 meals from menu 1, 19 from menu 2, 43 from menu 3, and 38 from menu 4.

a) Write the information in the table as a 4×5 matrix **M**.

b) Write a row matrix **N** that represents the number of each menu ordered.

c) Find the product **NM**.

d) State what the entries of **NM** represent.

Find A^{-1}, *if it exists.*

36. $A = \begin{bmatrix} -2 & 0 \\ 1 & 3 \end{bmatrix}$

37. $A = \begin{bmatrix} 0 & 0 & 3 \\ 0 & -2 & 0 \\ 4 & 0 & 0 \end{bmatrix}$

38. $A = \begin{bmatrix} 1 & 0 & 0 & 0 \\ 0 & 4 & -5 & 0 \\ 0 & 2 & 2 & 0 \\ 0 & 0 & 0 & 1 \end{bmatrix}$

39. Write a matrix equation equivalent to this system of equations:

$$\begin{aligned} 3x - 2y + 4z &= 13, \\ x + 5y - 3z &= 7, \\ 2x - 3y + 7z &= -8. \end{aligned}$$

Solve the system of equations using the inverse of the coefficient matrix of the equivalent matrix equation.

40. $\begin{aligned} 2x + 3y &= 5, \\ 3x + 5y &= 11 \end{aligned}$

41. $\begin{aligned} 5x - y + 2z &= 17, \\ 3x + 2y - 3z &= -16, \\ 4x - 3y - z &= 5 \end{aligned}$

42. $\begin{aligned} w - x - y + z &= -1, \\ 2w + 3x - 2y - z &= 2, \\ -w + 5x + 4y - 2z &= 3, \\ 3w - 2x + 5y + 3z &= 4 \end{aligned}$

Evaluate the determinant.

43. $\begin{vmatrix} 1 & -2 \\ 3 & 4 \end{vmatrix}$ **44.** $\begin{vmatrix} \sqrt{3} & -5 \\ -3 & -\sqrt{3} \end{vmatrix}$

45. $\begin{vmatrix} 2 & -1 & 1 \\ 1 & 2 & -1 \\ 3 & 4 & -3 \end{vmatrix}$ **46.** $\begin{vmatrix} 1 & -1 & 2 \\ -1 & 2 & 0 \\ -1 & 3 & 1 \end{vmatrix}$

Solve using Cramer's rule.

47. $\begin{aligned} 5x - 2y &= 19, \\ 7x + 3y &= 15 \end{aligned}$ **48.** $\begin{aligned} x + y &= 4, \\ 4x + 3y &= 11 \end{aligned}$

49. $3x - 2y + z = 5,$
$4x - 5y - z = -1,$
$3x + 2y - z = 4$

50. $2x - y - z = 2,$
$3x + 2y + 2z = 10,$
$x - 5y - 3z = -2$

Graph.

51. $y \leq 3x + 6$

52. $4x - 3y \geq 12$

53. Graph this system of inequalities and find the coordinates of any vertices formed.

$$2x + y \geq 9,$$
$$4x + 3y \geq 23,$$
$$x + 3y \geq 8,$$
$$x \geq 0,$$
$$y \geq 0$$

54. Find the maximum and minimum values of $T = 6x + 10y$ subject to

$$x + y \leq 10,$$
$$5x + 10y \geq 50,$$
$$x \geq 2,$$
$$y \geq 0.$$

55. *Maximizing a Test Score.* Marita is taking a test that contains questions in group A worth 7 points each and questions in group B worth 12 points each. The total number of questions answered must be at least 8. If Marita knows that group A questions take 8 min each and group B questions take 10 min each and the maximum time for the test is 80 min, how many questions from each group must she answer correctly in order to maximize her score? What is the maximum score?

Decompose into partial fractions.

56. $\dfrac{5}{(x + 2)^2(x + 1)}$

57. $\dfrac{-8x + 23}{2x^2 + 5x - 12}$

Collaborative Discussion and Writing

58. Write a problem for a classmate to solve that can be translated to a system of equations. Devise the problem so that the solution is "The caterer sold 20 cheese trays and 35 seafood trays."

59. For square matrices **A** and **B**, is it true, in general, that $(\mathbf{AB})^2 = \mathbf{A}^2\mathbf{B}^2$? Explain.

Synthesis

60. One year, Don invested a total of $40,000, part at 12%, part at 13%, and the rest at $14\frac{1}{2}$%. The total amount of interest received on the investments was $5370. The interest received on the $14\frac{1}{2}$% investment was $1050 more than the interest received on the 13% investment. How much was invested at each rate?

Solve.

61. $\dfrac{2}{3x} + \dfrac{4}{5y} = 8,$
$\dfrac{5}{4x} - \dfrac{3}{2y} = -6$

62. $\dfrac{3}{x} - \dfrac{4}{y} + \dfrac{1}{z} = -2,$
$\dfrac{5}{x} + \dfrac{1}{y} - \dfrac{2}{z} = 1,$
$\dfrac{7}{x} + \dfrac{3}{y} + \dfrac{2}{z} = 19$

Graph.

63. $|x| - |y| \leq 1$

64. $|xy| > 1$

Chapter Test

Solve. Use any method.

1. $3x + 2y = 1,$
$2x - y = -11$

2. $2x - 3y = 8,$
$5x - 2y = 9$

3. $4x + 2y + z = 4,$
$3x - y + 5z = 4,$
$5x + 3y - 3z = -2$

4. Classify the system of equations in Exercise 1 as consistent or inconsistent.

5. Classify the system of equations in Exercise 2 as dependent or independent.

6. *Ticket Sales.* One evening 750 tickets were sold for Shortridge Community College's spring musical. Tickets cost $3 for students and $5 for nonstudents. Total receipts were $3066. How many of each type of ticket were sold?

For Exercises 7–12, let

$$A = \begin{bmatrix} 1 & -1 & 3 \\ -2 & 5 & 2 \end{bmatrix}, \quad B = \begin{bmatrix} -5 & 1 \\ -2 & 4 \end{bmatrix}, \quad \text{and}$$

$$C = \begin{bmatrix} 3 & -4 \\ -1 & 0 \end{bmatrix}.$$

Find each of the following, if possible.

7. $B + C$

8. $A - C$

9. CB

10. AB

11. $2A$

12. C^{-1}

13. *Food Service Management.* The table below shows the cost per serving, in cents, for items on three lunch menus served at a senior citizens' center.

MENU	MAIN DISH	SIDE DISH	DESSERT
1	49	10	13
2	43	12	11
3	51	8	12

On a particular day, 26 Menu 1 meals, 18 Menu 2 meals, and 23 Menu 3 meals are served.

a) Write the information in the table as a 3 × 3 matrix **M**.

b) Write a row matrix **N** that represents the number of each menu served.

c) Find the product **NM**.

d) State what the entries of **NM** represent.

14. Write a matrix equation equivalent to the system of equations

$$3x - 4y + 2z = -8,$$
$$2x + 3y + z = 7,$$
$$x - 5y - 3z = 3.$$

15. Evaluate the determinant

$$\begin{vmatrix} 2 & -1 & 4 \\ -3 & 1 & -2 \\ 5 & 3 & -1 \end{vmatrix}.$$

16. Solve using Cramer's rule. Show your work.

$$5x + 2y = -1,$$
$$7x + 6y = 1$$

17. Graph $3x + 4y \le -12$.

18. Find the maximum and minimum values of $Q = 2x + 3y$ subject to

$$x + y \le 6,$$
$$2x - 3y \ge -3,$$
$$x \ge 1,$$
$$y \ge 0.$$

19. *Maximizing Profit.* Casey's Cakes prepares pound cakes and carrot cakes. In a given week, at most 100 cakes can be prepared, of which 25 pound cakes and 15 carrot cakes are required by regular customers. The profit from each pound cake is $3 and the profit from each carrot cake is $4. How many of each type of cake should be prepared in order to maximize the profit?

20. Decompose into partial fractions:

$$\frac{3x - 11}{x^2 + 2x - 3}.$$

Synthesis

21. Three solutions of the equation $Ax - By = Cz - 8$ are $(2, -2, 2)$, $(-3, -1, 1)$, and $(4, 2, 9)$. Find A, B, and C.

Conic Sections 9

*I*n this chapter, we study *conic sections*. These curves are formed by the intersection of a cone and a plane. Conic sections and their properties were first studied by the Greeks. We also study systems of equations in which at least one equation is nonlinear. Both conic sections and nonlinear systems of equations have many applications, as we will see.

APPLICATION

For a student recreation building at Southport Community College, an architect wants to lay out a rectangular piece of land that has a perimeter of 204 m and an area of 2565 m². Determine the dimensions of the piece of land.

This problem appears as Example 5 in Section 9.4.

9.1

The Parabola

• *Given an equation of a parabola, complete the square, if necessary, and then find the vertex, the focus, and the directrix and graph the parabola.*

A **conic section** is formed when a right circular cone with two parts, called *nappes*, is intersected by a plane. One of four types of curves can be formed: a parabola, a circle, an ellipse, or a hyperbola.

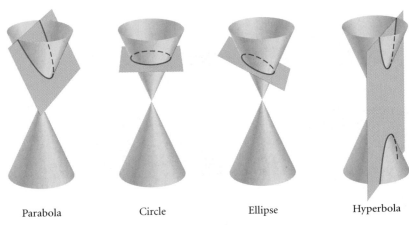

Parabola   Circle   Ellipse   Hyperbola

Conic Sections

Parabolas

Conic sections can be defined algebraically using second-degree equations of the form $Ax^2 + Bxy + Cy^2 + Dx + Ey + F = 0$. In addition, they can be defined geometrically as a set of points that satisfy certain conditions.

In Section 2.4, we saw that the graph of the quadratic function $f(x) = ax^2 + bx + c$, $a \neq 0$, is a parabola. A parabola can be defined geometrically.

Parabola

A **parabola** is the set of all points in a plane equidistant from a fixed line (the **directrix**) and a fixed point not on the line (the **focus**).

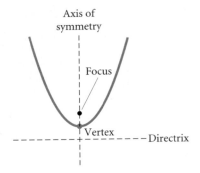

The line that is perpendicular to the directrix and contains the focus is the **axis of symmetry.** The **vertex** is the midpoint of the segment between the focus and the directrix. (See the figure at left.)

Let's derive the standard equation of a parabola with vertex $(0, 0)$ and directrix $y = -p$, where $p > 0$. We place the coordinate axes as shown in the figure at the top of the following page. The y-axis is the axis of symmetry and contains the focus *F*. The distance from the focus to the vertex is the same as the distance from the vertex to the directrix. Thus the coordinates of *F* are (0, *p*).

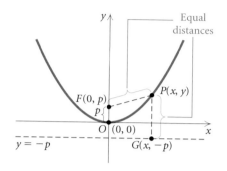

Let *P*(*x*, *y*) be any point of the parabola and consider $\overline{PG}$ perpendicular to the line *y* = −*p*. The coordinates of *G* are (*x*, −*p*). By the definition of a parabola,

$$PF = PG.$$ The distance from *P* to the focus is the same as the distance from *P* to the directrix.

Then using the distance formula, we have

$$\sqrt{(x - 0)^2 + (y - p)^2} = \sqrt{(x - x)^2 + [y - (-p)]^2}$$
$$x^2 + y^2 - 2py + p^2 = y^2 + 2py + p^2$$ Squaring both sides and squaring the binomials
$$x^2 = 4py.$$

We have shown that if *P*(*x*, *y*) is on the parabola shown above, then its coordinates satisfy this equation. The converse is also true, but we will not prove it here.

Note that if *p* > 0, as above, the graph opens up. If *p* < 0, the graph opens down.

The equation of a parabola with vertex (0, 0) and directrix *x* = −*p* is derived similarly. Such a parabola opens either right (*p* > 0) or left (*p* < 0).

Standard Equation of a Parabola with Vertex at the Origin

The standard equation of a parabola with vertex (0, 0) and directrix *y* = −*p* is

$$x^2 = 4py.$$

The focus is (0, *p*) and the *y*-axis is the axis of symmetry.

The standard equation of a parabola with vertex (0, 0) and directrix *x* = −*p* is

$$y^2 = 4px.$$

The focus is (*p*, 0) and the *x*-axis is the axis of symmetry.

EXAMPLE 1 Find the focus and the directrix of the parabola $y = -\frac{1}{12}x^2$. Then graph the parabola.

Solution We write $y = -\frac{1}{12}x^2$ in the form $x^2 = 4py$:

$$-\frac{1}{12}x^2 = y \qquad\qquad \text{Given equation}$$
$$x^2 = -12y \qquad\qquad \text{Multiplying by } -12 \text{ on both sides}$$
$$x^2 = 4(-3)y. \qquad\quad \text{Standard form}$$

Thus, $p = -3$, so the focus is $(0, p)$, or $(0, -3)$. The directrix is $y = -p = -(-3) = 3$.

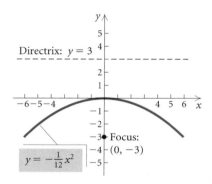

Technology Connection

We can use a graphing calculator to graph parabolas. Consider the parabola in Example 2, $y^2 = 20x$. It might be necessary to solve this equation for y before entering it in the calculator:

$$y^2 = 20x$$
$$y = \pm\sqrt{20x}.$$

We now graph $y_1 = \sqrt{20x}$ and $y_2 = -\sqrt{20x}$ in a squared viewing window. On some graphing calculators, it is possible to graph $y_1 = \sqrt{20x}$ and $y_2 = -y_1$ by using the Y-VARS menu.

EXAMPLE 2 Find an equation of the parabola with vertex $(0, 0)$ and focus $(5, 0)$. Then graph the parabola.

Solution The focus is on the x-axis so the line of symmetry is the x-axis. Thus the equation is of the type

$$y^2 = 4px.$$

Since the focus is 5 units to the right of the vertex, $p = 5$ and the equation is

$$y^2 = 4(5)x, \quad \text{or} \quad y^2 = 20x.$$

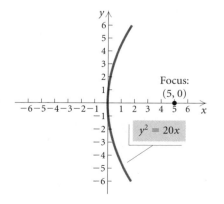

Finding Standard Form by Completing the Square

If a parabola with vertex at the origin is translated horizontally $|h|$ units and vertically $|k|$ units, it has an equation as follows.

Standard Equation of a Parabola with Vertex (h, k) and Vertical Axis of Symmetry

The standard equation of a parabola with vertex (h, k) and vertical axis of symmetry is

$$(x - h)^2 = 4p(y - k),$$

where the vertex is (h, k), the focus is $(h, k + p)$, and the directrix is $y = k - p$.

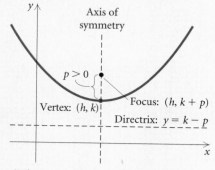

(When $p < 0$, the parabola opens down.)

Standard Equation of a Parabola with Vertex (h, k) and Horizontal Axis of Symmetry

The standard equation of a parabola with vertex (h, k) and horizontal axis of symmetry is

$$(y - k)^2 = 4p(x - h),$$

where the vertex is (h, k), the focus is $(h + p, k)$, and the directrix is $x = h - p.$

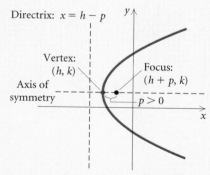

(When $p < 0$, the parabola opens left.)

COMPLETING THE SQUARE

REVIEW SECTION 2.3.

We can complete the square on equations of the form

$$y = ax^2 + bx + c \quad \text{or} \quad x = ay^2 + by + c$$

in order to write them in standard form.

EXAMPLE 3 For the parabola

$$x^2 + 6x + 4y + 5 = 0,$$

find the vertex, the focus, and the directrix. Then draw the graph.

Solution We first complete the square:

$$x^2 + 6x + 4y + 5 = 0$$

$$x^2 + 6x \qquad = -4y - 5 \qquad \text{Subtracting } 4y \text{ and } 5 \text{ on both sides}$$

$$x^2 + 6x + 9 = -4y - 5 + 9 \qquad \text{Adding } 9 \text{ on both sides to complete the square on the left side}$$

$$x^2 + 6x + 9 = -4y + 4$$

$$(x + 3)^2 = -4(y - 1) \qquad \text{Factoring}$$

$$[(x - (-3)]^2 = 4(-1)(y - 1). \qquad \text{Writing standard form: } (x - h)^2 = 4p(y - k)$$

We now have the following:

Vertex (h, k): $\qquad (-3, 1);$

Focus $(h, k + p)$: $\qquad (-3, 1 + (-1))$, or $(-3, 0);$

Directrix $y = k - p$: $\quad y = 1 - (-1)$, or $y = 2.$

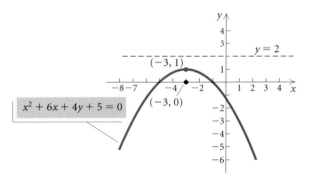

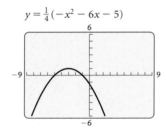

$y = \frac{1}{4}(-x^2 - 6x - 5)$

Technology Connection

We can check the graph in Example 3 on a graphing calculator using a squared viewing window. It might be necessary to solve for y first:

$$x^2 + 6x + 4y + 5 = 0$$

$$4y = -x^2 - 6x - 5$$

$$y = \tfrac{1}{4}(-x^2 - 6x - 5).$$

EXAMPLE 4 For the parabola

$$y^2 - 2y - 8x - 31 = 0,$$

find the vertex, the focus, and the directrix. Then draw the graph.

Solution We first complete the square:

$$y^2 - 2y - 8x - 31 = 0$$
$$y^2 - 2y \qquad\quad = 8x + 31 \qquad\qquad \text{Adding } 8x \text{ and } 31 \text{ on both sides}$$
$$y^2 - 2y + \ 1 = 8x + 31 + 1 \qquad \text{Adding 1 on both sides to complete the square on the left side}$$

$$y^2 - 2y + 1 = 8x + 32$$
$$(y - 1)^2 = 8(x + 4)$$
$$(y - 1)^2 = 4(2)[x - (-4)]. \qquad \text{Writing standard form: } (y - k)^2 = 4p(x - h)$$

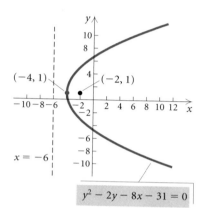

$x = -6$

$(-4, 1)$ $(-2, 1)$

$y^2 - 2y - 8x - 31 = 0$

We now have the following:

Vertex (h, k): $(-4, 1)$;

Focus $(h + p, k)$: $(-4 + 2, 1)$, or $(-2, 1)$;

Directrix $x = h - p$: $x = -4 - 2$, or $x = -6$.

Technology Connection

We can check the graph in Example 4 on a graphing calculator using a squared viewing window. We solve the original equation for y using the quadratic formula:

$$y^2 - 2y - 8x - 31 = 0$$
$$y^2 - 2y + (-8x - 31) = 0$$
$$a = 1, \quad b = -2, \quad c = -8x - 31$$
$$y = \frac{-(-2) \pm \sqrt{(-2)^2 - 4 \cdot 1(-8x - 31)}}{2 \cdot 1}$$
$$y = \frac{2 \pm \sqrt{32x + 128}}{2}.$$

We now graph

$$y_1 = \frac{2 + \sqrt{32x + 128}}{2} \quad \text{and} \quad y_2 = \frac{2 - \sqrt{32x + 128}}{2}.$$

$y^2 - 2y - 8x - 31 = 0$

$$y_1 = \frac{2 + \sqrt{32x + 128}}{2},$$
$$y_2 = \frac{2 - \sqrt{32x + 128}}{2}$$

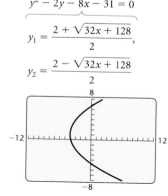

Applications

Parabolas have many applications. For example, cross sections of car headlights, flashlights, and searchlights are parabolas. The bulb is located at the focus and light from that point is reflected outward parallel to the axis of symmetry. Satellite dishes and field microphones used at sporting events often have parabolic cross sections. Incoming radio waves or sound waves parallel to the axis are reflected into the focus. Cables hung between structures in suspension bridges, such as the Golden Gate Bridge, form parabolas. When a cable supports only its own weight, however, it forms a curve called a *catenary* rather than a parabola.

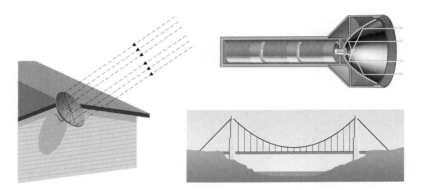

Exercise Set 9.1

In Exercises 1–6, match the equation with one of the graphs (a)–(f), which follow.

a)

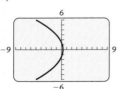

b)

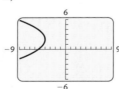

c)

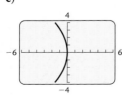

d)

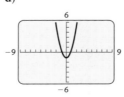

e)

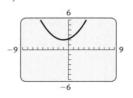

f)

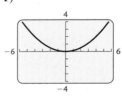

1. $x^2 = 8y$ **2.** $y^2 = -10x$

3. $(y - 2)^2 = -3(x + 4)$ **4.** $(x + 1)^2 = 5(y - 2)$

5. $13x^2 - 8y - 9 = 0$ **6.** $41x + 6y^2 = 12$

Find the vertex, the focus, and the directrix. Then draw the graph.

7. $x^2 = 20y$ **8.** $x^2 = 16y$

9. $y^2 = -6x$ **10.** $y^2 = -2x$

11. $x^2 - 4y = 0$

12. $y^2 + 4x = 0$

13. $x = 2y^2$

14. $y = \frac{1}{2}x^2$

Find an equation of a parabola satisfying the given conditions.

15. Focus $(4, 0)$, directrix $x = -4$

16. Focus $\left(0, \frac{1}{4}\right)$, directrix $y = -\frac{1}{4}$

17. Focus $(0, -\pi)$, directrix $y = \pi$

18. Focus $(-\sqrt{2}, 0)$, directrix $x = \sqrt{2}$

19. Focus $(3, 2)$, directrix $x = -4$

20. Focus $(-2, 3)$, directrix $y = -3$

Find the vertex, the focus, and the directrix. Then draw the graph.

21. $(x + 2)^2 = -6(y - 1)$

22. $(y - 3)^2 = -20(x + 2)$

23. $x^2 + 2x + 2y + 7 = 0$

24. $y^2 + 6y - x + 16 = 0$

25. $x^2 - y - 2 = 0$

26. $x^2 - 4x - 2y = 0$

27. $y = x^2 + 4x + 3$

28. $y = x^2 + 6x + 10$

29. $y^2 - y - x + 6 = 0$

30. $y^2 + y - x - 4 = 0$

31. *Satellite Dish.* An engineer designs a satellite dish with a parabolic cross section. The dish is 15 ft wide at the opening and the focus is placed 4 ft from the vertex.

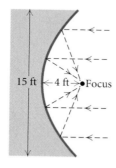

15 ft ← 4 ft →●Focus

a) Position a coordinate system with the origin at the vertex and the x-axis on the parabola's axis of symmetry and find an equation of the parabola.

b) Find the depth of the satellite dish at the vertex.

32. *Headlight Mirror.* A car headlight mirror has a parabolic cross section with diameter 6 in. and depth 1 in.

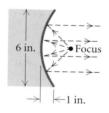

6 in. ●Focus

← 1 in.

a) Position a coordinate system with the origin at the vertex and the x-axis on the parabola's axis of symmetry and find an equation of the parabola.

b) How far from the vertex should the bulb be positioned if it is to be placed at the focus?

33. *Spotlight.* A spotlight has a parabolic cross section that is 4 ft wide at the opening and 1.5 ft deep at the vertex. How far from the vertex is the focus?

34. *Field Microphone.* A field microphone used at a football game has a parabolic cross section and is 18 in. deep. The focus is 4 in. from the vertex. Find the width of the microphone at the opening.

Technology Connection

Use a graphing calculator to find the vertex, the focus, and the directrix of each of the following.

35. $4.5x^2 - 7.8x + 9.7y = 0$

36. $134.1y^2 + 43.4x - 316.6y - 122.4 = 0$

Collaborative Discussion and Writing

37. Is a parabola always the graph of a function? Why or why not?

38. Explain how the distance formula is used to find the standard equation of a parabola.

Skill Maintenance

Complete the square and write the result as the square of a binomial.

39. $x^2 + 10x$

40. $y^2 - 9y$

Find the center and the radius of the circle.

41. $(x - 1)^2 + (y + 2)^2 = 9$

42. $(x + 3)^2 + (y - 5)^2 = 36$

Synthesis

43. Find an equation of the parabola with a vertical axis of symmetry and vertex $(-1, 2)$ and containing the point $(-3, 1)$.

44. Find an equation of a parabola with a horizontal axis of symmetry and vertex $(-2, 1)$ and containing the point $(-3, 5)$.

45. *Suspension Bridge.* The cables of a suspension bridge are 50 ft above the roadbed at the ends of the bridge and 10 ft above it in the center of the bridge. The roadbed is 200 ft long. Vertical cables are to be spaced every 20 ft along the bridge. Calculate the lengths of these vertical cables.

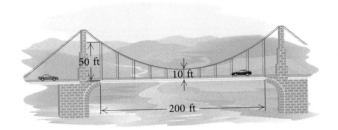

9.2

The Circle and the Ellipse

- *Given an equation of a circle, complete the square, if necessary, and then find the center and the radius and graph the circle.*
- *Given an equation of an ellipse, complete the square, if necessary, and then find the center, the vertices, and the foci and graph the ellipse.*

Circles

We can define a circle geometrically.

Circle

A **circle** is the set of all points in a plane that are at a fixed distance from a fixed point (the **center**) in the plane.

CIRCLES

REVIEW SECTION 1.1.

Circles were introduced in Section 1.1. Recall the standard equation of a circle with center (h, k) and radius r.

Standard Equation of a Circle

The standard equation of a circle with center (h, k) and radius r is

$$(x - h)^2 + (y - k)^2 = r^2.$$

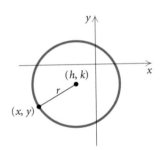

EXAMPLE 1 For the circle

$$x^2 + y^2 - 16x + 14y + 32 = 0,$$

find the center and the radius. Then graph the circle.

Solution   First, we complete the square twice:

$$x^2 + y^2 - 16x + 14y + 32 = 0$$

$$x^2 - 16x \qquad + y^2 + 14y \qquad = -32$$

$$x^2 - 16x + 64 + y^2 + 14y + 49 = -32 + 64 + 49$$

> Adding 64 and 49 on both sides to complete the square twice on the left side

$$(x - 8)^2 + (y + 7)^2 = 81$$

$$(x - 8)^2 + [y - (-7)]^2 = 9^2. \qquad \text{Writing standard form}$$

The center is $(8, -7)$ and the radius is 9. We graph the circle.

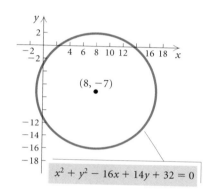

Technology Connection

To use a graphing calculator to graph the circle in Example 1, it might be necessary to solve for y first. The original equation can be solved using the quadratic formula, or the standard form of the equation can be solved using the principle of square roots. The second alternative is illustrated here:

$$(x - 8)^2 + (y + 7)^2 = 81$$

$$(y + 7)^2 = 81 - (x - 8)^2$$

$$y + 7 = \pm\sqrt{81 - (x - 8)^2} \qquad \text{Using the principle of square roots}$$

$$y = -7 \pm \sqrt{81 - (x - 8)^2}.$$

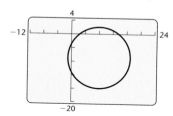

Then we graph

$$y_1 = -7 + \sqrt{81 - (x - 8)^2}$$

and

$$y_2 = -7 - \sqrt{81 - (x - 8)^2}$$

in a squared viewing window.

Some graphing calculators have a DRAW feature that provides a quick way to graph a circle when the center and the radius are known. This feature is described on p. 52.

Ellipses

We have studied two conic sections, the parabola and the circle. Now we turn our attention to a third, the *ellipse.*

Ellipse

An **ellipse** is the set of all points in a plane, the sum of whose distances from two fixed points (the **foci**) is constant. The **center** of an ellipse is the midpoint of the segment between the foci.

We can draw an ellipse by first placing two thumbtacks in a piece of cardboard. These are the foci (singular, *focus*). We then attach a piece of string to the tacks. Its length is the constant sum of the distances $d_1 + d_2$ from the foci to any point on the ellipse. Next, we trace a curve with a pencil held tight against the string. The figure traced is an ellipse.

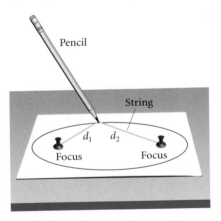

Let's first consider the ellipse shown below with center at the origin. The points F_1 and F_2 are the foci. The segment $\overline{A'A}$ is the **major axis,** and the points A' and A are the **vertices**. The segment $\overline{B'B}$ is the **minor axis,** and the points B and B' are the **y-intercepts**. Note that the major axis of an ellipse is longer than the minor axis.

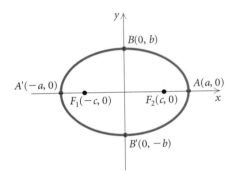

Standard Equation of an Ellipse with Center at the Origin

Major Axis Horizontal

$$\frac{x^2}{a^2} + \frac{y^2}{b^2} = 1, \ a > b > 0$$

Vertices: $(-a, 0)$, $(a, 0)$

y-intercepts:
$(0, -b)$, $(0, b)$

Foci: $(-c, 0)$, $(c, 0)$,
where $c^2 = a^2 - b^2$

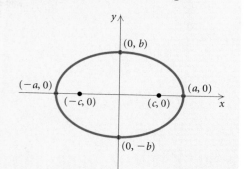

Major Axis Vertical

$$\frac{x^2}{b^2} + \frac{y^2}{a^2} = 1, \ a > b > 0$$

Vertices: $(0, -a)$, $(0, a)$

x-intercepts:
$(-b, 0)$, $(b, 0)$

Foci: $(0, -c)$, $(0, c)$,
where $c^2 = a^2 - b^2$

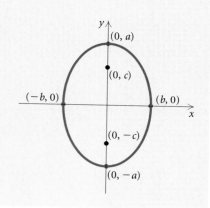

EXAMPLE 2 Find the standard equation of the ellipse with vertices $(-5, 0)$ and $(5, 0)$ and foci $(-3, 0)$ and $(3, 0)$. Then graph the ellipse.

Solution Since the foci are on the x-axis and the origin is the midpoint of the segment between them, the major axis is horizontal and $(0, 0)$ is the center of the ellipse. Thus the equation is of the form

$$\frac{x^2}{a^2} + \frac{y^2}{b^2} = 1.$$

Since the vertices are $(-5, 0)$ and $(5, 0)$ and the foci are $(-3, 0)$ and $(3, 0)$, we know that $a = 5$ and $c = 3$. These values can be used to find b^2:

$$c^2 = a^2 - b^2$$
$$3^2 = 5^2 - b^2$$
$$9 = 25 - b^2$$
$$b^2 = 16.$$

Thus the equation of the ellipse is

$$\frac{x^2}{25} + \frac{y^2}{16} = 1.$$

STUDY TIP

If you are finding it difficult to master a particular topic or concept, talk about it with a classmate. Verbalizing your questions about the material might help clarify it for you. If your classmate is also finding the material difficult, it is possible that the majority of the students in your class are confused, and you can ask your instructor to explain the concept again.

To graph the ellipse, we plot the vertices $(-5, 0)$ and $(5, 0)$. Since $b^2 = 16$, we know that $b = 4$ and the y-intercepts are $(0, -4)$ and $(0, 4)$. We plot these points as well and connect the four points we have plotted with a smooth curve.

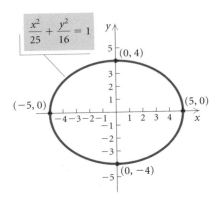

Technology Connection

To draw the graph in Example 2 using a graphing calculator, it might be necessary to solve for y first:

$$y = \pm \sqrt{\frac{400 - 16x^2}{25}}.$$

Then we graph

$$y_1 = -\sqrt{\frac{400 - 16x^2}{25}} \quad \text{and} \quad y_2 = \sqrt{\frac{400 - 16x^2}{25}}$$

or

$$y_1 = -\sqrt{\frac{400 - 16x^2}{25}} \quad \text{and} \quad y_2 = -y_1$$

in a squared viewing window.

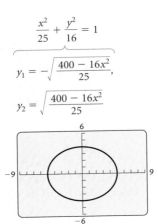

EXAMPLE 3   For the ellipse

$$9x^2 + 4y^2 = 36,$$

find the vertices and the foci. Then draw the graph.

Solution   We first find standard form:

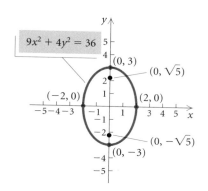

$$9x^2 + 4y^2 = 36$$

$$\frac{9x^2}{36} + \frac{4y^2}{36} = \frac{36}{36} \qquad \text{Dividing by 36 on both sides to get 1 on the right side}$$

$$\frac{x^2}{4} + \frac{y^2}{9} = 1$$

$$\frac{x^2}{2^2} + \frac{y^2}{3^2} = 1. \qquad \text{Writing standard form}$$

Thus, $a = 3$ and $b = 2$. The major axis is vertical, so the vertices are $(0, -3)$ and $(0, 3)$. Since we know that $c^2 = a^2 - b^2$, we have $c^2 = 3^2 - 2^2 = 5$, so $c = \sqrt{5}$ and the foci are $(0, -\sqrt{5})$ and $(0, \sqrt{5})$.

To graph the ellipse, we plot the vertices. Note also that since $b = 2$, the x-intercepts are $(-2, 0)$ and $(2, 0)$. We plot these points as well and connect the four points we have plotted with a smooth curve. ▬

If the center of an ellipse is not at the origin but at some point (h, k), then we can think of an ellipse with center at the origin being translated horizontally $|h|$ units and vertically $|k|$ units.

Standard Equation of an Ellipse with Center at (h, k)

Major Axis Horizontal

$$\frac{(x - h)^2}{a^2} + \frac{(y - k)^2}{b^2} = 1, \quad a > b > 0$$

Vertices: $(h - a, k)$, $(h + a, k)$

Length of minor axis: $2b$

Foci: $(h - c, k)$, $(h + c, k)$, where $c^2 = a^2 - b^2$

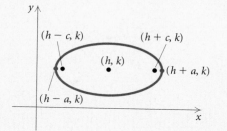

Major Axis Vertical

$$\frac{(x - h)^2}{b^2} + \frac{(y - k)^2}{a^2} = 1, \quad a > b > 0$$

Vertices: $(h, k - a)$, $(h, k + a)$

Length of minor axis: $2b$

Foci: $(h, k - c)$, $(h, k + c)$, where $c^2 = a^2 - b^2$

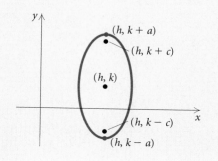

EXAMPLE 4   For the ellipse

$$4x^2 + y^2 + 24x - 2y + 21 = 0,$$

find the center, the vertices, and the foci. Then draw the graph.

Solution   First, we complete the square twice to get standard form:

$$4x^2 + y^2 + 24x - 2y + 21 = 0$$

$$4(x^2 + 6x \quad) + (y^2 - 2y \quad) = -21$$

$$4(x^2 + 6x + 9) + (y^2 - 2y + 1) = -21 + 4 \cdot 9 + 1$$

> Completing the square twice by adding $4 \cdot 9$ and 1 on both sides

$$4(x + 3)^2 + (y - 1)^2 = 16$$

$$\frac{1}{16}[4(x + 3)^2 + (y - 1)^2] = \frac{1}{16} \cdot 16$$

$$\frac{(x + 3)^2}{4} + \frac{(y - 1)^2}{16} = 1$$

$$\frac{[x - (-3)]^2}{2^2} + \frac{(y - 1)^2}{4^2} = 1.$$

> Writing standard form: $\dfrac{(x - h)^2}{b^2} + \dfrac{(y - k)^2}{a^2} = 1$

The center is $(-3, 1)$. Note that $a = 4$ and $b = 2$. The major axis is vertical, so the vertices are 4 units above and below the center:

$$(-3, 1 + 4) \text{ and } (-3, 1 - 4), \quad \text{or} \quad (-3, 5) \text{ and } (-3, -3).$$

We know that $c^2 = a^2 - b^2$, so $c^2 = 4^2 - 2^2 = 12$ and $c = \sqrt{12}$, or $2\sqrt{3}$. Then the foci are $2\sqrt{3}$ units above and below the center:

$$(-3, 1 + 2\sqrt{3}) \quad \text{and} \quad (-3, 1 - 2\sqrt{3}).$$

To graph the ellipse, we plot the vertices. Note also that since $b = 2$, two other points on the graph are the endpoints of the minor axis, 2 units right and left of the center:

$$(-3 + 2, 1) \text{ and } (-3 - 2, 1),$$

or

$$(-1, 1) \text{ and } (-5, 1).$$

We plot these points as well and connect the four points with a smooth curve.

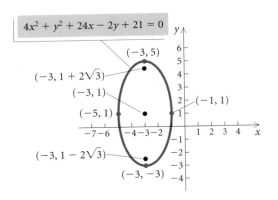

Applications

One of the most exciting recent applications of an ellipse is a medical device called a *lithotripter*. This machine uses underwater shock waves to crush kidney stones. The waves originate at one focus of an ellipse and are reflected to the kidney stone, which is positioned at the other focus. Recovery time following the use of this technique is much shorter than with conventional surgery and the mortality rate is far lower.

A room with an ellipsoidal ceiling is known as a *whispering gallery*. In such a room, a word whispered at one focus can be clearly heard at the other. Whispering galleries are found in the rotunda of the Capitol Building in Washington, D.C., and in the Mormon Tabernacle in Salt Lake City.

Ellipses have many other applications. Planets travel around the sun in elliptical orbits with the sun at one focus, for example, and satellites travel around the earth in elliptical orbits as well.

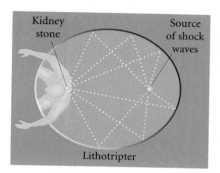

Lithotripter

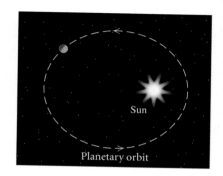

Planetary orbit

Exercise Set 9.2

In Exercises 1–6, match the equation with one of the graphs (a)–(f), which follow.

a)

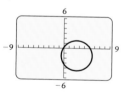

b)

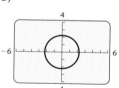

c)

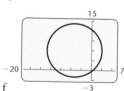

f

d)

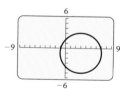

e)

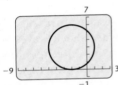

f)

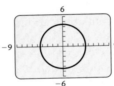

1. $x^2 + y^2 = 5$

2. $y^2 = 20 - x^2$

3. $x^2 + y^2 - 6x + 2y = 6$

4. $x^2 + y^2 + 10x - 12y = 3$

5. $x^2 + y^2 - 5x + 3y = 0$

6. $x^2 + 4x - 2 = 6y - y^2 - 6$

Find the center and the radius of the circle with the given equation. Then draw the graph.

7. $x^2 + y^2 - 14x + 4y = 11$

8. $x^2 + y^2 + 2x - 6y = -6$

9. $x^2 + y^2 + 4x - 6y - 12 = 0$

10. $x^2 + y^2 - 8x - 2y - 19 = 0$

11. $x^2 + y^2 + 6x - 10y = 0$

12. $x^2 + y^2 - 7x - 2y = 0$

13. $x^2 + y^2 - 9x = 7 - 4y$

14. $y^2 - 6y - 1 = 8x - x^2 + 3$

In Exercises 15–18, match the equation with one of the graphs (a)–(d), which follow.

a)

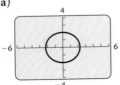

b)

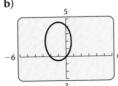

c)

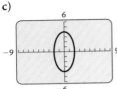

d)

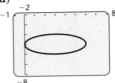

15. $16x^2 + 4y^2 = 64$

16. $4x^2 + 5y^2 = 20$

17. $x^2 + 9y^2 - 6x + 90y = -225$

18. $9x^2 + 4y^2 + 18x - 16y = 11$

Find the vertices and the foci of the ellipse with the given equation. Then draw the graph.

19. $\dfrac{x^2}{4} + \dfrac{y^2}{1} = 1$

20. $\dfrac{x^2}{25} + \dfrac{y^2}{36} = 1$

21. $16x^2 + 9y^2 = 144$

22. $9x^2 + 4y^2 = 36$

23. $2x^2 + 3y^2 = 6$

24. $5x^2 + 7y^2 = 35$

25. $4x^2 + 9y^2 = 1$

26. $25x^2 + 16y^2 = 1$

Find an equation of an ellipse satisfying the given conditions.

27. Vertices: $(-7, 0)$ and $(7, 0)$;
foci: $(-3, 0)$ and $(3, 0)$

28. Vertices: $(0, -6)$ and $(0, 6)$;
foci: $(0, -4)$ and $(0, 4)$

29. Vertices: $(0, -8)$ and $(0, 8)$;
length of minor axis: 10

30. Vertices: $(-5, 0)$ and $(5, 0)$;
length of minor axis: 6

31. Foci: $(-2, 0)$ and $(2, 0)$;
length of major axis: 6

32. Foci: $(0, -3)$ and $(0, 3)$;
length of major axis: 10

Find the center, the vertices, and the foci of the ellipse. Then draw the graph.

33. $\dfrac{(x - 1)^2}{9} + \dfrac{(y - 2)^2}{4} = 1$

34. $\dfrac{(x - 1)^2}{1} + \dfrac{(y - 2)^2}{4} = 1$

35. $\dfrac{(x + 3)^2}{25} + \dfrac{(y - 5)^2}{36} = 1$

36. $\dfrac{(x - 2)^2}{16} + \dfrac{(y + 3)^2}{25} = 1$

37. $3(x + 2)^2 + 4(y - 1)^2 = 192$

38. $4(x - 5)^2 + 3(y - 4)^2 = 48$

39. $4x^2 + 9y^2 - 16x + 18y - 11 = 0$

40. $x^2 + 2y^2 - 10x + 8y + 29 = 0$

41. $4x^2 + y^2 - 8x - 2y + 1 = 0$

42. $9x^2 + 4y^2 + 54x - 8y + 49 = 0$

*The **eccentricity** of an ellipse is defined as $e = c/a$. For an ellipse, $0 < c < a$, so $0 < e < 1$. When e is close to 0, an ellipse appears to be nearly circular. When e is close to 1, an ellipse is very flat.*

43. Observe the shapes of the ellipses in Examples 2 and 4. Which ellipse has the smaller eccentricity? Confirm your answer by computing the eccentricity of each ellipse.

44. Which ellipse below has the smaller eccentricity? (Assume that the coordinate systems have the same scale.)

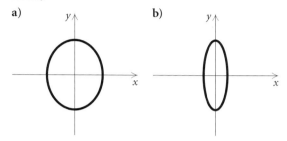

a) **b)**

45. Find an equation of an ellipse with vertices $(0, -4)$ and $(0, 4)$ and $e = \frac{1}{4}$.

46. Find an equation of an ellipse with vertices $(-3, 0)$ and $(3, 0)$ and $e = \frac{7}{10}$.

47. *Bridge Supports.* The bridge support shown in the figure below is the top half of an ellipse. Assuming that a coordinate system is superimposed on the drawing in such a way that the center of the ellipse is at point Q, find an equation of the ellipse.

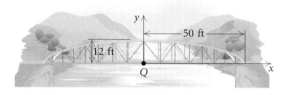

48. *The Ellipse.* In Washington, D.C., there is a large grassy area south of the White House known as the Ellipse. It is actually an ellipse with major axis of length 1048 ft and minor axis of length 898 ft. Assuming that a coordinate system is superimposed on the area in such a way that the center is at the origin and the major and minor axes are on the x- and y-axes of the coordinate system, respectively, find an equation of the ellipse.

49. *The Earth's Orbit.* The maximum distance of the earth from the sun is 9.3×10^7 miles. The

minimum distance is 9.1×10^7 miles. The sun is at one focus of the elliptical orbit. Find the distance from the sun to the other focus.

50. *Carpentry.* A carpenter is cutting a 3-ft by 4-ft elliptical sign from a 3-ft by 4-ft piece of plywood. The ellipse will be drawn using a string attached to the board at the foci of the ellipse.

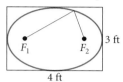

a) How far from the ends of the board should the string be attached?
b) How long should the string be?

Technology Connection

Use a graphing calculator to find the center and the vertices of each of the following.

51. $4x^2 + 9y^2 - 16.025x + 18.0927y - 11.346 = 0$

52. $9x^2 + 4y^2 + 54.063x - 8.016y + 49.872 = 0$

Collaborative Discussion and Writing

53. Explain why function notation is not used in this section.

54. Is the center of an ellipse part of the graph of the ellipse? Why or why not?

Skill Maintenance

In Exercises 55–58, given the function:

a) *Determine whether it is one-to-one.*
b) *If it is one-to-one, find a formula for the inverse.*

55. $f(x) = 2x - 3$ **56.** $f(x) = x^3 + 2$

57. $f(x) = \dfrac{5}{x - 1}$ **58.** $f(x) = \sqrt{x + 4}$

Synthesis

Find an equation of an ellipse satisfying the given conditions.

59. Vertices: $(3, -4)$, $(3, 6)$;
endpoints of minor axis: $(1, 1)$, $(5, 1)$

60. Vertices: $(-1, -1)$, $(-1, 5)$;
endpoints of minor axis: $(-3, 2)$, $(1, 2)$

61. Vertices: $(-3, 0)$ and $(3, 0)$; passing through $\left(2, \frac{22}{3}\right)$

62. Center: $(-2, 3)$; major axis vertical;
length of major axis: 4;
length of minor axis: 1

63. *Bridge Arch.*   A bridge with a semielliptical arch
spans a river as shown below. What is the clearance
6 ft from the riverbank?

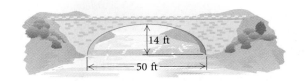

14 ft

50 ft

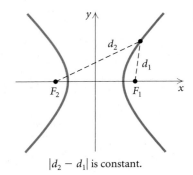

d_2

d_1

F_2

F_1

$|d_2 - d_1|$ is constant.

9.3

The Hyperbola

• *Given an equation of a hyperbola, complete the square, if
necessary, and then find the center, the vertices, and the foci and
graph the hyperbola.*

The last type of conic section that we will study is the *hyperbola*.

Hyperbola

A **hyperbola** is the set of all points in a plane for which the
absolute value of the difference of the distances from two fixed
points (the **foci**) is constant. The midpoint of the segment
between the foci is the **center** of the hyperbola.

Standard Equations of Hyperbolas

We first consider the equation of a hyperbola with center at the origin. In
the figure below, F_1 and F_2 are the foci. The segment $\overline{V_2 V_1}$ is the **transverse axis** and the points V_2 and V_1 are the **vertices**.

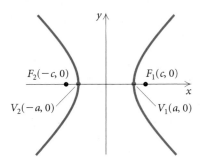

$F_2(-c, 0)$

$F_1(c, 0)$

$V_2(-a, 0)$

$V_1(a, 0)$

Standard Equation of a Hyperbola with Center at the Origin

Transverse Axis Horizontal

$$\frac{x^2}{a^2} - \frac{y^2}{b^2} = 1$$

Vertices: $(-a, 0)$, $(a, 0)$

Foci: $(-c, 0)$, $(c, 0)$,
where $c^2 = a^2 + b^2$

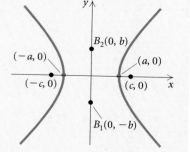

Transverse Axis Vertical

$$\frac{y^2}{a^2} - \frac{x^2}{b^2} = 1$$

Vertices: $(0, -a)$, $(0, a)$

Foci: $(0, -c)$, $(0, c)$,
where $c^2 = a^2 + b^2$

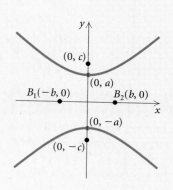

The segment $\overline{B_1 B_2}$ is the **conjugate axis** of the hyperbola.

To graph a hyperbola with a horizontal transverse axis, it is helpful to begin by graphing the lines $y = -(b/a)x$ and $y = (b/a)x$. These are the **asymptotes** of the hyperbola. For a hyperbola with a vertical transverse axis, the asymptotes are $y = -(a/b)x$ and $y = (a/b)x$. As $|x|$ gets larger and larger, the graph of the hyperbola gets closer and closer to the asymptotes.

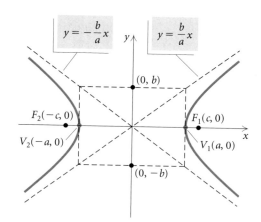

EXAMPLE 1   Find an equation of the hyperbola with vertices $(0, -4)$ and $(0, 4)$ and foci $(0, -6)$ and $(0, 6)$.

Solution   We know that $a = 4$ and $c = 6$. We find b^2.

$$c^2 = a^2 + b^2$$
$$6^2 = 4^2 + b^2$$
$$36 = 16 + b^2$$
$$20 = b^2$$

Since the vertices and the foci are on the y-axis, we know that the transverse axis is vertical. We can now write the equation of the hyperbola:

$$\frac{y^2}{a^2} - \frac{x^2}{b^2} = 1$$

$$\frac{y^2}{16} - \frac{x^2}{20} = 1.$$

EXAMPLE 2   For the hyperbola given by

$$9x^2 - 16y^2 = 144,$$

find the vertices, the foci, and the asymptotes. Then graph the hyperbola.

Solution   First, we find standard form:

$$9x^2 - 16y^2 = 144$$

$$\frac{1}{144}(9x^2 - 16y^2) = \frac{1}{144} \cdot 144 \qquad \text{Multiplying by } \tfrac{1}{144} \text{ to get 1 on the right side}$$

$$\frac{x^2}{16} - \frac{y^2}{9} = 1$$

$$\frac{x^2}{4^2} - \frac{y^2}{3^2} = 1. \qquad \text{Writing standard form}$$

The hyperbola has a horizontal transverse axis, so the vertices are $(-a, 0)$ and $(a, 0)$, or $(-4, 0)$ and $(4, 0)$. From the standard form of the equation, we know that $a^2 = 4^2$, or 16, and $b^2 = 3^2$, or 9. We find the foci:

$$c^2 = a^2 + b^2$$
$$c^2 = 16 + 9$$
$$c^2 = 25$$
$$c = 5.$$

Thus the foci are $(-5, 0)$ and $(5, 0)$.
Next, we find the asymptotes:

$$y = \frac{b}{a}x = \frac{3}{4}x$$

and

$$y = -\frac{b}{a}x = -\frac{3}{4}x.$$

To draw the graph, we sketch the asymptotes first. This is easily done by drawing the rectangle with horizontal sides passing through $(0, 3)$ and $(0, -3)$ and vertical sides through $(4, 0)$ and $(-4, 0)$. Then we draw and extend the diagonals of this rectangle. The two extended diagonals are the asymptotes of the hyperbola. Next, we plot the vertices and draw the branches of the hyperbola outward from the vertices toward the asymptotes.

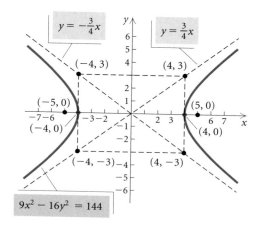

Technology Connection

To graph the hyperbola in Example 2 on a graphing calculator, it might be necessary to solve for y first and then graph the top and bottom halves of the hyperbola in the same squared viewing window.

$$9x^2 - 16y^2 = 144$$

$$y_1 = \sqrt{\frac{9x^2 - 144}{16}}, \quad y_2 = -\sqrt{\frac{9x^2 - 144}{16}}$$

If a hyperbola with center at the origin is translated horizontally $|h|$ units and vertically $|k|$ units, the center is at the point (h, k).

Standard Equation of a Hyperbola with Center (h, k)

Transverse Axis Horizontal

$$\frac{(x - h)^2}{a^2} - \frac{(y - k)^2}{b^2} = 1$$

Vertices: $(h - a, k)$, $(h + a, k)$

Asymptotes: $y - k = \dfrac{b}{a}(x - h)$, $y - k = -\dfrac{b}{a}(x - h)$

Foci: $(h - c, k)$, $(h + c, k)$, where $c^2 = a^2 + b^2$

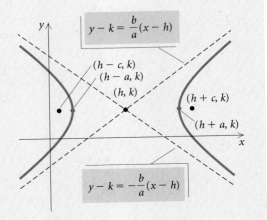

Transverse Axis Vertical

$$\frac{(y - k)^2}{a^2} - \frac{(x - h)^2}{b^2} = 1$$

Vertices: $(h, k - a)$, $(h, k + a)$

Asymptotes: $y - k = \dfrac{a}{b}(x - h)$, $y - k = -\dfrac{a}{b}(x - h)$

Foci: $(h, k - c)$, $(h, k + c)$, where $c^2 = a^2 + b^2$

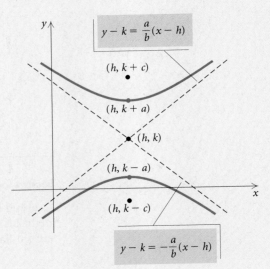

EXAMPLE 3   For the hyperbola given by
$$4y^2 - x^2 + 24y + 4x + 28 = 0,$$
find the center, the vertices, and the foci. Then draw the graph.

Solution   First, we complete the square to get standard form:

$$4y^2 - x^2 + 24y + 4x + 28 = 0$$
$$4(y^2 + 6y \qquad) - (x^2 - 4x \qquad) = -28$$
$$4(y^2 + 6y + 9 - 9) - (x^2 - 4x + 4 - 4) = -28$$
$$4(y^2 + 6y + 9) - (x^2 - 4x + 4) = -28 + 36 - 4.$$

Then

$$4(y + 3)^2 - (x - 2)^2 = 4$$

$$\frac{(y + 3)^2}{1} - \frac{(x - 2)^2}{4} = 1 \qquad \text{Dividing by 4}$$

$$\frac{[y - (-3)]^2}{1^2} - \frac{(x - 2)^2}{2^2} = 1. \qquad \text{Standard form}$$

The center is $(2, -3)$. Note that $a = 1$ and $b = 2$. The transverse axis is vertical, so the vertices are 1 unit below and above the center:

$$(2, -3 - 1) \text{ and } (2, -3 + 1), \quad \text{or} \quad (2, -4) \text{ and } (2, -2).$$

We know that $c^2 = a^2 + b^2$, so $c^2 = 1^2 + 2^2 = 1 + 4 = 5$ and $c = \sqrt{5}$. Thus the foci are $\sqrt{5}$ units below and above the center:

$$(2, -3 - \sqrt{5}) \quad \text{and} \quad (2, -3 + \sqrt{5}).$$

The asymptotes are

$$y - (-3) = \frac{1}{2}(x - 2) \quad \text{and} \quad y - (-3) = -\frac{1}{2}(x - 2),$$

or

$$y + 3 = \frac{1}{2}(x - 2) \quad \text{and} \quad y + 3 = -\frac{1}{2}(x - 2).$$

We sketch the asymptotes, plot the vertices, and draw the graph.

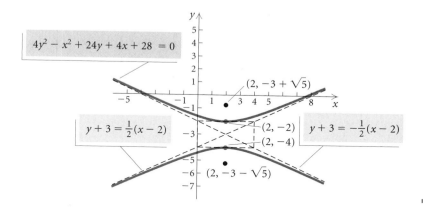

CONNECTING THE CONCEPTS

CLASSIFYING EQUATIONS OF CONIC SECTIONS

EQUATION	TYPE OF CONIC SECTION	GRAPH
$x - 4 + 4y = y^2$	Only one variable is squared, so this cannot be a circle, an ellipse, or a hyperbola. Find an equivalent equation: $$x = (y - 2)^2.$$ This is an equation of a parabola.	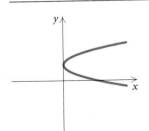
$3x^2 + 3y^2 = 75$	Both variables are squared, so this cannot be a parabola. The squared terms are added, so this cannot be a hyperbola. Divide by 3 on both sides to find an equivalent equation: $$x^2 + y^2 = 25.$$ This is an equation of a circle.	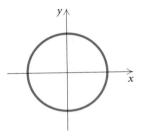
$y^2 = 16 - 4x^2$	Both variables are squared, so this cannot be a parabola. Add $4x^2$ on both sides to find an equivalent equation: $4x^2 + y^2 = 16$. The squared terms are added, so this cannot be a hyperbola. The coefficients of x^2 and y^2 are not the same, so this is not a circle. Divide by 16 on both sides to find an equivalent equation: $$\frac{x^2}{4} + \frac{y^2}{16} = 1.$$ This is an equation of an ellipse.	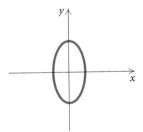
$x^2 = 4y^2 + 36$	Both variables are squared, so this cannot be a parabola. Subtract $4y^2$ on both sides to find an equivalent equation: $x^2 - 4y^2 = 36$. The squared terms are not added, so this cannot be a circle or an ellipse. Divide by 36 on both sides to find an equivalent equation: $$\frac{x^2}{36} - \frac{y^2}{9} = 1.$$ This is an equation of a hyperbola.	

Applications

Some comets travel in hyperbolic paths with the sun at one focus. Such comets pass by the sun only one time, unlike those with elliptical orbits, which reappear at intervals. A cross section of an amphitheater might be one branch of a hyperbola. A cross section of a nuclear cooling tower might also be a hyperbola.

One other application of hyperbolas is in the long-range navigation system LORAN. This system uses transmitting stations in three locations to send out simultaneous signals to a ship or aircraft. The difference in the arrival times of the signals from one pair of transmitters is recorded on the ship or aircraft. This difference is also recorded for signals from another pair of transmitters. For each pair, a computation is performed to determine the difference in the distances from each member of the pair to the ship or aircraft. If each pair of differences is kept constant, two hyperbolas can be drawn. Each has one of the pairs of transmitters as foci, and the ship or aircraft lies on the intersection of two of their branches.

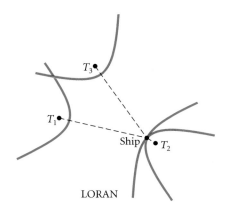

LORAN

Exercise Set 9.3

In Exercises 1–6, match the equation with one of the graphs (a)–(f), which follow.

a)
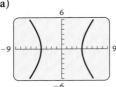

b)
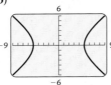

c)
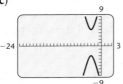

d)
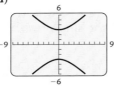

e)
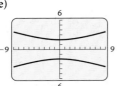

f)
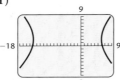

1. $\dfrac{x^2}{25} - \dfrac{y^2}{9} = 1$

2. $\dfrac{y^2}{4} - \dfrac{x^2}{36} = 1$

3. $\dfrac{(y-1)^2}{16} - \dfrac{(x+3)^2}{1} = 1$

4. $\dfrac{(x+4)^2}{100} - \dfrac{(y-2)^2}{81} = 1$

5. $25x^2 - 16y^2 = 400$

6. $y^2 - x^2 = 9$

Find an equation of a hyperbola satisfying the given conditions.

7. Vertices at $(0, 3)$ and $(0, -3)$; foci at $(0, 5)$ and $(0, -5)$

8. Vertices at $(1, 0)$ and $(-1, 0)$; foci at $(2, 0)$ and $(-2, 0)$

9. Asymptotes $y = \frac{3}{2}x$, $y = -\frac{3}{2}x$; one vertex $(2, 0)$

10. Asymptotes $y = \frac{5}{4}x$, $y = -\frac{5}{4}x$; one vertex $(0, 3)$

Find the center, the vertices, the foci, and the asymptotes. Then draw the graph.

11. $\dfrac{x^2}{4} - \dfrac{y^2}{4} = 1$

12. $\dfrac{x^2}{1} - \dfrac{y^2}{9} = 1$

13. $\dfrac{(x-2)^2}{9} - \dfrac{(y+5)^2}{1} = 1$

14. $\dfrac{(x-5)^2}{16} - \dfrac{(y+2)^2}{9} = 1$

15. $\dfrac{(y+3)^2}{4} - \dfrac{(x+1)^2}{16} = 1$

16. $\dfrac{(y+4)^2}{25} - \dfrac{(x+2)^2}{16} = 1$

17. $x^2 - 4y^2 = 4$ **18.** $4x^2 - y^2 = 16$

19. $9y^2 - x^2 = 81$ **20.** $y^2 - 4x^2 = 4$

21. $x^2 - y^2 = 2$ **22.** $x^2 - y^2 = 3$

23. $y^2 - x^2 = \frac{1}{4}$ **24.** $y^2 - x^2 = \frac{1}{9}$

Find the center, the vertices, the foci, and the asymptotes of the hyperbola. Then draw the graph.

25. $x^2 - y^2 - 2x - 4y - 4 = 0$

26. $4x^2 - y^2 + 8x - 4y - 4 = 0$

27. $36x^2 - y^2 - 24x + 6y - 41 = 0$

28. $9x^2 - 4y^2 + 54x + 8y + 41 = 0$

29. $9y^2 - 4x^2 - 18y + 24x - 63 = 0$

30. $x^2 - 25y^2 + 6x - 50y = 41$

31. $x^2 - y^2 - 2x - 4y = 4$

32. $9y^2 - 4x^2 - 54y - 8x + 41 = 0$

33. $y^2 - x^2 - 6x - 8y - 29 = 0$

34. $x^2 - y^2 = 8x - 2y - 13$

*The **eccentricity** of a hyperbola is defined as $e = c/a$. For a hyperbola, $c > a > 0$, so $e > 1$. When e is close to 1, a hyperbola appears to be very narrow. As the eccentricity increases, the hyperbola becomes "wider."*

35. Observe the shapes of the hyperbolas in Examples 2 and 3. Which hyperbola has the larger eccentricity?

Confirm your answer by computing the eccentricity of each hyperbola.

36. Which hyperbola below has the larger eccentricity? (Assume that the coordinate systems have the same scale.)

a)

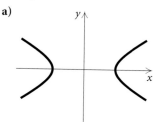

b)

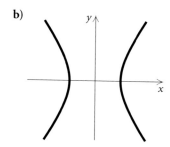

37. Find an equation of a hyperbola with vertices $(3, 7)$ and $(-3, 7)$ and $e = \frac{5}{3}$.

38. Find an equation of a hyperbola with vertices $(-1, 3)$ and $(-1, 7)$ and $e = 4$.

39. *Hyperbolic Mirror.* Certain telescopes contain both a parabolic and a hyperbolic mirror. In the telescope shown in the figure below, the parabola and the hyperbola share focus F_1, which is 14 m above the vertex of the parabola. The hyperbola's second focus F_2 is 2 m above the parabola's vertex. The vertex of the hyperbolic mirror is 1 m below F_1. Position a coordinate system with the origin at the center of the hyperbola and with the foci on the y-axis. Then find the equation of the hyperbola.

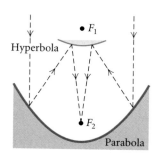

40. *Nuclear Cooling Tower.* A cross section of a nuclear cooling tower is a hyperbola with equation

$$\frac{x^2}{90^2} - \frac{y^2}{130^2} = 1.$$

The tower is 450 ft tall and the distance from the top of the tower to the center of the hyperbola is half the distance from the base of the tower to the center of the hyperbola. Find the diameter of the top and the base of the tower.

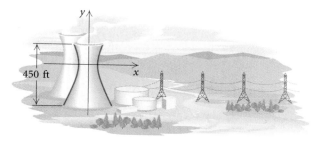

Technology Connection

Use a graphing calculator to find the center, the vertices, and the asymptotes.

41. $5x^2 - 3.5y^2 + 14.6x - 6.7y + 3.4 = 0$

42. $x^2 - y^2 - 2.046x - 4.088y - 4.228 = 0$

Collaborative Discussion and Writing

43. How does the graph of a parabola differ from the graph of one branch of a hyperbola?

44. Are the asymptotes of a hyperbola part of the graph of the hyperbola? Why or why not?

Skill Maintenance

Solve.

45. $x + y = 5,$
$x - y = 7$

46. $3x - 2y = 5,$
$5x + 2y = 3$

47. $2x - 3y = 7,$
$3x + 5y = 1$

48. $3x + 2y = -1,$
$2x + 3y = 6$

Synthesis

Find an equation of a hyperbola satisfying the given conditions.

49. Vertices at $(3, -8)$ and $(3, -2)$;
asymptotes $y = 3x - 14$, $y = -3x + 4$

50. Vertices at $(-9, 4)$ and $(-5, 4)$;
asymptotes $y = 3x + 25$, $y = -3x - 17$

51. *Navigation.* Two radio transmitters positioned 300 mi apart along the shore send simultaneous signals to a ship that is 200 mi offshore, sailing parallel to the shoreline. The signal from transmitter S reaches the ship 200 microseconds later than the signal from transmitter T. The signals travel at a speed of 186,000 miles per second, or 0.186 mile per microsecond. Find the equation of the hyperbola with foci S and T on which the ship is located. (*Hint*: For any point on the hyperbola, the absolute value of the difference of its distances from the foci is $2a$.)

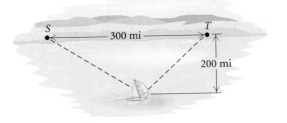

9.4

Nonlinear Systems of Equations

• Solve a nonlinear system of equations.
• Use nonlinear systems of equations to solve applied problems.

The systems of equations that we have studied so far have been composed of linear equations. Now we consider systems of two equations in two variables in which at least one equation is not linear.

Nonlinear Systems of Equations

The graphs of the equations in a nonlinear system of equations can have no point of intersection or one or more points of intersection. The coordinates of each point of intersection represent a solution of the system of equations. When no point of intersection exists, the system of equations has no real-number solution.

Solutions of nonlinear systems of equations can be found using the substitution or elimination method. The substitution method is preferable for a system consisting of one linear and one nonlinear equation. The elimination method is preferable in most, but not all, cases when both equations are nonlinear.

EXAMPLE 1   Solve the following system of equations:

$$x^2 + y^2 = 25, \quad (1) \qquad \text{(The graph is a circle.)}$$
$$3x - 4y = 0. \quad (2) \qquad \text{(The graph is a line.)}$$

Algebraic Solution

We use the substitution method. First, we solve equation (2) for x:

$$x = \tfrac{4}{3}y. \qquad (3) \qquad \text{We could have solved for } y \text{ instead.}$$

Next, we substitute $\tfrac{4}{3}y$ for x in equation (1) and solve for y:

$$\left(\tfrac{4}{3}y\right)^2 + y^2 = 25$$
$$\tfrac{16}{9}y^2 + y^2 = 25$$
$$\tfrac{25}{9}y^2 = 25$$
$$y^2 = 9 \qquad \text{Multiplying by } \tfrac{9}{25}$$
$$y = \pm 3.$$

Now we substitute these numbers for y in equation (3) and solve for x:

$$x = \tfrac{4}{3}(3) = 4, \qquad \text{The pair (4, 3) appears to be a solution.}$$
$$x = \tfrac{4}{3}(-3) = -4. \qquad \text{The pair } (-4, -3) \text{ appears to be a solution.}$$

CHECK: For (4, 3):

$$\begin{array}{c|c} x^2 + y^2 = 25 \\ \hline 4^2 + 3^2 \ ? \ 25 \\ 16 + 9 \\ 25 \ | \ 25 \quad \text{TRUE} \end{array} \qquad \begin{array}{c|c} 3x - 4y = 0 \\ \hline 3(4) - 4(3) \ ? \ 0 \\ 12 - 12 \\ 0 \ | \ 0 \quad \text{TRUE} \end{array}$$

For $(-4, -3)$:

$$\begin{array}{c|c} x^2 + y^2 = 25 \\ \hline (-4)^2 + (-3)^2 \ ? \ 25 \\ 16 + 9 \\ 25 \ | \ 25 \quad \text{TRUE} \end{array} \qquad \begin{array}{c|c} 3x - 4y = 0 \\ \hline 3(-4) - 4(-3) \ ? \ 0 \\ -12 + 12 \\ 0 \ | \ 0 \quad \text{TRUE} \end{array}$$

The pairs (4, 3) and $(-4, -3)$ check, so they are the solutions.

Visualizing the Solution

The ordered pairs corresponding to the points of intersection of the graphs of the equations are the solutions of the system of equations.

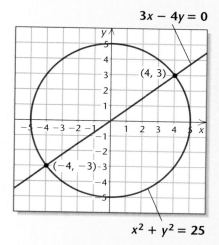

We see that the solutions are (4, 3) and $(-4, -3)$.

Technology Connection

To solve the system of equations in Example 1 on a graphing calculator, we graph both equations in the same viewing window. Note that there are two points of intersection. We can find their coordinates using the INTERSECT feature.

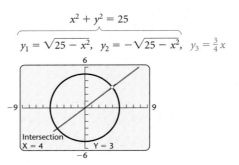

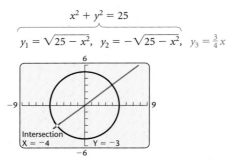

The solutions are $(4, 3)$ and $(-4, -3)$.

The graph can also be used to check the solutions that we found algebraically.

In the solution in Example 1, suppose that to find x, we had substituted 3 and -3 in equation (1) rather than equation (3). If $y = 3$, $y^2 = 9$, and if $y = -3$, $y^2 = 9$, so both substitutions can be performed at the same time:

$$x^2 + y^2 = 25 \qquad (1)$$
$$x^2 + (\pm 3)^2 = 25$$
$$x^2 + 9 = 25$$
$$x^2 = 16$$
$$x = \pm 4.$$

Thus, if $y = 3$, $x = 4$ or $x = -4$, and if $y = -3$, $x = 4$ or $x = -4$. The possible solutions are $(4, 3)$, $(-4, 3)$, $(4, -3)$, and $(-4, -3)$. A check reveals that $(4, -3)$ and $(-4, 3)$ are not solutions of equation (2). Since a circle and a line can intersect in at most two points, it is clear that there can be at most two real-number solutions.

EXAMPLE 2   Solve the following system of equations:

$$x + y = 5, \qquad (1) \qquad \text{(The graph is a line.)}$$
$$y = 3 - x^2. \qquad (2) \qquad \text{(The graph is a parabola.)}$$

Algebraic Solution

We use the substitution method, substituting $3 - x^2$ for y in equation (1):

$$x + 3 - x^2 = 5$$
$$-x^2 + x - 2 = 0 \qquad \text{Subtracting 5 and rearranging}$$
$$x^2 - x + 2 = 0. \qquad \text{Multiplying by } -1$$

Next, we use the quadratic formula:

$$x = \frac{-b \pm \sqrt{b^2 - 4ac}}{2a} = \frac{-(-1) \pm \sqrt{(-1)^2 - 4(1)(2)}}{2(1)}$$
$$= \frac{1 \pm \sqrt{1 - 8}}{2} = \frac{1 \pm \sqrt{-7}}{2} = \frac{1 \pm i\sqrt{7}}{2} = \frac{1}{2} \pm \frac{\sqrt{7}}{2}i.$$

Now, we substitute these values for x in equation (1) and solve for y:

$$\frac{1}{2} + \frac{\sqrt{7}}{2}i + y = 5$$
$$y = 5 - \frac{1}{2} - \frac{\sqrt{7}}{2}i = \frac{9}{2} - \frac{\sqrt{7}}{2}i$$

and $$\frac{1}{2} - \frac{\sqrt{7}}{2}i + y = 5$$
$$y = 5 - \frac{1}{2} + \frac{\sqrt{7}}{2}i = \frac{9}{2} + \frac{\sqrt{7}}{2}i.$$

The solutions are

$$\left(\frac{1}{2} + \frac{\sqrt{7}}{2}i, \frac{9}{2} - \frac{\sqrt{7}}{2}i\right) \quad \text{and} \quad \left(\frac{1}{2} - \frac{\sqrt{7}}{2}i, \frac{9}{2} + \frac{\sqrt{7}}{2}i\right).$$

There are no real-number solutions.

Visualizing the Solution

We graph the equations, as shown below.

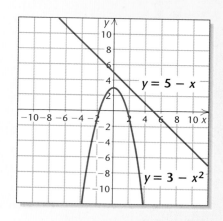

Note that there are no points of intersection. This indicates that there are no real-number solutions.

EXAMPLE 3 Solve the following system of equations:

$$2x^2 + 5y^2 = 39, \quad (1) \qquad \text{(The graph is an ellipse.)}$$
$$3x^2 - y^2 = -1. \quad (2) \qquad \text{(The graph is a hyperbola.)}$$

Algebraic Solution

We use the elimination method. First, we multiply equation (2) by 5 and add to eliminate the y^2-term:

$$2x^2 + 5y^2 = 39 \quad (1)$$
$$\underline{15x^2 - 5y^2 = -5} \quad \text{Multiplying (2) by 5}$$
$$17x^2 \qquad = 34 \quad \text{Adding}$$
$$x^2 = 2$$
$$x = \pm\sqrt{2}.$$

If $x = \sqrt{2}$, $x^2 = 2$, and if $x = -\sqrt{2}$, $x^2 = 2$. Thus substituting $\sqrt{2}$ or $-\sqrt{2}$ for x in equation (2) gives us

$$3(\pm\sqrt{2})^2 - y^2 = -1$$
$$3 \cdot 2 - y^2 = -1$$
$$6 - y^2 = -1$$
$$-y^2 = -7$$
$$y^2 = 7$$
$$y = \pm\sqrt{7}.$$

Thus, for $x = \sqrt{2}$, we have $y = \sqrt{7}$ or $y = -\sqrt{7}$, and for $x = -\sqrt{2}$, we have $y = \sqrt{7}$ or $y = -\sqrt{7}$. The possible solutions are $(\sqrt{2}, \sqrt{7})$, $(\sqrt{2}, -\sqrt{7})$, $(-\sqrt{2}, \sqrt{7})$, and $(-\sqrt{2}, -\sqrt{7})$. All four pairs check, so they are the solutions.

Visualizing the Solution

We graph the equations and observe that there are four points of intersection.

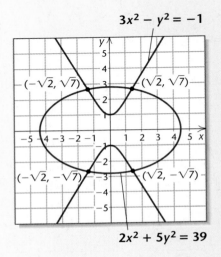

The coordinates of the points of intersection are $(\sqrt{2}, \sqrt{7})$, $(\sqrt{2}, -\sqrt{7})$, $(-\sqrt{2}, \sqrt{7})$, and $(-\sqrt{2}, -\sqrt{7})$. These are the solutions of the system of equations.

Technology Connection

To solve the system of equations in Example 3 graphically on a graphing calculator, we first graph both equations in the same viewing window. There are four points of intersection. We can use the INTERSECT feature to find their coordinates.

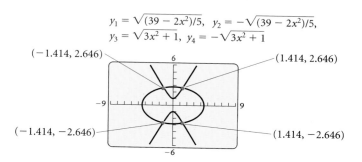

$$y_1 = \sqrt{(39 - 2x^2)/5}, \quad y_2 = -\sqrt{(39 - 2x^2)/5},$$
$$y_3 = \sqrt{3x^2 + 1}, \quad y_4 = -\sqrt{3x^2 + 1}$$

$(-1.414, 2.646)$ $(1.414, 2.646)$

$(-1.414, -2.646)$ $(1.414, -2.646)$

Note that the algebraic method yields exact solutions, whereas the graphical method yields decimal approximations of the solutions on most graphing calculators.

The solutions are approximately $(1.414, 2.646)$, $(1.414, -2.646)$, $(-1.414, 2.646)$, and $(-1.414, -2.646)$.

EXAMPLE 4 Solve the following system of equations:

$$x^2 - 3y^2 = 6, \quad (1)$$
$$xy = 3. \quad (2)$$

Algebraic Solution

We use the substitution method. First, we solve equation (2) for y:

$$xy = 3$$
$$y = \frac{3}{x}. \quad (3)$$

Next, we substitute $3/x$ for y in equation (1) and solve for x:

$$x^2 - 3\left(\frac{3}{x}\right)^2 = 6$$

$$x^2 - 3 \cdot \frac{9}{x^2} = 6$$

$$x^2 - \frac{27}{x^2} = 6$$

$x^4 - 27 = 6x^2$	Multiplying by x^2
$x^4 - 6x^2 - 27 = 0$	
$u^2 - 6u - 27 = 0$	Letting $u = x^2$
$(u - 9)(u + 3) = 0$	Factoring
$u = 9 \quad or \quad u = -3$	Principle of zero products
$x^2 = 9 \quad or \quad x^2 = -3$	Substituting x^2 for u
$x = \pm 3 \quad or \quad x = \pm i\sqrt{3}.$	

Since $y = 3/x$,

when $x = 3$, $\qquad y = \dfrac{3}{3} = 1$;

when $x = -3$, $\qquad y = \dfrac{3}{-3} = -1$;

when $x = i\sqrt{3}$, $\quad y = \dfrac{3}{i\sqrt{3}} = \dfrac{3}{i\sqrt{3}} \cdot \dfrac{-i\sqrt{3}}{-i\sqrt{3}} = -i\sqrt{3}$;

when $x = -i\sqrt{3}$, $\quad y = \dfrac{3}{-i\sqrt{3}} = \dfrac{3}{-i\sqrt{3}} \cdot \dfrac{i\sqrt{3}}{i\sqrt{3}} = i\sqrt{3}$.

The pairs $(3, 1)$, $(-3, -1)$, $(i\sqrt{3}, -i\sqrt{3})$, and $(-i\sqrt{3}, i\sqrt{3})$ check, so they are the solutions.

Visualizing the Solution

The coordinates of the points of intersection of the graphs of the equations give us the real-number solutions of the system of equations. These graphs do not show us the imaginary-number solutions.

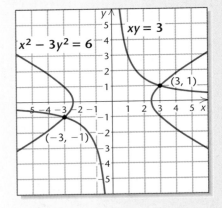

Modeling and Problem Solving

EXAMPLE 5 *Dimensions of a Piece of Land.* For a student recreation building at Southport Community College, an architect wants to lay out a rectangular piece of land that has a perimeter of 204 m and an area of 2565 m². Find the dimensions of the piece of land.

Solution

1. **Familiarize.** We make a drawing and label it, letting l = the length of the piece of land, in meters, and w = the width, in meters.

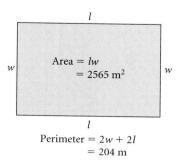

$$l$$

$$w \qquad \text{Area} = lw \qquad w$$
$$= 2565 \text{ m}^2$$

$$l$$

$$\text{Perimeter} = 2w + 2l$$
$$= 204 \text{ m}$$

2. **Translate.** We now have the following:

Perimeter: $2w + 2l = 204,$ (1)

Area: $lw = 2565.$ (2)

3. **Carry out.** We solve the system of equations.

$$2w + 2l = 204,$$
$$lw = 2565.$$

Solving the second equation for l gives us $l = 2565/w$. We then substitute $2565/w$ for l in equation (1) and solve for w:

$$2w + 2\left(\frac{2565}{w}\right) = 204$$

$$2w^2 + 2(2565) = 204w \qquad \text{Multiplying by } w$$

$$2w^2 - 204w + 2(2565) = 0$$

$$w^2 - 102w + 2565 = 0 \qquad \text{Multiplying by } \tfrac{1}{2}$$

$$(w - 57)(w - 45) = 0$$

$$w = 57 \quad or \quad w = 45. \qquad \text{Principle of zero products}$$

If $w = 57$, then $l = 2565/w = 2565/57 = 45$. If $w = 45$, then $l = 2565/w = 2565/45 = 57$. Since length is generally considered to be longer than width, we have the solution $l = 57$ and $w = 45$, or $(57, 45)$.

4. **Check.** If $l = 57$ and $w = 45$, the perimeter is $2 \cdot 45 + 2 \cdot 57$, or 204. The area is $57 \cdot 45$, or 2565. The numbers check.

5. **State.** The length of the piece of land is 57 m and the width is 45 m.

Exercise Set 9.4

In Exercises 1–6, match the system of equations with one of the graphs (a)–(f), which follow.

a)

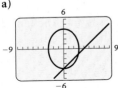

b)

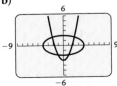

c)

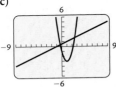

d)

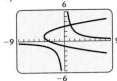

e)

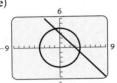

f)

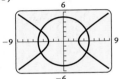

1. $x^2 + y^2 = 16,$
$x + y = 3$

2. $16x^2 + 9y^2 = 144,$
$x - y = 4$

3. $y = x^2 - 4x - 2,$
$2y - x = 1$

4. $4x^2 - 9y^2 = 36,$
$x^2 + y^2 = 25$

5. $y = x^2 - 3,$
$x^2 + 4y^2 = 16$

6. $y^2 - 2y = x + 3,$
$xy = 4$

Solve.

7. $x^2 + y^2 = 25,$
$y - x = 1$

8. $x^2 + y^2 = 100,$
$y - x = 2$

9. $4x^2 + 9y^2 = 36,$
$3y + 2x = 6$

10. $9x^2 + 4y^2 = 36,$
$3x + 2y = 6$

11. $x^2 + y^2 = 25,$
$y^2 = x + 5$

12. $y = x^2,$
$x = y^2$

13. $x^2 + y^2 = 9,$
$x^2 - y^2 = 9$

14. $y^2 - 4x^2 = 4,$
$4x^2 + y^2 = 4$

15. $y^2 - x^2 = 9,$
$2x - 3 = y$

16. $x + y = -6,$
$xy = -7$

17. $y^2 = x + 3,$
$2y = x + 4$

18. $y = x^2,$
$3x = y + 2$

19. $x^2 + y^2 = 25,$
$xy = 12$

20. $x^2 - y^2 = 16,$
$x + y^2 = 4$

21. $x^2 + y^2 = 4,$
$16x^2 + 9y^2 = 144$

22. $x^2 + y^2 = 25,$
$25x^2 + 16y^2 = 400$

23. $x^2 + 4y^2 = 25,$
$x + 2y = 7$

24. $y^2 - x^2 = 16,$
$2x - y = 1$

25. $x^2 - xy + 3y^2 = 27,$
$x - y = 2$

26. $2y^2 + xy + x^2 = 7,$
$x - 2y = 5$

27. $x^2 + y^2 = 16,$
$y^2 - 2x^2 = 10$

28. $x^2 + y^2 = 14,$
$x^2 - y^2 = 4$

29. $x^2 + y^2 = 5,$
$xy = 2$

30. $x^2 + y^2 = 20,$
$xy = 8$

31. $3x + y = 7,$
$4x^2 + 5y = 56$

32. $2y^2 + xy = 5,$
$4y + x = 7$

33. $a + b = 7,$
$ab = 4$

34. $p + q = -4,$
$pq = -5$

35. $x^2 + y^2 = 13,$
$xy = 6$

36. $x^2 + 4y^2 = 20,$
$xy = 4$

37. $x^2 + y^2 + 6y + 5 = 0,$
$x^2 + y^2 - 2x - 8 = 0$

38. $2xy + 3y^2 = 7,$
$3xy - 2y^2 = 4$

39. $2a + b = 1,$
$b = 4 - a^2$

40. $4x^2 + 9y^2 = 36,$
$x + 3y = 3$

41. $a^2 + b^2 = 89,$
$a - b = 3$

42. $xy = 4,$
$x + y = 5$

43. $xy - y^2 = 2,$
$2xy - 3y^2 = 0$

44. $4a^2 - 25b^2 = 0,$
$2a^2 - 10b^2 = 3b + 4$

45. $m^2 - 3mn + n^2 + 1 = 0,$
$3m^2 - mn + 3n^2 = 13$

46. $ab - b^2 = -4,$
$ab - 2b^2 = -6$

47. $x^2 + y^2 = 5,$
$x - y = 8$

48. $4x^2 + 9y^2 = 36,$
$y - x = 8$

49. $a^2 + b^2 = 14,$
$ab = 3\sqrt{5}$

50. $x^2 + xy = 5,$
$2x^2 + xy = 2$

51. $x^2 + y^2 = 25,$
$9x^2 + 4y^2 = 36$

52. $x^2 + y^2 = 1,$
$9x^2 - 16y^2 = 144$

53. $5y^2 - x^2 = 1,$
$xy = 2$

54. $x^2 - 7y^2 = 6,$
$xy = 1$

55. *Picture Frame Dimensions.*   Frank's Frame Shop is building a frame for a rectangular oil painting with a perimeter of 28 cm and a diagonal of 10 cm. Find the dimensions of the painting.

56. *Landscaping.*   Green Leaf Landscaping is planting a rectangular wildflower garden with a perimeter of 6 m and a diagonal of $\sqrt{5}$ m. Find the dimensions of the garden.

57. *Pamphlet Design.*   A graphic artist is designing a rectangular advertising pamphlet that is to have an area of 20 in^2 and a perimeter of 18 in. Find the dimensions of the pamphlet.

58. *Sign Dimensions.*   Peden's Advertising is building a rectangular sign with an area of 2 yd^2 and a perimeter of 6 yd. Find the dimensions of the sign.

59. *Banner Design.*   A rectangular banner with an area of $\sqrt{3}$ m^2 is being designed to advertise an exhibit at the Madison Art League. The length of a diagonal is 2 m. Find the dimensions of the banner.

Area $= \sqrt{3}$ m^2

60. *Carpentry.*   A carpenter wants to make a rectangular tabletop with an area of $\sqrt{2}$ m^2 and a diagonal of $\sqrt{3}$ m. Find the dimensions of the tabletop.

61. *Fencing.*   It will take 210 yd of fencing to enclose a rectangular dog run. The area of the run is 2250 yd^2. What are the dimensions of the run?

62. *Office Dimensions.*   The diagonal of the floor of a rectangular office cubicle is 1 ft longer than the length of the cubicle and 3 ft longer than twice the width. Find the dimensions of the cubicle.

63. *Seed Test Plots.*   The Burton Seed Company has two square test plots. The sum of their areas is 832 ft^2 and the difference of their areas is 320 ft^2. Find the length of a side of each plot.

64. *Investment.*   Jenna made an investment for 1 yr that earned $7.50 simple interest. If the principal had been $25 more and the interest rate 1% less, the interest would have been the same. Find the principal and the rate.

Technology Connection

65. *Social Security Trust Fund.*   Income and expenditures of the Social Security Trust Fund are shown in the table below.

YEAR	INCOME (IN BILLIONS)	EXPENDITURES (IN BILLIONS)
1985	$203.5	$190.6
1990	315.4	253.1
1991	329.7	274.2
1992	342.6	291.9
1993	355.6	308.8
1994	381.1	323.0
1995	399.7	339.8
1996	424.5	353.6
1997	457.7	369.1
1998	484.3	382.9
1999	503.7	396.3

Source: Social Security Administration

a) Find a linear function $I(x)$ that models the income data. Let x represent the number of years after 1985.

b) Find an exponential function $E(x)$ that models the expenditure data. Let x represent the number of years after 1985.

c) Use the functions found in parts (a) and (b) to determine when income will equal expenditures.

66. *Higher Education.* The percent of associate's, bachelor's, master's, first professional, and doctor's degrees conferred on men and women in the United States in recent years are shown in the table below.

YEAR	PERCENT MALE	PERCENT FEMALE
1960	65.8	34.2
1965	61.5	38.5
1970	59.2	40.8
1975	56.0	44.0
1980	51.1	48.9
1985	49.3	50.7
1990	46.6	53.4
1995	44.9	55.1

Source: U.S. National Center for Education Statistics

a) Find an exponential function $M(x)$ that models the percent of degrees that were earned by men. Let x represent the number of years after 1960.
b) Find a linear function $F(x)$ that models the percent of degrees that were earned by women. Let x represent the number of years after 1960.
c) Use the functions found in parts (a) and (b) to determine when equal percents of men and women earned degrees.

Solve using a graphing calculator.

67. $y - \ln x = 2,$
$y = x^2$

68. $y = \ln (x + 4),$
$x^2 + y^2 = 6$

69. $e^x - y = 1,$
$3x + y = 4$

70. $y - e^{-x} = 1,$
$y = 2x + 5$

71. $y = e^x,$
$x - y = -2$

72. $y = e^{-x},$
$x + y = 3$

73. $x^2 + y^2 = 19{,}380{,}510.36,$
$27{,}942.25x - 6.125y = 0$

74. $2x + 2y = 1660,$
$xy = 35{,}325$

75. $14.5x^2 - 13.5y^2 - 64.5 = 0,$
$5.5x - 6.3y - 12.3 = 0$

76. $13.5xy + 15.6 = 0,$
$5.6x - 6.7y - 42.3 = 0$

77. $0.319x^2 + 2688.7y^2 = 56{,}548,$
$0.306x^2 - 2688.7y^2 = 43{,}452$

78. $18.465x^2 + 788.723y^2 = 6408,$
$106.535x^2 - 788.723y^2 = 2692$

Collaborative Discussion and Writing

79. Which conic sections, if any, have equations that can be expressed using function notation? Explain why you answered as you did.

80. Write a problem that can be translated to a nonlinear system of equations, and ask a classmate to solve it. Devise the problem so that the solution is "The dimensions of the rectangle are 6 ft by 8 ft."

Skill Maintenance

Solve.

81. $2^{3x} = 64$

82. $5^x = 27$

83. $\log_3 x = 4$

84. $\log (x - 3) + \log x = 1$

Synthesis

85. Find an equation of the circle that passes through the points $(2, 4)$ and $(3, 3)$ and whose center is on the line $3x - y = 3$.

86. Find an equation of the circle that passes through the points $(-2, 3)$ and $(-4, 1)$ and whose center is on the line $5x + 8y = -2$.

87. Find an equation of an ellipse centered at the origin that passes through the points $(1, \sqrt{3}/2)$ and $(\sqrt{3}, 1/2)$.

88. Find an equation of a hyperbola of the type

$$\frac{x^2}{b^2} - \frac{y^2}{a^2} = 1$$

that passes through the points $(-3, -3\sqrt{5}/2)$ and $(-3/2, 0)$.

89. Find an equation of the circle that passes through the points $(4, 6)$, $(-6, 2)$, and $(1, -3)$.

90. Find an equation of the circle that passes through the points $(2, 3)$, $(4, 5)$, and $(0, -3)$.

91. Show that a hyperbola does not intersect its asymptotes. That is, solve the system of equations

$$\frac{x^2}{a^2} - \frac{y^2}{b^2} = 1,$$

$$y = \frac{b}{a}x \quad \left(\text{or } y = -\frac{b}{a}x\right).$$

92. *Numerical Relationship.* Find two numbers whose product is 2 and the sum of whose reciprocals is $\frac{33}{8}$.

93. *Numerical Relationship.* The square of a number exceeds twice the square of another number by $\frac{1}{8}$. The sum of their squares is $\frac{5}{16}$. Find the numbers.

94. *Box Dimensions.* Four squares with sides 5 in. long are cut from the corners of a rectangular metal sheet that has an area of 340 in^2. The edges are bent up to form an open box with a volume of 350 in^3. Find the dimensions of the box.

95. *Numerical Relationship.* The sum of two numbers is 1, and their product is 1. Find the sum of their cubes. There is a method to solve this problem that is easier than solving a nonlinear system of equations. Can you discover it?

96. Solve for x and y:

$$x^2 - y^2 = a^2 - b^2,$$

$$x - y = a - b.$$

Solve.

97. $x^3 + y^3 = 72,$
    $x + y = 6$

98. $a + b = \frac{5}{6},$

    $\frac{a}{b} + \frac{b}{a} = \frac{13}{6}$

99. $p^2 + q^2 = 13,$

    $\frac{1}{pq} = -\frac{1}{6}$

100. $x^2 + y^2 = 4,$
     $(x - 1)^2 + y^2 = 4$

101. $5^{x+y} = 100,$
     $3^{2x-y} = 1000$

102. $e^x - e^{x+y} = 0,$
     $e^y - e^{x-y} = 0$

9 | Chapter Summary and Review

Important Properties and Formulas

Standard Equation of a Parabola with Vertex at the Origin

The standard equation of a parabola with vertex $(0, 0)$ and directrix $y = -p$ is

$$x^2 = 4py.$$

The focus is $(0, p)$ and the y-axis is the axis of symmetry.

The standard equation of a parabola with vertex $(0, 0)$ and directrix $x = -p$ is

$$y^2 = 4px.$$

The focus is $(p, 0)$ and the x-axis is the axis of symmetry.

Standard Equation of a Parabola with Vertex (h, k) and Vertical Axis of Symmetry

The standard equation of a parabola with vertex (h, k) and vertical axis of symmetry is

$$(x - h)^2 = 4p(y - k),$$

where the vertex is (h, k), the focus is $(h, k + p)$, and the directrix is $y = k - p$.

(continued)

Standard Equation of a Parabola with Vertex (h, k) and Horizontal Axis of Symmetry

The standard equation of a parabola with vertex (h, k) and horizontal axis of symmetry is

$$(y - k)^2 = 4p(x - h),$$

where the vertex is (h, k), the focus is $(h + p, k)$, and the directrix is $x = h - p$.

Standard Equation of a Circle

The standard equation of a circle with center (h, k) and radius r is

$$(x - h)^2 + (y - k)^2 = r^2.$$

Standard Equation of an Ellipse with Center at the Origin

Major axis horizontal

$$\frac{x^2}{a^2} + \frac{y^2}{b^2} = 1, \ a > b > 0$$

Vertices: $(-a, 0), (a, 0)$

y-intercepts: $(0, -b), (0, b)$

Foci: $(-c, 0), (c, 0)$, where $c^2 = a^2 - b^2$

Major axis vertical

$$\frac{x^2}{b^2} + \frac{y^2}{a^2} = 1, \ a > b > 0$$

Vertices: $(0, -a), (0, a)$

x-intercepts: $(-b, 0), (b, 0)$

Foci: $(0, -c), (0, c)$, where $c^2 = a^2 - b^2$

Standard Equation of an Ellipse with Center at (h, k)

Major axis horizontal

$$\frac{(x - h)^2}{a^2} + \frac{(y - k)^2}{b^2} = 1, \ a > b > 0$$

Vertices: $(h - a, k), (h + a, k)$

Length of minor axis: $2b$

Foci: $(h - c, k), (h + c, k)$, where $c^2 = a^2 - b^2$

Major axis vertical

$$\frac{(x - h)^2}{b^2} + \frac{(y - k)^2}{a^2} = 1, \ a > b > 0$$

Vertices: $(h, k - a), (h, k + a)$

Length of minor axis: $2b$

Foci: $(h, k - c), (h, k + c)$, where $c^2 = a^2 - b^2$

Standard Equation of a Hyperbola with Center at the Origin

Transverse axis horizontal

$$\frac{x^2}{a^2} - \frac{y^2}{b^2} = 1$$

Vertices: $(-a, 0), (a, 0)$

Asymptotes: $y = -\frac{b}{a}x, \ y = \frac{b}{a}x$

Foci: $(-c, 0), (c, 0)$, where $c^2 = a^2 + b^2$

Transverse axis vertical

$$\frac{y^2}{a^2} - \frac{x^2}{b^2} = 1$$

Vertices: $(0, -a), (0, a)$

Asymptotes: $y = -\frac{a}{b}x, \ y = \frac{a}{b}x$

Foci: $(0, -c), (0, c)$, where $c^2 = a^2 + b^2$

Standard Equation of a Hyperbola with Center at (h, k)

Transverse axis horizontal

$$\frac{(x-h)^2}{a^2} - \frac{(y-k)^2}{b^2} = 1$$

Vertices: $(h-a, k), (h+a, k)$

Asymptotes: $y - k = \frac{b}{a}(x-h),$

$$y - k = -\frac{b}{a}(x-h)$$

Foci: $(h-c, k), (h+c, k)$, where
$c^2 = a^2 + b^2$

Transverse axis vertical

$$\frac{(y-k)^2}{a^2} - \frac{(x-h)^2}{b^2} = 1$$

Vertices: $(h, k-a), (h, k+a)$

Asymptotes: $y - k = \frac{a}{b}(x-h),$

$$y - k = -\frac{a}{b}(x-h)$$

Foci: $(h, k-c), (h, k+c)$, where
$c^2 = a^2 + b^2$

REVIEW EXERCISES

In Exercises 1–8, match the equation with one of the graphs (a)–(h), which follow.

a)
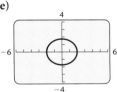

b)
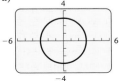

c)
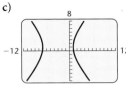

d)
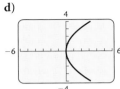

e)
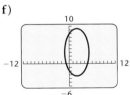

f)
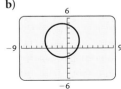

g)
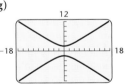

h)
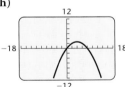

1. $y^2 = 5x$

2. $y^2 = 9 - x^2$

3. $3x^2 + 4y^2 = 12$

4. $9y^2 - 4x^2 = 36$

5. $x^2 + y^2 + 2x - 3y = 8$

6. $4x^2 + y^2 - 16x - 6y = 15$

7. $x^2 - 8x + 6y = 0$

8. $\dfrac{(x+3)^2}{16} - \dfrac{(y-1)^2}{25} = 1$

9. Find an equation of the parabola with directrix $y = \frac{3}{2}$ and focus $\left(0, -\frac{3}{2}\right)$.

10. Find the focus, the vertex, and the directrix of the parabola given by
$$y^2 = -12x.$$

11. Find the vertex, the focus, and the directrix of the parabola given by
$$x^2 + 10x + 2y + 9 = 0.$$

12. Find the center, the vertices, and the foci of the ellipse given by
$$16x^2 + 25y^2 - 64x + 50y - 311 = 0.$$
Then draw the graph.

13. Find an equation of the ellipse having vertices $(0, -4)$ and $(0, 4)$ with minor axis of length 6.

14. Find the center, the vertices, the foci, and the asymptotes of the hyperbola given by
$$x^2 - 2y^2 + 4x + y - \tfrac{1}{8} = 0.$$

15. *Spotlight.* A spotlight has a parabolic cross section that is 2 ft wide at the opening and 1.5 ft deep at the vertex. How far from the vertex is the focus?

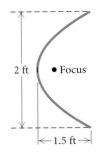

2 ft   • Focus

←1.5 ft→

Solve.

16. $x^2 - 16y = 0,$
$x^2 - y^2 = 64$

17. $4x^2 + 4y^2 = 65,$
$6x^2 - 4y^2 = 25$

18. $x^2 - y^2 = 33,$
$x + y = 11$

19. $x^2 - 2x + 2y^2 = 8,$
$2x + y = 6$

20. $x^2 - y = 3,$
$2x - y = 3$

21. $x^2 + y^2 = 25,$
$x^2 - y^2 = 7$

22. $x^2 - y^2 = 3,$
$y = x^2 - 3$

23. $x^2 + y^2 = 18,$
$2x + y = 3$

24. $x^2 + y^2 = 100,$
$2x^2 - 3y^2 = -120$

25. $x^2 + 2y^2 = 12,$
$xy = 4$

26. *Numerical Relationship.* The sum of two numbers is 11 and the sum of their squares is 65. Find the numbers.

27. *Dimensions of a Rectangle.* A rectangle has a perimeter of 38 m and an area of 84 m². What are the dimensions of the rectangle?

28. *Numerical Relationship.* Find two positive integers whose sum is 12 and the sum of whose reciprocals is $\tfrac{3}{8}$.

29. *Perimeter.* The perimeter of a square is 12 cm more than the perimeter of another square. The area of the first square exceeds the area of the other by 39 cm². Find the perimeter of each square.

30. *Radius of a Circle.* The sum of the areas of two circles is 130π ft². The difference of the areas is 112π ft². Find the radius of each circle.

Collaborative Discussion and Writing

31. What would you say to a classmate who tells you that it is always possible to visualize all of the solutions of a nonlinear system of equations?

32. Is a circle a special type of ellipse? Why or why not?

Synthesis

33. Find an equation of the ellipse containing the point $(-1/2, 3\sqrt{3}/2)$ and with vertices $(0, -3)$ and $(0, 3)$.

34. Find two numbers whose product is 4 and the sum of whose reciprocals is $\tfrac{65}{56}$.

35. Find an equation of the circle that passes through the points $(10, 7)$, $(-6, 7)$, and $(-8, 1)$.

36. *Navigation.* Two radio transmitters positioned 400 mi apart along the shore send simultaneous signals to a ship that is 250 mi offshore, sailing parallel to the shoreline. The signal from transmitter A reaches the ship 300 microseconds before the signal from transmitter B. The signals travel at a speed of 186,000 miles per second, or 0.186 mile per microsecond. Find the equation of the hyperbola with foci A and B on which the ship is located. (*Hint:* For any point on the hyperbola, the absolute value of the difference of its distances from the foci is $2a$.)

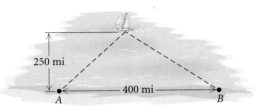

250 mi

A   ← 400 mi →   B

9 Chapter Test

In Exercises 1–4, match the equation with one of the graphs (a)–(d), which follow.

a)
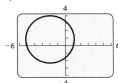

b)
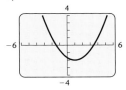

c)
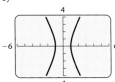

d)
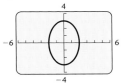

1. $4x^2 - y^2 = 4$ **2.** $x^2 - 2x - 3y = 5$

3. $x^2 + 4x + y^2 - 2y - 4 = 0$

4. $9x^2 + 4y^2 = 36$

5. Find an equation of the parabola with focus $(0, 2)$ and directrix $y = -2$.

6. Find the vertex of the parabola given by $y^2 - 8y + 2x - 4 = 0$.

7. Find the center and the radius of the circle given by $x^2 + y^2 + 2x - 6y - 15 = 0$.

8. Find an equation of the ellipse having vertices $(0, -5)$ and $(0, 5)$ and with minor axis of length 4.

9. Find the asymptotes of the hyperbola given by $2y^2 - x^2 = 18$.

10. *Satellite Dish.*   A satellite dish has a parabolic cross section that is 18 in. wide at the opening and 6 in. deep at the vertex. How far from the vertex is the focus?

Solve.

11. $2x^2 - 3y^2 = -10,$
      $x^2 + 2y^2 = 9$

12. $x^2 + y^2 = 13,$
      $x + y = 1$

13. $x + y = 5,$
      $xy = 6$

14. *Landscaping.*   Leisurescape is planting a rectangular flower garden with a perimeter of 18 ft and a diagonal of $\sqrt{41}$ ft. Find the dimensions of the garden.

Synthesis

15. Find an equation of the circle satisfying the following conditions: endpoints of a diameter: $(1, 1)$ and $(5, -3)$; center on the line $x - y = 4$.

Sequences, Series, and Combinatorics 10

We begin this chapter with a study of a special type of function whose domain is the set of positive integers, a *sequence*. When we add the terms of a sequence, we get a *series*. We also study a method of proof known as *mathematical induction* and we find a method for expanding binomials $(a + b)^n$, where n is a natural number. In addition, we then study *combinatorics*, methods for counting the arrangements and combinations of objects in a set and, finally, we calculate the *probability* that an event will occur.

APPLICATION

To create a college fund, a parent makes a sequence of 18 equal yearly deposits of $1000 in a savings account on which interest is compounded annually at 6.2%. Find the amount of the annuity.

This problem appears as Exercise 55 in Section 10.3.

10.1

Sequences and Series

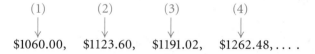

- *Find terms of sequences given the nth term.*
- *Look for a pattern in a sequence and try to determine a general term.*
- *Convert between sigma notation and other notation for a series.*
- *Construct the terms of a recursively defined sequence.*

In this section, we discuss sets or lists of numbers, considered in order, and their sums.

Sequences

Suppose that $1000 is invested at 6%, compounded annually. The amounts to which the account will grow after 1 yr, 2 yr, 3 yr, 4 yr, and so on, form the following sequence of numbers:

$$
\begin{array}{cccc}
(1) & (2) & (3) & (4) \\
\downarrow & \downarrow & \downarrow & \downarrow \\
\$1060.00, & \$1123.60, & \$1191.02, & \$1262.48, \ldots\,.
\end{array}
$$

We can think of this as a function that pairs 1 with $1060.00, 2 with $1123.60, 3 with $1191.02, and so on. A **sequence** is thus a *function,* where the domain is a set of consecutive positive integers beginning with 1.

If we continue to compute the amounts of money in the account forever, we obtain an **infinite sequence** with function values

$1060.00, $1123.60, $1191.02, $1262.48, $1338.23, $1418.52, ...

The dots at the end "..." indicate that the sequence goes on without stopping. If we stop after a certain number of years, we obtain a **finite sequence:**

$1060.00, $1123.60, $1191.02, $1262.48.

Sequences

An **infinite sequence** is a function having for its domain the set of positive integers, $\{1, 2, 3, 4, 5, \ldots\}$.

A **finite sequence** is a function having for its domain a set of positive integers, $\{1, 2, 3, 4, 5, \ldots, n\}$, for some positive integer n.

Consider the sequence given by the formula

$$a(n) = 2^n, \quad \text{or} \quad a_n = 2^n.$$

Some of the function values, also known as the **terms** of the sequence, follow:

$$a_1 = 2^1 = 2,$$
$$a_2 = 2^2 = 4,$$
$$a_3 = 2^3 = 8,$$
$$a_4 = 2^4 = 16,$$
$$a_5 = 2^5 = 32.$$

The first term of the sequence is denoted as a_1, the fifth term as a_5, and the nth term, or **general term,** as a_n. This sequence can also be denoted as

$$2, 4, 8, \ldots, \quad \text{or as} \quad 2, 4, 8, \ldots, 2^n, \ldots.$$

EXAMPLE 1 Find the first 4 terms and the 23rd term of the sequence whose general term is given by $a_n = (-1)^n n^2$.

Solution We have $a_n = (-1)^n n^2$, so

$$a_1 = (-1)^1 \cdot 1^2 = -1,$$
$$a_2 = (-1)^2 \cdot 2^2 = 4,$$
$$a_3 = (-1)^3 \cdot 3^2 = -9,$$
$$a_4 = (-1)^4 \cdot 4^2 = 16,$$
$$a_{23} = (-1)^{23} \cdot 23^2 = -529.$$

Note in Example 1 that the power $(-1)^n$ causes the signs of the terms to alternate between positive and negative, depending on whether n is even or odd. This kind of sequence is called an **alternating sequence.**

Technology Connection

We can use a graphing calculator to find the desired terms of the sequence in Example 1. We enter $y_1 = (-1)^x x^2$. We then set up a table in ASK mode and enter 1, 2, 3, 4, and 23 as values for x.

We can also use the SEQ feature to find the terms of a sequence. Suppose, for example, that we want to find the first 5 terms of the sequence whose general term is given by $a_n = n/(n+1)$. We select SEQ from the LIST OPS menu and enter the general term, the variable, and the numbers of the first and last terms desired. The calculator will write the terms horizontally as a list. The list can also be written in fractional notation.

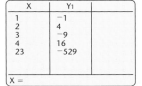

X	Y₁
1	−1
2	4
3	−9
4	16
23	−529

X =

```
seq(X/(X+1),X,1,5)▶Frac
{1/2 2/3 3/4 4/...
```

```
seq(X/(X+1),X,1,5)▶Frac
.../3 3/4 4/5 5/6}
```

We use the ▷ key to view the two items that do not initially appear on the screen. The first 5 terms of the sequence are 1/2, 2/3, 3/4, 4/5, and 5/6.

STUDY TIP

Refer to the *Graphing Calculator Manual* that accompanies this text to find the keystrokes for finding the terms of a sequence.

Finding the General Term

When only the first few terms of a sequence are known, we do not know for sure what the general term is, but we might be able to make a prediction by looking for a pattern.

EXAMPLE 2 For each of the following sequences, predict the general term.

a) $1, \sqrt{2}, \sqrt{3}, 2, \ldots$

b) $-1, 3, -9, 27, -81, \ldots$

c) $2, 4, 8, \ldots$

Solution

a) These are square roots of consecutive integers, so the general term might be $\sqrt{n}$.

b) These are powers of 3 with alternating signs, so the general term might be $(-1)^n 3^{n-1}$.

c) If we see the pattern of powers of 2, we will see 16 as the next term and guess 2^n for the general term. Then the sequence could be written with more terms as

$$2, 4, 8, 16, 32, 64, 128, \ldots .$$

If we see that we can get the second term by adding 2, the third term by adding 4, and the next term by adding 6, and so on, we will see 14 as the next term. A general term for the sequence is $n^2 - n + 2$, and the sequence can be written with more terms as

$$2, 4, 8, 14, 22, 32, 44, 58, \ldots .$$

Example 2(c) illustrates that, in fact, you can never be certain about the general term. The fewer the given terms, the greater the uncertainty.

Sums and Series

> ### Series
> Given the infinite sequence
>
> $$a_1, a_2, a_3, a_4, \ldots, a_n, \ldots,$$
>
> the sum of the terms
>
> $$a_1 + a_2 + a_3 + \cdots + a_n + \cdots$$
>
> is called an **infinite series.** A **partial sum** is the sum of the first n terms:
>
> $$a_1 + a_2 + a_3 + \cdots + a_n.$$
>
> A partial sum is also called a **finite series,** or **nth partial sum,** and is denoted S_n.

EXAMPLE 3   For the sequence $-2, 4, -6, 8, -10, 12, -14, \ldots$, find each of the following.

a) S_1 **b)** S_4 **c)** S_5

Solution

a) $S_1 = -2$

b) $S_4 = -2 + 4 + (-6) + 8 = 4$

c) $S_5 = -2 + 4 + (-6) + 8 + (-10) = -6$

Technology Connection

We can use a graphing calculator to find partial sums of a sequence when a formula for the general term is known. Suppose, for example, that we want to find S_1, S_2, S_3, and S_4 for the sequence whose general term is given by $a_n = n^2 - 3$. We can use the CUMSUM feature from the LIST OPS menu. The calculator will write the partial sums as a list. (Note that the calculator can be set in either FUNCTION mode or SEQUENCE mode. Here we show SEQUENCE mode.)

```
cumSum(seq(n²−3,n,1,4))
              {−2 −1 5 18}
```

We have $S_1 = -2$, $S_2 = -1$, $S_3 = 5$, and $S_4 = 18$.

Sigma Notation

The Greek letter Σ (sigma) can be used to simplify notation when the general term of a sequence is a formula. For example, the sum of the first four terms of the sequence $3, 5, 7, 9, \ldots, 2k + 1, \ldots$ can be named as follows, using what is called **sigma notation,** or **summation notation:**

$$\sum_{k=1}^{4} (2k + 1).$$

This is read "the sum as k goes from 1 to 4 of $(2k + 1)$." The letter k is called the **index of summation.** Sometimes the index of summation starts at a number other than 1, and sometimes letters other than k are used.

EXAMPLE 4   Find and evaluate each of the following sums.

a) $\displaystyle\sum_{k=1}^{5} k^3$ **b)** $\displaystyle\sum_{k=0}^{4} (-1)^k 5^k$ **c)** $\displaystyle\sum_{i=8}^{11} \left(2 + \frac{1}{i}\right)$

Technology Connection

We can combine the SUM and
SEQ features on a graphing
calculator to add the terms of
a sequence. We find the sum
in Example 4(a) as shown
below.

sum(seq(n^3,n,1,5))	
	225

Solution

a) We replace k with 1, 2, 3, 4, and 5. Then we add the results.

$$\sum_{k=1}^{5} k^3 = 1^3 + 2^3 + 3^3 + 4^3 + 5^3$$
$$= 1 + 8 + 27 + 64 + 125$$
$$= 225$$

b) $\displaystyle\sum_{k=0}^{4} (-1)^k 5^k = (-1)^0 5^0 + (-1)^1 5^1 + (-1)^2 5^2 + (-1)^3 5^3 + (-1)^4 5^4$

$$= 1 - 5 + 25 - 125 + 625 = 521$$

c) $\displaystyle\sum_{i=8}^{11} \left(2 + \frac{1}{i}\right) = \left(2 + \frac{1}{8}\right) + \left(2 + \frac{1}{9}\right) + \left(2 + \frac{1}{10}\right) + \left(2 + \frac{1}{11}\right)$

$$= 8\frac{1691}{3960}$$

EXAMPLE 5   Write sigma notation for each sum.

a) $1 + 2 + 4 + 8 + 16 + 32 + 64$

b) $-2 + 4 - 6 + 8 - 10$

c) $x + \dfrac{x^2}{2} + \dfrac{x^3}{3} + \dfrac{x^4}{4} + \cdots$

Solution

a) $1 + 2 + 4 + 8 + 16 + 32 + 64$

This is a sum of powers of 2, beginning with 2^0, or 1, and ending with 2^6, or 64. Sigma notation is $\sum_{k=0}^{6} 2^k$.

b) $-2 + 4 - 6 + 8 - 10$

Disregarding the alternating signs, we see that this is the sum of the first 5 even integers. Note that $2k$ is a formula for the kth positive even integer, and $(-1)^k = -1$ when k is odd and $(-1)^k = 1$ when k is even. Thus the general term is $(-1)^k(2k)$. The sum begins with $k = 1$ and ends with $k = 5$, so sigma notation is $\sum_{k=1}^{5} (-1)^k(2k)$.

c) $x + \dfrac{x^2}{2} + \dfrac{x^3}{3} + \dfrac{x^4}{4} + \cdots$

The general term is x^k/k, beginning with $k = 1$. This is also an infinite series. We use the symbol ∞ for infinity and write the series using sigma notation: $\sum_{k=1}^{\infty} (x^k/k)$.

Recursive Definitions

A sequence may be defined **recursively** or by using a **recursion formula.** Such a definition lists the first term, or the first few terms, and then describes how to determine the remaining terms from the given terms.

EXAMPLE 6 Find the first 5 terms of the sequence defined by

$$a_1 = 5, \qquad a_{k+1} = 2a_k - 3, \quad \text{for } k \geq 1.$$

Solution

$$a_1 = 5,$$

$$a_2 = 2a_1 - 3 = 2 \cdot 5 - 3 = 7,$$

$$a_3 = 2a_2 - 3 = 2 \cdot 7 - 3 = 11,$$

$$a_4 = 2a_3 - 3 = 2 \cdot 11 - 3 = 19,$$

$$a_5 = 2a_4 - 3 = 2 \cdot 19 - 3 = 35.$$

Technology Connection

Many graphing calculators have the capability to work with recursively defined sequences when they are set in SEQUENCE mode. In Example 6, for instance, the function could be entered as $u(n) = 2 * u(n - 1) - 3$ with $u(n\text{Min}) = 5$. We can read the terms of the sequence from a table.

Plot1	Plot2	Plot3
nMin=1		
\u(n)◼2*u(n−1)−3		
u(nMin)◼{5}		
\v(n)=		
v(nMin)=		
\w(n)=		

n	$u(n)$	
1	5	
2	7	
3	11	
4	19	
5	35	
6	67	
7	131	
$n = 1$		

Exercise Set

In each of the following, the nth term of a sequence is given. Find the first 4 terms, a_{10}, and a_{15}.

1. $a_n = 4n - 1$

2. $a_n = (n - 1)(n - 2)(n - 3)$

3. $a_n = \dfrac{n}{n - 1}, \ n \geq 2$

4. $a_n = n^2 - 1, \ n \geq 3$

5. $a_n = \dfrac{n^2 - 1}{n^2 + 1}$

6. $a_n = \left(-\dfrac{1}{2}\right)^{n-1}$

7. $a_n = (-1)^n n^2$

8. $a_n = (-1)^{n-1}(3n - 5)$

9. $a_n = 5 + \dfrac{(-2)^{n+1}}{2^n}$

10. $a_n = \dfrac{2n - 1}{n^2 + 2n}$

Find the indicated term of the given sequence.

11. $a_n = 5n - 6$; a_8

12. $a_n = (3n - 4)(2n + 5)$; a_7

13. $a_n = (2n + 3)^2$; a_6

14. $a_n = (-1)^{n-1}(4.6n - 18.3)$; a_{12}

15. $a_n = 5n^2(4n - 100)$; a_{11}

16. $a_n = \left(1 + \dfrac{1}{n}\right)^2$; a_{80}

17. $a_n = \ln e^n$; a_{67}

18. $a_n = 2 - \dfrac{1000}{n}$; a_{100}

Predict the general, or nth term, a_n, of the sequence.
Answers may vary.

19. 2, 4, 6, 8, 10, ...

20. 3, 9, 27, 81, 243, ...

21. −2, 6, −18, 54, ...

22. −2, 3, 8, 13, 18, ...

23. $\frac{2}{3}, \frac{3}{4}, \frac{4}{5}, \frac{5}{6}, \frac{6}{7}, \ldots$

24. $\sqrt{2}, 2, \sqrt{6}, 2\sqrt{2}, \sqrt{10}, \ldots$

25. $1 \cdot 2, 2 \cdot 3, 3 \cdot 4, 4 \cdot 5, \ldots$

26. −1, −4, −7, −10, −13, ...

27. 0, log 10, log 100, log 1000, ...

28. $\ln e^2$, $\ln e^3$, $\ln e^4$, $\ln e^5$, ...

Find the indicated partial sums for the sequence.

29. 1, 2, 3, 4, 5, 6, 7, ...; S_3 and S_7

30. 1, −3, 5, −7, 9, −11, ...; S_2 and S_5

31. 2, 4, 6, 8, ...; S_4 and S_5

32. $1, \frac{1}{4}, \frac{1}{9}, \frac{1}{16}, \frac{1}{25}, \ldots$; S_1 and S_5

Find and evaluate the sum.

33. $\displaystyle\sum_{k=1}^{5} \dfrac{1}{2k}$

34. $\displaystyle\sum_{i=1}^{6} \dfrac{1}{2i + 1}$

35. $\displaystyle\sum_{i=0}^{6} 2^i$

36. $\displaystyle\sum_{k=4}^{7} \sqrt{2k - 1}$

37. $\displaystyle\sum_{k=7}^{10} \ln k$

38. $\displaystyle\sum_{k=1}^{4} \pi k$

39. $\displaystyle\sum_{k=1}^{8} \dfrac{k}{k + 1}$

40. $\displaystyle\sum_{i=1}^{5} \dfrac{i - 1}{i + 3}$

41. $\displaystyle\sum_{i=1}^{5} (-1)^i$

42. $\displaystyle\sum_{k=0}^{5} (-1)^{k+1}$

43. $\displaystyle\sum_{k=1}^{8} (-1)^{k+1}3k$

44. $\displaystyle\sum_{k=0}^{7} (-1)^k 4^{k+1}$

45. $\displaystyle\sum_{k=0}^{6} \dfrac{2}{k^2 + 1}$

46. $\displaystyle\sum_{i=1}^{10} i(i + 1)$

47. $\displaystyle\sum_{k=0}^{5} (k^2 - 2k + 3)$

48. $\displaystyle\sum_{k=1}^{10} \dfrac{1}{k(k + 1)}$

49. $\displaystyle\sum_{i=0}^{10} \dfrac{2^i}{2^i + 1}$

50. $\displaystyle\sum_{k=0}^{3} (-2)^{2k}$

Write sigma notation.

51. $5 + 10 + 15 + 20 + 25 + \cdots$

52. $7 + 14 + 21 + 28 + 35 + \cdots$

53. $2 - 4 + 8 - 16 + 32 - 64$

54. $3 + 6 + 9 + 12 + 15$

55. $-\dfrac{1}{2} + \dfrac{2}{3} - \dfrac{3}{4} + \dfrac{4}{5} - \dfrac{5}{6} + \dfrac{6}{7}$

56. $\dfrac{1}{1^2} + \dfrac{1}{2^2} + \dfrac{1}{3^2} + \dfrac{1}{4^2} + \dfrac{1}{5^2}$

57. $4 - 9 + 16 - 25 + \cdots + (-1)^n n^2$

58. $9 - 16 + 25 + \cdots + (-1)^{n+1} n^2$

59. $\dfrac{1}{1 \cdot 2} + \dfrac{1}{2 \cdot 3} + \dfrac{1}{3 \cdot 4} + \dfrac{1}{4 \cdot 5} + \cdots$

60. $\dfrac{1}{1 \cdot 2^2} + \dfrac{1}{2 \cdot 3^2} + \dfrac{1}{3 \cdot 4^2} + \dfrac{1}{4 \cdot 5^2} + \cdots$

Find the first 4 terms of the recursively defined sequence.

61. $a_1 = 4$, $a_{k+1} = 1 + \dfrac{1}{a_k}$

62. $a_1 = 256$, $a_{k+1} = \sqrt{a_k}$

63. $a_1 = 6561$, $a_{k+1} = (-1)^k \sqrt{a_k}$

64. $a_1 = e^Q$, $a_{k+1} = \ln a_k$

65. $a_1 = 2$, $a_2 = 3$, $a_{k+1} = a_k + a_{k-1}$

66. $a_1 = -10$, $a_2 = 8$, $a_{k+1} = a_k - a_{k-1}$

67. *Compound Interest.* Suppose that $1000 is invested at 6.2%, compounded annually. The value of the investment after n years is given by the sequence model

$$a_n = \$1000(1.062)^n, \quad n = 1, 2, 3, \ldots.$$

a) Find the first 10 terms of the sequence.
b) Find the value of the investment after 20 yr.

68. *Salvage Value.* The value of an office machine is $5200. Its salvage value each year is 75% of its value the year before. Give a sequence that lists the salvage value of the machine for each year of a 10-yr period.

69. *Bacteria Growth.* A single cell of bacteria divides into two every 15 min. Suppose that the same rate of division is maintained for 4 hr. Give a sequence that lists the number of cells after successive 15-min periods.

70. *Salary Sequence.* Torrey is paid $6.30 per hour for working at Red Freight Limited. Each year he receives a $0.40 hourly raise. Give a sequence that lists Torrey's hourly salary over a 10-yr period.

71. *Fibonacci Sequence: Rabbit Population Growth.* One of the most famous recursively defined sequences is the *Fibonacci sequence*. In 1202, the Italian mathematician Leonardo Fibonacci (also called Leonardo da Pisa) proposed the following model for rabbit population growth. Start with one pair of rabbits, one female and one male. These rabbits mature and produce a new pair, again one female and one male. The first pair lives until each pair can reproduce, and then each produces a new pair. This pattern continues. The population of mature rabbits can be modeled by the following recursively defined sequence:

$$a_1 = 1, \quad a_2 = 1, \quad a_{k+1} = a_k + a_{k-1}, \quad \text{for } k \geq 2,$$

where a_k is the total number of pairs of rabbits after $k - 2$ reproductions. Find the first 7 terms of the Fibonacci sequence.

Technology Connection

Use a graphing calculator to construct a table of values for the first 10 terms of the sequence.

72. $a_n = \sqrt{n+1} - \sqrt{n}$

73. $a_n = \left(1 + \dfrac{1}{n}\right)^n$

74. $a_1 = 2, \quad a_{k+1} = \dfrac{1}{2}\left(a_k + \dfrac{2}{a_k}\right)$

75. $a_1 = 2, \quad a_{k+1} = \sqrt{1 + \sqrt{a_k}}$

76. *Household Discretionary Income by Age.* The following table contains data relating household discretionary income to age.

AGE, x	HOUSEHOLD DISCRETIONARY INCOME, y
25	$10,223
34	15,607
54	21,173
64	18,257
70	13,747

Source: U.S. Bureau of the Census

a) Use a graphing calculator to fit a quadratic sequence regression function

$$a_n = an^2 + bn + c$$

to the data.

b) Find the household discretionary income of people whose age is 16 yr, 20 yr, 40 yr, 60 yr, and 75 yr. Round to the nearest dollar.

77. *Driver Fatalities by Age.* The number of licensed drivers per 100,000 who died in motor vehicle accidents at age n can be estimated by the sequence model

$$a_n = 0.018n^2 - 1.7n + 48.7, \quad n = 15, 16, 17, \ldots, 90.$$

a) Find the first 10 terms of the sequence.

b) Find the number of fatalities per 100,000 of those drivers whose age is 16 yr, 23 yr, 50 yr, 75 yr, and 85 yr.

Collaborative Discussion and Writing

78. a) Find the first few terms of the sequence $a_n = n^2 - n + 41$ and describe the pattern you observe.

b) Does the pattern you found in part (a) hold for all choices of n? Why or why not?

79. The Fibonacci sequence has intrigued mathematicians for centuries. In fact, a journal called the *Fibonacci Quarterly* is devoted to publishing the results pertaining to such sequences. Do some research on the connection of the Fibonacci sequence to the idea of the "Golden Section."

Skill Maintenance

Solve.

80. $3x - 2y = 3,$
$\quad 2x + 3y = -11$

81. *Research and Development.* Sun Microsystems, Inc., spent a total of $1.19 billion on original research and development and on the purchase of in-process research and development in 1998. The amount spent on original research and development was $0.837 billion more than the amount spent on purchased research and development. (*Source*: Sun Microsystems, Inc., 1998 Annual Report) Find the amount spent on each category.

Find the center and the radius of the circle with the given equation.

82. $x^2 + y^2 - 6x + 4y = 3$

83. $x^2 + y^2 + 5x - 8y = 2$

Synthesis

Find the first 5 terms of the sequence, and then find S_5.

84. $a_n = \dfrac{1}{2^n} \log 1000^n$

85. $a_n = i^n, \ i = \sqrt{-1}$

86. $a_n = \ln (1 \cdot 2 \cdot 3 \cdots n)$

For each sequence, find a formula for S_n.

87. $a_n = \ln n$

88. $a_n = \dfrac{1}{n} - \dfrac{1}{n+1}$

10.2

Arithmetic Sequences and Series

- *For any arithmetic sequence, find the nth term when n is given and n when the nth term is given, and given two terms, find the common difference and construct the sequence.*
- *Find the sum of the first n terms of an arithmetic sequence.*

A sequence in which each term after the first is found by adding the same number to the preceding term is an **arithmetic sequence.**

Arithmetic Sequences

The sequence 2, 5, 8, 11, 14, 17, ... is arithmetic because adding 3 to any term produces the next term. In other words, the difference between any term and the preceding one is 3. Arithmetic sequences are also called *arithmetic progressions.*

Arithmetic Sequence

A sequence is **arithmetic** if there exists a number d, called the **common difference,** such that $a_{n+1} = a_n + d$ for any integer $n \geq 1$.

EXAMPLE 1 For each of the following arithmetic sequences, identify the first term, a_1, and the common difference, d.

a) 4, 9, 14, 19, 24, ...

b) 34, 27, 20, 13, 6, −1, −8, ...

c) 2, $2\frac{1}{2}$, 3, $3\frac{1}{2}$, 4, $4\frac{1}{2}$, ...

Solution The first term, a_1, is the first term listed. To find the common difference, d, we choose any term beyond the first and subtract the preceding term from it.

SEQUENCE	FIRST TERM, a_1	COMMON DIFFERENCE, d
a) 4, 9, 14, 19, 24, ...	4	5 $(9 - 4 = 5)$
b) 34, 27, 20, 13, 6, -1, -8, ...	34	-7 $(27 - 34 = -7)$
c) 2, $2\frac{1}{2}$, 3, $3\frac{1}{2}$, 4, $4\frac{1}{2}$, ...	2	$\frac{1}{2}$ $\left(2\frac{1}{2} - 2 = \frac{1}{2}\right)$

We obtained the common difference by subtracting a_1 from a_2. Had we subtracted a_2 from a_3 or a_3 from a_4, we would have obtained the same values for d. Thus we can check by adding d to each term in a sequence to see if we progress correctly to the next term.

CHECK:

a) $4 + 5 = 9$, $9 + 5 = 14$, $14 + 5 = 19$, $19 + 5 = 24$

b) $34 + (-7) = 27$, $27 + (-7) = 20$, $20 + (-7) = 13$,
$13 + (-7) = 6$, $6 + (-7) = -1$, $-1 + (-7) = -8$

c) $2 + \frac{1}{2} = 2\frac{1}{2}$, $2\frac{1}{2} + \frac{1}{2} = 3$, $3 + \frac{1}{2} = 3\frac{1}{2}$, $3\frac{1}{2} + \frac{1}{2} = 4$,
$4 + \frac{1}{2} = 4\frac{1}{2}$

To find a formula for the general, or nth, term of any arithmetic sequence, we denote the common difference by d, write out the first few terms, and look for a pattern:

a_1,

$a_2 = a_1 + d$,

$a_3 = a_2 + d = (a_1 + d) + d = a_1 + 2d$ Substituting for a_2

$a_4 = a_3 + d = (a_1 + 2d) + d = a_1 + 3d$ Substituting for a_3

Note that the coefficient of d in each case is 1 less than the subscript.

Generalizing, we obtain the following formula.

nth Term of an Arithmetic Sequence

The **nth term** of an arithmetic sequence is given by
$a_n = a_1 + (n - 1)d$, for any integer $n \geq 1$.

EXAMPLE 2 Find the 14th term of the arithmetic sequence 4, 7, 10, 13, ...

Solution We first note that $a_1 = 4$, $d = 3$, and $n = 14$. Then using the formula for the nth term, we obtain

$$a_n = a_1 + (n - 1)d$$
$$a_{14} = 4 + (14 - 1) \cdot 3 = 4 + 13 \cdot 3 = 4 + 39 = 43.$$

The 14th term is 43.

EXAMPLE 3 In the sequence of Example 2, which term is 301? That is, find n if $a_n = 301$.

Solution We substitute 301 for a_n, 4 for a_1, and 3 for d in the formula for the nth term and solve for n:

$$a_n = a_1 + (n - 1)d$$
$$301 = 4 + (n - 1) \cdot 3 \qquad \text{Substituting}$$
$$301 = 4 + 3n - 3$$
$$301 = 3n + 1$$
$$300 = 3n \qquad\qquad \text{Solving for } n$$
$$100 = n.$$

The term 301 is the 100th term of the sequence. ▬

Given two terms and their places in an arithmetic sequence, we can construct the sequence.

EXAMPLE 4 The 3rd term of an arithmetic sequence is 8, and the 16th term is 47. Find a_1 and d and construct the sequence.

Solution We know that $a_3 = 8$ and $a_{16} = 47$. Thus we would have to add d 13 times to get 47 from 8. That is,

$$8 + 13d = 47. \qquad a_3 \text{ and } a_{16} \text{ are 13 terms apart.}$$

Solving $8 + 13d = 47$, we obtain

$$13d = 39$$
$$d = 3.$$

Since $a_3 = 8$, we subtract d twice to get a_1. Thus,

$$a_1 = 8 - 2 \cdot 3 = 2. \qquad a_1 \text{ and } a_3 \text{ are 2 terms apart.}$$

The sequence is 2, 5, 8, 11, ... Note that we could also subtract d 15 times from a_{16} in order to find a_1. ▬

In general, d should be subtracted $n - 1$ times from a_n in order to find a_1.

Sum of the First n Terms of an Arithmetic Sequence

Consider the arithmetic sequence

$$3, 5, 7, 9, \ldots .$$

When we add the first 4 terms of the sequence, we get S_4, which is

$$3 + 5 + 7 + 9, \quad \text{or} \quad 24.$$

This sum is called an **arithmetic series.** To find a formula for the sum of the first n terms, S_n, of an arithmetic sequence, we first denote an arithmetic sequence, as follows:

> This term is two terms back from the last. If you add d to this term, the result is the next-to-last term, $a_n - d$.

$$a_1, \quad (a_1 + d), \quad (a_1 + 2d), \quad \ldots, \quad (a_n - 2d), \quad (a_n - d), \quad a_n.$$

> This is the next-to-last term. If you add d to this term, the result is a_n.

Then S_n is given by

$$S_n = a_1 + (a_1 + d) + (a_1 + 2d) + \cdots + (a_n - 2d)$$
$$+ (a_n - d) + a_n. \tag{1}$$

Reversing the order of the addition gives us

$$S_n = a_n + (a_n - d) + (a_n - 2d) + \cdots + (a_1 + 2d)$$
$$+ (a_1 + d) + a_1. \tag{2}$$

If we add corresponding terms of each side of equations (1) and (2), we get

$$2S_n = [a_1 + a_n] + [(a_1 + d) + (a_n - d)] + [(a_1 + 2d) + (a_n - 2d)]$$
$$+ \cdots + [(a_n - 2d) + (a_1 + 2d)]$$
$$+ [(a_n - d) + (a_1 + d)] + [a_n + a_1]. \qquad \text{There are } n \text{ pairs of square brackets.}$$

This simplifies to

$$2S_n = [a_1 + a_n] + [a_1 + a_n] + [a_1 + a_n] + \cdots + [a_n + a_1]$$
$$+ [a_n + a_1] + [a_n + a_1].$$

Since $a_1 + a_n$ is being added n times, it follows that

$$2S_n = n(a_1 + a_n),$$

from which we get the following formula.

Sum of the First n Terms

The sum of the first n terms of an arithmetic sequence is given by

$$S_n = \frac{n}{2}(a_1 + a_n).$$

EXAMPLE 5   Find the sum of the first 100 natural numbers.

Solution   The sum is

$$1 + 2 + 3 + \cdots + 99 + 100.$$

This is the sum of the first 100 terms of the arithmetic sequence for which

$$a_1 = 1, \qquad a_n = 100, \quad \text{and} \quad n = 100.$$

Thus substituting into the formula

$$S_n = \frac{n}{2}(a_1 + a_n),$$

we get

$$S_{100} = \frac{100}{2}(1 + 100) = 50(101) = 5050.$$

The sum of the first 100 natural numbers is 5050.   ▬

EXAMPLE 6   Find the sum of the first 15 terms of the arithmetic sequence 4, 7, 10, 13, ...

Solution   Note that $a_1 = 4$, $d = 3$, and $n = 15$. Before using the formula

$$S_n = \frac{n}{2}(a_1 + a_n),$$

we find the last term, a_{15}:

$$a_{15} = 4 + (15 - 1)3 \qquad \text{Substituting into the formula } a_n = a_1 + (n - 1)d$$
$$= 4 + 14 \cdot 3 = 46.$$

Thus,

$$S_{15} = \frac{15}{2}(4 + 46) = \frac{15}{2}(50) = 375.$$

The sum of the first 15 terms is 375.   ▬

EXAMPLE 7   Find the sum: $\displaystyle\sum_{k=1}^{130}(4k + 5)$.

Solution   It is helpful to first write out a few terms:

$$9 + 13 + 17 + \cdots .$$

It appears that this is an arithmetic series coming from an arithmetic sequence with $a_1 = 9$, $d = 4$, and $n = 130$. Before using the formula

$$S_n = \frac{n}{2}(a_1 + a_n),$$

we find the last term, a_{130}:

$$a_{130} = 4 \cdot 130 + 5 \qquad \text{The } k\text{th term is } 4k + 5.$$
$$= 520 + 5$$
$$= 525.$$

Thus,

$$S_{130} = \frac{130}{2}(9 + 525) \qquad \text{Substituting into } S_n = \frac{n}{2}(a_1 + a_n)$$
$$= 34{,}710.$$

Applications

The translation of some applications and problem-solving situations may involve arithmetic sequences or series. We consider some examples.

EXAMPLE 8 *Hourly Wages.* Gloria accepts a job, starting with an hourly wage of $14.25, and is promised a raise of 15¢ per hour every 2 months for 5 yr. At the end of 5 yr, what will Gloria's hourly wage be?

Solution It helps to first write down the hourly wage for several 2-month time periods:

Beginning:	$14.25,
After two months:	14.40,
After four months:	14.55,
and so on.	

What appears is a sequence of numbers: 14.25, 14.40, 14.55, ... This sequence is arithmetic, because adding $0.15 each time gives us the next term.

We want to find the last term of an arithmetic sequence, so we use the formula $a_n = a_1 + (n - 1)d$. We know that $a_1 = 14.25$ and $d = 0.15$, but what is n? That is, how many terms are in the sequence? Each year there are 6 raises, since Gloria gets a raise every 2 months. There are 5 yr, so the total number of raises will be $5 \cdot 6$, or 30. Thus there will be 31 terms: the original wage and 30 increased rates.

Substituting in the formula $a_n = a_1 + (n - 1)d$ gives us

$$a_{31} = 14.25 + (31 - 1) \cdot 0.15$$
$$= \$18.75.$$

Thus, at the end of 5 yr, Gloria's hourly wage will be $18.75.

The calculations in Example 8 could be done in a number of ways. There is often a variety of ways in which a problem can be solved. In this chapter, we will concentrate on the use of sequences and series and their related formulas.

EXAMPLE 9   *Total in a Stack.* A stack of telephone poles has 30 poles in the bottom row. There are 29 poles in the second row, 28 in the next row, and so on. How many poles are in the stack if there are 5 poles in the top row?

Solution   A picture will help in this case. The following figure shows the ends of the poles and the way in which they stack.

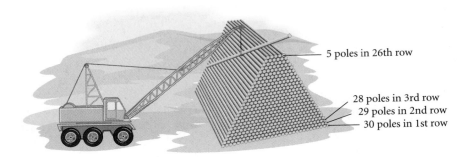

5 poles in 26th row

28 poles in 3rd row
29 poles in 2nd row
30 poles in 1st row

Since the number of poles goes from 30 in a row up to 5 in the top row, there must be 26 rows. We want the sum

$$30 + 29 + 28 + \cdots + 5.$$

Thus we have an arithmetic series. We use the formula

$$S_n = \frac{n}{2}(a_1 + a_n),$$

with $n = 26$, $a_1 = 30$, and $a_{26} = 5$.
    Substituting, we get

$$S_{26} = \frac{26}{2}(30 + 5) = 455.$$

There are 455 poles in the stack.

Exercise Set

Find the first term and the common difference.

1. 3, 8, 13, 18, ...

2. $1.08, $1.16, $1.24, $1.32, ...

3. 9, 5, 1, −3, ...

4. −8, −5, −2, 1, 4, ...

5. $\frac{3}{2}, \frac{9}{4}, 3, \frac{15}{4}, \ldots$

6. $\frac{3}{5}, \frac{1}{10}, -\frac{2}{5}, \ldots$

7. $316, $313, $310, $307, ...

8. Find the 11th term of the arithmetic sequence 0.07, 0.12, 0.17, ...

9. Find the 12th term of the arithmetic sequence 2, 6, 10, ...

10. Find the 17th term of the arithmetic sequence 7, 4, 1, ...

11. Find the 14th term of the arithmetic sequence $3, \frac{7}{3}, \frac{5}{3}, \ldots$.

12. Find the 13th term of the arithmetic sequence $1200, \$964.32, \$728.64, \ldots$.

13. Find the 10th term of the arithmetic sequence $2345.78, \$2967.54, \$3589.30, \ldots$.

14. In the sequence of Exercise 9, what term is the number 106?

15. In the sequence of Exercise 8, what term is the number 1.67?

16. In the sequence of Exercise 10, what term is -296?

17. In the sequence of Exercise 11, what term is -27?

18. Find a_{20} when $a_1 = 14$ and $d = -3$.

19. Find a_1 when $d = 4$ and $a_8 = 33$.

20. Find d when $a_1 = 8$ and $a_{11} = 26$.

21. Find n when $a_1 = 25$, $d = -14$, and $a_n = -507$.

22. In an arithmetic sequence, $a_{17} = -40$ and $a_{28} = -73$. Find a_1 and d. Write the first 5 terms of the sequence.

23. In an arithmetic sequence, $a_{17} = \frac{25}{3}$ and $a_{32} = \frac{95}{6}$. Find a_1 and d. Write the first 5 terms of the sequence.

24. Find the sum of the first 14 terms of the series $11 + 7 + 3 + \cdots$.

25. Find the sum of the first 20 terms of the series $5 + 8 + 11 + 14 + \cdots$.

26. Find the sum of the first 300 natural numbers.

27. Find the sum of the first 400 even natural numbers.

28. Find the sum of the odd numbers 1 to 199, inclusive.

29. Find the sum of the multiples of 7 from 7 to 98, inclusive.

30. Find the sum of all multiples of 4 that are between 14 and 523.

31. If an arithmetic series has $a_1 = 2$, $d = 5$, and $n = 20$, what is S_n?

32. If an arithmetic series has $a_1 = 7$, $d = -3$, and $n = 32$, what is S_n?

Find the sum.

33. $\displaystyle\sum_{k=1}^{40} (2k + 3)$

34. $\displaystyle\sum_{k=5}^{20} 8k$

35. $\displaystyle\sum_{k=0}^{19} \frac{k - 3}{4}$

36. $\displaystyle\sum_{k=2}^{50} (2000 - 3k)$

37. $\displaystyle\sum_{k=12}^{57} \frac{7 - 4k}{13}$

38. $\displaystyle\sum_{k=101}^{200} (1.14k - 2.8) - \sum_{k=1}^{5} \left(\frac{k + 4}{10} \right)$

39. *Pole Stacking.* How many poles will be in a stack of telephone poles if there are 50 in the first layer, 49 in the second, and so on, with 6 in the top layer?

40. *Investment Return.* Max is an investment counselor. He sets up an investment situation for a client that will return $5000 the first year, \$6125 the second year, \$7250 the third year, and so on, for 25 yr. How much is received from the investment altogether?

41. *Garden Plantings.* A gardener is making a planting in the shape of a trapezoid. It will have 35 plants in the front row, 31 in the second row, 27 in the third row, and so on. If the pattern is consistent, how many plants will there be in the last row? How many plants are there altogether?

42. *Band Formation.* A formation of a marching band has 14 marchers in the front row, 16 in the second row, 18 in the third row, and so on, for 25 rows. How many marchers are in the last row? How many marchers are there altogether?

43. *Total Savings.* If 10¢ is saved on October 1, 20¢ is saved on October 2, 30¢ on October 3, and so on, how much is saved during the 31 days of October?

44. *Parachutist Free Fall.* It has been found that when a parachutist jumps from an airplane, the distances, in feet, which the parachutist falls in each successive second before pulling the ripcord to release the parachute are as follows:

$$16, 48, 80, 112, 144, \ldots .$$

Is this sequence arithmetic? What is the common difference? What is the total distance fallen after 10 sec?

45. *Theater Seating.*   Theaters are often built with more seats per row as the rows move toward the back. Suppose that the first balcony of a theater has 28 seats in the first row, 32 in the second, 36 in the third, and so on, for 20 rows. How many seats are in the first balcony altogether?

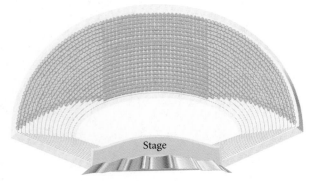

46. *Small Group Interaction.*   In a social science study, Stephan found the following data regarding an interaction measurement r_n for groups of size n.

n	r_n
3	0.5908
4	0.6080
5	0.6252
6	0.6424
7	0.6596
8	0.6768
9	0.6940
10	0.7112

Source: American Sociological Review, 17 (1952)

Is this sequence arithmetic? What is the common difference?

47. *Raw Material Production.*   In an industrial situation, it took 3 units of raw materials to produce 1 unit of a product. The raw material needs thus formed the sequence

$$3, 6, 9, \ldots, 3n, \ldots.$$

Is this sequence arithmetic? What is the common difference?

Collaborative Discussion and Writing

48. The sum of the first n terms of an arithmetic sequence can be given by

$$S_n = \frac{n}{2}[2a_1 + (n-1)d].$$

Compare this formula to

$$S_n = \frac{n}{2}(a_1 + a_n).$$

Discuss the reasons for the use of one formula over the other.

49. It is said that as a young child, the mathematician Karl F. Gauss (1777–1855) was able to compute the sum $1 + 2 + 3 + \cdots + 100$ very quickly in his head to the amazement of a teacher. Explain how Gauss might have done this had he possessed some knowledge of arithmetic sequences and series. Then give a formula for the sum of the first n natural numbers.

Skill Maintenance

Solve.

50. $7x - 2y = 4,$
$\quad\; x + 3y = 17$

51. $2x + y + 3z = 12,$
$\quad\; x - 3y + 2z = -1,$
$\quad\; 5x + 2y - 4z = -4$

52. Find the vertices and the foci of the ellipse with the equation $9x^2 + 16y^2 = 144$.

53. Find an equation of the ellipse with vertices $(0, -5)$ and $(0, 5)$ and minor axis of length 4.

Synthesis

54. Find three numbers in an arithmetic sequence such that the sum of the first and third is 10 and the product of the first and second is 15.

55. Find a formula for the sum of the first n odd natural numbers:

$$1 + 3 + 5 + \cdots + (2n - 1).$$

56. Find the first 10 terms of the arithmetic sequence for which

$$a_1 = \$8760 \quad \text{and} \quad d = -\$798.23.$$

Then find the sum of the first 10 terms.

57. Find the first term and the common difference for the arithmetic sequence for which

$$a_2 = 40 - 3q \quad \text{and} \quad a_4 = 10p + q.$$

58. The zeros of this polynomial function form an arithmetic sequence. Find them.

$$f(x) = x^4 + 4x^3 - 84x^2 - 176x + 640$$

*If p, m, and q form an arithmetic sequence, it can be shown that m = (p + q)/2. (See Exercise 65.) The number m is the **arithmetic mean,** or **average**, of p and q. Given two numbers p and q, if we find k other numbers m_1, m_2, ..., m_k such that*

$$p, m_1, m_2, \ldots, m_k, q$$

forms an arithmetic sequence, we say that we have "inserted k arithmetic means between p and q."

59. Insert three arithmetic means between 4 and 12.

60. Insert three arithmetic means between −3 and 5.

61. Insert four arithmetic means between 4 and 13.

62. Insert ten arithmetic means between 27 and 300.

63. Insert enough arithmetic means between 1 and 50 so that the sum of the resulting series will be 459.

64. *Straight-Line Depreciation.*   A company buys an office machine for $5200 on January 1 of a given year. The machine is expected to last for 8 yr, at the end of which time its **trade-in value,** or **salvage**

value, will be $1100. If the company's accountant figures the decline in value to be the same each year, then its **book values,** or **salvage values,** after t years, $0 \leq t \leq 8$, form an arithmetic sequence given by

$$a_t = C - t\left(\frac{C - S}{N}\right),$$

where C is the original cost of the item ($5200), N is the number of years of expected life (8), and S is the salvage value ($1100).

a) Find the formula for a_t for the straight-line depreciation of the office machine.

b) Find the salvage value after 0 yr, 1 yr, 2 yr, 3 yr, 4 yr, 7 yr, and 8 yr.

65. Prove that if p, m, and q form an arithmetic sequence, then

$$m = \frac{p + q}{2}.$$

10.3

Geometric Sequences and Series

• *Identify the common ratio of a geometric sequence, and find a given term and the sum of the first n terms.*
• *Find the sum of an infinite geometric series, if it exists.*

A sequence in which each term after the first is found by multiplying the preceding term by the same number is a **geometric sequence.**

Geometric Sequences

Consider this sequence:

$$2, 6, 18, 54, 162, \ldots .$$

Note that multiplying each term by 3 produces the next term. We call the number 3 the **common ratio** because it can be found by dividing any term by the preceding term. A geometric sequence is also called a *geometric progression.*

> ### Geometric Sequence
>
> A sequence is **geometric** if there is a number r, called the **common ratio,** such that
>
> $$\frac{a_{n+1}}{a_n} = r, \quad \text{or} \quad a_{n+1} = a_n r, \quad \text{for any integer } n \geq 1.$$

EXAMPLE 1 For each of the following geometric sequences, identify the common ratio.

a) 3, 6, 12, 24, 48, . . .

b) $1, -\dfrac{1}{2}, \dfrac{1}{4}, -\dfrac{1}{8}, \ldots$

c) $5200, $3900, $2925, $2193.75, . . .

d) $1000, $1060, $1123.60, . . .

Solution

SEQUENCE	COMMON RATIO
a) 3, 6, 12, 24, 48, . . .	$2 \quad \left(\frac{6}{3} = 2, \frac{12}{6} = 2, \text{ and so on}\right)$
b) $1, -\dfrac{1}{2}, \dfrac{1}{4}, -\dfrac{1}{8}, \ldots$	$-\dfrac{1}{2} \quad \left(\dfrac{-\frac{1}{2}}{1} = -\dfrac{1}{2}, \dfrac{\frac{1}{4}}{-\frac{1}{2}} = -\dfrac{1}{2}, \text{ and so on}\right)$
c) $5200, $3900, $2925, $2193.75, . . .	$0.75 \quad \left(\dfrac{\$3900}{\$5200} = 0.75, \dfrac{\$2925}{\$3900} = 0.75, \text{ and so on}\right)$
d) $1000, $1060, $1123.60, . . .	$1.06 \quad \left(\dfrac{\$1060}{\$1000} = 1.06, \dfrac{\$1123.60}{\$1060} = 1.06, \text{ and so on}\right)$

We now find a formula for the general, or nth, term of a geometric sequence. Let a_1 be the first term and r the common ratio. The first few terms are as follows:

$$a_1,$$
$$a_2 = a_1 r,$$
$$a_3 = a_2 r = (a_1 r)r = a_1 r^2, \qquad \text{Substituting } a_1 r \text{ for } a_2$$
$$a_4 = a_3 r = (a_1 r^2)r = a_1 r^3. \qquad \text{Substituting } a_1 r^2 \text{ for } a_3$$

Note that the exponent is 1 less than the subscript.

Generalizing, we obtain the following.

> ### nth Term of a Geometric Sequence
>
> The **nth term** of a geometric sequence is given by
>
> $$a_n = a_1 r^{n-1}, \quad \text{for any integer } n \geq 1.$$

EXAMPLE 2 Find the 7th term of the geometric sequence 4, 20, 100, ...

Solution We first note that

$$a_1 = 4 \quad \text{and} \quad n = 7.$$

To find the common ratio, we can divide any term (other than the first) by the preceding term. Since the second term is 20 and the first is 4, we get

$$r = \frac{20}{4}, \quad \text{or} \quad 5.$$

Then using the formula $a_n = a_1 r^{n-1}$, we have

$$a_7 = 4 \cdot 5^{7-1} = 4 \cdot 5^6 = 4 \cdot 15{,}625 = 62{,}500.$$

Thus the 7th term is 62,500.

EXAMPLE 3 Find the 10th term of the geometric sequence 64, -32, 16, -8, ...

Solution We first note that

$$a_1 = 64, \qquad n = 10, \quad \text{and} \quad r = \frac{-32}{64}, \quad \text{or} \quad -\frac{1}{2}.$$

Then using the formula $a_n = a_1 r^{n-1}$, we have

$$a_{10} = 64 \cdot \left(-\frac{1}{2}\right)^{10-1} = 64 \cdot \left(-\frac{1}{2}\right)^{9} = 2^6 \cdot \left(-\frac{1}{2^9}\right) = -\frac{1}{8}.$$

Thus the 10th term is $-\frac{1}{8}$.

Sum of the First *n* Terms of a Geometric Sequence

Next, we develop a formula for the sum S_n of the first *n* terms of a geometric sequence:

$$a_1, a_1 r, a_1 r^2, a_1 r^3, \ldots, a_1 r^{n-1}, \ldots .$$

The associated **geometric series** is given by

$$S_n = a_1 + a_1 r + a_1 r^2 + a_1 r^3 + \cdots + a_1 r^{n-1}. \tag{1}$$

We want to find a formula for this sum. If we multiply on both sides of equation (1) by *r*, we have

$$r S_n = a_1 r + a_1 r^2 + a_1 r^3 + a_1 r^4 + \cdots + a_1 r^{n}. \tag{2}$$

Subtracting equation (2) from equation (1), we see that the differences of the red terms are 0, leaving

$$S_n - r S_n = a_1 - a_1 r^n,$$

or

$$S_n(1 - r) = a_1(1 - r^n). \qquad \text{Factoring}$$

Dividing on both sides by $1 - r$ gives us the following formula.

Sum of the First n Terms

The sum of the first n terms of a geometric sequence is given by

$$S_n = \frac{a_1(1 - r^n)}{1 - r}, \quad \text{for any } r \neq 1.$$

EXAMPLE 4 Find the sum of the first 7 terms of the geometric sequence 3, 15, 75, 375, ...

Solution We first note that

$$a_1 = 3, \quad n = 7, \quad \text{and} \quad r = \tfrac{15}{3}, \text{ or } 5.$$

Then using the formula

$$S_n = \frac{a_1(1 - r^n)}{1 - r},$$

we have

$$S_7 = \frac{3(1 - 5^7)}{1 - 5}$$

$$= \frac{3(1 - 78,125)}{-4}$$

$$= 58,593.$$

Thus the sum of the first 7 terms is 58,593. ▬

EXAMPLE 5 Find the sum: $\displaystyle\sum_{k=1}^{11} (0.3)^k$.

Solution This is a geometric series with $a_1 = 0.3$, $r = 0.3$, and $n = 11$. Thus,

$$S_{11} = \frac{0.3(1 - 0.3^{11})}{1 - 0.3}$$

$$\approx 0.42857.$$ ▬

Infinite Geometric Series

The sum of the terms of an infinite geometric sequence is an **infinite geometric series.** For some geometric sequences, S_n gets close to a specific number as n gets large. For example, consider the infinite series

$$\frac{1}{2} + \frac{1}{4} + \frac{1}{8} + \frac{1}{16} + \cdots + \frac{1}{2^n} + \cdots.$$

We examine some partial sums. Note that each of the partial sums is less than 1, but S_n gets very close to 1 as n gets large. We say that 1 is the **limit** of S_n and also that 1 is the **sum of the infinite geometric sequence.**

The sum of an infinite geometric sequence is denoted S_∞. In this case, $S_\infty = 1$.

n	S_n
1	0.5
5	0.96875
10	0.9990234375
20	0.9999990463
30	0.9999999991

Some infinite sequences do not have sums. Consider the infinite geometric series

$$2 + 4 + 8 + 16 + \cdots + 2^n + \cdots.$$

We again examine some partial sums. Note that as n gets large, S_n gets large without bound. This sequence does not have a sum.

n	S_n
1	2
5	62
10	2,046
20	2,097,150
30	2,147,483,646

It can be shown (but we will not do so here) that the sum of the terms of an infinite geometric series exists if and only if $|r| < 1$ (that is, the absolute value of the common ratio is less than 1).

To find a formula for the sum of an infinite geometric series, we first consider the sum of the first n terms:

$$S_n = \frac{a_1(1 - r^n)}{1 - r} = \frac{a_1 - a_1 r^n}{1 - r}. \qquad \text{Using the distributive law}$$

For $|r| < 1$, values of r^n get close to 0 as n gets large. As r^n gets close to 0, so does $a_1 r^n$. Thus, S_n gets close to $a_1/(1 - r)$.

Limit or Sum of an Infinite Geometric Series

When $|r| < 1$, the limit or sum of an infinite geometric series is given by

$$S_\infty = \frac{a_1}{1 - r}.$$

EXAMPLE 6   Determine whether each of the following infinite geometric series has a limit. If a limit exists, find it.

a) $1 + 3 + 9 + 27 + \cdots$
b) $-2 + 1 - \frac{1}{2} + \frac{1}{4} - \frac{1}{8} + \cdots$

Solution
a) Here $r = 3$, so $|r| = |3| = 3$. Since $|r| > 1$, the series *does not* have a limit.
b) Here $r = -\frac{1}{2}$, so $|r| = \left|-\frac{1}{2}\right| = \frac{1}{2}$. Since $|r| < 1$, the series *does* have a limit. We find the limit:

$$S_\infty = \frac{a_1}{1 - r} = \frac{-2}{1 - \left(-\frac{1}{2}\right)} = \frac{-2}{\frac{3}{2}} = -\frac{4}{3}.$$ ▬

EXAMPLE 7   Find fractional notation for $0.78787878\ldots$, or $0.\overline{78}$.

Solution   We can express this as

$$0.78 + 0.0078 + 0.000078 + \cdots .$$

Then we see that this is an infinite geometric series, where $a_1 = 0.78$ and $r = 0.01$. Since $|r| < 1$, this series has a limit:

$$S_\infty = \frac{a_1}{1 - r} = \frac{0.78}{1 - 0.01} = \frac{0.78}{0.99} = \frac{78}{99}, \quad \text{or} \quad \frac{26}{33}.$$

Thus fractional notation for $0.78787878\ldots$ is $\frac{26}{33}$. You can check this on your calculator.   ▬

Applications

The translation of some applications and problem-solving situations may involve geometric sequences or series. Examples 9 and 10 in particular show applications in business and economics.

EXAMPLE 8   *A Daily Doubling Salary.* Suppose someone offered you a job for the month of September (30 days) under the following conditions. You will be paid $0.01 for the first day, $0.02 for the second, $0.04 for the third, and so on, doubling your previous day's salary each day. How much would you earn? (Would you take the job? Make a conjecture before reading further.)

Solution   You earn $0.01 the first day, $0.01(2) the second day, $0.01(2)(2) the third day, and so on. The amount earned is the geometric series

$$\$0.01 + \$0.01(2) + \$0.01(2^2) + \$0.01(2^3) + \cdots + \$0.01(2^{29}),$$

where $a_1 = \$0.01$, $r = 2$, and $n = 30$. Using the formula

$$S_n = \frac{a_1(1 - r^n)}{1 - r},$$

we have

$$S_{30} = \frac{\$0.01(1 - 2^{30})}{1 - 2} = \$10,737,418.23.$$

The pay exceeds $10.7 million for the month. (Most people would probably take the job!)

EXAMPLE 9 *The Amount of an Annuity.* An **annuity** is a sequence of equal payments made at equal time intervals that earn interest. Fixed deposits in a savings account are an example of an annuity. Suppose that to save money to buy a car, Andrea deposits $1000 at the *end* of each of 5 yr in an account that pays 8% interest, compounded annually. The total amount in the account at the end of 5 yr is called the **amount of the annuity.** Find that amount.

Solution The following time diagram can help visualize the problem. Note that no deposit is made until the end of the first year.

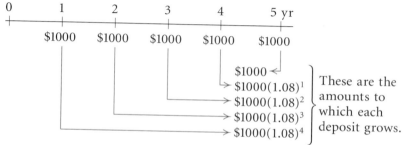

The amount of the annuity is the geometric series

$$\$1000 + \$1000(1.08)^1 + \$1000(1.08)^2 + \$1000(1.08)^3 + \$1000(1.08)^4,$$

where $a_1 = \$1000$, $n = 5$, and $r = 1.08$. Using the formula

$$S_n = \frac{a_1(1 - r^n)}{1 - r},$$

we have

$$S_5 = \frac{\$1000(1 - 1.08^5)}{1 - 1.08} \approx \$5866.60.$$

The amount of the annuity is $5866.60.

EXAMPLE 10 *The Economic Multiplier.* The finals of the NCAA men's basketball tournament have a significant effect on the economy of the host city. Suppose that 45,000 people visit the city and spend $600 each while there. Then assume that 80% of that money is spent again in the city, and then 80% of that money is spent again, and so on. This is known as the *economic multiplier effect.* Find the total effect on the economy.

Solution   The initial effect is $45,000 \cdot \$600$, or $27,000,000$. The total effect is given by the infinite series

$$\$27,000,000 + \$27,000,000(0.80) + \$27,000,000(0.80^2) + \cdots.$$

Using the formula for the sum of an infinite geometric series, we have

$$S_\infty = \frac{a_1}{1-r} = \frac{27,000,000}{1-0.80} = \$135,000,000.$$

Thus the total effect on the economy of the spending from the tournament is $135,000,000. —

Exercise Set 10.3

Find the common ratio.

1. 2, 4, 8, 16, ...

2. 18, −6, 2, −$\frac{2}{3}$, ...

3. −1, 1, −1, 1, ...

4. −8, −0.8, −0.08, −0.008, ...

5. $\frac{2}{3}$, −$\frac{4}{3}$, $\frac{8}{3}$, −$\frac{16}{3}$, ...

6. 75, 15, 3, $\frac{3}{5}$, ...

7. 6.275, 0.6275, 0.06275, ...

8. $\frac{1}{x}$, $\frac{1}{x^2}$, $\frac{1}{x^3}$, ...

9. 5, $\frac{5a}{2}$, $\frac{5a^2}{4}$, $\frac{5a^3}{8}$, ...

10. $780, $858, $943.80, $1038.18, ...

Find the indicated term.

11. 2, 4, 8, 16, ...; the 7th term

12. 2, −10, 50, −250, ...; the 9th term

13. 2, 2$\sqrt{3}$, 6, ...; the 9th term

14. 1, −1, 1, −1, ...; the 57th term

15. $\frac{7}{625}$, −$\frac{7}{25}$, ...; the 23rd term

16. $1000, $1060, $1123.60, ...; the 5th term

Find the nth, or general, term.

17. 1, 3, 9, ...

18. 25, 5, 1, ...

19. 1, −1, 1, −1, ...

20. −2, 4, −8, ...

21. $\frac{1}{x}$, $\frac{1}{x^2}$, $\frac{1}{x^3}$, ...

22. 5, $\frac{5a}{2}$, $\frac{5a^2}{4}$, $\frac{5a^3}{8}$, ...

23. Find the sum of the first 7 terms of the geometric series

$$6 + 12 + 24 + \cdots.$$

24. Find the sum of the first 10 terms of the geometric series

$$16 - 8 + 4 - \cdots.$$

25. Find the sum of the first 9 terms of the geometric series

$$\tfrac{1}{18} - \tfrac{1}{6} + \tfrac{1}{2} - \cdots.$$

26. Find the sum of the geometric series

$$-8 + 4 + (-2) + \cdots + \left(-\tfrac{1}{32}\right).$$

Find the sum, if it exists.

27. $4 + 2 + 1 + \cdots$

28. $7 + 3 + \frac{9}{7} + \cdots$

29. $25 + 20 + 16 + \cdots$

30. $100 - 10 + 1 - \frac{1}{10} + \cdots$

31. $8 + 40 + 200 + \cdots$

32. $-6 + 3 - \frac{3}{2} + \frac{3}{4} - \cdots$

33. $0.6 + 0.06 + 0.006 + \cdots$

34. $\displaystyle\sum_{k=0}^{10} 3^k$

35. $\displaystyle\sum_{k=1}^{11} 15\left(\frac{2}{3}\right)^k$

36. $\displaystyle\sum_{k=0}^{50} 200(1.08)^k$

37. $\displaystyle\sum_{k=1}^{\infty} \left(\frac{1}{2}\right)^{k-1}$

38. $\displaystyle\sum_{k=1}^{\infty} 2^k$

39. $\displaystyle\sum_{k=1}^{\infty} 12.5^k$

40. $\displaystyle\sum_{k=1}^{\infty} 400(1.0625)^k$

41. $\displaystyle\sum_{k=1}^{\infty} \$500(1.11)^{-k}$

42. $\displaystyle\sum_{k=1}^{\infty} \$1000(1.06)^{-k}$

43. $\displaystyle\sum_{k=1}^{\infty} 16(0.1)^{k-1}$

44. $\displaystyle\sum_{k=1}^{\infty} \frac{8}{3}\left(\frac{1}{2}\right)^{k-1}$

Find fractional notation.

45. $0.131313\ldots$, or $0.\overline{13}$

46. $0.2222\ldots$, or $0.\overline{2}$

47. $8.9999\overline{9}$

48. $6.1616\overline{16}$

49. $3.4125\overline{125}$

50. $12.7809\overline{809}$

51. *Bouncing Ping-Pong Ball.* A ping-pong ball is dropped from a height of 16 ft and always rebounds $\frac{1}{4}$ of the distance fallen.

   a) How high does it rebound the 6th time?

   b) Find the total sum of the rebound heights of the ball.

52. *Daily Doubling Salary.* Suppose someone offered you a job for the month of February (28 days) under the following conditions. You will be paid $0.01 the 1st day, $0.02 the 2nd, $0.04 the 3rd, and so on, doubling your previous day's salary each day. How much would you earn altogether?

53. *Bungee Jumping.* A bungee jumper rebounds 60% of the height jumped. A bungee jump is made using a cord that stretches to 200 ft.

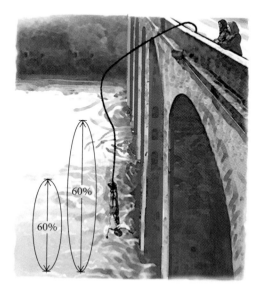

   a) After jumping and then rebounding 9 times, how far has a bungee jumper traveled upward (the total rebound distance)?

   b) About how far will a jumper have traveled upward (bounced) before coming to rest?

54. *Population Growth.* Gaintown has a present population of 100,000, and the population is increasing by 3% each year.

   a) What will the population be in 15 yr?

   b) How long will it take for the population to double?

55. *Amount of an Annuity.* To create a college fund, a parent makes a sequence of 18 yearly deposits of $1000 each in a savings account on which interest is compounded annually at 6.2%. Find the amount of the annuity.

56. *Amount of an Annuity.* A sequence of yearly payments of P dollars is invested at the end of each of N years at interest rate i, compounded annually. The total amount in the account, or the amount of the annuity, is V.

   a) Show that

$$V = \frac{P[(1+i)^N - 1]}{i}.$$

   b) Suppose that interest is compounded n times per year and deposits are made every compounding period. Show that the formula for V is then given by

$$V = \frac{P\left[\left(1 + \dfrac{i}{n}\right)^{nN} - 1\right]}{i/n}.$$

57. *Loan Repayment.* A family borrows $120,000. The loan is to be repaid in 13 yr at 12% interest, compounded annually. How much will be repaid at the end of 13 yr?

58. *Doubling Paper Folds.* A piece of paper is 0.01 in. thick. It is cut and stacked repeatedly in such a way that its thickness is doubled each time for 20 times. How thick is the result?

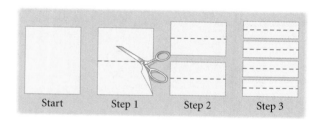

Start   Step 1   Step 2   Step 3

59. *The Economic Multiplier.*   The government is making a \$13,000,000,000 expenditure for educational improvement. If 85% of this is spent again, and so on, what is the total effect on the economy?

60. *Advertising Effect.*   The cereal company Box-o-Vim is about to market a new low-fat great-tasting cereal in a city of 5,000,000 people. They plan an advertising campaign that they think will induce 30% of the people to buy the product. They estimate that if those people like the product, they will induce 30% · 30% · 5,000,000 more to buy the product, and those will induce 30% · 30% · 30% · 5,000,000, and so on. In all, how many people will buy the product as a result of the advertising campaign? What percentage of the population is this?

Technology Connection

61. Use a graphing calculator to find the sum in Exercise 35.

62. Use a graphing calculator to find the sum in Exercise 34.

Collaborative Discussion and Writing

63. Write a problem for a classmate to solve. Devise the problem so that a geometric series is involved and the solution is "The total amount in the bank is \$900(1.08)^{40}, or about \$19,552."

64. The infinite series

$$S_\infty = 2 + \frac{1}{2} + \frac{1}{2 \cdot 3} + \frac{1}{2 \cdot 3 \cdot 4} + \frac{1}{2 \cdot 3 \cdot 4 \cdot 5} + \cdots$$

is not geometric, but it does have a sum. Consider $S_1, S_2, S_3, S_4, S_5,$ and S_6. Construct a table and a graph of the sequence. Expand the sequence of sums, if needed. Make a conjecture about the value of S_∞ and explain your reasoning.

Skill Maintenance

For each pair of functions, find $(f \circ g)(x)$ and $(g \circ f)(x)$.

65. $f(x) = x^2$, $g(x) = 4x + 5$

66. $f(x) = x - 1$, $g(x) = x^2 + x + 3$

Solve.

67. $5^x = 35$

68. $\log_2 x = -4$

Synthesis

69. Prove that

$$\sqrt{3} - \sqrt{2}, \qquad 4 - \sqrt{6}, \quad \text{and} \quad 6\sqrt{3} - 2\sqrt{2}$$

form a geometric sequence.

70. Consider the sequence

$$4, \ 20.4, \ 104.04, \ 531.6444, \ \dots .$$

What is the error in using $a_{277} = 4(5.1)^{276}$ to find the 277th term?

71. Consider the sequence

$$x + 3, \ x + 7, \ 4x - 2, \ \dots .$$

a) If the sequence is arithmetic, find x and then determine each of the 3 terms and the 4th term.
b) If the sequence is geometric, find x and then determine each of the 3 terms and the 4th term.

72. Find the sum of the first n terms of

$$1 + x + x^2 + \cdots .$$

73. Find the sum of the first n terms of

$$x^2 - x^3 + x^4 - x^5 + \cdots .$$

In Exercises 74 and 75, assume that $a_1, a_2, a_3, \dots$ is a geometric sequence.

74. Prove that $a_1^2, a_2^2, a_3^2, \dots$, is a geometric sequence.

75. Prove that $\ln a_1, \ln a_2, \ln a_3, \dots$, is an arithmetic sequence.

76. Prove that $5^{a_1}, 5^{a_2}, 5^{a_3}, \dots$, is a geometric sequence, if $a_1, a_2, a_3, \dots$, is an arithmetic sequence.

77. The sides of a square are 16 cm long. A second square is inscribed by joining the midpoints of the sides, successively. In the second square, we repeat the process, inscribing a third square. If this process is continued indefinitely, what is the sum of all the areas of all the squares? (*Hint:* Use an infinite geometric series.)

10.4

Mathematical Induction

- *List the statements of an infinite sequence that is defined by a formula.*
- *Do proofs by mathematical induction.*

In this section we learn to prove a sequence of mathematical statements using a procedure called *mathematical induction*.

Sequences of Statements

Infinite sequences of statements occur often in mathematics. In an infinite sequence of statements, there is a statement for each natural number. For example, consider the sequence of statements represented by the following:

"For each x between 0 and 1, $0 < x^n < 1$."

Let's think of this as $S(n)$, or S_n. Substituting natural numbers for n gives a sequence of statements. We list a few of them.

Statement 1, S_1:   For x between 0 and 1, $0 < x^1 < 1$.
Statement 2, S_2:   For x between 0 and 1, $0 < x^2 < 1$.
Statement 3, S_3:   For x between 0 and 1, $0 < x^3 < 1$.
Statement 4, S_4:   For x between 0 and 1, $0 < x^4 < 1$.

In this context, the symbols S_1, S_2, S_3, and so on, do not represent sums.

EXAMPLE 1   List the first four statements in the sequence obtainable from each of the following.

a) $\log n < n$
b) $1 + 3 + 5 + \cdots + (2n - 1) = n^2$

Solution

a) This time, S_n is "$\log n < n$."

S_1: $\log 1 < 1$
S_2: $\log 2 < 2$
S_3: $\log 3 < 3$
S_4: $\log 4 < 4$

b) This time, S_n is "$1 + 3 + 5 + \cdots + (2n - 1) = n^2$."

S_1: $1 = 1^2$
S_2: $1 + 3 = 2^2$
S_3: $1 + 3 + 5 = 3^2$
S_4: $1 + 3 + 5 + 7 = 4^2$

Proving Infinite Sequences of Statements

We now develop a method of proof, called **mathematical induction,** which we can use to try to prove that all statements in an infinite sequence of statements are true. The statements usually have the form:

"For all natural numbers n, S_n",

where S_n is some mathematical sentence such as those of the preceding examples. Of course, we cannot prove each statement of an infinite sequence individually. Instead, we try to show that "whenever S_k holds, then S_{k+1} must hold." We abbreviate this as $S_k \rightarrow S_{k+1}$. (This is also read "*If S_k, then S_{k+1},*" or "*S_k implies S_{k+1}.*") Suppose that we could somehow establish that this holds for all natural numbers k. Then we would have the following:

$S_1 \longrightarrow S_2$ meaning "if S_1 is true, then S_2 is true";

$S_2 \longrightarrow S_3$ meaning "if S_2 is true, then S_3 is true";

$S_3 \longrightarrow S_4$ meaning "if S_3 is true, then S_4 is true";

and so on, indefinitely.

Even knowing that $S_k \rightarrow S_{k+1}$, we would still not be certain whether there is *any* k for which S_k is true. All we would know is that "if S_k is true, then S_{k+1} is true." Suppose now that S_k is true for some k, say, $k = 1$. We then must have the following.

S_1 is true.	We have verified, or proved, this.
$S_1 \longrightarrow S_2$	This means that whenever S_1 is true, S_2 is true.
Therefore, S_2 is true.	
$S_2 \longrightarrow S_3$	This means that whenever S_2 is true, S_3 is true.
Therefore, S_3 is true.	
and so on.	

We conclude that S_n is true for all natural numbers n.

This leads us to the principle of mathematical induction, which we use to prove the types of statements considered here.

The Principle of Mathematical Induction

We can prove an infinite sequence of statements S_n by showing the following.

(1) *Basis step.* S_1 is true.

(2) *Induction step.* For all natural numbers k, $S_k \rightarrow S_{k+1}$.

Mathematical induction is analogous to lining up a sequence of dominoes. The induction step tells us that if any one domino is knocked over, then the one next to it will be hit and knocked over. The basis step tells us that the first domino can indeed be knocked over. Note that in order for all dominoes to fall, *both* conditions must be satisfied.

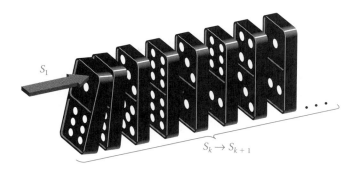

When you are learning to do proofs by mathematical induction, it is helpful to first write out S_n, S_1, S_k, and S_{k+1}. This helps to identify what is to be assumed and what is to be deduced.

EXAMPLE 2   Prove: For every natural number n,

$$1 + 3 + 5 + \cdots + (2n - 1) = n^2.$$

Proof   We first list S_n, S_1, S_k, and S_{k+1}.

S_n: $1 + 3 + 5 + \cdots + (2n - 1) = n^2$
S_1: $1 = 1^2$
S_k: $1 + 3 + 5 + \cdots + (2k - 1) = k^2$
S_{k+1}: $1 + 3 + 5 + \cdots + (2k - 1) + [2(k + 1) - 1] = (k + 1)^2$

(1) *Basis step.* S_1, as listed, is true.

(2) *Induction step.* We let k be any natural number. We assume S_k to be true and try to show that it implies that S_{k+1} is true. Now S_k is

$$1 + 3 + 5 + \cdots + (2k - 1) = k^2.$$

Starting with the left side of S_{k+1} and substituting k^2 for $1 + 3 + 5 + \cdots + (2k - 1)$, we have

$$\underbrace{1 + 3 + \cdots + (2k - 1)}_{\downarrow} + [2(k + 1) - 1]$$

$$= k^2 + [2(k + 1) - 1] = k^2 + 2k + 1 = (k + 1)^2.$$

We have derived S_{k+1} from S_k. Thus we have shown that for all natural numbers k, $S_k \rightarrow S_{k+1}$. This completes the induction step. It and the basis step tell us that the proof is complete.

EXAMPLE 3 Prove: For every natural number n,

$$\frac{1}{2} + \frac{1}{4} + \frac{1}{8} + \cdots + \frac{1}{2^n} = \frac{2^n - 1}{2^n}.$$

Proof   We first list S_n, S_1, S_k, and S_{k+1}.

S_n: $\dfrac{1}{2} + \dfrac{1}{4} + \dfrac{1}{8} + \cdots + \dfrac{1}{2^n} = \dfrac{2^n - 1}{2^n}$

S_1: $\dfrac{1}{2} = \dfrac{2^1 - 1}{2^1}$

S_k: $\dfrac{1}{2} + \dfrac{1}{4} + \dfrac{1}{8} + \cdots + \dfrac{1}{2^k} = \dfrac{2^k - 1}{2^k}$

S_{k+1}: $\dfrac{1}{2} + \dfrac{1}{4} + \dfrac{1}{8} + \cdots + \dfrac{1}{2^k} + \dfrac{1}{2^{k+1}} = \dfrac{2^{k+1} - 1}{2^{k+1}}$

(1) *Basis step.* We show S_1 to be true as follows:

$$\frac{2^1 - 1}{2^1} = \frac{2 - 1}{2} = \frac{1}{2}.$$

(2) *Induction step.* We let k be any natural number. We assume S_k to be true and try to show that it implies that S_{k+1} is true. Now S_k is

$$\frac{1}{2} + \frac{1}{4} + \frac{1}{8} + \cdots + \frac{1}{2^k} = \frac{2^k - 1}{2^k}.$$

Starting with the left side of S_{k+1} and substituting

$$\frac{2^k - 1}{2^k} \quad \text{for} \quad \frac{1}{2} + \frac{1}{4} + \cdots + \frac{1}{2^k},$$

we have

$$\underbrace{\frac{1}{2} + \frac{1}{4} + \frac{1}{8} + \cdots + \frac{1}{2^k}} + \frac{1}{2^{k+1}}$$

$$= \frac{2^k - 1}{2^k} + \frac{1}{2^{k+1}} = \frac{2^k - 1}{2^k} \cdot \frac{2}{2} + \frac{1}{2^{k+1}} = \frac{(2^k - 1) \cdot 2 + 1}{2^{k+1}}$$

$$= \frac{2^{k+1} - 2 + 1}{2^{k+1}} = \frac{2^{k+1} - 1}{2^{k+1}}.$$

We have derived S_{k+1} from S_k. Thus we have shown that for all natural numbers k, $S_k \rightarrow S_{k+1}$. This completes the induction step. It and the basis step tell us that the proof is complete.

EXAMPLE 4 Prove: For every natural number n, $n < 2^n$.

Proof We first list S_n, S_1, S_k, and S_{k+1}.

$$S_n: \quad n < 2^n$$
$$S_1: \quad 1 < 2^1$$
$$S_k: \quad k < 2^k$$
$$S_{k+1}: \quad k + 1 < 2^{k+1}$$

(1) *Basis step*. S_1, as listed, is true since $2^1 = 2$ and $1 < 2$.

(2) *Induction step*. We let k be any natural number. We assume S_k to be true and try to show that it implies that S_{k+1} is true. Now

$$k < 2^k \qquad \text{This is } S_k.$$
$$2k < 2 \cdot 2^k \qquad \text{Multiplying by 2 on both sides}$$
$$2k < 2^{k+1} \qquad \text{Adding exponents on the right}$$
$$k + k < 2^{k+1}. \qquad \text{Rewriting } 2k \text{ as } k + k$$

Since k is any natural number, we know that $1 \le k$. Thus,

$$k + 1 \le k + k. \qquad \text{Adding } k \text{ on both sides}$$

Putting the results $k + 1 \le k + k$ and $k + k < 2^{k+1}$ together gives us

$$k + 1 < 2^{k+1}. \qquad \text{This is } S_{k+1}.$$

We have derived S_{k+1} from S_k. Thus we have shown that for all natural numbers k, $S_k \to S_{k+1}$. This completes the induction step. It and the basis step tell us that the proof is complete. ▬

Exercise Set   10.4

List the first five statements in the sequence obtainable from each of the following. Determine whether each of the statements is true or false.

1. $n^2 < n^3$

2. $n^2 - n + 41$ is prime. Find a value for n for which the statement is false.

3. A polygon of n sides has $[n(n - 3)]/2$ diagonals.

4. The sum of the angles of a polygon of n sides is $(n - 2) \cdot 180°$.

Use mathematical induction to prove each of the following.

5. $2 + 4 + 6 + \cdots + 2n = n(n + 1)$

6. $4 + 8 + 12 + \cdots + 4n = 2n(n + 1)$

7. $1 + 5 + 9 + \cdots + (4n - 3) = n(2n - 1)$

8. $3 + 6 + 9 + \cdots + 3n = \dfrac{3n(n + 1)}{2}$

9. $2 + 4 + 8 + \cdots + 2^n = 2(2^n - 1)$

10. $2 \le 2^n$

11. $n < n + 1$

12. $3^n < 3^{n+1}$

13. $2n \le 2^n$

14. $\dfrac{1}{1 \cdot 2} + \dfrac{1}{2 \cdot 3} + \cdots + \dfrac{1}{n(n + 1)} = \dfrac{n}{n + 1}$

15. $\dfrac{1}{1 \cdot 2 \cdot 3} + \dfrac{1}{2 \cdot 3 \cdot 4} + \dfrac{1}{3 \cdot 4 \cdot 5} + \cdots$

$$+ \dfrac{1}{n(n+1)(n+2)} = \dfrac{n(n+3)}{4(n+1)(n+2)}$$

16. If x is any real number greater than 1, then for any natural number n, $x \leq x^n$.

The following formulas can be used to find sums of powers of natural numbers. Use mathematical induction to prove each of the following.

17. $1 + 2 + 3 + \cdots + n = \dfrac{n(n+1)}{2}$

18. $1^2 + 2^2 + 3^2 + \cdots + n^2 = \dfrac{n(n+1)(2n+1)}{6}$

19. $1^3 + 2^3 + 3^3 + \cdots + n^3 = \dfrac{n^2(n+1)^2}{4}$

20. $1^4 + 2^4 + 3^4 + \cdots + n^4$

$$= \dfrac{n(n+1)(2n+1)(3n^2+3n-1)}{30}$$

21. $1^5 + 2^5 + 3^5 + \cdots + n^5$

$$= \dfrac{n^2(n+1)^2(2n^2+2n-1)}{12}$$

Use mathematical induction to prove each of the following.

22. $\displaystyle\sum_{i=1}^{n} (3i - 1) = \dfrac{n(3n+1)}{2}$

23. $\displaystyle\sum_{i=1}^{n} i(i + 1) = \dfrac{n(n+1)(n+2)}{3}$

24. $\left(1 + \dfrac{1}{1}\right)\left(1 + \dfrac{1}{2}\right)\left(1 + \dfrac{1}{3}\right) \cdots \left(1 + \dfrac{1}{n}\right) = n + 1$

25. The sum of n terms of an arithmetic sequence:

$a_1 + (a_1 + d) + (a_1 + 2d) + \cdots + [a_1 + (n-1)d]$

$$= \dfrac{n}{2}[2a_1 + (n-1)d]$$

Collaborative Discussion and Writing

26. Write an explanation of the idea behind mathematical induction for a fellow student.

27. Find two statements not considered in this section that are not true for all natural numbers. Then try to find where a proof by mathematical induction fails.

Skill Maintenance

Solve.

28. $2x - 3y = 1,$
$3x - 4y = 3$

29. $x + y + z = 3,$
$2x - 3y - 2z = 5,$
$3x + 2y + 2z = 8$

30. *e-Commerce.* ebooks.com ran a one-day promotion offering a hardback title for $24.95 and a paperback title for $9.95. A total of 80 books were sold and $1546 was taken in. How many of each type of book were sold?

31. *Investment.* Martin received $376 in simple interest one year from three investments. Part is invested at 6%, part at 8%, and part at 10%. The amount invested at 8% is twice the amount invested at 6%. There is $400 more invested at 10% than at 8%. Find the amount invested at each rate.

Synthesis

Use mathematical induction to prove each of the following.

32. The sum of n terms of a geometric sequence:

$$a_1 + a_1 r + a_1 r^2 + \cdots + a_1 r^{n-1} = \dfrac{a_1 - a_1 r^n}{1 - r}$$

33. $x + y$ is a factor of $x^{2n} - y^{2n}$

Prove each of the following using mathematical induction. Do the basis step for $n = 2$.

34. For every natural number $n \geq 2$,

$$2n + 1 < 3^n.$$

35. For every natural number $n \geq 2$,

$\log_a (b_1 b_2 \cdots b_n)$
$$= \log_a b_1 + \log_a b_2 + \cdots + \log_a b_n.$$

Prove each of the following for any complex numbers $z_1, z_2, \ldots, z_n$, where $i^2 = -1$ and $\bar{z}$ is the conjugate of z (see Section 2.2).

36. $\overline{z^n} = \bar{z}^n$

37. $\overline{z_1 + z_2 + \cdots + z_n} = \overline{z_1} + \overline{z_2} + \cdots + \overline{z_n}$

38. $\overline{z_1 z_2 \cdots z_n} = \overline{z_1} \cdot \overline{z_2} \cdots \overline{z_n}$

39. i^n is either 1, -1, i, or $-i$.

For any integers *a* and *b*, *b* is a factor of *a* if there exists an integer *c* such that $a = bc$. Prove each of the following for any natural number *n*.

40. 2 is a factor of $n^2 + n$.

41. 3 is a factor of $n^3 + 2n$.

42. *The Tower of Hanoi Problem.* There are three pegs on a board. On one peg are *n* disks, each smaller than the one on which it rests. The problem is to move this pile of disks to another peg. The final order must be the same, but you can move only one disk at a time and can never place a larger disk on a smaller one.

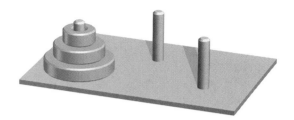

a) What is the *smallest* number of moves needed to move 3 disks? 4 disks? 2 disks? 1 disk?

b) Conjecture a formula for the *smallest* number of moves needed to move *n* disks. Prove it by mathematical induction.

Combinatorics: Permutations

- *Evaluate factorial and permutation notation and solve related applied problems.*

In order to study probability, it is first necessary to learn about **combinatorics**, the theory of counting.

Permutations

In this section, we will consider the part of combinatorics called *permutations*.

> The study of permutations involves *order* and *arrangements*.

EXAMPLE 1 How many 3-letter code symbols can be formed with the letters A, B, C *without* repetition (that is, using each letter only once)?

Solution Consider placing the letters in these boxes.

We can select any of the 3 letters for the first letter in the symbol. Once this letter has been selected, the second must be selected from the 2 remaining letters. After this, the third letter is already determined, since only 1 possibility is left. That is, we can place any of the 3 letters in the first box, either of the remaining 2 letters in the second box, and the only remaining letter in the third box. The possibilities can be arrived at using a **tree diagram,** as shown below.

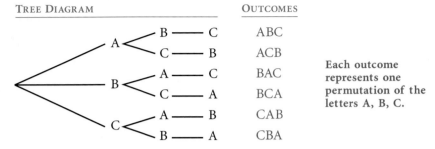

TREE DIAGRAM   OUTCOMES

A —— B —— C   ABC
  —— C —— B   ACB
B —— A —— C   BAC
  —— C —— A   BCA
C —— A —— B   CAB
  —— B —— A   CBA

Each outcome represents one permutation of the letters A, B, C.

We see that there are 6 possibilities. The set of all the possibilities is

{ABC, ACB, BAC, BCA, CAB, CBA}.

Suppose that we perform an experiment such as selecting letters (as in the preceding example), flipping a coin, or drawing a card. The results are called **outcomes.** An **event** is a set of outcomes. The following principle pertains to the counting of actions that occur together, or are combined to form an event.

The Fundamental Counting Principle

Given a combined action, or *event,* in which the first action can be performed in n_1 ways, the second action can be performed in n_2 ways, and so on, the total number of ways in which the combined action can be performed is the product

$$n_1 \cdot n_2 \cdot n_3 \cdot \cdots \cdot n_k.$$

Thus, in Example 1, there are 3 choices for the first letter, 2 for the second letter, and 1 for the third letter, making a total of $3 \cdot 2 \cdot 1$, or 6 possibilities.

EXAMPLE 2 How many 3-letter code symbols can be formed with the letters A, B, C, D, and E *with* repetition (that is, allowing letters to be repeated)?

Solution Since repetition is allowed, there are 5 choices for the first letter, 5 choices for the second, and 5 for the third. Thus, by the fundamental counting principle, there are $5 \cdot 5 \cdot 5$, or 125 code symbols.

Permutation

A **permutation** of a set of n objects is an ordered arrangement of all n objects.

Consider, for example, a set of 4 objects

{A, B, C, D}.

To find the number of ordered arrangements of the set, we select a first letter: There are 4 choices. Then we select a second letter: There are 3 choices. Then we select a third letter: There are 2 choices. Finally, there is 1 choice for the last selection. Thus, by the fundamental counting principle, there are $4 \cdot 3 \cdot 2 \cdot 1$, or 24, permutations of a set of 4 objects.

We can find a formula for the total number of permutations of all objects in a set of n objects. We have n choices for the first selection, $n - 1$ choices for the second, $n - 2$ for the third, and so on. For the nth selection, there is only 1 choice.

The Total Number of Permutations of n Objects

The total number of permutations of n objects, denoted $_nP_n$, is given by

$$_nP_n = n(n - 1)(n - 2) \cdots 3 \cdot 2 \cdot 1.$$

Technology Connection

We can find the total number of permutations of n objects, as in Example 3, using the $_nP_r$ operation from the MATH PRB (probability) menu on a graphing calculator.

```
4 nPr 4
              24
7 nPr 7
            5040
```

EXAMPLE 3 Find each of the following.

a) $_4P_4$ **b)** $_7P_7$

Solution

Start with 4.

a) $_4P_4 = \underbrace{4 \cdot 3 \cdot 2 \cdot 1}_{\text{4 factors}} = 24$

b) $_7P_7 = 7 \cdot 6 \cdot 5 \cdot 4 \cdot 3 \cdot 2 \cdot 1 = 5040$

EXAMPLE 4 In how many different ways can 9 packages be placed in 9 mailboxes, one package in a box?

Solution We have

$$_9P_9 = 9 \cdot 8 \cdot 7 \cdot 6 \cdot 5 \cdot 4 \cdot 3 \cdot 2 \cdot 1 = 362,880.$$

Factorial Notation

We will use products such as $7 \cdot 6 \cdot 5 \cdot 4 \cdot 3 \cdot 2 \cdot 1$ so often that it is convenient to adopt a notation for them. For the product

$$7 \cdot 6 \cdot 5 \cdot 4 \cdot 3 \cdot 2 \cdot 1,$$

we write 7!, read "7 factorial."

We now define factorial notation for natural numbers and for 0.

> *Factorial Notation*
>
> For any natural number n,
>
> $$n! = n(n-1)(n-2) \cdots 3 \cdot 2 \cdot 1.$$
>
> For the number 0,
>
> $$0! = 1.$$

We define 0! as 1 so that certain formulas can be stated concisely and with a consistent pattern.

Here are some examples.

$$
\begin{aligned}
7! &= 7 \cdot 6 \cdot 5 \cdot 4 \cdot 3 \cdot 2 \cdot 1 = 5040 \\
6! &= \phantom{7 \cdot{}} 6 \cdot 5 \cdot 4 \cdot 3 \cdot 2 \cdot 1 = 720 \\
5! &= \phantom{7 \cdot 6 \cdot{}} 5 \cdot 4 \cdot 3 \cdot 2 \cdot 1 = 120 \\
4! &= \phantom{7 \cdot 6 \cdot 5 \cdot{}} 4 \cdot 3 \cdot 2 \cdot 1 = 24 \\
3! &= \phantom{7 \cdot 6 \cdot 5 \cdot 4 \cdot{}} 3 \cdot 2 \cdot 1 = 6 \\
2! &= \phantom{7 \cdot 6 \cdot 5 \cdot 4 \cdot 3 \cdot{}} 2 \cdot 1 = 2 \\
1! &= \phantom{7 \cdot 6 \cdot 5 \cdot 4 \cdot 3 \cdot 2 \cdot{}} 1 = 1 \\
0! &= \phantom{7 \cdot 6 \cdot 5 \cdot 4 \cdot 3 \cdot 2 \cdot{}} 1 = 1
\end{aligned}
$$

We now see that the following statement is true.

$$_nP_n = n!$$

We will often need to manipulate factorial notation. For example, note that

$$
\begin{aligned}
8! &= 8 \cdot 7 \cdot 6 \cdot 5 \cdot 4 \cdot 3 \cdot 2 \cdot 1 \\
&= 8 \cdot (7 \cdot 6 \cdot 5 \cdot 4 \cdot 3 \cdot 2 \cdot 1) = 8 \cdot 7!.
\end{aligned}
$$

Generalizing, we get the following.

For any natural number n, $n! = n(n-1)!$.

By using this result repeatedly, we can further manipulate factorial notation.

EXAMPLE 5   Rewrite 7! with a factor of 5!.

Solution   We have

$$7! = 7 \cdot 6! = 7 \cdot 6 \cdot 5!.$$

In general, we have the following.

Technology Connection

We can evaluate factorial notation using the ! operation from the MATH PRB (probability) menu.

7!	
	5040
6!	
	720
5!	
	120

> For any natural numbers k and n, with $k < n$,
> $$n! = \underbrace{n(n-1)(n-2)\cdots[n-(k-1)]}_{k \text{ factors}} \cdot \underbrace{(n-k)!}_{n-k \text{ factors}}$$

Permutations of n Objects Taken k at a Time

Consider a set of 5 objects

$$\{A, B, C, D, E\}.$$

How many ordered arrangements can be formed using 3 objects without repetition? Examples of such an arrangement are EBA, CAB, and BCD. There are 5 choices for the first object, 4 choices for the second, and 3 choices for the third. By the fundamental counting principle, there are

$$5 \cdot 4 \cdot 3,$$

or

60 *permutations* of a set of 5 objects taken 3 at a time.

Note that

$$5 \cdot 4 \cdot 3 = \frac{5 \cdot 4 \cdot 3 \cdot 2 \cdot 1}{2 \cdot 1}, \quad \text{or} \quad \frac{5!}{2!}.$$

Permutation of n Objects Taken k at a Time

A **permutation** of a set of n objects taken k at a time is an ordered arrangement of k objects taken from the set.

Consider a set of n objects and the selection of an ordered arrangement of k of them. There would be n choices for the first object. Then there would remain $n-1$ choices for the second, $n-2$ choices for the third, and so on. We make k choices in all, so there are k factors in the product. By the fundamental counting principle, the total number of permutations is

$$\underbrace{n(n-1)(n-2)\cdots[n-(k-1)]}_{k \text{ factors}}.$$

We can express this in another way by multiplying by 1, as follows:

$$n(n-1)(n-2)\cdots[n-(k-1)] \cdot \frac{(n-k)!}{(n-k)!}$$

$$= \frac{n(n-1)(n-2)\cdots[n-(k-1)](n-k)!}{(n-k)!}$$

$$= \frac{n!}{(n-k)!}.$$

This gives us the following.

The Number of Permutations of n Objects Taken k at a Time

The number of permutations of a set of n objects taken k at a time, denoted $_nP_k$, is given by

$$_nP_k = \underbrace{n(n-1)(n-2) \cdots [n-(k-1)]}_{k \text{ factors}} \qquad (1)$$

$$= \frac{n!}{(n-k)!}. \qquad (2)$$

EXAMPLE 6   Compute $_8P_4$ using both forms of the formula.

SOLUTION   Using form (1), we have

$$_8P_4 = \underbrace{8 \cdot 7 \cdot 6 \cdot 5}_{} = 1680.$$

The 8 tells where to start.

The 4 tells how many factors.

Using form (2), we have

$$_8P_4 = \frac{8!}{(8-4)!}$$

$$= \frac{8!}{4!}$$

$$= \frac{8 \cdot 7 \cdot 6 \cdot 5 \cdot 4 \cdot 3 \cdot 2 \cdot 1}{4 \cdot 3 \cdot 2 \cdot 1}$$

$$= 8 \cdot 7 \cdot 6 \cdot 5 = 1680.$$

Technology Connection

We can do computations like the one in Example 6 using the $_nP_r$ operation from the MATH PRB menu on a graphing calculator.

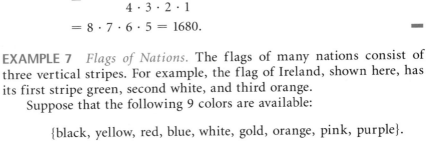

EXAMPLE 7   *Flags of Nations.* The flags of many nations consist of three vertical stripes. For example, the flag of Ireland, shown here, has its first stripe green, second white, and third orange.

Suppose that the following 9 colors are available:

{black, yellow, red, blue, white, gold, orange, pink, purple}.

How many different flags of 3 colors can be made without repetition of colors? This assumes that the order in which the stripes appear is considered.

Solution   We are determining the number of permutations of 9 objects taken 3 at a time. There is no repetition of colors. Using form (1), we get

$$_9P_3 = 9 \cdot 8 \cdot 7 = 504.$$

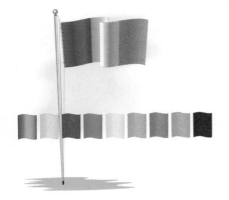

EXAMPLE 8 *Batting Orders.* A baseball manager arranges the batting order as follows: The 4 infielders will bat first. Then the 3 outfielders, the catcher, and the pitcher will follow, not necessarily in that order. How many different batting orders are possible?

Solution The infielders can bat in $_4P_4$ different ways, the rest in $_5P_5$ different ways. Then by the fundamental counting principle, we have

$$_4P_4 \cdot {_5P_5} = 4! \cdot 5!, \quad \text{or} \quad 2880 \text{ possible batting orders.} \quad \blacksquare$$

If we allow repetition, a situation like the following can occur.

EXAMPLE 9 How many 5-letter code symbols can be formed with the letters A, B, C, and D if we allow a letter to occur more than once?

Solution We can select each of the 5 letters in 4 ways. That is, we can select the first letter in 4 ways, the second in 4 ways, and so on. Thus there are 4^5, or 1024 arrangements. $\quad \blacksquare$

> The number of distinct arrangements of n objects taken k at a time, allowing repetition, is n^k.

Permutations of Sets with Nondistinguishable Objects

Consider a set of 7 marbles, 4 of which are blue and 3 of which are red. Although the marbles are all different, when they are lined up, one red marble will look just like any other red marble. In this sense, we say that the red marbles are nondistinguishable and, similarly, the blue marbles are nondistinguishable.

We know that there are 7! permutations of this set. Many of them will look alike, however. We develop a formula for finding the number of distinguishable permutations.

Consider a set of n objects in which n_1 are of one kind, n_2 are of a second kind, ..., n_k are of a kth kind. The total number of permutations of the set is $n!$, but this includes many that are nondistinguishable. Let N be the total number of distinguishable permutations. For each of these N permutations, there are $n_1!$ actual, nondistinguishable permutations, obtained by permuting the objects of the first kind. For each of these $N \cdot n_1!$ permutations, there are $n_2!$ nondistinguishable permutations, obtained by permuting the objects of the second kind, and so on. By the fundamental counting principle, the total number of permutations, including those that are nondistinguishable, is

$$N \cdot n_1! \cdot n_2! \cdot \cdots \cdot n_k!.$$

Then we have $N \cdot n_1! \cdot n_2! \cdot \cdots \cdot n_k! = n!$. Solving for N, we obtain

$$N = \frac{n!}{n_1! \cdot n_2! \cdot \cdots \cdot n_k!}.$$

Now, to finish our problem with the marbles, we have

$$N = \frac{7!}{4!\,3!} = \frac{7 \cdot 6 \cdot 5 \cdot 4 \cdot 3 \cdot 2 \cdot 1}{4 \cdot 3 \cdot 2 \cdot 1 \cdot 3 \cdot 2 \cdot 1}$$

$$= \frac{7 \cdot 5}{1}, \quad \text{or} \quad 35$$

distinguishable permutations of the marbles.
   In general:

For a set of n objects in which n_1 are of one kind, n_2 are of another kind, ..., n_k are of a kth kind, the number of distinguishable permutations is

$$\frac{n!}{n_1! \cdot n_2! \cdot \cdots \cdot n_k!}.$$

EXAMPLE 10   In how many distinguishable ways can the letters of the word CINCINNATI be arranged?

Solution   There are 2 C's, 3 I's, 3 N's, 1 A, and 1 T for a total of 10 letters. Thus,

$$N = \frac{10!}{2! \cdot 3! \cdot 3! \cdot 1! \cdot 1!}, \quad \text{or} \quad 50,400.$$

The letters of the word CINCINNATI can be arranged in 50,400 distinguishable ways.   ▄

Exercise Set

Evaluate.

1. $_6P_6$

2. $_4P_3$

3. $_{10}P_7$

4. $_{10}P_3$

5. $5!$

6. $7!$

7. $0!$

8. $1!$

9. $\dfrac{9!}{5!}$

10. $\dfrac{9!}{4!}$

11. $(8 - 3)!$

12. $(8 - 5)!$

13. $\dfrac{10!}{7!\,3!}$

14. $\dfrac{7!}{(7 - 2)!}$

15. $_8P_0$

16. $_{13}P_1$

17. $_{52}P_4$

18. $_{52}P_5$

19. $_nP_3$

20. $_nP_2$

21. $_nP_1$

22. $_nP_0$

In each of the following exercises, give your answer using permutation notation, factorial notation, or other operations. Then evaluate.

How many permutations are there of the letters in each of the following words, if all the letters are used without repetition?

23. MARVIN

24. JUDY

25. UNDERMOST

26. COMBINES

27. How many permutations are there of the letters of the word UNDERMOST if the letters are taken 4 at a time?

28. How many permutations are there of the letters of the word COMBINES if the letters are taken 5 at a time?

29. How many 5-digit numbers can be formed using the digits 2, 4, 6, 8, and 9 without repetition? with repetition?

30. In how many ways can 7 athletes be arranged in a straight line?

31. How many distinguishable code symbols can be formed from the letters of the word BUSINESS? BIOLOGY? MATHEMATICS?

32. A professor is going to grade her 24 students on a curve. She will give 3 A's, 5 B's, 9 C's, 4 D's, and 3 F's. In how many ways can she do this?

33. *Phone Numbers.* How many 7-digit phone numbers can be formed with the digits 0, 1, 2, 3, 4, 5, 6, 7, 8, and 9, assuming that the first number cannot be 0 or 1? Accordingly, how many telephone numbers can there be within a given area code, before the area needs to be split with a new area code?

34. *Program Planning.* A program is planned to have 5 rock numbers and 4 speeches. In how many ways can this be done if a rock number and a speech are to alternate and a rock number is to come first?

35. Suppose the expression $a^2b^3c^4$ is rewritten without exponents. In how many ways can this be done?

36. *Coin Arrangements.* A penny, a nickel, a dime, and a quarter are arranged in a straight line.

a) Considering just the coins, in how many ways can they be lined up?

b) Considering the coins and heads and tails, in how many ways can they be lined up?

37. How many code symbols can be formed using 5 out of 6 letters of A, B, C, D, E, F if the letters:

a) are not repeated?

b) can be repeated?

c) are not repeated but must begin with D?

d) are not repeated but must begin with DE?

38. *License Plates.* A state forms its license plates by first listing a number that corresponds to the county in which the owner of the car resides (the names of the counties are alphabetized and the number is its location in that order). Then the plate lists a letter of the alphabet, and this is followed by a number from 1 to 9999. How many such plates are possible if there are 80 counties?

39. *Zip Codes.* A U.S. postal zip code is a five-digit number.

a) How many zip codes are possible if any of the digits 0 to 9 can be used?

b) If each post office has its own zip code, how many possible post offices can there be?

40. *Zip-Plus-4 Codes.* A zip-plus-4 postal code uses a 9-digit number like 75247-5456. How many 9-digit zip-plus-4 postal codes are possible?

41. *Social Security Numbers.* A social security number is a 9-digit number like 243-47-0825.

a) How many different social security numbers can there be?

b) There are about 284 million people in the United States. Can each person have a unique social security number?

Collaborative Discussion and Writing

42. How "long" is 15!? You own 15 different books and decide to make up all the possible arrangements of the books on a shelf. About how long, in years, would it take you if you make one arrangement per second? Write out the reasoning you used for this problem in the form of a paragraph.

43. *Circular Arrangements.* In how many ways can the numbers on a clock face be arranged? See if you can derive a formula for the number of distinct circular arrangements of n objects. Explain your reasoning.

Skill Maintenance

Find the zero(s) of the function.

44. $f(x) = 4x - 9$

45. $f(x) = x^2 + x - 6$

46. $f(x) = 2x^2 - 3x - 1$

47. $f(x) = x^3 - 4x^2 - 7x + 10$

Synthesis

Solve for n.

48. $_nP_5 = 7 \cdot {}_nP_4$ **49.** $_nP_4 = 8 \cdot {}_{n-1}P_3$

50. $_nP_5 = 9 \cdot {}_{n-1}P_4$ **51.** $_nP_4 = 8 \cdot {}_nP_3$

52. Show that $n! = n(n - 1)(n - 2)(n - 3)!$.

53. *Single-Elimination Tournaments.*   In a single-elimination sports tournament consisting of n teams, a team is eliminated when it loses one game. How many games are required to complete the tournament?

54. *Double-Elimination Tournaments.*   In a double-elimination softball tournament consisting of n teams, a team is eliminated when it loses two games. At most, how many games are required to complete the tournament?

10.6

Combinatorics: Combinations

• *Evaluate combination notation and solve related applied problems.*

We now consider counting techniques in which order is not considered.

Combinations

We sometimes make a selection from a set *without regard to order.* Such a selection is called a *combination.* If you play cards, for example, you know that in most situations the *order* in which you hold cards is not important. That is,

The hand

is "equivalent" to these hands.

Each hand contains the same combination of three cards.

EXAMPLE 1   Find all the combinations of 3 letters taken from the set of 5 letters {A, B, C, D, E}.

Solution The combinations are

$$\{A, B, C\}, \quad \{A, B, D\},$$
$$\{A, B, E\}, \quad \{A, C, D\},$$
$$\{A, C, E\}, \quad \{A, D, E\},$$
$$\{B, C, D\}, \quad \{B, C, E\},$$
$$\{B, D, E\}, \quad \{C, D, E\}.$$

There are 10 combinations of the 5 letters taken 3 at a time.

When we find all the combinations from a set of 5 objects taken 3 at a time, we are finding all the 3-element subsets. When a set is named, the order of the listing is *not* considered. Thus,

$$\{A, C, B\} \quad \text{names the same set as} \quad \{A, B, C\}.$$

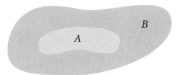

Subset

Set A is a subset of set B, denoted $A \subseteq B$, if every element of A is an element of B.

The elements of a subset are not ordered. When thinking of *combinations*, do *not* think about order!

Combination

A **combination** containing k objects is a subset containing k objects.

We want to develop a formula for computing the number of combinations of n objects taken k at a time without actually listing the combinations or subsets.

Combination Notation

The number of combinations of n objects taken k at a time is denoted $_nC_k$.

We call $_nC_k$ **combination notation.** We want to derive a general formula for $_nC_k$ for any $k \leq n$. First, it is true that $_nC_n = 1$, because a set with n objects has only 1 subset with n objects, the set itself. Second, $_nC_1 = n$, because a set with n objects has n subsets with 1 element each. Finally, $_nC_0 = 1$, because a set with n objects has only one subset with 0 elements, namely the empty set $\varnothing$. To consider other possibilities, let's return to Example 1 and compare the number of combinations with the number of permutations.

COMBINATIONS		PERMUTATIONS				

$_5C_3$ of these $\left\{\begin{array}{l} \\ \\ \\ \\ \\ \\ \\ \\ \\ \\ \end{array}\right.$

$\{A, B, C\} \longrightarrow$	ABC	BCA	CAB	CBA	BAC	ACB
$\{A, B, D\} \longrightarrow$	ABD	BDA	DAB	DBA	BAD	ADB
$\{A, B, E\} \longrightarrow$	ABE	BEA	EAB	EBA	BAE	AEB
$\{A, C, D\} \longrightarrow$	ACD	CDA	DAC	DCA	CAD	ADC
$\{A, C, E\} \longrightarrow$	ACE	CEA	EAC	ECA	CAE	AEC
$\{A, D, E\} \longrightarrow$	ADE	DEA	EAD	EDA	DAE	AED
$\{B, C, D\} \longrightarrow$	BCD	CDB	DBC	DCB	CBD	BDC
$\{B, C, E\} \longrightarrow$	BCE	CEB	EBC	ECB	CBE	BEC
$\{B, D, E\} \longrightarrow$	BDE	DEB	EBD	EDB	DBE	BED
$\{C, D, E\} \longrightarrow$	CDE	DEC	ECD	EDC	DCE	CED

$\left.\begin{array}{r} \\ \\ \\ \\ \\ \\ \\ \\ \\ \\ \end{array}\right\}$ $3! \cdot {_5C_3}$ of these

Note that each combination of 3 objects yields 6, or 3!, permutations.

$$3! \cdot {_5C_3} = 60 = {_5P_3} = 5 \cdot 4 \cdot 3,$$

so

$${_5C_3} = \frac{{_5P_3}}{3!} = \frac{5 \cdot 4 \cdot 3}{3 \cdot 2 \cdot 1} = 10.$$

In general, the number of combinations of n objects taken k at a time, $_nC_k$, times the number of permutations of these objects, $k!$, must equal the number of permutations of n objects taken k at a time:

$$k! \cdot {_nC_k} = {_nP_k}$$

$${_nC_k} = \frac{{_nP_k}}{k!}$$

$$= \frac{1}{k!} \cdot {_nP_k}$$

$$= \frac{1}{k!} \cdot \frac{n!}{(n-k)!} = \frac{n!}{k!(n-k)!}.$$

Combinations of n Objects Taken k at a Time

The total number of combinations of n objects taken k at a time, denoted $_nC_k$, is given by

$${_nC_k} = \frac{n!}{k!(n-k)!}, \tag{1}$$

or

$${_nC_k} = \frac{{_nP_k}}{k!} = \frac{n(n-1)(n-2) \cdots [n-(k-1)]}{k!}. \tag{2}$$

Another kind of notation for $_nC_k$ is **binomial coefficient notation.** The reason for such terminology will be seen later.

Binomial Coefficient Notation

$$\binom{n}{k} = {}_nC_k$$

You should be able to use either notation and either form of the formula.

EXAMPLE 2 Evaluate $\binom{7}{5}$, using forms (1) and (2).

Solution

a) By form (1),

$$\binom{7}{5} = \frac{7!}{5!\,2!} = \frac{7 \cdot 6 \cdot 5 \cdot 4 \cdot 3 \cdot 2 \cdot 1}{5 \cdot 4 \cdot 3 \cdot 2 \cdot 1 \cdot 2 \cdot 1} = \frac{7 \cdot 6}{2 \cdot 1} = 21.$$

b) By form (2),

The 7 tells where to start.

$$\binom{7}{5} = \frac{7 \cdot 6 \cdot 5 \cdot 4 \cdot 3}{5 \cdot 4 \cdot 3 \cdot 2 \cdot 1} = \frac{7 \cdot 6}{2 \cdot 1} = 21.$$

The 5 tells how many factors there are in both the numerator and the denominator and where to start the denominator.

Be sure to keep in mind that $\binom{n}{k}$ does not mean $n \div k$, or n/k.

EXAMPLE 3 Evaluate $\binom{n}{0}$ and $\binom{n}{2}$.

Solution We use form (1) for the first expression and form (2) for the second. Then

$$\binom{n}{0} = \frac{n!}{0!\,(n-0)!} = \frac{n!}{1 \cdot n!} = 1,$$

using form (1), and

$$\binom{n}{2} = \frac{n(n-1)}{2!} = \frac{n(n-1)}{2}, \quad \text{or} \quad \frac{n^2-n}{2},$$

using form (2).

Note that

$$\binom{7}{2} = \frac{7 \cdot 6}{2 \cdot 1} = 21,$$

so that using the result of Example 2 gives us

$$\binom{7}{5} = \binom{7}{2}.$$

This says that the number of 5-element subsets of a set of 7 objects is the same as the number of 2-element subsets of a set of 7 objects. When 5 elements are chosen from a set, one also chooses *not* to include 2 elements. To see this, consider such a set:

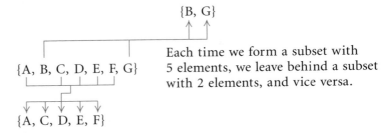

In general, we have the following.

Subsets of Size k and of Size n − k

$$\binom{n}{k} = \binom{n}{n-k} \quad \text{and} \quad {}_nC_k = {}_nC_{n-k}$$

The number of subsets of size k of a set with n objects is the same as the number of subsets of size $n - k$. The number of combinations of n objects taken k at a time is the same as the number of combinations of n objects taken $n - k$ at a time.

This result provides an alternative way to compute combinations. We now solve problems involving combinations.

EXAMPLE 4 *Michigan Lotto.* The state of Michigan runs a 6-out-of-49-number lotto twice a week that pays at least $2 million. You purchase a card for $1 and pick any 6 numbers from 1 to 49. If your numbers match those that the state draws, you win.

a) How many 6-number combinations are there for drawing?

b) Suppose that it takes 10 min to pick your numbers and buy a ticket. How many tickets can you buy in 4 days?

c) How many people would you have to hire to buy tickets with all the possible combinations and ensure that you win?

Solution

a) No order is implied here. You pick any 6 numbers from 1 to 49. Thus the number of combinations is

$$_{49}C_6 = \binom{49}{6} = \frac{49!}{6!\,43!}$$
$$= \frac{49 \cdot 48 \cdot 47 \cdot 46 \cdot 45 \cdot 44}{6 \cdot 5 \cdot 4 \cdot 3 \cdot 2 \cdot 1}$$
$$= 13{,}983{,}816.$$

b) In 4 days, there are $4 \cdot 24 \cdot 60$, or 5760 min, so you could buy 5760/10, or 576 tickets in that entire time period.

c) You would need to hire 13,983,816/576, or about 24,278 people to buy tickets with all the possible combinations and ensure a win. (This presumes lottery tickets can be bought 24 hours a day.) ▬

EXAMPLE 5   How many committees can be formed from a group of 5 governors and 7 senators if each committee consists of 3 governors and 4 senators?

Solution   The 3 governors can be selected in $_5C_3$ ways and the 4 senators can be selected in $_7C_4$ ways. If we use the fundamental counting principle, it follows that the number of possible committees is

$$_5C_3 \cdot {_7C_4} = 10 \cdot 35 = 350.$$ ▬

CONNECTING THE CONCEPTS

PERMUTATIONS AND COMBINATIONS

PERMUTATIONS

Permutations involve order and arrangements of objects.

Given 5 books, we can arrange 3 of them on a shelf in $_5P_3$, or 60 ways.

Placing the books in different orders produces different arrangements.

COMBINATIONS

Combinations do not involve order or arrangements of objects.

Given 5 books, we can select 3 of them in $_5C_3$, or 10 ways.

The order in which the books are chosen does not matter.

Exercise Set 10.6

Evaluate.

1. $_{13}C_2$

2. $_9C_6$

3. $\begin{pmatrix} 13 \\ 11 \end{pmatrix}$

4. $\begin{pmatrix} 9 \\ 3 \end{pmatrix}$

5. $\begin{pmatrix} 7 \\ 1 \end{pmatrix}$

6. $\begin{pmatrix} 8 \\ 8 \end{pmatrix}$

7. $\dfrac{_5P_3}{3!}$

8. $\dfrac{_{10}P_5}{5!}$

9. $\begin{pmatrix} 6 \\ 0 \end{pmatrix}$

10. $\begin{pmatrix} 6 \\ 1 \end{pmatrix}$

11. $\begin{pmatrix} 6 \\ 2 \end{pmatrix}$

12. $\begin{pmatrix} 6 \\ 3 \end{pmatrix}$

13. $\begin{pmatrix} 7 \\ 0 \end{pmatrix} + \begin{pmatrix} 7 \\ 1 \end{pmatrix} + \begin{pmatrix} 7 \\ 2 \end{pmatrix} + \begin{pmatrix} 7 \\ 3 \end{pmatrix} + \begin{pmatrix} 7 \\ 4 \end{pmatrix} + \begin{pmatrix} 7 \\ 5 \end{pmatrix}$
$+ \begin{pmatrix} 7 \\ 6 \end{pmatrix} + \begin{pmatrix} 7 \\ 7 \end{pmatrix}$

14. $\begin{pmatrix} 6 \\ 0 \end{pmatrix} + \begin{pmatrix} 6 \\ 1 \end{pmatrix} + \begin{pmatrix} 6 \\ 2 \end{pmatrix} + \begin{pmatrix} 6 \\ 3 \end{pmatrix} + \begin{pmatrix} 6 \\ 4 \end{pmatrix} + \begin{pmatrix} 6 \\ 5 \end{pmatrix} + \begin{pmatrix} 6 \\ 6 \end{pmatrix}$

15. $_{52}C_4$

16. $_{52}C_5$

17. $\begin{pmatrix} 27 \\ 11 \end{pmatrix}$

18. $\begin{pmatrix} 37 \\ 8 \end{pmatrix}$

19. $\begin{pmatrix} n \\ 1 \end{pmatrix}$

20. $\begin{pmatrix} n \\ 3 \end{pmatrix}$

21. $\begin{pmatrix} m \\ m \end{pmatrix}$

22. $\begin{pmatrix} t \\ 4 \end{pmatrix}$

In each of the following exercises, give an expression for the answer using permutation notation, combination notation, factorial notation, or other operations. Then evaluate.

23. *Fraternity Officers.* There are 23 students in a fraternity. How many sets of 4 officers can be selected?

24. *League Games.* How many games can be played in a 9-team sports league if each team plays all other teams once? twice?

25. *Test Options.* On a test, a student is to select 10 out of 13 questions. In how many ways can this be done?

26. *Test Options.* Of the first 10 questions on a test, a student must answer 7. Of the second 5 questions, the student must answer 3. In how many ways can this be done?

27. *Lines and Triangles from Points.* How many lines are determined by 8 points, no 3 of which are collinear? How many triangles are determined by the same points?

28. *Senate Committees.* Suppose the Senate of the United States consists of 58 Republicans and 42 Democrats. How many committees can be formed consisting of 6 Republicans and 4 Democrats?

29. *Poker Hands.* How many 5-card poker hands are possible with a 52-card deck?

30. *Bridge Hands.* How many 13-card bridge hands are possible with a 52-card deck?

31. *Baskin-Robbins Ice Cream.* Baskin-Robbins, a national firm, sells ice cream in 31 flavors.

  a) How many 2-dip cones are possible if order of flavors is to be considered and no flavor is repeated?

  b) How many 2-dip cones are possible if order is to be considered and a flavor can be repeated?

  c) How many 2-dip cones are possible if order is not considered and no flavor is repeated?

Collaborative Discussion and Writing

32. Explain why a "combination" lock should really be called a "permutation" lock.

33. Give an explanation that you might use with a fellow student to explain that

$$\binom{n}{k} = \binom{n}{n-k}.$$

Skill Maintenance

Solve.

34. $3x - 7 = 5x + 10$ **35.** $2x^2 - x = 3$

36. $x^2 + 5x + 1 = 0$ **37.** $x^3 + 3x^2 - 10x = 24$

Synthesis

38. *Full House.* A full house in poker consists of a pair (two of a kind) and three of a kind. How many full houses are there that consist of 3 aces and 2 queens? (See Section 7.8 for a description of a 52-card deck.)

39. *Flush.* A flush in poker consists of a 5-card hand with all cards of the same suit. How many 5-card hands (flushes) are there that consist of all diamonds?

40. There are n points on a circle. How many quadrilaterals can be inscribed with these points as vertices?

41. *League Games.* How many games are played in a league with n teams if each team plays each other team once? twice?

Solve for n.

42. $\binom{n+1}{3} = 2 \cdot \binom{n}{2}$ **43.** $\binom{n}{n-2} = 6$

44. $\binom{n}{3} = 2 \cdot \binom{n-1}{2}$ **45.** $\binom{n+2}{4} = 6 \cdot \binom{n}{2}$

46. How many line segments are determined by the n vertices of an n-agon? Of these, how many are diagonals? Use mathematical induction to prove the result for the diagonals.

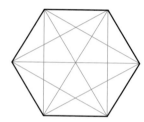

47. Prove that

$$\binom{n}{k-1} + \binom{n}{k} = \binom{n+1}{k}$$

for any natural numbers n and k, $k \le n$.

10.7

The Binomial Theorem

- Expand a power of a binomial using Pascal's triangle or factorial notation.
- Find a specific term of a binomial expansion.
- Find the total number of subsets of a set of n objects.

In this section, we consider ways of expanding a binomial $(a + b)^n$.

Binomial Expansions Using Pascal's Triangle

Consider the following expanded powers of $(a + b)^n$, where $a + b$ is any binomial and n is a whole number. Look for patterns.

$$(a + b)^0 = 1$$
$$(a + b)^1 = a + b$$
$$(a + b)^2 = a^2 + 2ab + b^2$$
$$(a + b)^3 = a^3 + 3a^2b + 3ab^2 + b^3$$
$$(a + b)^4 = a^4 + 4a^3b + 6a^2b^2 + 4ab^3 + b^4$$
$$(a + b)^5 = a^5 + 5a^4b + 10a^3b^2 + 10a^2b^3 + 5ab^4 + b^5$$

Each expansion is a polynomial. There are some patterns to be noted.

1. There is one more term than the power of the exponent, n. That is, there are $n + 1$ terms in the expansion of $(a + b)^n$.

2. In each term, the sum of the exponents is n, the power to which the binomial is raised.

3. The exponents of a start with n, the power of the binomial, and decrease to 0. The last term has no factor of a. The first term has no factor of b, so powers of b start with 0 and increase to n.

4. The coefficients start at 1 and increase through certain values about "half"-way and then decrease through these same values back to 1.

Let's explore the coefficients further. Suppose that we want to find an expansion of $(a + b)^6$. The patterns we noted above indicate that there are 7 terms in the expansion:

$$a^6 + c_1a^5b + c_2a^4b^2 + c_3a^3b^3 + c_4a^2b^4 + c_5ab^5 + b^6.$$

How can we determine the value of each coefficient, c_i? We can do so in two different ways. The first method involves writing the coefficients in a triangular array, as follows. This is known as **Pascal's triangle:**

$$
\begin{array}{lccccccccccc}
(a + b)^0: & & & & & & 1 & & & & & \\
(a + b)^1: & & & & & 1 & & 1 & & & & \\
(a + b)^2: & & & & 1 & & 2 & & 1 & & & \\
(a + b)^3: & & & 1 & & 3 & & 3 & & 1 & & \\
(a + b)^4: & & 1 & & 4 & & 6 & & 4 & & 1 & \\
(a + b)^5: & 1 & & 5 & & 10 & & 10 & & 5 & & 1 \\
\end{array}
$$

There are many patterns in the triangle. Find as many as you can.

Perhaps you discovered a way to write the next row of numbers, given the numbers in the row above it. There are always 1's on the outside. Each remaining number is the sum of the two numbers above it. Let's try to find an expansion for $(a + b)^6$ by adding another row using the patterns we have discovered:

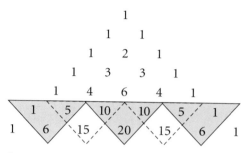

We see that in the last row

the 1st and last numbers are **1**;

the 2nd number is 1 + 5, or **6**;

the 3rd number is 5 + 10, or **15**;

the 4th number is 10 + 10, or **20**;

the 5th number is 10 + 5, or **15**; and

the 6th number is 5 + 1, or **6**.

Thus the expansion for $(a + b)^6$ is

$$(a + b)^6 = 1a^6 + 6a^5b + 15a^4b^2 + 20a^3b^3 + 15a^2b^4 + 6ab^5 + 1b^6.$$

To find an expansion for $(a + b)^8$, we complete two more rows of Pascal's triangle:

$$
\begin{array}{ccccccccccccccc}
&&&&&&&& 1 \\
&&&&&&& 1 && 1 \\
&&&&&& 1 && 2 && 1 \\
&&&&& 1 && 3 && 3 && 1 \\
&&&& 1 && 4 && 6 && 4 && 1 \\
&&& 1 && 5 && 10 && 10 && 5 && 1 \\
&& 1 && 6 && 15 && 20 && 15 && 6 && 1 \\
& 1 && 7 && 21 && 35 && 35 && 21 && 7 && 1 \\
1 && 8 && 28 && 56 && 70 && 56 && 28 && 8 && 1
\end{array}
$$

Thus the expansion of $(a + b)^8$ is

$$(a + b)^8 = a^8 + 8a^7b^1 + 28a^6b^2 + 56a^5b^3 + 70a^4b^4 + 56a^3b^5$$
$$+ 28a^2b^6 + 8a^1b^7 + b^8.$$

We can generalize our results as follows.

The Binomial Theorem Using Pascal's Triangle

For any binomial $a + b$ and any natural number n,

$$(a + b)^n = c_0 a^n b^0 + c_1 a^{n-1} b^1 + c_2 a^{n-2} b^2 + \cdots + c_{n-1} a^1 b^{n-1}$$
$$+ c_n a^0 b^n,$$

where the numbers $c_0, c_1, c_2, \ldots, c_{n-1}, c_n$ are from the $(n + 1)$st row of Pascal's triangle.

EXAMPLE 1   Expand: $(u - v)^5$.

Solution   We have $(a + b)^n$, where $a = u$, $b = -v$, and $n = 5$. We use the 6th row of Pascal's triangle:

$$1 \quad 5 \quad 10 \quad 10 \quad 5 \quad 1$$

Then we have

$$(u - v)^5 = [u + (-v)]^5$$
$$= 1(u)^5 + 5(u)^4(-v)^1 + 10(u)^3(-v)^2 + 10(u)^2(-v)^3$$
$$+ 5(u)(-v)^4 + 1(-v)^5$$
$$= u^5 - 5u^4 v + 10u^3 v^2 - 10u^2 v^3 + 5uv^4 - v^5.$$

Note that the signs of the terms alternate between $+$ and $-$. When the power of $-v$ is odd, the sign is $-$.   ▬

EXAMPLE 2   Expand: $\left(2t + \dfrac{3}{t} \right)^4$.

Solution   We have $(a + b)^n$, where $a = 2t$, $b = 3/t$, and $n = 4$. We use the 5th row of Pascal's triangle:

$$1 \quad 4 \quad 6 \quad 4 \quad 1$$

Then we have

$$\left(2t + \frac{3}{t} \right)^4 = (2t)^4 + 4(2t)^3 \left(\frac{3}{t} \right)^1 + 6(2t)^2 \left(\frac{3}{t} \right)^2 + 4(2t)^1 \left(\frac{3}{t} \right)^3 + \left(\frac{3}{t} \right)^4$$

$$= 16t^4 + 4(8t^3) \left(\frac{3}{t} \right) + 6(4t^2) \left(\frac{9}{t^2} \right) + 4(2t) \left(\frac{27}{t^3} \right) + \frac{81}{t^4}$$

$$= 16t^4 + 96t^2 + 216 + 216t^{-2} + 81t^{-4}.$$ ▬

Binomial Expansion Using Factorial Notation

Suppose that we want to find the expansion of $(a + b)^{11}$. The disadvantage in using Pascal's triangle is that we must compute all the preceding rows of the triangle to obtain the row needed for the expansion. The following method avoids this. It also enables us to find a specific term—

say, the 8th term—without computing all the other terms of the expansion. This method is useful in such courses as finite mathematics, calculus, and statistics, and it uses the *binomial coefficient notation* $\binom{n}{k}$ developed in Section 10.6.

We can restate the binomial theorem as follows.

The Binomial Theorem Using Factorial Notation

For any binomial $a + b$ and any natural number n,

$$(a + b)^n = \binom{n}{0}a^n b^0 + \binom{n}{1}a^{n-1}b^1 + \binom{n}{2}a^{n-2}b^2 + \cdots$$

$$+ \binom{n}{n-1}a^1 b^{n-1} + \binom{n}{n}a^0 b^n$$

$$= \sum_{k=0}^{n} \binom{n}{k} a^{n-k}b^k.$$

The binomial theorem can be proved by mathematical induction, but we will not do so here. This form shows why $\binom{n}{k}$ is called a *binomial coefficient*.

EXAMPLE 3 Expand: $(x^2 - 2y)^5$.

Solution We have $(a + b)^n$, where $a = x^2$, $b = -2y$, and $n = 5$. Then using the binomial theorem, we have

$$(x^2 - 2y)^5 = \binom{5}{0}(x^2)^5 + \binom{5}{1}(x^2)^4(-2y) + \binom{5}{2}(x^2)^3(-2y)^2$$

$$+ \binom{5}{3}(x^2)^2(-2y)^3 + \binom{5}{4}x^2(-2y)^4 + \binom{5}{5}(-2y)^5$$

$$= \frac{5!}{0!\,5!}x^{10} + \frac{5!}{1!\,4!}x^8(-2y) + \frac{5!}{2!\,3!}x^6(4y^2) + \frac{5!}{3!\,2!}x^4(-8y^3)$$

$$+ \frac{5!}{4!\,1!}x^2(16y^4) + \frac{5!}{5!\,0!}(-32y^5)$$

$$= x^{10} - 10x^8 y + 40x^6 y^2 - 80x^4 y^3 + 80x^2 y^4 - 32y^5. \qquad \blacksquare$$

EXAMPLE 4 Expand: $\left(\dfrac{2}{x} + 3\sqrt{x}\right)^4$.

Solution   We have $(a + b)^n$, where $a = 2/x$, $b = 3\sqrt{x}$, and $n = 4$. Then using the binomial theorem, we have

$$\left(\frac{2}{x} + 3\sqrt{x}\right)^4 = \binom{4}{0}\left(\frac{2}{x}\right)^4 + \binom{4}{1}\left(\frac{2}{x}\right)^3(3\sqrt{x}) + \binom{4}{2}\left(\frac{2}{x}\right)^2(3\sqrt{x})^2$$

$$+ \binom{4}{3}\left(\frac{2}{x}\right)(3\sqrt{x})^3 + \binom{4}{4}(3\sqrt{x})^4$$

$$= \frac{4!}{0!\,4!}\left(\frac{2}{x}\right)^4 + \frac{4!}{1!\,3!}\left(\frac{2}{x}\right)^3(3\sqrt{x}) + \frac{4!}{2!\,2!}\left(\frac{2}{x}\right)^2(3\sqrt{x})^2$$

$$+ \frac{4!}{3!\,1!}\left(\frac{2}{x}\right)(3\sqrt{x})^3 + \frac{4!}{4!\,0!}(3\sqrt{x})^4$$

$$= \frac{16}{x^4} + \frac{96}{x^{5/2}} + \frac{216}{x} + 216\sqrt{x} + 81x^2.$$

Finding a Specific Term

Suppose that we want to determine only a particular term of an expansion. The method we have developed will allow us to find such a term without computing all the rows of Pascal's triangle or all the preceding coefficients.

Note that in the binomial theorem, $\binom{n}{0}a^n b^0$ gives us the 1st term, $\binom{n}{1}a^{n-1}b^1$ gives us the 2nd term, $\binom{n}{2}a^{n-2}b^2$ gives us the 3rd term, and so on. This can be generalized as follows.

Finding the $(k + 1)$st Term

The $(k + 1)$st term of $(a + b)^n$ is $\binom{n}{k}a^{n-k}b^k$.

EXAMPLE 5   Find the 5th term in the expansion of $(2x - 5y)^6$.

Solution   First, we note that $5 = 4 + 1$. Thus, $k = 4$, $a = 2x$, $b = -5y$, and $n = 6$. Then the 5th term of the expansion is

$$\binom{6}{4}(2x)^{6-4}(-5y)^4, \quad \text{or} \quad \frac{6!}{4!\,2!}(2x)^2(-5y)^4, \quad \text{or} \quad 37{,}500x^2y^4.$$

EXAMPLE 6   Find the 8th term in the expansion of $(3x - 2)^{10}$.

Solution   First, we note that $8 = 7 + 1$. Thus, $k = 7$, $a = 3x$, $b = -2$, and $n = 10$. Then the 8th term of the expansion is

$$\binom{10}{7}(3x)^{10-7}(-2)^7, \quad \text{or} \quad \frac{10!}{7!\,3!}(3x)^3(-2)^7, \quad \text{or} \quad -414{,}720x^3.$$

Total Number of Subsets

Suppose that a set has n objects. The number of subsets containing k elements is $\binom{n}{k}$ by a result of Section 10.6. The total number of subsets of a set is the number of subsets with 0 elements, plus the number of subsets with 1 element, plus the number of subsets with 2 elements, and so on. The total number of subsets of a set with n elements is

$$\binom{n}{0} + \binom{n}{1} + \binom{n}{2} + \cdots + \binom{n}{n}.$$

Now consider the expansion of $(1 + 1)^n$:

$$(1 + 1)^n = \binom{n}{0} \cdot 1^n + \binom{n}{1} \cdot 1^{n-1} \cdot 1^1 + \binom{n}{2} \cdot 1^{n-2} \cdot 1^2$$

$$+ \cdots + \binom{n}{n} \cdot 1^n$$

$$= \binom{n}{0} + \binom{n}{1} + \binom{n}{2} + \cdots + \binom{n}{n}.$$

Thus the total number of subsets is $(1 + 1)^n$, or 2^n. We have proved the following.

Total Number of Subsets

The total number of subsets of a set with n elements is 2^n.

EXAMPLE 7 The set {A, B, C, D, E} has how many subsets?

Solution The set has 5 elements, so the number of subsets is 2^5, or 32.

EXAMPLE 8 Wendy's, a national restaurant firm, offers the following condiments for its hamburgers:

{*catsup, mustard, mayonnaise, tomato,*

lettuce, onions, pickle, relish, cheese}.

How many different kinds of hamburgers can Wendy's serve, excluding size of hamburger or number of patties?

Solution The condiments on each hamburger are the elements of a subset of the set of all possible condiments, the empty set being a plain hamburger. The total number of possible hamburgers is

$$\binom{9}{0} + \binom{9}{1} + \binom{9}{2} + \cdots + \binom{9}{9} = 2^9 = 512.$$

Thus Wendy's serves hamburgers in 512 different ways.

Exercise Set 10.7

Expand.

1. $(x + 5)^4$

2. $(x - 1)^4$

3. $(x - 3)^5$

4. $(x + 2)^9$

5. $(x - y)^5$

6. $(x + y)^8$

7. $(5x + 4y)^6$

8. $(2x - 3y)^5$

9. $\left(2t + \dfrac{1}{t}\right)^7$

10. $\left(3y - \dfrac{1}{y}\right)^4$

11. $(x^2 - 1)^5$

12. $(1 + 2q^3)^8$

13. $(\sqrt{5} + t)^6$

14. $(x - \sqrt{2})^6$

15. $\left(a - \dfrac{2}{a}\right)^9$

16. $(1 + 3)^n$

17. $(\sqrt{2} + 1)^6 - (\sqrt{2} - 1)^6$

18. $(1 - \sqrt{2})^4 + (1 + \sqrt{2})^4$

19. $(x^{-2} + x^2)^4$

20. $\left(\dfrac{1}{\sqrt{x}} - \sqrt{x}\right)^6$

Find the indicated term of the binomial expansion.

21. 3rd; $(a + b)^7$

22. 6th; $(x + y)^8$

23. 6th; $(x - y)^{10}$

24. 5th; $(p - 2q)^9$

25. 12th; $(a - 2)^{14}$

26. 11th; $(x - 3)^{12}$

27. 5th; $(2x^3 - \sqrt{y})^8$

28. 4th; $\left(\dfrac{1}{b^2} + \dfrac{b}{3}\right)^7$

29. Middle; $(2u - 3v^2)^{10}$

30. Middle two; $(\sqrt{x} + \sqrt{3})^5$

Determine the number of subsets of each of the following.

31. A set of 7 elements

32. A set of 6 members

33. The set of letters of the Greek alphabet, which contains 24 letters

34. The set of letters of the English alphabet, which contains 26 letters

35. What is the degree of $(x^5 + 3)^4$?

36. What is the degree of $(2 - 5x^3)^7$?

Expand each of the following, where $i^2 = -1$.

37. $(3 + i)^5$

38. $(1 + i)^6$

39. $(\sqrt{2} - i)^4$

40. $\left(\dfrac{\sqrt{3}}{2} - \dfrac{1}{2}i\right)^{11}$

41. Find a formula for $(a - b)^n$. Use sigma notation.

42. Expand and simplify:
$$\dfrac{(x + h)^{13} - x^{13}}{h}.$$

43. Expand and simplify:
$$\dfrac{(x + h)^n - x^n}{h}.$$
Use sigma notation.

Collaborative Discussion and Writing

44. Discuss the pros and cons of each method of finding a binomial expansion. Give examples of when you might use one method rather than the other.

45. Blaise Pascal (1623–1662) was a French scientist and philosopher who founded the modern theory of probability. Do some research on Pascal and see if you can find out how he discovered his famous "triangle of numbers."

Skill Maintenance

Given that $f(x) = x^2 + 1$ and $g(x) = 2x - 3$, find each of the following.

46. $(f + g)(x)$

47. $(fg)(x)$

48. $(f \circ g)(x)$

49. $(g \circ f)(x)$

Synthesis

Solve for x.

50. $\displaystyle\sum_{k=0}^{8} \binom{8}{k} x^{8-k} 3^k = 0$

51. $\displaystyle\sum_{k=0}^{4} \binom{4}{k} 5^{4-k} x^k = 64$

52. $\displaystyle\sum_{k=0}^{5} \binom{5}{k} (-1)^k x^{5-k} 3^k = 32$

53. $\sum_{k=0}^{4} \binom{4}{k}(-1)^k x^{4-k} 6^k = 81$

54. Find the term of

$$\left(\frac{3x^2}{2} - \frac{1}{3x}\right)^{12}$$

that does not contain x.

55. Find the middle term of $(x^2 - 6y^{3/2})^6$.

56. Find the ratio of the 4th term of

$$\left(p^2 - \frac{1}{2}p\sqrt[3]{q}\right)^5$$

to the 3rd term.

57. Find the term of

$$\left(\sqrt[3]{x} - \frac{1}{\sqrt{x}}\right)^7$$

containing $1/x^{1/6}$.

58. *Money Combinations.*   A money clip contains one each of the following bills: $1, $2, $5, $10, $20, $50,

and $100. How many different sums of money can be formed using the bills?

Find the sum.

59. $_{100}C_0 + {}_{100}C_1 + \cdots + {}_{100}C_{100}$

60. $_nC_0 + {}_nC_1 + \cdots + {}_nC_n$

Simplify.

61. $\sum_{k=0}^{23} \binom{23}{k}(\log_a x)^{23-k}(\log_a t)^k$

62. $\sum_{k=0}^{15} \binom{15}{k}i^{30-2k}$

63. Use mathematical induction and the property

$$\binom{n}{r-1} + \binom{n}{r} = \binom{n+1}{r}$$

to prove the binomial theorem.

Probability

- *Compute the probability of a simple event.*

When a coin is tossed, we can reason that the chance, or likelihood, that it will fall heads is 1 out of 2, or the **probability** that it will fall heads is $\frac{1}{2}$. Of course, this does not mean that if a coin is tossed 10 times it will necessarily fall heads 5 times. If the coin is a "fair coin" and it is tossed a great many times, however, it will fall heads very nearly half of the time. Here we give an introduction to two kinds of probability, **experimental** and **theoretical**.

Experimental and Theoretical Probability

If we toss a coin a great number of times—say, 1000—and count the number of times it falls heads, we can determine the probability that it will fall heads. If it falls heads 503 times, we would calculate the probability of its falling heads to be

$$\frac{503}{1000}, \quad \text{or} \quad 0.503.$$

This is an **experimental** determination of probability. Such a determination of probability is discovered by the observation and study of data and is quite common and very useful. Here, for example, are some probabilities that have been determined *experimentally:*

STUDY TIP

It is always best to study for a final exam over a period of at least two weeks. If you have only one or two days of study time, however, begin by reviewing each chapter, studying the formulas, theorems, properties, and procedures in the sections and in the Chapter Summary and Review. Then do the exercises in each Chapter Summary and Review and on the Chapter Test. Also attend a review session if one is available.

1. The probability that a woman will get breast cancer in her lifetime is $\frac{1}{11}$.
2. If you kiss someone who has a cold, the probability of your catching a cold is 0.07.
3. A person who has just been released from prison has an 80% probability of returning.

If we consider a coin and reason that it is just as likely to fall heads as tails, we would calculate the probability that it will fall heads to be $\frac{1}{2}$. This is a **theoretical** determination of probability. Here are some other probabilities that have been determined *theoretically*, using mathematics:

1. If there are 30 people in a room, the probability that two of them have the same birthday (excluding year) is 0.706.
2. While on a trip, you meet someone and, after a period of conversation, discover that you have a common acquaintance. The typical reaction, "It's a small world!", is actually not appropriate, because the probability of such an occurrence is quite high—just over 22%.

In summary, experimental probabilities are determined by making observations and gathering data. Theoretical probabilities are determined by reasoning mathematically. Examples of experimental and theoretical probability like those above, especially those we do not expect, lead us to see the value of a study of probability. You might ask, "What is the *true* probability?" In fact, there is none. Experimentally, we can determine probabilities within certain limits. These may or may not agree with the probabilities that we obtain theoretically. There are situations in which it is much easier to determine one of these types of probabilities than the other. For example, it would be quite difficult to arrive at the probability of catching a cold using theoretical probability.

Computing Experimental Probabilities

We first consider experimental determination of probability. The basic principle we use in computing such probabilities is as follows.

Principle P (Experimental)

An experiment is performed in which n observations are made. If a situation, or event, E occurs m times out of n observations, then we say that the *experimental probability* of the event, $P(E)$, is given by

$$P(E) = \frac{m}{n}.$$

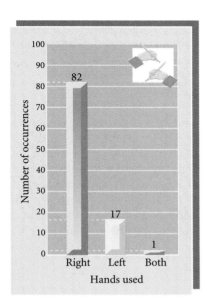

EXAMPLE 1 *Sociological Survey.* The authors of this text conducted an experiment to determine the number of people who are left-handed, right-handed, or both. The results are shown in the graph at left.

a) Determine the probability that a person is right-handed.

b) Determine the probability that a person is left-handed.

c) Determine the probability that a person is ambidextrous (uses both hands with equal ability).

d) For most tournaments held by the Professional Bowlers Association, there are 120 bowlers. On the basis of the data in this experiment, how many of the bowlers would you expect to be left-handed?

Solution

a) The number of people who are right-handed is 82, the number who are left-handed is 17, and the number who are ambidextrous is 1. The total number of observations is 82 + 17 + 1, or 100. Thus the probability that a person is right-handed is P, where

$$P = \frac{82}{100}, \quad \text{or} \quad 0.82, \quad \text{or} \quad 82\%.$$

b) The probability that a person is left-handed is P, where

$$P = \frac{17}{100}, \quad \text{or} \quad 0.17, \quad \text{or} \quad 17\%.$$

c) The probability that a person is ambidextrous is P, where

$$P = \frac{1}{100}, \quad \text{or} \quad 0.01, \quad \text{or} \quad 1\%.$$

d) There are 120 bowlers, and from part (b) we can expect 17% to be left-handed. Since

$$17\% \text{ of } 120 = 0.17 \cdot 120 = 20.4,$$

we can expect that about 20 of the bowlers will be left-handed. ▬

EXAMPLE 2 *Quality Control.* It is very important for a manufacturer to maintain the quality of its products. In fact, companies hire quality control inspectors to ensure this process. The goal is to produce as few defective products as possible. But since a company is producing thousands of products every day, it cannot afford to check every product to see if it is defective. To find out what percentage of its products are defective, the company checks a smaller sample.

The U.S. Department of Agriculture requires that 80% of the seeds that a company produces must sprout. To find out about the quality of the seeds it produces, a company takes 500 seeds from those it has produced and plants them. It finds that 417 of the seeds sprout.

a) What is the probability that a seed will sprout?

b) Did the seeds pass government standards?

Solution

a) We know that 500 seeds were planted and 417 sprouted. The probability of a seed sprouting is P, where

$$P = \frac{417}{500} = 0.834, \quad \text{or} \quad 83.4\%.$$

b) Since the percentage of seeds that sprouted exceeded the 80% requirement, the company determines that it is producing quality seeds. ▬

EXAMPLE 3   *Television Ratings.* Television networks are always concerned about the percentage of homes that have TVs and are watching their programs. A sample of the homes are contacted by attaching an electronic device to the TVs of about 1400 homes across the country. Viewing information is then fed into a computer. The following are the results of a recent survey.

NETWORK	ABC	CBS	NBC	Fox	Other, or not watching
NUMBER OF HOMES WATCHING	118	134	143	99	906

What is the probability that a home was tuned to NBC during the time period? to Fox?

Solution   The probability that a home was tuned to NBC is P, where

$$P = \frac{143}{1400} \approx 0.102 \approx 10.2\%.$$

The probability that a home was tuned to Fox is P, where

$$P = \frac{99}{1400} \approx 0.071 \approx 7.1\%.$$

▬

The percentages found in Example 3 are called *ratings*.

Theoretical Probability

Suppose that we perform an experiment such as flipping a coin, throwing a dart, drawing a card from a deck, or checking an item off an assembly line for quality. The results of such an experiment are called **outcomes**. The set of all possible outcomes is called the **sample space.** An **event** is a set of outcomes, that is, a subset of the sample space.

EXAMPLE 4   *Dart Throwing.* Consider this dartboard. Assume that the experiment is "throwing a dart" and that the dart hits the board. Find each of the following.

a) The outcomes

b) The sample space

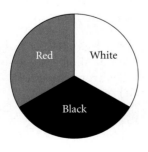

Solution

a) The outcomes are *hitting black* (B), *hitting red* (R), and *hitting white* (W).

b) The sample space is {*hitting black, hitting red, hitting white*}, which can be simply stated as {B, R, W}.

EXAMPLE 5 *Die Rolling.* A die (pl., dice) is a cube, with six faces, each containing a number of dots from 1 to 6 on each side.

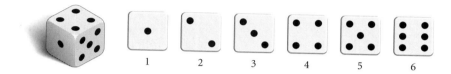

A die is rolled. Find each of the following.

a) The outcomes

b) The sample space

Solution

a) The outcomes are 1, 2, 3, 4, 5, 6.

b) The sample space is {1, 2, 3, 4, 5, 6}.

We denote the probability that an event E occurs as $P(E)$. For example, "a coin falling heads" may be denoted H. Then $P(H)$ represents the probability of the coin falling heads. When all the outcomes of an experiment have the same probability of occurring, we say that they are *equally likely*. To see the distinction between events that are equally likely and those that are not, consider the dartboards shown below.

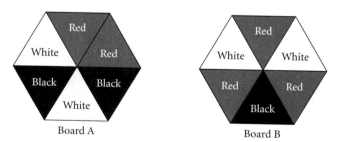

Board A   Board B

For board A, the events *hitting black*, *hitting red*, and *hitting white* are equally likely, because the black, red, and white areas are the same. However, for board B the areas are not the same so these events are not equally likely.

Principle P (Theoretical)

If an event E can occur m ways out of n possible equally likely outcomes of a sample space S, then the **theoretical probability** of the event, $P(E)$, is given by

$$P(E) = \frac{m}{n}.$$

EXAMPLE 6 What is the probability of rolling a 3 on a die?

Solution On a fair die, there are 6 equally likely outcomes and there is 1 way to roll a 3. By Principle P, $P(3) = \frac{1}{6}$.   —

EXAMPLE 7 What is the probability of rolling an even number on a die?

Solution The event is rolling an *even* number. It can occur 3 ways (getting 2, 4, or 6). The number of equally likely outcomes is 6. By Principle P, $P(\text{even}) = \frac{3}{6}$, or $\frac{1}{2}$.   —

We now use a number of examples related to a standard bridge deck of 52 cards. Such a deck is made up as shown in the following figure.

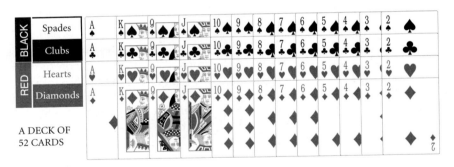

A DECK OF
52 CARDS

EXAMPLE 8 What is the probability of drawing an ace from a well-shuffled deck of cards?

Solution There are 52 outcomes (the number of cards in the deck), they are equally likely (from a well-shuffled deck), and there are 4 ways to obtain an ace, so by Principle P, we have

$$P(\text{drawing an ace}) = \frac{4}{52}, \quad \text{or} \quad \frac{1}{13}.$$ —

EXAMPLE 9 Suppose that we select, without looking, one marble from a bag containing 3 red marbles and 4 green marbles. What is the probability of selecting a red marble?

Solution   There are 7 equally likely ways of selecting any marble, and since the number of ways of getting a red marble is 3, we have

$$P(\text{selecting a red marble}) = \frac{3}{7}.$$

The following are some results that follow from Principle *P*.

Probability Properties

a) If an event *E* cannot occur, then $P(E) = 0$.
b) If an event *E* is certain to occur, then $P(E) = 1$.
c) The probability that an event *E* will occur is a number from 0 to 1: $0 \le P(E) \le 1$.

For example, in coin tossing, the event that a coin will land on its edge has probability 0. The event that a coin falls either heads or tails has probability 1.

In the following examples, we use the combinatorics that we studied in Sections 10.5 and 10.6 to calculate theoretical probabilities.

EXAMPLE 10   Suppose that 2 cards are drawn from a well-shuffled deck of 52 cards. What is the probability that both of them are spades?

Solution   The number of ways *n* of drawing 2 cards from a well-shuffled deck of 52 is $_{52}C_2$. Since 13 of the 52 cards are spades, the number of ways *m* of drawing 2 spades is $_{13}C_2$. Thus,

$$P(\text{getting 2 spades}) = \frac{m}{n} = \frac{_{13}C_2}{_{52}C_2} = \frac{78}{1326} = \frac{1}{17}.$$

EXAMPLE 11   Suppose that 3 people are selected at random from a group that consists of 6 men and 4 women. What is the probability that 1 man and 2 women are selected?

Solution   The number of ways of selecting 3 people from a group of 10 is $_{10}C_3$. One man can be selected in $_6C_1$ ways, and 2 women can be selected in $_4C_2$ ways. By the fundamental counting principle, the number of ways of selecting 1 man and 2 women is $_6C_1 \cdot {_4C_2}$. Thus the probability that 1 man and 2 women are selected is

$$P = \frac{_6C_1 \cdot {_4C_2}}{_{10}C_3} = \frac{3}{10}.$$

EXAMPLE 12   *Rolling Two Dice.* What is the probability of getting a total of 8 on a roll of a pair of dice?

Solution   On each die, there are 6 possible outcomes. The outcomes are paired so there are $6 \cdot 6$, or 36, possible ways in which the two can fall.

Technology Connection

We can use the $_nC_r$ operation from the MATH PRB menu and the ►Frac operation from the MATH MATH menu to compute the probabilities in Examples 10 and 11 on a graphing calculator.

```
13 nCr 2/52 nCr 2►Frac
                    1/17
```

```
6 nCr 1*4 nCr 2/10 nCr
3►Frac
                   3/10
```

(Assuming that the dice are different—say, one red and one blue—can help in visualizing this.)

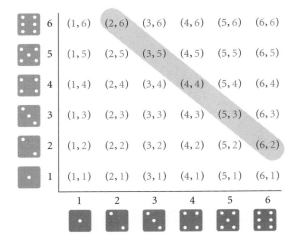

The pairs that total 8 are as shown in the figure above. There are 5 possible ways of getting a total of 8, so the probability is $\frac{5}{36}$.   ▬

Exercise Set

1. *Select a Number.* In a survey conducted by the authors, 100 people were polled and asked to select a number from 1 to 5. The results are shown in the following table.

NUMBER OF CHOICES	1	2	3	4	5
NUMBER WHO CHOSE THAT NUMBER	18	24	23	23	12

a) What is the probability that the number chosen is 1? 2? 3? 4? 5?

b) What general conclusion might a psychologist make from the experiment?

2. *Mason Dots®.* Made by the Tootsie Industries of Chicago, Illinois, Mason Dots® is a gumdrop candy. A box was opened by the authors and was found to contain the following number of gumdrops:

Orange	9
Lemon	8
Strawberry	7
Grape	6
Lime	5
Cherry	4

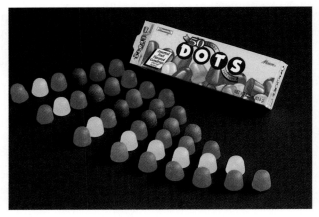

If we take one gumdrop out of the box, what is the probability of getting a lemon? lime? orange? grape? strawberry? licorice?

3. *Junk Mail.*   In experimental studies, the U.S. Postal Service has found that the probability that a piece of advertising is opened and read is 78%. A business sends out 15,000 pieces of advertising. How many of these can the company expect to be opened and read?

4. *Linguistics.*   An experiment was conducted by the authors to determine the relative occurrence of various letters of the English alphabet. The front page of a newspaper was considered. In all, there were 9136 letters. The number of occurrences of each letter of the alphabet is listed in the following table.

LETTER	NUMBER OF OCCURRENCES	PROBABILITY
A	853	$853/9136 \approx 9.3\%$
B	136	
C	273	
D	286	
E	1229	
F	173	
G	190	
H	399	
I	539	
J	21	
K	57	
L	417	
M	231	
N	597	
O	705	
P	238	
Q	4	
R	609	
S	745	
T	789	
U	240	
V	113	
W	127	
X	20	
Y	124	
Z	21	$21/9136 \approx 0.2\%$

a) Complete the table of probabilities with the percentage, to the nearest tenth of a percent, of the occurrence of each letter.

b) What is the probability of a vowel occurring?

c) What is the probability of a consonant occurring?

5. *Wheel of Fortune®.*   The results of the experiment in Exercise 4 can be quite useful to a person playing the popular television game show *Wheel of Fortune*. Players guess letters in order to spell out a phrase, a person, or a thing.

a) What 5 consonants have the greatest probability of occurring?

b) What vowel has the greatest probability of occurring?

c) The winner of the main part of the show plays for a grand prize and at one time was allowed to guess 5 consonants and a vowel in order to discover the secret wording. The 5 consonants R, S, T, L, N, and the vowel E seemed to be chosen most often. Do the results in parts (a) and (b) support such a choice?

6. *Card Drawing.*   Suppose we draw a card from a well-shuffled deck of 52 cards.

a) How many equally likely outcomes are there? What is the probability of drawing each of the following?

b) A queen   **c)** A heart

d) A 7   **e)** A red card

f) A 9 or a king   **g)** A black ace

7. *Marbles.*   Suppose we select, without looking, one marble from a bag containing 4 red marbles and 10 green marbles. What is the probability of selecting each of the followng?

a) A red marble

b) A green marble

c) A purple marble

d) A red or a green marble

8. *Production Unit.*   The sales force of a business consists of 10 men and 10 women. A production unit of 4 people is set up at random. What is the probability that 2 men and 2 women are chosen?

9. *Coin Drawing.*   A sack contains 7 dimes, 5 nickels, and 10 quarters. Eight coins are drawn at random. What is the probability of getting 4 dimes, 3 nickels, and 1 quarter?

10. *Michigan Lotto.*   Twice a week, the state of Michigan runs a 6-out-of-49-number lotto that pays at least $2 million. You purchase a card for $1 and pick any 6 numbers from 1 to 49. If your numbers match those that the state draws, you win.

a) How many 6-number combinations are there for drawing?

b) You buy 1 lottery ticket. What is your probability of winning?

Five-Card Poker Hands.   Suppose that 5 cards are drawn from a deck of 52 cards. What is the probability of drawing each of the following?

11. 3 sevens and 2 kings

12. 5 aces

13. 5 spades

14. 4 aces and 1 five

15. *Tossing Three Coins.*   Three coins are flipped. An outcome might be HTH.

a) Find the sample space.

What is the probability of getting each of the following?

b) Exactly one head
c) At most two tails
d) At least one head
e) Exactly two tails

Roulette.   An American roulette wheel contains 38 slots numbered 00, 0, 1, 2, 3, . . . , 35, 36. Eighteen of the slots numbered 1–36 are colored red and 18 are colored black. The 00 and 0 slots are considered to be uncolored. The wheel is spun, and a ball is rolled around the rim until it falls into a slot. What is the probability that the ball falls in each of the following?

16. A red slot   **17.** A black slot

18. The 00 slot   **19.** A red or a black slot

20. Either the 00 or the 0 slot (in this case, the house always wins)

21. The 0 slot

22. The number 24

23. An odd-numbered slot

24. *Dartboard.*   The figure below shows a dartboard. A dart is thrown and hits the board. Find the probabilities

$$P(\text{red}), P(\text{green}), P(\text{blue}), P(\text{yellow}).$$

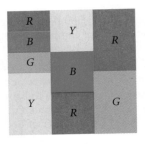

Technology Connection

25. *Random-Number Generator.*   Many graphing calculators have a **random-number generator.** This feature produces a random number in the interval [0, 1]. (Consult your user's manual.) We can use such a feature to simulate coin flipping. A number r such that $0 \le r \le 0.5$ would indicate heads, H. A number r such that $0.5 < r \le 1.0$ would indicate tails, T. Use a random-number generator 100 times.

a) What is the experimental probability of getting heads?

b) What is the experimental probability of getting tails?

Collaborative Discussion and Writing

26. *Random Best-Selling Novels.*   Sir Arthur Stanley Eddington, an astronomer, once wrote in a satirical essay that if a monkey were left alone long enough with a typewriter and typed randomly, any great novel could be replicated.

What is the probability that the following passage could have been written by a monkey? Ignore capital letters and punctuation and consider only letters and spaces.

"*It was the best of times, it was the worst of times, . . .*" (Charles Dickens, 1859). Explain your answer.

27. Find at least one use of probability in today's newspaper. Make a report.

Skill Maintenance

Solve.

28. $3x^2 - 4x = 3$

29. $2x^3 + 5x^2 - 4x - 3 = 0$

30. $x - y = 1,$
$2x - 3y = 6$

31. $2x + y - 3z = 5,$
$3x + 3y - 5z = 4,$
$x - 2y + 2z = 11$

Synthesis

Five-Card Poker Hands. *Suppose that 5 cards are drawn from a deck of 52 cards. For the following exercises, give both a reasoned expression and an answer.*

32. *Royal Flush.* A *royal flush* consists of a 5-card hand with A-K-Q-J-10 of the same suit.

a) How many royal flushes are there?
b) What is the probability of getting a royal flush?

33. *Straight Flush.* A *straight flush* consists of 5 cards in sequence in the same suit, but excludes royal flushes. An ace can be used low, before a two, or high, following a king.

a) How many straight flushes are there?
b) What is the probability of getting a straight flush?

34. *Four of a Kind.* A *four-of-a-kind* is a 5-card hand in which 4 of the cards are of the same denomination, such as J-J-J-J-6, 7-7-7-7-A, or 2-2-2-2-5.

a) How many four-of-a-kind hands are there?
b) What is the probability of getting four of a kind?

35. *Full House.* A *full house* consists of a pair and 3 of a kind, such as Q-Q-Q-4-4.

a) How many full houses are there?
b) What is the probability of getting a full house?

36. *Three of a Kind.* A *three-of-a-kind* is a 5-card hand in which exactly 3 of the cards are of the same denomination and the other 2 are *not*, such as Q-Q-Q-10-7.

a) How many three-of-a-kind hands are there?
b) What is the probability of getting three of a kind?

37. *Flush.* An ordinary *flush* is a 5-card hand in which all the cards are of the same suit, but not all in sequence (not a straight flush or royal flush).

a) How many flushes are there?
b) What is the probability of getting a flush?

38. *Two Pairs.* A hand with *two pairs* is a hand like Q-Q-3-3-A.

a) How many are there?
b) What is the probability of getting two pairs?

39. *Straight.* An ordinary *straight* is any 5 cards in sequence, but not of the same suit—for example, 4 of spades, 5 of hearts, 6 of diamonds, 7 of hearts, and 8 of clubs.

a) How many straights are there?
b) What is the probability of getting a straight?

Chapter Summary and Review

Important Properties and Formulas

Arithmetic Sequences and Series

General term: $a_n = a_{n-1} + d$

$a_n = a_1 + (n-1)d$

Common difference: d

Sum of the first n terms: $S_n = \dfrac{n}{2}(a_1 + a_n)$

Geometric Sequences and Series

General term: $a_{n+1} = a_n r$

$a_n = a_1 r^{n-1}$

Common ratio: r

Sum of the first n terms: $S_n = \dfrac{a_1(1 - r^n)}{1 - r}$

Sum of an infinite geometric series:

$S_\infty = \dfrac{a_1}{1 - r}, \quad |r| < 1$

The Principle of Mathematical Induction

(1) *Basis step*: Prove S_1 is true.

(2) *Induction step*: Prove for all numbers k, $S_k \to S_{k+1}$.

The Fundamental Counting Principle

The total number of ways in which k actions can be performed together is $n_1 \cdot n_2 \cdot n_3 \cdots n_k$.

Permutations of n Objects Taken n at a Time

$_nP_n = n! = n(n-1)(n-2) \cdots 3 \cdot 2 \cdot 1$

Permutations of n Objects Taken k at a Time

$_nP_k = \underbrace{n(n-1)(n-2) \cdots [n - (k-1)]}_{k \text{ factors}}$

$= \dfrac{n!}{(n-k)!}$

Permutations of Sets with Some Nondistinguishable Objects

$\dfrac{n!}{n_1! \cdot n_2! \cdots n_k!}$

Combinations of n Objects Taken k at a Time

$_nC_k = \dbinom{n}{k} = \dfrac{_nP_k}{k!} = \dfrac{n!}{k!\,(n-k)!}$

$= \dfrac{n(n-1)(n-2) \cdots [n - (k-1)]}{k!}$

The Binomial Theorem

$(a + b)^n = \displaystyle\sum_{k=0}^{n} \binom{n}{k} a^{n-k} b^k$

The (k + 1)st Term of Binomial Expansion

The $(k+1)$st term of $(a+b)^n$ is $\dbinom{n}{k} a^{n-k} b^k$.

The total number of subsets of a set with n elements is 2^n.

Probability Principle P

$P(E) = \dfrac{m}{n}$

REVIEW EXERCISES

1. Find the first 4 terms, a_{11}, and a_{23}:

$$a_n = (-1)^n\left(\frac{n^2}{n^4 + 1}\right).$$

2. Predict the general, or nth, term. Answers may vary.

$$2, -5, 10, -17, 26, \ldots$$

3. Find and evaluate:

$$\sum_{k=1}^{4} \frac{(-1)^{k+1}3^k}{3^k - 1}.$$

4. Write sigma notation:

$$0 + 3 + 8 + 15 + 24 + 35 + 48.$$

5. Find the 10th term of the arithmetic sequence

$$\frac{3}{4}, \frac{13}{12}, \frac{17}{12}, \ldots.$$

6. Find the 6th term of the arithmetic sequence

$$a - b, a, a + b, \ldots.$$

7. Find the sum of the first 18 terms of the arithmetic sequence

$$4, 7, 10, \ldots.$$

8. Find the sum of the first 200 natural numbers.

9. The 1st term in an arithmetic sequence is 5, and the 17th term is 53. Find the 3rd term.

10. The common difference in an arithmetic sequence is 3. The 10th term is 23. Find the first term.

11. For a geometric sequence, $a_1 = -2$, $r = 2$, and $a_n = -64$. Find n and S_n.

12. For a geometric sequence, $r = \frac{1}{2}$, $n = 5$, and $S_n = \frac{31}{2}$. Find a_1 and a_n.

Find the sum of each infinite geometric series, if it exists.

13. $25 + 27.5 + 30.25 + 33.275 + \cdots$

14. $0.27 + 0.0027 + 0.000027 + \cdots$

15. $\frac{1}{2} - \frac{1}{6} + \frac{1}{18} - \cdots$

16. Find fractional notation for $2.\overline{43}$.

17. Insert four arithmetic means between 5 and 9.

18. *Bouncing Golfball.* A golfball is dropped from a height of 30 ft to the pavement. It always rebounds three fourths of the distance that it drops. How far (up and down) will the ball have traveled when it hits the pavement for the 6th time?

19. *The Amount of an Annuity.* To create a college fund, a parent makes a sequence of 18 yearly deposits of $2000 each in a savings account on which interest is compounded annually at 5.8%. Find the amount of the annuity.

20. *Total Gift.* You receive 10¢ on the first day of the year, 12¢ on the 2nd day, 14¢ on the 3rd day, and so on.

a) How much will you receive on the 365th day?
b) What is the sum of all these 365 gifts?

21. *The Economic Multiplier.* The government is making a $24,000,000,000 expenditure for travel to Mars. If 73% of this amount is spent again, and so on, what is the total effect on the economy?

Use mathematical induction to prove each of the following.

22. For every natural number n,

$$1 + 4 + 7 + \cdots + (3n - 2) = \frac{n(3n - 1)}{2}.$$

23. For every natural number n,

$$1 + 3 + 3^2 + \cdots + 3^{n-1} = \frac{3^n - 1}{2}.$$

24. For every natural number $n \geq 2$,

$$\left(1 - \frac{1}{2}\right)\left(1 - \frac{1}{3}\right)\cdots\left(1 - \frac{1}{n}\right) = \frac{1}{n}.$$

25. *Book Arrangements.* In how many ways can 6 books be arranged on a shelf?

26. *Flag Displays.* If 9 different signal flags are available, how many different displays are possible using 4 flags in a row?

27. *Prize Choices.* The winner of a contest can choose any 8 of 15 prizes. How many different sets of prizes can be chosen?

28. *Fraternity–Sorority Names.* The Greek alphabet contains 24 letters. How many fraternity or sorority names can be formed using 3 different letters?

29. *Letter Arrangements.* In how many distinguishable ways can the letters of the word TENNESSEE be arranged?

30. *Floor Plans.* A manufacturer of houses has 1 floor plan but achieves variety by having 3 different roofs, 4 different ways of attaching the garage, and 3 different types of entrance. Find the number of different houses that can be produced.

31. *Code Symbols.* How many code symbols can be formed using 5 out of 6 of the letters of G, H, I, J, K, L if the letters:

a) cannot be repeated?
b) can be repeated?
c) cannot be repeated but must begin with K?
d) cannot be repeated but must end with IGH?

32. Determine the number of subsets of a set containing 8 members.

Expand.

33. $(m + n)^7$

34. $(x - \sqrt{2})^5$

35. $(x^2 - 3y)^4$

36. $\left(a + \dfrac{1}{a}\right)^8$

37. $(1 + 5i)^6$, where $i^2 = -1$

38. Find the 4th term of $(a + x)^{12}$.

39. Find the 12th term of $(2a - b)^{18}$. Do not multiply out the factorials.

40. *Election Poll.* Before an election, a poll was conducted to see which candidate was favored. Three people were running for a particular office. During the polling, 86 favored candidate A, 97 favored B, and 23 favored C. Assuming that the poll is a valid indicator of the election, what is the probability that the election will be won by A? B? C?

41. *Rolling Dice.* What is the probability of getting a 10 on a roll of a pair of dice? on a roll of 1 die?

42. *Drawing a Card.* From a deck of 52 cards, 1 card is drawn at random. What is the probability that it is a club?

43. *Drawing Three Cards.* From a deck of 52 cards, 3 are drawn at random without replacement. What is the probability that 2 are aces and 1 is a king?

Technology Connection

44. Use a graphing calculator to construct a table of values for the first 10 terms of this sequence.

$$a_1 = 0.3, \quad a_{k+1} = 5a_k + 1$$

45. *Video Sales.* The table at the top of the next column shows the revenue from video sales in the United States in several recent years.

YEAR	VIDEO SALES REVENUE (IN BILLIONS)
1985	$0.86
1990	3.18
1997	8.24
2000	9.76

Source: Paul Kagan Associates, Inc.

a) Find a linear sequence function $a_n = an + b$ that models the data. Let n represent the number of years after 1985.
b) Use the sequence found in part (a) to estimate the revenue in 2005.

Collaborative Discussion and Writing

46. Write an exercise for a classmate to solve. Design it so that the solution is $_9C_4$.

47. *Chain Business Deals.* Chain letters have been outlawed by the U.S. government. Nevertheless, "chain" business deals still exist and they can be fraudulent. Suppose that a salesperson is charged with the task of hiring 4 new salespersons. Each of them gives half of their profits to the person who hires them. Each of these people hires 4 new salespersons. Each of these gives half of his or her profits to the person who hired them. Half of these profits then go back to the original hiring person. Explain the lure of this business to someone who has managed several sequences of hirings. Explain the fallacy of such a business as well. Keep in mind that there are only 284 million people in the United States.

Synthesis

48. Explain why the following cannot be proved by mathematical induction: For every natural number n,

a) $3 + 5 + \cdots + (2n + 1) = (n + 1)^2$.
b) $1 + 3 + \cdots + (2n - 1) = n^2 + 3$.

49. Suppose that $a_1, a_2, \ldots, a_n$ and $b_1, b_2, \ldots, b_n$ are geometric sequences. Prove that $c_1, c_2, \ldots, c_n$ is a geometric sequence, where $c_n = a_n b_n$.

50. Suppose that $a_1, a_2, \ldots, a_n$ is an arithmetic sequence. Is $b_1, b_2, \ldots, b_n$ an arithmetic sequence if:

a) $b_n = |a_n|$?   b) $b_n = a_n + 8$?

c) $b_n = 7a_n$?   d) $b_n = \dfrac{1}{a_n}$?

e) $b_n = \log a_n$?   f) $b_n = a_n^3$?

51. The zeros of this polynomial function form an arithmetic sequence. Find them.
$$f(x) = x^4 - 4x^3 - 4x^2 + 16x$$

52. Write the first 3 terms of the infinite geometric series with $r = -\frac{1}{3}$ and $S_\infty = \frac{3}{8}$.

53. Simplify:
$$\sum_{k=0}^{10} (-1)^k \binom{10}{k} (\log x)^{10-k} (\log y)^k.$$

Solve for n.

54. $\binom{n}{6} = 3 \cdot \binom{n-1}{5}$

55. $\binom{n}{n-1} = 36$

56. Solve for a:
$$\sum_{k=0}^{5} \binom{5}{k} 9^{5-k} a^k = 0.$$

Chapter Test

1. For the sequence whose nth term is $a_n = (-1)^n(2n + 1)$, find a_{21}.

2. Find the first 5 terms of the sequence with general term
$$a_n = \frac{n+1}{n+2}.$$

3. Find and evaluate:
$$\sum_{k=1}^{4} (k^2 + 1).$$

4. Find the 15th term of the arithmetic sequence 2, 5, 8, ...

5. The 1st term of an arithmetic sequence is 8 and the 21st term is 108. Find the 7th term.

6. Find the sum of the first 20 terms of the series $17 + 13 + 9 + \cdots$.

7. For a geometric sequence, $r = 0.2$ and $S_4 = 1248$. Find a_1.

8. Find the sum, if it exists: $18 + 6 + 2 + \cdots$.

9. Find fractional notation for $0.\overline{56}$.

10. *Amount of an Annuity.* To create a college fund, a parent makes a sequence of 18 equal yearly deposits of $2500 in a savings account on which interest is compounded annually at 5.6%. Find the amount of the annuity.

11. Use mathematical induction to prove that, for every natural number n,
$$2 + 5 + 8 + \cdots + (3n - 1) = \frac{n(3n + 1)}{2}.$$

Evaluate.

12. $_{15}P_6$

13. $_{21}C_{10}$

14. $\binom{n}{4}$

15. How many 4-digit numbers can be formed using the digits 1, 3, 5, 6, 7, and 9 without repetition?

16. *Test Options.* On a test with 20 questions, a student must answer 8 of the first 12 questions and 4 of the last 8. In how many ways can this be done?

17. Expand: $(x + 1)^5$.

18. Determine the number of subsets of a set containing 9 members.

19. *Marbles.* Suppose we select, without looking, one marble from a bag containing 6 red marbles and 8 blue marbles. What is the probability of selecting a blue marble?

Synthesis

20. Solve for n: $_nP_7 = 9 \cdot {_nP_6}$.

Appendix

Parametric Equations

- *Graph parametric equations.*
- *Determine an equivalent rectangular equation for parametric equations.*
- *Determine parametric equations for a rectangular equation.*
- *Determine the location of a moving object at a specific time.*

Graphing Parametric Equations

We have graphed *plane curves* that are composed of sets of ordered pairs (x, y) in the rectangular coordinate plane. Now we discuss a way to represent plane curves in which x and y are functions of a third variable t.

EXAMPLE 1   Graph the curve represented by the equations

$$x = \tfrac{1}{2}t, \quad y = t^2 - 3, \quad -3 \le t \le 3.$$

Solution   We can choose values for t between -3 and 3 and find the corresponding values of x and y. When $t = -3$, we have

$$x = \tfrac{1}{2}(-3) = -\tfrac{3}{2}, \qquad y = (-3)^2 - 3 = 6.$$

The table below lists other ordered pairs. We plot these points and then draw the curve.

t	x	y	(x, y)
-3	$-\tfrac{3}{2}$	6	$\left(-\tfrac{3}{2}, 6\right)$
-2	-1	1	$(-1, 1)$
-1	$-\tfrac{1}{2}$	-2	$\left(-\tfrac{1}{2}, -2\right)$
0	0	-3	$(0, -3)$
1	$\tfrac{1}{2}$	-2	$\left(\tfrac{1}{2}, -2\right)$
2	1	1	$(1, 1)$
3	$\tfrac{3}{2}$	6	$\left(\tfrac{3}{2}, 6\right)$

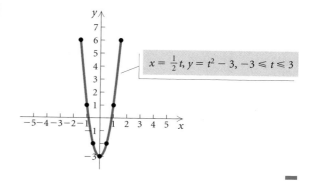

The curve above appears to be part of a parabola. Let's verify this by finding the equivalent rectangular equation. Solving $x = \frac{1}{2}t$ for t, we get $t = 2x$. Substituting $2x$ for t in $y = t^2 - 3$, we have

$$y = (2x)^2 - 3 = 4x^2 - 3.$$

This is a quadratic equation. Hence its graph is a parabola. Thus the curve is part of the parabola $y = 4x^2 - 3$. Since $-3 \leq t \leq 3$ and $x = \frac{1}{2}t$, we must include the restriction $-\frac{3}{2} \leq x \leq \frac{3}{2}$ when we write the equivalent rectangular equation:

$$y = 4x^2 - 3, \quad -\frac{3}{2} \leq x \leq \frac{3}{2}.$$

The equations $x = \frac{1}{2}t$ and $y = t^2 - 3$ are **parametric equations** for the curve. The variable t is the **parameter**.

Parametric Equations

If f and g are continuous functions of t on an interval I, then the set of ordered pairs (x, y) such that $x = f(t)$ and $y = g(t)$ is a **plane curve**. The equations $x = f(t)$ and $y = g(t)$ are **parametric equations** for the curve. The variable t is the **parameter**.

Determining a Rectangular Equation for Given Parametric Equations

EXAMPLE 2 Find a rectangular equation equivalent to each pair of parametric equations.

a) $x = t^2, y = t - 1; -1 \leq t \leq 4$
b) $x = \sqrt{t}, y = 2t + 3; 0 \leq t \leq 3$

Solution

a) We can first solve either equation for t. We choose the equation $y = t - 1$:

$$y = t - 1$$
$$y + 1 = t.$$

We then substitute $y + 1$ for t in $x = t^2$:

$$x = t^2$$
$$x = (y + 1)^2. \quad \text{Substituting}$$

This is an equation of a parabola that opens to the right. Given that $-1 \leq t \leq 4$, we have the corresponding restrictions on x and y: $0 \leq x \leq 16$ and $-2 \leq y \leq 3$. Thus the equivalent rectangular equation is

$$x = (y + 1)^2; \quad 0 \leq x \leq 16.$$

b) We first solve $x = \sqrt{t}$ for t:

$$x = \sqrt{t}$$
$$x^2 = t.$$

Then we substitute x^2 for t in $y = 2t + 3$:

$$y = 2t + 3$$
$$y = 2x^2 + 3. \qquad \text{Substituting}$$

When $0 \leq t \leq 3$, we have $0 \leq x \leq \sqrt{3}$. The equivalent rectangular equation is

$$y = 2x^2 + 3; \quad 0 \leq x \leq \sqrt{3}.$$

Technology Connection

We can graph parametric equations on a graphing calculator set in PARAMETRIC mode. The window dimensions include minimum and maximum values for x, y, and t. For the parametric equations in Example 2(b), we can use the settings shown below.

WINDOW
  Tmin = 0
  Tmax = 3
  Tstep = .1
  Xmin = −3
  Xmax = 3
  Xscl = 1
  Ymin = −2
  Ymax = 10
  Yscl = 1

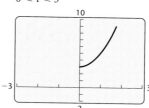

EXAMPLE 3 Find a rectangular equation equivalent to each pair of parametric equations.

a) $x = \cos t$, $y = \sin t$, $0 \leq t \leq 2\pi$

b) $x = 5 \cos t$, $y = 3 \sin t$; $0 \leq t \leq 2\pi$

Solution

a) First, we square both sides of each parametric equation:

$$x^2 = \cos^2 t \quad \text{and} \quad y^2 = \sin^2 t.$$

This allows us to use the trigonometric identity $\sin^2 \theta + \cos^2 \theta = 1$. Substituting, we get

$$x^2 + y^2 = 1.$$

This is the equation of a circle with center $(0, 0)$ and radius 1.

b) First, we solve for cos t and sin t in the parametric equations:

$$x = 5 \cos t \qquad y = 3 \sin t$$

$$\frac{x}{5} = \cos t, \qquad \frac{y}{3} = \sin t.$$

Using the identity $\sin^2 \theta + \cos^2 \theta = 1$, we can substitute to eliminate the parameter:

$$\sin^2 t + \cos^2 t = 1$$

$$\left(\frac{y}{3}\right)^2 + \left(\frac{x}{5}\right)^2 = 1 \qquad \text{Substituting}$$

$$\frac{x^2}{25} + \frac{y^2}{9} = 1.$$

This is the equation of an ellipse centered at the origin with vertices at $(5, 0)$ and $(-5, 0)$.   ▬

Determining Parametric Equations for a Given Rectangular Equation

Many sets of parametric equations can represent the same plane curve. In fact, there are infinitely many such equations.

EXAMPLE 4   Find three sets of parametric equations for the parabola

$$y = 4 - (x + 3)^2.$$

Solution

If $x = t$, then $y = 4 - (t + 3)^2$, or $-5 - 6t - t^2$.

If $x = t - 3$, then $y = 4 - (t - 3 + 3)^2$, or $4 - t^2$.

If $x = \frac{t}{3}$, then $y = 4 - \left(\frac{t}{3} + 3\right)^2$, or $-\frac{t^2}{9} - 2t - 5$.   ▬

Applications

The motion of an object that is propelled upward can be described with parametric equations. Such motion is called **projectile motion.** It can be shown using more advanced mathematics that, neglecting air resistance, the following equations describe the path of a projectile propelled upward at an angle θ with the horizontal from a height h, in feet, at an initial speed v_0, in feet per second:

$$x = (v_0 \cos \theta)t, \qquad y = h + (v_0 \sin \theta)t - 16t^2.$$

We can use these equations to determine the location of the object at time t, in seconds.

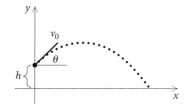

EXAMPLE 5 *Projectile Motion.* A baseball is thrown from a height of 6 ft with an initial speed of 100 ft/sec at an angle of 45° with the horizontal.

a) Find parametric equations that give the position of the ball at time t, in seconds.

b) Find the height of the ball after 1 sec, 2 sec, and 3 sec.

c) Determine how long the ball is in the air.

d) Determine the horizontal distance that the ball travels.

e) Find the maximum height of the ball.

Solution

a) We substitute 6 for h, 100 for v_0, and 45° for θ in the equations above:

$$x = (v_0 \cos \theta)t$$
$$= (100 \cos 45°)t$$
$$= \left(100 \cdot \frac{\sqrt{2}}{2}\right)t = 50\sqrt{2}t;$$

$$y = h + (v_0 \sin \theta)t - 16t^2$$
$$= 6 + (100 \sin 45°)t - 16t^2$$
$$= 6 + \left(100 \cdot \frac{\sqrt{2}}{2}\right)t - 16t^2$$
$$= 6 + 50\sqrt{2}t - 16t^2.$$

b) The height of the ball at time t is represented by y.

When $t = 1$, $y = 6 + 50\sqrt{2}(1) - 16(1)^2 \approx 60.7$ ft.

When $t = 2$, $y = 6 + 50\sqrt{2}(2) - 16(2)^2 \approx 83.4$ ft.

When $t = 3$, $y = 6 + 50\sqrt{2}(3) - 16(3)^2 \approx 74.1$ ft.

c) The ball hits the ground when $y = 0$. Thus, in order to determine how long the ball is in the air, we solve the equation $y = 0$:

$$6 + 50\sqrt{2}t - 16t^2 = 0$$
$$-16t^2 + 50\sqrt{2}t + 6 = 0 \qquad \text{Standard form}$$
$$t = \frac{-50\sqrt{2} \pm \sqrt{(50\sqrt{2})^2 - 4(-16)(6)}}{2(-16)}$$

Using the quadratic formula

$$t \approx -0.1 \quad \text{or} \quad t \approx 4.5.$$

The negative value for t has no meaning in this application. Thus we determine that the ball is in the air about 4.5 sec.

d) Since the ball is in the air for about 4.5 sec, the horizontal distance that it travels is given by

$$x = 50\sqrt{2}(4.5) \approx 318.2 \text{ ft.}$$

e) To find the maximum height of the ball, we find the maximum value of y. This occurs at the vertex of the quadratic function represented by y. At the vertex, we have

$$t = -\frac{b}{2a} = -\frac{50\sqrt{2}}{2(-16)} \approx 2.2.$$

When $t = 2.2$,

$$y = 6 + 50\sqrt{2}(2.2) - 16(2.2)^2 \approx 84.1 \text{ ft.}$$

The path of a fixed point on the circumference of a circle as it rolls along a line is called a **cycloid**. For example, a point on the rim of a bicycle wheel traces a cycloid curve.

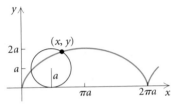

The parametric equations of a cycloid are

$$x = a(t - \sin t), \qquad y = a(1 - \cos t),$$

where a is the radius of the circle that traces the curve and t is in radian measure. The graph of the cycloid described by the parametric equations

$$x = 3(t - \sin t), \qquad y = 3(1 - \cos t); \quad 0 \leq t \leq 6\pi$$

is shown below.

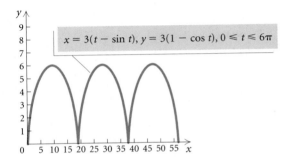

Exercise Set

Graph the plane curve given by the parametric equations. Then find an equivalent rectangular equation.

1. $x = \frac{1}{2}t,\ y = 6t - 7;\ -1 \le t \le 6$

2. $x = t,\ y = 5 - t;\ -2 \le t \le 3$

3. $x = 4t^2,\ y = 2t;\ -1 \le t \le 1$

4. $x = \sqrt{t},\ y = 2t + 3;\ 0 \le t \le 8$

Find a rectangular equation equivalent to the given pair of parametric equations.

5. $x = t^2,\ y = \sqrt{t};\ 0 \le t \le 4$

6. $x = t^3 + 1,\ y = t;\ -3 \le t \le 3$

7. $x = t + 3,\ y = \dfrac{1}{t + 3};\ -2 \le t \le 2$

8. $x = 2t^3 + 1,\ y = 2t^3 - 1;\ -4 \le t \le 4$

9. $x = 2t - 1,\ y = t^2;\ -3 \le t \le 3$

10. $x = \frac{1}{3}t,\ y = t;\ -5 \le t \le 5$

11. $x = e^{-t},\ y = e^t;\ -\infty < t < \infty$

12. $x = 2 \ln t,\ y = t^2;\ 0 < t < \infty$

13. $x = 3 \cos t,\ y = 3 \sin t;\ 0 \le t \le 2\pi$

14. $x = 2 \cos t,\ y = 4 \sin t;\ 0 \le t \le 2\pi$

15. $x = \cos t,\ y = 2 \sin t;\ 0 \le t \le 2\pi$

16. $x = 2 \cos t,\ y = 2 \sin t;\ 0 \le t \le 2\pi$

17. $x = \sec t,\ y = \cos t;\ -\dfrac{\pi}{2} < t < \dfrac{\pi}{2}$

18. $x = \sin t,\ y = \csc t;\ 0 < t < \pi$

19. $x = 1 + 2 \cos t,\ y = 2 + 2 \sin t;\ 0 \le t \le 2\pi$

20. $x = 2 + \sec t,\ y = 1 + 3 \tan t;\ 0 < t < \dfrac{\pi}{2}$

Find two sets of parametric equations for the rectangular equation.

21. $y = 4x - 3$

22. $y = x^2 - 1$

23. $y = (x - 2)^2 - 6x$

24. $y = x^3 + 3$

25. *Projectile Motion.* A ball is thrown from a height of 7 ft with an initial speed of 80 ft/sec at an angle of 30° with the horizontal.

  a) Find parametric equations that give the position of the ball at time t, in seconds.

  b) Find the height of the ball after 1 sec and 2 sec.

  c) Determine how long the ball is in the air.

  d) Determine the horizontal distance that the ball travels.

  e) Find the maximum height of the ball.

26. *Projectile Motion.* A projectile is launched from the ground with an initial speed of 200 ft/sec at an angle of 60° with the horizontal.

  a) Find parametric equations that give the position of the projectile at time t, in seconds.

  b) Find the height of the projectile after 4 sec and 8 sec.

  c) Determine how long the projectile is in the air.

  d) Determine the horizontal distance that the projectile travels.

  e) Find the maximum height of the projectile.

Synthesis

27. Consider the curve described by

$$x = 3 \cos t,\quad y = 3 \sin t;\quad 0 \le t \le 2\pi.$$

As t increases, the curve is traced in the counterclockwise direction. How can the equations be changed so that the curve is traced in the clockwise direction?

28. Find an equivalent rectangular equation for the curve described by

$$x = \cos^3 t,\quad y = \sin^3 t;\quad 0 \le t \le 2\pi.$$

Answers

Exercise Set R.1

1. $\sqrt[3]{8}$, 0, 9, $\sqrt{25}$
3. $\sqrt{7}$, $5.242242224\ldots$, $-\sqrt{14}$, $\sqrt[5]{5}$, $\sqrt[3]{4}$
5. -12, $5.\overline{3}$, $-\frac{7}{3}$, $\sqrt[3]{8}$, 0, -1.96, 9, $4\frac{2}{3}$, $\sqrt{25}$, $\frac{5}{7}$
7. Answers may vary; $\frac{3}{4}$ **9.** Answers may vary; -5
11. $[-3, 3]$ **13.** $[-14, -11]$ **15.** $(-\infty, -4]$
17. $(-\infty, 3.8)$ **19.** $(7, \infty)$ **21.** $(0, 5)$
23. $[-9, -4)$ **25.** $[x, x + h]$ **27.** (p, ∞) **29.** True
31. False   **33.** True   **35.** False   **37.** False   **39.** True
41. True   **43.** True   **45.** False   **47.** Commutative property of multiplication
49. Multiplicative identity property
51. Associative property of multiplication
53. Commutative property of multiplication
55. Commutative property of addition
57. Multiplicative inverse property
59. 7.1 **61.** $\frac{5}{4}$ **63.** 11 **65.** 6 **67.** 5.4
69. Discussion and Writing
71. Answers may vary; $0.124124412444\ldots$
73. Answers may vary; -0.00999
75.

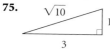

Exercise Set R.2

1. 1 **3.** 5^2, or 25 **5.** 1 **7.** 7^{-1}, or $\frac{1}{7}$ **9.** $6x^5$
11. $15a^{-1}b^5$, or $\dfrac{15b^5}{a}$ **13.** $72x^5$ **15.** b^3

17. x^3y^{-3}, or $\dfrac{x^3}{y^3}$ **19.** $8xy^{-5}$, or $\dfrac{8x}{y^5}$ **21.** $8a^3b^6$
23. $16x^{12}$ **25.** $\dfrac{c^2d^4}{25}$ **27.** $\dfrac{8^5a^{20}c^{10}}{b^{25}}$ **29.** 4.05×10^5
31. 3.9×10^{-7} **33.** 1.6×10^{-5} **35.** 0.000083
37. $20{,}700{,}000$ **39.** $11{,}358{,}000{,}000$ **41.** 1.395×10^3
43. 8×10^{-14} **45.** $\$1.19 \times 10^7$ **47.** 1.332×10^{14}
49. 2 **51.** 2048 **53.** 5 **55.** $\$2883.67$
57. $\$8763.54$ **59.** Discussion and Writing   **61.** x^{8t}
63. t^{8x} **65.** $9x^{2a}y^{2b}$ **67.** $\$660.29$ **69.** $\$942.54$

Exercise Set R.3

1. $-5y^4$, $3y^3$, $7y^2$, $-y$, -4; 4 **3.** $3a^4b$, $-7a^3b^3$, $5ab$, -2; 6
5. $3x^2y - 5xy^2 + 7xy + 2$ **7.** $3x + 2y - 2z - 3$
9. $-2x^2 + 6x - 2$ **11.** $x^4 - 3x^3 - 4x^2 + 9x - 3$
13. $2a^4 - 2a^3b - a^2b + 4ab^2 - 3b^3$ **15.** $x^2 + 2x - 15$
17. $x^2 + 10x + 24$ **19.** $2a^2 + 13a + 15$
21. $4x^2 + 8xy + 3y^2$ **23.** $y^2 + 10y + 25$
25. $x^2 - 8x + 16$ **27.** $25x^2 - 30x + 9$
29. $4x^2 + 12xy + 9y^2$ **31.** $4x^4 - 12x^2y + 9y^2$
33. $a^2 - 9$ **35.** $4x^2 - 25$ **37.** $9x^2 - 4y^2$
39. $4x^2 + 12xy + 9y^2 - 16$ **41.** $x^4 - 1$
43. Discussion and Writing   **45.** $a^{2n} - b^{2n}$
47. $a^{2n} + 2a^nb^n + b^{2n}$ **49.** $x^6 - 1$ **51.** $x^{a^2-b^2}$
53. $a^2 + b^2 + c^2 + 2ab + 2ac + 2bc$

Exercise Set R.4

1. $2(x - 5)$ **3.** $3x^2(x^2 - 3)$ **5.** $4(a^2 - 3a + 4)$
7. $(b - 2)(a + c)$ **9.** $(x + 3)(x^2 + 6)$
11. $(y - 3)(y + 2)(y - 2)$ **13.** $(x - 1)(x^2 + 1)$
15. $(p + 2)(p + 4)$ **17.** $(2n - 7)(n + 8)$
19. $(3x + 2)(4x + 1)$ **21.** $(y^2 - 7)(y^2 + 3)$

23. $(m + 2)(m - 2)$ **25.** $(3x + 5)(3x - 5)$
27. $4x(y^2 + z)(y^2 - z)$ **29.** $(y - 3)^2$
31. $(2z + 3)^2$ **33.** $(1 - 4x)^2$
35. $(x + 2)(x^2 - 2x + 4)$ **37.** $(m - 1)(m^2 + m + 1)$
39. $3ab(6a - 5b)$ **41.** $(x - 4)(x^2 + 5)$
43. $8(x + 2)(x - 2)$ **45.** Prime
47. $(m + 3n)(m - 3n)$
49. $(x + 4)(x + 5)$ **51.** $(y - 5)(y - 1)$
53. $(2a + 1)(a + 4)$ **55.** $(3x - 1)(2x + 3)$
57. $(y - 9)^2$ **59.** $(3z - 4)^2$ **61.** $(xy - 7)^2$
63. $4a(x + 7)(x - 2)$ **65.** $3(z - 2)(z^2 + 2z + 4)$
67. $2ab(2a^2 + 3b^2)(4a^4 - 6a^2b^2 + 9b^4)$
69. Discussion and Writing   **71.** $(y^2 + 12)(y^2 - 7)$
73. $\left(y + \frac{4}{7}\right)\left(y - \frac{2}{7}\right)$ **75.** $h(3x^2 + 3xh + h^2)$
77. $(y + 4)(y - 7)$ **79.** $(x^n + 8)(x^n - 3)$
81. $(x + a)(x + b)$ **83.** $(5y^m + x^n - 1)(5y^m - x^n + 1)$
85. $y(y - 1)^2(y - 2)$

Exercise Set R.5

1. $\{x \mid x \text{ is a real number}\}$
3. $\{x \mid x \text{ is a real number } and \ x \neq 0 \text{ and } x \neq 1\}$
5. $\{x \mid x \text{ is a real number } and \ x \neq -5 \text{ and } x \neq 1\}$
7. $\{x \mid x \text{ is a real number } and \ x \neq -2 \text{ and } x \neq 2 \text{ and } x \neq -5\}$
9. $\frac{1}{x - y}$ **11.** $\frac{(x + 5)(2x + 3)}{7x}$ **13.** $\frac{a + 2}{a - 5}$
15. $m + n$ **17.** $\frac{3(x - 4)}{2(x + 4)}$ **19.** $\frac{1}{x + y}$
21. $\frac{x - y - z}{x + y + z}$ **23.** 1 **25.** $\frac{y - 2}{y - 1}$ **27.** $\frac{x + y}{2x - 3y}$
29. $\frac{3x - 4}{(x + 2)(x - 2)}$ **31.** $\frac{-y + 10}{(y + 4)(y - 5)}$
33. $\frac{4x - 8y}{(x + y)(x - y)}$ **35.** $\frac{3x - 4}{(x - 2)(x - 1)}$
37. $\frac{5a^2 + 10ab - 4b^2}{(a + b)(a - b)}$
39. $\frac{11x^2 - 18x + 8}{(2 + x)(2 - x)^2}$, or $\frac{11x^2 - 18x + 8}{(x + 2)(x - 2)^2}$ **41.** 0
43. $\frac{x + y}{x}$ **45.** $x - y$ **47.** $\frac{c^2 - 2c + 4}{c}$ **49.** $\frac{xy}{x - y}$
51. $\frac{a^2 - 1}{a^2 + 1}$ **53.** $\frac{3(x - 1)^2(x + 2)}{(x - 3)(x + 3)(-x + 10)}$
55. $\frac{1 + a}{1 - a}$ **57.** $\frac{b + a}{b - a}$ **59.** Discussion and Writing
61. $2x + h$ **63.** $3x^2 + 3xh + h^2$ **65.** x^5
67. $\frac{(n + 1)(n + 2)(n + 3)}{2 \cdot 3}$ **69.** $\frac{x^3 + 2x^2 + 11x + 20}{2(x + 1)(2 + x)}$

Exercise Set R.6

1. 11 **3.** $4|y|$ **5.** $|b + 1|$ **7.** $-3x$ **9.** $|x - 2|$
11. 2 **13.** $6\sqrt{5}$ **15.** $3\sqrt[3]{2}$ **17.** $8\sqrt{2}|c|d^2$
19. $2|x||y|\sqrt[4]{3x^2}$ **21.** $2x^2y\sqrt{6}$ **23.** $3x\sqrt[3]{4y}$
25. $2(x + 4)\sqrt[3]{(x + 4)^2}$ **27.** $\frac{m^2n^4}{2}$ **29.** 2
31. $\frac{1}{2x}$ **33.** $\frac{4a\sqrt[3]{a}}{3b}$ **35.** $\frac{x\sqrt{7x}}{6y^3}$ **37.** $51\sqrt{2}$
39. $-2x\sqrt{2} - 12\sqrt{5x}$ **41.** 1 **43.** $4 + 2\sqrt{3}$
45. $-9 - 5\sqrt{15}$ **47.** About 13,709.5 ft
49. (a) $h = \frac{a}{2}\sqrt{3}$; (b) $A = \frac{a^2}{4}\sqrt{3}$ **51.** 8
53. $\frac{\sqrt{6}}{3}$ **55.** $\frac{\sqrt[3]{10}}{2}$
57. $\frac{2\sqrt[3]{6}}{3}$ **59.** $\frac{9 - 3\sqrt{5}}{2}$ **61.** $\frac{6\sqrt{m} + 6\sqrt{n}}{m - n}$
63. $\frac{6}{5\sqrt{3}}$ **65.** $\frac{7}{\sqrt[3]{98}}$ **67.** $\frac{11}{\sqrt{33}}$
69. $\frac{76}{27 + 3\sqrt{5} - 9\sqrt{3} - \sqrt{15}}$ **71.** $\frac{a - b}{3a\sqrt{a} - 3a\sqrt{b}}$
73. $\sqrt[4]{x^3}$ **75.** 8 **77.** $\frac{1}{5}$ **79.** $\frac{a\sqrt[4]{a}}{\sqrt[4]{b^3}}$, or $a\sqrt[4]{\frac{a}{b^3}}$
81. $13\sqrt[4]{13}$ **83.** $20^{2/3}$ **85.** $11^{1/6}$ **87.** $5^{5/6}$
89. 4 **91.** $8a^2$ **93.** $\frac{x^3}{3b^{-2}}$, or $\frac{x^3b^2}{3}$ **95.** $x\sqrt[3]{y}$
97. $\sqrt[6]{288}$ **99.** $\sqrt[12]{x^{11}y^7}$ **101.** $a\sqrt[6]{a^5}$
103. $(a + x)\sqrt[12]{(a + x)^{11}}$ **105.** Discussion and Writing
107. $\frac{(2 + x^2)\sqrt{1 + x^2}}{1 + x^2}$ **109.** $a^{a/2}$

Review Exercises, Chapter R

1. [R.1] $12, -3, -1, -19, 31, 0$ **2.** [R.1] $12, 31$
3. [R.1] $-43.89, 12, -3, -\frac{1}{5}, -1, -\frac{4}{3}, 7\frac{2}{3}, -19, 31, 0$
4. [R.1] All of them   **5.** [R.1] $\sqrt{7}, \sqrt[3]{10}$
6. [R.1] $12, 31, 0$ **7.** [R.1] $[-3, 5)$ **8.** [R.1] 3.5
9. [R.1] 16 **10.** [R.1] 10 **11.** [R.2] 117 **12.** [R.2] -10
13. [R.2] $3,261,000$ **14.** [R.2] 0.00041
15. [R.2] 1.432×10^{-2} **16.** [R.2] 4.321×10^4
17. [R.2] 7.8125×10^{-22} **18.** [R.2] 5.46×10^{-32}
19. [R.2] $-14a^{-2}b^7$, or $\frac{-14b^7}{a^2}$
20. [R.2] $6x^9y^{-6}z^6$, or $\frac{6x^9z^6}{y^6}$ **21.** [R.6] 3 **22.** [R.6] -2
23. [R.5] $\frac{b}{a}$ **24.** [R.5] $\frac{x + y}{xy}$ **25.** [R.6] -4
26. [R.6] $25x^4 - 10\sqrt{2}x^2 + 2$ **27.** [R.6] $13\sqrt{5}$
28. [R.3] $x^3 + t^3$ **29.** [R.3] $10a^2 - 7ab - 12b^2$
30. [R.3] $8xy^4 - 9xy^2 + 4x^2 + 2y - 7$

31. [R.4] $(x + 2)(x^2 - 3)$
32. [R.4] $3a(2a + 3b^2)(2a - 3b^2)$ **33.** [R.4] $(x + 12)^2$
34. [R.4] $x(9x - 1)(x + 4)$
35. [R.4] $(2x - 1)(4x^2 + 2x + 1)$
36. [R.4] $(3x^2 + 5y^2)(9x^4 - 15x^2y^2 + 25y^4)$
37. [R.4] $6(x + 2)(x^2 - 2x + 4)$
38. [R.4] $(x - 1)(2x + 3)(2x - 3)$
39. [R.4] $(3x - 5)^2$ **40.** [R.4] $3(6x^2 - x + 2)$
41. [R.4] $(ab - 3)(ab + 2)$ **42.** [R.5] 3
43. [R.5] $\dfrac{x - 5}{(x + 5)(x + 3)}$ **44.** [R.6] $y^3\sqrt[6]{y}$
45. [R.6] $\sqrt[3]{(a + b)^2}$ **46.** [R.6] $b\sqrt[5]{b^2}$ **47.** [R.6] $\dfrac{m^4n^2}{3}$
48. [R.6] $\dfrac{x - 2\sqrt{xy} + y}{x - y}$ **49.** [R.6] About 18.8 ft
50. Discussion and Writing [R.2] Anya is not following the rules for order of operations. She is subtracting 6 from 15 first, then dividing the difference by 3, and finally multiplying the quotient by 4. The correct answer is 7.
51. Discussion and Writing [R.2] When the number 4 is raised to an integer power greater than 1, the last digit of the result is 4 or 6. Since the calculator returns $4.398046511 \times 10^{12}$, or $4,398,046,511,000$, we can conclude that this result is an approximation.
52. [R.3] $x^{2n} + 6x^n - 40$ **53.** [R.3] $t^{2a} + 2 + t^{-2a}$
54. [R.3] $y^{2b} - z^{2c}$
55. [R.3] $a^{3n} - 3a^{2n}b^n + 3a^nb^{2n} - b^{3n}$
56. [R.4] $(y^n + 8)^2$ **57.** [R.4] $(x^t - 7)(x^t + 4)$
58. [R.4] $m^{3n}(m^n - 1)(m^{2n} + m^n + 1)$

Test, Chapter R

1. [R.1] **(a)** $-8, 0, 36$; **(b)** $-8, \frac{11}{3}, 0, -5.49, 36, 10\frac{1}{6}$
2. [R.1] $1.2|x||y|$ **3.** [R.1] $(-3, 6]$ **4.** [R.2] -5
5. [R.2] 7.5×10^6 **6.** [R.2] $-15a^4b^{-1}$, or $-\dfrac{15a^4}{b}$
7. [R.3] $3x^4 - 5x^3 + x^2 + 5x$ **8.** [R.3] $2x^2 + x - 15$
9. [R.3] $4y^2 - 4y + 1$ **10.** [R.6] $21\sqrt{3}$
11. [R.5] $\dfrac{x-y}{xy}$ **12.** [R.4] $(2n - 3)(n + 4)$
13. [R.4] $2(2x + 3)(2x - 3)$
14. [R.4] $(m - 2)(m^2 + 2m + 4)$
15. [R.5] $\dfrac{x - 5}{x - 2}$ **16.** [R.5] $\dfrac{x + 3}{(x + 1)(x + 5)}$
17. [R.6] $\sqrt[7]{t^5}$ **18.** [R.6] $\dfrac{35 + 5\sqrt{3}}{46}$
19. [R.6] 13 ft   **20.** [R.3] $x^2 - 2xy + y^2 - 2x + 2y + 1$

Chapter 1

Exercise Set 1.1

1.

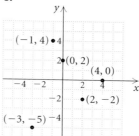

3.

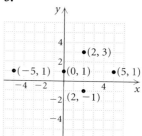

5. Yes; no   **7.** Yes; no   **9.** No; yes   **11.** No; yes

13.

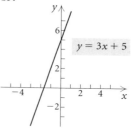

15.

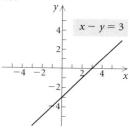

17.

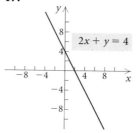

19.

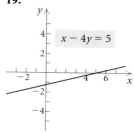

21.

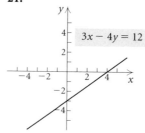

23.

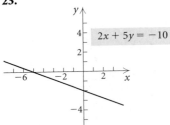

$2x + 5y = -10$

25.

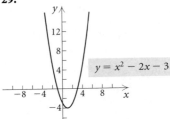

$y = -x^2$

27.

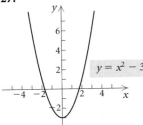

$y = x^2 - 3$

29.

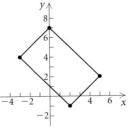

$y = x^2 - 2x - 3$

31. $\sqrt{10}$, 3.162   **33.** $\sqrt{45}$, 6.708   **35.** $\sqrt{14 + 6\sqrt{2}}$, 4.742
37. 6.5   **39.** Yes   **41.** The distances between the points
are $\sqrt{10}$, 10, and $\sqrt{90}$. Since $(\sqrt{10})^2 + (\sqrt{90})^2 = 10^2$, the
three points are vertices of a right triangle.
43. $\left(\dfrac{9}{2}, \dfrac{15}{2}\right)$ **45.** $\left(\dfrac{15}{2}, 2\right)$ **47.** $\left(\dfrac{\sqrt{3} - \sqrt{6}}{2}, -\dfrac{\sqrt{5}}{2}\right)$

49.

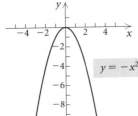

$\left(-\dfrac{1}{2}, \dfrac{3}{2}\right), \left(\dfrac{7}{2}, \dfrac{1}{2}\right), \left(\dfrac{5}{2}, \dfrac{9}{2}\right), \left(-\dfrac{3}{2}, \dfrac{11}{2}\right)$; no

51. $\left(\dfrac{\sqrt{7} + \sqrt{2}}{2}, -\dfrac{1}{2}\right)$ **53.** $(x - 2)^2 + (y - 3)^2 = \dfrac{25}{9}$

55. $(x + 1)^2 + (y - 4)^2 = 25$
57. $(x - 2)^2 + (y - 1)^2 = 169$
59. $(x + 2)^2 + (y - 3)^2 = 4$
61. $(0, 0)$; 2;

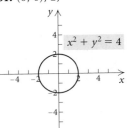

$x^2 + y^2 = 4$

63. $(0, 3)$; 4;

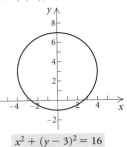

$x^2 + (y - 3)^2 = 16$

65. $(1, 5)$; 6;

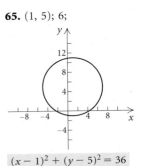

$(x - 1)^2 + (y - 5)^2 = 36$

67. $(-4, -5)$; 3;

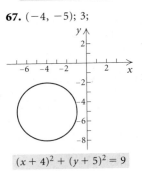

$(x + 4)^2 + (y + 5)^2 = 9$

69. (b)   **71.** (a)

73.

$y = 2x + 1$

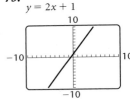

75.

$4x + y = 7$

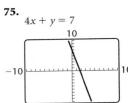

77.

$y = \frac{1}{3}x + 2$

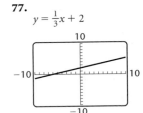

79.

$2x + 3y = -5$

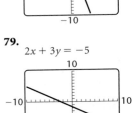

81.

$y = x^2 + 6$

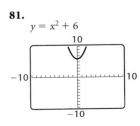

83.

$y = 2 - x^2$

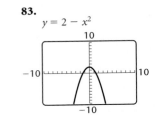

85.

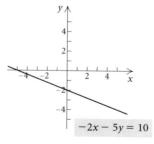

$$y = x^2 + 4x - 2$$

87. $[-25, 25, -25, 25]$ **89.** Standard window
91. $[-1, 1, -0.3, 0.3]$
93. Square the window; for example, $[-12, 9, -4, 10]$.
95.–101. Left to the student
103. Discussion and Writing
105. $\sqrt{h^2 + h + 2a - 2\sqrt{a^2 + ah}}$,
$\left(\dfrac{2a + h}{2}, \dfrac{\sqrt{a} + \sqrt{a + h}}{2}\right)$
107. $(x - 2)^2 + (y + 7)^2 = 36$
109. $a_1 \approx 2.7$ ft, $a_2 \approx 37.3$ ft   **111.** Yes
113. Yes   **115.** $(0, 4)$
117. Let $P_1 = (x_1, y_1)$, $P_2 = (x_2, y_2)$, and
$M = \left(\dfrac{x_1 + x_2}{2}, \dfrac{y_1 + y_2}{2}\right)$. Let $d(AB)$ denote the distance
from point A to point B.

(a) $d(P_1 M) = \sqrt{\left(\dfrac{x_1 + x_2}{2} - x_1\right)^2 + \left(\dfrac{y_1 + y_2}{2} - y_1\right)^2}$

$= \dfrac{1}{2}\sqrt{(x_2 - x_1)^2 + (y_2 - y_1)^2};$

$d(P_2 M) = \sqrt{\left(\dfrac{x_1 + x_2}{2} - x_2\right)^2 + \left(\dfrac{y_1 + y_2}{2} - y_2\right)^2}$

$= \dfrac{1}{2}\sqrt{(x_1 - x_2)^2 + (y_1 - y_2)^2}$

$= \dfrac{1}{2}\sqrt{(x_2 - x_1)^2 + (y_2 - y_1)^2} = d(P_1 M).$

(b) $d(P_1 M) + d(P_2 M) = \dfrac{1}{2}\sqrt{(x_2 - x_1)^2 + (y_2 - y_1)^2}$

$+ \dfrac{1}{2}\sqrt{(x_2 - x_1)^2 + (y_2 - y_1)^2}$

$= \sqrt{(x_2 - x_1)^2 + (y_2 - y_1)^2}$

$= d(P_1 P_2).$

Exercise Set 1.2

1. Yes   **3.** Yes   **5.** No   **7.** Yes
9. Yes   **11.** Yes   **13.** No
15. Function; domain: $\{2, 3, 4\}$; range: $\{10, 15, 20\}$
17. Not a function; domain: $\{-7, -2, 0\}$;
range: $\{3, 1, 4, 7\}$
19. Function; domain: $\{-2, 0, 2, 4, -3\}$; range: $\{1\}$
21. $f(-1) = 2$; $f(0) = 0$; $f(1) = -2$

23. (a) 1; **(b)** 6; **(c)** 22; **(d)** $3x^2 + 2x + 1$; **(e)** $3t^2 - 4t + 2$
25. (a) 8; **(b)** -8; **(c)** $-x^3$; **(d)** $27y^3$; **(e)** $8 + 12h + 6h^2 + h^3$
27. (a) $\dfrac{1}{8}$; **(b)** 0; **(c)** does not exist; **(d)** $\dfrac{81}{53}$, or
approximately 1.5283; **(e)** $\dfrac{x + h - 4}{x + h + 3}$
29. 0; does not exist; does not exist as a real number; $\dfrac{1}{\sqrt{3}}$,
or $\dfrac{\sqrt{3}}{3}$ **31.** All real numbers, or $(-\infty, \infty)$
33. $\{x \mid x \neq 0\}$, or $(-\infty, 0) \cup (0, \infty)$
35. $\{x \mid x \neq 2\}$, or $(-\infty, 2) \cup (2, \infty)$
37. $\{x \mid x \neq -1 \text{ and } x \neq 5\}$, or $(-\infty, -1) \cup (-1, 5) \cup$
$(5, \infty)$ **39.** Domain: all real numbers; range: $[0, \infty)$
41. Domain: $[-3, 3]$; range: $[0, 3]$
43. Domain: all real numbers; range: all real numbers
45. Domain: $(-\infty, 7]$; range: $[0, \infty)$
47. Domain: all real numbers; range: $(-\infty, 3]$
49. No   **51.** Yes   **53.** Yes   **55.** No
57. Function; domain: $[0, 5]$; range: $[0, 3]$
59. Function; domain: $[-2\pi, 2\pi]$; range: $[-1, 1]$
61. 645 m; 0 m   **63.** 0.4 acre; 20.4 acres; 50.6 acres; 416.9
acres; 1033.6 acres
65. $g(-2.1) \approx -21.8$; $g(5.08) \approx -130.4$; $g(10.003) \approx$
-468.3 **67.** Discussion and Writing
69. [1.1] $(0, -7)$, no; $(8, 11)$, yes
70. [1.1] $\left(\frac{4}{5}, -2\right)$, yes; $\left(\frac{11}{5}, \frac{1}{10}\right)$, yes
71. [1.1]   **72.** [1.1]

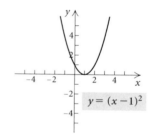
$$y = (x - 1)^2$$

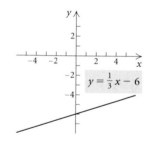
$$y = \frac{1}{3}x - 6$$

73. [1.1]

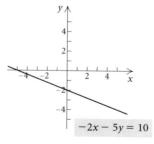
$$-2x - 5y = 10$$

75. $f(x) = x$, $g(x) = x + 1$ **77.** 35

Exercise Set 1.3

1. (a) Yes; **(b)** yes; **(c)** yes   **3. (a)** Yes; **(b)** no; **(c)** no
5. $-\frac{3}{5}$ **7.** 0   **9.** $\frac{1}{5}$ **11.** $-\frac{5}{3}$ **13.** Not defined
15. $2a + h$ **17.** 6.7% grade; $y = 0.067x$
19. $100 per year   **21.** $\frac{1}{6}$ km per minute
23. (a) $W(h) = 3.5h - 110$; **(b)** 107 lb; **(c)** $\{h \mid h > 31.43\}$, or $(31.43, \infty)$
25. (a) 115, 75, 135, 179; **(b)** Below $-57.5°$, stopping distance is negative; above 32°, ice doesn't form.
27. (a) $\frac{11}{10}$. For each mile per hour faster that the car travels, it takes $\frac{11}{10}$ ft longer to stop;
(b) 6, 11.5, 22.5, 55.5, 72; **(c)** $\{r \mid r > 0\}$, or $(0, \infty)$. If r is allowed to be 0, the function says that a stopped car travels $\frac{1}{2}$ ft before stopping.

29. $\dfrac{\text{Height of corn (in feet)}}{\text{Number of weeks after planting}}$
31. $C(t) = 60 + 40t$; $C(6) = 300
33. $C(x) = 800 + 3x$; $C(75) = 1025
35. Left to the student   **37.** Discussion and Writing
38. [1.2] 10   **39.** [1.2] 40   **40.** [1.2] $a^2 + 3a$
41. [1.2] $a^2 + 2ah + h^2 - 3a - 3h$
43. False   **45.** False   **47.** $f(x) = x + b$

Exercise Set 1.4

1. $\frac{3}{5}$; $(0, -7)$ **3.** $-\frac{1}{2}$; $(0, 5)$ **5.** $-\frac{3}{2}$; $(0, 5)$
7. $\frac{4}{3}$; $(0, -7)$ **9.** $y = \frac{2}{9}x + 4$ **11.** $y = -4x - 7$
13. $y = -4.2x + \frac{3}{4}$ **15.** $y = \frac{2}{9}x + \frac{19}{3}$ **17.** $y = 3x - 5$
19. $y = -\frac{3}{5}x - \frac{17}{5}$ **21.** $y = -3x + 2$
23. $y = -\frac{1}{2}x + \frac{7}{2}$ **25.** Parallel   **27.** Perpendicular
29. $y = \frac{2}{7}x + \frac{29}{7}$; $y = -\frac{7}{2}x + \frac{31}{2}$
31. $y = -0.3x - 2.1$; $y = \frac{10}{3}x + \frac{70}{3}$
33. $y = -\frac{3}{4}x + \frac{1}{4}$; $y = \frac{4}{3}x - 6$
35. $x = 3$; $y = -3$
37. (a) Model 1, using (10, 121) and (30, 205): $y = \frac{21}{5}x + 79$; model 2, using (0, 88) and (37, 217): $y = \frac{129}{37}x + 88$; **(b)** model 1: 268 million tons; model 2: about 245 million tons; **(c)** model 2
39. (a) $M = 0.2H + 156$; **(b)** 164, 169, 171, 173; **(c)** $r = 1$; the regression line fits the data perfectly and should be a good predictor.
41. (a) $y = 3.650586114x + 85.7786294$; **(b)** 250 million tons. This value is only 5 million more than the value found with model 2. **(c)** $r \approx 0.9932$. The line fits the data well.
43. Discussion and Writing   **45.** [1.2] -1
46. [1.2] Not defined
47. [1.1] $(x + 7)^2 + (y + 1)^2 = \frac{81}{25}$
48. [1.1] $x^2 + (y - 3)^2 = 6.25$ **49.** -7.75

Exercise Set 1.5

1. (a) $(-5, 1)$; **(b)** $(3, 5)$; **(c)** $(1, 3)$
3. (a) $(-3, -1)$, $(3, 5)$; **(b)** $(1, 3)$; **(c)** $(-5, -3)$
5. Increasing: $(0, \infty)$; decreasing: $(-\infty, 0)$; relative minimum: 0 at $x = 0$
7. Increasing: $(-\infty, 0)$; decreasing: $(0, \infty)$; relative maximum: 5 at $x = 0$
9. Increasing: $(3, \infty)$; decreasing: $(-\infty, 3)$; relative minimum: 1 at $x = 3$
11. $d(t) = \sqrt{(120t)^2 + (400)^2}$

13. $A(x) = 24x - x^2$ **15.** $A(w) = 10w - \dfrac{w^2}{2}$

17. $d(s) = \dfrac{14}{s}$

19. (a) $A(x) = x(30 - x)$, or $30x - x^2$;
(b) $\{x \mid 0 < x < 30\}$; **(c)** 15 ft by 15 ft
21. (a) $V(x) = x(12 - 2x)(12 - 2x)$, or $4x(6 - x)^2$;
(b) $\{x \mid 0 < x < 6\}$; **(c)** 8 cm by 8 cm by 2 cm
23.
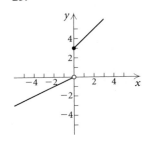

25.
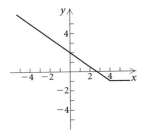

27.
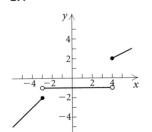

29.
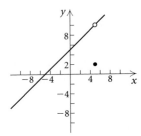

31.
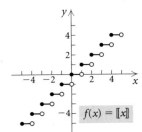

33.
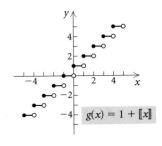

35. 33   **37.** −1   **39.** 5   **41.** 2   **43.** 0   **45.** 0

47. (a) Domain of f, all real numbers; domain of g, $[-4, \infty)$; domain of $f + g$, $f - g$, and fg, $[-4, \infty)$; domain of ff, all real numbers; domain of f/g, $(-4, \infty)$; domain of g/f, $[-4, 3) \cup (3, \infty)$; **(b)** $(f + g)(x) = x - 3 + \sqrt{x + 4}$; $(f - g)(x) = x - 3 - \sqrt{x + 4}$; $(fg)(x) = (x - 3)\sqrt{x + 4}$; $(ff)(x) = x^2 - 6x + 9$; $(f/g)(x) = (x - 3)/\sqrt{x + 4}$; $(g/f)(x) = \sqrt{x + 4}/(x - 3)$

49. (a) Domain of f, g, $f + g$, $f - g$, fg, and ff, all real numbers; domain of f/g, all real numbers except -3 and $\frac{1}{2}$; domain of g/f, all real numbers except 0; **(b)** $(f + g)(x) = x^3 + 2x^2 + 5x - 3$; $(f - g)(x) = x^3 - 2x^2 - 5x + 3$; $(fg)(x) = 2x^5 + 5x^4 - 3x^3$; $(ff)(x) = x^6$; $(f/g)(x) = x^3/(2x^2 + 5x - 3)$; $(g/f)(x) = (2x^2 + 5x - 3)/x^3$

51. $(f + g)(x) = 2x^2 - 2$; $(f - g)(x) = -6$; $(fg)(x) = x^4 - 2x^2 - 8$; $(f/g)(x) = (x^2 - 4)/(x^2 + 2)$

53. $(f + g)(x) = \sqrt{x - 7} + x^2 - 25$; $(f - g)(x) = \sqrt{x - 7} - x^2 + 25$; $(fg)(x) = \sqrt{x - 7}\,(x^2 - 25)$; $(f/g)(x) = \sqrt{x - 7}/(x^2 - 25)$

55. Domain of F: $[0, 9]$; domain of G: $[3, 10]$; domain of $F + G$: $[3, 9]$ **57.** $[3, 6) \cup (6, 8) \cup (8, 9]$

59.

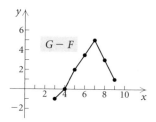

61. (a) $P(x) = -0.4x^2 + 57x - 13$; **(b)** $R(100) = 2000$; $C(100) = 313$; $P(100) = 1687$

63. $6x + 3h - 2$ **65.** $3x^2 + 3xh + h^2$

67. $\dfrac{7}{(x + h + 3)(x + 3)}$

69. (a)

$y = -0.1x^2 + 1.2x + 98.6$

(b) 102.2; **(c)** 6 days after the illness began, the temperature is 102.2°F.

71. Increasing: $(-1, 1)$; decreasing: $(-\infty, -1)$, $(1, \infty)$

73. Increasing: $(-1.414, 1.414)$; decreasing: $(-2, -1.414)$, $(1.414, 2)$ **75. (a)** $A(x) = x\sqrt{256 - x^2}$;

(b) $\{x \mid 0 < x < 16\}$;

(c)

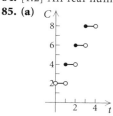

$y = x\sqrt{256 - x^2}$

(d) 11.314 ft by 11.314 ft

77. Left to the student   **79.** Discussion and Writing

81. [1.2] All real numbers, or $(-\infty, \infty)$

82. [1.2] $\{x \mid x \neq 3\}$ **83.** [1.2] $\{x \mid x \neq -5\}$

84. [1.2] All real numbers, or $(-\infty, \infty)$

85. (a)

(b) $C(t) = 2(\llbracket t \rrbracket + 1)$, $t > 0$

87. $\{x \mid -5 \leq x < -4 \ or \ 5 \leq x < 6\}$

89. (a) $h(r) = \dfrac{30 - 5r}{3}$; **(b)** $V(r) = \pi r^2 \left(\dfrac{30 - 5r}{3}\right)$;

(c) $V(h) = \pi h \left(\dfrac{30 - 3h}{5}\right)^2$

Exercise Set 1.6

1. x-axis, no; y-axis, yes; origin, no

3. x-axis, yes; y-axis, no; origin, no

5. x-axis, no; y-axis, no; origin, yes

7. x-axis, no; y-axis, no; origin, yes

9. x-axis, yes; y-axis, yes; origin, yes

11. x-axis, no; y-axis, yes; origin, no

13. x-axis, yes; y-axis, yes; origin, yes

15. x-axis, no; y-axis, no; origin, no

17. x-axis, no; y-axis, no; origin, yes

19. Even   **21.** Odd   **23.** Neither   **25.** Odd

27. Even   **29.** Odd   **31.** Even   **33.** Even

35. Start with the graph of $g(x) = x^2$. Shift it up 1 unit.

37. Start with the graph of $f(x) = x^3$. Shift it left 5 units.

39. Start with the graph of $g(x) = x^2$. Reflect it across the x-axis.

41. Start with the graph of $g(x) = \sqrt{x}$. Stretch it vertically by multiplying each y-coordinate by 3, then shift it down 5 units.

43. Start with the graph of $f(x) = |x|$. Stretch it horizontally by multiplying each x-coordinate by 3, then shift it down 4 units.
45. Start with the graph of $g(x) = x^2$. Shift it left 5 units, then down 4 units.
47. Start with the graph of $g(x) = x^2$. Shift it right 5 units, shrink it vertically by multiplying each y-coordinate by $\frac{1}{4}$, and reflect it across the x-axis.
49. Start with the graph of $g(x) = 1/x$. Shift it left 3 units, then up 2 units.
51. $f(x) = -(x - 8)^2$ **53.** $f(x) = |x + 7| + 2$
55. $f(x) = \dfrac{1}{2x} - 3$ **57.** $f(x) = -(x - 3)^2 + 4$
59. $f(x) = \sqrt{-(x + 2)} - 1$
61. $f(x) = 0.83(x + 4)^3$
63.

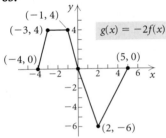

65.

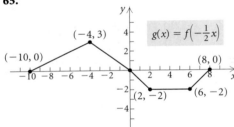

67.

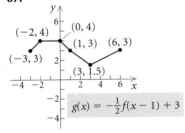

69. (f)   **71.** (f)   **73.** (d)   **75.** (c)
77. $f(-x) = 2(-x)^4 - 35(-x)^3 + 3(-x) - 5 = 2x^4 + 35x^3 - 3x - 5 = g(x)$
79. $g(x) = x^3 - 3x^2 + 2$
81. $k(x) = (x + 1)^3 - 3(x + 1)^2$

83. Discussion and Writing   **85.** Discussion and Writing
86. [1.2] **(a)** 22; **(b)** -22; **(c)** $4a^3 - 5a$; **(d)** $-4a^3 + 5a$
87. [1.2] **(a)** 38; **(b)** 38; **(c)** $5a^2 - 7$; **(d)** $5a^2 - 7$
88. [1.3] Slope is $\frac{2}{9}$; y-intercept is $\left(0, \frac{1}{9}\right)$.
89. [1.3] $y = -\frac{1}{8}x + \frac{7}{8}$ **91.** Odd
93. x-axis, yes; y-axis, yes; origin, yes
95. x-axis, yes; y-axis, no; origin, no
97. 5   **99.** True
101. $E(-x) = \dfrac{f(-x) + f(-(-x))}{2} = \dfrac{f(-x) + f(x)}{2} = E(x)$
103. **(a)** $E(x) + O(x) = \dfrac{f(x) + f(-x)}{2} + \dfrac{f(x) - f(-x)}{2} = \dfrac{2f(x)}{2} = f(x)$; **(b)** $f(x) = \dfrac{-22x^2 + \sqrt{x} + \sqrt{-x} - 20}{2} + \dfrac{8x^3 + \sqrt{x} - \sqrt{-x}}{2}$

Review Exercises, Chapter 1

1. [1.1] Yes; no   **2.** [1.1] Yes; no
3. [1.1]   **4.** [1.1]

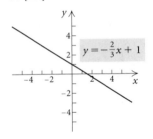

5. [1.1] $\sqrt{34} \approx 5.831$ **6.** [1.1] $\left(\frac{1}{2}, \frac{11}{2}\right)$
7. [1.1] $(x + 2)^2 + (y - 6)^2 = 13$ **8.** [1.1] $(-1, 3)$; 4
9. [1.1] $(x - 2)^2 + (y - 4)^2 = 26$
10. [1.2] Not a function; domain: $\{3, 5, 7\}$; range: $\{1, 3, 5, 7\}$
11. [1.2] Function; domain: $\{-2, 0, 1, 2, 7\}$; range: $\{-7, -4, -2, 2, 7\}$ **12.** [1.2] No   **13.** [1.2] Yes
14. [1.2] No   **15.** [1.2] Yes   **16.** [1.2] All real numbers   **17.** [1.2] $\{x \mid x \neq 0\}$ **18.** [1.2] $\{x \mid x \neq 5$ and $x \neq 1\}$ **19.** [1.2] $\{x \mid x \neq -4$ and $x \neq 4\}$
20. [1.2] Domain: $[-4, 4]$; range: $[0, 4]$
21. [1.2] Domain: all real numbers; range: $[-0, \infty)$
22. [1.2] Domain: all real numbers; range: all real numbers
23. [1.2] Domain: all real numbers; range: $[0, \infty)$
24. [1.2] **(a)** -3; **(b)** 9; **(c)** $a^2 - 3a - 1$; **(d)** $2x + h - 1$
25. [1.3] **(a)** Yes; **(b)** no; **(c)** no, strictly speaking, but data might be modeled by a linear regression function.
26. [1.3] **(a)** Yes; **(b)** yes; **(c)** yes   **27.** [1.3] $\frac{5}{3}$ **28.** [1.3] 0
29. [1.3] Not defined   **30.** [1.3] -2; $(0, -7)$
31. [1.4] $y + 4 = -\frac{2}{3}(x - 0)$
32. [1.4] $y + 1 = 3(x + 2)$

33. [1.4] $y - 1 = \frac{1}{3}(x - 4)$, or $y + 1 = \frac{1}{3}(x + 2)$
34. [1.4] Parallel   **35.** [1.4] Neither
36. [1.4] Perpendicular   **37.** [1.4] $y = -\frac{2}{3}x - \frac{1}{3}$
38. [1.4] $y = \frac{3}{2}x - \frac{5}{2}$
39. [1.3] $C(t) = 25 + 20t$; \$145
40. [1.3] **(a)** 70°C, 220°C, 10,020°C; **(b)** [0, 5600]
41. [1.4] Using (10, 16.5) and (40, 48.6), $y = 1.07x + 5.8$;
70 million. Answers may vary.
42. [1.5]   **43.** [1.5]

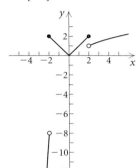

44. [1.5]   **45.** [1.5]

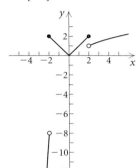

46. [1.5] $A(x) = 2x\sqrt{4 - x^2}$

47. [1.5] **(a)** $A(x) = x^2 + \dfrac{432}{x}$; **(b)** $(0, \infty)$;

(c) $x = 6$ in.; height $= 3$ in.
48. [1.6] x-axis, yes; y-axis, yes; origin, yes
49. [1.6] x-axis, yes; y-axis, yes; origin, yes
50. [1.6] x-axis, no; y-axis, no; origin, no
51. [1.6] x-axis, no; y-axis, yes; origin, no
52. [1.6] x-axis, no; y-axis, no; origin, yes
53. [1.6] x-axis, no; y-axis, yes; origin, no
54. [1.6] Even   **55.** [1.6] Even   **56.** [1.6] Odd
57. [1.6] Even   **58.** [1.6] Even   **59.** [1.6] Neither
60. [1.6] Odd   **61.** [1.6] Even   **62.** [1.6] Even
63. [1.6] Odd   **64.** [1.6] $f(x) = (x + 3)^2$
65. [1.6] $f(x) = -\sqrt{x - 3} + 4$
66. [1.6] $f(x) = 2|x - 3|$

67. [1.6]   **68.** [1.6]

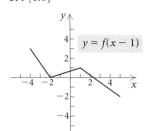
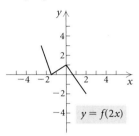

69. [1.6]   **70.** [1.6]

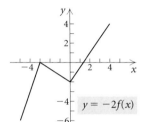
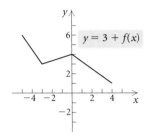

71. [1.5] **(a)** Domain of f, $\{x \,|\, x \neq 0\}$; domain of g, all real numbers; domain of $f + g$, $f - g$, and fg, $\{x \,|\, x \neq 0\}$;
domain of f/g, $\left\{ x \,\middle|\, x \neq 0 \text{ and } x \neq \dfrac{3}{2} \right\}$

(b) $(f + g)(x) = \dfrac{4}{x^2} + 3 - 2x$; $(f - g)(x) = \dfrac{4}{x^2} - 3 + 2x$;

$fg(x) = \dfrac{12}{x^2} - \dfrac{8}{x}$; $(f/g)(x) = \dfrac{4}{x^2(3 - 2x)}$

72. [1.5] **(a)** Domain of f, g, $f + g$, $f - g$, and fg, all real numbers; domain of f/g, $\left\{ x \,\middle|\, x \neq \dfrac{1}{2} \right\}$

(b) $(f + g)(x) = 3x^2 + 6x - 1$;
$(f - g)(x) = 3x^2 + 2x + 1$; $fg(x) = 6x^3 + 5x^2 - 4x$;
$(f/g)(x) = \dfrac{3x^2 + 4x}{2x - 1}$

73. [1.5] $P(x) = -0.5x^2 + 105x - 6$
74. [1.4] **(a)** $y = 0.9362739726x + 9.656712329$;
65.8 million; **(b)** $r = 0.9946$, a good fit
75. Discussion and Writing [1.2], [1.6] **(a)** $4x^3 - 2x + 9$;
(b) $4x^3 + 24x^2 + 46x + 35$; **(c)** $4x^3 - 2x + 42$. **(a)** Adds
2 to each function value; **(b)** adds 2 to each input before
finding a function value; **(c)** adds the output for 2 to the
output for x
76. Discussion and Writing [1.6] In the graph of
$y = f(cx)$, the constant c stretches or shrinks the graph of
$y = f(x)$ horizontally. The constant c in $y = cf(x)$
stretches or shrinks the graph of $y = f(x)$ vertically. For
$y = f(cx)$, the x-coordinates of $y = f(x)$ are divided by c;
for $y = cf(x)$, the y-coordinates of $y = f(x)$ are multiplied
by c.

77. Discussion and Writing [1.6] **(a)** To draw the graph of y_2 from the graph of y_1, reflect across the x-axis the portions of the graph for which the y-coordinates are negative. **(b)** To draw the graph of y_2 from the graph of y_1, draw the portion of the graph of y_1 to the right of the y-axis; then draw its reflection across the y-axis.
78. [1.2] $\{x \mid x < 0\}$
79. [1.2] $\{x \mid x \neq -3 \text{ and } x \neq 0 \text{ and } x \neq 3\}$
80. [1.5], [1.6] Let $f(x)$ and $g(x)$ be odd functions. Then by definition, $f(-x) = -f(x)$, or $f(x) = -f(-x)$, and $g(-x) = -g(x)$, or $g(x) = -g(-x)$. Thus, $(f + g)(x) = f(x) + g(x) = -f(-x) + [-g(-x)] = -[f(-x) + g(-x)] = -(f + g)(-x)$ and $f + g$ is odd.
81. [1.6] Reflect the graph of $y = f(x)$ across the x-axis and then across the y-axis.

Test, Chapter 1

1. [1.1]

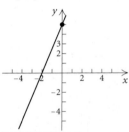

$5x - 2y = -10$

2. [1.1] $\sqrt{45} \approx 6.708$ **3.** [1.1] $\left(-3, \frac{9}{2}\right)$
4. [1.1] $(x + 1)^2 + (y - 2)^2 = 5$
5. [1.2] **(a)** Yes; **(b)** $\{-4, 3, 1, 0\}$; **(c)** $\{7, 0, 5\}$
6. [1.2] **(a)** 8; **(b)** $2a^2 + 7a + 11$
7. [1.2] **(a)**   **(b)** $(-\infty, \infty)$; **(c)** $[3, \infty)$

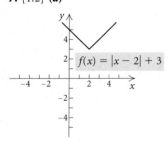

$f(x) = |x - 2| + 3$

8. [1.2] **(a)** No; **(b)** yes

9. [1.5]

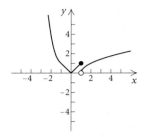

10. [1.3] Slope: $\frac{3}{2}$; y-intercept: $\left(0, \frac{5}{2}\right)$
11. [1.4] $y - 4 = -\frac{3}{4}(x - (-5))$, or $y - (-2) = -\frac{3}{4}(x - 3)$, or $y = -\frac{3}{4}x + \frac{1}{4}$
12. [1.4] $y - 3 = -\frac{1}{2}(x + 1)$, or $y = -\frac{1}{2}x + \frac{5}{2}$
13. [1.4] Using $(0, 8.4)$ and $(2, 8.0)$, $y = -0.2x + 8.4$; 6.0 gal   **14.** [1.6] x-axis: no; y-axis: yes; origin: no
15. [1.6] Odd   **16.** [1.6] $f(x) = (x - 2)^2 - 1$
17. [1.6] $f(x) = (x + 2)^2 - 3$
18. [1.6]

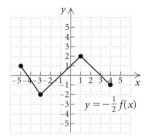

$y = -\frac{1}{2}f(x)$

19. [1.5] **(a)** $(-\infty, \infty)$; **(b)** $[3, \infty)$; **(c)** $(f - g)(x) = x^2 - \sqrt{x - 3}$; **(d)** $(fg)(x) = x^2\sqrt{x - 3}$; **(e)** $(3, \infty)$
20. [1.6] $(-1, 1)$

Chapter 2

Exercise Set 2.1

1. 4   **3.** $-\frac{3}{4}$ **5.** 8   **7.** -4 **9.** 6   **11.** $\frac{4}{5}$ **13.** $-\frac{3}{2}$
15. $-\frac{2}{3}$ **17.** $\frac{1}{2}$ **19.** $b = \dfrac{2A}{h}$ **21.** $w = \dfrac{P - 2l}{2}$
23. $h = \dfrac{2A}{b_1 + b_2}$ **25.** $\pi = \dfrac{3V}{4r^3}$ **27.** $C = \frac{5}{9}(F - 32)$
29. -5 **31.** 18   **33.** 16   **35.** -12 **37.** 6   **39.** 20
41. 6   **43.** 15   **45.** **(a)** $(4, 0)$; **(b)** 4   **47.** 47.1 million
49. 125   **51.** 1.93 million
53. Length: 93 m; width: 68 m   **55.** $26°$, $130°$, $24°$
57. About 280,526
59. Freight: 66 mph; passenger: 80 mph   **61.** 67.5 lb
63. About 5.5 million   **65.** 23   **67.** 660 yr

69. Left to the student   **71.** Discussion and Writing
73. [1.4] $y = -\frac{3}{4}x + \frac{13}{4}$ **74.** [1.4] $y = -\frac{3}{4}x + \frac{1}{4}$
75. [1.5] $x^2 + 2x - 6$ **76.** [1.5] 12   **77.** $-\frac{2}{3}$
79. Yes   **81.** No

Exercise Set 2.2

1. $2 + 11i$ **3.** $5 - 12i$ **5.** $5 + 9i$ **7.** $5 + 4i$
9. $5 + 7i$ **11.** $13 - i$ **13.** $-11 + 16i$ **15.** $35 + 14i$
17. $-14 + 23i$ **19.** 41   **21.** $12 + 16i$ **23.** $-45 - 28i$
25. $\dfrac{-4\sqrt{3} + 10}{41} + \dfrac{5\sqrt{3} + 8}{41}i$ **27.** $-\dfrac{14}{13} + \dfrac{5}{13}i$
29. $\frac{15}{146} + \frac{33}{146}i$ **31.** $-\frac{1}{2} + \frac{1}{2}i$ **33.** $-\frac{1}{2} - \frac{13}{2}i$
35. $-i$ **37.** $-i$ **39.** 1   **41.** i **43.** 625
45. Left to the student   **47.** Discussion and Writing
49. [1.4] $y = -2x + 1$ **50.** [1.5] $x^2 - 3x - 1$
51. [1.5] $\frac{8}{11}$ **52.** [1.5] $2x + h - 3$ **53.** True
55. True   **57.** $a^2 + b^2$

Exercise Set 2.3

1. $\frac{2}{3}, \frac{3}{2}$ **3.** $-1, \frac{2}{3}$ **5.** $-\sqrt{3}, \sqrt{3}$ **7.** $0, 3$ **9.** $-\frac{1}{3}, 0, 2$
11. $-1, -\frac{1}{7}, 1$ **13.** (a) $(-4, 0), (2, 0)$; (b) $-4, 2$
15. $-7, 1$ **17.** $4 \pm \sqrt{7}$ **19.** $-4 \pm 3i$ **21.** $-2, \frac{1}{3}$
23. $-3, 5$ **25.** $-1, \frac{2}{5}$ **27.** $\dfrac{5 \pm \sqrt{7}}{3}$ **29.** $-\dfrac{1}{2} \pm \dfrac{\sqrt{7}}{2}i$
31. $\dfrac{4 \pm \sqrt{31}}{5}$ **33.** $\dfrac{5}{6} \pm \dfrac{\sqrt{23}}{6}i$ **35.** No   **37.** Yes
39. No   **41.** $-5, -1$ **43.** $\dfrac{3 \pm \sqrt{21}}{2}$ **45.** $\dfrac{5 \pm \sqrt{21}}{2}$
47. $-1 \pm \sqrt{6}$ **49.** $\pm 1, \pm \sqrt{2}$ **51.** 16   **53.** $-8, 64$
55. $\frac{5}{2}, 3$ **57.** $-\frac{3}{2}, -1, \frac{1}{2}, 1$ **59.** About 11.5 sec
61. Length: 4 ft; width: 3 ft
63. 4 and 9; -9 and -4 **65.** 2 cm   **67.** Linear
69. Quadratic   **71.** Linear   **73.** 2, 6   **75.** 0.143, 6
77. $-0.151, 1.651$ **79.** $-0.637, 3.137$ **81.** $-1.535, 0.869$
83. $-0.347, 1.181$ **85.** Discussion and Writing
87. [1.6] x-axis: yes; y-axis: yes; origin: yes
88. [1.6] x-axis: no; y-axis: yes; origin: no
89. [1.6] Odd   **90.** [1.6] Neither   **91.** 1
93. $-\sqrt{7}, -\frac{3}{2}, 0, \frac{1}{3}, \sqrt{7}$ **95.** $\dfrac{-1 \pm \sqrt{1 + 4\sqrt{2}}}{2}$
97. $3 \pm \sqrt{5}$ **99.** 19   **101.** $-2 \pm \sqrt{2}, \dfrac{1}{2} \pm \dfrac{\sqrt{7}}{2}i$
103. $t = \dfrac{-v_0 \pm \sqrt{v_0^2 - 2ax_0}}{a}$

Exercise Set 2.4

1. (a) $\left(-\frac{1}{2}, -\frac{9}{4}\right)$; (b) $x = -\frac{1}{2}$; (c) minimum: $-\frac{9}{4}$

3. (a) $(4, -4)$; (b) $x = 4$; (c) minimum: -4
5. (a) $\left(\frac{7}{2}, -\frac{1}{4}\right)$; (b) $x = \frac{7}{2}$; (c) minimum: $-\frac{1}{4}$
7. (a) $\left(-\frac{3}{2}, \frac{7}{2}\right)$; (b) $x = -\frac{3}{2}$; (c) minimum: $\frac{7}{2}$
9. (a) $\left(\frac{1}{2}, \frac{3}{2}\right)$; (b) $x = \frac{1}{2}$; (c) maximum: $\frac{3}{2}$
11. (f)   **13.** (b)   **15.** (h)   **17.** (c)
19. (a) $(3, -4)$; (b) minimum: -4; (c) $[-4, \infty)$;
(d) increasing: $(3, \infty)$; decreasing: $(-\infty, 3)$
21. (a) $(-1, -18)$; (b) minimum: -18; (c) $[-18, \infty)$;
(d) increasing: $(-1, \infty)$; decreasing: $(-\infty, -1)$
23. (a) $\left(5, \frac{9}{2}\right)$; (b) maximum: $\frac{9}{2}$; (c) $\left(-\infty, \frac{9}{2}\right]$;
(d) increasing: $(-\infty, 5)$; decreasing: $(5, \infty)$
25. (a) $(-1, 2)$; (b) minimum: 2; (c) $[2, \infty)$;
(d) increasing: $(-1, \infty)$; decreasing: $(-\infty, -1)$
27. 4.5 in.   **29.** Base: 10 cm; height: 10 cm
31. 350.6 ft   **33.** 350   **35.** $797; 40
37. 4800 yd^2 **39.** Left to the student
41. Left to the student   **43.** Discussion and Writing
45. Discussion and Writing
46. [1.5] 3   **47.** [1.5] $4x + 2h - 1$
48. [1.6]   **49.** [1.6]

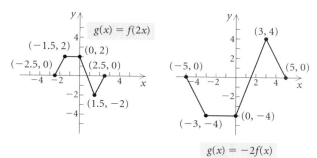

$g(x) = -2f(x)$

51. (a) 2; (b) $\frac{11}{2}$ **53.** (a) 2; (b) $1 - i$ **55.** -236.25
57.

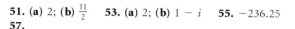

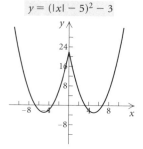

Exercise Set 2.5

1. $\frac{20}{9}$ **3.** 286   **5.** 6   **7.** No solution
9. No solution   **11.** No solution
13. $\{x \mid x \text{ is a real number } and \ x \neq 0 \ and \ x \neq 6\}$
15. 2   **17.** $\frac{5}{3}$ **19.** $\pm\sqrt{2}$ **21.** No solution   **23.** -6

25. 7   **27.** No solution   **29.** 1   **31.** 3, 7   **33.** −8
35. 81   **37.** −7, 7   **39.** No solution   **41.** −3, 5
43. $-\frac{1}{3}, \frac{1}{3}$ **45.** 0   **47.** −1, $-\frac{1}{3}$ **49.** −24, 44
51. −2, 4   **53.** $T_1 = \dfrac{P_1 V_1 T_2}{P_2 V_2}$ **55.** $R_2 = \dfrac{RR_1}{R_1 - R}$
57. $p = \dfrac{Fm}{m - F}$ **59.** Left to the student
61. Discussion and Writing   **62.** [2.1] 3   **63.** [2.1] 7.5
64. [2.1] 10.4%   **65.** [2.1] 481,000   **67.** $3 \pm 2\sqrt{2}$
69. −1

Exercise Set 2.6

1. $\{x \mid x > 3\}$, or $(3, \infty)$ **3.** $\left\{x \mid x \geq -\frac{5}{12}\right\}$, or $\left[-\frac{5}{12}, \infty\right)$
5. $\left\{y \mid y \geq \frac{22}{13}\right\}$, or $\left[\frac{22}{13}, \infty\right)$ **7.** $\left\{x \mid x \leq \frac{15}{34}\right\}$, or $\left(-\infty, \frac{15}{34}\right]$
9. $\{x \mid x < 1\}$, or $(-\infty, 1)$ **11.** $[-3, 3)$ **13.** $[8, 10]$
15. $[-7, -1]$ **17.** $\left(-\frac{3}{2}, 2\right)$ **19.** $(1, 5]$ **21.** $\left(-\frac{11}{3}, \frac{13}{3}\right]$
23. $(-\infty, -2] \cup (1, \infty)$ **25.** $\left(-\infty, -\frac{7}{2}\right] \cup \left[\frac{1}{2}, \infty\right)$
27. $(-\infty, 9.6) \cup (10.4, \infty)$ **29.** $\left(-\infty, -\frac{57}{4}\right] \cup \left[-\frac{55}{4}, \infty\right)$
31. $(-7, 7)$;

33. $(-\infty, -4.5] \cup [4.5, \infty)$;

35. $(-17, 1)$ **37.** $(-\infty, -17] \cup [1, \infty)$ **39.** $\left(-\frac{1}{4}, \frac{3}{4}\right)$
41. $\left(-\frac{1}{3}, \frac{1}{3}\right)$ **43.** $[-6, 3]$ **45.** $(-\infty, 4.9) \cup (5.1, \infty)$
47. $\left[-\frac{1}{2}, \frac{7}{2}\right]$ **49.** $\left[-\frac{7}{3}, 1\right]$ **51.** $(-\infty, -8) \cup (7, \infty)$
53. No solution   **55.** Left to the student
57. Discussion and Writing   **59.** [2.2] $10 - 7i$
60. [2.2] $2 + i$ **61.** [2.2] $\frac{7}{25} + \frac{26}{25}i$ **62.** [2.2] $13 - 13i$
63. $\left(-\frac{1}{4}, \frac{5}{9}\right)$ **65.** $\left(-\infty, \frac{1}{2}\right)$ **67.** No solution
69. $\left(-\infty, -\frac{8}{3}\right) \cup (-2, \infty)$

Review Exercises, Chapter 2

1. [2.1] $\frac{3}{2}$ **2.** [2.1] −6   **3.** [2.1] −1   **4.** [2.1] −21
5. [2.3] $-\frac{5}{2}, \frac{1}{3}$ **6.** [2.3] −5, 1   **7.** [2.3] $-2, \frac{4}{3}$
8. [2.3] $-\sqrt{3}, \sqrt{3}$ **9.** [2.3] $-\sqrt{10}, \sqrt{10}$
10. [2.1] 3   **11.** [2.1] 4   **12.** [2.1] 0.2, or $\frac{1}{5}$ **13.** [2.1] 4
14. [2.3] 1   **15.** [2.3] −5, 3   **16.** [2.3] $\dfrac{1 \pm \sqrt{41}}{4}$
17. [2.3] $\dfrac{-1 \pm \sqrt{10}}{3}$ **18.** [2.5] $\frac{27}{7}$ **19.** [2.5] $-\frac{1}{2}, \frac{9}{4}$
20. [2.5] 0, 3   **21.** [2.5] 5   **22.** [2.5] 1, 7
23. [2.5] −8, 1   **24.** [2.6] $\left[-\frac{4}{3}, \frac{4}{3}\right]$ **25.** [2.6] $\left(\frac{2}{5}, 2\right]$
26. [2.6] $(-\infty, \infty)$ **27.** [2.6] $\left(-\infty, -\frac{5}{3}\right] \cup [1, \infty)$
28. [2.6] $\left(-\frac{2}{3}, 1\right)$ **29.** [2.6] $(-\infty, -6] \cup [-2, \infty)$
30. [2.1] $h = \dfrac{V}{lw}$ **31.** [2.1] $s = \dfrac{M - n}{0.3}$ **32.** [2.5] $h = \dfrac{v^2}{2g}$

33. [2.5] $t = \dfrac{ab}{a + b}$ **34.** [2.2] $-2\sqrt{10}i$
35. [2.2] $-4\sqrt{15}$ **36.** [2.2] $-\frac{7}{8}$ **37.** [2.2] $-18 - 26i$
38. [2.2] $\frac{11}{10} + \frac{3}{10}i$ **39.** [2.2] $1 - 4i$ **40.** [2.2] $2 - i$
41. [2.2] $-i$ **42.** [2.2] 3^{28}
43. [2.3] $x^2 - 3x + \frac{9}{4} = 18 + \frac{9}{4}; \left(x - \frac{3}{2}\right)^2 = \frac{81}{4}; x = \frac{3}{2} \pm \frac{9}{2};$
$-3, 6$ **44.** [2.3] $x^2 - 4x = 2; x^2 - 4x + 4 = 2 + 4;$
$(x - 2)^2 = 6; x = 2 \pm \sqrt{6}; 2 - \sqrt{6}, 2 + \sqrt{6}$
45. [2.3] $-4, \frac{2}{3}$ **46.** [2.3] $1 - 3i, 1 + 3i$
47. [2.3] $-3, 6$ **48.** [2.3] 1
49. [2.3] $\pm\sqrt{\dfrac{3 \pm \sqrt{5}}{2}}$ **50.** [2.3] $-\sqrt{3}, 0, \sqrt{3}$
51. [2.3] $-2, -\frac{2}{3}, 3$ **52.** [2.3] $-5, -2, 2$
53. [2.4] **(a)** $\left(\frac{3}{8}, -\frac{7}{16}\right)$; **(b)** $x = \frac{3}{8}$; **(c)** maximum: $-\frac{7}{16}$
54. [2.4] **(a)** $(1, -2)$; **(b)** $x = 1$; **(c)** minimum: -2
55. [2.4] (d)   **56.** [2.4] (c)   **57.** [2.4] (b)   **58.** [2.4] (a)
59. [2.3] 30 ft, 40 ft   **60.** [2.5] 6 mph
61. [2.3] 80 km/h   **62.** [2.3] $35 - 5\sqrt{33}$ ft,
or about 6.3 ft   **63.** [2.3] 6 ft by 6 ft
64. [2.3] $\dfrac{15 - \sqrt{115}}{2}$ cm, or about 2.1 cm
65. [2.3] 2.1 cm   **66.** Left to the student
67. Left to the student   **68.** Left to the student
69. Left to the student
70. Discussion and Writing [2.1] If an equation contains no fractions, using the addition principle before using the multiplication principle eliminates the need to add or subtract fractions.
71. Discussion and Writing [2.4] You can conclude that $|a_1| = |a_2|$ since these constants determine how wide the parabolas are. Nothing can be concluded about the h's and the k's.
72. [2.5] 256   **73.** [2.5] −7, 9   **74.** [2.5] $4 \pm \sqrt[4]{243}$, or 0.052, 7.948   **75.** [2.3] −1   **76.** [2.3] $-\frac{1}{4}, 2$
77. [2.4] ± 6 **78.** [2.3] 9%

Test, Chapter 2

1. [2.1] −1   **2.** [2.3] $\frac{1}{2}, -5$ **3.** [2.3] −1, 3
4. [2.3] $-\sqrt{6}, \sqrt{6}$ **5.** [2.3] $\dfrac{3}{4} \pm \dfrac{\sqrt{23}}{4}i$ **6.** [2.5] 16
7. [2.5] $-1, \frac{13}{6}$ **8.** [2.5] 5   **9.** [2.5] $-\frac{1}{2}, 2$
10. [2.6] $(-5, 3)$ **11.** [2.6] $(-\infty, -7) \cup (-3, \infty)$
12. [2.1] $h = \dfrac{3V}{2\pi r^2}$ **13.** [2.5] $n = \dfrac{R^2}{3p}$ **14.** [2.1] −3
15. [2.3] $-\frac{1}{4}, 3$ **16.** [2.3] $\dfrac{1 \pm \sqrt{57}}{4}$ **17.** [2.3] $-2 \pm \sqrt{5}$
18. [2.2] $\sqrt{43}\,i$ **19.** [2.2] $-5i$ **20.** [2.2] $10 + 5i$
21. [2.2] i **22.** [2.4] **(a)** $(1, 9)$; **(b)** $x = 1$;
(c) maximum: 9; **(d)** $(-\infty, 9]$ **23.** [2.4] 15 ft by 30 ft
24. [2.4] $\frac{1}{6}$

Chapter 3

Exercise Set 3.1

1. Cubic; $\frac{1}{2}x^3$; $\frac{1}{2}$; 3   **3.** Linear; $0.9x$; 0.9; 1
5. Quartic; $305x^4$; 305; 4   **7.** (d)   **9.** (f)   **11.** (b)
13. $f(-5) = -18$ and $f(-4) = 7$. By the intermediate value theorem, since $f(-5)$ and $f(-4)$ have opposite signs, then $f(x)$ has a zero between -5 and -4.   **15.** $f(-3) = 22$ and $f(-2) = 5$. Both $f(-3)$ and $f(-2)$ are positive. We cannot use the intermediate value theorem to determine if there is a zero between -3 and -2.
17. $f(2) = 2$ and $f(3) = 57$. Both $f(2)$ and $f(3)$ are positive. We cannot use the intermediate value theorem to determine if there is a zero between 2 and 3.
19. 0.866 sec   **21.** \$3240   **23.** (a) 18.75%; (b) 10%
25. Relative maximum: 1.506 at $x = -0.632$, relative minimum: 0.494 at $x = 0.632$; $(-\infty, \infty)$ **27.** Relative minimum: -3.8 at $x = 0$, no relative maxima; $[-3.8, \infty)$
29. Relative maximum: 11.012 at $x = 1.258$, relative minimum: -8.183 at $x = -1.116$; $(-\infty, \infty)$
31. (a) linear: $y = 4.717948718x + 37.93076923$; quadratic: $y = 0.0281175273x^2 + 1.711977591x + 96.76946268$; cubic: $y = -0.0016285158x^3 + 0.2897410836x^2 - 10.07476028x + 227.1101167$; power: $y = 22.9451257x^{0.6396947491}$, where x is the number of years after 1900; (b) linear; 510, 557; answers may vary
33. (a) linear: $y = 62.43636364x + 668.6363636$; quadratic: $y = 7.465909091x^2 - 4.756818182x + 758.2272727$; cubic: $y = -1.431429681x^3 + 26.79020979x^2 - 70.7457265x + 794.2993007$; quartic: $y = -0.6566142191x^4 + 10.38762626x^3 - 39.52782634x^2 + 47.44483294x + 765.9335664$, where x is the number of years after 1989; (b) quadratic; \$2155 billion; answers may vary
35. Discussion and Writing   **37.** [1.6] 5
38. [1.6] $6\sqrt{2}$ **39.** [1.1] Center: $(3, -5)$; radius 7
40. [1.1] Center: $(-4, 3)$; radius $2\sqrt{2}$ **41.** 7%

Exercise Set 3.2

1. Yes; no; no   **3.** (a)
5. $P(x) = (x + 2)(x^2 - 2x + 4) - 16$
7. $P(x) = (x^2 + 4)(x^2 + 5) + 0$
9. $Q(x) = 2x^3 + x^2 - 3x + 10$, $R(x) = -42$
11. $Q(x) = x^2 - 4x + 8$, $R(x) = -24$
13. $Q(x) = x^3 + x^2 + x + 1$, $R(x) = 0$
15. $Q(x) = 2x^3 + x^2 + \frac{7}{2}x + \frac{7}{4}$, $R(x) = -\frac{1}{8}$
17. $Q(x) = x^3 + x^2y + xy^2 + y^3$, $R(x) = 0$
19. 0; -60; 0   **21.** 5,935,988; -772
23. 0; 0; 65; $1 - 12\sqrt{2}$ **25.** Yes; no   **27.** No; no
29. $f(x) = (x - 1)(x + 2)(x + 3)$; 1, -2, -3

31. $f(x) = (x - 2)(x - 5)(x + 1)$; 2, 5, -1
33. $f(x) = (x - 2)(x - 3)(x + 4)$; 2, 3, -4
35. $f(x) = (x - 1)(x - 2)(x - 3)(x + 5)$; 1, 2, 3, -5
37. Left to the student   **39.** Discussion and Writing
40. [2.1] -5 **41.** [2.3] $\dfrac{5 \pm i\sqrt{71}}{4}$ **42.** [2.3] -1, $\dfrac{3}{7}$
43. [2.3] $b = 15$ in., $h = 15$ in.
45. (a) $x + 4$, $x + 3$, $x - 2$, $x - 5$; (b) $P(x) = (x + 4)(x + 3)(x - 2)(x - 5)$; (c) yes; two examples are $f(x) = c \cdot P(x)$ for any nonzero constant c; and $g(x) = (x - a)P(x)$; (d) no   **47.** $\frac{14}{3}$ **49.** $-1 \pm \sqrt{7}$
51. Answers can vary. One possibility is $P(x) = x^{15} - x^{14}$.
53. $x - 3 + i$, R $6 - 3i$

Exercise Set 3.3

1. -3, multiplicity 2; 1, multiplicity 1
3. 0, multiplicity 3; 1, multiplicity 2; -4, multiplicity 1
5. $\pm\sqrt{3}$, ± 1, each has multiplicity 1
7. -3, -1, 1, each has multiplicity 1
9. $f(x) = x^3 - 6x^2 - x + 30$
11. $f(x) = x^3 + 3x^2 + 4x + 12$
13. $f(x) = x^3 - \sqrt{3}x^2 - 2x + 2\sqrt{3}$
15. $f(x) = x^5 + 2x^4 - 2x^2 - x$
17. $-3 - 4i$, $4 + \sqrt{5}$
19. $f(x) = x^3 - 4x^2 + 6x - 4$
21. $f(x) = x^3 - 5x^2 + 16x - 80$
23. $f(x) = x^4 + 4x^2 - 45$
25. i, 2, 3   **27.** $1 + 2i$, $1 - 2i$ **29.** ± 1
31. ± 1, ± 2, $\pm\frac{1}{3}$, $\pm\frac{1}{5}$, $\pm\frac{2}{3}$, $\pm\frac{2}{5}$, $\pm\frac{1}{15}$, $\pm\frac{2}{15}$
33. (a) Rational: -3; other: $\pm\sqrt{2}$;
(b) $f(x) = (x + 3)(x + \sqrt{2})(x - \sqrt{2})$
35. (a) Rational: -2, 1; other: none;
(b) $f(x) = (x + 2)(x - 1)^2$
37. (a) Rational: -1; other: $3 \pm 2\sqrt{2}i$;
(b) $f(x) = (x + 1)(x - 3 - 2\sqrt{2}i)(x - 3 + 2\sqrt{2}i)$
39. (a) Rational: $-\frac{1}{5}$, 1; other: $\pm 2i$;
(b) $f(x) = (5x + 1)(x - 1)(x + 2i)(x - 2i)$
41. (a) Rational: -2, -1; other: $3 \pm \sqrt{13}$;
(b) $f(x) = (x + 2)(x + 1)(x - 3 - \sqrt{13})(x - 3 + \sqrt{13})$
43. (a) Rational: 2; other: $1 \pm \sqrt{3}$;
(b) $f(x) = (x - 2)(x - 1 - \sqrt{3})(x - 1 + \sqrt{3})$
45. (a) Rational: -2; other: $1 \pm \sqrt{3}i$;
(b) $f(x) = (x + 2)(x - 1 - \sqrt{3}i)(x - 1 + \sqrt{3}i)$
47. (a) Rational: $\dfrac{1}{2}$; other: $\dfrac{1 \pm \sqrt{5}}{2}$;
(b) $f(x) = \dfrac{1}{3}\left(x - \dfrac{1}{2}\right)\left(x - \dfrac{1 + \sqrt{5}}{2}\right)\left(x - \dfrac{1 - \sqrt{5}}{2}\right)$
49. No rational zeros   **51.** No rational zeros
53. No rational zeros   **55.** -2, 1, 2   **57.** 3 or 1; 0
59. 0; 3 or 1   **61.** 2 or 0; 2 or 0   **63.** 1; 1   **65.** 1; 0

67. 2 or 0; 2 or 0   **69.** 3 or 1; 1   **71.** 1; 1   **73.** Left to the student   **75.** Discussion and Writing
76. [2.4] **(a)** $(1, -4)$; **(b)** $x = 1$; **(c)** minimum: -4 at $x = 1$ **77.** [2.4] **(a)** $(4, -6)$; **(b)** $x = 4$; **(c)** minimum: -6 at $x = 4$ **78.** [2.3] $-3, 11$ **79.** [2.1] 10
81. **(a)** $-1, \frac{1}{2}, 3$; **(b)** $0, \frac{3}{2}, 4$; **(c)** $-3, -\frac{3}{2}, 1$; **(d)** $-\frac{1}{2}, \frac{1}{4}, \frac{3}{2}$
83. $-8, -\frac{3}{2}, 4, 7, 15$

Exercise Set 3.4

1. (d); $x = 2, x = -2, y = 0$
3. (e); $x = 2, x = -2, y = 0$
5. (c); $x = 2, x = -2, y = 8x$
7. No x-intercepts, y-intercept: $\left(0, \frac{1}{3}\right)$;

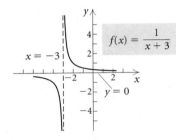

9. No x-intercepts, y-intercept: $\left(0, \frac{2}{5}\right)$;

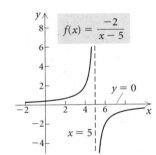

11. x-intercept: $\left(-\frac{1}{2}, 0\right)$, no y-intercept;

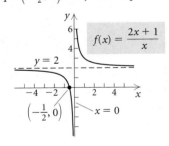

13. No x-intercepts, y-intercept: $\left(0, \frac{1}{4}\right)$;

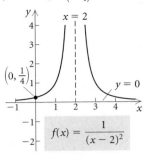

15. No x-intercepts, no y-intercept;

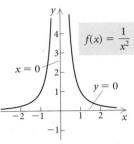

17. No x-intercepts, y-intercept: $\left(0, \frac{1}{3}\right)$;

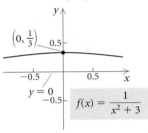

19. x-intercept: $(-2, 0)$, y-intercept: $(0, 2)$;

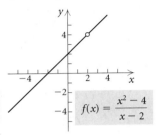

21. x-intercept: $(1, 0)$, y-intercept: $\left(0, -\frac{1}{2}\right)$;

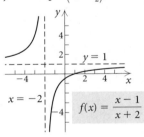

23. x-intercept: $(-3, 0)$, y-intercept: $(0, -1)$;

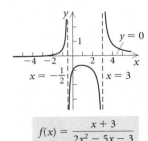

25. x-intercepts: $(-3, 0)$ and $(3, 0)$, y-intercept: $(0, -9)$;

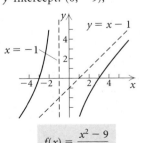

27. x-intercepts: $(-2, 0)$ and $(1, 0)$, y-intercept: $(0, -2)$;

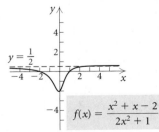

29. x-intercept: $\left(-\frac{2}{3}, 0\right)$, y-intercept: $(0, 2)$;

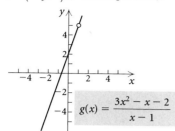

$$g(x) = \frac{3x^2 - x - 2}{x - 1}$$

31. x-intercept: $(1, 0)$, y-intercept: $\left(0, \frac{1}{3}\right)$;

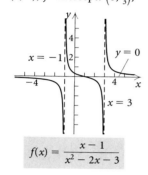

$$f(x) = \frac{x - 1}{x^2 - 2x - 3}$$

33. x-intercept: $(3, 0)$, y-intercept: $(0, -3)$;

35. x-intercept: $(-1, 0)$, no y-intercept;

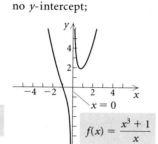

$$f(x) = \frac{x - 3}{(x + 1)^3}$$

$$f(x) = \frac{x^3 + 1}{x}$$

37. x-intercepts: $(-5, 0)$, $(0, 0)$, and $(3, 0)$, y-intercept: $(0, 0)$;

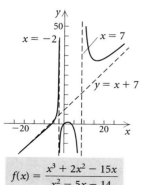

$$f(x) = \frac{x^3 + 2x^2 - 15x}{x^2 - 5x - 14}$$

39. x-intercept: $(0, 0)$, y-intercept: $(0, 0)$;

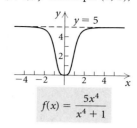

$$f(x) = \frac{5x^4}{x^4 + 1}$$

41. x-intercept: $(0, 0)$, y-intercept: $(0, 0)$;

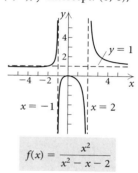

$$f(x) = \frac{x^2}{x^2 - x - 2}$$

43. $f(x) = \dfrac{1}{x^2 - x - 20}$　**45.** $f(x) = \dfrac{3x^2 + 12x + 12}{2x^2 - 2x - 40}$

47. (a) $N(t) \to 0.16$ as $t \to \infty$; **(b)** The medication never completely disappears from the body; a trace amount remains.

49. (a) $P(0) = 0$; $P(1) = 45{,}455$; $P(3) = 55{,}556$; $P(8) = 29{,}197$; **(b)** $P(t) \to 0$ as $t \to \infty$; **(c)** In time, no one lives in Lordsburg.

51. 58,926 at $t \approx 2.12$ months

53. Discussion and Writing

55. [2.6] $\{y \mid y \geq 3\}$, or $[3, \infty)$　**56.** [2.6] $\left\{x \mid x > \frac{5}{3}\right\}$, or $\left(\frac{5}{3}, \infty\right)$　**57.** [2.6] $\{x \mid x \leq -13 \ or \ x \geq 1\}$, or $(-\infty, -13] \cup [1, \infty)$　**58.** [2.6] $\left\{x \mid -\frac{11}{12} \leq x \leq \frac{5}{12}\right\}$, or $\left[-\frac{11}{12}, \frac{5}{12}\right]$　**59.** $y = x^3 + 4$

61.

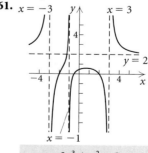

$$f(x) = \frac{2x^3 + x^2 - 8x - 4}{x^3 + x^2 - 9x - 9}$$

63. $(-\infty, -3) \cup (7, \infty)$

Exercise Set 3.5

1. $(-\infty, -5) \cup (-3, 2)$
3. $(-2, 0] \cup (2, \infty)$ **5.** $(-4, 1)$ **7.** $(-\infty, -2] \cup [4, \infty)$
9. $(-\infty, -2) \cup (1, \infty)$ **11.** $(-\infty, -5) \cup (5, \infty)$
13. $(-\infty, -2] \cup [2, \infty)$ **15.** $(-\infty, 3) \cup (3, \infty)$
17. $\varnothing$ **19.** $\left(-\infty, -\frac{5}{4}\right] \cup [0, 3]$ **21.** $[-3, -1] \cup [1, \infty)$
23. $(-\infty, -2) \cup (1, 3)$ **25.** $[-\sqrt{2}, -1] \cup [\sqrt{2}, \infty)$
27. $(-\infty, -1] \cup \left[\frac{3}{2}, 2\right]$ **29.** $(-\infty, 5]$
31. $(-4, \infty)$ **33.** $\left(-\frac{5}{2}, \infty\right)$ **35.** $\left(-3, -\frac{1}{5}\right] \cup (1, \infty)$
37. $(-\infty, -3) \cup \left[\dfrac{5 - \sqrt{105}}{10}, -\dfrac{1}{3}\right) \cup \left[\dfrac{5 + \sqrt{105}}{10}, \infty\right)$
39. $\left(2, \frac{7}{2}\right]$ **41.** $(1 - \sqrt{2}, 0) \cup (1 + \sqrt{2}, \infty)$
43. $(-\infty, -3) \cup (1, 3) \cup \left[\frac{11}{3}, \infty\right)$ **45.** $(-\infty, \infty)$
47. $\left(-3, \dfrac{1 - \sqrt{61}}{6}\right) \cup \left(-\dfrac{1}{2}, 0\right) \cup \left(\dfrac{1 + \sqrt{61}}{6}, \infty\right)$
49. $(-1, 0) \cup \left(\frac{2}{7}, \frac{7}{2}\right)$
51. $[-6 - \sqrt{33}, -5) \cup [-6 + \sqrt{33}, 1) \cup (5, \infty)$
53. $(0.408, 2.449)$
55. (a) $(10, 200)$; (b) $(0, 10) \cup (200, \infty)$
57. $\{n \mid 9 \le n \le 23\}$
59. Left to the student **61.** Discussion and Writing
62. [1.1] $x^2 + (y + 3)^2 = \frac{49}{16}$
63. [1.1] $(x + 2)^2 + (y - 4)^2 = 9$
64. [2.4] (a) $(5, -23)$; (b) minimum: -23 when $x = 5$;
(c) $[-23, \infty)$ **65.** [2.4] (a) $\left(\frac{3}{4}, -\frac{55}{8}\right)$; (b) maximum: $-\frac{55}{8}$
when $x = \frac{3}{4}$; (c) $\left(-\infty, -\frac{55}{8}\right]$
67. $\{3\}$ **69.** $(-\infty, \infty)$ **71.** $[-\sqrt{5}, \sqrt{5}]$ **73.** $\left[-\frac{3}{2}, \frac{3}{2}\right]$
75. $\left(-\infty, -\frac{1}{4}\right) \cup \left(\frac{1}{2}, \infty\right)$ **77.** $(-4, -2) \cup (-1, 1)$
79. $x^2 + x - 12 < 0$; answers may vary

Exercise Set 3.6

1. 4.5; $y = 4.5x$ **3.** 36; $y = \dfrac{36}{x}$ **5.** 4; $y = 4x$
7. 4; $y = \dfrac{4}{x}$ **9.** $\dfrac{3}{8}$; $y = \dfrac{3}{8}x$ **11.** 0.54; $y = \dfrac{0.54}{x}$
13. 3.5 hr **15.** 90 g **17.** About 686 kg **19.** 40 lb
21. $66\frac{2}{3}$ cm **23.** 1.92 ft **25.** $y = \dfrac{0.0015}{x^2}$
27. $y = 15x^2$ **29.** $y = xz$ **31.** $y = \dfrac{3}{10}xz^2$
33. $y = \dfrac{xz}{5wp}$ **35.** 2.5 m **37.** 36 mph **39.** About 106
41. Discussion and Writing

43. [1.5]

44. [1.6] x-axis, no; y-axis, yes; origin, no
45. [1.6] x-axis, yes; y-axis, no; origin, no
46. [1.6] x-axis, no; y-axis, no; origin, yes

47. $\$7.20$ **49.** $\dfrac{\pi}{4}$

Review Exercises, Chapter 3

1. [3.1] Quartic, $0.45x^4$, 0.45, 4 **2.** [3.1] Constant, -25,
-25, 0 **3.** [3.1] Linear, $-0.5x$, -0.5, 1 **4.** [3.1] Cubic,
$\frac{1}{3}x^3$, $\frac{1}{3}$, 3 **5.** [3.1] $f(1) = -4$ and $f(2) = 3$. Since $f(1)$ and
$f(2)$ have opposite signs, $f(x)$ has a zero between 1 and 2.
6. [3.1] $f(0) = 2$ and $f(1) = -0.5$. Since $f(0)$ and $f(1)$ have
opposite signs, $f(x)$ has a zero between 0 and 1.
7. [3.1] (a) 4%; (b) 5%
8. [3.2] $4x^2 - \frac{14}{5}x - \frac{17}{25}$, R $284/25$
9. [3.2] $2x^3 + x + 1$, R $2x + 6$ **10.** [3.2] $\{-5, 1, 2\}$
11. [3.2] $\{-2 - \sqrt{11}, -5, -2 + \sqrt{11}, 4\}$
12. [3.2] $x^2 + 7x + 22$, R 120
13. [3.2] $x^3 + x^2 + x + 1$, R 0 **14.** [3.2] 120
15. [3.2] 0
16. [3.2] -3 is a zero; $P(x) = (x + 5)(x + 3)(x - 1)$;
other zeros: -5, 1
17. [3.2] -2 is a zero; $P(x) = (x + 2)(x + 1) \cdot$
$[x - (-3 + \sqrt{10})][x - (-3 - \sqrt{10})]$;
other zeros: -1, $-3 \pm \sqrt{10}$
18. [3.3] $x^3 - x^2 - 10x - 8 = 0$; answers may vary
19. [3.3] $x^4 + 3x^3 - 7x^2 - 9x + 12 = 0$; answers may vary
20. [3.3] $x^6 - 13x^5 + 62x^4 - 132x^3 - 45x^2 + 745x - $
$858 = 0$; answers may vary
21. [3.3] $\{-1, 2, 3\}$ **22.** [3.3] $\{-5, -3, 1\}$
23. [3.3] $\{-3, 2 - \sqrt{11}, 2 + \sqrt{11}, 3\}$
24. [3.3] (a) $\{-10, 1\}$; (b) $P(x) = (x - 1)^2(x + 10)$
25. [3.3] (a) $\{-4, 0, 3, 4\}$;
(b) $P(x) = x^2(x + 4)^2(x - 3)(x - 4)$
26. [3.3] 3 or 1; 0 **27.** [3.3] 4 or 2 or 0; 2 or 0
28. [3.3] 3 or 1; 0

29. [3.4]

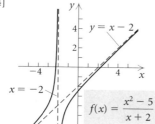

30. [3.4]

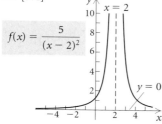

31. [3.4]   **32.** [3.4]

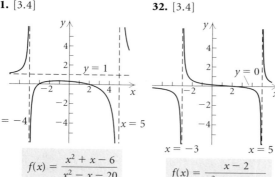

33. [3.4] $f(x) = \dfrac{1}{x^2 - x - 6}$

34. [3.4] $f(x) = \dfrac{4x^2 + 12x}{x^2 - x - 6}$

35. [3.4] **(a)** $N(t) \to 0.0875$ as $t \to \infty$; **(b)** The medication never completely disappears from the body; a trace amount remains.   **36.** [3.5] $(-3, 3)$

37. [3.5] $\left(-\infty, -\frac{1}{2}\right) \cup (2, \infty)$

38. [3.5] $[-4, 1] \cup [2, \infty)$

39. [3.5] $\left(-\infty, -\frac{14}{3}\right) \cup (-3, \infty)$

40. [3.5] $(-5, -3.386] \cup (-2, 2) \cup (4, \infty)$

41. [3.1], [3.5] **(a)** $t = 7$; **(b)** $(2, 3)$

42. [3.5] $\left[\dfrac{5 - \sqrt{15}}{2}, \dfrac{5 + \sqrt{15}}{2}\right]$

43. [3.6] $y = 4x$ **44.** [3.6] $y = \dfrac{2500}{x}$ **45.** [3.6] 20 min
46. [3.6] 500 watts   **47.** [3.6] About 78
48. [3.1] **(a)** $-2.637, 1.137$; **(b)** relative maximum: 7.125 at $x = -0.75$; **(c)** none; **(d)** domain: all real numbers; range: $(-\infty, 7.125]$
49. [3.1] **(a)** $-3, -1.414, 1.414$; **(b)** relative maximum: 2.303 at $x = -2.291$; **(c)** relative minimum: -6.303 at $x = 0.291$; **(d)** domain: all real numbers; range: all real numbers
50. [3.1] **(a)** $0, 1, 2$; **(b)** relative maximum: 0.202 at $x = 0.610$; **(c)** relative minima: 0 at $x = 0$, -0.620 at $x = 1.640$; **(d)** domain: all real numbers; range: $[-0.620, \infty)$
51. [3.1] **(a)** Linear: $f(x) = 0.5408695652x - 30.30434783$; quadratic: $f(x) = 0.0030322581x^2 - 0.5764516129x + 57.53225806$; cubic: $f(x) = 0.0000247619x^3 - 0.0112857143x^2 + 2.002380952x - 82.14285714$; **(b)** the cubic function; **(c)** 298, 498
52. Discussion and Writing [3.1], [3.4] A polynomial function is a function that can be defined by a polynomial expression. A rational function is a function that can be defined as a quotient of two polynomials.
53. Discussion and Writing [3.4] Vertical asymptotes occur at any x-values that make the denominator zero. The graph of a rational function does not cross any vertical asymptotes. Horizontal asymptotes occur when the degree of the numerator is less than or equal to the degree of the denominator. Oblique asymptotes occur when the degree of the numerator is 1 greater than the degree of the denominator. Graphs of rational functions may cross horizontal or oblique asymptotes.
54. [3.1] 9%   **55.** [3.5] $(-\infty, -1 - \sqrt{6}] \cup [-1 + \sqrt{6}, \infty)$
56. [3.5] $\left(-\infty, -\frac{1}{2}\right) \cup \left(\frac{1}{2}, \infty\right)$
57. [3.3] $\{1 + i, 1 - i, i, -i\}$ **58.** [3.5] $(-\infty, 2)$
59. [3.3] $(x - 1)\left(x + \dfrac{1}{2} - \dfrac{\sqrt{3}}{2}i\right)\left(x + \dfrac{1}{2} + \dfrac{\sqrt{3}}{2}i\right)$
60. [3.2] 7   **61.** [3.2] -4 **62.** [3.5] $(-\infty, -5] \cup [2, \infty)$
63. [3.5] $(-\infty, 1.1] \cup [2, \infty)$ **64.** [3.5] $\left(-1, \frac{3}{7}\right)$

Test, Chapter 3

1. [3.1] Quartic, $-x^4$, $-1, 4$ **2.** [3.1] Linear, $-4.7x$, $-4.7, 1$ **3.** [3.1] $f(0) = 3$ and $f(2) = -17$. Since $f(0)$ and $f(2)$ have opposite signs, $f(x)$ has a zero between 0 and 2.
4. [3.1] $g(-2) = 5$ and $g(-1) = 1$. Both $g(-2)$ and $g(-1)$ are positive. We cannot use the intermediate value theorem to determine if there is a zero between -2 and -1.
5. [3.1] 5.5%   **6.** [3.2] $x^2 + 3x + 1 + \dfrac{5x - 4}{x^2 - 1}$
7. [3.2] $3x^2 + 15x + 63$, R 322   **8.** [3.2] Yes
9. [3.2] -115 **10.** [3.3] $-\sqrt{3}, 2 + i$

11. [3.3] **(a)** $1, -2, \dfrac{-1 \pm \sqrt{13}}{2}$; **(b)** $P(x) =$

$(x - 1)(x + 2)\left(x - \dfrac{-1 + \sqrt{13}}{2}\right)\left(x - \dfrac{-1 - \sqrt{13}}{2}\right)$

12. [3.3] 2 or 0; 2 or 0
13. [3.4]

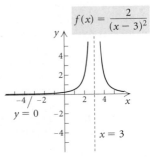

$$f(x) = \dfrac{2}{(x-3)^2}$$

14. [3.4] Answers may vary; $f(x) = \dfrac{x + 4}{x^2 - x - 2}$

15. [3.5] $\left(-\infty, -\frac{1}{2}\right) \cup (3, \infty)$
16. [3.5] $(-\infty, 4) \cup \left[\frac{13}{2}, \infty\right)$

17. [3.4] **(a)** 6 sec; **(b)** (1, 3)   **18.** [3.6] $y = \dfrac{30}{x}$

19. [3.6] 50 ft   **20.** [3.1] $(-\infty, -4] \cup [3, \infty)$

Chapter 4

Exercise Set 4.1

1. $(f \circ g)(x) = (g \circ f)(x) = x$
3. $(f \circ g)(x) = (g \circ f)(x) = x$
5. $(f \circ g)(x) = 20;\ (g \circ f)(x) = 0.05$
7. $(f \circ g)(x) = |x|;\ (g \circ f)(x) = x$
9. $(f \circ g)(x) = (g \circ f)(x) = x$
11. $(f \circ g)(x) = x^3 - 2x^2 - 4x + 6;$
$(g \circ f)(x) = x^3 - 5x^2 + 3x + 8$
13. $f(x) = x^5;\ g(x) = 4 + 3x$
15. $f(x) = \dfrac{1}{x};\ g(x) = (x - 2)^4$
17. $f(x) = \dfrac{x - 1}{x + 1};\ g(x) = x^3$
19. $f(x) = x^6;\ g(x) = \dfrac{2 + x^3}{2 - x^3}$
21. $f(x) = \sqrt{x};\ g(x) = \dfrac{x - 5}{x + 2}$
23. $f(x) = x^3 - 5x^2 + 3x - 1;\ g(x) = x + 2$
25. $f(x) = 2(x - 20)$
27. $\{(8, 7), (8, -2), (-4, 3), (-8, 8)\}$

29. $\{(-1, -1), (4, -3)\}$
31. $x = 4y - 5$
33. $y^3x = -5$
35.

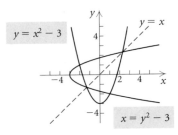

37.

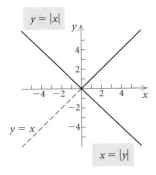

39. Yes   **41.** No   **43.** No   **45.** Yes   **47.** Yes   **49.** No
51. No   **53.** Yes
55. **(a)** One-to-one; **(b)** $f^{-1}(x) = x - 4$
57. **(a)** One-to-one; **(b)** $f^{-1}(x) = \dfrac{x + 1}{2}$
59. **(a)** One-to-one; **(b)** $f^{-1}(x) = \dfrac{4}{x} - 7$
61. **(a)** One-to-one; **(b)** $f^{-1}(x) = \dfrac{3x + 4}{x - 1}$
63. **(a)** One-to-one; **(b)** $f^{-1}(x) = \sqrt[3]{x + 1}$
65. **(a)** Not one-to-one; **(b)** does not have an inverse that is a function
67. **(a)** One-to-one; **(b)** $f^{-1}(x) = \sqrt{\dfrac{x + 2}{5}},\ x \geq -2$
69. **(a)** One-to-one; **(b)** $f^{-1}(x) = x^2 - 1,\ x \geq 0$
71. $\frac{1}{3}x$ **73.** $-x$ **75.** $x^3 + 5$
77. $(f^{-1} \circ f)(x) = f^{-1}(f(x))$
$$= f^{-1}\!\left(\tfrac{7}{8}x\right)$$
$$= \tfrac{8}{7}\!\left(\tfrac{7}{8}x\right) = x;$$
$(f \circ f^{-1})(x) = f(f^{-1}(x))$
$$= f\!\left(\tfrac{8}{7}x\right)$$
$$= \tfrac{7}{8}\!\left(\tfrac{8}{7}x\right) = x$$

79. $(f^{-1} \circ f)(x) = f^{-1}\left(\dfrac{1-x}{x}\right)$

$$= \dfrac{1}{\dfrac{1-x}{x} + 1}$$

$$= \dfrac{1}{\dfrac{1}{x}} = x;$$

$$(f \circ f^{-1})(x) = f\left(\dfrac{1}{x+1}\right)$$

$$= \dfrac{1 - \dfrac{1}{x+1}}{\dfrac{1}{x+1}}$$

$$= \dfrac{\dfrac{x+1-1}{x+1}}{\dfrac{1}{x+1}} = x$$

81. 5; a

83. (a) 36, 40, 44, 52, 60; **(b)** $g^{-1}(x) = \dfrac{x}{2} - 12$;

(c) 6, 8, 10, 14, 18

85.

$$y_1 = 0.8x + 1.7,$$
$$y_2 = \dfrac{x - 1.7}{0.8}$$

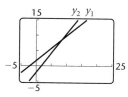

Domain and range of both f and f^{-1}: all real numbers

87.

$$y_1 = \tfrac{1}{2}x - 4,$$
$$y_2 = 2x + 8$$

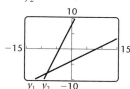

Domain and range of both f and f^{-1}: all real numbers

89.

$$y_1 = \sqrt{x - 3},$$
$$y_2 = x^2 + 3, \, x \geqslant 0$$

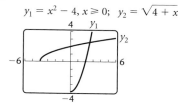

Domain of f: $[3, \infty]$, range of f: $[0, \infty)$; domain of f^{-1}: $[0, \infty)$, range of f^{-1}: $[3, \infty)$

91.

$$y_1 = x^2 - 4, \, x \geqslant 0; \quad y_2 = \sqrt{4 + x}$$

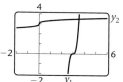

Domain of f: $[0, \infty)$, range of f: $[-4, \infty)$; domain of f^{-1}: $[-4, \infty)$, range of f^{-1}: $[0, \infty)$

93.

$$y_1 = (3x - 9)^3, \quad y_2 = \dfrac{\sqrt[3]{x} + 9}{3}$$

Domain and range of both f and f^{-1}: all real numbers

95. (a) 0.5, 11.5, 22.5, 55.5, 72;

(b) $D^{-1}(r) = \dfrac{10r - 5}{11}$; the speed, in miles per hour, that the car is traveling when the reaction distance is r feet;

(c)

$$y_1 = \dfrac{11x + 5}{10}, \quad y_2 = \dfrac{10x - 5}{11}$$

97. Discussion and Writing

99. [1.1]

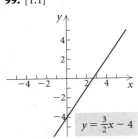

$$y = \frac{3}{2}x - 4$$

100. [1.1]

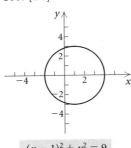

$$(x-1)^2 + y^2 = 9$$

13.

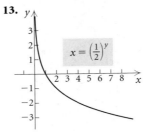

$$x = \left(\frac{1}{2}\right)^y$$

15.

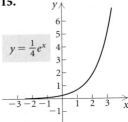

$$y = \frac{1}{4}e^x$$

17.

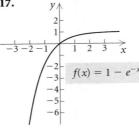

$$f(x) = 1 - e^{-x}$$

19. Shift the graph of $y = 2^x$ left 1 unit.

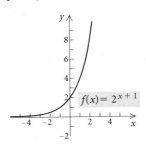

$$f(x) = 2^{x+1}$$

101. [3.4]

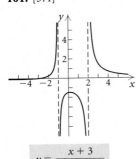

$$y = \frac{x + 3}{x^2 - x - 2}$$

102. [2.4]

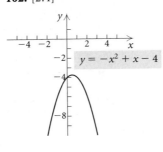

$$y = -x^2 + x - 4$$

103. $f(x) = x^2 - 3$, for inputs $x \geq 0$; $f^{-1}(x) = \sqrt{x + 3}$, for inputs $x \geq -3$ **105.** Answers may vary. $f(x) = 3/x$, $f(x) = 1 - x$, $f(x) = x$

Exercise Set 4.2

1. 54.5982 **3.** 0.0856

5.

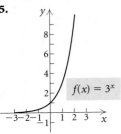

$$f(x) = 3^x$$

7.

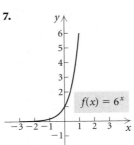

$$f(x) = 6^x$$

9.

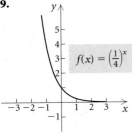

$$f(x) = \left(\frac{1}{4}\right)^x$$

11.

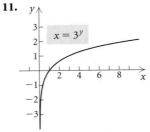

$$x = 3^y$$

21. Shift the graph of $y = 2^x$ down 3 units.

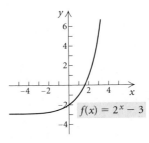

$$f(x) = 2^x - 3$$

23. Reflect the graph of $y = 3^x$ across the y-axis, then across the x-axis, and then shift it up 4 units.

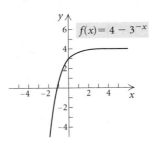

$$f(x) = 4 - 3^{-x}$$

25. Shift the graph of $y = \left(\frac{3}{2}\right)^x$ right 1 unit.

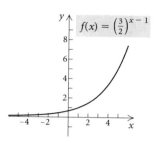

$f(x) = \left(\frac{3}{2}\right)^{x-1}$

27. Shift the graph of $y = 2^x$ left 3 units, and then down 5 units.

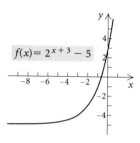

$f(x) = 2^{x+3} - 5$

29. Shrink the graph of $y = e^x$ horizontally.

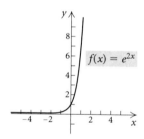

$f(x) = e^{2x}$

31. Shift the graph of $y = e^x$ left 1 unit and reflect it across the y-axis.

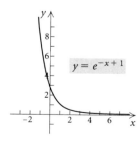

$y = e^{-x+1}$

33. Reflect the graph of $y = e^x$ across the y-axis, then across the x-axis, then shift it up 1 unit, and then stretch it vertically.

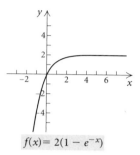

$f(x) = 2(1 - e^{-x})$

35. (a) 799,053; **(b)** 1,363,576
37. (a) 350,000; 233,333; 69,136; 6070;
(b)

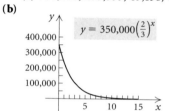

$y = 350,000\left(\frac{2}{3}\right)^x$

39. $5800, $4640, $3712, $1900.54, $622.77
41. About 71.5 billion ft^3, 78.2 billion ft^3
43. About 63%   **45.** (c)   **47.** (a)   **49.** (l)   **51.** (g)
53. (i)   **55.** (k)   **57.** (m)   **59.** Discussion and Writing
61. Discussion and Writing   **63.** [3.1] -1, 0, 1
64. [3.1] -4, 0, 3   **65.** [3.1] -8, 0, 2
66. [3.1] $\dfrac{5 \pm \sqrt{97}}{6}$ **67.** π^7; 70^{80}

Exercise Set 4.3

1.

3.

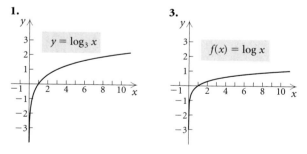

$y = \log_3 x$

$f(x) = \log x$

5. 4   **7.** 3   **9.** -3 **11.** -2 **13.** 0   **15.** 1
17. $\log_{10} 1000 = 3$ **19.** $\log_8 2 = \frac{1}{3}$ **21.** $\log_e t = 3$
23. $\log_e 7.3891 = 2$ **25.** $\log_p 3 = k$ **27.** $5^1 = 5$
29. $10^{-2} = 0.01$ **31.** $e^{3.4012} = 30$ **33.** $a^{-x} = M$
35. $a^x = T^3$ **37.** 0.4771   **39.** 2.7259   **41.** -0.2441
43. Does not exist   **45.** 0.6931   **47.** 6.6962
49. Does not exist   **51.** 3.3219   **53.** -0.2614

55. 0.7384
57. Shift the graph of $y = \log_2 x$ left 3 units. Domain: $(-3, \infty)$; vertical asymptote: $x = -3$;

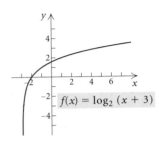

$f(x) = \log_2 (x + 3)$

59. Shift the graph of $y = \log_3 x$ down 1 unit. Domain: $(0, \infty)$; vertical asymptote: $x = 0$;

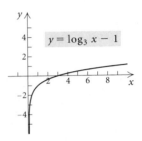

$y = \log_3 x - 1$

61. Stretch the graph of $y = \ln x$ vertically. Domain: $(0, \infty)$; vertical asymptote: $x = 0$;

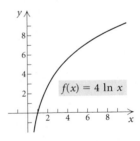

$f(x) = 4 \ln x$

63. Reflect the graph of $y = \ln x$ across the x-axis and shift it up 2 units. Domain: $(0, \infty)$; vertical asymptote: $x = 0$;

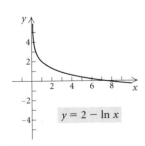

$y = 2 - \ln x$

65.

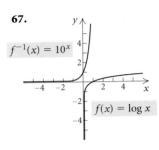

$f(x) = 3^x$
$f^{-1}(x) = \log_3 x$

67.

$f^{-1}(x) = 10^x$
$f(x) = \log x$

69. (a) 2.3 ft/sec; **(b)** 3.0 ft/sec; **(c)** 2.2 ft/sec; **(d)** 1.9 ft/sec; **(e)** 1.7 ft/sec
71. (a) 78%; **(b)** 67.5%, 57%
73. (a) 10^{-7}; **(b)** 4.0×10^{-6}; **(c)** 6.3×10^{-4}; **(d)** 1.6×10^{-5}
75. (a) 34 decibels; **(b)** 64 decibels; **(c)** 60 decibels; **(d)** 90 decibels   **77.** Discussion and Writing
79. [3.2] -280 **80.** [3.2] -4 **81.** [3.3] $f(x) = x^3 - 7x$
82. [3.3] $f(x) = x^3 - x^2 + 16x - 16$ **83.** 3   **85.** $(0, \infty)$
87. $(-\infty, 0) \cup (0, \infty)$ **89.** $\left(-\frac{5}{2}, -2\right)$ **91.** (d)   **93.** (b)

Exercise Set 4.4

1. $\log_3 81 + \log_3 27$ **3.** $\log_5 5 + \log_5 125$
5. $\log_t 8 + \log_t Y$ **7.** $3 \log_b t$ **9.** $8 \log y$
11. $-6 \log_c K$ **13.** $\log_t M - \log_t 8$
15. $\log_a x - \log_a y$
17. $\log_a 6 + \log_a x + 5 \log_a y + 4 \log_a z$
19. $2 \log_b p + 5 \log_b q - 4 \log_b m - 9$
21. $3 \log_a x - \frac{5}{2} \log_a p - 4 \log_a q$
23. $2 \log_a m + 3 \log_a n - \frac{3}{4} - \frac{5}{4} \log_a b$
25. $\log_a 150$ **27.** $\log 100 = 2$
29. $\log_a x^{-5/2} y^4$, or $\log_a \dfrac{y^4}{x^{5/2}}$ **31.** $\ln x$ **33.** $\ln (x - 2)$
35. $\ln \dfrac{x}{(x^2 - 25)^3}$ **37.** $\ln \dfrac{2^{11/5} x^9}{y^8}$ **39.** -0.5108
41. -1.6094 **43.** $\frac{1}{2}$ **45.** 2.6094 **47.** 4.3174
49. 3   **51.** $|x - 4|$ **53.** $4x$ **55.** w **57.** $8t$
59. Discussion and Writing   **61.** [2.2] $31 - 22i$
62. [2.2] $\frac{1}{2} - \frac{1}{2} i$ **63.** [2.3] $\left(-\frac{1}{2}, 0\right)$, $(7, 0)$; $-\frac{1}{2}$, 7
64. [3.3] $(1, 0)$; 1   **65.** 4   **67.** $\log_a (x^3 - y^3)$
69. $\frac{1}{2} \log_a (x - y) - \frac{1}{2} \log_a (x + y)$
71. 7   **73.** True   **75.** True   **77.** True   **79.** -2
81. 3   **83.** $\log_a \left(\dfrac{x + \sqrt{x^2 - 5}}{5} \cdot \dfrac{x - \sqrt{x^2 - 5}}{x - \sqrt{x^2 - 5}} \right)$

$$= \log_a \dfrac{5}{5(x - \sqrt{x^2 - 5})}$$
$$= -\log_a (x - \sqrt{x^2 - 5})$$

Exercise Set 4.5

1. 4   **3.** $\frac{3}{2}$ **5.** 5.044   **7.** $\frac{5}{2}$ **9.** $-3, \frac{1}{2}$ **11.** 0.959
13. 6.908   **15.** 84.191   **17.** -1.710 **19.** 2.844
21. $-1.567, 1.567$ **23.** 0.347   **25.** 625   **27.** 0.0001
29. e **31.** $\frac{22}{3}$ **33.** 10   **35.** $\frac{1}{63}$ **37.** 5   **39.** $\frac{21}{8}$
41. 0.367   **43.** 0.621   **45.** -1.532 **47.** 7.062
49. 2.444   **51.** (4.093, 0.786)   **53.** (7.586, 6.684)
55. Discussion and Writing
56. [2.4] **(a)** (3, 1); **(b)** $x = 3$; **(c)** maximum: 1 when $x = 3$
57. [2.4] **(a)** (0, -6); **(b)** $x = 0$; **(c)** minimum: -6 when $x = 0$ **58.** [2.4] **(a)** (2, 4); **(b)** $x = 2$; **(c)** minimum: 4 when $x = 2$ **59.** [2.4] **(a)** ($-1, -5$); **(b)** $x = -1$; **(c)** maximum: -5 when $x = -1$ **61.** 10
63. 1, e^4 or 1, 54.598   **65.** $\frac{1}{3}$, 27   **67.** 1, e^2 or 1, 7.389
69. 0, 0.431   **71.** $-9, 9$ **73.** e^{-2}, e^2 or 0.135, 7.389
75. $\frac{7}{4}$ **77.** 5   **79.** $a = \frac{2}{3}b$ **81.** 88

Exercise Set 4.6

1. (a) $P(t) = 6.0e^{0.013t}$; **(b)** 6.5 billion, 6.9 billion; **(c)** in 22.1 yr; **(d)** 53.3 yr
3. (a) 36.5 yr; **(b)** 0.2% per year; **(c)** 21.0 yr; **(d)** 138.6 yr; **(e)** 3.4% per year; **(f)** 2.2% per year   **5.** About 322 yr
7. (a) $P(t) = 10,000e^{0.054t}$; **(b)** \$10,555; \$11,140; \$13,100; \$17,160; **(c)** 12.8 yr   **9.** About 5135 yr
11. (a) 23.1% per minute; **(b)** 3.15% per year; **(c)** 7.2 days; **(d)** 11 yr; **(e)** 2.8% per year; **(f)** 0.015% per year; **(g)** 0.003% per year
13. (a) $k \approx 0.016$; $P(t) = 80e^{-0.016t}$; **(b)** about 61 lb per person; **(c)** 86.6 yr
15. (a) 96; **(b)** 286, 1005, 1719, 1971, 1997; **(c)** as $t \to \infty$, $N(t) \to 2000$; the number approaches 2000 but never actually reaches it.
17. 400, 520, 1214, 2059, 2396, 2478
19. (a) $y = 4.195491964(1.025306189)^x$, or $y = 4.195491964e^{0.024991289x}$; since $r \approx 0.9954$, the function is a good fit.
(b)

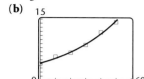

(c) 4.8 million, 7.8 million, 51.1 million
21. (a)

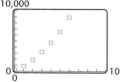

(b) linear: $y = 1429.214286x - 530.0714286$, $r^2 = 0.9641$; quadratic: $y = 158.0952381x^2 + 480.6428571x + 260.4047619$, $r^2 = 0.9995$; exponential: $y = 445.8787388(1.736315606)^x$, $r^2 = 0.9361$; according to r^2, the quadratic has the best fit;
(c)

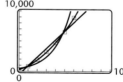

(d) linear: \$19,479 million, or \$19.479 billion; quadratic: \$37,976 million, or \$37.976 billion; exponential: \$1,009,295 million, or \$1009.295 billion; the revenue from the quadratic model seems most realistic. Answers may vary.
23. Discussion and Writing   **25.** [1.2] 0; (0, 6)
26. [1.2] $\frac{3}{10}$; $\left(0, -\frac{7}{5}\right)$ **27.** [1.2] 2; $\left(0, -\frac{3}{13}\right)$
28. [1.2] Slope not defined; no y-intercept
29. \$14,182.70   **31.** \$166.16   **33.** 46.7°F
35. $t = -\frac{L}{R}\left[\ln\left(1 - \frac{iR}{V}\right)\right]$ **37.** Linear

Review Exercises, Chapter 4

1. [4.1] **(a)** Domain of $f \circ g$, $\left\{x \,\middle|\, x \neq \frac{3}{2}\right\}$; domain of $g \circ f$, $\{x \mid x \neq 0\}$; **(b)** $(f \circ g)(x) = \frac{4}{(3 - 2x)^2}$; $(g \circ f)(x) = 3 - \frac{8}{x^2}$
2. [4.1] **(a)** Domain of $f \circ g$ and $g \circ f$, all real numbers; **(b)** $(f \circ g)(x) = 12x^2 - 4x - 1$; $(g \circ f)(x) = 6x^2 + 8x - 1$
3. [4.1] $f(x) = \sqrt{x}$, $g(x) = 5x + 2$. Answers may vary.
4. [4.1] $f(x) = 4x^2 + 9$, $g(x) = 5x - 1$. Answers may vary.
5. [4.1] $\{(-2.7, 1.3), (-3, 8), (3, -5), (-3, 6), (-5, 7)\}$
6. [4.1] **(a)** $x = 3y^2 + 2y - 1$; **(b)** $0.8y^3 - 5.4x^2 = 3y$
7. [4.1] **(a)** Yes; **(b)** $f^{-1}(x) = x^2 + 6$, $x \geq 0$
8. [4.1] **(a)** Yes; **(b)** $f^{-1}(x) = \sqrt[3]{x + 8}$ **9.** [4.1] **(a)** No
10. [4.1] **(a)** Yes; **(b)** $f^{-1}(x) = \ln x$ **11.** [4.1] 657
12. [4.2] (c)   **13.** [4.3] (a)   **14.** [4.3] (b)   **15.** [4.2] (f)
16. [4.2] (e)   **17.** [4.3] (d)   **18.** [4.3] $4^2 = x$
19. [4.3] $\log_e 80 = x$ **20.** [4.5] 16   **21.** [4.5] $\frac{1}{5}$
22. [4.5] 4.382   **23.** [4.5] 2   **24.** [4.5] $\frac{1}{2}$
25. [4.5] 5   **26.** [4.5] 4   **27.** [4.5] 9   **28.** [4.5] 1
29. [4.5] 3.912   **30.** [4.4] $\log_b \frac{x^3\sqrt{z}}{y^4}$
31. [4.4] $\ln(x^2 - 4)$ **32.** [4.4] $\frac{1}{4}\ln w + \frac{1}{2}\ln r$
33. [4.4] $\frac{2}{3}\log M - \frac{1}{3}\log N$ **34.** [4.4] 0.477
35. [4.4] 1.699   **36.** [4.4] -0.699 **37.** [4.4] 0.233
38. [4.4] $-5k$ **39.** [4.4] $-6t$ **40.** [4.6] 8.1 yr
41. [4.6] 2.3%   **42.** [4.6] About 2623 yr   **43.** [4.3] 5.6
44. [4.3] 30 decibels

45. [4.6] **(a)** $P(t) = 11e^{0.012t}$; **(b)** 11.8 million, 14.3 million;
(c) in 68.4 yr; **(d)** 57.8 yr
46. [4.6] **(a)** $k = 0.304$; **(b)** $P(t) = 7e^{0.304t}$;
(c) 292.4 billion, 727.9 billion; **(d)** 2006
47. [4.3] **(a)** 2.7 ft/sec; **(b)** 8,553,143
48. [4.6] **(a)** $y = 11.96557466(1.00877703)^x$;
(b)

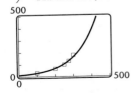

$y = 11.96557466(1.00877703)^x$

(c) 44, 255, 394; **(d)** Comparing the coefficient of
correlation values (power: 0.9739; exponential: 0.9955), we
see that the exponential function is the better fit.
49. [4.1] No
50. [4.2], [4.3] **(a)**

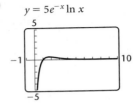

$y = 5e^{-x} \ln x$

(b) relative maximum: 0.486 at $x = 1.763$
51. Discussion and Writing [4.4] By the product rule,
$\log_2 x + \log_2 5 = \log_2 5x$, not $\log_2 (x + 5)$. Also,
substituting various numbers for x shows that both sides
of the inequality are indeed unequal. You could also graph
each side and show that the graphs do not coincide.
52. Discussion and Writing [4.1] The inverse of a function
$f(x)$ is written $f^{-1}(x)$, whereas $[f(x)]^{-1}$ means $\dfrac{1}{f(x)}$.
53. [4.5] $\frac{1}{64}$, 64   **54.** [4.5] 1   **55.** [4.5] 16
56. [4.3] $(1, \infty)$

Test, Chapter 4

1. [4.1] $(f \circ g)(x) = x^2 - 4$; $(g \circ f)(x) = x^2 - 10x + 26$
2. [4.1] $\{(5, -2), (3, 4), (-1, 0), (-3, -6)\}$ **3.** [4.1] No
4. [4.1] $f^{-1}(x) = \sqrt[3]{x - 1}$
5. [4.2]
6. [4.3]

$f(x) = e^x - 3$

$f(x) = \ln(x + 2)$

7. [4.3] $x = e^4$ **8.** [4.3] $x = \log_3 5.4$ **9.** [4.5] $\frac{1}{2}$
10. [4.5] 1   **11.** [4.5] 1   **12.** [4.5] 4.174
13. [4.4] $\frac{2}{5} \ln x + \frac{1}{5} \ln y$ **14.** [4.4] 0.656   **15.** [4.4] $-4t$
16. [4.6] 0.0154   **17.** [4.6] **(a)** 4.5%; **(b)** $P(t) = 1000e^{0.045t}$;
(c) \$1433.33; **(d)** 15.4 yr   **18.** [4.5] $\frac{27}{8}$

Chapter 5

Exercise Set 5.1

1. $\sin \phi = \dfrac{15}{17}$, $\cos \phi = \dfrac{8}{17}$, $\tan \phi = \dfrac{15}{8}$, $\csc \phi = \dfrac{17}{15}$,
$\sec \phi = \dfrac{17}{8}$, $\cot \phi = \dfrac{8}{15}$

3. $\sin \alpha = \dfrac{\sqrt{3}}{2}$, $\cos \alpha = \dfrac{1}{2}$, $\tan \alpha = \sqrt{3}$, $\csc \alpha = \dfrac{2\sqrt{3}}{3}$,
$\sec \alpha = 2$, $\cot \alpha = \dfrac{\sqrt{3}}{3}$

5. $\sin \phi = \dfrac{7\sqrt{65}}{65}$, $\cos \phi = \dfrac{4\sqrt{65}}{65}$, $\tan \phi = \dfrac{7}{4}$,
$\csc \phi = \dfrac{\sqrt{65}}{7}$, $\sec \phi = \dfrac{\sqrt{65}}{4}$, $\cot \phi = \dfrac{4}{7}$

7. $\cos \theta = \dfrac{7}{25}$, $\tan \theta = \dfrac{24}{7}$, $\csc \theta = \dfrac{25}{24}$, $\sec \theta = \dfrac{25}{7}$,
$\cot \theta = \dfrac{7}{24}$

9. $\sin \phi = \dfrac{2\sqrt{5}}{5}$, $\cos \phi = \dfrac{\sqrt{5}}{5}$, $\csc \phi = \dfrac{\sqrt{5}}{2}$, $\sec \phi = \sqrt{5}$,
$\cot \phi = \dfrac{1}{2}$

11. $\sin \theta = \dfrac{2}{3}$, $\cos \theta = \dfrac{\sqrt{5}}{3}$, $\tan \theta = \dfrac{2\sqrt{5}}{5}$,
$\sec \theta = \dfrac{3\sqrt{5}}{5}$, $\cot \theta = \dfrac{\sqrt{5}}{2}$
13. $\dfrac{\sqrt{2}}{2}$ **15.** 2   **17.** $\dfrac{\sqrt{3}}{3}$ **19.** $\dfrac{1}{2}$
21. 127.3 ft   **23.** 9.72°   **25.** 35.01°   **27.** 3.03°
29. 49.65°   **31.** 17°36′   **33.** 83°1′30″   **35.** 11°45′
37. 47°49′36″   **39.** 0.6293   **41.** 0.0737
43. 1.2765   **45.** 0.7621   **47.** 0.9336
49. 12.4288   **51.** 1.0000   **53.** 1.7032   **55.** 30.8°
57. 12.5°   **59.** 64.4°   **61.** 46.5°   **63.** 25.2°   **65.** 38.6°
67. 45°   **69.** 60°   **71.** 45°
73. $\cos 20° = \sin 70° = \dfrac{1}{\sec 20°}$
75. $\tan 52° = \cot 38° = \dfrac{1}{\cot 52°}$
77. $\sin 25° \approx 0.4226$, $\cos 25° \approx 0.9063$, $\tan 25° \approx 0.4663$,
$\csc 25° \approx 2.366$, $\sec 25° \approx 1.103$, $\cot 25° \approx 2.145$

79. $\sin 8° = q$, $\cos 8° = p$, $\tan 8° = \dfrac{1}{r}$, $\csc 8° = \dfrac{1}{q}$,

$\sec 8° = \dfrac{1}{p}$, $\cot 8° = r$ **81.** Discussion and Writing

83. [4.2]

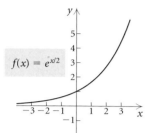

$f(x) = e^{x/2}$

84. [4.2]

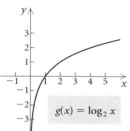

$f(x) = 2^{-x}$

85. [4.3]

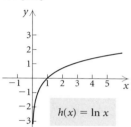

$h(x) = \ln x$

86. [4.3]

$g(x) = \log_2 x$

87. 0.6534
89. Area $= \frac{1}{2} ab$. But $a = c \sin A$, so Area $= \frac{1}{2} bc \sin A$.

Exercise Set 5.2

1. $F = 60°$, $d = 3$, $f \approx 5.2$
3. $A = 22.7°$, $a \approx 52.7$, $c \approx 136.6$
5. $P = 47°38'$, $n \approx 34.4$, $p \approx 25.4$
7. $B = 2°17'$, $b \approx 0.39$, $c \approx 9.74$
9. $A \approx 77.2°$, $B \approx 12.8°$, $a \approx 439$
11. $B = 42.42°$, $a \approx 35.7$, $b \approx 32.6$
13. $B = 55°$, $a \approx 28.0$, $c \approx 48.8$
15. $A \approx 62.4°$, $B \approx 27.6°$, $a \approx 3.56$
17. About 62.2 ft   **19.** About 2.5 ft   **21.** About 606 ft
23. About 96.7 cm   **25.** About 599 ft   **27.** About 8 km
29. About 275 ft   **31.** About 24 km
33. Discussion and Writing
35. [1.1] $10\sqrt{2}$, or about 14.142   **36.** [1.1] $3\sqrt{10}$, or about
9.487   **37.** [4.3] $10^{-3} = 0.001$ **38.** [4.3] $\ln t = 4$
39. 3.3   **41.** Cut so that $\theta = 79.38°$ **43.** $\theta \approx 27°$

Exercise Set 5.3

1. III   **3.** III   **5.** I   **7.** III   **9.** II   **11.** II
13. 434°, 794°, −286°, −646°
15. 475.3°, 835.3°, −244.7°, −604.7°

17. 180°, 540°, −540°, −900°   **19.** 72.89°, 162.89°
21. 77°56'46", 167°56'46"   **23.** 44.8°, 134.8°
25. $\sin \beta = \dfrac{5}{13}$, $\cos \beta = -\dfrac{12}{13}$, $\tan \beta = -\dfrac{5}{12}$, $\csc \beta = \dfrac{13}{5}$,

$\sec \beta = -\dfrac{13}{12}$, $\cot \beta = -\dfrac{12}{5}$

27. $\sin \phi = -\dfrac{2\sqrt{7}}{7}$, $\cos \phi = -\dfrac{\sqrt{21}}{7}$, $\tan \phi = \dfrac{2\sqrt{3}}{3}$,

$\csc \phi = -\dfrac{\sqrt{7}}{2}$, $\sec \phi = -\dfrac{\sqrt{21}}{3}$, $\cot \phi = \dfrac{\sqrt{3}}{2}$

29. $\sin \theta = -\dfrac{2\sqrt{13}}{13}$, $\cos \theta = \dfrac{3\sqrt{13}}{13}$, $\tan \theta = -\dfrac{2}{3}$

31. $\sin \theta = \dfrac{5\sqrt{41}}{41}$, $\cos \theta = \dfrac{4\sqrt{41}}{41}$, $\tan \theta = \dfrac{5}{4}$

33. $\cos \theta = -\dfrac{2\sqrt{2}}{3}$, $\tan \theta = \dfrac{\sqrt{2}}{4}$, $\csc \theta = -3$,

$\sec \theta = -\dfrac{3\sqrt{2}}{4}$, $\cot \theta = 2\sqrt{2}$

35. $\sin \theta = -\dfrac{\sqrt{5}}{5}$, $\cos \theta = \dfrac{2\sqrt{5}}{5}$, $\tan \theta = -\dfrac{1}{2}$,

$\csc \theta = -\sqrt{5}$, $\sec \theta = \dfrac{\sqrt{5}}{2}$

37. $\sin \phi = -\dfrac{4}{5}$, $\tan \phi = -\dfrac{4}{3}$, $\csc \phi = -\dfrac{5}{4}$,

$\sec \phi = \dfrac{5}{3}$, $\cot \phi = -\dfrac{3}{4}$

39. 30°; $-\dfrac{\sqrt{3}}{2}$ **41.** 45°; 1   **43.** 0   **45.** 45°; $-\dfrac{\sqrt{2}}{2}$

47. 30°; 2   **49.** 30°; $\sqrt{3}$ **51.** 30°; $-\dfrac{\sqrt{3}}{3}$

53. Undefined   **55.** −1   **57.** 60°; $\sqrt{3}$ **59.** 45°; $\dfrac{\sqrt{2}}{2}$

61. 45°; $-\sqrt{2}$ **63.** 1   **65.** 0   **67.** 0   **69.** 0
71. Positive: cos, sec; negative: sin, csc, tan, cot
73. Positive: tan, cot; negative: sin, csc, cos, sec
75. Positive: sin, csc; negative: cos, sec, tan, cot
77. Positive: all
79. $\sin 319° = -0.6561$, $\cos 319° = 0.7547$,
$\tan 319° = -0.8693$, $\csc 319° \approx -1.5242$,
$\sec 319° \approx 1.3250$, $\cot 319° \approx -1.1504$
81. $\sin 115° = 0.9063$, $\cos 115° = -0.4226$,
$\tan 115° = -2.1445$, $\csc 115° \approx 1.1034$,
$\sec 115° \approx -2.3663$, $\cot 115° \approx -0.4663$
83. East: about 130 km; south: 75 km
85. About 223 km   **87.** −1.1585   **89.** −1.4910
91. 0.8771   **93.** 0.4352   **95.** 0.9563   **97.** 2.9238
99. 275.4°   **101.** 200.1°   **103.** 288.1°   **105.** 72.6°
107. Discussion and Writing

109. [3.4]

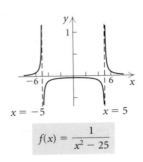

$$f(x) = \frac{1}{x^2 - 25}$$

110. [3.1]

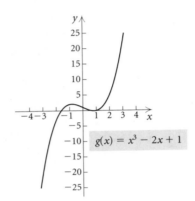

$$g(x) = x^3 - 2x + 1$$

111. [1.2], [3.4] Domain: $\{x \mid x \neq -2\}$; range: $\{x \mid x \neq 1\}$
112. [1.2], [3.1] Domain: $\left\{x \mid x \neq -\frac{3}{2} \text{ and } x \neq 5\right\}$; range:
all real numbers   **113.** 19.625 in.

Exercise Set 5.4

1.   **3.**

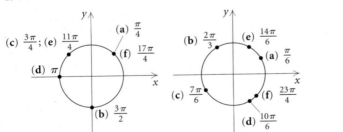

5. $M: \dfrac{2\pi}{3}, -\dfrac{4\pi}{3}$; $N: \dfrac{3\pi}{2}, -\dfrac{\pi}{2}$; $P: \dfrac{5\pi}{4}, -\dfrac{3\pi}{4}$; $Q: \dfrac{11\pi}{6}$,
$-\dfrac{\pi}{6}$

7.

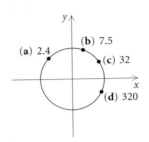

9. $\dfrac{9\pi}{4}, -\dfrac{7\pi}{4}$ **11.** $\dfrac{19\pi}{6}, -\dfrac{5\pi}{6}$

13. Complement: $\dfrac{\pi}{6}$; supplement: $\dfrac{2\pi}{3}$

15. Complement: $\dfrac{\pi}{8}$; supplement: $\dfrac{5\pi}{8}$

17. $\dfrac{5\pi}{12}$ **19.** $\dfrac{10\pi}{9}$ **21.** $-\dfrac{214.6\pi}{180}$ **23.** $-\pi$

25. 4.19   **27.** -1.05 **29.** 2.06   **31.** 0.02   **33.** $-135°$
35. 1440°   **37.** 57.30°   **39.** 134.47°

41. $0° = 0$ radian, $30° = \dfrac{\pi}{6}$, $45° = \dfrac{\pi}{4}$, $60° = \dfrac{\pi}{3}$,

$90° = \dfrac{\pi}{2}$, $135° = \dfrac{3\pi}{4}$, $180° = \pi$, $225° = \dfrac{5\pi}{4}$, $270° = \dfrac{3\pi}{2}$,

$315° = \dfrac{7\pi}{4}$, $360° = 2\pi$ **43.** 1.1, 63°   **45.** 3.2 yd

47. $\dfrac{5\pi}{3}$, or about 5.24   **49.** 3150 $\dfrac{\text{cm}}{\text{min}}$

51. About 18,852 revolutions per hour   **53.** 1047 mph
55. 10 mph   **57.** About 202   **59.** Left to the student
61. Discussion and Writing   **63.** [4.5] 4   **64.** [4.5] 9.21
65. [4.5] 343   **66.** [4.5] $\frac{101}{97}$ **67.** 111.7 km; 69.8 mi
69. (a) 5°37′30″; (b) 19°41′15″   **71.** 1.676 radians/sec
73. 1.46 nautical miles

Exercise Set 5.5

1. (a) $\left(-\dfrac{3}{4}, -\dfrac{\sqrt{7}}{4}\right)$; (b) $\left(\dfrac{3}{4}, \dfrac{\sqrt{7}}{4}\right)$; (c) $\left(\dfrac{3}{4}, -\dfrac{\sqrt{7}}{4}\right)$

3. (a) $\left(\dfrac{2}{5}, \dfrac{\sqrt{21}}{5}\right)$; (b) $\left(-\dfrac{2}{5}, -\dfrac{\sqrt{21}}{5}\right)$; (c) $\left(-\dfrac{2}{5}, \dfrac{\sqrt{21}}{5}\right)$

5. $\left(\dfrac{\sqrt{2}}{2}, -\dfrac{\sqrt{2}}{2}\right)$ **7.** 0   **9.** $\sqrt{3}$ **11.** 0   **13.** $-\dfrac{\sqrt{3}}{2}$

15. 1   **17.** $\dfrac{\sqrt{3}}{2}$ **19.** $-\dfrac{\sqrt{2}}{2}$ **21.** 0   **23.** 0

25. 0.4816   **27.** 1.3065   **29.** -2.1599 **31.** 1
33. -1.1747 **35.** -1 **37.** -0.7071 **39.** 0
41. 0.8391

43. (a)

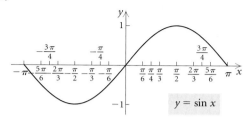

$y = \sin x$

(b)

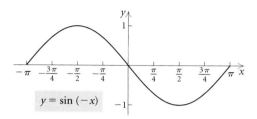

$y = \sin(-x)$

(c) same as (b); **(d)** the same
45. (a) See Exercise 43(a);
(b)

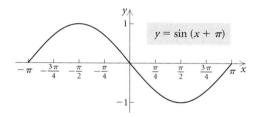

$y = \sin(x + \pi)$

(c) same as (b); **(d)** the same
47. (a)

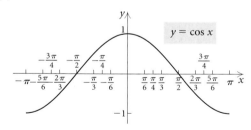

$y = \cos x$

(b)

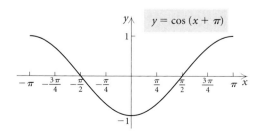

$y = \cos(x + \pi)$

(c) same as (b); **(d)** the same
49. Even: cosine, secant; odd: sine, tangent, cosecant, cotangent   **51.** Positive: I, III; negative: II, IV
53. Positive: I, IV; negative: II, III
55. Domain: $(-\infty, \infty)$; range: $[0, 1]$; period: π; amplitude: $\dfrac{1}{2}$

57. 1   **58.** $\left(-\dfrac{3\pi}{4} + 2k\pi, \dfrac{\pi}{4} + 2k\pi \right), k \in \mathbb{Z}$

59. Discussion and Writing
61. [1.6]

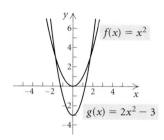

$f(x) = x^2$

$g(x) = 2x^2 - 3$

Stretch the graph of f vertically, then shift it down 3 units.
62. [1.6]

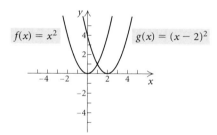

$f(x) = x^2$

$g(x) = (x - 2)^2$

Shift the graph of f right 2 units.
63. [1.6]

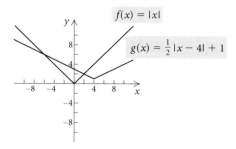

$f(x) = |x|$

$g(x) = \dfrac{1}{2}|x - 4| + 1$

Shift the graph of f to the right 4 units, shrink it vertically, then shift it up 1 unit.

64. [1.6]

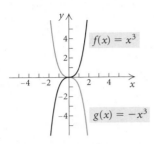

$f(x) = x^3$

$g(x) = -x^3$

Reflect the graph of f across the x-axis.

65. [1.6] $y = -(x - 2)^3 - 1$ **66.** [1.6] $y = \dfrac{1}{4x} + 3$

67. $\cos x$ **69.** $\sin x$ **71.** $\sin x$ **73.** $-\cos x$

75. $-\sin x$ **77.** (a) $\dfrac{\pi}{2} + 2k\pi,\ k \in \mathbb{Z}$; (b) $\pi + 2k\pi,$

$k \in \mathbb{Z}$; (c) $k\pi,\ k \in \mathbb{Z}$

79. $\left[-\dfrac{\pi}{2} + 2k\pi, \dfrac{\pi}{2} + 2k\pi\right],\ k \in \mathbb{Z}$

81. $\left\{x \mid x \neq \dfrac{\pi}{2} + k\pi,\ k \in \mathbb{Z}\right\}$

83.

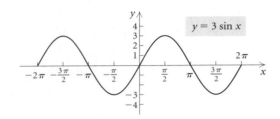

$y = 3 \sin x$

85.

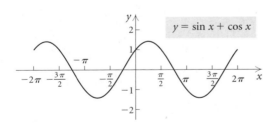

$y = \sin x + \cos x$

87. (a) $\triangle OPA \sim \triangle ODB$;   (b) $\triangle OPA \sim \triangle ODB$;

Thus, $\dfrac{AP}{OA} = \dfrac{BD}{OB}$ $\dfrac{OD}{OP} = \dfrac{OB}{OA}$

$\dfrac{\sin \theta}{\cos \theta} = \dfrac{BD}{1}$ $\dfrac{OD}{1} = \dfrac{1}{\cos \theta}$

$\tan \theta = BD$ $OD = \sec \theta$

(c) $\triangle OAP \sim \triangle ECO$;   (d) $\triangle OAP \sim \triangle ECO$

$\dfrac{OE}{PO} = \dfrac{CO}{AP}$ $\dfrac{CE}{AO} = \dfrac{CO}{AP}$

$\dfrac{OE}{1} = \dfrac{1}{\sin \theta}$ $\dfrac{CE}{\cos \theta} = \dfrac{1}{\sin \theta}$

$OE = \csc \theta$ $CE = \dfrac{\cos \theta}{\sin \theta}$

$CE = \cot \theta$

Exercise Set 5.6

1. Amplitude: 1; period: 2π; phase shift: 0

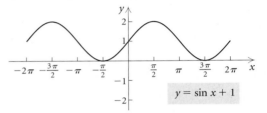

$y = \sin x + 1$

3. Amplitude: 3; period: 2π; phase shift: 0

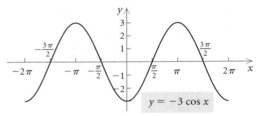

$y = -3 \cos x$

5. Amplitude: 2; period: 4π; phase shift: 0

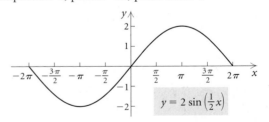

$y = 2 \sin \left(\dfrac{1}{2}x\right)$

7. Amplitude: $\dfrac{1}{2}$; period: 2π; phase shift: $-\dfrac{\pi}{2}$

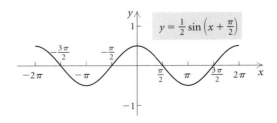

$y = \dfrac{1}{2} \sin \left(x + \dfrac{\pi}{2}\right)$

9. Amplitude: 3; period: 2π; phase shift: π

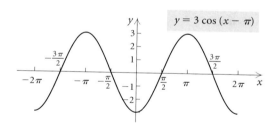

$y = 3 \cos (x - \pi)$

11. Amplitude: $\dfrac{1}{3}$; period: 2π; phase shift: 0

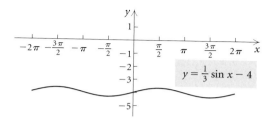

$y = \dfrac{1}{3} \sin x - 4$

13. Amplitude: 1; period: 2π; phase shift: 0

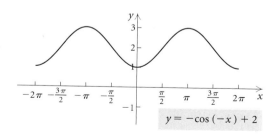

$y = -\cos (-x) + 2$

15. Amplitude: 2; period: 4π; phase shift: π

17. Amplitude: $\dfrac{1}{2}$; period: π; phase shift: $-\dfrac{\pi}{4}$

19. Amplitude: 3; period: 2; phase shift: $\dfrac{3}{\pi}$

21. Amplitude: $\dfrac{1}{2}$; period: 1; phase shift: 0

23. Amplitude: 1; period: 4π; phase shift: π

25. Amplitude: 1; period: 1; phase shift: 0

27. Amplitude: $\dfrac{1}{4}$; period: 2; phase shift: $\dfrac{4}{\pi}$

29. (b)   **31.** (h)   **33.** (a)   **35.** (f)

37. $y = \dfrac{1}{2} \cos x + 1$ **39.** $y = \cos \left(x + \dfrac{\pi}{2} \right) - 2$

41.

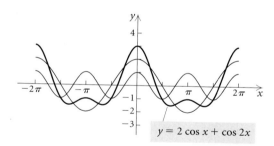

$y = 2 \cos x + \cos 2x$

43.

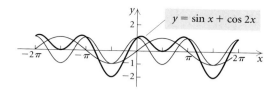

$y = \sin x + \cos 2x$

45.

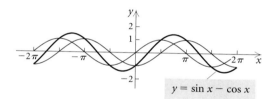

$y = \sin x - \cos x$

47.

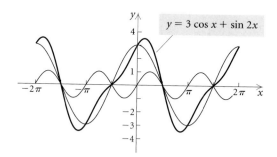

$y = 3 \cos x + \sin 2x$

49.

$y = x + \sin x$

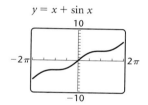

51.

$y = \cos x - x$

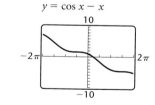

53.

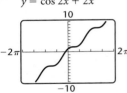

$y = \cos 2x + 2x$

55. $y = 4 \cos 2x - 2 \sin x$

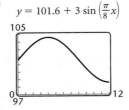

57. $-9.42, -6.28, -3.14, 3.14, 6.28, 9.42$
59. $-3.14, 0, 3.14$
61. (a) $y = 101.6 + 3 \sin \left(\frac{\pi}{8} x \right)$

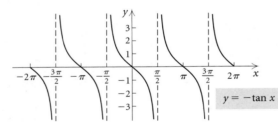

(b) $104.6°, 98.6°$
63. Discussion and Writing **65.** [2.2] 12
66. [2.3] $-2, 3$ **67.** [2.1] (12, 0)
68. [2.3] $(-2, 0), (3, 0)$
69.

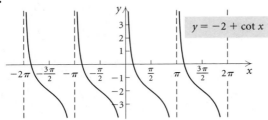

$y = -\tan x$

71.

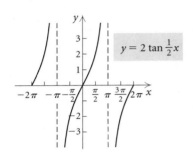

$y = -2 + \cot x$

73.

$y = 2 \tan \frac{1}{2} x$

75.

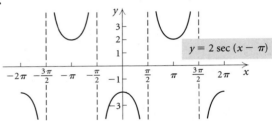

$y = 2 \sec (x - \pi)$

77.

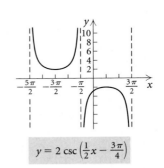

$y = 2 \csc \left(\frac{1}{2} x - \frac{3\pi}{4} \right)$

79. Amplitude: 3000; period: 90; phase shift: 10
81. 4 in.

Review Exercises, Chapter 5

1. [5.1] $\sin \theta = \frac{3\sqrt{73}}{73}$, $\cos \theta = \frac{8\sqrt{73}}{73}$, $\tan \theta = \frac{3}{8}$,

$\csc \theta = \frac{\sqrt{73}}{3}$, $\sec \theta = \frac{\sqrt{73}}{8}$, $\cot \theta = \frac{8}{3}$ **2.** [5.1] $\frac{\sqrt{2}}{2}$

3. [5.1] $\frac{\sqrt{3}}{3}$ **4.** [5.3] $-\frac{\sqrt{2}}{2}$ **5.** [5.3] $\frac{1}{2}$

6. [5.3] Undefined **7.** [5.3] $-\sqrt{3}$ **8.** [5.1] $22°16'12''$
9. [5.1] $47.56°$ **10.** [5.3] 0.4452 **11.** [5.3] 1.1315
12. [5.3] 0.9498 **13.** [5.3] -0.9092 **14.** [5.3] -1.5282
15. [5.3] -0.2778 **16.** [5.3] $205.3°$ **17.** [5.3] $47.2°$
18. [5.1] $60°$ **19.** [5.1] $60°$
20. [5.1] $\sin 30.9° \approx 0.5135$, $\cos 30.9° \approx 0.8581$,
$\tan 30.9° \approx 0.5985$, $\csc 30.9° \approx 1.9474$, $\sec 30.9° \approx 1.1654$,
$\cot 30.9° \approx 1.6709$
21. [5.2] $b \approx 4.5$, $A \approx 58.1°$, $B \approx 31.9°$
22. [5.2] $A = 38.83°$, $b \approx 37.9$, $c \approx 48.6$ **23.** [5.2] 1748 m

24. [5.2] 14 ft **25.** [5.3] $425°, -295°$ **26.** [5.4] $\frac{\pi}{3}, -\frac{5\pi}{3}$

27. [5.3] Complement: $76.6°$; supplement: $166.6°$

28. [5.4] Complement: $\frac{\pi}{3}$; supplement: $\frac{5\pi}{6}$

29. [5.3] $\sin \theta = \frac{3\sqrt{13}}{13}$, $\cos \theta = \frac{-2\sqrt{13}}{13}$, $\tan \theta = -\frac{3}{2}$,

$\csc \theta = \frac{\sqrt{13}}{3}$, $\sec \theta = -\frac{\sqrt{13}}{2}$, $\cot \theta = -\frac{2}{3}$

30. [5.3] $\sin\theta = -\dfrac{2}{3}$, $\cos\theta = -\dfrac{\sqrt{5}}{3}$, $\cot\theta = \dfrac{\sqrt{5}}{2}$,

$\sec\theta = -\dfrac{3\sqrt{5}}{5}$, $\csc\theta = -\dfrac{3}{2}$ **31.** [5.3] About 1743 mi

32. [5.4]

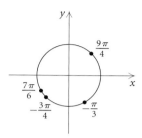

33. [5.4] II, $\dfrac{121}{150}\pi$, 2.534   **34.** [5.4] IV, $-\dfrac{\pi}{6}$, -0.524

35. [5.4] 270°   **36.** [5.4] 171.89°   **37.** [5.4] $\dfrac{7\pi}{4}$, or 5.5 cm

38. [5.4] 2.25, 129°   **39.** [5.4] About 37.9 ft/min
40. [5.4] 497,829 radians/hr

41. [5.5] $\left(\dfrac{3}{5},\dfrac{4}{5}\right),\left(-\dfrac{3}{5},-\dfrac{4}{5}\right),\left(-\dfrac{3}{5},\dfrac{4}{5}\right)$

42. [5.5] -1 **43.** [5.5] 1   **44.** [5.5] $-\dfrac{\sqrt{3}}{2}$ **45.** [5.5] $\dfrac{1}{2}$

46. [5.5] $\dfrac{\sqrt{3}}{3}$ **47.** [5.5] -1 **48.** [5.5] -0.9056

49. [5.5] 0.9218   **50.** [5.5] Undefined   **51.** [5.5] 4.3813
52. [5.5] -6.1685 **53.** [5.5] 0.8090
54. [5.5]

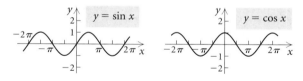

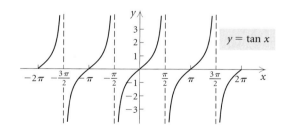

55. [5.5] Period of sin, cos, sec, csc: 2π; period of tan, cot: π
56. [5.5]

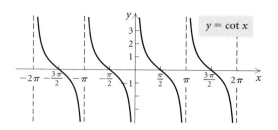

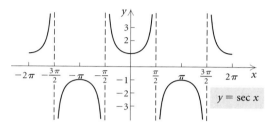

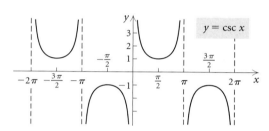

FUNCTION	DOMAIN	RANGE
Sine	$(-\infty,\infty)$	$[-1,1]$
Cosine	$(-\infty,\infty)$	$[-1,1]$
Tangent	$\left\{x\,\middle|\,x\neq\dfrac{\pi}{2}+k\pi,\,k\in\mathbb{Z}\right\}$	$(-\infty,\infty)$

57. [5.3]

FUNCTION	I	II	III	IV
Sine	+	+	−	−
Cosine	+	−	−	+
Tangent	+	−	+	−

58. [5.6] Amplitude: 1; period: 2π; phase shift: $-\dfrac{\pi}{2}$

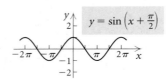

59. [5.6] Amplitude: $\dfrac{1}{2}$; period: π; phase shift: $\dfrac{\pi}{4}$

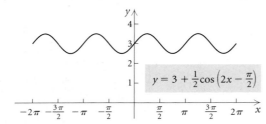

60. [5.6] (d)   **61.** [5.6] (a)   **62.** [5.6] (c)   **63.** [5.6] (b)
64. [5.6]

$y = 3 \cos x + \sin x$

65. Discussion and Writing [5.1], [5.4] Both degrees and radians are units of angle measure. A degree is defined to be $\frac{1}{360}$ of one complete positive revolution. Degree notation has been in use since Babylonian times. Radians are defined in terms of intercepted arc length on a circle, with one radian being the measure of the angle for which the arc length equals the radius. There are 2π radians in one complete revolution.
66. Discussion and Writing [5.5] The graph of the cosine function is shaped like a continuous wave, with "high" points at $y = 1$ and "low" points at $y = -1$. The maximum value of the cosine function is 1, and it occurs at all points where $x = 2k\pi$, $k \in \mathbb{Z}$.
67. Discussion and Writing [5.5] No; $\sin x$ is never greater than 1.

68. [5.6] Domain $(-\infty, \infty)$; range $[-3, 3]$; period 4π

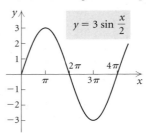

69. [5.6] $y_2 = 2 \sin\left(x + \dfrac{\pi}{2}\right) - 2$
70. [5.6] The domain consists of the intervals
$\left(-\dfrac{\pi}{2} + 2k\pi, \dfrac{\pi}{2} + 2k\pi\right)$, $k \in \mathbb{Z}$.
71. [5.3] $\cos x = -0.7890$, $\tan x = -0.7787$,
$\cot x = -1.2842$, $\sec x = -1.2674$, $\csc x = 1.6276$

Test, Chapter 5

1. [5.1] $\sin\theta = \dfrac{4}{\sqrt{65}}$, or $\dfrac{4\sqrt{65}}{65}$; $\cos\theta = \dfrac{7}{\sqrt{65}}$, or $\dfrac{7\sqrt{65}}{65}$;
$\tan\theta = \dfrac{4}{7}$; $\csc\theta = \dfrac{\sqrt{65}}{4}$; $\sec\theta = \dfrac{\sqrt{65}}{7}$; $\cot\theta = \dfrac{7}{4}$
2. [5.3] $\dfrac{\sqrt{3}}{2}$ **3.** [5.3] -1 **4.** [5.4] -1 **5.** [5.4] $-\sqrt{2}$
6. [5.1] $38.47°$ **7.** [5.3] -0.2419 **8.** [5.3] -0.2079
9. [5.4] -5.7588 **10.** [5.4] 0.7827 **11.** [5.1] $30°$
12. [5.1] $\sin 61.6° \approx 0.8796$; $\cos 61.6° \approx 0.4756$;
$\tan 61.6° \approx 1.8495$; $\csc 61.6° \approx 1.1369$; $\sec 61.6° \approx 2.1026$;
$\cot 61.6° \approx 0.5407$
13. [5.2] $B = 54.1°$, $a \approx 32.6$, $c \approx 55.7$
14. [5.3] Answers may vary; $472°$, $-248°$ **15.** [5.4] $\dfrac{\pi}{6}$
16. [5.3] $\cos\theta = \dfrac{5}{\sqrt{41}}$; $\tan\theta = -\dfrac{4}{5}$; $\csc\theta = -\dfrac{\sqrt{41}}{4}$;
$\sec\theta = \dfrac{\sqrt{41}}{5}$; $\cot\theta = -\dfrac{5}{4}$ **17.** [5.4] $\dfrac{7\pi}{6}$ **18.** [5.4] $135°$
19. [5.4] $\dfrac{16\pi}{3} \approx 16.755$ cm   **20.** [5.5] 1 **21.** [5.5] 2π
22. [5.5] $\dfrac{\pi}{2}$ **23.** [5.6] (c)   **24.** [5.2] About 444 ft
25. [5.2] About 272 mi   **26.** [5.4] $18\pi \approx 56.55$ m/min
27. [5.5] $\left\{x \,\middle|\, -\dfrac{\pi}{2} + 2k\pi < x < \dfrac{\pi}{2} + 2k\pi, k \text{ an integer}\right\}$

Chapter 6

Exercise Set 6.1

1. $\sin^2 x - \cos^2 x$ **3.** $\sin y + \cos y$
5. $1 - 2 \sin \phi \cos \phi$ **7.** $\sin^3 x + \csc^3 x$
9. $\cos x (\sin x + \cos x)$
11. $(\sin x + \cos x)(\sin x - \cos x)$
13. $(2 \cos x + 3)(\cos x - 1)$
15. $(\sin x + 3)(\sin^2 x - 3 \sin x + 9)$ **17.** $\tan x$
19. $\sin x + 1$ **21.** $\dfrac{2 \tan t + 1}{3 \tan t + 1}$ **23.** 1
25. $\dfrac{5 \cot \phi}{\sin \phi + \cos \phi}$ **27.** $\dfrac{1 + 2 \sin s + 2 \cos s}{\sin^2 s - \cos^2 s}$
29. $\dfrac{5(\sin \theta - 3)}{3}$ **31.** $\sin x \cos x$
33. $\sqrt{\cos \alpha}\,(\sin \alpha - \cos \alpha)$ **35.** $1 - \sin y$
37. $\dfrac{\sqrt{\sin x \cos x}}{\cos x}$ **39.** $\dfrac{\sqrt{2} \cot y}{2}$
41. $\dfrac{\cos x}{\sqrt{\sin x \cos x}}$ **43.** $\dfrac{1 + \sin y}{\cos y}$
45. $\cos \theta = \dfrac{\sqrt{a^2 - x^2}}{a}$, $\tan \theta = \dfrac{x}{\sqrt{a^2 - x^2}}$
47. $\sin \theta = \dfrac{\sqrt{x^2 - 9}}{x}$, $\cos \theta = \dfrac{3}{x}$
49. $\sin \theta \tan \theta$ **51.** $\dfrac{\sqrt{6} - \sqrt{2}}{4}$
53. $\dfrac{\sqrt{3} + 1}{1 - \sqrt{3}}$, or $-2 - \sqrt{3}$ **55.** $\dfrac{\sqrt{6} + \sqrt{2}}{4}$
57. $\sin 59° \approx 0.8572$ **59.** $\tan 52° \approx 1.2799$

61. $\tan (\mu + \nu) = \dfrac{\sin (\mu + \nu)}{\cos (\mu + \nu)}$

$= \dfrac{\sin \mu \cos \nu + \cos \mu \sin \nu}{\cos \mu \cos \nu - \sin \mu \sin \nu}$

$= \dfrac{\sin \mu \cos \nu + \cos \mu \sin \nu}{\cos \mu \cos \nu - \sin \mu \sin \nu} \cdot \dfrac{\dfrac{1}{\cos \mu \cos \nu}}{\dfrac{1}{\cos \mu \cos \nu}}$

$= \dfrac{\dfrac{\sin \mu}{\cos \mu} + \dfrac{\sin \nu}{\cos \nu}}{1 - \dfrac{\sin \mu \sin \nu}{\cos \mu \cos \nu}}$

$= \dfrac{\tan \mu + \tan \nu}{1 - \tan \mu \tan \nu}$

63. 0 **65.** $-\dfrac{7}{25}$ **67.** -1.5789

69. $2 \sin \alpha \cos \beta$ **71.** $\cos u$ **73.** Left to the student
75. Discussion and Writing   **77.** [2.1] All real numbers
78. [2.1] No solution   **79.** [5.1] 1.9417   **80.** [5.1] 1.6645
81. $0°$; the lines are parallel   **83.** $\dfrac{3\pi}{4}$, or 135°   **85.** 22.83°
87. $\dfrac{\cos (x + h) - \cos x}{h}$

$= \dfrac{\cos x \cos h - \sin x \sin h - \cos x}{h}$

$= \dfrac{\cos x \cos h - \cos x}{h} - \dfrac{\sin x \sin h}{h}$

$= \cos x \left(\dfrac{\cos h - 1}{h} \right) - \sin x \left(\dfrac{\sin h}{h} \right)$

89. Let $x = \dfrac{\pi}{5}$. Then $\dfrac{\sin 5x}{x} = \dfrac{\sin \pi}{\pi/5} = 0 \ne \sin 5$.
Answers may vary.

91. Let $\alpha = \dfrac{\pi}{4}$. Then $\cos (2\alpha) = \cos \dfrac{\pi}{2} = 0$, but

$2 \cos \alpha = 2 \cos \dfrac{\pi}{4} = \sqrt{2}$. Answer may vary.

93. Let $x = \dfrac{\pi}{6}$. Then $\dfrac{\cos 6x}{\cos x} = \dfrac{\cos \pi}{\cos \dfrac{\pi}{6}} = \dfrac{-1}{\sqrt{3}/2} \ne 6$.

Answer may vary.   **95.** $\dfrac{6 - 3\sqrt{3}}{9 + 2\sqrt{3}} \approx 0.0645$

97. 168.7°   **99.** $\cos 2\theta = \cos^2 \theta - \sin^2 \theta$, or $1 - 2 \sin^2 \theta$,
or $2 \cos^2 \theta - 1$

101. $\tan \left(x + \dfrac{\pi}{4} \right) = \dfrac{\tan x + \tan \dfrac{\pi}{4}}{1 - \tan x \tan \dfrac{\pi}{4}} = \dfrac{1 + \tan x}{1 - \tan x}$

103. $\sin (\alpha + \beta) + \sin (\alpha - \beta) = \sin \alpha \cos \beta + \cos \alpha \sin \beta + \sin \alpha \cos \beta - \cos \alpha \sin \beta = 2 \sin \alpha \cos \beta$

Exercise Set 6.2

1. (a) $\tan \dfrac{3\pi}{10} \approx 1.3763$, $\csc \dfrac{3\pi}{10} \approx 1.2361$, $\sec \dfrac{3\pi}{10} \approx 1.7013$,

$\cot \dfrac{3\pi}{10} \approx 0.7266$; **(b)** $\sin \dfrac{\pi}{5} \approx 0.5878$, $\cos \dfrac{\pi}{5} \approx 0.8090$,

$\tan \dfrac{\pi}{5} \approx 0.7266$, $\csc \dfrac{\pi}{5} \approx 1.7013$, $\sec \dfrac{\pi}{5} \approx 1.2361$,

$\cot \dfrac{\pi}{5} \approx 1.3763$

3. (a) $\cos \theta = -\dfrac{2\sqrt{2}}{3}$, $\tan \theta = -\dfrac{\sqrt{2}}{4}$, $\csc \theta = 3$,

$\sec \theta = -\dfrac{3\sqrt{2}}{4}$, $\cot \theta = -2\sqrt{2}$; **(b)** $\sin\left(\dfrac{\pi}{2} - \theta\right) = -\dfrac{2\sqrt{2}}{3}$,

$\cos\left(\dfrac{\pi}{2} - \theta\right) = \dfrac{1}{3}$, $\tan\left(\dfrac{\pi}{2} - \theta\right) = -2\sqrt{2}$,

$\csc\left(\dfrac{\pi}{2} - \theta\right) = -\dfrac{3\sqrt{2}}{4}$, $\sec\left(\dfrac{\pi}{2} - \theta\right) = 3$,

$\cot\left(\dfrac{\pi}{2} - \theta\right) = -\dfrac{\sqrt{2}}{4}$; **(c)** $\sin\left(\theta - \dfrac{\pi}{2}\right) = \dfrac{2\sqrt{2}}{3}$,

$\cos\left(\theta - \dfrac{\pi}{2}\right) = \dfrac{1}{3}$, $\tan\left(\theta - \dfrac{\pi}{2}\right) = 2\sqrt{2}$,

$\csc\left(\theta - \dfrac{\pi}{2}\right) = \dfrac{3\sqrt{2}}{4}$, $\sec\left(\theta - \dfrac{\pi}{2}\right) = 3$,

$\cot\left(\theta - \dfrac{\pi}{2}\right) = \dfrac{\sqrt{2}}{4}$

5. $\sec\left(x + \dfrac{\pi}{2}\right) = -\csc x$ **7.** $\tan\left(x - \dfrac{\pi}{2}\right) = -\cot x$

9. $\sin 2\theta = \dfrac{24}{25}$, $\cos 2\theta = -\dfrac{7}{25}$, $\tan 2\theta = -\dfrac{24}{7}$; II

11. $\sin 2\theta = \dfrac{24}{25}$, $\cos 2\theta = -\dfrac{7}{25}$, $\tan 2\theta = -\dfrac{24}{7}$; II

13. $\sin 2\theta = -\dfrac{120}{169}$, $\cos 2\theta = \dfrac{119}{169}$, $\tan 2\theta = -\dfrac{120}{119}$; IV

15. $\cos 4x = 1 - 8 \sin^2 x \cos^2 x$, or $\cos^4 x -$
$6 \sin^2 x \cos^2 x + \sin^4 x$, or $8 \cos^4 x - 8 \cos^2 x + 1$

17. $\dfrac{\sqrt{2 + \sqrt{3}}}{2}$ **19.** $\dfrac{\sqrt{2 + \sqrt{2}}}{2}$ **21.** $2 + \sqrt{3}$

23. 0.6421   **25.** 0.1735   **27.** $\cos x$ **29.** 1

31. $\cos 2x$ **33.** 8

35. (d); $\dfrac{\cos 2x}{\cos x - \sin x} = \dfrac{\cos^2 x - \sin^2 x}{\cos x - \sin x}$

$ = \dfrac{(\cos x + \sin x)(\cos x - \sin x)}{\cos x - \sin x}$

$ = \cos x + \sin x$

$ = \dfrac{\sin x}{\sin x}(\cos x + \sin x)$

$ = \sin x \left(\dfrac{\cos x}{\sin x} + \dfrac{\sin x}{\sin x}\right)$

$ = \sin x (\cot x + 1)$

37. (d); $\dfrac{\sin 2x}{2 \cos x} = \dfrac{2 \sin x \cos x}{2 \cos x} = \sin x$

39. Discussion and Writing   **41.** [6.1] $\sin^2 x$ **42.** [6.1] 1

43. [6.1] $-\cos^2 x$ **44.** [6.1] $\csc^2 x$

45. $\sin 141° \approx 0.6293$, $\cos 141° \approx -0.7772$,
$\tan 141° \approx -0.8097$, $\csc 141° \approx 1.5891$,
$\sec 141° \approx -1.2867$, $\cot 141° \approx -1.2350$

47. $-\cos x(1 + \cot x)$ **49.** $\cot^2 y$

51. $\sin \theta = -\dfrac{15}{17}$, $\cos \theta = -\dfrac{8}{17}$, $\tan \theta = \dfrac{15}{8}$

53. (a) 9.80359 m/sec²; **(b)** 9.80180 m/sec²;
(c) $g = 9.78049(1 + 0.005264 \sin^2 \phi + 0.000024 \sin^4 \phi)$.

Exercise Set 6.3

1.

$$\begin{array}{c|c}
\sec x - \sin x \tan x & \cos x \\ \hline
\dfrac{1}{\cos x} - \sin x \cdot \dfrac{\sin x}{\cos x} & \\
\dfrac{1 - \sin^2 x}{\cos x} & \\
\dfrac{\cos^2 x}{\cos x} & \\
\cos x &
\end{array}$$

3.

$$\begin{array}{c|c}
\dfrac{1 - \cos x}{\sin x} & \dfrac{\sin x}{1 + \cos x} \\ \hline
 & \dfrac{\sin x}{1 + \cos x} \cdot \dfrac{1 - \cos x}{1 - \cos x} \\
 & \dfrac{\sin x (1 - \cos x)}{1 - \cos^2 x} \\
 & \dfrac{\sin x (1 - \cos x)}{\sin^2 x} \\
 & \dfrac{1 - \cos x}{\sin x}
\end{array}$$

5.

$$\begin{array}{c|c}
\dfrac{1 + \tan \theta}{1 - \tan \theta} + \dfrac{1 + \cot \theta}{1 - \cot \theta} & 0 \\ \hline
\dfrac{1 + \dfrac{\sin \theta}{\cos \theta}}{1 - \dfrac{\sin \theta}{\cos \theta}} + \dfrac{1 + \dfrac{\cos \theta}{\sin \theta}}{1 - \dfrac{\cos \theta}{\sin \theta}} & \\
\dfrac{\dfrac{\cos \theta + \sin \theta}{\cos \theta}}{\dfrac{\cos \theta - \sin \theta}{\cos \theta}} + \dfrac{\dfrac{\sin \theta + \cos \theta}{\sin \theta}}{\dfrac{\sin \theta - \cos \theta}{\sin \theta}} & \\
\dfrac{\cos \theta + \sin \theta}{\cos \theta} \cdot \dfrac{\cos \theta}{\cos \theta - \sin \theta} + & \\
\dfrac{\sin \theta + \cos \theta}{\sin \theta} \cdot \dfrac{\sin \theta}{\sin \theta - \cos \theta} & \\
\dfrac{\cos \theta + \sin \theta}{\cos \theta - \sin \theta} + \dfrac{\sin \theta + \cos \theta}{\sin \theta - \cos \theta} & \\
\dfrac{\cos \theta + \sin \theta}{\cos \theta - \sin \theta} - \dfrac{\cos \theta + \sin \theta}{\cos \theta - \sin \theta} & \\
0 &
\end{array}$$

7.

$$\dfrac{\cos^2 \alpha + \cot \alpha}{\cos^2 \alpha - \cot \alpha} \quad \bigg| \quad \dfrac{\cos^2 \alpha \tan \alpha + 1}{\cos^2 \alpha \tan \alpha - 1}$$

$$\dfrac{\cos^2 \alpha + \dfrac{\cos \alpha}{\sin \alpha}}{\cos^2 \alpha - \dfrac{\cos \alpha}{\sin \alpha}} \quad \bigg| \quad \dfrac{\cos^2 \alpha \dfrac{\sin \alpha}{\cos \alpha} + 1}{\cos^2 \alpha \dfrac{\sin \alpha}{\cos \alpha} - 1}$$

$$\dfrac{\cos \alpha \left(\cos \alpha + \dfrac{1}{\sin \alpha}\right)}{\cos \alpha \left(\cos \alpha - \dfrac{1}{\sin \alpha}\right)} \quad \bigg| \quad \dfrac{\sin \alpha \cos \alpha + 1}{\sin \alpha \cos \alpha - 1}$$

$$\dfrac{\cos \alpha + \dfrac{1}{\sin \alpha}}{\cos \alpha - \dfrac{1}{\sin \alpha}}$$

$$\dfrac{\dfrac{\sin \alpha \cos \alpha + 1}{\sin \alpha}}{\dfrac{\sin \alpha \cos \alpha - 1}{\sin \alpha}}$$

$$\dfrac{\sin \alpha \cos \alpha + 1}{\sin \alpha \cos \alpha - 1}$$

9.

$$\dfrac{2 \tan \theta}{1 + \tan^2 \theta} \quad \bigg| \quad \sin 2\theta$$

$$\dfrac{2 \tan \theta}{\sec^2 \theta} \quad \bigg| \quad 2 \sin \theta \cos \theta$$

$$\dfrac{2 \sin \theta}{\cos \theta} \cdot \dfrac{\cos^2 \theta}{1}$$

$$2 \sin \theta \cos \theta \quad \bigg|$$

11.

$$1 - \cos 5\theta \cos 3\theta - \sin 5\theta \sin 3\theta \quad \bigg| \quad 2 \sin^2 \theta$$

$$1 - [\cos 5\theta \cos 3\theta + \sin 5\theta \sin 3\theta] \quad \bigg| \quad 1 - \cos 2\theta$$

$$1 - \cos (5\theta - 3\theta)$$

$$1 - \cos 2\theta \quad \bigg|$$

13.

$$2 \sin \theta \cos^3 \theta + 2 \sin^3 \theta \cos \theta \quad \bigg| \quad \sin 2\theta$$

$$2 \sin \theta \cos \theta (\cos^2 \theta + \sin^2 \theta) \quad \bigg| \quad 2 \sin \theta \cos \theta$$

$$2 \sin \theta \cos \theta \quad \bigg|$$

15.

$$\dfrac{\tan x - \sin x}{2 \tan x} \quad \bigg| \quad \sin^2 \dfrac{x}{2}$$

$$\dfrac{1}{2}\left[\dfrac{\dfrac{\sin x}{\cos x} - \sin x}{\dfrac{\sin x}{\cos x}}\right] \quad \bigg| \quad \dfrac{1 - \cos x}{2}$$

$$\dfrac{1}{2}\dfrac{\dfrac{\sin x - \sin x \cos x}{\cos x}}{\dfrac{\sin x}{\cos x}} \cdot \dfrac{\cos x}{\sin x}$$

$$\dfrac{1 - \cos x}{2} \quad \bigg|$$

17.

$$\sin (\alpha + \beta) \sin (\alpha - \beta) \quad \bigg| \quad \sin^2 \alpha - \sin^2 \beta$$

$$\left(\begin{array}{c}\sin \alpha \cos \beta + \\ \cos \alpha \sin \beta\end{array}\right)\left(\begin{array}{c}\sin \alpha \cos \beta - \\ \cos \alpha \sin \beta\end{array}\right) \quad \bigg| \quad \begin{array}{c}1 - \cos^2 \alpha - \\ (1 - \cos^2 \beta)\end{array}$$

$$\sin^2 \alpha \cos^2 \beta - \cos^2 \alpha \sin^2 \beta \quad \bigg| \quad \cos^2 \beta - \cos^2 \alpha$$

$$\begin{array}{c}\cos^2 \beta (1 - \cos^2 \alpha) - \\ \cos^2 \alpha (1 - \cos^2 \beta)\end{array}$$

$$\begin{array}{c}\cos^2 \beta - \cos^2 \alpha \cos^2 \beta - \\ \cos^2 \alpha + \cos^2 \alpha \cos^2 \beta\end{array}$$

$$\cos^2 \beta - \cos^2 \alpha \quad \bigg|$$

19.

$$\tan \theta (\tan \theta + \cot \theta) \quad \bigg| \quad \sec^2 \theta$$

$$\tan^2 \theta + \tan \theta \cot \theta$$

$$\tan^2 \theta + 1$$

$$\sec^2 \theta \quad \bigg|$$

21.

$$\dfrac{1 + \cos^2 x}{\sin^2 x} \quad \bigg| \quad 2 \csc^2 x - 1$$

$$\dfrac{1}{\sin^2 x} + \dfrac{\cos^2 x}{\sin^2 x}$$

$$\csc^2 x + \cot^2 x$$

$$\csc^2 x + \csc^2 x - 1$$

$$2 \csc^2 x - 1 \quad \bigg|$$

23.

$$\dfrac{1 + \sin x}{1 - \sin x} + \dfrac{\sin x - 1}{1 + \sin x} \quad \bigg| \quad 4 \sec x \tan x$$

$$\dfrac{(1 + \sin x)^2 - (1 - \sin x)^2}{1 - \sin^2 x} \quad \bigg| \quad 4 \cdot \dfrac{1}{\cos x} \cdot \dfrac{\sin x}{\cos x}$$

$$\dfrac{(1 + 2 \sin x + \sin^2 x) - (1 - 2 \sin x + \sin^2 x)}{\cos^2 x} \quad \bigg| \quad \dfrac{4 \sin x}{\cos^2 x}$$

$$\dfrac{4 \sin x}{\cos^2 x} \quad \bigg|$$

25.

| $\begin{array}{c} \cos^2 \alpha \cot^2 \alpha \\ (1 - \sin^2 \alpha) \cot^2 \alpha \\ \cot^2 \alpha - \sin^2 \alpha \cdot \dfrac{\cos^2 \alpha}{\sin^2 \alpha} \\ \cot^2 \alpha - \cos^2 \alpha \end{array}$ | $\cot^2 \alpha - \cos^2 \alpha$ |

27.

| $\begin{array}{c} 2 \sin^2 \theta \cos^2 \theta + \cos^4 \theta \\ \cos^2 \theta \, (2 \sin^2 \theta + \cos^2 \theta) \\ \cos^2 \theta \, (\sin^2 \theta + \sin^2 \theta + \cos^2 \theta) \\ \cos^2 \theta \, (\sin^2 \theta + 1) \end{array}$ | $\begin{array}{c} 1 - \sin^4 \theta \\ (1 + \sin^2 \theta)(1 - \sin^2 \theta) \\ (1 + \sin^2 \theta)(\cos^2 \theta) \end{array}$ |

29.

$\dfrac{1 + \sin x}{1 - \sin x}$	$(\sec x + \tan x)^2$
$\dfrac{1 + \sin x}{1 - \sin x} \cdot \dfrac{1 + \sin x}{1 + \sin x}$	$\left(\dfrac{1}{\cos x} + \dfrac{\sin x}{\cos x} \right)^2$
$\dfrac{(1 + \sin x)^2}{1 - \sin^2 x}$	$\dfrac{(1 + \sin x)^2}{\cos^2 x}$
$\dfrac{(1 + \sin x)^2}{\cos^2 x}$	

31. B;

$\dfrac{\cos x + \cot x}{1 + \csc x}$	$\cos x$
$\dfrac{\dfrac{\cos x}{1} + \dfrac{\cos x}{\sin x}}{1 + \dfrac{1}{\sin x}}$	
$\dfrac{\sin x \cos x + \cos x}{\sin x} \cdot \dfrac{\sin x}{\sin x + 1}$	
$\dfrac{\cos x \, (\sin x + 1)}{\sin x + 1}$	
$\cos x$	

33. A;

| $\sin x \cos x + 1$ | $\begin{array}{c} \dfrac{\sin^3 x - \cos^3 x}{\sin x - \cos x} \\ \dfrac{(\sin x - \cos x)(\sin^2 x + \sin x \cos x + \cos^2 x)}{\sin x - \cos x} \\ \sin^2 x + \sin x \cos x + \cos^2 x \\ \sin x \cos x + 1 \end{array}$ |

35. C;

$\dfrac{1}{\cot x \sin^2 x}$	$\tan x + \cot x$
$\dfrac{1}{\dfrac{\cos x}{\sin x} \cdot \sin^2 x}$	$\dfrac{\sin x}{\cos x} + \dfrac{\cos x}{\sin x}$
$\dfrac{1}{\cos x \sin x}$	$\dfrac{\sin^2 x + \cos^2 x}{\cos x \sin x}$
	$\dfrac{1}{\cos x \sin x}$

37. Discussion and Writing

39. [4.1] **(a)**, **(d)**

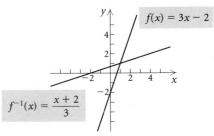

(b) yes; **(c)** $f^{-1}(x) = \dfrac{x + 2}{3}$

40. [4.1] **(a)**, **(d)**

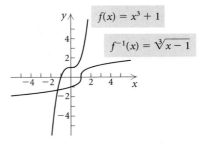

(b) yes; **(c)** $f^{-1}(x) = \sqrt[3]{x - 1}$

41. [4.1] **(a)**, **(d)**

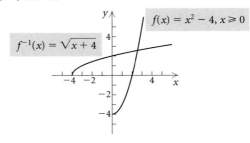

(b) yes; **(c)** $f^{-1}(x) = \sqrt{x + 4}$

42. [4.1] **(a)**, **(d)**

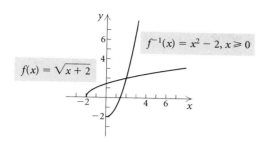

(b) yes; **(c)** $f^{-1}(x) = x^2 - 2,\ x \geq 0$

43.

$\ln |\tan x|$	$-\ln |\cot x|$
$\ln \left\| \dfrac{1}{\cot x} \right\|$	
$\ln |1| - \ln |\cot x|$	
$0 - \ln |\cot x|$	
$-\ln |\cot x|$	

45. $\log (\cos x - \sin x) + \log (\cos x + \sin x)$
$$= \log [(\cos x - \sin x)(\cos x + \sin x)]$$
$$= \log (\cos^2 x - \sin^2 x) = \log \cos 2x$$

47. $\dfrac{1}{\omega C(\tan \theta + \tan \phi)} = \dfrac{1}{\omega C\left(\dfrac{\sin \theta}{\cos \theta} + \dfrac{\sin \phi}{\cos \phi}\right)}$

$$= \dfrac{1}{\omega C\left(\dfrac{\sin \theta \cos \phi + \sin \phi \cos \theta}{\cos \theta \cos \phi}\right)}$$

$$= \dfrac{\cos \theta \cos \phi}{\omega C \sin (\theta + \phi)}$$

Exercise Set 6.4

1. $-\dfrac{\pi}{3}$, $-60°$ **3.** $\dfrac{\pi}{4}$, $45°$ **5.** $\dfrac{\pi}{4}$, $45°$ **7.** 0, $0°$

9. $\dfrac{\pi}{6}$, $30°$ **11.** $\dfrac{\pi}{6}$, $30°$ **13.** $-\dfrac{\pi}{6}$, $-30°$

15. $-\dfrac{\pi}{6}$, $-30°$ **17.** $\dfrac{\pi}{2}$, $90°$ **19.** $\dfrac{\pi}{3}$, $60°$

21. 0.3520, $20.2°$ **23.** 1.2917, $74.0°$ **25.** 2.9463, $168.8°$
27. -0.1600, $-9.2°$ **29.** 0.8289, $47.5°$
31. -0.9600, $-55.0°$
33. $\sin^{-1}$: $[-1, 1]$; $\cos^{-1}$: $[-1, 1]$; $\tan^{-1}$: $(-\infty, \infty)$

35. $\theta = \sin^{-1}\left(\dfrac{2000}{d}\right)$ **37.** 0.3 **39.** $\dfrac{\pi}{4}$ **41.** $\dfrac{\pi}{5}$

43. $-\dfrac{\pi}{3}$ **45.** $\dfrac{1}{2}$ **47.** 1 **49.** $\dfrac{\pi}{3}$ **51.** $\dfrac{\sqrt{11}}{33}$

53. $-\dfrac{\pi}{6}$ **55.** $\dfrac{a}{\sqrt{a^2 + 9}}$ **57.** $\dfrac{\sqrt{q^2 - p^2}}{p}$ **59.** $\dfrac{p}{3}$

61. $\dfrac{\sqrt{3}}{2}$ **63.** $-\dfrac{\sqrt{2}}{10}$ **65.** $xy + \sqrt{(1 - x^2)(1 - y^2)}$

67. 0.9861 **69.** Discussion and Writing

71. Discussion and Writing   **72.** [2.3] -4, $\dfrac{7}{3}$

73. [2.3] 0, $\dfrac{5}{2}$ **74.** [2.3] $5 \pm 2\sqrt{6}$ **75.** [2.3] ± 2, $\pm 3i$

76. [2.5] 9 **77.** [2.5] 27

79.

$\sin^{-1} x + \cos^{-1} x$	$\dfrac{\pi}{2}$
$\sin (\sin^{-1} x + \cos^{-1} x)$	$\sin \dfrac{\pi}{2}$
$[\sin (\sin^{-1} x)][\cos (\cos^{-1} x)] +$ $[\cos (\sin^{-1} x)][\sin (\cos^{-1} x)]$ $x \cdot x + \sqrt{1 - x^2} \cdot \sqrt{1 - x^2}$ $x^2 + 1 - x^2$	1
1	

81.

$\sin^{-1} x$	$\tan^{-1} \dfrac{x}{\sqrt{1 - x^2}}$
$\sin (\sin^{-1} x)$	$\sin \left(\tan^{-1} \dfrac{x}{\sqrt{1 - x^2}}\right)$
x	x

83.

$\arcsin x$	$\arccos \sqrt{1 - x^2}$
$\sin (\arcsin x)$	$\sin (\arccos \sqrt{1 - x^2})$
x	x

85. $\theta = \arctan \dfrac{y + h}{x} - \arctan \dfrac{y}{x}$; $38.7°$

Exercise Set 6.5

1. $\dfrac{\pi}{6} + 2k\pi$, $\dfrac{11\pi}{6} + 2k\pi$, or $30° + k \cdot 360°$, $330° + k \cdot 360°$

3. $\dfrac{2\pi}{3} + k\pi$, or $120° + k \cdot 180°$

5. $98.09°$, $261.91°$ **7.** $\dfrac{4\pi}{3}$, $\dfrac{5\pi}{3}$ **9.** $\dfrac{\pi}{4}$, $\dfrac{3\pi}{4}$, $\dfrac{5\pi}{4}$, $\dfrac{7\pi}{4}$

11. $\dfrac{\pi}{6}$, $\dfrac{5\pi}{6}$, $\dfrac{3\pi}{2}$ **13.** $\dfrac{\pi}{6}$, $\dfrac{\pi}{2}$, $\dfrac{3\pi}{2}$, $\dfrac{11\pi}{6}$

15. $109.47°$, $120°$, $240°$, $250.53°$

17. 0, $\dfrac{\pi}{4}$, $\dfrac{3\pi}{4}$, π, $\dfrac{5\pi}{4}$, $\dfrac{7\pi}{4}$ **19.** $139.81°$, $220.19°$

21. $37.22°$, $169.35°$, $217.22°$, $349.35°$ **23.** 0, π, $\dfrac{7\pi}{6}$, $\dfrac{11\pi}{6}$

25. 0, $\dfrac{\pi}{2}$, π, $\dfrac{3\pi}{2}$ **27.** 0, π **29.** $\dfrac{3\pi}{4}$, $\dfrac{7\pi}{4}$

31. $\dfrac{2\pi}{3}$, $\dfrac{4\pi}{3}$, $\dfrac{3\pi}{2}$ **33.** $\dfrac{\pi}{4}$, $\dfrac{3\pi}{4}$, $\dfrac{5\pi}{4}$, $\dfrac{7\pi}{4}$

35. $\dfrac{\pi}{12}, \dfrac{5\pi}{12}$ **37.** 0.967, 1.853, 4.109, 4.994

39. $\dfrac{2\pi}{3}, \dfrac{4\pi}{3}$ **41.** Left to the student   **43.** 1.114, 2.773

45. 0.515   **47.** 0.422, 1.756

49. (a) $y = 7 \sin(-2.6180x + 0.5236) + 7$; **(b)** $10,500$, $13,062

51. Discussion and Writing

53. [5.2] $B = 35°$, $b \approx 140.7$, $c \approx 245.4$

54. [5.2] $R \approx 15.5°$, $T \approx 74.5°$, $t \approx 13.7$ **55.** [2.1] 36

56. [2.1] 14   **57.** $\dfrac{\pi}{3}, \dfrac{2\pi}{3}, \dfrac{4\pi}{3}, \dfrac{5\pi}{3}$ **59.** $\dfrac{\pi}{3}, \dfrac{4\pi}{3}$

61. 0   **63.** $e^{3\pi/2 + 2k\pi}$, where k (an integer) ≤ -1

65. 1.24 days, 6.76 days   **67.** 16.5°N

69. 1   **71.** 0.1923

Review Exercises, Chapter 6

1. [6.1] $\csc^2 x$ **2.** [6.1] 1   **3.** [6.1] $\tan^2 y - \cot^2 y$

4. [6.1] $\dfrac{(\cos^2 x + 1)^2}{\cos^2 x}$ **5.** [6.1] $\csc x (\sec x - \csc x)$

6. [6.1] $(3 \sin y + 5)(\sin y - 4)$

7. [6.1] $(10 - \cos u)(100 + 10 \cos u + \cos^2 u)$ **8.** [6.1] 1

9. [6.1] $\dfrac{1}{2} \sec x$ **10.** [6.1] $\dfrac{3 \tan x}{\sin x - \cos x}$

11. [6.1] $\dfrac{3 \cos y + 3 \sin y + 2}{\cos^2 y - \sin^2 y}$ **12.** [6.1] 1

13. [6.1] $\dfrac{1}{4} \cot x$ **14.** [6.1] $\sin x + \cos x$

15. [6.1] $\dfrac{\cos x}{1 - \sin x}$ **16.** [6.1] $\dfrac{\cos x}{\sqrt{\sin x}}$ **17.** [6.1] $3 \sec \theta$

18. [6.1] $\cos x \cos \dfrac{3\pi}{2} - \sin x \sin \dfrac{3\pi}{2}$

19. [6.1] $\dfrac{\tan 45° - \tan 30°}{1 + \tan 45° \tan 30°}$

20. [6.1] $\cos(27° - 16°)$, or $\cos 11°$ **21.** [6.1] $\dfrac{-\sqrt{6} - \sqrt{2}}{4}$

22. [6.1] $2 - \sqrt{3}$ **23.** [6.1] -0.3745 **24.** [6.2] $-\sin x$

25. [6.2] $\sin x$ **26.** [6.2] $-\cos x$

27. [6.2] **(a)** $\sin \alpha = -\dfrac{4}{5}$, $\tan \alpha = \dfrac{4}{3}$, $\cot \alpha = \dfrac{3}{4}$,

$\sec \alpha = -\dfrac{5}{3}$, $\csc \alpha = -\dfrac{5}{4}$; **(b)** $\sin\left(\dfrac{\pi}{2} - \alpha\right) = -\dfrac{3}{5}$,

$\cos\left(\dfrac{\pi}{2} - \alpha\right) = -\dfrac{4}{5}$, $\tan\left(\dfrac{\pi}{2} - \alpha\right) = \dfrac{3}{4}$,

$\cot\left(\dfrac{\pi}{2} - \alpha\right) = \dfrac{4}{3}$, $\sec\left(\dfrac{\pi}{2} - \alpha\right) = -\dfrac{5}{4}$,

$\csc\left(\dfrac{\pi}{2} - \alpha\right) = -\dfrac{5}{3}$; **(c)** $\sin\left(\alpha + \dfrac{\pi}{2}\right) = -\dfrac{3}{5}$,

$\cos\left(\alpha + \dfrac{\pi}{2}\right) = \dfrac{4}{5}$, $\tan\left(\alpha + \dfrac{\pi}{2}\right) = -\dfrac{3}{4}$,

$\cot\left(\alpha + \dfrac{\pi}{2}\right) = -\dfrac{4}{3}$, $\sec\left(\alpha + \dfrac{\pi}{2}\right) = \dfrac{5}{4}$,

$\csc\left(\alpha + \dfrac{\pi}{2}\right) = -\dfrac{5}{3}$ **28.** [6.2] $-\sec x$

29. [6.2] $\tan 2\theta = \dfrac{24}{7}$, $\cos 2\theta = \dfrac{7}{25}$, $\sin 2\theta = \dfrac{24}{25}$; I

30. [6.2] $\dfrac{\sqrt{2 - \sqrt{2}}}{2}$

31. [6.2] $\sin 2\beta = 0.4261$, $\cos \dfrac{\beta}{2} = 0.9940$,

$\cos 4\beta = 0.6369$ **32.** [6.2] $\cos x$ **33.** [6.2] 1

34. [6.2] $\sin 2x$ **35.** [6.2] $\tan 2x$

36. [6.3]

$\dfrac{1 - \sin x}{\cos x}$	$\dfrac{\cos x}{1 + \sin x}$
$\dfrac{1 - \sin x}{\cos x} \cdot \dfrac{\cos x}{\cos x}$	$\dfrac{\cos x}{1 + \sin x} \cdot \dfrac{1 - \sin x}{1 - \sin x}$
$\dfrac{\cos x - \sin x \cos x}{\cos^2 x}$	$\dfrac{\cos x - \sin x \cos x}{1 - \sin^2 x}$
	$\dfrac{\cos x - \sin x \cos x}{\cos^2 x}$

37. [6.3]

$\dfrac{1 + \cos 2\theta}{\sin 2\theta}$	$\cot \theta$
$\dfrac{1 + 2 \cos^2 \theta - 1}{2 \sin \theta \cos \theta}$	$\dfrac{\cos \theta}{\sin \theta}$
$\dfrac{\cos \theta}{\sin \theta}$	

38. [6.3]

$\dfrac{\tan y + \sin y}{2 \tan y}$	$\cos^2 \dfrac{y}{2}$
$\dfrac{1}{2}\left[\dfrac{\dfrac{\sin y + \sin y \cos y}{\cos y}}{\dfrac{\sin y}{\cos y}}\right]$	$\dfrac{1 + \cos y}{2}$
$\dfrac{1}{2}\left[\dfrac{\sin y (1 + \cos y)}{\cos y} \cdot \dfrac{\cos y}{\sin y}\right]$	
$\dfrac{1 + \cos y}{2}$	

39. [6.3]

$\dfrac{\sin x - \cos x}{\cos^2 x}$	$\begin{array}{c} \tan^2 x - 1 \\[4pt] \hline \sin x + \cos x \\[6pt] \dfrac{\frac{\sin^2 x}{\cos^2 x} - 1}{\sin x + \cos x} \\[10pt] \dfrac{\sin^2 x - \cos^2 x}{\cos^2 x} \cdot \dfrac{1}{\sin x + \cos x} \\[10pt] \dfrac{\sin x - \cos x}{\cos^2 x} \end{array}$

40. [6.4] $-\dfrac{\pi}{6},\ -30°$ **41.** [6.4] $\dfrac{\pi}{6},\ 30°$

42. [6.4] $\dfrac{\pi}{4},\ 45°$ **43.** [6.4] $0,\ 0°$ **44.** [6.4] $1.7920,\ 102.7°$

45. [6.4] $0.3976,\ 22.8°$ **46.** [6.4] $\dfrac{1}{2}$ **47.** [6.4] $\dfrac{\sqrt{3}}{3}$

48. [6.4] $\dfrac{\pi}{7}$ **49.** [6.4] $\dfrac{\sqrt{2}}{2}$ **50.** [6.4] $\dfrac{3}{\sqrt{b^2 + 9}}$

51. [6.4] $-\dfrac{7}{25}$

52. [6.5] $\dfrac{3\pi}{4} + 2k\pi,\ \dfrac{5\pi}{4} + 2k\pi$, or $135° + k \cdot 360°,$

$225° + k \cdot 360°$ **53.** [6.5] $\dfrac{\pi}{3} + k\pi$, or $60° + k \cdot 180°$

54. [6.5] $\dfrac{\pi}{6},\ \dfrac{5\pi}{6},\ \dfrac{7\pi}{6},\ \dfrac{11\pi}{6}$

55. [6.5] $\dfrac{\pi}{4},\ \dfrac{\pi}{2},\ \dfrac{3\pi}{4},\ \dfrac{5\pi}{4},\ \dfrac{3\pi}{2},\ \dfrac{7\pi}{4}$ **56.** [6.5] $\dfrac{2\pi}{3},\ \pi,\ \dfrac{4\pi}{3}$

57. [6.5] $0,\ \pi$ **58.** [6.5] $\dfrac{\pi}{4},\ \dfrac{3\pi}{4},\ \dfrac{5\pi}{4},\ \dfrac{7\pi}{4}$

59. [6.5] $0,\ \dfrac{\pi}{2},\ \pi,\ \dfrac{3\pi}{2}$ **60.** [6.5] $\dfrac{7\pi}{12},\ \dfrac{23\pi}{12}$

61. [6.5] $0.864,\ 2.972,\ 4.006,\ 6.114$

62. [6.3] B;

$\csc x - \cos x \cot x$	$\sin x$
$\dfrac{1}{\sin x} - \cos x \dfrac{\cos x}{\sin x}$	
$\dfrac{1 - \cos^2 x}{\sin x}$	
$\dfrac{\sin^2 x}{\sin x}$	
$\sin x$	

63. [6.3] D;

$\dfrac{1}{\sin x \cos x} - \dfrac{\cos x}{\sin x}$	$\dfrac{\sin x \cos x}{1 - \sin^2 x}$
$\dfrac{1}{\sin x \cos x} - \dfrac{\cos^2 x}{\sin x \cos x}$	$\dfrac{\sin x \cos x}{\cos^2 x}$
$\dfrac{1 - \cos^2 x}{\sin x \cos x}$	$\dfrac{\sin x}{\cos x}$
$\dfrac{\sin^2 x}{\sin x \cos x}$	
$\dfrac{\sin x}{\cos x}$	

64. [6.3] A;

$\dfrac{\cot x - 1}{1 - \tan x}$	$\dfrac{\csc x}{\sec x}$
$\dfrac{\frac{\cos x}{\sin x} - \frac{\sin x}{\sin x}}{\frac{\sin x}{\cos x} - \frac{\sin x}{\cos x}}$	$\dfrac{\frac{1}{\sin x}}{\frac{1}{\cos x}}$
$\dfrac{\cos x - \sin x}{\sin x} \cdot \dfrac{\cos x}{\cos x - \sin x}$	$\dfrac{1}{\sin x} \cdot \dfrac{\cos x}{1}$
$\dfrac{\cos x}{\sin x}$	$\dfrac{\cos x}{\sin x}$

65. [6.3] C;

$\dfrac{\cos x + 1}{\sin x} + \dfrac{\sin x}{\cos x + 1}$	$\dfrac{2}{\sin x}$
$\dfrac{(\cos x + 1)^2 + \sin^2 x}{\sin x (\cos x + 1)}$	
$\dfrac{\cos^2 x + 2\cos x + 1 + \sin^2 x}{\sin x (\cos x + 1)}$	
$\dfrac{2\cos x + 2}{\sin x (\cos x + 1)}$	
$\dfrac{2(\cos x + 1)}{\sin x (\cos x + 1)}$	
$\dfrac{2}{\sin x}$	

66. [6.5] 4.917 **67.** [6.5] No solution in $[0, 2\pi)$

68. [6.3] Discussion and Writing

(a) $2\cos^2 x - 1 = \cos 2x = \cos^2 x - \sin^2 x$
$= 1 \cdot (\cos^2 x - \sin^2 x)$
$= (\cos^2 x + \sin^2 x)(\cos^2 x - \sin^2 x)$
$= \cos^4 x - \sin^4 x;$

(b) $\cos^4 x - \sin^4 x = (\cos^2 x + \sin^2 x)(\cos^2 x - \sin^2 x)$
$= 1 \cdot (\cos^2 x - \sin^2 x)$
$= \cos^2 x - \sin^2 x = \cos 2x$
$= 2\cos^2 x - 1;$

(c)

$$\begin{array}{c|l} \hline 2\cos^2 x - 1 & \cos^4 x - \sin^4 x \\ \cos 2x & (\cos^2 x + \sin^2 x)(\cos^2 x - \sin^2 x) \\ & 1 \cdot (\cos^2 x - \sin^2 x) \\ & \cos^2 x - \sin^2 x \\ & \cos 2x \\ \hline \end{array}$$

Answer may vary. Method 2 may be the more efficient because it involves straightforward factorization and simplification. Method 1(a) requires a "trick" such as multiplying by a particular expression equivalent to 1.
69. Discussion and Writing [6.4] The ranges of the inverse trigonometric functions are restricted in order that they might be functions.
70. [6.1] 108.4°
71. [6.1]
$$\cos(u + v) = \cos u \cos v - \sin u \sin v$$
$$= \cos u \cos v - \cos\left(\frac{\pi}{2} - u\right)\cos\left(\frac{\pi}{2} - v\right)$$
72. [6.2] $\cos^2 x$
73. [6.2] $\sin\theta = \sqrt{\dfrac{1}{2} + \dfrac{\sqrt{6}}{5}}$; $\cos\theta = \sqrt{\dfrac{1}{2} - \dfrac{\sqrt{6}}{5}}$;
$$\tan\theta = \sqrt{\frac{5 + 2\sqrt{6}}{5 - 2\sqrt{6}}}$$
74. [6.3] $\ln e^{\sin t} = \log_e e^{\sin t} = \sin t$
75. [6.4]

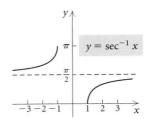

$y = \sec^{-1} x$

76. [6.4] Let $x = \dfrac{\sqrt{2}}{2}$. Then $\tan^{-1}\dfrac{\sqrt{2}}{2} \approx 0.6155$ and
$$\frac{\sin^{-1}\dfrac{\sqrt{2}}{2}}{\cos^{-1}\dfrac{\sqrt{2}}{2}} = \frac{\dfrac{\pi}{4}}{\dfrac{\pi}{4}} = 1. \quad \textbf{77.}\ [6.5]\ \frac{\pi}{2}, \frac{3\pi}{2}$$

Test, Chapter 6

1. [6.1] $2\cos x + 1$ **2.** [6.1] $2\cos\theta$ **3.** [6.1] $\dfrac{\sqrt{2} + \sqrt{6}}{4}$
4. [6.2] $\dfrac{\sqrt{5}}{3}$ **5.** [6.2] $\frac{24}{25}$ **6.** [6.2] $\dfrac{\sqrt{2} + \sqrt{3}}{2}$

7. [6.3] $\csc x - \cos x \cot x = \dfrac{1}{\sin x} - \cos x \cdot \dfrac{\cos x}{\sin x}$
$$= \frac{1 - \cos^2 x}{\sin x}$$
$$= \frac{\sin^2 x}{\sin x}$$
$$= \sin x$$
8. [6.3] $(\sin x + \cos x)^2 = \sin^2 x + 2\sin x \cos x + \cos^2 x$
$$= 1 + 2\sin x \cos x$$
$$= 1 + \sin 2x$$
9. [6.4] $-45°$ **10.** [6.4] $\dfrac{\pi}{3}$ **11.** [6.4] 2.3072
12. [6.4] $\dfrac{\sqrt{3}}{2}$ **13.** [6.5] $\dfrac{\pi}{6}, \dfrac{5\pi}{6}, \dfrac{7\pi}{6}, \dfrac{11\pi}{6}$
14. [6.5] $0, \dfrac{\pi}{4}, \dfrac{3\pi}{4}, \pi$ **15.** [6.2] $-\sqrt{\frac{11}{12}}$

Chapter 7

Exercise Set 7.1

1. $A = 121°$, $a \approx 33$, $c \approx 14$ **3.** $B \approx 57.4°$, $C \approx 86.1°$, $c \approx 40$, or $B \approx 122.6°$, $C \approx 20.9°$, $c \approx 14$
5. $B \approx 44°24'$, $A \approx 74°26'$, $a \approx 33.3$
7. $A = 110.36°$, $a \approx 5$ mi, $b \approx 3$ mi
9. $B \approx 83.78°$, $A \approx 12.44°$, $a \approx 12.30$ yd
11. $B \approx 14.7°$, $C \approx 135.0°$, $c \approx 28.04$ cm
13. No solution   **15.** $B = 125.27°$, $b \approx 302$ m, $c \approx 138$ m
17. 8.2 ft²   **19.** 12 yd²   **21.** 596.98 ft²   **23.** 787 ft²
25. About 12.86 ft, or 12 ft, 10 in.   **27.** About 51 ft
29. From A: about 35 mi; from B: about 66 mi
31. About 60 mi   **33.** Discussion and Writing
35. [5.1] 1.348, 77.2°   **36.** [5.1] No angle
37. [5.1] 18.24°   **38.** [5.1] 125.06°
39. Use the formula for the area of a triangle and the law of sines.
$$K = \frac{1}{2}bc\sin A \quad \text{and} \quad b = \frac{c\sin B}{\sin C},$$
$$\text{so} \quad K = \frac{c^2 \sin A \sin B}{2\sin C}.$$
$$K = \frac{1}{2}ab\sin C \quad \text{and} \quad b = \frac{a\sin B}{\sin A},$$
$$\text{so} \quad K = \frac{a^2 \sin B \sin C}{2\sin A}.$$
$$K = \frac{1}{2}bc\sin A \quad \text{and} \quad c = \frac{b\sin C}{\sin B},$$
$$\text{so} \quad K = \frac{b^2 \sin A \sin C}{2\sin B}.$$

41.

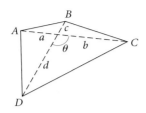

For the quadrilateral $ABCD$, we have

$$\text{Area} = \frac{1}{2} bd \sin \theta + \frac{1}{2} ac \sin \theta$$

$$+ \frac{1}{2} ad(\sin 180° - \theta) + \frac{1}{2} bc \sin (180° - \theta)$$

Note: $\sin \theta = \sin(180° - \theta)$.

$$= \frac{1}{2}(bd + ac + ad + bc) \sin \theta$$

$$= \frac{1}{2}(a + b)(c + d) \sin \theta$$

$$= \frac{1}{2} d_1 d_2 \sin \theta,$$

where $d_1 = a + b$ and $d_2 = c + d$.

Exercise Set 7.2

1. $a \approx 15$, $B \approx 24°$, $C \approx 126°$
3. $A \approx 36.18°$, $B \approx 43.53°$, $C \approx 100.29°$
5. $b \approx 75$ m, $A \approx 94°51'$, $C \approx 12°29'$
7. $A \approx 24.15°$, $B \approx 30.75°$, $C \approx 125.10°$
9. No solution
11. $A \approx 79.93°$, $B \approx 53.55°$, $C \approx 46.52°$
13. $c \approx 45.17$ mi, $A \approx 89.3°$, $B \approx 42.0°$
15. $a \approx 13.9$ in., $B \approx 36.127°$, $C \approx 90.417°$
17. Law of sines; $C = 98°$, $a \approx 96.7$, $c \approx 101.9$
19. Law of cosines; $A \approx 73.71°$, $B \approx 51.75°$, $C \approx 54.54°$
21. Cannot be solved
23. Law of cosines; $A \approx 33.71°$, $B \approx 107.08°$, $C \approx 39.21°$
25. About 367 ft   **27.** About 1.5 mi
29. About 37 nautical mi   **31.** About 912 km
33. (a) About 16 ft; **(b)** about 122 ft^2 **35.** About 4.7 cm
37. Discussion and Writing
39. [R.1] 5   **40.** [5.3] $\dfrac{\sqrt{3}}{2}$ **41.** [5.3] $\dfrac{\sqrt{2}}{2}$
42. [5.3] $-\dfrac{\sqrt{3}}{2}$ **43.** [5.3] $-\dfrac{1}{2}$ **44.** [2.2] 2
45. About 9386 ft
47. $A = \dfrac{1}{2} a^2 \sin \theta$; when $\theta = 90°$

Exercise Set 7.3

1. 5;

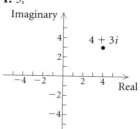

3. 1;

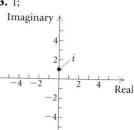

5. $\sqrt{17}$;

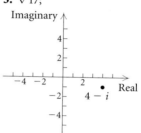

7. 3;

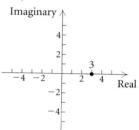

9. $3 - 3i$; $3\sqrt{2}\left(\cos \dfrac{7\pi}{4} + i \sin \dfrac{7\pi}{4}\right)$, or $3\sqrt{2}(\cos 315° + i \sin 315°)$

11. $4i$; $4\left(\cos \dfrac{\pi}{2} + i \sin \dfrac{\pi}{2}\right)$, or $4(\cos 90° + i \sin 90°)$

13. $\sqrt{2}\left(\cos \dfrac{7\pi}{4} + i \sin \dfrac{7\pi}{4}\right)$, or $\sqrt{2}(\cos 315° + i \sin 315°)$

15. $3\left(\cos \dfrac{3\pi}{2} + i \sin \dfrac{3\pi}{2}\right)$, or $3(\cos 270° + i \sin 270°)$

17. $2\left(\cos \dfrac{\pi}{6} + i \sin \dfrac{\pi}{6}\right)$, or $2(\cos 30° + i \sin 30°)$

19. $\dfrac{2}{5}(\cos 0 + i \sin 0)$, or $\dfrac{2}{5}(\cos 0° + i \sin 0°)$

21. $\dfrac{3\sqrt{3}}{2} + \dfrac{3}{2} i$ **23.** $-10i$ **25.** $2 + 2i$ **27.** $2i$

29. $4(\cos 42° + i \sin 42°)$ **31.** $11.25(\cos 56° + i \sin 56°)$

33. 4   **35.** $-i$ **37.** $6 + 6\sqrt{3}i$ **39.** $-2i$

41. $8(\cos \pi + i \sin \pi)$ **43.** $8\left(\cos \dfrac{3\pi}{2} + i \sin \dfrac{3\pi}{2}\right)$

45. $\dfrac{27}{2} + \dfrac{27\sqrt{3}}{2} i$ **47.** $-4 + 4i$ **49.** -1

51. $-\dfrac{\sqrt{2}}{2} + \dfrac{\sqrt{2}}{2} i$, $\dfrac{\sqrt{2}}{2} - \dfrac{\sqrt{2}}{2} i$

53. $2(\cos 157.5° + i \sin 157.5°)$, $2(\cos 337.5° + i \sin 337.5°)$

55. $\dfrac{\sqrt{3}}{2} + \dfrac{1}{2} i$, $-\dfrac{\sqrt{3}}{2} + \dfrac{1}{2} i$, $-i$

57. $\sqrt[3]{4}(\cos 110° + i \sin 110°)$, $\sqrt[3]{4}(\cos 230° + i \sin 230°)$, $\sqrt[3]{4}(\cos 350° + i \sin 350°)$

59. $2, 2i, -2, -2i$;

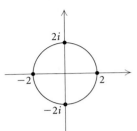

61. $\cos 36° + i \sin 36°$, $\cos 108° + i \sin 108°$, -1, $\cos 252° + i \sin 252°$, $\cos 324° + i \sin 324°$;

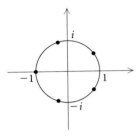

63. $\sqrt[10]{8}$, $\sqrt[10]{8}(\cos 36° + i \sin 36°)$, $\sqrt[10]{8}(\cos 72° + i \sin 72°)$, $\sqrt[10]{8}(\cos 108° + i \sin 108°)$, $\sqrt[10]{8}(\cos 144° + i \sin 144°)$, $-\sqrt[10]{8}$, $\sqrt[10]{8}(\cos 216° + i \sin 216°)$, $\sqrt[10]{8}(\cos 252° + i \sin 252°)$, $\sqrt[10]{8}(\cos 288° + i \sin 288°)$, $\sqrt[10]{8}(\cos 324° + i \sin 324°)$

65. $\frac{\sqrt{3}}{2} + \frac{1}{2}i$, i, $-\frac{\sqrt{3}}{2} + \frac{1}{2}i$, $-\frac{\sqrt{3}}{2} - \frac{1}{2}i$, $-i$, $\frac{\sqrt{3}}{2} - \frac{1}{2}i$

67. $1, -\frac{1}{2} + \frac{\sqrt{3}}{2}i, -\frac{1}{2} - \frac{\sqrt{3}}{2}i$

69. $\cos 67.5° + i \sin 67.5°$, $\cos 157.5° + i \sin 157.5°$, $\cos 247.5° + i \sin 247.5°$, $\cos 337.5° + i \sin 337.5°$

71. $\sqrt{3} + i, 2i, -\sqrt{3} + i, -\sqrt{3} - i, -2i, \sqrt{3} - i$

73. Left to the student   **75.** Discussion and Writing

77. [5.4] $15°$ **78.** [5.4] $540°$ **79.** [5.4] $\frac{11\pi}{6}$

80. [5.4] $-\frac{5\pi}{4}$ **81.** [R.6] $3\sqrt{5}$

82. [1.1]

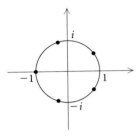

83. $-\frac{1 + \sqrt{3}}{2} + \frac{1 + \sqrt{3}}{2}i$, $-\frac{1 - \sqrt{3}}{2} + \frac{1 - \sqrt{3}}{2}i$

85. $\cos \theta - i \sin \theta$

87. $z = a + bi$, $|z| = \sqrt{a^2 + b^2}$; $\bar{z} = a - bi$, $|\bar{z}| = \sqrt{a^2 + (-b)^2} = \sqrt{a^2 + b^2}$, $\therefore |z| = |\bar{z}|$

89. $|(a + bi)^2| = |a^2 - b^2 + 2abi| = \sqrt{(a^2 - b^2)^2 + 4a^2b^2}$ $= \sqrt{a^4 + 2a^2b^2 + b^4} = a^2 + b^2$,

$|a + bi|^2 = (\sqrt{a^2 + b^2})^2 = a^2 + b^2$

91. $\dfrac{z}{w} = \dfrac{r_1(\cos \theta_1 + i \sin \theta_1)}{r_2(\cos \theta_2 + i \sin \theta_2)}$

$= \dfrac{r_1}{r_2}(\cos (\theta_1 - \theta_2) + i \sin (\theta_1 - \theta_2))$,

$\left|\dfrac{z}{w}\right| = \sqrt{\left[\dfrac{r_1}{r_2}\cos (\theta_1 - \theta_2)\right]^2 + \left[\dfrac{r_1}{r_2}\sin (\theta_1 - \theta_2)\right]^2}$

$= \sqrt{\dfrac{r_1^2}{r_2^2}} = \dfrac{|r_1|}{|r_2|}$;

$|z| = \sqrt{(r_1 \cos \theta_1)^2 + (r_1 \sin \theta_1)^2} = \sqrt{r_1^2} = |r_1|$;

$|w| = \sqrt{(r_2 \cos \theta_2)^2 + (r_2 \sin \theta_2)^2} = \sqrt{r_2^2} = |r_2|$;

Then $\left|\dfrac{z}{w}\right| = \dfrac{|r_1|}{|r_2|} = \dfrac{|z|}{|w|}$.

93.

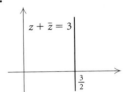

Exercise Set 7.4

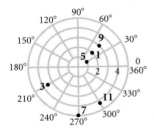

13. A: $(4, 30°)$, $(4, 390°)$, $(-4, 210°)$; B: $(5, 300°)$, $(5, -60°)$, $(-5, 120°)$; C: $(2, 150°)$, $(2, 510°)$, $(-2, 330°)$; D: $(3, 225°)$, $(3, -135°)$, $(-3, 45°)$; answers may vary

15. $(3, 270°)$, $\left(3, \dfrac{3\pi}{2}\right)$ **17.** $(6, 300°)$, $\left(6, \dfrac{5\pi}{3}\right)$

19. $(8, 330°)$, $\left(8, \dfrac{11\pi}{6}\right)$ **21.** $(2, 225°)$, $\left(2, \dfrac{5\pi}{4}\right)$

23. $\left(\dfrac{5}{2}, \dfrac{5\sqrt{3}}{2}\right)$ **25.** $\left(-\dfrac{3\sqrt{2}}{2}, -\dfrac{3\sqrt{2}}{2}\right)$

27. $\left(-\dfrac{3}{2}, -\dfrac{3\sqrt{3}}{2}\right)$ **29.** $(-1, \sqrt{3})$

31. $r(3\cos\theta + 4\sin\theta) = 5$ **33.** $r\cos\theta = 5$
35. $r = 6$ **37.** $r^2\cos^2\theta = 25r\sin\theta$
39. $r^2\sin^2\theta - 5r\cos\theta - 25 = 0$ **41.** $x^2 + y^2 = 25$
43. $y = 2$ **45.** $y^2 = -6x + 9$
47. $x^2 - 9x + y^2 - 7y = 0$ **49.** $x = 5$
51. **53.**

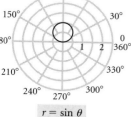

$r = \sin\theta$

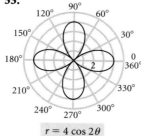

$r = 4\cos 2\theta$

55.

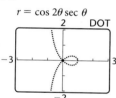

$r = \cos\theta$

57. $(7.616, 66.8°)$, $(7.616, 1.166)$
59. $(4.643, 132.9°)$, $(4.643, 2.320)$ **61.** $(2.19, -2.05)$
63. $(1.30, -3.99)$ **65.** (d) **67.** (g) **69.** (j) **71.** (b)
73. (e) **75.** (k)
77. **79.**

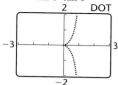

$r = \sin\theta\tan\theta$

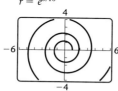

$r = e^{\theta/10}$

81. **83.**

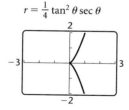

$r = \cos 2\theta\sec\theta$

$r = \frac{1}{4}\tan^2\theta\sec\theta$

85. Discussion and Writing
87. [2.1] 12 **88.** [2.1] $\dfrac{1}{5}$
89. [1.3] **90.** [1.3]

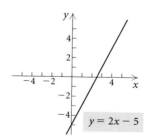

$y = 2x - 5$

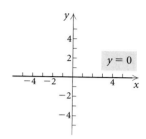

$4x - y = 6$

91. [1.3] **92.** [1.3]

$x = -3$

$y = 0$

93. $y^2 = -4x + 4$

Exercise Set 7.5

1. Yes **3.** No **5.** Yes **7.** No **9.** No **11.** Yes
13. 55 N, 55° **15.** 929 N, 19° **17.** 57.0, 38°
19. 18.4, 37° **21.** 20.9, 58° **23.** 68.3, 18°
25. 11 ft/sec, 63° **27.** 174 nautical mi, S15°E
29. 60° **31.** 70.7 east; 70.7 south
33. Horizontal: 215.17 mph forward; vertical: 65.78 mph up
35. Horizontal: 390 lb forward; vertical: 675.5 lb up
37. Northerly: 115 km/h; westerly: 164 km/h
39. Perpendicular: 90.6 lb; parallel: 42.3 lb
41. 48.1 lb **43.** Discussion and Writing
45. [5.5] $\dfrac{\sqrt{3}}{2}$ **46.** [5.5] $\dfrac{\sqrt{3}}{2}$ **47.** [5.5] $\dfrac{\sqrt{2}}{2}$
48. [5.5] $\dfrac{1}{2}$ **49.** **(a)** $(4.950, 4.950)$; **(b)** $(0.950, -1.978)$

Exercise Set 7.6

1. $\langle -9, 5\rangle$; $\sqrt{106}$ **3.** $\langle -3, 6\rangle$; $3\sqrt{5}$ **5.** $\langle 4, 0\rangle$; 4
7. $\sqrt{37}$ **9.** $\langle 4, -5\rangle$ **11.** $\sqrt{257}$ **13.** $\langle -9, 9\rangle$
15. $\langle 41, -38\rangle$ **17.** $\sqrt{261} - \sqrt{65}$ **19.** $\langle -1, -1\rangle$
21. $\langle -8, 14\rangle$ **23.** 1 **25.** -34

27.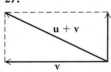

29.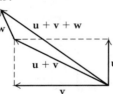

31. (a) $\mathbf{w} = \mathbf{u} + \mathbf{v}$; **(b)** $\mathbf{v} = \mathbf{w} - \mathbf{u}$

33. $\left\langle -\dfrac{5}{13}, \dfrac{12}{13} \right\rangle$ **35.** $\left\langle \dfrac{1}{\sqrt{101}}, -\dfrac{10}{\sqrt{101}} \right\rangle$

37. $\left\langle -\dfrac{1}{\sqrt{17}}, -\dfrac{4}{\sqrt{17}} \right\rangle$ **39.** $\mathbf{w} = -4\mathbf{i} + 6\mathbf{j}$

41. $\mathbf{s} = 2\mathbf{i} + 5\mathbf{j}$ **43.** $-7\mathbf{i} + 5\mathbf{j}$

45. (a) $3\mathbf{i} + 29\mathbf{j}$; **(b)** $\langle 3, 29 \rangle$ **47. (a)** $4\mathbf{i} + 16\mathbf{j}$; **(b)** $\langle 4, 16 \rangle$

49. $\mathbf{j}$, or $\langle 0, 1 \rangle$

51. $-\dfrac{1}{2}\mathbf{i} - \dfrac{\sqrt{3}}{2}\mathbf{j}$, or $\left\langle -\dfrac{1}{2}, -\dfrac{\sqrt{3}}{2} \right\rangle$ **53.** $248°$

55. $63°$ **57.** $50°$ **59.** $|\mathbf{u}| = 3$; $\theta = 45°$

61. 1; $120°$ **63.** $144.2°$ **65.** $14.0°$ **67.** $101.3°$

69.

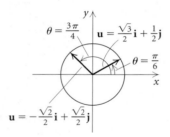

71. $\mathbf{u} = -\dfrac{\sqrt{2}}{2}\mathbf{i} - \dfrac{\sqrt{2}}{2}\mathbf{j}$

73. $\mathbf{u} = -\dfrac{\sqrt{10}}{10}\mathbf{i} + \dfrac{3\sqrt{10}}{10}\mathbf{j}$ **75.** $\sqrt{13}\left(\dfrac{2\sqrt{13}}{13}\mathbf{i} - \dfrac{3\sqrt{13}}{13}\mathbf{j} \right)$

77.

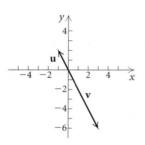

79. 174 nautical mi, S17°E   **81.** $60°$

83. 500 lb on left, 866 lb on right

85. Cable: 224-lb tension, boom: 167-lb compression

87. $\mathbf{u} + \mathbf{v} = \langle u_1, u_2 \rangle + \langle v_1, v_2 \rangle$
$= \langle u_1 + v_1, u_2 + v_2 \rangle$
$= \langle v_1 + u_1, v_2 + u_2 \rangle$
$= \langle v_1, v_2 \rangle + \langle u_1, u_2 \rangle$
$= \mathbf{v} + \mathbf{u}$

89. Discussion and Writing

91. [1.3] $-\dfrac{1}{5}$; $(0, -15)$ **92.** [1.3] 0; $(0, 7)$

93. [3.1] $0, 4$ **94.** [2.3] $-\dfrac{11}{3}, \dfrac{5}{2}$

95. (a) $\cos \theta = \dfrac{\mathbf{u} \cdot \mathbf{v}}{|\mathbf{u}| \, |\mathbf{v}|} = \dfrac{0}{|\mathbf{u}| \, |\mathbf{v}|}$, $\therefore \cos \theta = 0$ and $\theta = 90°$.
(b) Answers may vary. $\mathbf{u} = \langle 2, -3 \rangle$ and $\mathbf{v} = \langle -3, -2 \rangle$;
$\mathbf{u} \cdot \mathbf{v} = 2(-3) + (-3)(-2) = 0$

97. $\dfrac{3}{5}\mathbf{i} - \dfrac{4}{5}\mathbf{j}, \ -\dfrac{3}{5}\mathbf{i} + \dfrac{4}{5}\mathbf{j}$ **99.** $(5, 8)$

Review Exercises, Chapter 7

1. [7.2] $A \approx 153°, B \approx 18°, C \approx 9°$
2. [7.1] $A = 118°, a \approx 37$ in., $c \approx 24$ in.
3. [7.1] $B = 14°50', a \approx 2523$ m, $c \approx 1827$ m
4. [7.1] No solution   **5.** [7.1] 33 m^2 **6.** [7.1] 13.72 ft^2
7. [7.1] 63 ft^2 **8.** [7.2] $92°, 33°, 55°$
9. [7.1] 419 ft   **10.** [7.2] About 650 km
11. [7.3] $\sqrt{29}$;   **12.** [7.3] 4;

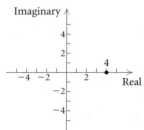

13. [7.3] 2;   **14.** [7.3] $\sqrt{10}$;

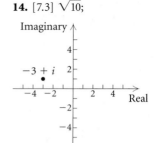

15. [7.3] $\sqrt{2}\left(\cos \dfrac{\pi}{4} + i \sin \dfrac{\pi}{4} \right)$, or
$\sqrt{2}(\cos 45° + i \sin 45°)$

16. [7.3] $4\left(\cos \dfrac{3\pi}{2} + i \sin \dfrac{3\pi}{2} \right)$, or $4(\cos 270° + i \sin 270°)$

17. [7.3] $10\left(\cos \dfrac{5\pi}{6} + i \sin \dfrac{5\pi}{6}\right)$, or $10(\cos 150° + i \sin 150°)$

18. [7.3] $\dfrac{3}{4}(\cos 0 + i \sin 0)$, or $\dfrac{3}{4}(\cos 0° + i \sin 0°)$

19. [7.3] $2 + 2\sqrt{3}i$ **20.** [7.3] 7 **21.** [7.3] $-\dfrac{5}{2} + \dfrac{5\sqrt{3}}{2}i$

22. [7.3] $1 - \sqrt{3}i$ **23.** [7.3] $1 + \sqrt{3} + (-1 + \sqrt{3})i$

24. [7.3] $-i$ **25.** [7.3] $2i$ **26.** [7.3] $3\sqrt{3} + 3i$

27. [7.3] $8(\cos 180° + i \sin 180°)$

28. [7.3] $4(\cos 7\pi + i \sin 7\pi)$ **29.** [7.3] $-8i$

30. [7.3] $-\dfrac{1}{2} - \dfrac{\sqrt{3}}{2}i$

31. [7.3] $\sqrt[4]{2}\left(\cos \dfrac{3\pi}{8} + i \sin \dfrac{3\pi}{8}\right)$,

$\sqrt[4]{2}\left(\cos \dfrac{11\pi}{8} + i \sin \dfrac{11\pi}{8}\right)$

32. [7.3] $\sqrt[3]{6}(\cos 110° + i \sin 110°)$,

$\sqrt[3]{6}(\cos 230° + i \sin 230°)$, $\sqrt[3]{6}(\cos 350° + i \sin 350°)$

33. [7.3] $3, 3i, -3, -3i$

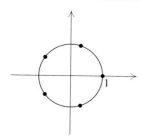

34. [7.3] $1, \cos 72° + i \sin 72°, \cos 144° + i \sin 144°$,
$\cos 216° + i \sin 216°, \cos 288° + i \sin 288°$

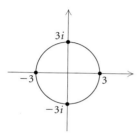

35. [7.3] $\cos 22.5° + i \sin 22.5°, \cos 112.5° + i \sin 112.5°$,
$\cos 202.5° + i \sin 202.5°, \cos 292.5° + i \sin 292.5°$

36. [7.3] $\dfrac{1}{2} + \dfrac{\sqrt{3}}{2}i, -1, \dfrac{1}{2} - \dfrac{\sqrt{3}}{2}i$

37. [7.4] A: $(5, 120°)$, $(5, 480°)$, $(-5, 300°)$; B: $(3, 210°)$,
$(-3, 30°)$, $(-3, 390°)$; C: $(4, 60°)$, $(4, 420°)$, $(-4, 240°)$;
D: $(1, 300°)$, $(1, -60°)$, $(-1, 120°)$; answers may vary

38. [7.4] $(8, 135°)$, $\left(8, \dfrac{3\pi}{4}\right)$ **39.** [7.4] $(5, 270°)$, $\left(5, \dfrac{3\pi}{2}\right)$

40. [7.4] $\left(\dfrac{3\sqrt{2}}{2}, \dfrac{3\sqrt{2}}{2}\right)$ **41.** [7.4] $(3, 3\sqrt{3})$

42. [7.4] $r(5 \cos \theta - 2 \sin \theta) = 6$ **43.** [7.4] $r \sin \theta = 3$

44. [7.4] $r = 3$ **45.** [7.4] $r^2 \sin^2 \theta - 4r \cos \theta - 16 = 0$

46. [7.4] $x^2 + y^2 = 36$ **47.** [7.4] $x^2 + 2y = 1$

48. [7.4] $y^2 - 6x = 9$ **49.** [7.4] $x^2 - 2x + y^2 - 3y = 0$

50. [7.4] (b)   **51.** [7.4] (d)   **52.** [7.4] (a)   **53.** [7.4] (c)

54. [7.5] 13.7, 71°   **55.** [7.5] 98.7, 15°

56. [7.5]   **57.** [7.5]

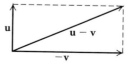

58. [7.5] 666.7 N, 36°   **59.** [7.5] 29 km/h, 329°

60. [7.5] 102.4 nautical mi, S43°E   **61.** [7.6] $\langle -4, 3\rangle$

62. [7.6] $\langle 2, -6\rangle$ **63.** [7.6] $\sqrt{61}$ **64.** [7.6] $\langle 10, -21\rangle$

65. [7.6] $\langle 14, -64\rangle$ **66.** [7.6] $5 + \sqrt{116}$ **67.** [7.6] 14

68. [7.6] $\left\langle -\dfrac{3}{\sqrt{10}}, -\dfrac{1}{\sqrt{10}}\right\rangle$ **69.** [7.6] $-9\mathbf{i} + 4\mathbf{j}$

70. [7.6] 194.0°   **71.** [7.6] $\sqrt{34}$; $\theta = 211.0°$

72. [7.6] 111.8°   **73.** [7.6] 85.1°   **74.** [7.6] $34\mathbf{i} - 55\mathbf{j}$

75. [7.6] $\mathbf{i} - 12\mathbf{j}$ **76.** [7.6] $5\sqrt{2}$

77. [7.6] $3\sqrt{65} + \sqrt{109}$ **78.** [7.6] $-5\mathbf{i} + 5\mathbf{j}$

79. [7.6]

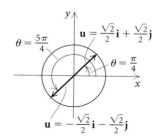

80. [7.6]

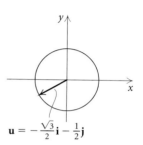

81. [7.6] $\sqrt{10}\left(\dfrac{3\sqrt{10}}{10}\mathbf{i} - \dfrac{\sqrt{10}}{10}\mathbf{j}\right)$

82. Discussion and Writing [7.1] When only the lengths of the sides of a triangle are known, the law of sines cannot be used to find the first angle because there will be two unknowns in the equation. To use the law of sines, you must know the measure of at least one angle and the length of the side opposite it.
83. Discussion and Writing [7.1], [7.2] A triangle has no solution when a sine or cosine value found is less than -1 or greater than 1. A triangle also has no solution if the sum of the angle measures calculated is greater than 180°. A triangle has only one solution if only one possible answer is found, or if one of the possible answers has an angle sum greater than 180°. A triangle has two solutions when two possible answers are found and neither results in an angle sum greater than 180°.
84. Discussion and Writing [7.2] For 150-yd shot, about 13.1 yd; for 300-yd shot, about 26.2 yd
85. [7.6] $\frac{36}{13}\mathbf{i} + \frac{15}{13}\mathbf{j}$ **86.** [7.1] 50.52°, 129.48°

Test, Chapter 7

1. [7.1] $A = 83°$, $b \approx 14.7$ ft, $c \approx 12.4$ ft
2. [7.1] $A \approx 73.9°$, $B \approx 70.1°$, $a \approx 8.2$ m, or $A \approx 34.1°$, $B \approx 109.9°$, $a \approx 4.8$ m
3. [7.2] $A \approx 99.9°$, $B \approx 36.8°$, $C \approx 43.3°$
4. [7.1] About 43.6 cm² **5.** [7.1] About 77 m
6. [7.5] About 930 km
7. [7.3]

Imaginary

$-4 + i$

Real

8. [7.3] $\sqrt{13}$ **9.** [7.3] $3\sqrt{2}(\cos 315° + i \sin 315°)$
10. [7.3] $\frac{1}{4}i$ **11.** [7.3] 16
12. [7.4] $2(\cos 120° + i \sin 120°)$
13. [7.4] $\left(\frac{1}{2}, -\frac{\sqrt{3}}{2}\right)$ **14.** [7.4] $r = \sqrt{10}$
15. [7.4]

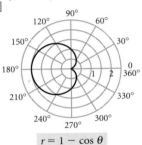

$r = 1 - \cos \theta$

16. [7.4] (a)   **17.** [7.5] Magnitude: 11.2, direction: 23.4°
18. [7.6] $-11i - 17j$ **19.** [7.6] $-\frac{4}{5}i + \frac{3}{5}j$
20. [7.1] 28.9°, 151.1°

Chapter 8

Exercise Set 8.1

1. (c)   **3.** (f)   **5.** (b)   **7.** $(-1, 3)$ **9.** $(-1, 1)$
11. No solution   **13.** $(5, 4)$ **15.** $(1, -3)$ **17.** $(2, -2)$
19. $\left(\frac{39}{11}, -\frac{1}{11}\right)$ **21.** $(1, -1)$
23. $(1, 3)$; consistent, independent
25. $(-4, -2)$; consistent, independent
27. $(-3, 0)$; consistent, independent
29. $(4y + 2, y)$ or $\left(x, \frac{1}{4}x - \frac{1}{2}\right)$; consistent, dependent
31. $(10, 8)$; consistent, independent
33. $(1, 1)$; consistent, independent
35. 10 free rentals, 38 boxes of popcorn
37. Adult: $6, child: $3   **39.** (15, $100)
41. 140   **43.** 6000
45. $1\frac{1}{2}$ servings of spaghetti, 2 servings of lettuce
47. Boat: 20 km/h, stream: 3 km/h
49. $6000 at 7%, $9000 at 9%
51. 6 lb of French roast, 4 lb of Kenyan
53. $21,428.57   **55.** Left to the student
57. **(a)** $t(x) = 1.69541779x + 44.15363881$, $r(x) = -0.4528301887x + 68.83018868$; **(b)** 11.5 yr
59. Discussion and Writing   **61.** [2.1] $\frac{5}{2}$, or 2.5
62. [2.1] 2   **63.** [2.3] 1, 3   **64.** [2.3] $\dfrac{1 \pm \sqrt{61}}{6}$
65. 4 km   **67.** First train: 36 km/h, second train: 54 km/h   **69.** $A = \frac{1}{10}$, $B = -\frac{7}{10}$
71. City: 209; highway: 196

Exercise Set 8.2

1. $(3, -2, 1)$ **3.** $(-3, 2, 1)$ **5.** $\left(2, \frac{1}{2}, -2\right)$
7. No solution   **9.** $\left(\dfrac{11y + 19}{5}, y, \dfrac{9y + 11}{5}\right)$
11. $\left(\frac{1}{2}, \frac{2}{3}, -\frac{5}{6}\right)$ **13.** $(1, -2, 4, -1)$
15. Under 10 lb: 60; 10 lb up to 15 lb: 70; 15 lb or more: 20
17. $1\frac{1}{4}$ servings of beef, 1 baked potato, $\frac{3}{4}$ serving strawberries   **19.** 4%: $1300; 6%: $900; 7%: $2800
21. Orange juice: $1; bagel: $1.25; coffee: $0.75
23. Airways: 351 billion; bus: 32 billion; railroad: 22 billion   **25.** Par-3: 4; par-4: 10; par-5: 4
27. **(a)** $f(x) = -\frac{31}{60}x^2 + \frac{233}{30}x + 211$;
(b) about 82 thousand

29. (a) $f(x) = \frac{69}{110}x^2 - \frac{939}{110}x + 594$; **(b)** 674 lb
31. (a) $f(x) = 0.1729980538x^2 - 10.70013387x +$
465.7917864; **(b)** 788; 844   **33.** Discussion and Writing
35. [2.2] $5 - 3i$ **36.** [2.2] $6 - 2i$ **37.** [2.2] $10 - 10i$
38. [2.2] $\frac{9}{25} + \frac{13}{25}i$ **39.** $\left(-1, \frac{1}{5}, -\frac{1}{2}\right)$
41. $180°$ **43.** $3x + 4y + 2z = 12$
45. $y = -4x^3 + 5x^2 - 3x + 1$
47. Adults: 5; students: 1; children: 94

Exercise Set 8.3

1. 3×2 **3.** 1×4 **5.** 3×3 **7.** $\begin{bmatrix} 2 & -1 & 7 \\ 1 & 4 & -5 \end{bmatrix}$

9. $\begin{bmatrix} 1 & -2 & 3 & 12 \\ 2 & 0 & -4 & 8 \\ 0 & 3 & 1 & 7 \end{bmatrix}$

11. $3x - 5y = 1$,
$\quad x + 4y = -2$
13. $2x + y - 4z = 12$,
$\quad 3x \quad + 5z = -1$,
$\quad x - y + z = 2$
15. $\left(\frac{3}{2}, \frac{5}{2}\right)$ **17.** $\left(-\frac{63}{29}, -\frac{114}{29}\right)$ **19.** $\left(-1, \frac{5}{2}\right)$
21. $(0, 3)$ **23.** No solution   **25.** $(3y - 2, y)$
27. $(-1, 2, -2)$ **29.** $\left(\frac{3}{2}, -4, 3\right)$ **31.** $(-1, 6, 3)$
33. $\left(\frac{1}{2}z + \frac{1}{2}, -\frac{1}{2}z - \frac{1}{2}, z\right)$ **35.** $(r - 2, -2r + 3, r)$
37. No solution   **39.** $(1, -3, -2, -1)$ **41.** 1:00 A.M.
43. \$8000 at 8%; \$12,000 at 10%; \$10,000 at 12%
45. Discussion and Writing   **47.** [2.3] $\dfrac{-1 \pm \sqrt{57}}{4}$
48. [2.5] $-3, -2$ **49.** [2.5] 4   **50.** [2.5] 9
51. $y = 3x^2 + \frac{5}{2}x - \frac{15}{2}$ **53.** $\begin{bmatrix} 1 & 5 \\ 0 & 1 \end{bmatrix}, \begin{bmatrix} 1 & 0 \\ 0 & 1 \end{bmatrix}$
55. $\left(-\frac{4}{3}, -\frac{1}{3}, 1\right)$ **57.** $\left(-\frac{14}{13}z - 1, \frac{3}{13}z - 2, z\right)$
59. $(-3, 3)$

Exercise Set 8.4

1. $x = -3, y = 5$ **3.** $x = -1, y = 1$

5. $\begin{bmatrix} -2 & 7 \\ 6 & 2 \end{bmatrix}$ **7.** $\begin{bmatrix} 1 & 3 \\ 2 & 6 \end{bmatrix}$ **9.** $\begin{bmatrix} 9 & 9 \\ -3 & -3 \end{bmatrix}$

11. $\begin{bmatrix} 11 & 13 \\ 5 & 3 \end{bmatrix}$ **13.** $\begin{bmatrix} -4 & 3 \\ -2 & -4 \end{bmatrix}$ **15.** $\begin{bmatrix} 17 & 9 \\ -2 & 1 \end{bmatrix}$

17. $\begin{bmatrix} 0 & 0 \\ 0 & 0 \end{bmatrix}$ **19.** $\begin{bmatrix} 1 & 2 \\ 4 & 3 \end{bmatrix}$ **21.** $\begin{bmatrix} 1 \\ 40 \end{bmatrix}$

23. $\begin{bmatrix} -10 & 28 \\ 14 & -26 \\ 0 & -6 \end{bmatrix}$ **25.** Not defined   **27.** $\begin{bmatrix} 3 & 16 & 3 \\ 0 & -32 & 0 \\ -6 & 4 & 5 \end{bmatrix}$

29. (a) $[150 \quad 80 \quad 40]$; **(b)** $[157.5 \quad 84 \quad 42]$;
(c) $[307.5 \quad 164 \quad 82]$, the total budget for each area in June
and July   **31. (a)** $\mathbf{C} = [140 \quad 27 \quad 3 \quad 13 \quad 64]$,
$\mathbf{P} = [180 \quad 4 \quad 11 \quad 24 \quad 662]$, $\mathbf{B} = [50 \quad 5 \quad 1 \quad 82 \quad 20]$;
(b) $[650 \quad 50 \quad 28 \quad 307 \quad 1448]$, the total nutritional values
of a meal of 1 serving of chicken, 1 cup of potato salad,
and 3 broccoli spears

33. (a) $\begin{bmatrix} 45.29 & 6.63 & 10.94 & 7.42 & 8.01 \\ 53.78 & 4.95 & 9.83 & 6.16 & 12.56 \\ 47.13 & 8.47 & 12.66 & 8.29 & 9.43 \\ 51.64 & 7.12 & 11.57 & 9.35 & 10.72 \end{bmatrix}$;

(b) $[65 \quad 48 \quad 93 \quad 57]$;
(c) $[12{,}851.86 \quad 1862.1 \quad 3019.81 \quad 2081.9 \quad 2611.56]$;
(d) the total cost, in cents, for each item for the day's
meals

35. (a) $\begin{bmatrix} 8 & 15 \\ 6 & 10 \\ 4 & 3 \end{bmatrix}$; **(b)** $[3 \quad 1.50 \quad 2]$; **(c)** $[41 \quad 66]$;

(d) the total cost, in dollars, of ingredients for each coffee
shop
37. (a) $[6 \quad 4.50 \quad 5.20]$; **(b)** $\mathbf{PS} = [95.80 \quad 150.60]$

39. $\begin{bmatrix} 2 & -3 \\ 1 & 5 \end{bmatrix}\begin{bmatrix} x \\ y \end{bmatrix} = \begin{bmatrix} 7 \\ -6 \end{bmatrix}$

41. $\begin{bmatrix} 1 & 1 & -2 \\ 3 & -1 & 1 \\ 2 & 5 & -3 \end{bmatrix}\begin{bmatrix} x \\ y \\ z \end{bmatrix} = \begin{bmatrix} 6 \\ 7 \\ 8 \end{bmatrix}$

43. $\begin{bmatrix} 3 & -2 & 4 \\ 2 & 1 & -5 \end{bmatrix}\begin{bmatrix} x \\ y \\ z \end{bmatrix} = \begin{bmatrix} 17 \\ 13 \end{bmatrix}$

45. $\begin{bmatrix} -4 & 1 & -1 & 2 \\ 1 & 2 & -1 & -1 \\ -1 & 1 & 4 & -3 \\ 2 & 3 & 5 & -7 \end{bmatrix}\begin{bmatrix} w \\ x \\ y \\ z \end{bmatrix} = \begin{bmatrix} 12 \\ 0 \\ 1 \\ 9 \end{bmatrix}$

47. Left to the student   **49.** Discussion and Writing
51. [2.4] **(a)** $\left(\frac{3}{2}, -\frac{49}{4}\right)$; **(b)** $x = \frac{3}{2}$; **(c)** minimum: $-\frac{49}{4}$
52. [2.4] **(a)** $\left(\frac{5}{4}, -\frac{49}{8}\right)$; **(b)** $x = \frac{5}{4}$; **(c)** minimum: $-\frac{49}{8}$
53. [2.4] **(a)** $\left(-\frac{3}{2}, \frac{29}{4}\right)$; **(b)** $x = -\frac{3}{2}$; **(c)** maximum: $\frac{29}{4}$
54. [2.4] **(a)** $\left(\frac{2}{3}, \frac{7}{3}\right)$; **(b)** $x = \frac{2}{3}$; **(c)** maximum: $\frac{7}{3}$
55. $(\mathbf{A} + \mathbf{B})(\mathbf{A} - \mathbf{B}) = \begin{bmatrix} -2 & 1 \\ 2 & -1 \end{bmatrix}$; $\mathbf{A}^2 - \mathbf{B}^2 = \begin{bmatrix} 0 & 3 \\ 0 & -3 \end{bmatrix}$

57. $(A + B)(A - B) = \begin{bmatrix} -2 & 1 \\ 2 & -1 \end{bmatrix}$

$$= A^2 + BA - AB - B^2$$

59. $A + B =$

$$\begin{bmatrix} a_{11} + b_{11} & a_{12} + b_{12} & a_{13} + b_{13} & \cdots & a_{1n} + b_{1n} \\ a_{21} + b_{21} & a_{22} + b_{22} & a_{23} + b_{23} & \cdots & a_{2n} + b_{2n} \\ a_{31} + b_{31} & a_{32} + b_{32} & a_{33} + b_{33} & \cdots & a_{3n} + b_{3n} \\ \vdots & \vdots & \vdots & & \vdots \\ a_{m1} + b_{m1} & a_{m2} + b_{m2} & a_{m3} + b_{m3} & \cdots & a_{mn} + b_{mn} \end{bmatrix}$$

$$=$$

$$\begin{bmatrix} b_{11} + a_{11} & b_{12} + a_{12} & b_{13} + a_{13} & \cdots & b_{1n} + a_{1n} \\ b_{21} + a_{21} & b_{22} + a_{22} & b_{23} + a_{23} & \cdots & b_{2n} + a_{2n} \\ b_{31} + a_{31} & b_{32} + a_{32} & b_{33} + a_{33} & \cdots & b_{3n} + a_{3n} \\ \vdots & \vdots & \vdots & & \vdots \\ b_{m1} + a_{m1} & b_{m2} + a_{m2} & b_{m3} + a_{m3} & \cdots & b_{mn} + a_{mn} \end{bmatrix}$$

$$= B + A$$

61. $(kl)A =$

$$\begin{bmatrix} (kl)a_{11} & (kl)a_{12} & (kl)a_{13} & \cdots & (kl)a_{1n} \\ (kl)a_{21} & (kl)a_{22} & (kl)a_{23} & \cdots & (kl)a_{2n} \\ (kl)a_{31} & (kl)a_{32} & (kl)a_{33} & \cdots & (kl)a_{3n} \\ \vdots & \vdots & \vdots & & \vdots \\ (kl)a_{m1} & (kl)a_{m2} & (kl)a_{m3} & \cdots & (kl)a_{mn} \end{bmatrix}$$

$$=$$

$$\begin{bmatrix} k(la_{11}) & k(la_{12}) & k(la_{13}) & \cdots & k(la_{1n}) \\ k(la_{21}) & k(la_{22}) & k(la_{23}) & \cdots & k(la_{2n}) \\ k(la_{31}) & k(la_{32}) & k(la_{33}) & \cdots & k(la_{3n}) \\ \vdots & \vdots & \vdots & & \vdots \\ k(la_{m1}) & k(la_{m2}) & k(la_{m3}) & \cdots & k(la_{mn}) \end{bmatrix}$$

$$= k \begin{bmatrix} la_{11} & la_{12} & la_{13} & \cdots & la_{1n} \\ la_{21} & la_{22} & la_{23} & \cdots & la_{2n} \\ la_{31} & la_{32} & la_{33} & \cdots & la_{3n} \\ \vdots & \vdots & \vdots & & \vdots \\ la_{m1} & la_{m2} & la_{m3} & \cdots & la_{mn} \end{bmatrix}$$

$$= k(lA)$$

63. $(k + l)A =$

$$\begin{bmatrix} (k + l)a_{11} & (k + l)a_{12} & (k + l)a_{13} & \cdots & (k + l)a_{1n} \\ (k + l)a_{21} & (k + l)a_{22} & (k + l)a_{23} & \cdots & (k + l)a_{2n} \\ (k + l)a_{31} & (k + l)a_{32} & (k + l)a_{33} & \cdots & (k + l)a_{3n} \\ \vdots & \vdots & \vdots & & \vdots \\ (k + l)a_{m1} & (k + l)a_{m2} & (k + l)a_{m3} & \cdots & (k + l)a_{mn} \end{bmatrix}$$

$$=$$

$$\begin{bmatrix} ka_{11} + la_{11} & ka_{12} + la_{12} & ka_{13} + la_{13} & \cdots & ka_{1n} + la_{1n} \\ ka_{21} + la_{21} & ka_{22} + la_{22} & ka_{23} + la_{23} & \cdots & ka_{2n} + la_{2n} \\ ka_{31} + la_{31} & ka_{32} + la_{32} & ka_{33} + la_{33} & \cdots & ka_{3n} + la_{3n} \\ \vdots & \vdots & \vdots & & \vdots \\ ka_{m1} + la_{m1} & ka_{m2} + la_{m2} & ka_{m3} + la_{m3} & \cdots & ka_{mn} + la_{mn} \end{bmatrix}$$

$$= kA + lA$$

Exercise Set 8.5

1. Yes   **3.** No   **5.** $\begin{bmatrix} -3 & 2 \\ 5 & -3 \end{bmatrix}$ **7.** Does not exist

9. $\begin{bmatrix} \frac{2}{5} & -\frac{3}{5} \\ \frac{1}{5} & -\frac{4}{5} \end{bmatrix}$ **11.** $\begin{bmatrix} \frac{3}{8} & -\frac{1}{4} & \frac{1}{8} \\ -\frac{1}{8} & \frac{3}{4} & -\frac{3}{8} \\ -\frac{1}{4} & \frac{1}{2} & \frac{1}{4} \end{bmatrix}$ **13.** Does not exist

15. $\begin{bmatrix} -1 & -1 & -6 \\ 1 & 0 & 2 \\ 0 & 1 & 3 \end{bmatrix}$ **17.** $\begin{bmatrix} 1 & 1 & 2 \\ 1 & 1 & 1 \\ 2 & 3 & 4 \end{bmatrix}$

19. Does not exist   **21.** $\begin{bmatrix} 1 & -2 & 3 & 8 \\ 0 & 1 & -3 & 1 \\ 0 & 0 & 1 & -2 \\ 0 & 0 & 0 & -1 \end{bmatrix}$

23. $\begin{bmatrix} 0.25 & 0.25 & 1.25 & -0.25 \\ 0.5 & 1.25 & 1.75 & -1 \\ -0.25 & -0.25 & -0.75 & 0.75 \\ 0.25 & 0.5 & 0.75 & -0.5 \end{bmatrix}$

25. $(-23, 83)$ **27.** $(-1, 5, 1)$ **29.** $(2, -2)$ **31.** $(0, 2)$
33. $(3, -3, -2)$ **35.** $(-1, 0, 1)$ **37.** $(1, -1, 0, 1)$
39. 50 sausages, 95 hot dogs
41. Topsoil: \$239; mulch: \$179; pea gravel: \$222
43. Left to the student   **45.** Left to the student
47. Discussion and Writing
49. [3.2] -48 **50.** [3.2] 194
51. [3.2] $(x + 2)(x - 1)(x - 4)$
52. [3.2] $(x + 5)(x + 1)(x - 1)(x - 3)$

53. A^{-1} exists if and only if $x \neq 0$. $A^{-1} = \begin{bmatrix} \dfrac{1}{x} \end{bmatrix}$

55. $\mathbf{A}^{-1}$ exists if and only if $xyz \neq 0$. $\mathbf{A}^{-1} = \begin{bmatrix} 0 & 0 & \dfrac{1}{z} \\ 0 & \dfrac{1}{y} & 0 \\ \dfrac{1}{x} & 0 & 0 \end{bmatrix}$

Exercise Set 8.6

1. -11 **3.** $x^3 - 4x$ **5.** -109 **7.** $-x^4 + x^2 - 5x$
9. $M_{11} = 6$, $M_{32} = -9$, $M_{22} = -29$
11. $A_{11} = 6$, $A_{32} = 9$, $A_{22} = -29$
13. -10 **15.** -10 **17.** $M_{41} = -14$, $M_{33} = 20$
19. $A_{24} = 15$, $A_{43} = 30$ **21.** 110 **23.** $\left(-\frac{25}{2}, -\frac{11}{2}\right)$
25. $(3, 1)$ **27.** $\left(\frac{1}{2}, -\frac{1}{3}\right)$ **29.** $(1, 1)$ **31.** $\left(\frac{3}{2}, \frac{13}{14}, \frac{33}{14}\right)$
33. $(3, -2, 1)$ **35.** $(1, 3, -2)$ **37.** $\left(\frac{1}{2}, \frac{2}{3}, -\frac{5}{6}\right)$
39. Left to the student   **41.** 110 **43.** Left to the student
45. Discussion and Writing   **47.** [4.1] $f^{-1}(x) = \dfrac{x-2}{3}$
48. [4.1] Not one-to-one   **49.** [4.1] Not one-to-one
50. [4.1] $f^{-1}(x) = (x-1)^3$
51. ± 2 **53.** $(-\infty, -\sqrt{3}] \cup [\sqrt{3}, \infty)$ **55.** -34
57. 4
59. Answers may vary.   **61.** Answers may vary.
$\begin{vmatrix} L & -W \\ 2 & 2 \end{vmatrix}$ $\begin{vmatrix} a & b \\ -b & a \end{vmatrix}$
63. Answers may vary.
$\begin{vmatrix} 2\pi r & 2\pi r \\ -h & r \end{vmatrix}$

Exercise Set 8.7

1. (f)   **3.** (h)   **5.** (g)   **7.** (b)
9.
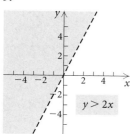
$y > 2x$

11.
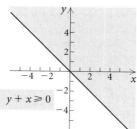
$y + x \geqslant 0$

13.
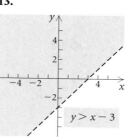
$y > x - 3$

15.
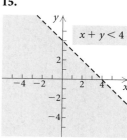
$x + y < 4$

17.
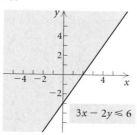
$3x - 2y \leqslant 6$

19.
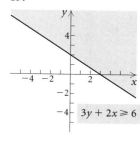
$3y + 2x \geqslant 6$

21.
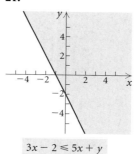
$3x - 2 \leqslant 5x + y$

23.
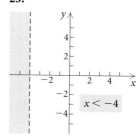
$x < -4$

25.
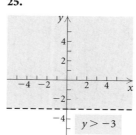
$y > -3$

27.
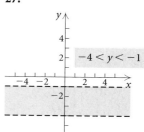
$-4 < y < -1$

29.

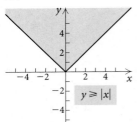

$y \geq |x|$

31. (f)   **33.** (a)   **35.** (b)

37.

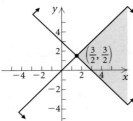

$\left(\frac{3}{2}, \frac{3}{2}\right)$

39.

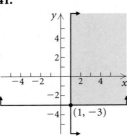

41.

$(1, -3)$

43.

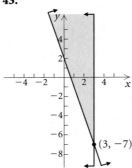

$(3, -7)$

45.

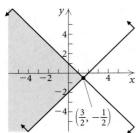

$\left(\frac{3}{2}, -\frac{1}{2}\right)$

47.

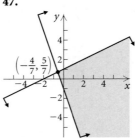

$\left(-\frac{4}{7}, \frac{5}{7}\right)$

49.

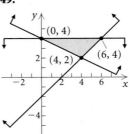

$(0, 4)$ $(4, 2)$ $(6, 4)$

51.

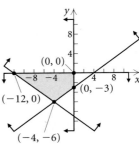

$(0, 0)$ $(-12, 0)$ $(0, -3)$ $(-4, -6)$

53.

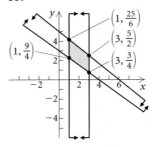

$\left(1, \frac{25}{6}\right)$ $\left(3, \frac{5}{2}\right)$ $\left(1, \frac{9}{4}\right)$ $\left(3, \frac{3}{4}\right)$

55. Maximum: 179 when $x = 7$ and $y = 0$; minimum: 48 when $x = 0$ and $y = 4$

57. Maximum: 216 when $x = 0$ and $y = 6$; minimum: 0 when $x = 0$ and $y = 0$

59. Maximum income of \$18 is achieved when 100 of each type of biscuit is made.

61. Maximum profit of \$11,000 is achieved by producing 100 units of lumber and 300 units of plywood.

63. Minimum cost of \$36$\frac{12}{13}$ is achieved by using $1\frac{11}{13}$ sacks of soybean meal and $1\frac{11}{13}$ sacks of oats.

65. Maximum income of \$3110 is achieved when \$22,000 is invested in corporate bonds and \$18,000 is invested in municipal bonds.

67. Minimum cost of \$460 thousand is achieved using 30 P_1's and 10 P_2's.

69. Maximum profit per day of \$192 is achieved when 2 knit suits and 4 worsted suits are made.

71. Minimum weekly cost of \$19.05 is achieved when 1.5 lb of meat and 3 lb of cheese are used.

73. Maximum total number of 800 is achieved when there are 550 of A and 250 of B.

75. Left to the student   **77.** Discussion and Writing

79. [2.6] $\{x \mid -7 \leq x < 2\}$, or $[-7, 2)$

80. [2.6] $\{x \mid x \leq 1 \ or \ x \geq 5\}$, or $(-\infty, 1] \cup [5, \infty)$

81. [3.5] $\{x \mid -1 \leq x \leq 3\}$, or $[-1, 3]$

82. [3.5] $\{x \mid -3 < x < -2\}$, or $(-3, -2)$

83.

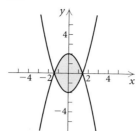

85.

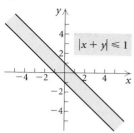

$|x + y| \leq 1$

87.

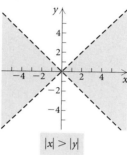

$|x| > |y|$

89. Maximum income of $28,500 is achieved by making 30 less expensive assemblies and 30 more expensive assemblies.

Exercise Set 8.8

1. $\dfrac{2}{x - 3} - \dfrac{1}{x + 2}$ **3.** $-\dfrac{4}{3x - 1} + \dfrac{5}{2x - 1}$

5. $-\dfrac{3}{x - 2} + \dfrac{2}{x + 2} + \dfrac{4}{x + 1}$

7. $-\dfrac{3}{(x + 2)^2} - \dfrac{1}{x + 2} + \dfrac{1}{x - 1}$ **9.** $\dfrac{3}{x - 1} - \dfrac{4}{2x - 1}$

11. $x - 2 - \dfrac{\frac{11}{4}}{(x + 1)^2} + \dfrac{\frac{17}{16}}{x + 1} - \dfrac{\frac{17}{16}}{x - 3}$

13. $\dfrac{3x + 5}{x^2 + 2} - \dfrac{4}{x - 1}$

15. $-\dfrac{2}{x + 2} + \dfrac{10}{(x + 2)^2} + \dfrac{3}{2x - 1}$

17. $3x + 1 + \dfrac{2}{2x - 1} + \dfrac{3}{x + 1}$

19. $-\dfrac{1}{x - 3} + \dfrac{3x}{x^2 + 2x - 5}$

21. $\dfrac{5}{3x + 5} - \dfrac{3}{x + 1} + \dfrac{4}{(x + 1)^2}$ **23.** $\dfrac{8}{4x - 5} + \dfrac{3}{3x + 2}$

25. $\dfrac{2x - 5}{3x^2 + 1} - \dfrac{2}{x - 2}$ **27.** Discussion and Writing

29. Discussion and Writing   **30.** [3.3] 3, $\pm i$

31. [3.3] $-2, \dfrac{1 \pm \sqrt{5}}{2}$ **32.** [3.3] $-2, 3, \pm i$

33. [3.3] $-3, -1 \pm \sqrt{2}$

35. $-\dfrac{\frac{1}{2a^2}x}{x^2 + a^2} + \dfrac{\frac{1}{4a^2}}{x - a} + \dfrac{\frac{1}{4a^2}}{x + a}$

37. $-\dfrac{3}{25(\ln x + 2)} + \dfrac{3}{25(\ln x - 3)} + \dfrac{7}{5(\ln x - 3)^2}$

Review Exercises, Chapter 8

1. [8.1] (a)   **2.** [8.1] (e)   **3.** [8.1] (h)   **4.** [8.1] (d)
5. [8.6] (b)   **6.** [8.6] (g)   **7.** [8.6] (c)   **8.** [8.6] (f)
9. [8.1] $(-2, -2)$ **10.** [8.1] $(-5, 4)$
11. [8.1] No solution   **12.** [8.1] $(y - 2, y)$, or $(x, x + 2)$
13. [8.2] No solution   **14.** [8.2] $(0, 0, 0)$
15. [8.2] $(-5, 13, 8, 2)$
16. [8.1], [8.2] Consistent: 9, 10, 12, 14, 15; the others are inconsistent.
17. [8.1], [8.2] Dependent: 12; the others are independent.
18. [8.3] $(1, 2)$ **19.** [8.3] $(-3, 4, -2)$
20. [8.3] $\left(\dfrac{z}{2}, -\dfrac{z}{2}, z\right)$ **21.** [8.3] $(-4, 1, -2, 3)$
22. [8.1] 31 nickels, 44 dimes
23. [8.1] $1600 at 10%, $3400 at 10.5%
24. [8.2] 1 bagel, $\frac{1}{2}$ serving cream cheese, 2 bananas
25. [8.2] 74.5, 68.5, 82
26. [8.2] **(a)** $f(x) = \frac{1}{12}x^2 - \frac{31}{60}x + \frac{77}{10}$; **(b)** 18.7 lb

27. [8.4] $\begin{bmatrix} 0 & -1 & 6 \\ 3 & 1 & -2 \\ -2 & 1 & -2 \end{bmatrix}$ **28.** [8.4] $\begin{bmatrix} -3 & 3 & 0 \\ -6 & -9 & 6 \\ 6 & 0 & -3 \end{bmatrix}$

29. [8.4] $\begin{bmatrix} -1 & 1 & 0 \\ -2 & -3 & 2 \\ 2 & 0 & -1 \end{bmatrix}$ **30.** [8.4] $\begin{bmatrix} -2 & 2 & 6 \\ 1 & -8 & 18 \\ 2 & 1 & -15 \end{bmatrix}$

31. [8.4] Not possible   **32.** [8.4] $\begin{bmatrix} 2 & -1 & -6 \\ 1 & 5 & -2 \\ -2 & -1 & 4 \end{bmatrix}$

33. [8.4] $\begin{bmatrix} 3 & -2 & -6 \\ 3 & 8 & -4 \\ -4 & -1 & 5 \end{bmatrix}$ **34.** [8.4] $\begin{bmatrix} -2 & -1 & 18 \\ 5 & -3 & -2 \\ -2 & 3 & -8 \end{bmatrix}$

35. [8.4] **(a)** $\begin{bmatrix} 46.1 & 5.9 & 10.1 & 8.5 & 11.4 \\ 54.6 & 4.6 & 9.6 & 7.6 & 10.6 \\ 48.9 & 5.5 & 12.7 & 9.4 & 9.3 \\ 51.3 & 4.8 & 11.3 & 6.9 & 12.7 \end{bmatrix}$

(b) $[32 \quad 19 \quad 43 \quad 38]$;
(c) $[6564.7 \quad 695.1 \quad 1481.1 \quad 1082.8 \quad 1448.7]$;
(d) the total cost, in cents, for each item for the day's meal

36. [8.5] $\begin{bmatrix} -\frac{1}{2} & 0 \\ \frac{1}{6} & \frac{1}{3} \end{bmatrix}$ **37.** [8.5] $\begin{bmatrix} 0 & 0 & \frac{1}{4} \\ 0 & -\frac{1}{2} & 0 \\ \frac{1}{3} & 0 & 0 \end{bmatrix}$

38. [8.5] $\begin{bmatrix} 1 & 0 & 0 & 0 \\ 0 & \frac{1}{9} & \frac{5}{18} & 0 \\ 0 & -\frac{1}{9} & \frac{2}{9} & 0 \\ 0 & 0 & 0 & 1 \end{bmatrix}$

39. [8.4] $\begin{bmatrix} 3 & -2 & 4 \\ 1 & 5 & -3 \\ 2 & -3 & 7 \end{bmatrix} \begin{bmatrix} x \\ y \\ z \end{bmatrix} = \begin{bmatrix} 13 \\ 7 \\ -8 \end{bmatrix}$ **40.** [8.5] $(-8, 7)$

41. [8.5] $(1, -2, 5)$ **42.** [8.5] $(2, -1, 1, -3)$
43. [8.6] 10 **44.** [8.6] -18 **45.** [8.6] -6
46. [8.6] -1 **47.** [8.6] $(3, -2)$ **48.** [8.6] $(-1, 5)$
49. [8.6] $\left(\frac{3}{2}, \frac{13}{14}, \frac{33}{14}\right)$ **50.** [8.6] $(2, -1, 3)$
51. [8.7]

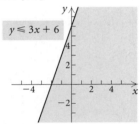

$y \le 3x + 6$

52. [8.7]

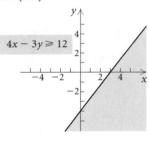

$4x - 3y \ge 12$

53. [8.7]

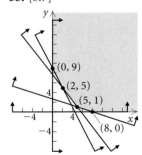

$(0, 9)$
$(2, 5)$
$(5, 1)$
$(8, 0)$

54. [8.7] Minimum $= 52$ at $(2, 4)$; maximum $= 92$ at $(2, 8)$
55. [8.7] Maximum score of 96 when 0 group A questions and 8 group B questions are answered

56. [8.8] $\dfrac{5}{x+1} - \dfrac{5}{x+2} - \dfrac{5}{(x+2)^2}$

57. [8.8] $\dfrac{2}{2x-3} - \dfrac{5}{x+4}$

58. Discussion and Writing [8.1] During a holiday season, a caterer sold a total of 55 food trays. She sold 15 more seafood trays than cheese trays. How many of each were sold?

59. Discussion and Writing [8.4] In general, $(\mathbf{AB})^2 \neq \mathbf{A}^2\mathbf{B}^2$. $(\mathbf{AB})^2 = \mathbf{ABAB}$ and $\mathbf{A}^2\mathbf{B}^2 = \mathbf{AABB}$. Since matrix multiplication is not commutative, $\mathbf{BA} \neq \mathbf{AB}$, so $(\mathbf{AB})^2 \neq \mathbf{A}^2\mathbf{B}^2$.
60. [8.2] 12%: \$10,000; 13%: \$12,000; $14\frac{1}{2}$%: \$18,000
61. [8.1] $\left(\frac{5}{18}, \frac{1}{7}\right)$ **62.** [8.2] $\left(1, \frac{1}{2}, \frac{1}{3}\right)$
63. [8.7]   **64.** [8.7]

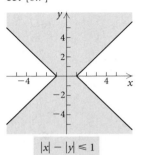   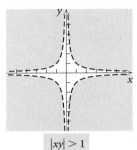

$|x| - |y| \le 1$ $|xy| > 1$

Test, Chapter 8

1. [8.1] $(-3, 5)$ **2.** [8.1] $(1, -2)$
3. [8.2] $(-1, 3, 2)$ **4.** [8.1] Consistent
5. [8.1] Independent
6. [8.1] Student: 342; nonstudent: 408

7. [8.4] $\begin{bmatrix} -2 & -3 \\ -3 & 4 \end{bmatrix}$ **8.** [8.4] Not defined

9. [8.4] $\begin{bmatrix} -7 & -13 \\ 5 & -1 \end{bmatrix}$ **10.** [8.4] Not defined

11. [8.4] $\begin{bmatrix} 2 & -2 & 6 \\ -4 & 10 & 4 \end{bmatrix}$ **12.** [8.5] $\begin{bmatrix} 0 & -1 \\ -\frac{1}{4} & -\frac{3}{4} \end{bmatrix}$

13. [8.4] **(a)** $\begin{bmatrix} 49 & 10 & 13 \\ 43 & 12 & 11 \\ 51 & 8 & 12 \end{bmatrix}$; **(b)** $[26 \quad 18 \quad 23]$;

(c) $[3221 \quad 660 \quad 812]$; **(d)** the total cost, in cents, for each type of menu item served on the given day

14. [8.4] $\begin{bmatrix} 3 & -4 & 2 \\ 2 & 3 & 1 \\ 1 & -5 & -3 \end{bmatrix} \begin{bmatrix} x \\ y \\ z \end{bmatrix} = \begin{bmatrix} -8 \\ 7 \\ 3 \end{bmatrix}$

15. [8.6] -33 **16.** [8.6] $\left(-\frac{1}{2}, \frac{3}{4}\right)$
17. [8.7]

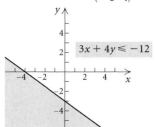

$3x + 4y \le -12$

18. [8.7] Maximum: 15 at (3, 3); minimum: 2 at (1, 0)
19. [8.7] Pound cakes: 25; carrot cakes: 75
20. [8.8] $-\dfrac{2}{x-1} + \dfrac{5}{x+3}$
21. [8.2] $A = 1,\ B = -3,\ C = 2$

Chapter 9

Exercise Set 9.1

1. (f)   **3.** (b)   **5.** (d)

7. V: (0, 0); F: (0, 5); D: $y = -5$

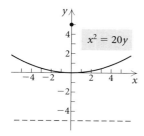

$x^2 = 20y$

9. V: (0, 0); F: $\left(-\dfrac{3}{2}, 0\right)$; D: $x = \dfrac{3}{2}$

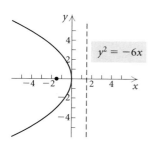

$y^2 = -6x$

11. V: (0, 0); F: (0, 1); D: $y = -1$

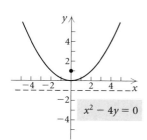

$x^2 - 4y = 0$

13. V: (0, 0); F: $\left(\dfrac{1}{8}, 0\right)$; D: $x = -\dfrac{1}{8}$

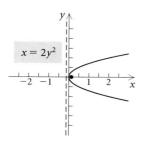

$x = 2y^2$

15. $y^2 = 16x$ **17.** $x^2 = -4\pi y$

19. $(y - 2)^2 = 14\left(x + \dfrac{1}{2}\right)$

21. V: (-2, 1); F: $\left(-2, -\dfrac{1}{2}\right)$; D: $y = \dfrac{5}{2}$

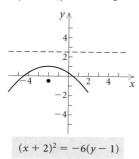

$(x + 2)^2 = -6(y - 1)$

23. V: (-1, -3); F: $\left(-1, -\dfrac{7}{2}\right)$; D: $y = -\dfrac{5}{2}$

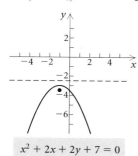

$x^2 + 2x + 2y + 7 = 0$

25. V: (0, -2); F: $\left(0, -1\dfrac{3}{4}\right)$; D: $y = -2\dfrac{1}{4}$

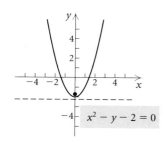

$x^2 - y - 2 = 0$

27. V: (-2, -1); F: $\left(-2, -\dfrac{3}{4}\right)$; D: $y = -1\dfrac{1}{4}$

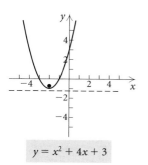

$y = x^2 + 4x + 3$

29. V: $\left(5\frac{3}{4}, \frac{1}{2}\right)$; F: $\left(6, \frac{1}{2}\right)$; D: $x = 5\frac{1}{2}$

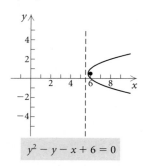

$$y^2 - y - x + 6 = 0$$

31. (a) $y^2 = 16x$; **(b)** $3\frac{33}{64}$ ft   **33.** $\frac{2}{3}$ ft, or 8 in.

35. V: $(0.867, 0.348)$; F: $(0.867, -0.190)$; D: $y = 0.887$
37. Discussion and Writing   **39.** [2.3] $(x + 5)^2$
40. [2.3] $\left(y - \frac{9}{2}\right)^2$ **41.** [1.1] $(1, -2)$; 3
42. [1.1] $(-3, 5)$; 6   **43.** $(x + 1)^2 = -4(y - 2)$
45. 10 ft, 11.6 ft, 16.4 ft, 24.4 ft, 35.6 ft, 50 ft

Exercise Set 9.2

1. (b)   **3.** (d)   **5.** (a)
7. $(7, -2)$; 8   **9.** $(-2, 3)$; 5

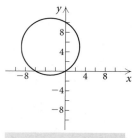

$$x^2 + y^2 - 14x + 4y = 11$$

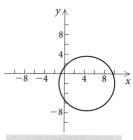

$$x^2 + y^2 + 4x - 6y - 12 = 0$$

11. $(-3, 5)$; $\sqrt{34}$ **13.** $\left(\frac{9}{2}, -2\right)$; $\frac{5\sqrt{5}}{2}$

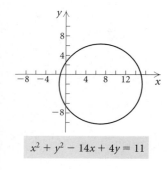

$$x^2 + y^2 + 6x - 10y = 0$$

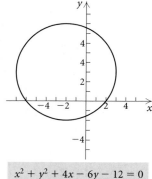

$$x^2 + y^2 - 9x = 7 - 4y$$

15. (c)   **17.** (d)
19. V: $(2, 0)$, $(-2, 0)$;   **21.** V: $(0, 4)$, $(0, -4)$;
   F: $(\sqrt{3}, 0)$, $(-\sqrt{3}, 0)$ F: $(0, \sqrt{7})$, $(0, -\sqrt{7})$

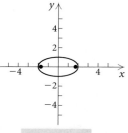

$$\frac{x^2}{4} + \frac{y^2}{1} = 1$$

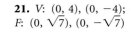

$$16x^2 + 9y^2 = 144$$

23. V: $(-\sqrt{3}, 0)$, $(\sqrt{3}, 0)$;
   F: $(-1, 0)$, $(1, 0)$

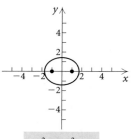

$$2x^2 + 3y^2 = 6$$

25. V: $\left(-\frac{1}{2}, 0\right)$, $\left(\frac{1}{2}, 0\right)$;

   F: $\left(-\frac{\sqrt{5}}{6}, 0\right)$, $\left(\frac{\sqrt{5}}{6}, 0\right)$

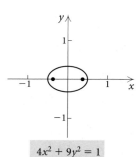

$$4x^2 + 9y^2 = 1$$

27. $\frac{x^2}{49} + \frac{y^2}{40} = 1$ **29.** $\frac{x^2}{25} + \frac{y^2}{64} = 1$ **31.** $\frac{x^2}{9} + \frac{y^2}{5} = 1$
33. C: $(1, 2)$; V: $(4, 2)$, $(-2, 2)$; F: $(1 + \sqrt{5}, 2)$,
   $(1 - \sqrt{5}, 2)$

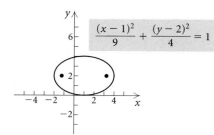

$$\frac{(x - 1)^2}{9} + \frac{(y - 2)^2}{4} = 1$$

35. C: $(-3, 5)$; V: $(-3, 11)$, $(-3, -1)$; F: $(-3, 5 + \sqrt{11})$, $(-3, 5 - \sqrt{11})$

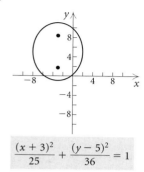

$$\frac{(x + 3)^2}{25} + \frac{(y - 5)^2}{36} = 1$$

37. C: $(-2, 1)$; V: $(-10, 1)$, $(6, 1)$; F: $(-6, 1)$, $(2, 1)$

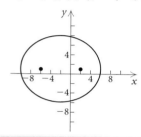

$$3(x + 2)^2 + 4(y - 1)^2 = 192$$

39. C: $(2, -1)$; V: $(-1, -1)$, $(5, -1)$; F: $(2 + \sqrt{5}, -1)$, $(2 - \sqrt{5}, -1)$

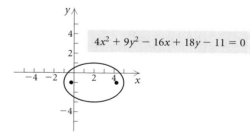

$$4x^2 + 9y^2 - 16x + 18y - 11 = 0$$

41. C: $(1, 1)$; V: $(1, 3)$, $(1, -1)$; F: $(1, 1 + \sqrt{3})$, $(1, 1 - \sqrt{3})$

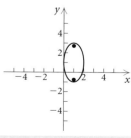

$$4x^2 + y^2 - 8x - 2y + 1 = 0$$

43. Example 2; $\dfrac{3}{5} < \dfrac{\sqrt{12}}{4}$ **45.** $\dfrac{x^2}{15} + \dfrac{y^2}{16} = 1$

47. $\dfrac{x^2}{2500} + \dfrac{y^2}{144} = 1$ **49.** 2×10^6 mi

51. C: $(2.003, -1.005)$; V: $(-1.017, -1.005)$, $(5.023, -1.005)$ **53.** Discussion and Writing

55. [4.1] **(a)** Yes; **(b)** $f^{-1}(x) = \dfrac{x + 3}{2}$

56. [4.1] **(a)** Yes; **(b)** $f^{-1}(x) = \sqrt[3]{x - 2}$

57. [4.1] **(a)** Yes; **(b)** $f^{-1}(x) = \dfrac{5}{x} + 1$, or $\dfrac{5 + x}{x}$

58. [4.1] **(a)** Yes; **(b)** $f^{-1}(x) = x^2 - 4$, $x \geq 0$

59. $\dfrac{(x - 3)^2}{4} + \dfrac{(y - 1)^2}{25} = 1$ **61.** $\dfrac{x^2}{9} + \dfrac{y^2}{484/5} = 1$

63. About 9.1 ft

Exercise Set 9.3

1. (b)   **3.** (c)   **5.** (a)   **7.** $\dfrac{y^2}{9} - \dfrac{x^2}{16} = 1$

9. $\dfrac{x^2}{4} - \dfrac{y^2}{9} = 1$ **11.** C: $(0, 0)$; V: $(2, 0)$, $(-2, 0)$; F: $(2\sqrt{2}, 0)$, $(-2\sqrt{2}, 0)$; A: $y = x$, $y = -x$

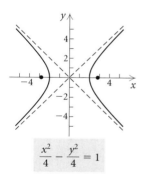

$$\frac{x^2}{4} - \frac{y^2}{4} = 1$$

13. C: $(2, -5)$; V: $(-1, -5)$, $(5, -5)$; F: $(2 - \sqrt{10}, -5)$, $(2 + \sqrt{10}, -5)$; A: $y = -\dfrac{x}{3} - \dfrac{13}{3}$, $y = \dfrac{x}{3} - \dfrac{17}{3}$

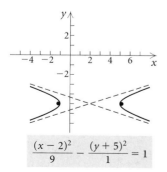

$$\frac{(x - 2)^2}{9} - \frac{(y + 5)^2}{1} = 1$$

15. C: $(-1, -3)$; V: $(-1, -1)$, $(-1, -5)$;
F: $(-1, -3 + 2\sqrt{5})$, $(-1, -3 - 2\sqrt{5})$; A: $y = \frac{1}{2}x - \frac{5}{2}$,
$y = -\frac{1}{2}x - \frac{7}{2}$

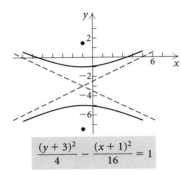

$$\frac{(y+3)^2}{4} - \frac{(x+1)^2}{16} = 1$$

17. C: $(0, 0)$; V: $(-2, 0)$, $(2, 0)$; F: $(-\sqrt{5}, 0)$, $(\sqrt{5}, 0)$;
A: $y = -\frac{1}{2}x$, $y = \frac{1}{2}x$

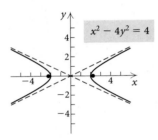

$x^2 - 4y^2 = 4$

19. C: $(0, 0)$; V: $(0, -3)$, $(0, 3)$; F: $(0, -3\sqrt{10})$, $(0, 3\sqrt{10})$;
A: $y = \frac{1}{3}x$, $y = -\frac{1}{3}x$

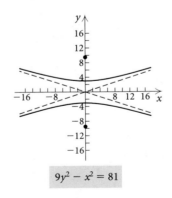

$9y^2 - x^2 = 81$

21. C: $(0, 0)$; V: $(-\sqrt{2}, 0)$, $(\sqrt{2}, 0)$; F: $(-2, 0)$, $(2, 0)$;
A: $y = x$, $y = -x$

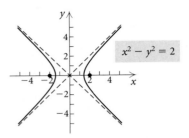

$x^2 - y^2 = 2$

23. C: $(0, 0)$; V: $\left(0, -\frac{1}{2}\right)$, $\left(0, \frac{1}{2}\right)$; F: $\left(0, -\frac{\sqrt{2}}{2}\right)$,
$\left(0, \frac{\sqrt{2}}{2}\right)$; A: $y = x$, $y = -x$

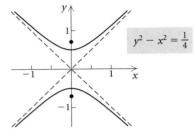

$y^2 - x^2 = \frac{1}{4}$

25. C: $(1, -2)$; V: $(0, -2)$, $(2, -2)$; F: $(1 - \sqrt{2}, -2)$,
$(1 + \sqrt{2}, -2)$; A: $y = -x - 1$, $y = x - 3$

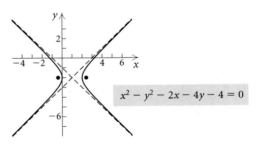

$x^2 - y^2 - 2x - 4y - 4 = 0$

27. C: $\left(\frac{1}{3}, 3\right)$; V: $\left(-\frac{2}{3}, 3\right)$, $\left(\frac{4}{3}, 3\right)$; F: $\left(\frac{1}{3} - \sqrt{37}, 3\right)$,
$\left(\frac{1}{3} + \sqrt{37}, 3\right)$; A: $y = 6x + 1$, $y = -6x + 5$

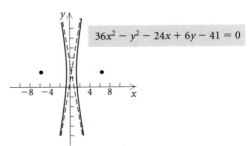

$36x^2 - y^2 - 24x + 6y - 41 = 0$

29. C: $(3, 1)$; V: $(3, 3)$, $(3, -1)$; F: $(3, 1 + \sqrt{13})$, $(3, 1 - \sqrt{13})$; A: $y = \frac{2}{3}x - 1$, $y = -\frac{2}{3}x + 3$

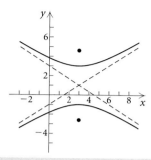

$$9y^2 - 4x^2 - 18y + 24x - 63 = 0$$

31. C: $(1, -2)$; V: $(2, -2)$, $(0, -2)$; F: $(1 + \sqrt{2}, -2)$, $(1 - \sqrt{2}, -2)$; A: $y = x - 3$, $y = -x - 1$

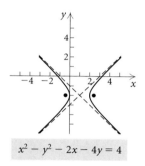

$$x^2 - y^2 - 2x - 4y = 4$$

33. C: $(-3, 4)$; V: $(-3, 10)$, $(-3, -2)$; F: $(-3, 4 + 6\sqrt{2})$, $(-3, 4 - 6\sqrt{2})$; A: $y = x + 7$, $y = -x + 1$

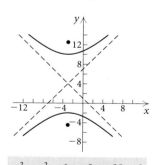

$$y^2 - x^2 - 6x - 8y - 29 = 0$$

35. Example 3; $\dfrac{\sqrt{5}}{1} > \dfrac{5}{4}$ **37.** $\dfrac{x^2}{9} - \dfrac{(y - 7)^2}{16} = 1$

39. $\dfrac{y^2}{25} - \dfrac{x^2}{11} = 1$

41. C: $(-1.460, -0.957)$; V: $(-2.360, -0.957)$, $(-0.560, -0.957)$; A: $y = -1.429x - 3.043$, $y = 1.429x + 1.129$

43. Discussion and Writing
45. [8.1], [8.3], [8.5], [8.6] $(6, -1)$
46. [8.1], [8.3], [8.5], [8.6] $(1, -1)$
47. [8.1], [8.3], [8.5], [8.6] $(2, -1)$
48. [8.1], [8.3], [8.5], [8.6] $(-3, 4)$
49. $\dfrac{(y + 5)^2}{9} - (x - 3)^2 = 1$

51. $\dfrac{x^2}{345.96} - \dfrac{y^2}{22,154.04} = 1$

Exercise Set 9.4

1. (e)   **3.** (c)   **5.** (b)   **7.** $(-4, -3)$, $(3, 4)$
9. $(0, 2)$, $(3, 0)$ **11.** $(-5, 0)$, $(4, 3)$, $(4, -3)$
13. $(3, 0)$, $(-3, 0)$ **15.** $(0, -3)$, $(4, 5)$ **17.** $(-2, 1)$
19. $(3, 4)$, $(-3, -4)$, $(4, 3)$, $(-4, -3)$
21. $\left(\dfrac{6\sqrt{21}}{7}, \dfrac{4i\sqrt{35}}{7}\right)$, $\left(\dfrac{6\sqrt{21}}{7}, -\dfrac{4i\sqrt{35}}{7}\right)$,

$\left(-\dfrac{6\sqrt{21}}{7}, \dfrac{4i\sqrt{35}}{7}\right)$, $\left(-\dfrac{6\sqrt{21}}{7}, -\dfrac{4i\sqrt{35}}{7}\right)$

23. $(3, 2)$, $\left(4, \dfrac{3}{2}\right)$

25. $\left(\dfrac{5 + \sqrt{70}}{3}, \dfrac{-1 + \sqrt{70}}{3}\right)$, $\left(\dfrac{5 - \sqrt{70}}{3}, \dfrac{-1 - \sqrt{70}}{3}\right)$

27. $(\sqrt{2}, \sqrt{14})$, $(-\sqrt{2}, \sqrt{14})$, $(\sqrt{2}, -\sqrt{14})$, $(-\sqrt{2}, -\sqrt{14})$ **29.** $(1, 2)$, $(-1, -2)$, $(2, 1)$, $(-2, -1)$

31. $\left(\dfrac{15 + \sqrt{561}}{8}, \dfrac{11 - 3\sqrt{561}}{8}\right)$,

$\left(\dfrac{15 - \sqrt{561}}{8}, \dfrac{11 + 3\sqrt{561}}{8}\right)$

33. $\left(\dfrac{7 - \sqrt{33}}{2}, \dfrac{7 + \sqrt{33}}{2}\right)$, $\left(\dfrac{7 + \sqrt{33}}{2}, \dfrac{7 - \sqrt{33}}{2}\right)$

35. $(3, 2)$, $(-3, -2)$, $(2, 3)$, $(-2, -3)$

37. $\left(\dfrac{5 - 9\sqrt{15}}{20}, \dfrac{-45 + 3\sqrt{15}}{20}\right)$,

$\left(\dfrac{5 + 9\sqrt{15}}{20}, \dfrac{-45 - 3\sqrt{15}}{20}\right)$

39. $(3, -5)$, $(-1, 3)$ **41.** $(8, 5)$, $(-5, -8)$ **43.** $(3, 2)$, $(-3, -2)$ **45.** $(2, 1)$, $(-2, -1)$, $(1, 2)$, $(-1, -2)$

47. $\left(4 + \dfrac{3i\sqrt{6}}{2}, -4 + \dfrac{3i\sqrt{6}}{2}\right)$, $\left(4 - \dfrac{3i\sqrt{6}}{2}, -4 - \dfrac{3i\sqrt{6}}{2}\right)$

49. $(3, \sqrt{5})$, $(-3, -\sqrt{5})$, $(\sqrt{5}, 3)$, $(-\sqrt{5}, -3)$

51. $\left(\dfrac{8\sqrt{5}}{5}i, \dfrac{3\sqrt{105}}{5}\right)$, $\left(\dfrac{8\sqrt{5}}{5}i, -\dfrac{3\sqrt{105}}{5}\right)$,

$\left(-\dfrac{8\sqrt{5}}{5}i, \dfrac{3\sqrt{105}}{5}\right)$, $\left(-\dfrac{8\sqrt{5}}{5}i, -\dfrac{3\sqrt{105}}{5}\right)$

53. $(2, 1)$, $(-2, -1)$, $\left(-i\sqrt{5}, \dfrac{2i\sqrt{5}}{5}\right)$, $\left(i\sqrt{5}, -\dfrac{2i\sqrt{5}}{5}\right)$

55. 6 cm by 8 cm   **57.** 4 in. by 5 in.
59. 1 m by $\sqrt{3}$ m   **61.** 30 yd by 75 yd   **63.** 16 ft, 24 ft
65. (a) $I(x) = 21.15508287x + 198.9151934$;
(b) $E(x) = 197.3380928 \cdot 1.053908223^x$; **(c)** about 25 yr
after 1985   **67.** (1.564, 2.448), (0.138, 0.019)
69. (0.871, 1.388)   **71.** (1.146, 3.146), (−1.841, 0.159)
73. (0.965, 4402.33), (−0.965, −4402.33)
75. (2.112, −0.109), (−13.041, −13.337)
77. (400, 1.431), (−400, 1.431), (400, −1.431),
(−400, −1.431)   **79.** Discussion and Writing
81. [4.5] 2   **82.** [4.5] 2.048   **83.** [4.5] 81   **84.** [4.5] 5
85. $(x - 2)^2 + (y - 3)^2 = 1$

87. $\dfrac{x^2}{4} + y^2 = 1$

89. $\left(x + \dfrac{5}{13}\right)^2 + \left(y - \dfrac{32}{13}\right)^2 = \dfrac{5365}{169}$

91. There is no number x such that $\dfrac{x^2}{a^2} - \dfrac{\left(\dfrac{b}{a}x\right)^2}{b^2} = 1$,

because the left side simplifies to $\dfrac{x^2}{a^2} - \dfrac{x^2}{a^2}$, which is 0.

93. $\left(\frac{1}{2}, \frac{1}{4}\right), \left(\frac{1}{2}, -\frac{1}{4}\right), \left(-\frac{1}{2}, \frac{1}{4}\right), \left(-\frac{1}{2}, -\frac{1}{4}\right)$
95. Factor: $x^3 + y^3 = (x + y)(x^2 - xy + y^2)$. We know
that $x + y = 1$, so $(x + y)^2 = x^2 + 2xy + y^2 = 1$, or
$x^2 + y^2 = 1 - 2xy$. We also know that $xy = 1$, so
$x^2 + y^2 = 1 - 2 \cdot 1 = -1$. Then
$x^3 + y^3 = 1 \cdot (-1 - 1) = -2$.   **97.** (2, 4), (4, 2)
99. (3, −2), (−3, 2), (2, −3), (−2, 3)

101. $\left(\dfrac{2 \log 3 + 3 \log 5}{3(\log 3 \cdot \log 5)}, \dfrac{4 \log 3 - 3 \log 5}{3(\log 3 \cdot \log 5)}\right)$

Review Exercises, Chapter 9

1. [9.1] (d)   **2.** [9.2] (a)   **3.** [9.2] (e)   **4.** [9.3] (g)
5. [9.2] (b)   **6.** [9.2] (f)   **7.** [9.1] (h)   **8.** [9.3] (c)
9. [9.1] $x^2 = -6y$ **10.** [9.1] F: (−3, 0); V: (0, 0);
D: $x = 3$ **11.** [9.1] V: (−5, 8); F: $\left(-5, \frac{15}{2}\right)$; D: $y = \frac{17}{2}$
12. [9.2] C: (2, −1); V: (−3, −1), (7, −1); F: (−1, −1),
(5, −1);

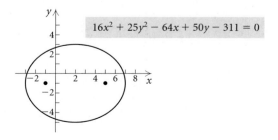

13. [9.2] $\dfrac{x^2}{9} + \dfrac{y^2}{16} = 1$

14. [9.3] C: $\left(-2, \dfrac{1}{4}\right)$; V: $\left(0, \dfrac{1}{4}\right), \left(-4, \dfrac{1}{4}\right)$;

F: $\left(-2 + \sqrt{6}, \dfrac{1}{4}\right), \left(-2 - \sqrt{6}, \dfrac{1}{4}\right)$;

A: $y - \dfrac{1}{4} = \dfrac{\sqrt{2}}{2}(x + 2), y - \dfrac{1}{4} = -\dfrac{\sqrt{2}}{2}(x + 2)$

15. [9.1] 0.167 ft   **16.** [9.4] (−8$\sqrt{2}$, 8), (8$\sqrt{2}$, 8)

17. [9.4] $\left(3, \dfrac{\sqrt{29}}{2}\right), \left(-3, \dfrac{\sqrt{29}}{2}\right), \left(3, -\dfrac{\sqrt{29}}{2}\right),$

$\left(-3, -\dfrac{\sqrt{29}}{2}\right)$ **18.** [9.4] (7, 4)

19. [9.4] (2, 2), $\left(\frac{32}{9}, -\frac{10}{9}\right)$ **20.** [9.4] (0, −3), (2, 1)
21. [9.4] (4, 3), (4, −3), (−4, 3), (−4, −3)
22. [9.4] (−$\sqrt{3}$, 0), ($\sqrt{3}$, 0), (−2, 1), (2, 1)
23. [9.4] $\left(-\frac{3}{5}, \frac{21}{5}\right)$, (3, −3)
24. [9.4] (6, 8), (6, −8), (−6, 8), (−6, −8)
25. [9.4] (2, 2), (−2, −2), (2$\sqrt{2}$, $\sqrt{2}$), (−2$\sqrt{2}$, −$\sqrt{2}$)
26. [9.4] 7, 4   **27.** [9.4] 7 m by 12 m   **28.** [9.4] 4, 8
29. [9.4] 32 cm, 20 cm   **30.** [9.4] 11 ft, 3 ft
31. Discussion and Writing [9.4] Although we can always
visualize the real-number solutions, we cannot visualize
the imaginary-number solutions.
32. Discussion and Writing [9.2] The equation of a circle
can be written as

$$\dfrac{(x - h)^2}{a^2} + \dfrac{(y - k)^2}{b^2} = 1,$$

where $a = b = r$, the radius of the circle. In an ellipse,
$a > b$, so a circle is not a special type of ellipse.

33. [9.2] $x^2 + \dfrac{y^2}{9} = 1$ **34.** [9.4] $\dfrac{8}{7}, \dfrac{7}{2}$

35. [9.2], [9.4] $(x - 2)^2 + (y - 1)^2 = 100$

36. [9.3] $\dfrac{x^2}{778.41} - \dfrac{y^2}{39,221.59} = 1$

Test, Chapter 9

1. [9.3] (c)   **2.** [9.1] (b)   **3.** [9.2] (a)   **4.** [9.2] (d)
5. [9.1] $x^2 = 8y$ **6.** [9.1] (10, 4)
7. [9.2] Center: (−1, 3); radius: 5
8. [9.2] $\dfrac{x^2}{4} + \dfrac{y^2}{25} = 1$ **9.** [9.3] $y = \dfrac{\sqrt{2}}{2}x, y = -\dfrac{\sqrt{2}}{2}x$
10. [9.1] $\frac{27}{8}$ in.
11. [9.4] (1, 2), (1, −2), (−1, 2), (−1, −2)
12. [9.4] (3, −2), (−2, 3)   **13.** [9.4] (2, 3), (3, 2)
14. [9.4] 5 ft by 4 ft
15. [9.2] $(x - 1)^2 + (y + 3)^2 = 16$

Chapter 10

Exercise Set 10.1

1. 3, 7, 11, 15; 39; 59   **3.** $2, \frac{3}{2}, \frac{4}{3}, \frac{5}{4}; \frac{10}{9}; \frac{15}{14}$

5. $0, \frac{3}{5}, \frac{4}{5}, \frac{15}{17}; \frac{99}{101}; \frac{112}{113}$ **7.** $-1, 4, -9, 16; 100; -225$

9. 7, 3, 7, 3; 3; 7   **11.** 34   **13.** 225   **15.** $-33{,}880$

17. 67   **19.** $2n$ **21.** $(-1)^n \cdot 2 \cdot 3^{n-1}$ **23.** $\dfrac{n+1}{n+2}$

25. $n(n+1)$ **27.** $\log 10^{n-1}$, or $n-1$ **29.** 6; 28

31. 20; 30   **33.** $\frac{1}{2} + \frac{1}{4} + \frac{1}{6} + \frac{1}{8} + \frac{1}{10} = \frac{137}{120}$

35. $1 + 2 + 4 + 8 + 16 + 32 + 64 = 127$

37. $\ln 7 + \ln 8 + \ln 9 + \ln 10 = \ln (7 \cdot 8 \cdot 9 \cdot 10) =$ $\ln 5040 \approx 8.5252$

39. $\frac{1}{2} + \frac{2}{3} + \frac{3}{4} + \frac{4}{5} + \frac{5}{6} + \frac{6}{7} + \frac{7}{8} + \frac{8}{9} = \frac{15{,}551}{2520}$

41. $-1 + 1 + 1 - 1 + 1 - 1 = -1$

43. $3 - 6 + 9 - 12 + 15 - 18 + 21 - 24 = -12$

45. $2 + 1 + \frac{2}{5} + \frac{1}{5} + \frac{2}{17} + \frac{1}{13} + \frac{2}{37} = \frac{157{,}351}{40{,}885}$

47. $3 + 2 + 3 + 6 + 11 + 18 = 43$

49. $\frac{1}{2} + \frac{2}{3} + \frac{4}{5} + \frac{8}{9} + \frac{16}{17} + \frac{32}{33} + \frac{64}{65} + \frac{128}{129} + \frac{256}{257} + \frac{512}{513} + \frac{1024}{1025} \approx 9.736$

51. $\displaystyle\sum_{k=1}^{\infty} 5k$ **53.** $\displaystyle\sum_{k=1}^{6} (-1)^{k+1} 2^k$ **55.** $\displaystyle\sum_{k=1}^{6} (-1)^k \frac{k}{k+1}$

57. $\displaystyle\sum_{k=2}^{n} (-1)^k k^2$ **59.** $\displaystyle\sum_{k=1}^{\infty} \frac{1}{k(k+1)}$

61. $4, 1\frac{1}{4}, 1\frac{4}{5}, 1\frac{5}{9}$ **63.** $6561, -81, 9i, -3\sqrt{i}$

65. 2, 3, 5, 8

67. (a) 1062, 1127.84, 1197.77, 1272.03, 1350.90, 1434.65, 1523.60, 1618.07, 1718.39, 1824.93; **(b)** $3330.35

69. 1, 2, 4, 8, 16, 32, 64, 128, 256, 512, 1024, 2048, 4096, 8192, 16,384, 32,768, 65,536

71. 1, 1, 2, 3, 5, 8, 13

73.

n	U_n
1	2
2	2.25
3	2.3704
4	2.4414
5	2.4883
6	2.5216
7	2.5465
8	2.5658
9	2.5812
10	2.5937

75.

n	U_n
1	2
2	1.5538
3	1.4988
4	1.4914
5	1.4904
6	1.4902
7	1.4902
8	1.4902
9	1.4902
10	1.4902

77. (a) 27.25, 26.108, 25.002, 23.932, 22.898, 21.9, 20.938, 20.012, 19.122, 18.268; **(b)** 26, 19, 9, 22, 34

79. Discussion and Writing
80. [8.1], [8.3], [8.5], [8.6] $(-1, -3)$
81. [8.1], [8.3], [8.5], [8.6] Original: $1.0135 billion; purchased: $0.1765 billion   **82.** [9.1] $(3, -2)$; 4
83. [9.1] $\left(-\dfrac{5}{2}, 4\right)$; $\dfrac{\sqrt{97}}{2}$ **85.** $i, -1, -i, 1, i; i$
87. $\ln (1 \cdot 2 \cdot 3 \cdot \cdots \cdot n)$

Exercise Set 10.2

1. $a_1 = 3, d = 5$ **3.** $a_1 = 9, d = -4$ **5.** $a_1 = \frac{3}{2}, d = \frac{3}{4}$

7. $a_1 = \$316, d = -\$3$ **9.** $a_{12} = 46$ **11.** $a_{14} = -\frac{17}{3}$

13. $a_{10} = \$7941.62$ **15.** 33rd   **17.** 46th   **19.** $a_1 = 5$

21. $n = 39$ **23.** $a_1 = \frac{1}{3}; d = \frac{1}{2}; \frac{1}{3}, \frac{5}{6}, \frac{4}{3}, \frac{11}{6}, \frac{7}{3}$ **25.** 670

27. 160,400   **29.** 735   **31.** 990   **33.** 1760   **35.** $\frac{65}{2}$

37. $-\frac{6026}{13}$ **39.** 1260   **41.** 3; 171   **43.** 4960¢, or $49.60

45. 1320   **47.** Yes; 3   **49.** Discussion and Writing

50. [8.1], [8.3], [8.5], [8.6] $(2, 5)$

51. [8.2], [8.3], [8.5], [8.6] $(2, -1, 3)$

52. [9.1] $(-4, 0), (4, 0); (-\sqrt{7}, 0), (\sqrt{7}, 0)$

53. [9.1] $\dfrac{x^2}{4} + \dfrac{y^2}{25} = 1$ **55.** n^2

57. $a_1 = 60 - 5p - 5q; d = 5p + 2q - 20$ **59.** 6, 8, 10

61. $5\frac{4}{5}, 7\frac{3}{5}, 9\frac{2}{5}, 11\frac{1}{5}$ **63.** Insert 16 arithmetic means between 1 and 50 with $d = \frac{49}{17}$.

65.
$$\begin{aligned} m &= p + d \\ m &= q - d \\ \hline 2m &= p + q \qquad \textbf{Adding} \\ m &= \frac{p+q}{2} \end{aligned}$$

Exercise Set 10.3

1. 2   **3.** -1 **5.** -2 **7.** 0.1   **9.** $\dfrac{a}{2}$ **11.** 128

13. 162   **15.** $7(5)^{40}$ **17.** 3^{n-1} **19.** $(-1)^{n-1}$

21. $\dfrac{1}{x^n}$ **23.** 762   **25.** $\frac{4921}{18}$ **27.** 8   **29.** 125

31. Does not exist   **33.** $\frac{2}{3}$ **35.** $29\frac{38{,}569}{59{,}049}$

37. 2   **39.** Does not exist   **41.** $4545.\overline{45}$

43. $\frac{160}{9}$ **45.** $\frac{13}{99}$ **47.** 9   **49.** $\frac{34{,}091}{9990}$ **51. (a)** $\frac{1}{256}$ ft;

(b) $5\frac{1}{3}$ ft   **53. (a)** About 297 ft; **(b)** 300 ft

55. $31,497.57   **57.** $523,619.17   **59.** $86,666,666,667

61. Left to the student   **63.** Discussion and Writing

65. [4.1] $(f \circ g)(x) = 16x^2 + 40x + 25$; $(g \circ f)(x) = 4x^2 + 5$ **66.** [4.1] $(f \circ g)(x) = x^2 + x + 2$; $(g \circ f)(x) = x^2 - x + 3$ **67.** [4.5] 2.209   **68.** [4.5] $\frac{1}{16}$

69. $(4 - \sqrt{6})/(\sqrt{3} - \sqrt{2}) = 2\sqrt{3} + \sqrt{2}$, $(6\sqrt{3} - 2\sqrt{2})/(4 - \sqrt{6}) = 2\sqrt{3} + \sqrt{2}$; there exists a common ratio, $2\sqrt{3} + \sqrt{2}$; thus the sequence is geometric.

71. (a) $\frac{13}{3}, \frac{22}{3}, \frac{34}{3}, \frac{46}{3}, \frac{58}{3}$; **(b)** $-\frac{11}{3}; -\frac{2}{3}, \frac{10}{3}, -\frac{50}{3}, \frac{250}{3}$ or 5; 8, 12, 18, 27   **73.** $S_n = \dfrac{x^2(1 - (-x)^n)}{x + 1}$

75. $\dfrac{a_{n+1}}{a_n} = r$, so $\ln \dfrac{a_{n+1}}{a_n} = \ln r$. But $\ln \dfrac{a_{n+1}}{a_n} = \ln a_{n+1} - \ln a_n = \ln r$. Thus, $\ln a_1, \ln a_2, \ldots$, is an arithmetic sequence with common difference $\ln r$.
77. 512 cm^2

Exercise Set 10.4

1. $1^2 < 1^3$, false; $2^2 < 2^3$, true; $3^2 < 3^3$, true; $4^2 < 4^3$, true; $5^2 < 5^3$, true

3. A polygon of 3 sides has $\dfrac{3(3 - 3)}{2}$ diagonals. True; A polygon of 4 sides has $\dfrac{4(4 - 3)}{2}$ diagonals. True; A polygon of 5 sides has $\dfrac{5(5 - 3)}{2}$ diagonals. True; A polygon of 6 sides has $\dfrac{6(6 - 3)}{2}$ diagonals. True; A polygon of 7 sides has $\dfrac{7(7 - 3)}{2}$ diagonals. True.

5. S_n: $2 + 4 + 6 + \cdots + 2n = n(n + 1)$
S_1: $2 = 1(1 + 1)$
S_k: $2 + 4 + 6 + \cdots + 2k = k(k + 1)$
S_{k+1}: $2 + 4 + 6 + \cdots + 2k + 2(k + 1)$
$= (k + 1)(k + 2)$
1. *Basis step*: S_1 true by substitution.
2. *Induction step*: Assume S_k. Deduce S_{k+1}.
   Starting with the left side of S_{k+1}, we have
   $2 + 4 + 6 + \cdots + 2k + 2(k + 1)$
   $= k(k + 1) + 2(k + 1)$ **By S_k**
   $= (k + 1)(k + 2)$.   **Distributive law**

7. S_n: $1 + 5 + 9 + \cdots + (4n - 3) = n(2n - 1)$
S_1: $1 = 1(2 \cdot 1 - 1)$
S_k: $1 + 5 + 9 + \cdots + (4k - 3) = k(2k - 1)$
S_{k+1}: $1 + 5 + 9 + \cdots + (4k - 3) + [4(k + 1) - 3]$
$= (k + 1)[2(k + 1) - 1]$
$= (k + 1)(2k + 1)$
1. *Basis step*: S_1 true by substitution.
2. *Induction step*: Assume S_k. Deduce S_{k+1}.
   Starting with the left side of S_{k+1}, we have
   $1 + 5 + 9 + \cdots + (4k - 3) + [4(k + 1) - 3]$
   $= k(2k - 1) + [4(k + 1) - 3]$ **By S_k**
   $= 2k^2 - k + 4k + 4 - 3$
   $= (k + 1)(2k + 1)$.

9. S_n: $2 + 4 + 8 + \cdots + 2^n = 2(2^n - 1)$
S_1: $2 = 2(2 - 1)$
S_k: $2 + 4 + 8 + \cdots + 2^k = 2(2^k - 1)$
S_{k+1}: $2 + 4 + 8 + \cdots + 2^k + 2^{k+1} = 2(2^{k+1} - 1)$
1. *Basis step*: S_1 is true by substitution.
2. *Induction step*: Assume S_k. Deduce S_{k+1}.
   Starting with the left side of S_{k+1}, we have
   $2 + 4 + 8 + \cdots + 2^k + 2^{k+1}$
   $= 2(2^k - 1) + 2^{k+1}$ **By S_k**
   $= 2^{k+1} - 2 + 2^{k+1}$
   $= 2 \cdot 2^{k+1} - 2$
   $= 2(2^{k+1} - 1)$.

11. 1. *Basis step*: Since $1 < 1 + 1$, S_1 is true.
2. *Induction step*: Assume S_k. Deduce S_{k+1}. Now
   $k < k + 1$ **By S_k**
   $k + 1 < k + 1 + 1$;   **Adding 1**
   $k + 1 < k + 2$.

13. 1. *Basis step*: Since $2 = 2$, S_1 is true.
2. *Induction step*: Let k be any natural number. Assume S_k. Deduce S_{k+1}.
   $2k \leq 2^k$ **By S_k**
   $2 \cdot 2k \leq 2 \cdot 2^k$ **Multiplying by 2**
   $4k \leq 2^{k+1}$
   Since $1 \leq k$, $k + 1 \leq k + k$, or $k + 1 \leq 2k$.
   Then $2(k + 1) \leq 4k$.
   Thus, $2(k + 1) \leq 4k \leq 2^{k+1}$, so $2(k + 1) \leq 2^{k+1}$.

15.
S_n: $\dfrac{1}{1 \cdot 2 \cdot 3} + \dfrac{1}{2 \cdot 3 \cdot 4} + \dfrac{1}{3 \cdot 4 \cdot 5} + \cdots$
$+ \dfrac{1}{n(n + 1)(n + 2)}$
$= \dfrac{n(n + 3)}{4(n + 1)(n + 2)}$

S_1: $\dfrac{1}{1 \cdot 2 \cdot 3} = \dfrac{1(1 + 3)}{4 \cdot 2 \cdot 3}$

S_k: $\dfrac{1}{1 \cdot 2 \cdot 3} + \dfrac{1}{2 \cdot 3 \cdot 4} + \cdots + \dfrac{1}{k(k + 1)(k + 2)}$
$= \dfrac{k(k + 3)}{4(k + 1)(k + 2)}$

S_{k+1}: $\dfrac{1}{1 \cdot 2 \cdot 3} + \dfrac{1}{2 \cdot 3 \cdot 4} + \cdots + \dfrac{1}{k(k + 1)(k + 2)}$
$+ \dfrac{1}{(k + 1)(k + 2)(k + 3)}$
$= \dfrac{(k + 1)(k + 1 + 3)}{4(k + 1 + 1)(k + 1 + 2)} = \dfrac{(k + 1)(k + 4)}{4(k + 2)(k + 3)}$

1. *Basis step*: Since $\dfrac{1}{1 \cdot 2 \cdot 3} = \dfrac{1}{6}$ and $\dfrac{1(1 + 3)}{4 \cdot 2 \cdot 3} = \dfrac{1 \cdot 4}{4 \cdot 2 \cdot 3} = \dfrac{1}{6}$, S_1 is true.
2. *Induction step*: Assume S_k. Deduce S_{k+1}.

Add $\dfrac{1}{(k + 1)(k + 2)(k + 3)}$ on both sides of S_k and simplify the right side. Only the right side is shown here.

$$\dfrac{k(k + 3)}{4(k + 1)(k + 2)} + \dfrac{1}{(k + 1)(k + 2)(k + 3)}$$

$$= \dfrac{k(k + 3)(k + 3) + 4}{4(k + 1)(k + 2)(k + 3)}$$

$$= \dfrac{k^3 + 6k^2 + 9k + 4}{4(k + 1)(k + 2)(k + 3)}$$

$$= \dfrac{(k + 1)^2(k + 4)}{4(k + 1)(k + 2)(k + 3)}$$

$$= \dfrac{(k + 1)(k + 4)}{4(k + 2)(k + 3)}$$

17.

S_n: $1 + 2 + 3 + \cdots + n = \dfrac{n(n + 1)}{2}$

S_1: $1 = \dfrac{1(1 + 1)}{2}$

S_k: $1 + 2 + 3 + \cdots + k = \dfrac{k(k + 1)}{2}$

S_{k+1}: $1 + 2 + 3 + \cdots + k + (k + 1) = \dfrac{(k + 1)(k + 2)}{2}$

1. *Basis step*: S_1 true by substitution.
2. *Induction step*: Assume S_k. Deduce S_{k+1}.
   Starting with the left side of S_{k+1}, we have

$$1 + 2 + 3 + \cdots + k + (k + 1)$$

$$= \dfrac{k(k + 1)}{2} + (k + 1) \qquad \textbf{By } S_k$$

$$= \dfrac{k(k + 1) + 2(k + 1)}{2} \qquad \textbf{Adding}$$

$$= \dfrac{(k + 1)(k + 2)}{2}. \qquad \textbf{Distributive law}$$

19. 1. *Basis step*. S_1: $1^3 = \dfrac{1^2(1 + 1)^2}{4} = 1$. True.

2. *Induction step*: Assume S_k. Deduce S_{k+1}.

S_k: $1^3 + 2^3 + \cdots + k^3 = \dfrac{k^2(k + 1)^2}{4}$

$1^3 + 2^3 + \cdots + (k + 1)^3 = \dfrac{k^2(k + 1)^2}{4} + (k + 1)^3 \qquad \textbf{By } S_k$

$$= \dfrac{(k + 1)^2}{4}[k^2 + 4(k + 1)]$$

$$= \dfrac{(k + 1)^2(k + 2)^2}{4}$$

21.

1. *Basis step*. S_1: $1^5 = \dfrac{1^2(1 + 1)^2(2 \cdot 1^2 + 2 \cdot 1 - 1)}{12}$.
True.

2. *Induction step*. Assume S_k:

$$1^5 + 2^5 + \cdots + k^5 = \dfrac{k^2(k + 1)^2(2k^2 + 2k - 1)}{12}.$$

Then $1^5 + 2^5 + \cdots + k^5 + (k + 1)^5$

$$= \dfrac{k^2(k + 1)^2(2k^2 + 2k - 1)}{12} + (k + 1)^5$$

$$= \dfrac{k^2(k + 1)^2(2k^2 + 2k - 1) + 12(k + 1)^5}{12}$$

$$= \dfrac{(k + 1)^2(2k^4 + 14k^3 + 35k^2 + 36k + 12)}{12}$$

$$= \dfrac{(k + 1)^2(k + 2)^2(2k^2 + 6k + 3)}{12}$$

$$= \dfrac{(k + 1)^2(k + 1 + 1)^2(2(k + 1)^2 + 2(k + 1) - 1)}{12}.$$

23.

1. *Basis step*. S_1: $1(1 + 1) = \dfrac{1(1 + 1)(1 + 2)}{3}$. True.

2. *Induction step*. Assume S_k:
   $1(1 + 1) + 2(2 + 1) + \cdots + k(k + 1)$
   $= \dfrac{k(k + 1)(k + 2)}{3}$.

Then $1(1 + 1) + 2(2 + 1) + \cdots + k(k + 1)$
$+ (k + 1)(k + 1 + 1)$

$$= \dfrac{k(k + 1)(k + 2)}{3} + (k + 1)(k + 2)$$

$$= \dfrac{(k + 1)(k + 2)(k + 3)}{3}$$

$$= \dfrac{(k + 1)(k + 1 + 1)(k + 1 + 2)}{3}.$$

25. 1. *Basis step*: Since $\frac{1}{2}[2a_1 + (1 - 1)d] =$
$\frac{1}{2} \cdot 2a_1 = a_1$, S_1 is true.

2. *Induction step*: Assume S_k. Deduce S_{k+1}. Starting with the left side of S_{k+1}, we have

$$a_1 + (a_1 + d) + \cdots + [a_1 + (k - 1)d] + [a_1 + kd]$$

$$= \dfrac{k}{2}[2a_1 + (k - 1)d] + [a_1 + kd] \qquad \textbf{By } S_k$$

$$= \dfrac{k[2a_1 + (k - 1)d]}{2} + \dfrac{2[a_1 + kd]}{2}$$

$$= \dfrac{2ka_1 + k(k - 1)d + 2a_1 + 2kd}{2}$$

$$= \dfrac{2a_1(k + 1) + k(k - 1)d + 2kd}{2}$$

$$= \dfrac{2a_1(k + 1) + (k - 1 + 2)kd}{2}$$

$$= \dfrac{2a_1(k + 1) + (k + 1)kd}{2}$$

$$= \dfrac{k + 1}{2}[2a_1 + kd].$$

27. Discussion and Writing
28. [8.1], [8.3], [8.5], [8.6] (5, 3)
29. [8.2], [8.3], [8.5], [8.6] (2, −3, 4)
30. [8.1], [8.3], [8.5], [8.6] 50 hardback, 30 paperback
31. [8.2], [8.3], [8.5], [8.6] $800 at 6%, $1600 at 8%, $2000 at 10%

33.
1. *Basis step.* S_1: $x + y$ is a factor of $x^2 - y^2$. True.
       S_2: $x + y$ is a factor of $x^4 - y^4$. True.
2. *Induction step.* Assume S_{k-1}: $x + y$ is a factor of $x^{2(k-1)} - y^{2(k-1)}$. Then $x^{2(k-1)} - y^{2(k-1)} = (x + y)Q(x)$ for some polynomial Q.
Assume S_k: $x + y$ is a factor of $x^{2k} - y^{2k}$. Then $x^{2k} - y^{2k} = (x + y)P(x)$ for some polynomial P.

$$x^{2(k+1)} - y^{2(k+1)}$$
$$= (x^{2k} - y^{2k})(x^2 + y^2) - (x^{2(k-1)} - y^{2(k-1)})(x^2 y^2)$$
$$= (x + y)P(x)(x^2 + y^2) - (x + y)Q(x)(x^2 y^2)$$
$$= (x + y)[P(x)(x^2 + y^2) - Q(x)(x^2 y^2)]$$

so $x + y$ is a factor of $x^{2(k+1)} - y^{2(k+1)}$.

35.
S_2: $\log_a (b_1 b_2) = \log_a b_1 + \log_a b_2$
S_k: $\log_a (b_1 b_2 \cdots b_k) = \log_a b_1 + \log_a b_2 + \cdots + \log_a b_k$
S_{k+1}: $\log_a (b_1 b_2 \cdots b_{k+1}) = \log_a b_1 + \log_a b_2 + \cdots + \log_a b_{k+1}$

1. *Basis step*: S_2 is true by the properties of logarithms.
2. *Induction step*: Let k be a natural number $k \geq 2$.
   Assume S_k. Deduce S_{k+1}.

$\log_a (b_1 b_2 \cdots b_{k+1})$ **Left side of S_{k+1}**
$\quad = \log_a (b_1 b_2 \cdots b_k) + \log_a b_{k+1}$ **By S_2**
$\quad = \log_a b_1 + \log_a b_2 + \cdots + \log_a b_k + \log_a b_{k+1}$

37. S_2: $\overline{z_1 + z_2} = \bar{z}_1 + \bar{z}_2$:
$\overline{(a + bi) + (c + di)} = \overline{(a + c) + (b + d)i}$
$\qquad\qquad\qquad = (a + c) - (b + d)i$
$\overline{(a + bi)} + \overline{(c + di)} = a - bi + c - di$
$\qquad\qquad\qquad = (a + c) - (b + d)i$.
S_k: $\overline{z_1 + z_2 + \cdots + z_k} = \bar{z}_1 + \bar{z}_2 + \cdots + \bar{z}_k$.
$\overline{(z_1 + z_2 + \cdots + z_k) + z_{k+1}}$
$\qquad = \overline{(z_1 + z_2 + \cdots + z_k)} + \overline{z_{k+1}}$ **By S_2**
$\qquad = \bar{z}_1 + \bar{z}_2 + \cdots + \bar{z}_k + \bar{z}_{k+1}$ **By S_k**

39. S_1: i is either i or -1 or $-i$ or 1.
   S_k: i^k is either i or -1 or $-i$ or 1.
   $i^{k+1} = i^k \cdot i$ is then $i \cdot i = -1$ or $-1 \cdot i = -i$ or $-i \cdot i = 1$ or $1 \cdot i = i$.

41. S_1:   3 is a factor of $1^3 + 2 \cdot 1$.
   S_k: 3 is a factor of $k^3 + 2k$, i.e., $k^3 + 2k = 3 \cdot m$.
   S_{k+1}: 3 is a factor of $(k + 1)^3 + 2(k + 1)$.

Consider
$$(k + 1)^3 + 2(k + 1) = k^3 + 3k^2 + 5k + 3$$
$$= (k^3 + 2k) + 3k^2 + 3k + 3$$
$$= 3m + 3(k^2 + k + 1).$$
A multiple of 3

Exercise Set 10.5

1. 720   **3.** 604,800   **5.** 120   **7.** 1   **9.** 3024   **11.** 120
13. 120   **15.** 1   **17.** 6,497,400   **19.** $n(n - 1)(n - 2)$
21. n **23.** 6! = 720   **25.** 9! = 362,880   **27.** $_9P_4 = 3024$
29. $_5P_5 = 120$; $5^5 = 3125$ **31.** $\dfrac{8!}{3!} = 6720$; $\dfrac{7!}{2!} = 2520$;
$\dfrac{11!}{2! \, 2! \, 2!} = 4,989,600$ **33.** $8 \cdot 10^6 = 8,000,000$; 8 million
35. $\dfrac{9!}{2! \, 3! \, 4!} = 1260$
37. (a) $_6P_5 = 720$; (b) $6^5 = 7776$; (c) $1 \cdot {}_5P_4 = 120$;
(d) $1 \cdot 1 \cdot {}_4P_3 = 24$ **39.** (a) 10^5, or 100,000; (b) 100,000
41. (a) $10^9 = 1,000,000,000$; (b) yes
43. Discussion and Writing
44. [2.1] $\dfrac{9}{4}$, or 2.25   **45.** [2.3] −3, 2   **46.** [2.3] $\dfrac{3 \pm \sqrt{17}}{4}$
47. [3.3] −2, 1, 5   **49.** 8   **51.** 11   **53.** $n - 1$

Exercise Set 10.6

1. 78   **3.** 78   **5.** 7   **7.** 10   **9.** 1   **11.** 15   **13.** 128
15. 270,725   **17.** 13,037,895   **19.** n **21.** 1
23. $_{23}C_4 = 8855$ **25.** $_{13}C_{10} = 286$
27. $\dbinom{8}{2} = 28$; $\dbinom{8}{3} = 56$ **29.** $\dbinom{52}{5} = 2,598,960$
31. (a) $_{31}P_2 = 930$; (b) $31^2 = 961$; (c) $_{31}C_2 = 465$
33. Discussion and Writing   **34.** [2.1] $-\dfrac{17}{2}$
35. [2.3] $-1, \dfrac{3}{2}$ **36.** [2.3] $\dfrac{-5 \pm \sqrt{21}}{2}$ **37.** [3.3] −4, −2, 3
39. $\dbinom{13}{5} = 1287$ **41.** $\dbinom{n}{2}$; $2\dbinom{n}{2}$ **43.** 4   **45.** 7
47. $\dbinom{n}{k - 1} + \dbinom{n}{k}$
$= \dfrac{n!}{(k - 1)!(n - k + 1)!} \cdot \dfrac{k}{k}$
$\quad + \dfrac{n!}{k!(n - k)!} \cdot \dfrac{(n - k + 1)}{(n - k + 1)}$
$= \dfrac{n!(k + (n - k + 1))}{k!(n - k + 1)!}$
$= \dfrac{(n + 1)!}{k!(n - k + 1)!} = \dbinom{n + 1}{k}$

Exercise Set 10.7

1. $x^4 + 20x^3 + 150x^2 + 500x + 625$
3. $x^5 - 15x^4 + 90x^3 - 270x^2 + 405x - 243$
5. $x^5 - 5x^4y + 10x^3y^2 - 10x^2y^3 + 5xy^4 - y^5$
7. $15{,}625x^6 + 75{,}000x^5y + 150{,}000x^4y^2 +$
$160{,}000x^3y^3 + 96{,}000x^2y^4 + 30{,}720xy^5 + 4096y^6$
9. $128t^7 + 448t^5 + 672t^3 + 560t + 280t^{-1} + 84t^{-3}$
$+ 14t^{-5} + t^{-7}$
11. $x^{10} - 5x^8 + 10x^6 - 10x^4 + 5x^2 - 1$
13. $125 + 150\sqrt{5}t + 375t^2 + 100\sqrt{5}t^3 + 75t^4 +$
$6\sqrt{5}t^5 + t^6$
15. $a^9 - 18a^7 + 144a^5 - 672a^3 + 2016a - 4032a^{-1} +$
$5376a^{-3} - 4608a^{-5} + 2304a^{-7} - 512a^{-9}$
17. $140\sqrt{2}$ **19.** $x^{-8} + 4x^{-4} + 6 + 4x^4 + x^8$
21. $21a^5b^2$ **23.** $-252x^5y^5$ **25.** $-745{,}472a^3$
27. $1120x^{12}y^2$ **29.** $-1{,}959{,}552u^5v^{10}$ **31.** 128
33. 2^{24}, or $16{,}777{,}216$ **35.** 20 **37.** $-12 + 316i$
39. $-7 - 4\sqrt{2}\,i$ **41.** $\sum_{k=0}^{n} \binom{n}{k}(-1)^k a^{n-k}b^k$
43. $\sum_{k=1}^{n} \binom{n}{k}x^{n-k}h^{k-1}$ **45.** Discussion and Writing
46. [1.5] $x^2 + 2x - 2$ **47.** [1.5] $2x^3 - 3x^2 + 2x - 3$
48. [4.1] $4x^2 - 12x + 10$ **49.** [4.1] $2x^2 - 1$
51. $-5 \pm 2\sqrt{2},\ -5 \pm 2\sqrt{2}\,i$ **53.** $3, 9, 6 \pm 3i$
55. $-4320x^6y^{9/2}$ **57.** $-\dfrac{35}{x^{1/6}}$ **59.** 2^{100} **61.** $[\log_a (xt)]^{23}$
63. 1. *Basis step*: Since $a + b = (a + b)^1$, S_1 is true.
2. *Induction step*: Let S_k be the statement of the
binomial theorem with n replaced by k. Multiply
both sides of S_k by $(a + b)$ to obtain
$(a + b)^{k+1}$
$$= \left[a^k + \cdots + \binom{k}{r-1}a^{k-(r-1)}b^{r-1} \right.$$
$$\left. + \binom{k}{r}a^{k-r}b^r + \cdots + b^k \right](a + b)$$
$$= a^{k+1} + \cdots + \left[\binom{k}{r-1} + \binom{k}{r} \right]a^{(k+1)-r}b^r$$
$$+ \cdots + b^{k+1}$$
$$= a^{k+1} + \cdots + \binom{k+1}{r}a^{(k+1)-r}b^r + \cdots + b^{k+1}.$$
This proves S_{k+1}, assuming S_k. Hence S_n is true for
$n = 1, 2, 3, \ldots$.

Exercise Set 10.8

1. (a) 0.18, 0.24, 0.23, 0.23, 0.12; **(b)** Opinions may vary,
but it seems that people tend not to pick the first or last
numbers.   **3.** 11,700

5. (a) T, S, R, N, L; **(b)** E; **(c)** yes
7. (a) $\frac{2}{7}$; **(b)** $\frac{5}{7}$; **(c)** 0; **(d)** 1   **9.** $\frac{350}{31{,}977}$ **11.** $\frac{1}{108{,}290}$
13. $\frac{33}{66{,}640}$ **15. (a)** HHH, HHT, HTH, HTT, THH, THT,
TTH, TTT; **(b)** $\frac{3}{8}$; **(c)** $\frac{7}{8}$; **(d)** $\frac{7}{8}$; **(e)** $\frac{3}{8}$ **17.** $\frac{9}{19}$ **19.** $\frac{18}{19}$
21. $\frac{1}{38}$ **23.** $\frac{9}{19}$
25. Answers will vary.   **27.** Discussion and Writing
28. [2.3] $\dfrac{2 \pm \sqrt{13}}{3}$ **29.** [3.3] $-3, -\frac{1}{2}$
30. [8.1], [8.3], [8.5], [8.6] $(-3, -4)$
31. [8.2], [8.3], [8.5], [8.6] $(5, -2, 1)$
33. (a) 36; **(b)** 1.39×10^{-5}
35. (a) $(13 \cdot {}_4C_3) \cdot (12 \cdot {}_4C_2) = 3744$; **(b)** 0.00144
37. (a) $4 \cdot \binom{13}{5} - 4 - 36 = 5108$; **(b)** 0.00197
39. (a) $\binom{10}{1}\binom{4}{1}\binom{4}{1}\binom{4}{1}\binom{4}{1}\binom{4}{1} - 4 - 36 = 10{,}200$;
(b) 0.00392

Review Exercises, Chapter 10

1. [10.1] $-\frac{1}{2}, \frac{4}{17}, -\frac{9}{82}, \frac{16}{257}; -\frac{121}{14{,}642}; -\frac{529}{279{,}842}$
2. [10.1] $(-1)^{n+1}(n^2 + 1)$
3. [10.1] $\frac{3}{2} - \frac{9}{8} + \frac{27}{26} - \frac{81}{80} = \frac{417}{1040}$
4. [10.1] $\sum_{k=1}^{7} k^2 - 1$ **5.** [10.2] $3\frac{3}{4}$ **6.** [10.2] $a + 4b$
7. [10.2] 531   **8.** [10.2] 20,100   **9.** [10.2] 11
10. [10.2] -4 **11.** [10.3] $n = 6, S_n = -126$
12. [10.3] $a_1 = 8, a_5 = \frac{1}{2}$ **13.** [10.3] Does not exist
14. [10.3] $\frac{3}{11}$ **15.** [10.3] $\frac{3}{8}$ **16.** [10.3] $\frac{241}{99}$
17. [10.2] $5\frac{4}{5}, 6\frac{3}{5}, 7\frac{2}{5}, 8\frac{1}{5}$ **18.** [10.3] 167.3 ft
19. [10.3] \$60,653.05   **20.** [10.2] **(a)** \$7.38; **(b)** \$1365.10
21. [10.3] \$88,888,888,889
22. [10.4] $S_n: 1 + 4 + 7 + \cdots + (3n - 2)$
$$= \frac{n(3n - 1)}{2}$$
$$S_1: \quad 1 = \frac{1(3 - 1)}{2}$$
$$S_k: \quad 1 + 4 + 7 + \cdots + (3k - 2) = \frac{k(3k - 1)}{2}$$
$$S_{k+1}: \quad 1 + 4 + 7 + \cdots + [3(k + 1) - 2]$$
$$= 1 + 4 + 7 + \cdots + (3k - 2) + (3k + 1)$$
$$= \frac{(k + 1)(3k + 2)}{2}$$
1. *Basis step*: $1 = \frac{2}{2} = \frac{1(3 - 1)}{2}$ is true.
2. *Induction step*: Assume S_k. Add $(3k + 1)$ to
both sides.

$1 + 4 + 7 + \cdots + (3k - 2) + (3k + 1)$

$$= \frac{k(3k - 1)}{2} + (3k + 1)$$

$$= \frac{k(3k - 1)}{2} + \frac{2(3k + 1)}{2}$$

$$= \frac{3k^2 - k + 6k + 2}{2}$$

$$= \frac{3k^2 + 5k + 2}{2}$$

$$= \frac{(k + 1)(3k + 2)}{2}$$

23. [10.4] S_1: $1 = \dfrac{3^1 - 1}{2}$; S_2: $1 + 3 = \dfrac{3^2 - 1}{2}$

S_k: $1 + 3 + 3^2 + \cdots + 3^{k-1} = \dfrac{3^k - 1}{2}$

2. *Induction step*: Assume S_k. Add 3^k on both sides.

$1 + 3 + \cdots + 3^{k-1} + 3^k$

$$= \frac{3^k - 1}{2} + 3^k = \frac{3^k - 1}{2} + 3^k \cdot \frac{2}{2}$$

$$= \frac{3 \cdot 3^k - 1}{2} = \frac{3^{k+1} - 1}{2}$$

24. [10.4]

S_n: $\left(1 - \dfrac{1}{2}\right)\left(1 - \dfrac{1}{3}\right) \cdots \left(1 - \dfrac{1}{n}\right) = \dfrac{1}{n}$

S_2: $\left(1 - \dfrac{1}{2}\right) = \dfrac{1}{2}$

S_k: $\left(1 - \dfrac{1}{2}\right)\left(1 - \dfrac{1}{3}\right) \cdots \left(1 - \dfrac{1}{k}\right) = \dfrac{1}{k}$

S_{k+1}: $\left(1 - \dfrac{1}{2}\right)\left(1 - \dfrac{1}{3}\right) \cdots \left(1 - \dfrac{1}{k}\right)\left(1 - \dfrac{1}{k+1}\right)$

$$= \frac{1}{k + 1}$$

1. *Basis step*: S_2 is true by substitution.
2. *Induction step*: Assume S_k. Deduce S_{k+1}.
   Starting with the left side of S_{k+1}, we have

$\underbrace{\left(1 - \dfrac{1}{2}\right)\left(1 - \dfrac{1}{3}\right) \cdots \left(1 - \dfrac{1}{k}\right)}\left(1 - \dfrac{1}{k+1}\right)$

$$= \frac{1}{k} \cdot \left(1 - \frac{1}{k + 1}\right). \quad \textbf{By } S_k$$

$$= \frac{1}{k} \cdot \left(\frac{k + 1 - 1}{k + 1}\right)$$

$$= \frac{1}{k} \cdot \frac{k}{k + 1}$$

$$= \frac{1}{k + 1}. \quad \textbf{Simplifying}$$

25. [10.5] $6! = 720$ **26.** [10.5] $9 \cdot 8 \cdot 7 \cdot 6 = 3024$

27. [10.6] $\dbinom{15}{8} = 6435$ **28.** [10.5] $24 \cdot 23 \cdot 22 = 12{,}144$

29. [10.5] $\dfrac{9!}{1! \, 4! \, 2! \, 2!} = 3780$ **30.** [10.5] 36

31. [10.5] **(a)** $_6P_5 = 720$; **(b)** $6^5 = 7776$; **(c)** $_5P_4 = 120$;
(d) $_3P_2 = 6$ **32.** [10.7] 256

33. [10.7] $m^7 + 7m^6n + 21m^5n^2 + 35m^4n^3 + 35m^3n^4 + 21m^2n^5 + 7mn^6 + n^7$

34. [10.7] $x^5 - 5\sqrt{2}\,x^4 + 20x^3 - 20\sqrt{2}\,x^2 + 20x - 4\sqrt{2}$

35. [10.7] $x^8 - 12x^6y + 54x^4y^2 - 108x^2y^3 + 81y^4$

36. [10.7] $a^8 + 8a^6 + 28a^4 + 56a^2 + 70 + 56a^{-2} + 28a^{-4} + 8a^{-6} + a^{-8}$ **37.** [10.7] $-6624 + 16{,}280i$

38. [10.7] $220a^9x^3$ **39.** [10.7] $-\dbinom{18}{11}128a^7b^{11}$

40. [10.8] $\frac{86}{206} \approx 0.42$, $\frac{97}{206} \approx 0.47$, $\frac{23}{206} \approx 0.11$

41. [10.8] $\frac{1}{12}$, 0   **42.** [10.8] $\frac{1}{4}$ **43.** [10.8] $\frac{6}{5525}$

44. [10.1]

n	U_n
1	0.3
2	2.5
3	13.5
4	68.5
5	343.5
6	1718.5
7	8593.5
8	42968.5
9	214843.5
10	1074218.5

45. [10.1] **(a)** $a_n = 0.6149275362n + 0.5905797101$;
(b) \$12.89 billion

46. Discussion and Writing [10.6] A list of 9 candidates for an office is to be narrowed down to 4 candidates. In how many ways can this be done?

47. Discussion and Writing [10.3] Someone who has managed several sequences of hiring has a considerable income from the sales of the people in the lower levels. However, with a finite population, it will not be long before the salespersons in the lowest level have no one to hire and no one to sell to.

48. [10.4] S_1 fails for both (a) and (b).

49. [10.3] $\dfrac{a_{k+1}}{a_k} = r_1$, $\dfrac{b_{k+1}}{b_k} = r_2$, so $\dfrac{a_{k+1}b_{k+1}}{a_kb_k} = r_1r_2$, a constant   **50.** [10.2] **(a)** No (unless a_n is all positive or all negative); **(b)** yes; **(c)** yes; **(d)** no (unless a_n is constant); **(e)** no (unless a_n is constant); **(f)** no (unless a_n is constant)   **51.** [10.2] $-2, 0, 2, 4$

52. [10.3] $\frac{1}{2}, -\frac{1}{6}, \frac{1}{18}$ **53.** [10.6] $\left(\log \dfrac{x}{y}\right)^{10}$

54. [10.6] 18   **55.** [10.6] 36   **56.** [10.7] -9

Test, Chapter 10

1. [10.1] -43 **2.** [10.1] $\frac{2}{3}, \frac{3}{4}, \frac{4}{5}, \frac{5}{6}; \frac{6}{7}$

3. [10.1] $2 + 5 + 10 + 17 = 34$ **4.** [10.2] 44

5. [10.2] 38   **6.** [10.2] -420 **7.** [10.3] 156,000

8. [10.3] 27   **9.** [10.3] $\frac{56}{99}$ **10.** [10.3] $74,399.77

11. [10.4]

$$S_1: \quad 2 = \frac{1(3 \cdot 1 + 1)}{2}$$

$$S_k: \quad 2 + 5 + 8 + \cdots + (3k - 1) = \frac{k(3k + 1)}{2}$$

$$S_{k+1}: \quad 2 + 5 + 8 + \cdots + (3k - 1) + [3(k + 1) - 1]$$
$$= \frac{(k + 1)[3(k + 1) + 1]}{2}$$

Basis step: $\dfrac{1(3 \cdot 1 + 1)}{2} = \dfrac{1 \cdot 4}{2} = 2$, so S_1 is true.

Induction step:

$$2 + 5 + 8 + \cdots + (3k - 1) + [3(k + 1) - 1]$$
$$= \frac{k(3k + 1)}{2} + [3k + 3 - 1]$$
$$= \frac{3k^2}{2} + \frac{k}{2} + 3k + 2$$
$$= \frac{3k^2}{2} + \frac{7k}{2} + 2$$
$$= \frac{3k^2 + 7k + 4}{2}$$
$$= \frac{(k + 1)(3k + 4)}{2}$$
$$= \frac{(k + 1)[3(k + 1) + 1]}{2}$$

12. [10.5] 3,603,600   **13.** [10.6] 352,716

14. [10.6] $\dfrac{n(n-1)(n - 2)(n - 3)}{24}$ **15.** [10.5] 360

16. [10.6] 34,650

17. [10.7] $x^5 + 5x^4 + 10x^3 + 10x^2 + 5x + 1$

18. [10.7] $2^9 = 512$ **19.** [10.8] $\frac{4}{7}$ **20.** [10.5] 15

Appendix

Exercise Set

1.

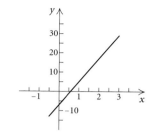

$$x = \tfrac{1}{2}t, \ y = 6t - 7, \ -1 \leq t \leq 6$$

$$y = 12x - 7, \ -\tfrac{1}{2} \leq x \leq 3$$

3.

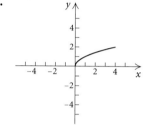

$$x = 4t^2, \ y = 2t, \ -1 \leq t \leq 1$$

$$x = y^2, \ 0 \leq x \leq 4$$

5. $x = y^4, \ 0 \leq x \leq 16$ **7.** $y = \dfrac{1}{x}, \ 1 \leq x \leq 5$

9. $y = \tfrac{1}{4}(x + 1)^2, \ -7 \leq x \leq 5$ **11.** $y = \dfrac{1}{x}, \ x > 0$

13. $x^2 + y^2 = 9, \ -3 \leq x \leq 3$

15. $x^2 + \dfrac{y^2}{4} = 1, \ -1 \leq x \leq 1$ **17.** $y = \dfrac{1}{x}, \ x \geq 1$

19. $(x - 1)^2 + (y - 2)^2 = 4, \ -1 \leq x \leq 3$

21. Answers may vary. $x = t, \ y = 4t - 3; \ x = \dfrac{t}{4} + 3,$
$y = t + 9$ **23.** Answers may vary. $x = t,$
$y = (t - 2)^2 - 6t; \ x = t + 2, \ y = t^2 - 6t - 12$
25. **(a)** $x = 40\sqrt{3}t, \ y = 7 + 40t - 16t^2;$ **(b)** 31 ft, 23 ft;
(c) about 2.7 sec; **(d)** about 187.1 ft; **(e)** 32 ft
27. $x = 3 \cos t, \ y = -3 \sin t$

Index

Index of Applications

Trigonometry

Trigonometric Functions

Acute Angles

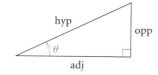

$$\sin \theta = \frac{\text{opp}}{\text{hyp}}, \quad \csc \theta = \frac{\text{hyp}}{\text{opp}},$$

$$\cos \theta = \frac{\text{adj}}{\text{hyp}}, \quad \sec \theta = \frac{\text{hyp}}{\text{adj}},$$

$$\tan \theta = \frac{\text{opp}}{\text{adj}}, \quad \cot \theta = \frac{\text{adj}}{\text{opp}}$$

Any Angle

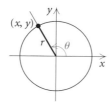

$$\sin \theta = \frac{y}{r}, \quad \csc \theta = \frac{r}{y},$$

$$\cos \theta = \frac{x}{r}, \quad \sec \theta = \frac{r}{x},$$

$$\tan \theta = \frac{y}{x}, \quad \cot \theta = \frac{x}{y}$$

Real Numbers

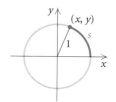

$$\sin s = y, \quad \csc s = \frac{1}{y},$$

$$\cos s = x, \quad \sec s = \frac{1}{x},$$

$$\tan s = \frac{y}{x}, \quad \cot s = \frac{x}{y}$$

Basic Trigonometric Identities

$$\sin (-x) = -\sin x,$$
$$\cos (-x) = \cos x,$$
$$\tan (-x) = -\tan x,$$

$$\tan x = \frac{\sin x}{\cos x},$$

$$\cot x = \frac{\cos x}{\sin x},$$

$$\csc x = \frac{1}{\sin x},$$

$$\sec x = \frac{1}{\cos x},$$

$$\cot x = \frac{1}{\tan x}$$

Pythagorean Identities

$$\sin^2 x + \cos^2 x = 1,$$
$$1 + \cot^2 x = \csc^2 x,$$
$$1 + \tan^2 x = \sec^2 x$$

Identities Involving $\pi/2$

$$\sin (\pi/2 - x) = \cos x,$$
$$\cos (\pi/2 - x) = \sin x, \qquad \sin (x \pm \pi/2) = \pm\cos x,$$
$$\tan (\pi/2 - x) = \cot x, \qquad \cos (x \pm \pi/2) = \mp\sin x$$

Sum and Difference Identities

$$\sin (u \pm v) = \sin u \cos v \pm \cos u \sin v,$$
$$\cos (u \pm v) = \cos u \cos v \mp \sin u \sin v,$$

$$\tan (u \pm v) = \frac{\tan u \pm \tan v}{1 \mp \tan u \tan v}$$

Double-Angle Identities

$$\sin 2x = 2 \sin x \cos x,$$
$$\cos 2x = \cos^2 x - \sin^2 x$$
$$= 1 - 2 \sin^2 x$$
$$= 2 \cos^2 x - 1,$$

$$\tan 2x = \frac{2 \tan x}{1 - \tan^2 x}$$

Half-Angle Identities

$$\sin \frac{x}{2} = \pm \sqrt{\frac{1 - \cos x}{2}},$$

$$\cos \frac{x}{2} = \pm \sqrt{\frac{1 + \cos x}{2}},$$

$$\tan \frac{x}{2} = \pm \sqrt{\frac{1 - \cos x}{1 + \cos x}} = \frac{\sin x}{1 + \cos x} = \frac{1 - \cos x}{\sin x}$$

(continued)

Trigonometry *(continued)*

The Law of Sines

In any $\triangle ABC$,

$$\frac{a}{\sin A} = \frac{b}{\sin B} = \frac{c}{\sin C}.$$

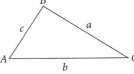

The Law of Cosines

In any $\triangle ABC$,

$$a^2 = b^2 + c^2 - 2bc \cos A,$$
$$b^2 = a^2 + c^2 - 2ac \cos B,$$
$$c^2 = a^2 + b^2 - 2ab \cos C.$$

Trigonometric Function Values of Special Angles

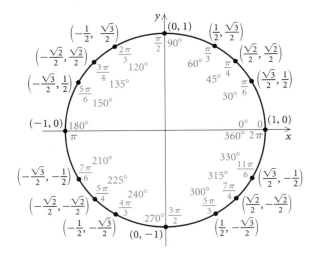

Graphs of Trigonometric Functions

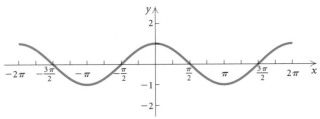

The sine function: $f(x) = \sin x$

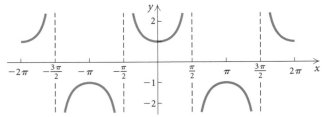

The cosecant function: $f(x) = \csc x$

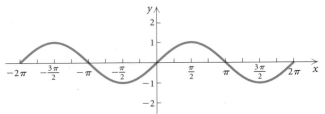

The cosine function: $f(x) = \cos x$

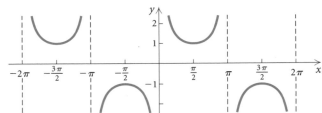

The secant function: $f(x) = \sec x$

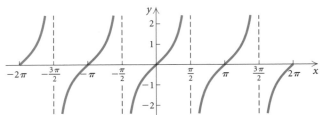

The tangent function: $f(x) = \tan x$

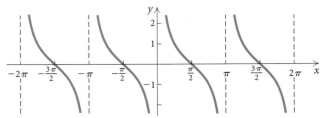

The cotangent function: $f(x) = \cot x$